TWO-DIMENSIONAL
SIGNAL AND IMAGE
PROCESSING

PRENTICE HALL SIGNAL PROCESSING SERIES

Alan V. Oppenheim, Editor

ANDREWS AND HUNT *Digital Image Restoration*
BRIGHAM *The Fast Fourier Transform*
BRIGHAM *The Fast Fourier Transform and Its Applications*
BURDIC *Underwater Acoustic System Analysis*
CASTLEMAN *Digital Image Processing*
COWAN AND GRANT *Adaptive Filters*
CROCHIERE AND RABINER *Multirate Digital Signal Processing*
DUDGEON AND MERSEREAU *Multidimensional Digital Signal Processing*
HAMMING *Digital Filters, 3/E*
HAYKIN, ED. *Array Signal Processing*
JAYANT AND NOLL *Digital Coding of Waveforms*
KAY *Modern Spectral Estimation*
KINO *Acoustic Waves: Devices, Imaging, and Analog Signal Processing*
LEA, ED. *Trends in Speech Recognition*
LIM *Two-Dimensional Signal and Image Processing*
LIM, ED. *Speech Enhancement*
LIM AND OPPENHEIM, EDS. *Advanced Topics in Signal Processing*
MARPLE *Digital Spectral Analysis with Applications*
MCCLELLAN AND RADER *Number Theory in Digital Signal Processing*
MENDEL *Lessons in Digital Estimation Theory*
OPPENHEIM, ED. *Applications of Digital Signal Processing*
OPPENHEIM, WILLSKY, WITH YOUNG *Signals and Systems*
OPPENHEIM AND SCHAFER *Digital Signal Processing*
OPPENHEIM AND SCHAFER *Discrete-Time Signal Processing*
QUACKENBUSH ET AL. *Objective Measures of Speech Quality*
RABINER AND GOLD *Theory and Applications of Digital Signal Processing*
RABINER AND SCHAFER *Digital Processing of Speech Signals*
ROBINSON AND TREITEL *Geophysical Signal Analysis*
STEARNS AND DAVID *Signal Processing Algorithms*
TRIBOLET *Seismic Applications of Homomorphic Signal Processing*
WIDROW AND STEARNS *Adaptive Signal Processing*

TWO-DIMENSIONAL
SIGNAL AND IMAGE
PROCESSING

JAE S. LIM

Department of Electrical Engineering
and Computer Science
Massachusetts Institute of Technology

P T R PRENTICE HALL, Englewood Cliffs, New Jersey 07632

Library of Congress Cataloging-in-Publication Data

Lim, Jae S.
    Two-dimensional signal and image processing / Jae S. Lim.
        p.   cm. — (Prentice Hall signal processing series)
    Bibliography: p.
    Includes index.
    ISBN 0-13-935322-4
    1. Signal processing—Digital techniques.   2. Image processing—
Digital techniques.   I. Title.   II. Series.
TK5102.5.L54   1990
621.382′2—dc20   89-33088
                                                                      CIP

Editorial/production supervision: *Raeia Maes*
Cover design: *Ben Santora*
Manufacturing buyer: *Mary Ann Gloriande*

 © 1990 by P T R Prentice Hall
Prentice-Hall, Inc.
A Paramount Communications Company
Englewood Cliffs, New Jersey 07632

Printed in the United States of America

10   9   8   7   6   5

ISBN   0-13-935322-4

Prentice-Hall International (UK) Limited, *London*
Prentice-Hall of Australia Pty. Limited, *Sydney*
Prentice-Hall Canada Inc., *Toronto*
Prentice-Hall Hispanoamericana, S.A., *Mexico*
Prentice-Hall of India Private Limited, *New Delhi*
Prentice-Hall of Japan, Inc., *Tokyo*
Simon & Schuster Asia Pte. Ltd., *Singapore*
Editora Prentice-Hall do Brasil, Ltda., *Rio de Janeiro*

**TO
KYUHO and TAEHO**

Contents

Contents

Contents

Preface

This book has grown out of the author's teaching and research activities in the field of two-dimensional signal and image processing. It is designed as a text for an upper-class undergraduate level or a graduate level course. The notes on which this book is based have been used since 1982 for a one-semester course in the Department of Electrical Engineering and Computer Science at M.I.T. and for a continuing education course at industries including Texas Instruments and Bell Laboratories.

In writing this book, the author has assumed that readers have prior exposure to fundamentals of one-dimensional digital signal processing, which are readily available in a variety of excellent text and reference books. Many two-dimensional signal processing theories are developed in the book by extension and generalization of one-dimensional signal processing theories.

This book consists of ten chapters. The first six chapters are devoted to fundamentals of two-dimensional digital signal processing. Chapter 1 is on signals, systems, and Fourier transform, which are the most basic concepts in signal processing and serve as a foundation for all other chapters. Chapter 2 is on z-transform representation and related topics including the difference equation and stability. Chapter 3 is on the discrete Fourier series, discrete Fourier transform, and fast Fourier transform. The chapter also covers the cosine and discrete cosine transforms which are closely related to Fourier and discrete Fourier transforms. Chapter 4 is on the design and implementation of finite impulse response filters. Chapter 5 is on the design and implementation of infinite impulse response filters. Chapter 6 is on random signals and spectral estimation. Throughout the first six chapters, the notation used and the theories developed are for two-dimensional signals and systems. Essentially all the results extend to more general multidimensional signals and systems in a straightforward manner.

The remaining four chapters are devoted to fundamentals of digital image processing. Chapter 7 is on the basics of image processing. Chapter 8 is on image enhancement including topics on contrast enhancement, noise smoothing, and use of color. The chapter also covers related topics on edge detection, image interpolation, and motion-compensated image processing. Chapter 9 is on image restoration and treats restoration of images degraded by both signal-independent and signal-dependent degradation. Chapter 10 is on image coding and related topics.

One goal of this book is to provide a single-volume text for a course that covers both two-dimensional signal processing and image processing. In a one-semester course at M.I.T., the author covered most topics in the book by treating some topics in reasonable depth and others with less emphasis. The book can also be used as a text for a course in which the primary emphasis is on either two-dimensional signal processing or image processing. A typical course with emphasis on two-dimensional signal processing, for example, would cover topics in Chapters 1 through 6 with reasonable depth and some selected topics from Chapters 7 and 9. A typical course with emphasis on image processing would cover topics in Chapters 1 and 3, Section 6.1, and Chapters 7 through 10. This book can also be used for a two-semester course, the first semester on two-dimensional signal processing and the second semester on image processing.

Many problems are included at the end of each chapter. These problems are, of course, intended to help the reader understand the basic concepts through drill and practice. The problems also extend some concepts presented previously and develop some new concepts.

The author is indebted to many students, friends, and colleagues for their assistance, support, and suggestions. The author was very fortunate to learn digital signal processing and image processing from Professor Alan Oppenheim, Professor Russell Mersereau, and Professor William Schreiber. Thrasyvoulos Pappas, Srinivasa Prasanna, Mike McIlrath, Matthew Bace, Roz Wright Picard, Dennis Martinez, and Giovanni Aliberti produced many figures. Many students and friends used the lecture notes from which this book originated and provided valuable comments and suggestions. Many friends and colleagues read drafts of this book, and their comments and suggestions have been incorporated. The book was edited by Beth Parkhurst and Patricia Johnson. Phyllis Eiro, Leslie Melcer, and Cindy LeBlanc typed many versions of the manuscript.

The author acknowledges the support of M.I.T. which provided an environment in which many ideas were developed and a major portion of the work was accomplished. The author is also grateful to the Woods Hole Oceanographic Institution and the Naval Postgraduate School where the author spent most of his sabbatical year completing the manuscript.

Jae S. Lim

Introduction

The fields of two-dimensional digital signal processing and digital image processing have maintained tremendous vitality over the past two decades and there is every indication that this trend will continue. Advances in hardware technology provide the capability in signal processing chips and microprocessors which were previously associated with mainframe computers. These advances allow sophisticated signal processing and image processing algorithms to be implemented in real time at a substantially reduced cost. New applications continue to be found and existing applications continue to expand in such diverse areas as communications, consumer electronics, medicine, defense, robotics, and geophysics. Along with advances in hardware technology and expansion in applications, new algorithms are developed and existing algorithms are better understood, which in turn lead to further expansion in applications and provide a strong incentive for further advances in hardware technology.

At a conceptual level, there is a great deal of similarity between one-dimensional signal processing and two-dimensional signal processing. In one-dimensional signal processing, the concepts discussed are filtering, Fourier transform, discrete Fourier transform, fast Fourier transform algorithms, and so on. In two-dimensional signal processing, we again are concerned with the same concepts. As a consequence, the general concepts that we develop in two-dimensional signal processing can be viewed as straightforward extensions of the results in one-dimensional signal processing.

At a more detailed level, however, considerable differences exist between one-dimensional and two-dimensional signal processing. For example, one major difference is the amount of data involved in typical applications. In speech processing, an important one-dimensional signal processing application, speech is typically sampled at a 10-kHz rate and we have 10,000 data points to process in a second. However, in video processing, where processing an image frame is an important two-dimensional signal processing application, we may have 30 frames per second, with each frame consisting of 500×500 pixels (picture elements). In this case, we would have 7.5 million data points to process per second, which is orders of magnitude greater than the case of speech processing. Due to this difference in data rate requirements, the computational efficiency of a signal processing algorithm plays a much more important role in two-dimensional signal processing, and advances in hardware technology will have a much greater impact on two-dimensional signal processing applications.

Another major difference comes from the fact that the mathematics used for one-dimensional signal processing is often simpler than that used for two-dimensional signal processing. For example, many one-dimensional systems are described by differential equations, while many two-dimensional systems are described by partial differential equations. It is generally much easier to solve differential equations than partial differential equations. Another example is the absence of the fundamental theorem of algebra for two-dimensional polynomials. For one-dimensional polynomials, the fundamental theorem of algebra states that any one-dimensional polynomial can be factored as a product of lower-order polynomials. This difference has a major impact on many results in signal processing. For example, an important structure for realizing a one-dimensional digital filter is the cascade structure. In the cascade structure, the z-transform of the digital filter's impulse response is factored as a product of lower-order polynomials and the realizations of these lower-order factors are cascaded. The z-transform of a two-dimensional digital filter's impulse response cannot, in general, be factored as a product of lower-order polynomials and the cascade structure therefore is not a general structure for a two-dimensional digital filter realization. Another consequence of the nonfactorability of a two-dimensional polynomial is the difficulty associated with issues related to system stability. In a one-dimensional system, the pole locations can be determined easily, and an unstable system can be stabilized without affecting the magnitude response by simple manipulation of pole locations. In a two-dimensional system, because poles are surfaces rather than points and there is no fundamental theorem of algebra, it is extremely difficult to determine the pole locations. As a result, checking the stability of a two-dimensional system and stabilizing an unstable two-dimensional system without affecting the magnitude response are extremely difficult.

As we have seen, there is considerable similarity and at the same time considerable difference between one-dimensional and two-dimensional signal processing. We will study the results in two-dimensional signal processing that are simple extensions of one-dimensional signal processing. Our discussion will rely heavily on the reader's knowledge of one-dimensional signal processing theories. We will also study, with much greater emphasis, the results in two-dimensional signal processing that are significantly different from those in one-dimensional signal processing. We will study what the differences are, where they come from, and what impacts they have on two-dimensional signal processing applications. Since we will study the similarities and differences of one-dimensional and two-dimensional signal processing and since one-dimensional signal processing is a special case of two-dimensional signal processing, this book will help us understand not only two-dimensional signal processing theories but also one-dimensional signal processing theories at a much deeper level.

An important application of two-dimensional signal processing theories is image processing. Image processing is closely tied to human vision, which is one of the most important means by which humans perceive the outside world. As a result, image processing has a large number of existing and potential applications and will play an increasingly important role in our everyday life.

Digital image processing can be classified broadly into four areas: image enhancement, restoration, coding, and understanding. In image enhancement, images either are processed for human viewers, as in television, or preprocessed to aid machine performance, as in object identification by machine. In image restoration, an image has been degraded in some manner and the objective is to reduce or eliminate the effect of degradation. Typical degradations that occur in practice include image blurring, additive random noise, quantization noise, multiplicative noise, and geometric distortion. The objective in image coding is to represent an image with as few bits as possible, preserving a certain level of image quality and intelligibility acceptable for a given application. Image coding can be used in reducing the bandwidth of a communication channel when an image is transmitted and in reducing the amount of required storage when an image needs to be retrieved at a future time. We study image enhancement, restoration, and coding in the latter part of the book.

The objective of image understanding is to symbolically represent the contents of an image. Applications of image understanding include computer vision and robotics. Image understanding differs from the other three areas in one major respect. In image enhancement, restoration, and coding, both the input and the output are images, and signal processing has been the backbone of many successful systems in these areas. In image understanding, the input is an image, but the output is symbolic representation of the contents of the image. Successful development of systems in this area involves not only signal processing but also other disciplines such as artificial intelligence. In a typical image understanding system, signal processing is used for such lower-level processing tasks as reduction of degradation and extraction of edges or other image features, and artificial intelligence is used for such higher-level processing tasks as symbol manipulation and knowledge base management. We treat some of the lower-level processing techniques useful in image understanding as part of our general discussion of image enhancement, restoration, and coding. A complete treatment of image understanding is outside the scope of this book.

Two-dimensional signal processing and image processing cover a large number of topics and areas, and a selection of topics was necessary due to space limitation. In addition, there are a variety of ways to present the material. The main objective of this book is to provide fundamentals of two-dimensional signal processing and image processing in a tutorial manner. We have selected the topics and chosen the style of presentation with this objective in mind. We hope that the fundamentals of two-dimensional signal processing and image processing covered in this book will form a foundation for additional reading of other books and articles in the field, application of theoretical results to real-world problems, and advancement of the field through research and development.

TWO-DIMENSIONAL
SIGNAL AND IMAGE
PROCESSING

1

Signals, Systems, and the Fourier Transform

1.0 INTRODUCTION

Most signals can be classified into three broad groups. One group, which consists of *analog* or *continuous-space* signals, is continuous in both space* and amplitude. In practice, a majority of signals falls into this group. Examples of analog signals include image, seismic, radar, and speech signals. Signals in the second group, *discrete-space* signals, are discrete in space and continuous in amplitude. A common way to generate discrete-space signals is by sampling analog signals. Signals in the third group, *digital* or *discrete* signals, are discrete in both space and amplitude. One way in which digital signals are created is by amplitude quantization of discrete-space signals. Discrete-space signals and digital signals are also referred to as *sequences*.

Digital systems and computers use only digital signals, which are discrete in both space and amplitude. The development of signal processing concepts based on digital signals, however, requires a detailed treatment of amplitude quantization, which is extremely difficult and tedious. Many useful insights would be lost in such a treatment because of its mathematical complexity. For this reason, most digital signal processing concepts have been developed based on discrete-space signals. Experience shows that theories based on discrete-space signals are often applicable to digital signals.

A system maps an input signal to an output signal. A major element in studying signal processing is the analysis, design, and implementation of a system that transforms an input signal to a more desirable output signal for a given application. When developing theoretical results about systems, we often impose

---

*Although we refer to "space," an analog signal can instead have a variable in time, as in the case of speech processing.

the constraints of linearity and shift invariance. Although these constraints are very restrictive, the theoretical results thus obtained apply in practice at least approximately to many systems. We will discuss signals and systems in Sections 1.1 and 1.2, respectively.

The Fourier transform representation of signals and systems plays a central role in both one-dimensional (1-D) and two-dimensional (2-D) signal processing. In Sections 1.3 and 1.4, the Fourier transform representation including some aspects that are specific to image processing applications is discussed. In Section 1.5, we discuss digital processing of analog signals. Many of the theoretical results, such as the 2-D sampling theorem summarized in that section, can be derived from the Fourier transform results.

Many of the theoretical results discussed in this chapter can be viewed as straightforward extensions of the one-dimensional case. Some, however, are unique to two-dimensional signal processing. Very naturally, we will place considerably more emphasis on these. We will now begin our journey with the discussion of signals.

1.1 SIGNALS

The signals we consider are discrete-space signals. A 2-D discrete-space signal (sequence) will be denoted by a function whose two arguments are integers. For example, $x(n_1, n_2)$ represents a sequence which is defined for all integer values of n_1 and n_2. Note that $x(n_1, n_2)$ for a noninteger n_1 or n_2 is not zero, but is undefined. The notation $x(n_1, n_2)$ may refer either to the discrete-space function x or to the value of the function x at a specific (n_1, n_2). The distinction between these two will be evident from the context.

An example of a 2-D sequence $x(n_1, n_2)$ is sketched in Figure 1.1. In the figure, the height at (n_1, n_2) represents the amplitude at (n_1, n_2). It is often tedious to sketch a 2-D sequence in the three-dimensional (3-D) perspective plot as shown

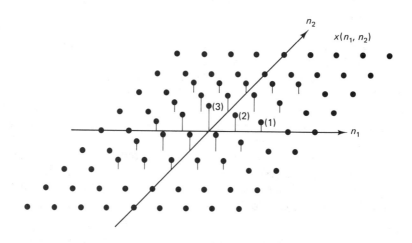

Figure 1.1   2-D sequence $x(n_1, n_2)$.

in Figure 1.1. An alternate way to sketch the 2-D sequence in Figure 1.1 is shown in Figure 1.2. In this figure, open circles represent amplitudes of 0 and filled-in circles represent nonzero amplitudes, with the values in parentheses representing the amplitudes. For example, $x(3, 0)$ is 0 and $x(1, 1)$ is 2.

Many sequences we use have amplitudes of 0 or 1 for large regions of (n_1, n_2). In such instances, the open circles and parentheses will be eliminated for convenience. If there is neither an open circle nor a filled-in circle at a particular (n_1, n_2), then the sequence has zero amplitude at that point. If there is a filled-in circle with no amplitude specification at a particular (n_1, n_2), then the sequence has an amplitude of 1 at that point. Figure 1.3 shows the result when this additional simplification is made to the sequence in Figure 1.2.

1.1.1 Examples of Sequences

Certain sequences and classes of sequences play a particularly important role in 2-D signal processing. These are impulses, step sequences, exponential sequences, separable sequences, and periodic sequences.

Impulses. The impulse or unit sample sequence, denoted by $\delta(n_1, n_2)$, is defined as

$$\delta(n_1, n_2) = \begin{cases} 1, & n_1 = n_2 = 0 \\ 0, & \text{otherwise.} \end{cases} \tag{1.1}$$

The sequence $\delta(n_1, n_2)$, sketched in Figure 1.4, plays a role similar to the impulse $\delta(n)$ in 1-D signal processing.

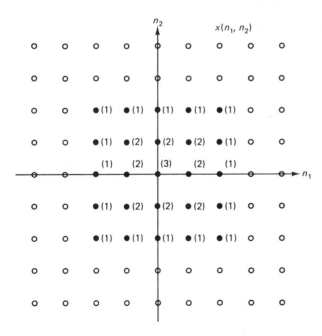

Figure 1.2 Alternate way to sketch the 2-D sequence in Figure 1.1. Open circles represent amplitudes of zero, and filled-in circles represent nonzero amplitudes, with values in parentheses representing the amplitude.

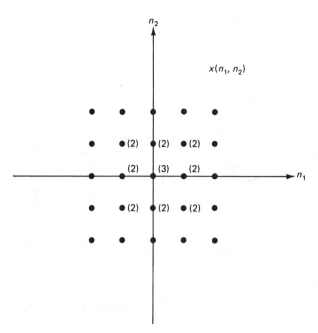

Figure 1.3 Sequence in Figure 1.2 sketched with some simplification. Open circles have been eliminated and filled-in circles with amplitude of 1 have no amplitude specifications.

Any sequence $x(n_1, n_2)$ can be represented as a linear combination of shifted impulses as follows:

$$x(n_1, n_2) = \cdots + x(-1, -1)\delta(n_1 + 1, n_2 + 1) + x(0, -1)\delta(n_1, n_2 + 1)$$

$$+ x(1, -1)\delta(n_1 - 1, n_2 + 1) + \cdots + x(-1, 0)\delta(n_1 + 1, n_2)$$

$$+ x(0, 0)\delta(n_1, n_2) + x(1, 0)\delta(n_1 - 1, n_2)$$

$$+ \cdots + x(-1, 1)\delta(n_1 + 1, n_2 - 1)$$

$$+ x(0, 1)\delta(n_1, n_2 - 1) + x(1, 1)\delta(n_1 - 1, n_2 - 1) + \cdots$$

$$= \sum_{k_1 = -\infty}^{\infty} \sum_{k_2 = -\infty}^{\infty} x(k_1, k_2)\delta(n_1 - k_1, n_2 - k_2). \tag{1.2}$$

The representation of $x(n_1, n_2)$ by (1.2) is very useful in system analysis.

Line impulses constitute a class of impulses which do not have any counterparts in 1-D. An example of a line impulse is the 2-D sequence $\delta_T(n_1)$, which is sketched in Figure 1.5 and is defined as

$$x(n_1, n_2) = \delta_T(n_1) = \begin{cases} 1, & n_1 = 0 \\ 0, & \text{otherwise.} \end{cases} \tag{1.3}$$

Other examples include $\delta_T(n_2)$ and $\delta_T(n_1 - n_2)$, which are defined similarly to $\delta_T(n_1)$. The subscript T in $\delta_T(n_1)$ indicates that $\delta_T(n_1)$ is a 2-D sequence. This notation is used to avoid confusion in cases where the 2-D sequence is a function of only one variable. For example, without the subscript T, $\delta_T(n_1)$ might be

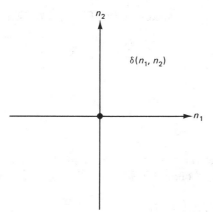

$\delta(n_1, n_2)$

Figure 1.4 Impulse $\delta(n_1, n_2)$.

confused with the 1-D impulse $\delta(n_1)$. For clarity, then, the subscript T will be used whenever a 2-D sequence is a function of one variable. The sequence $x_T(n_1)$ is thus a 2-D sequence, while $x(n_1)$ is a 1-D sequence.

Step sequences. The unit step sequence, denoted by $u(n_1, n_2)$, is defined as

$$u(n_1, n_2) = \begin{cases} 1, & n_1, n_2 \geq 0 \\ 0, & \text{otherwise.} \end{cases} \qquad (1.4)$$

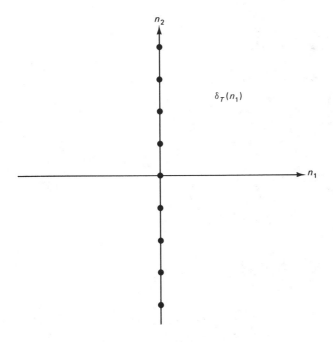

$\delta_T(n_1)$

Figure 1.5 Line impulse $\delta_T(n_1)$.

The sequence $u(n_1, n_2)$, which is sketched in Figure 1.6, is related to $\delta(n_1, n_2)$ as

$$u(n_1, n_2) = \sum_{k_1 = -\infty}^{n_1} \sum_{k_2 = -\infty}^{n_2} \delta(k_1, k_2) \tag{1.5a}$$

or

$$\delta(n_1, n_2) = u(n_1, n_2) - u(n_1 - 1, n_2) - u(n_1, n_2 - 1) + u(n_1 - 1, n_2 - 1). \tag{1.5b}$$

Some step sequences have no counterparts in 1-D. An example is the 2-D sequence $u_T(n_1)$, which is sketched in Figure 1.7 and is defined as

$$x(n_1, n_2) = u_T(n_1) = \begin{cases} 1, & n_1 \geq 0 \\ 0, & \text{otherwise.} \end{cases} \tag{1.6}$$

Other examples include $u_T(n_2)$ and $u_T(n_1 - n_2)$, which are defined similarly to $u_T(n_1)$.

Exponential sequences. Exponential sequences of the type $x(n_1, n_2) = A\alpha^{n_1}\beta^{n_2}$ are important for system analysis. As we shall see later, sequences of this class are eigenfunctions of linear shift-invariant (LSI) systems.

Separable sequences. A 2-D sequence $x(n_1, n_2)$ is said to be a separable sequence if it can be expressed as

$$x(n_1, n_2) = f(n_1)g(n_2) \tag{1.7}$$

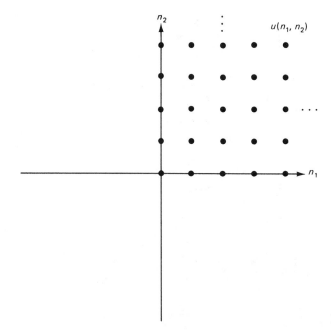

Figure 1.6 Unit step sequence $u(n_1, n_2)$.

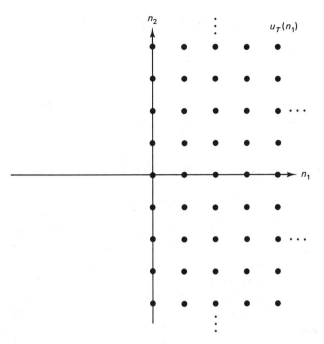

Figure 1.7 Step sequence $u_T(n_1)$.

where $f(n_1)$ is a function of only n_1 and $g(n_2)$ is a function of only n_2. Although it is possible to view $f(n_1)$ and $g(n_2)$ as 2-D sequences, it is more convenient to consider them to be 1-D sequences. For that reason, we use the notations $f(n_1)$ and $g(n_2)$ rather than $f_T(n_1)$ and $g_T(n_2)$.

The impulse $\delta(n_1, n_2)$ is a separable sequence since $\delta(n_1, n_2)$ can be expressed as

$$\delta(n_1, n_2) = \delta(n_1)\,\delta(n_2) \tag{1.8}$$

where $\delta(n_1)$ and $\delta(n_2)$ are 1-D impulses. The unit step sequence $u(n_1, n_2)$ is also a separable sequence since $u(n_1, n_2)$ can be expressed as

$$u(n_1, n_2) = u(n_1)u(n_2) \tag{1.9}$$

where $u(n_1)$ and $u(n_2)$ are 1-D unit step sequences. Another example of a separable sequence is $a^{n_1}b^{n_2} + b^{n_1+n_2}$, which can be written as $(a^{n_1} + b^{n_1})b^{n_2}$.

Separable sequences form a very special class of 2-D sequences. A typical 2-D sequence is not a separable sequence. As an illustration, consider a sequence $x(n_1, n_2)$ which is zero outside $0 \leq n_1 \leq N_1 - 1$ and $0 \leq n_2 \leq N_2 - 1$. A general sequence $x(n_1, n_2)$ of this type has $N_1 N_2$ degrees of freedom. If $x(n_1, n_2)$ is a separable sequence, $x(n_1, n_2)$ is completely specified by some $f(n_1)$ which is zero outside $0 \leq n_1 \leq N_1 - 1$ and some $g(n_2)$ which is zero outside $0 \leq n_2 \leq N_2 - 1$, and consequently has only $N_1 + N_2 - 1$ degrees of freedom.

Despite the fact that separable sequences constitute a very special class of 2-D sequences, they play an important role in 2-D signal processing. In those cases where the results that apply to 1-D sequences do not extend to general 2-D sequences in a straightforward manner, they often do for separable 2-D sequences.

In addition, the separability of the sequence can be exploited in order to reduce computation in various contexts, such as digital filtering and computation of the discrete Fourier transform. This will be discussed further in later sections.

Periodic sequences. A sequence $x(n_1, n_2)$ is said to be periodic with a period of $N_1 \times N_2$ if $x(n_1, n_2)$ satisfies the following condition:

$$x(n_1, n_2) = x(n_1 + N_1, n_2) = x(n_1, n_2 + N_2) \quad \text{for all } (n_1, n_2) \qquad (1.10)$$

where N_1 and N_2 are positive integers. For example, $\cos(\pi n_1 + (\pi/2)n_2)$ is a periodic sequence with a period of 2×4, since $\cos(\pi n_1 + (\pi/2)n_2) = \cos(\pi(n_1 + 2) + (\pi/2)n_2) = \cos(\pi n_1 + (\pi/2)(n_2 + 4))$ for all (n_1, n_2). The sequence $\cos(n_1 + n_2)$ is not periodic, however, since $\cos(n_1 + n_2)$ cannot be expressed as $\cos((n_1 + N_1) + n_2) = \cos(n_1 + (n_2 + N_2))$ for all (n_1, n_2) for any nonzero integers N_1 and N_2. A periodic sequence is often denoted by adding a "~" (tilde), for example, $\tilde{x}(n_1, n_2)$, to distinguish it from an aperiodic sequence.

Equation (1.10) is not the most general representation of a 2-D periodic sequence. As an illustration, consider the sequence $x(n_1, n_2)$ shown in Figure 1.8. Even though $x(n_1, n_2)$ can be considered a periodic sequence with a period of 3×2 it cannot be represented as such a sequence by using (1.10). Specifically, $x(n_1, n_2) \neq x(n_1 + 3, n_2)$ for all (n_1, n_2). It is possible to generalize (1.10) to incorporate cases such as that in Figure 1.8. However, in this text we will use (1.10) to define a periodic sequence, since it is sufficient for our purposes, and sequences such as that in Figure 1.8 can be represented by (1.10) by increasing N_1

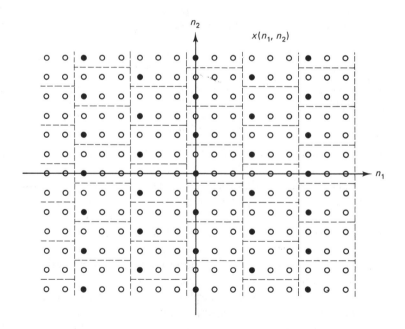

Figure 1.8 Periodic sequence with a period of 6×2.

and/or N_2. For example, the sequence in Figure 1.8 is periodic with a period of 6×2 using (1.10).

1.1.2 Digital Images

Many examples of sequences used in this book are digital images. A digital image, which can be denoted by $x(n_1, n_2)$, is typically obtained by sampling an analog image, for instance, an image on film. The amplitude of a digital image is often quantized to 256 levels (which can be represented by eight bits). Each level is commonly denoted by an integer, with 0 corresponding to the darkest level and 255 to the brightest. Each point (n_1, n_2) is called a pixel or pel (picture element). A digital image $x(n_1, n_2)$ of 512×512 pixels with each pixel represented by eight bits is shown in Figure 1.9. As we reduce the number of amplitude quantization levels, the signal-dependent quantization noise begins to appear as false contours. This is shown in Figure 1.10, where the image in Figure 1.9 is displayed with 64 levels (six bits), 16 levels (four bits), 4 levels (two bits), and 2 levels (one bit) of amplitude quantization. As we reduce the number of pixels in a digital image, the spatial resolution is decreased and the details in the image begin to disappear. This is shown in Figure 1.11, where the image in Figure 1.9 is displayed at a spatial resolution of 256×256 pixels, 128×128 pixels, 64×64 pixels, and 32×32 pixels. A digital image of 512×512 pixels has a spatial resolution similar to that seen in a television frame. To have a spatial resolution similar to that of an image on 35-mm film, we need a spatial resolution of 1024×1024 pixels in the digital image.

Figure 1.9   Digital image of 512×512 pixels quantized at 8 bits/pixel.

(a)

(b)

(c)

(d)

Figure 1.10   Image in Figure 1.9 with amplitude quantization at (a) 6 bits/pixel, (b) 4 bits/pixel, (c) 2 bits/pixel, and (d) 1 bit/pixel.

(a)

(b)

(c)

(d)

Figure 1.11   Image in Figure 1.9 with spatial resolution of (a) 256 × 256 pixels, (b) 128 × 128 pixels, (c) 64 × 64 pixels, and (d) 32 × 32 pixels.

1.2 SYSTEMS

1.2.1 Linear Systems and Shift-Invariant Systems

An input-output relationship is called a system if there is a unique output for any given input. A system T that relates an input $x(n_1, n_2)$ to an output $y(n_1, n_2)$ is represented by

$$y(n_1, n_2) = T[x(n_1, n_2)]. \tag{1.11}$$

This definition of a system is very broad. Without any restrictions, characterizing a system requires a complete input-output relationship. Knowing the output of a system to one set of inputs does not generally allow us to determine the output of the system to any other set of inputs. Two types of restriction which greatly simplify the characterization and analysis of a system are linearity and shift invariance. In practice, fortunately, many systems can be approximated to be linear and shift invariant.

The linearity of a system T is defined as

$$\text{Linearity} \iff T[ax_1(n_1, n_2) + bx_2(n_1, n_2)] = ay_1(n_1, n_2) + by_2(n_1, n_2) \tag{1.12}$$

where $T[x_1(n_1, n_2)] = y_1(n_1, n_2)$, $T[x_2(n_1, n_2)] = y_2(n_1, n_2)$, a and b are any scalar constants, and $A \iff B$ means that A implies B and B implies A. The condition in (1.12) is called the *principle of superposition*. To illustrate this concept, a linear system and a nonlinear system are shown in Figure 1.12. The linearity of the system in Figure 1.12(a) and the nonlinearity of the system in Figure 1.12(b) can be easily verified by using (1.12).

The shift invariance (SI) or space invariance of a system is defined as

$$\text{Shift invariance} \iff T[x(n_1 - m_1, n_2 - m_2)] = y(n_1 - m_1, n_2 - m_2) \tag{1.13}$$

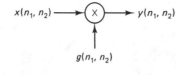

$x(n_1, n_2) \longrightarrow \boxed{\times} \longrightarrow y(n_1, n_2)$

$\uparrow$

$g(n_1, n_2)$

$y(n_1, n_2) = T[x(n_1, n_2)] = x(n_1, n_2)g(n_1, n_2)$

(a)

$x(n_1, n_2) \longrightarrow \boxed{(\,\cdot\,)^2} \longrightarrow y(n_1, n_2)$

$y(n_1, n_2) = T[x(n_1, n_2)] = x^2(n_1, n_2)$

(b)

Figure 1.12 (a) Example of a linear shift-variant system; (b) example of a nonlinear shift-invariant system.

where $y(n_1, n_2) = T[x(n_1, n_2)]$ and m_1 and m_2 are any integers. The system in Figure 1.12(a) is not shift invariant since $T[x(n_1 - m_1, n_2 - m_2)] = x(n_1 - m_1, n_2 - m_2)g(n_1, n_2)$ and $y(n_1 - m_1, n_2 - m_2) = x(n_1 - m_1, n_2 - m_2)g(n_1 - m_1, n_2 - m_2)$. The system in Figure 1.12(b), however, is shift invariant, since $T[x(n_1 - m_1, n_2 - m_2)] = x^2(n_1 - m_1, n_2 - m_2)$ and $y(n_1 - m_1, n_2 - m_2) = x^2(n_1 - m_1, n_2 - m_2)$.

Consider a linear system T. Using (1.2) and (1.12), we can express the output $y(n_1, n_2)$ for an input $x(n_1, n_2)$ as

$$y(n_1, n_2) = T[x(n_1, n_2)] = T\left[\sum_{k_1=-\infty}^{\infty}\sum_{k_2=-\infty}^{\infty} x(k_1, k_2)\delta(n_1 - k_1, n_2 - k_2)\right]$$

$$= \sum_{k_1=-\infty}^{\infty}\sum_{k_2=-\infty}^{\infty} x(k_1, k_2)T[\delta(n_1 - k_1, n_2 - k_2)]. \tag{1.14}$$

From (1.14), a linear system can be completely characterized by the response of the system to the impulse $\delta(n_1, n_2)$ and its shifts $\delta(n_1 - k_1, n_2 - k_2)$. If we know $T[\delta(n_1 - k_1, n_2 - k_2)]$ for all integer values of k_1 and k_2, the output of the linear system to any input $x(n_1, n_2)$ can be obtained from (1.14). For a nonlinear system, knowledge of $T[\delta(n_1 - k_1, n_2 - k_2)]$ for all integer values of k_1 and k_2 does not tell us the output of the system when the input $x(n_1, n_2)$ is $2\delta(n_1, n_2)$, $\delta(n_1, n_2) + \delta(n_1 - 1, n_2)$, or many other sequences.

System characterization is further simplified if we impose the additional restriction of shift invariance. Suppose we denote the response of a system T to an input $\delta(n_1, n_2)$ by $h(n_1, n_2)$;

$$h(n_1, n_2) = T[\delta(n_1, n_2)]. \tag{1.15}$$

From (1.13) and (1.15),

$$h(n_1 - k_1, n_2 - k_2) = T[\delta(n_1 - k_1, n_2 - k_2)] \tag{1.16}$$

for a shift-invariant system T. For a linear and shift-invariant (LSI) system, then, from (1.14) and (1.16), the input-output relation is given by

$$y(n_1, n_2) = T[x(n_1, n_2)] = \sum_{k_1=-\infty}^{\infty}\sum_{k_2=-\infty}^{\infty} x(k_1, k_2)h(n_1 - k_1, n_2 - k_2). \tag{1.17}$$

Equation (1.17) states that an LSI system is completely characterized by the impulse response $h(n_1, n_2)$. Specifically, for an LSI system, knowledge of $h(n_1, n_2)$ alone allows us to determine the output of the system to any input from (1.17). Equation (1.17) is referred to as *convolution*, and is denoted by the convolution operator "*" as follows:

For an LSI system,

$$y(n_1, n_2) = x(n_1, n_2) * h(n_1, n_2)$$
$$= \sum_{k_1=-\infty}^{\infty}\sum_{k_2=-\infty}^{\infty} x(k_1, k_2)h(n_1 - k_1, n_2 - k_2). \tag{1.18}$$

Note that the impulse response $h(n_1, n_2)$, which plays such an important role for an LSI system, loses its significance for a nonlinear or shift-variant system. Note also that an LSI system can be completely characterized by the system response to one of many other input sequences. The choice of $\delta(n_1, n_2)$ as the input in characterizing an LSI system is the simplest, both conceptually and in practice.

1.2.2 Convolution

The convolution operator in (1.18) has a number of properties that are straight-forward extensions of 1-D results. Some of the more important are listed below.

Commutativity

$$x(n_1, n_2) * y(n_1, n_2) = y(n_1, n_2) * x(n_1, n_2) \tag{1.19}$$

Associativity

$$(x(n_1, n_2) * y(n_1, n_2)) * z(n_1, n_2) = x(n_1, n_2) * (y(n_1, n_2) * z(n_1, n_2)) \tag{1.20}$$

Distributivity

$$x(n_1, n_2) * (y(n_1, n_2) + z(n_1, n_2))$$

$$= (x(n_1, n_2) * y(n_1, n_2)) + (x(n_1, n_2) * z(n_1, n_2)) \tag{1.21}$$

Convolution with Shifted Impulse

$$x(n_1, n_2) * \delta(n_1 - m_1, n_2 - m_2) = x(n_1 - m_1, n_2 - m_2) \tag{1.22}$$

The commutativity property states that the output of an LSI system is not affected when the input and the impulse response interchange roles. The associativity property states that a cascade of two LSI systems with impulse responses $h_1(n_1, n_2)$ and $h_2(n_1, n_2)$ has the same input-output relationship as one LSI system with impulse response $h_1(n_1, n_2) * h_2(n_1, n_2)$. The distributivity property states that a parallel combination of two LSI systems with impulse responses $h_1(n_1, n_2)$ and $h_2(n_1, n_2)$ has the same input-output relationship as one LSI system with impulse response given by $h_1(n_1, n_2) + h_2(n_1, n_2)$. In a special case of (1.22), when $m_1 = m_2 = 0$, we see that the impulse response of an identity system is $\delta(n_1, n_2)$.

The convolution of two sequences $x(n_1, n_2)$ and $h(n_1, n_2)$ can be obtained by explicitly evaluating (1.18). It is often simpler and more instructive, however, to evaluate (1.18) graphically. Specifically, the convolution sum in (1.18) can be interpreted as multiplying two sequences $x(k_1, k_2)$ and $h(n_1 - k_1, n_2 - k_2)$, which are functions of the variables k_1 and k_2, and summing the product over all integer values of k_1 and k_2. The output, which is a function of n_1 and n_2, is the result of convolving $x(n_1, n_2)$ and $h(n_1, n_2)$. To illustrate, consider the two sequences $x(n_1, n_2)$ and $h(n_1, n_2)$, shown in Figures 1.13(a) and (b). From $x(n_1, n_2)$ and $h(n_1, n_2)$, $x(k_1, k_2)$ and $h(n_1 - k_1, n_2 - k_2)$ as functions of k_1 and k_2 can be obtained, as shown in Figures 1.13(c)–(f). Note that $g(k_1 - n_1, k_2 - n_2)$ is

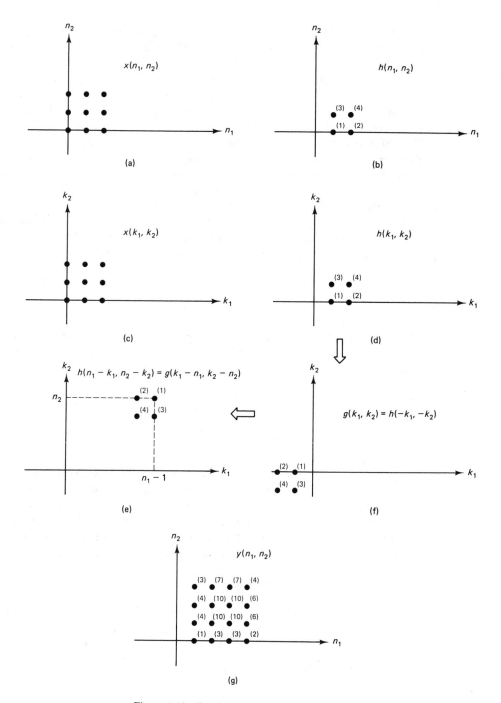

Figure 1.13 Example of convolving two sequences.

$g(k_1, k_2)$ shifted in the positive k_1 and k_2 directions by n_1 and n_2 points, respectively. Figures 1.13(d)–(f) show how to obtain $h(n_1 - k_1, n_2 - k_2)$ as a function of k_1 and k_2 from $h(n_1, n_2)$ in three steps. It is useful to remember how to obtain $h(n_1 - k_1, n_2 - k_2)$ directly from $h(n_1, n_2)$. One simple way is to first change the variables n_1 and n_2 to k_1 and k_2, flip the sequence with respect to the origin, and then shift the result in the positive k_1 and k_2 directions by n_1 and n_2 points, respectively. Once $x(k_1, k_2)$ and $h(n_1 - k_1, n_2 - k_2)$ are obtained, they can be multiplied and summed over k_1 and k_2 to produce the output at each different value of (n_1, n_2). The result is shown in Figure 1.13(g).

An LSI system is said to be *separable*, if its impulse response $h(n_1, n_2)$ is a separable sequence. For a separable system, it is possible to reduce the number of arithmetic operations required to compute the convolution sum. For large amounts of data, as typically found in images, the computational reduction can be considerable. To illustrate this, consider an input sequence $x(n_1, n_2)$ of $N \times N$ points and an impulse response $h(n_1, n_2)$ of $M \times M$ points:

$$x(n_1, n_2) = 0 \quad \text{outside} \quad 0 \le n_1 \le N - 1, \qquad 0 \le n_2 \le N - 1$$

$$\text{and} \quad h(n_1, n_2) = 0 \quad \text{outside} \quad 0 \le n_1 \le M - 1, \qquad 0 \le n_2 \le M - 1 \tag{1.23}$$

where $N \gg M$ in typical cases. The regions of (n_1, n_2) where $x(n_1, n_2)$ and $h(n_1, n_2)$ can have nonzero amplitudes are shown in Figures 1.14(a) and (b). The output of the system, $y(n_1, n_2)$, can be expressed as

$$y(n_1, n_2) = x(n_1, n_2) * h(n_1, n_2)$$

$$= \sum_{k_1 = -\infty}^{\infty} \sum_{k_2 = -\infty}^{\infty} x(k_1, k_2) h(n_1 - k_1, n_2 - k_2). \tag{1.24}$$

The region of (n_1, n_2) where $y(n_1, n_2)$ has nonzero amplitude is shown in Figure 1.14(c). If (1.24) is used directly to compute $y(n_1, n_2)$, approximately $(N + M - 1)^2 M^2$ arithmetic operations (one arithmetic operation = one multiplication and one addition) are required since the number of nonzero output points is $(N + M - 1)^2$ and computing each output point requires approximately M^2 arithmetic operations. If $h(n_1, n_2)$ is a separable sequence, it can be expressed as

$$h(n_1, n_2) = h_1(n_1) h_2(n_2)$$

$$h_1(n_1) = 0 \quad \text{outside} \quad 0 \le n_1 \le M - 1 \tag{1.25}$$

$$h_2(n_2) = 0 \quad \text{outside} \quad 0 \le n_2 \le M - 1.$$

From (1.24) and (1.25),

$$y(n_1, n_2) = \sum_{k_1 = -\infty}^{\infty} \sum_{k_2 = -\infty}^{\infty} x(k_1, k_2) h_1(n_1 - k_1) h_2(n_2 - k_2)$$

$$= \sum_{k_1 = -\infty}^{\infty} h_1(n_1 - k_1) \sum_{k_2 = -\infty}^{\infty} x(k_1, k_2) h_2(n_2 - k_2). \tag{1.26}$$

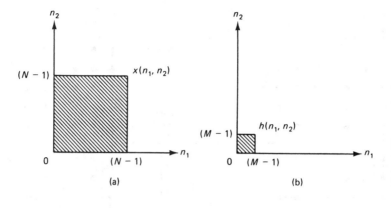

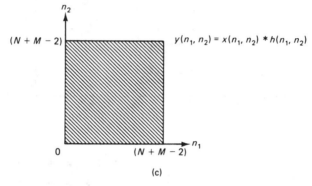

Figure 1.14 Regions of (n_1, n_2) where $x(n_1, n_2)$, $h(n_1, n_2)$, and $y(n_1, n_2) = x(n_1, n_2) * h(n_1, n_2)$ can have nonzero amplitude.

For a fixed k_1, $\sum_{k_2=-\infty}^{\infty} x(k_1, k_2)h_2(n_2 - k_2)$ in (1.26) is a 1-D convolution of $x(k_1, n_2)$ and $h_2(n_2)$. For example, using the notation

$$f(k_1, n_2) = \sum_{k_2=-\infty}^{\infty} x(k_1, k_2)h_2(n_2 - k_2), \tag{1.27}$$

$f(0, n_2)$ is the result of 1-D convolution of $x(0, n_2)$ with $h_2(n_2)$, as shown in Figure 1.15. Since there are N different values of k_1 for which $x(k_1, k_2)$ is nonzero, computing $f(k_1, n_2)$ requires N 1-D convolutions and therefore requires approximately $NM(N + M - 1)$ arithmetic operations. Once $f(k_1, n_2)$ is computed, $y(n_1, n_2)$ can be computed from (1.26) and (1.27) by

$$y(n_1, n_2) = \sum_{k_1=-\infty}^{\infty} h_1(n_1 - k_1)f(k_1, n_2). \tag{1.28}$$

From (1.28), for a fixed n_2, $y(n_1, n_2)$ is a 1-D convolution of $h_1(n_1)$ and $f(n_1, n_2)$. For example, $y(n_1, 1)$ is the result of a 1-D convolution of $f(n_1, 1)$ and $h_1(n_1)$, as shown in Figure 1.15, where $f(n_1, n_2)$ is obtained from $f(k_1, n_2)$ by a simple

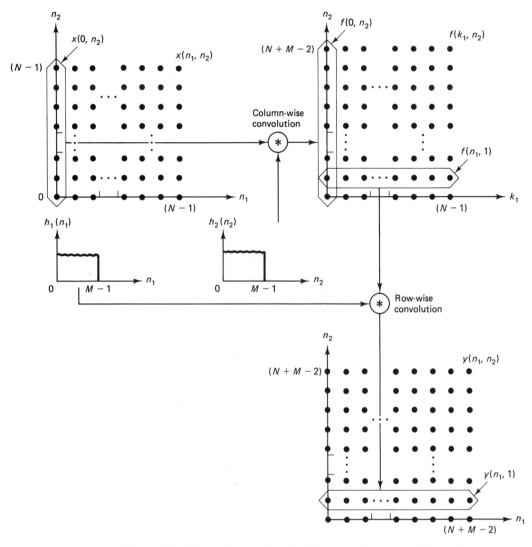

Figure 1.15 Convolution of $x(n_1, n_2)$ with a separable sequence $h(n_1, n_2)$.

change of variables. Since there are $N + M - 1$ different values of n_2, computing $y(n_1, n_2)$ from $f(k_1, n_2)$ requires $N + M - 1$ 1-D convolutions thus approximately $M(N + M - 1)^2$ arithmetic operations. Computing $y(n_1, n_2)$ from (1.27) and (1.28), exploiting the separability of $h(n_1, n_2)$, requires approximately $NM(N + M - 1) + M(N + M - 1)^2$ arithmetic operations. This can be a considerable computational saving over $(N + M - 1)^2 M^2$. If we assume $N \gg M$, exploiting the separability of $h(n_1, n_2)$ reduces the number of arithmetic operations by approximately a factor of $M/2$.

As an example, consider $x(n_1, n_2)$ and $h(n_1, n_2)$, shown in Figures 1.16(a) and (b). The sequence $h(n_1, n_2)$ can be expressed as $h_1(n_1)h_2(n_2)$, where $h_1(n_1)$ and

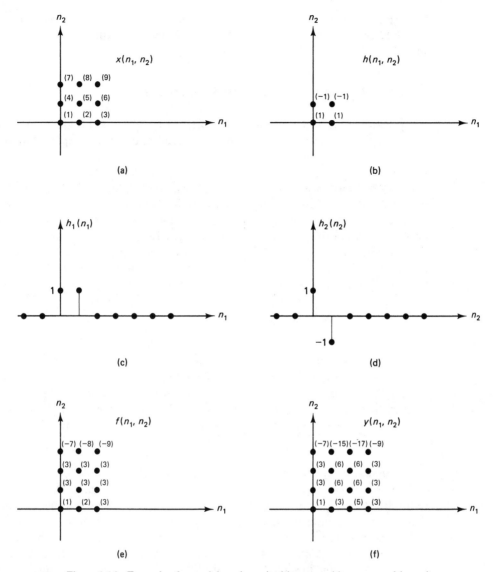

Figure 1.16 Example of convolving $x(n_1, n_2)$ with a separable sequence $h(n_1, n_2)$.

$h_2(n_2)$ are shown in Figures 1.16(c) and (d), respectively. The sequences $f(n_1, n_2)$ and $y(n_1, n_2)$ are shown in Figures 1.16(e) and (f).

In the above discussion, we performed a 1-D convolution first for each column of $x(n_1, n_2)$ with $h_2(n_2)$ and then a 1-D convolution for each row of $f(n_1, n_2)$ with $h_1(n_1)$. By changing the order of the two summations in (1.26) and following the same procedure, it is simple to show that $y(n_1, n_2)$ can be computed by performing a 1-D convolution first for each row of $x(n_1, n_2)$ with $h_1(n_1)$ and then a 1-D convolution for each column of the result with $h_2(n_2)$. In the above discussion, we have assumed that $x(n_1, n_2)$ and $h(n_1, n_2)$ are $N_1 \times N_2$-point and $M_1 \times M_2$-

point sequences respectively with $N_1 = N_2$ and $M_1 = M_2$. We note that the results discussed above can be generalized straightforwardly to the case when $N_1 \neq N_2$ and $M_1 \neq M_2$.

1.2.3 Stable Systems and Special Support Systems

For practical reasons, it is often appropriate to impose additional constraints on the class of systems we consider. Stable systems and special support systems have such constraints.

Stable systems. A system is considered stable in the bounded-input-bounded-output (BIBO) sense if and only if a bounded input always leads to a bounded output. Stability is often a desirable constraint to impose, since an unstable system can generate an unbounded output, which can cause system overload or other difficulties. From this definition and (1.18), it can be shown that a necessary and sufficient condition for an LSI system to be stable is that its impulse response $h(n_1, n_2)$ be absolutely summable:

$$\text{Stability of an LSI system} \iff \sum_{n_1 = -\infty}^{\infty} \sum_{n_2 = -\infty}^{\infty} |h(n_1, n_2)| < \infty. \qquad (1.29)$$

Although (1.29) is a straightforward extension of 1-D results, 2-D systems differ greatly from 1-D systems when a system's stability is tested. This will be discussed further in Section 2.3. Because of (1.29), an absolutely summable sequence is defined to be a *stable sequence*. Using this definition, a necessary and sufficient condition for an LSI system to be stable is that its impulse response be a stable sequence.

Special support systems. A 1-D system is said to be causal if and only if the current output $y(n)$ does not depend on any future values of the input, for example, $x(n + 1), x(n + 2), x(n + 3), \ldots$. Using this definition, we can show that a necessary and sufficient condition for a 1-D LSI system to be causal is that its impulse response $h(n)$ be zero for $n < 0$. Causality is often a desirable constraint to impose in designing 1-D systems. A noncausal system would require delay, which is undesirable in such applications as real time speech processing. In typical 2-D signal processing applications such as image processing, the causality constraint may not be necessary. At any given time, a complete frame of an image may be available for processing, and it may be processed from left to right, from top to bottom, or in any direction one chooses. Although the notion of causality may not be useful in 2-D signal processing, it is useful to extend the notion that a 1-D causal LSI system has an impulse response $h(n)$ whose nonzero values lie in a particular region. A 2-D LSI system whose impulse response $h(n_1, n_2)$ has all its nonzero values in a particular region is called a special support system.

A 2-D LSI system is said to be a *quadrant support system* when its impulse response $h(n_1, n_2)$ is a *quadrant support sequence*. A quadrant support sequence, or a quadrant sequence for short, is one which has all its nonzero values in one quadrant. An example of a first-quadrant support sequence is the unit step sequence $u(n_1, n_2)$.

A 2-D LSI system is said to be a *wedge support system* when its impulse response $h(n_1, n_2)$ is a *wedge support sequence*. Consider two lines emanating from the origin. If all the nonzero values in a sequence lie in the region bounded by these two lines, and the angle between the two lines is less than 180°, the sequence is called a wedge support sequence, or a wedge sequence for short. An example of a wedge support sequence $x(n_1, n_2)$ is shown in Figure 1.17.

Quadrant support sequences and wedge support sequences are closely related. A quadrant support sequence is always a wedge support sequence. In addition, it can be shown that any wedge support sequence can always be mapped to a first-quadrant support sequence by a linear mapping of variables without affecting its stability. To illustrate this, consider the wedge support sequence $x(n_1, n_2)$ shown in Figure 1.17. Suppose we obtain a new sequence $y(n_1, n_2)$ from $x(n_1, n_2)$ by the following linear mapping of variables:

$$y(n_1, n_2) = x(m_1, m_2)|_{m_1 = l_1 n_1 + l_2 n_2, m_2 = l_3 n_1 + l_4 n_2} \qquad (1.30)$$

where the integers l_1, l_2, l_3, and l_4 are chosen to be 1, 0, -1 and 1 respectively. The sequence $y(n_1, n_2)$ obtained by using (1.30) is shown in Figure 1.18, and is clearly a first-quadrant support sequence. In addition, the stability of $x(n_1, n_2)$ is equivalent to the stability of $y(n_1, n_2)$, since

$$\sum_{n_1 = -\infty}^{\infty} \sum_{n_2 = -\infty}^{\infty} |x(n_1, n_2)| = \sum_{n_1 = -\infty}^{\infty} \sum_{n_2 = -\infty}^{\infty} |y(n_1, n_2)|.$$

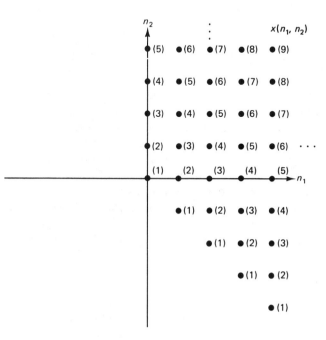

Figure 1.17 Example of a wedge support sequence.

The notion that a wedge support sequence can always be transformed to a first-quadrant support sequence by a simple linear mapping of variables without affecting its stability is very useful in studying the stability of a 2-D system. As we will discuss in Chapter 2, our primary concern in testing the stability of a 2-D system will be limited to a class of systems known as *recursively computable systems*. To test the stability of a recursively computable system, we need to test the stability of a wedge support sequence $h'(n_1, n_2)$. To accomplish this, we will transform $h'(n_1, n_2)$ to a first-quadrant support sequence $h''(n_1, n_2)$ by an appropriate linear mapping of variables and then check the stability of $h''(n_1, n_2)$. This approach exploits the fact that it is much easier to develop stability theorems for first-quadrant support sequences than for wedge support sequences. This will be discussed further in Section 2.3.

1.3 THE FOURIER TRANSFORM

1.3.1 The Fourier Transform Pair

It is a remarkable fact that any stable sequence $x(n_1, n_2)$ can be obtained by appropriately combining complex exponentials of the form $X(\omega_1, \omega_2)e^{j\omega_1 n_1}e^{j\omega_2 n_2}$. The function $X(\omega_1, \omega_2)$, which represents the amplitude associated with the complex exponential $e^{j\omega_1 n_1}e^{j\omega_2 n_2}$, can be obtained from $x(n_1, n_2)$. The relationships between $x(n_1, n_2)$ and $X(\omega_1, \omega_2)$ are given by

---

Discrete-Space Fourier Transform Pair

$$X(\omega_1, \omega_2) = \sum_{n_1=-\infty}^{\infty} \sum_{n_2=-\infty}^{\infty} x(n_1, n_2)e^{-j\omega_1 n_1}e^{-j\omega_2 n_2} \qquad (1.31a)$$

$$x(n_1, n_2) = \frac{1}{(2\pi)^2}\int_{\omega_1=-\pi}^{\pi}\int_{\omega_2=-\pi}^{\pi} X(\omega_1, \omega_2)e^{j\omega_1 n_1}e^{j\omega_2 n_2}\,d\omega_1\,d\omega_2 \qquad (1.31b)$$

Equation (1.31a) shows how the amplitude $X(\omega_1, \omega_2)$ associated with the exponential $e^{j\omega_1 n_1}e^{j\omega_2 n_2}$ can be determined from $x(n_1, n_2)$. The function $X(\omega_1, \omega_2)$ is called the *discrete-space Fourier transform*, or *Fourier transform* for short, of $x(n_1, n_2)$. Equation (1.31b) shows how complex exponentials $X(\omega_1, \omega_2)e^{j\omega_1 n_1}e^{j\omega_2 n_2}$ are specifically combined to form $x(n_1, n_2)$. The sequence $x(n_1, n_2)$ is called the *inverse discrete-space Fourier transform* or *inverse Fourier transform* of $X(\omega_1, \omega_2)$. The consistency of (1.31a) and (1.31b) can be easily shown by combining them.

From (1.31), it can be seen that $X(\omega_1, \omega_2)$ is in general complex, even though $x(n_1, n_2)$ may be real. It is often convenient to express $X(\omega_1, \omega_2)$ in terms of its magnitude $|X(\omega_1, \omega_2)|$ and phase $\theta_x(\omega_1, \omega_2)$ or in terms of its real part $X_R(\omega_1, \omega_2)$ and imaginary part $X_I(\omega_1, \omega_2)$ as

$$X(\omega_1, \omega_2) = |X(\omega_1, \omega_2)|e^{j\theta_x(\omega_1, \omega_2)} = X_R(\omega_1, \omega_2) + jX_I(\omega_1, \omega_2). \qquad (1.32)$$

From (1.31), it can also be seen that $X(\omega_1, \omega_2)$ is a function of continuous variables ω_1 and ω_2, although $x(n_1, n_2)$ is a function of discrete variables n_1 and n_2. In

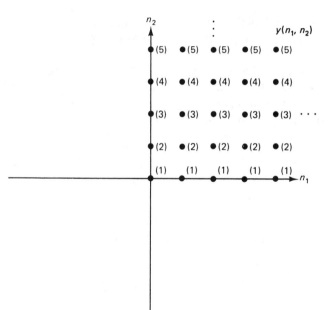

Figure 1.18 First-quadrant support sequence obtained from the wedge support sequence in Figure 1.17 by linear mapping of variables.

addition, $X(\omega_1, \omega_2)$ is always periodic with a period of $2\pi \times 2\pi$; that is, $X(\omega_1, \omega_2) = X(\omega_1 + 2\pi, \omega_2) = X(\omega_1, \omega_2 + 2\pi)$ for all ω_1 and ω_2. We can also show that the Fourier transform converges uniformly for stable sequences. The Fourier transform of $x(n_1, n_2)$ is said to *converge uniformly* when $X(\omega_1, \omega_2)$ is finite and

$$\lim_{N_1 \to \infty} \lim_{N_2 \to \infty} \sum_{n_1 = -N_1}^{N_1} \sum_{n_2 = -N_2}^{N_2} x(n_1, n_2)e^{-j\omega_1 n_1}e^{-j\omega_2 n_2} = X(\omega_1, \omega_2) \quad \text{for all } \omega_1 \text{ and } \omega_2.$$

(1.33)

When the Fourier transform of $x(n_1, n_2)$ converges uniformly, $X(\omega_1, \omega_2)$ is an analytic function and is infinitely differentiable with respect to ω_1 and ω_2.

A sequence $x(n_1, n_2)$ is said to be an eigenfunction of a system T if $T[x(n_1, n_2)] = kx(n_1, n_2)$ for some scalar k. Suppose we use a complex exponential $e^{j\omega_1 n_1}e^{j\omega_2 n_2}$ as an input $x(n_1, n_2)$ to an LSI system with impulse response $h(n_1, n_2)$. The output of the system $y(n_1, n_2)$ can be obtained as

$$y(n_1, n_2) = \sum_{k_1 = -\infty}^{\infty} \sum_{k_2 = -\infty}^{\infty} h(k_1, k_2)x(n_1 - k_1, n_2 - k_2)$$

$$= \sum_{k_1 = -\infty}^{\infty} \sum_{k_2 = -\infty}^{\infty} h(k_1, k_2)e^{j\omega_1(n_1 - k_1)}e^{j\omega_2(n_2 - k_2)}$$

$$= \sum_{k_1 = -\infty}^{\infty} \sum_{k_2 = -\infty}^{\infty} h(k_1, k_2)e^{-j\omega_1 k_1}e^{-j\omega_2 k_2}e^{j\omega_1 n_1}e^{j\omega_2 n_2}$$

$$= H(\omega_1, \omega_2)e^{j\omega_1 n_1}e^{j\omega_2 n_2}.$$

(1.34)

From (1.34), $e^{j\omega_1 n_1} e^{j\omega_2 n_2}$ is an eigenfunction of any LSI system for which $H(\omega_1, \omega_2)$ is well defined and $H(\omega_1, \omega_2)$ is the Fourier transform of $h(n_1, n_2)$. The function $H(\omega_1, \omega_2)$ is called the *frequency response* of the LSI system. The fact that $e^{j\omega_1 n_1} e^{j\omega_2 n_2}$ is an eigenfunction of an LSI system and that $H(\omega_1, \omega_2)$ is the scaling factor by which $e^{j\omega_1 n_1} e^{j\omega_2 n_2}$ is multiplied when it is an input to the LSI system simplifies system analysis for a sinusoidal input. For example, the output of an LSI system with frequency response $H(\omega_1, \omega_2)$ when the input is $\cos(\omega_1' n_1 + \omega_2' n_2)$ can be obtained as follows:

$$T[\cos(\omega_1' n_1 + \omega_2' n_2)] = T\left[\frac{e^{j\omega_1' n_1} e^{j\omega_2' n_2}}{2} + \frac{e^{-j\omega_1' n_1} e^{-j\omega_2' n_2}}{2}\right]$$

$$= \tfrac{1}{2} T[e^{j\omega_1' n_1} e^{j\omega_2' n_2}] + \tfrac{1}{2} T[e^{-j\omega_1' n_1} e^{-j\omega_2' n_2}] \qquad (1.35)$$

$$= \tfrac{1}{2} H(\omega_1', \omega_2') e^{j\omega_1' n_1} e^{j\omega_2' n_2} + \tfrac{1}{2} H(-\omega_1', -\omega_2') e^{-j\omega_1' n_1} e^{-j\omega_2' n_2}.$$

1.3.2 Properties

We can derive a number of useful properties from the Fourier transform pair in (1.31). Some of the more important properties, often useful in practice, are listed in Table 1.1. Most are essentially straightforward extensions of 1-D Fourier transform properties. The only exception is Property 4, which applies to separable sequences. If a 2-D sequence $x(n_1, n_2)$ can be written as $x_1(n_1) x_2(n_2)$, then its Fourier transform, $X(\omega_1, \omega_2)$, is given by $X_1(\omega_1) X_2(\omega_2)$, where $X_1(\omega_1)$ and $X_2(\omega_2)$ represent the 1-D Fourier transforms of $x_1(n_1)$ and $x_2(n_2)$, respectively. This property follows directly from the Fourier transform pair of (1.31). Note that this property is quite different from Property 3, the multiplication property. In the multiplication property, both $x(n_1, n_2)$ and $y(n_1, n_2)$ are 2-D sequences. In Property 4, $x_1(n_1)$ and $x_2(n_2)$ are 1-D sequences, and their product $x_1(n_1) x_2(n_2)$ forms a 2-D sequence.

1.3.3 Examples

Example 1

We wish to determine $H(\omega_1, \omega_2)$ for the sequence $h(n_1, n_2)$ shown in Figure 1.19(a). From (1.31),

$$H(\omega_1, \omega_2) = \sum_{n_1 = -\infty}^{\infty} \sum_{n_2 = -\infty}^{\infty} h(n_1, n_2) e^{-j\omega_1 n_1} e^{-j\omega_2 n_2}$$

$$= \tfrac{1}{3} + \tfrac{1}{6} e^{-j\omega_1} + \tfrac{1}{6} e^{-j\omega_2} + \tfrac{1}{6} e^{j\omega_1} + \tfrac{1}{6} e^{j\omega_2}$$

$$= \tfrac{1}{3} + \tfrac{1}{3} \cos \omega_1 + \tfrac{1}{3} \cos \omega_2.$$

The function $H(\omega_1, \omega_2)$ for this example is real and its magnitude is sketched in Figure 1.19(b). If $H(\omega_1, \omega_2)$ in Figure 1.19(b) is the frequency response of an LSI system, the system corresponds to a lowpass filter. The function $|H(\omega_1, \omega_2)|$ shows smaller values in frequency regions away from the origin. A lowpass filter applied to an

TABLE 1.1 PROPERTIES OF THE FOURIER TRANSFORM

$$x(n_1, n_2) \longleftrightarrow X(\omega_1, \omega_2)$$
$$y(n_1, n_2) \longleftrightarrow Y(\omega_1, \omega_2)$$

Property 1. *Linearity*
$$ax(n_1, n_2) + by(n_1, n_2) \longleftrightarrow aX(\omega_1, \omega_2) + bY(\omega_1, \omega_2)$$

Property 2. *Convolution*
$$x(n_1, n_2) * y(n_1, n_2) \longleftrightarrow X(\omega_1, \omega_2)Y(\omega_1, \omega_2)$$

Property 3. *Multiplication*
$$x(n_1, n_2)y(n_1, n_2) \longleftrightarrow X(\omega_1, \omega_2) \circledast Y(\omega_1, \omega_2)$$
$$= \frac{1}{(2\pi)^2} \int_{\theta_1 = -\pi}^{\pi} \int_{\theta_2 = -\pi}^{\pi} X(\theta_1, \theta_2)Y(\omega_1 - \theta_1, \omega_2 - \theta_2) \, d\theta_1 \, d\theta_2$$

Property 4. *Separable Sequence*
$$x(n_1, n_2) = x_1(n_1)x_2(n_2) \longleftrightarrow X(\omega_1, \omega_2) = X_1(\omega_1)X_2(\omega_2)$$

Property 5. *Shift of a Sequence and a Fourier Transform*
(a) $x(n_1 - m_1, n_2 - m_2) \longleftrightarrow X(\omega_1, \omega_2)e^{-j\omega_1 m_1}e^{-j\omega_2 m_2}$
(b) $e^{j v_1 n_1}e^{j v_2 n_2}x(n_1, n_2) \longleftrightarrow X(\omega_1 - v_1, \omega_2 - v_2)$

Property 6. *Differentiation*
(a) $-jn_1 x(n_1, n_2) \longleftrightarrow \dfrac{\partial X(\omega_1, \omega_2)}{\partial \omega_1}$

(b) $-jn_2 x(n_1, n_2) \longleftrightarrow \dfrac{\partial X(\omega_1, \omega_2)}{\partial \omega_2}$

Property 7. *Initial Value and DC Value Theorem*
(a) $x(0, 0) = \dfrac{1}{(2\pi)^2} \displaystyle\int_{\omega_1 = -\pi}^{\pi} \int_{\omega_2 = -\pi}^{\pi} X(\omega_1, \omega_2) \, d\omega_1 \, d\omega_2$

(b) $X(0, 0) = \displaystyle\sum_{n_1 = -\infty}^{\infty} \sum_{n_2 = -\infty}^{\infty} x(n_1, n_2)$

Property 8. *Parseval's Theorem*
(a) $\displaystyle\sum_{n_1 = -\infty}^{\infty} \sum_{n_2 = -\infty}^{\infty} x(n_1, n_2)y^*(n_1, n_2)$
$$= \frac{1}{(2\pi)^2} \int_{\omega_1 = -\pi}^{\pi} \int_{\omega_2 = -\pi}^{\pi} X(\omega_1, \omega_2)Y^*(\omega_1, \omega_2) \, d\omega_1 \, d\omega_2$$

(b) $\displaystyle\sum_{n_1 = -\infty}^{\infty} \sum_{n_2 = -\infty}^{\infty} |x(n_1, n_2)|^2 = \frac{1}{(2\pi)^2} \int_{\omega_1 = -\pi}^{\pi} \int_{\omega_2 = -\pi}^{\pi} |X(\omega_1, \omega_2)|^2 \, d\omega_1 \, d\omega_2$

Property 9. *Symmetry Properties*
(a) $x(-n_1, n_2) \longleftrightarrow X(-\omega_1, \omega_2)$
(b) $x(n_1, -n_2) \longleftrightarrow X(\omega_1, -\omega_2)$
(c) $x(-n_1, -n_2) \longleftrightarrow X(-\omega_1, -\omega_2)$
(d) $x^*(n_1, n_2) \longleftrightarrow X^*(-\omega_1, -\omega_2)$
(e) $x(n_1, n_2)$: real $\longleftrightarrow X(\omega_1, \omega_2) = X^*(-\omega_1, -\omega_2)$
$X_R(\omega_1, \omega_2)$, $|X(\omega_1, \omega_2)|$: even (symmetric with respect to the origin)
$X_I(\omega_1, \omega_2)$, $\theta_x(\omega_1, \omega_2)$: odd (antisymmetric with respect to the origin)
(f) $x(n_1, n_2)$: real and even $\longleftrightarrow X(\omega_1, \omega_2)$: real and even
(g) $x(n_1, n_2)$: real and odd $\longleftrightarrow X(\omega_1, \omega_2)$: pure imaginary and odd

Property 10. *Uniform Convergence*
For a stable $x(n_1, n_2)$, the Fourier transform of $x(n_1, n_2)$ uniformly converges.

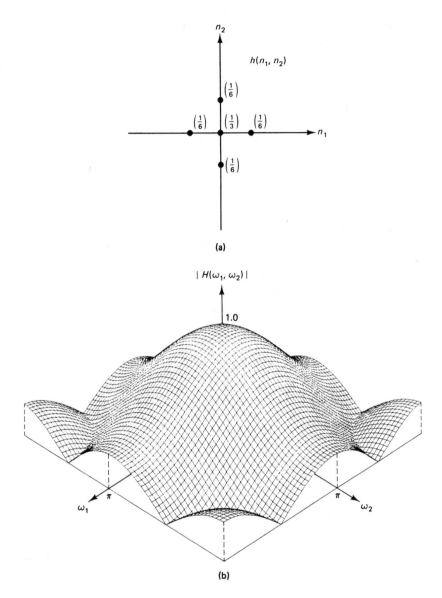

Figure 1.19 (a) 2-D sequence $h(n_1, n_2)$; (b) Fourier transform magnitude $|H(\omega_1, \omega_2)|$ of $h(n_1, n_2)$ in (a).

image blurs the image. The function $H(\omega_1, \omega_2)$ is 1 at $\omega_1 = \omega_2 = 0$, and therefore the average intensity of an image is not affected by the filter. A bright image will remain bright and a dark image will remain dark after processing with the filter. Figure 1.20(a) shows an image of 256×256 pixels. Figure 1.20(b) shows the image obtained by processing the image in Figure 1.20(a) with a lowpass filter whose impulse response is given by $h(n_1, n_2)$ in this example.

Signals, Systems, and the Fourier Transform   Chap. 1

(a)   (b)

Figure 1.20 (a) Image of 256 × 256 pixels; (b) image processed by filtering the image in (a) with a lowpass filter whose impulse response is given by $h(n_1, n_2)$ in Figure 1.19 (a).

Example 2

We wish to determine $H(\omega_1, \omega_2)$ for the sequence $h(n_1, n_2)$ shown in Figure 1.21(a). We can use (1.31) to determine $H(\omega_1, \omega_2)$, as in Example 1. Alternatively, we can use Property 4 in Table 1.1. The sequence $h(n_1, n_2)$ can be expressed as $h_1(n_1)h_2(n_2)$, where one possible choice of $h_1(n_1)$ and $h_2(n_2)$ is shown in Figure 1.21(b). Computing the 1-D Fourier transforms $H_1(\omega_1)$ and $H_2(\omega_2)$ and using Property 4 in Table 1.1, we have

$$H(\omega_1, \omega_2) = H_1(\omega_1)H_2(\omega_2) = (3 - 2\cos\omega_1)(3 - 2\cos\omega_2).$$

The function $H(\omega_1, \omega_2)$ is again real, and its magnitude is sketched in Figure 1.21(c). A system whose frequency response is given by the $H(\omega_1, \omega_2)$ above is a highpass filter. The function $|H(\omega_1, \omega_2)|$ has smaller values in frequency regions near the origin. A highpass filter applied to an image tends to accentuate image details or local contrast, and the processed image appears sharper. Figure 1.22(a) shows an original image of 256 × 256 pixels and Figure 1.22(b) shows the highpass filtered image using $h(n_1, n_2)$ in this example. When an image is processed, for instance by highpass filtering, the pixel intensities may no longer be integers between 0 and 255. They may be negative, noninteger, or above 255. In such instances, we typically add a bias and then scale and quantize the processed image so that all the pixel intensities are integers between 0 and 255. It is common practice to choose the bias and scaling factors such that the minimum intensity is mapped to 0 and the maximum intensity is mapped to 255.

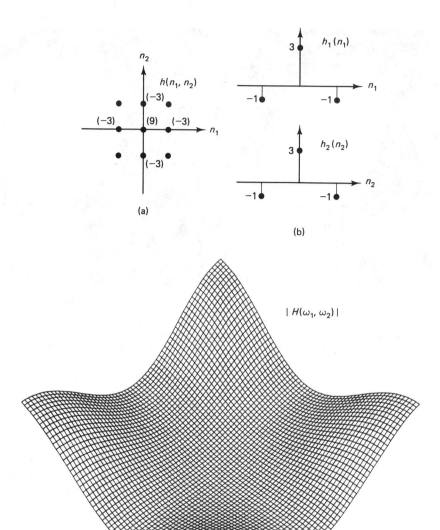

Figure 1.21 (a) 2-D sequence $h(n_1, n_2)$; (b) possible choice of $h_1(n_1)$ and $h_2(n_2)$ where $h(n_1, n_2) = h_1(n_1)h_2(n_2)$; (c) Fourier transform magnitude $|H(\omega_1, \omega_2)|$ of $h(n_1, n_2)$ in (a).

Example 3

We wish to determine $h(n_1, n_2)$ for the Fourier transform $H(\omega_1, \omega_2)$ shown in Figure 1.23. The function $H(\omega_1, \omega_2)$ is given by

$$H(\omega_1, \omega_2) = \begin{cases} 1, & |\omega_1| \le a \quad \text{and} \quad |\omega_2| \le b \text{ (shaded region)} \\ 0, & a < |\omega_1| \le \pi \quad \text{or} \quad b < |\omega_2| \le \pi \text{ (unshaded region).} \end{cases}$$

Signals, Systems, and the Fourier Transform   Chap. 1

(a)

(b)

Figure 1.22 (a) Image of 256 × 256 pixels; (b) image obtained from filtering the image in (a) with a highpass filter whose impulse response is given by $h(n_1, n_2)$ in Figure 1.21(a).

Since $H(\omega_1, \omega_2)$ is always periodic with a period of 2π along each of the two variables ω_1 and ω_2, $H(\omega_1, \omega_2)$ is shown only for $|\omega_1| \leq \pi$ and $|\omega_2| \leq \pi$. The function $H(\omega_1, \omega_2)$ can be expressed as $H_1(\omega_1)H_2(\omega_2)$, where one possible choice of $H_1(\omega_1)$ and $H_2(\omega_2)$ is also shown in Figure 1.23. When $H(\omega_1, \omega_2)$ above is the frequency response of a 2-D LSI system, the system is called a *separable ideal lowpass filter*. Computing the 1-D inverse Fourier transforms of $H_1(\omega_1)$ and $H_2(\omega_2)$ and using Property 4 in Table 1.1, we obtain

$$h(n_1, n_2) = h_1(n_1)h_2(n_2) = \frac{\sin an_1}{\pi n_1} \frac{\sin bn_2}{\pi n_2}$$

Example 4

We wish to determine $h(n_1, n_2)$ for the Fourier transform $H(\omega_1, \omega_2)$ shown in Figure 1.24. The function $H(\omega_1, \omega_2)$ is given by

$$H(\omega_1, \omega_2) = \begin{cases} 1, & \sqrt{\omega_1^2 + \omega_2^2} \leq \omega_C \quad \text{(shaded region)} \\ 0, & \omega_c < \sqrt{\omega_1^2 + \omega_2^2} \quad \text{and} \quad |\omega_1|, |\omega_2| \leq \pi \text{ (unshaded region).} \end{cases}$$

When $H(\omega_1, \omega_2)$ above is the frequency response of a 2-D LSI system, the system is called a *circularly symmetric ideal lowpass filter*, or an *ideal lowpass filter* for short. The inverse Fourier transform of $H(\omega_1, \omega_2)$ in this example requires a fair amount of algebra (see Problem 1.24). The result is

$$h(n_1, n_2) = \frac{\omega_C}{2\pi\sqrt{n_1^2 + n_2^2}} J_1(\omega_C \sqrt{n_1^2 + n_2^2}) \tag{1.36}$$

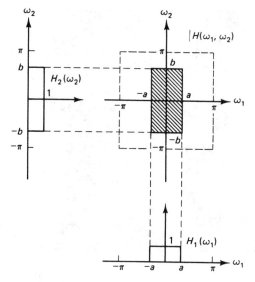

Figure 1.23 Separable Fourier transform $H(\omega_1, \omega_2)$ and one possible choice of $H_1(\omega_1)$ and $H_2(\omega_2)$ such that $H(\omega_1, \omega_2)$ $= H_1(\omega_1)H_2(\omega_2)$. The function $H(\omega_1, \omega_2)$ is 1 in the shaded region and 0 in the unshaded region.

where $J_1(\cdot)$ represents the Bessel function of the first kind and the first order and can be expanded in series form as

$$J_1(x) = \frac{x}{2} - \frac{x^3}{2^3 1!\, 2!} + \frac{x^5}{2^5 2!\, 3!} - \frac{x^7}{2^7 3!\, 4!} + \frac{x^9}{2^9 4!\, 5!} - \cdots. \qquad (1.37)$$

This example shows that 2-D Fourier transform or inverse Fourier transform operations can become much more algebraically complex than 1-D Fourier transform or inverse Fourier transform operations, despite the fact that the 2-D Fourier transform pair and many 2-D Fourier transform properties are straightforward extensions of 1-D results. From (1.36), we observe that the impulse response of a 2-D circularly symmetric ideal lowpass filter is also circularly symmetric, that is, it is a function of $n_1^2 + n_2^2$. This is a special case of a more general result. Specifically, if $H(\omega_1, \omega_2)$ is a function of $\omega_1^2 + \omega_2^2$ in the region $\sqrt{\omega_1^2 + \omega_2^2} \le \pi$ and is a constant outside the region, then the corresponding $h(n_1, n_2)$ is a function of $n_1^2 + n_2^2$. Note, however, that circular symmetry of $h(n_1, n_2)$ does not imply circular symmetry of $H(\omega_1, \omega_2)$. The function $J_1(x)/x$ is sketched in Figure 1.25. The sequence $h(n_1, n_2)$ in (1.36) is sketched in Figure 1.26 for the case $\omega_C = 0.4\pi$.

The impulse responses $h(n_1, n_2)$ obtained from the separable and circularly symmetric ideal lowpass filters in Examples 3 and 4 above are not absolutely sum-

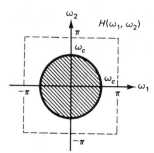

Figure 1.24 Frequency response of a circularly symmetric ideal lowpass filter.

Signals, Systems, and the Fourier Transform   Chap. 1

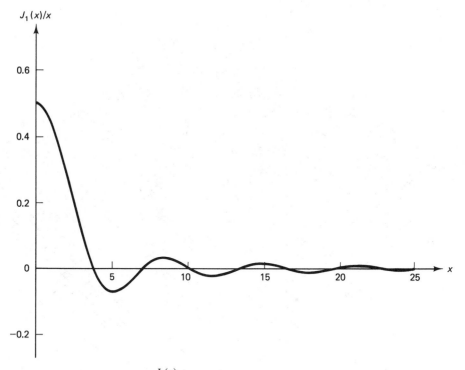

Figure 1.25 Sketch of $\dfrac{J_1(x)}{x}$, where $J_1(x)$ is the Bessel function of the first kind and first order.

mable, and their Fourier transforms do not converge uniformly to $H(\omega_1, \omega_2)$ used to obtain $h(n_1, n_2)$. This is evident from the observation that the two $H(\omega_1, \omega_2)$ contain discontinuities and are not analytic functions. Nevertheless, we will regard them as valid Fourier transform pairs, since they play an important role in digital filtering and the Fourier transforms of the two $h(n_1, n_2)$ converge to $H(\omega_1, \omega_2)$ in the mean square sense.*

1.4 ADDITIONAL PROPERTIES OF THE FOURIER TRANSFORM

1.4.1 Signal Synthesis and Reconstruction from Phase or Magnitude

The Fourier transform of a sequence is in general complex-valued, and the unique representation of a sequence in the Fourier transform domain requires both the

---

*The Fourier transform of $h(n_1, n_2)$ is said to converge to $H(\omega_1, \omega_2)$ in the mean square sense when

$$\lim_{N_1 \to \infty} \lim_{N_2 \to \infty} \int_{\omega_1 = -\pi}^{\pi} \int_{\omega_2 = -\pi}^{\pi} \left| \sum_{n_1 = -N_1}^{N_1} \sum_{n_2 = -N_2}^{N_2} h(n_1, n_2) e^{-j\omega_1 n_1} e^{-j\omega_2 n_2} - H(\omega_1, \omega_2) \right|^2 d\omega_1 \, d\omega_2 = 0.$$

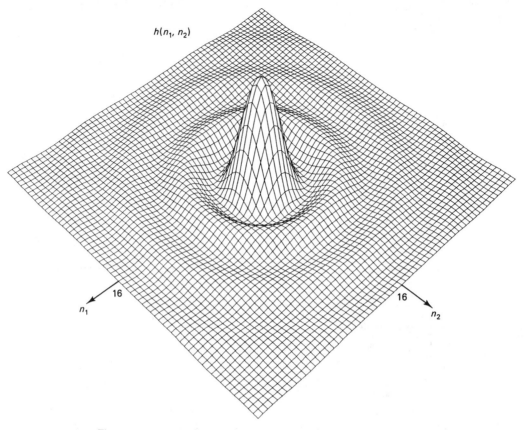

$h(n_1, n_2)$

16 16

n_1 n_2

Figure 1.26 Impulse response of a circularly symmetric ideal lowpass filter with $\omega_c = 0.4\pi$ in Equation (1.36). The value at the origin, $h(0, 0)$, is 0.126.

phase and magnitude of the Fourier transform. In various contexts, however, it is often desirable to synthesize or reconstruct a signal from only partial Fourier domain information [Saxton; Ramachandran and Srinivasan]. In this section, we discuss the problem of signal synthesis and reconstruction from the Fourier transform phase alone or from the Fourier transform magnitude alone.

Consider a 2-D sequence $x(n_1, n_2)$ with Fourier transform $X(\omega_1, \omega_2)$ so that

$$X(\omega_1, \omega_2) = F[x(n_1, n_2)] = |X(\omega_1, \omega_2)|e^{j\theta_x(\omega_1, \omega_2)}. \tag{1.38}$$

It has been observed that a straightforward signal synthesis from the Fourier transform phase $\theta_x(\omega_1, \omega_2)$ alone often captures most of the intelligibility of the original signal $x(n_1, n_2)$. A straightforward synthesis from the Fourier transform magnitude $|X(\omega_1, \omega_2)|$ alone, however, does not generally capture the original signal's intelligibility. To illustrate this, we synthesize the phase-only signal $x_p(n_1, n_2)$ and the magnitude-only signal $x_m(n_1, n_2)$ by

$$x_p(n_1, n_2) = F^{-1}[1e^{j\theta_x(\omega_1, \omega_2)}] \tag{1.39}$$

$$x_m(n_1, n_2) = F^{-1}[|X(\omega_1, \omega_2)|e^{j0}] \tag{1.40}$$

where $F^{-1}[\cdot]$ represents the inverse Fourier transform operation. In phase-only signal synthesis, the correct phase is combined with an arbitrary constant magnitude. In the magnitude-only signal synthesis, the correct magnitude is combined with an arbitrary constant phase. In this synthesis, $x_p(n_1, n_2)$ often preserves the intelligibility of $x(n_1, n_2)$, while $x_m(n_1, n_2)$ does not. An example of this is shown in Figure 1.27. Figure 1.27(a) shows an original image $x(n_1, n_2)$, and Figures 1.27(b) and (c) show $x_p(n_1, n_2)$ and $x_m(n_1, n_2)$, respectively.

(a)

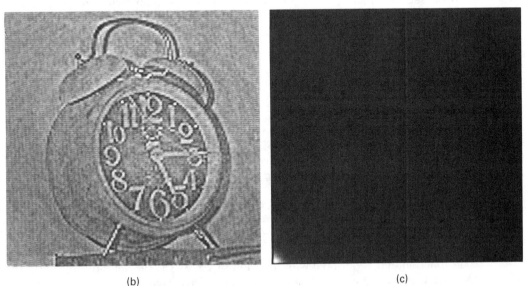

(b)   (c)

Figure 1.27 Example of phase-only and magnitude-only synthesis. (a) Original image of 128 × 128 pixels; (b) result of phase-only synthesis; (c) result of magnitude-only synthesis.

Sec. 1.4   Additional Properties of the Fourier Transform   **33**

An experiment which more dramatically illustrates the observation that phase-only signal synthesis captures more of the signal intelligibility than magnitude-only synthesis can be performed as follows. Consider two images $x(n_1, n_2)$ and $y(n_1, n_2)$. From these two images, we synthesize two other images $f(n_1, n_2)$ and $g(n_1, n_2)$ by

$$f(n_1, n_2) = F^{-1}[|Y(\omega_1, \omega_2)|e^{j\theta_x(\omega_1, \omega_2)}] \tag{1.41}$$

$$g(n_1, n_2) = F^{-1}[|X(\omega_1, \omega_2)|e^{j\theta_y(\omega_1, \omega_2)}]. \tag{1.42}$$

In this experiment, $f(n_1, n_2)$ captures the intelligibility of $x(n_1, n_2)$, while $g(n_1, n_2)$ captures the intelligibility of $y(n_1, n_2)$. An example is shown in Figure 1.28. Figures 1.28(a) and (b) show the two images $x(n_1, n_2)$ and $y(n_1, n_2)$ and Figures 1.28(c) and (d) show the two images $f(n_1, n_2)$ and $g(n_1, n_2)$.

The high intelligibility of phase-only synthesis raises the possibility of exactly reconstructing a signal $x(n_1, n_2)$ from its Fourier transform phase $\theta_x(\omega_1, \omega_2)$. This is known as the *magnitude-retrieval* problem. In fact, it has been shown [Hayes] that a sequence $x(n_1, n_2)$ is uniquely specified within a scale factor if $x(n_1, n_2)$ is real and has finite extent, and if its Fourier transform cannot be factored as a product of lower-order polynomials in $e^{j\omega_1}$ and $e^{j\omega_2}$. Typical images $x(n_1, n_2)$ are real and have finite regions of support. In addition, the fundamental theorem of algebra does not apply to 2-D polynomials, and their Fourier transforms cannot generally be factored as products of lower-order polynomials in $e^{j\omega_1}$ and $e^{j\omega_2}$. Typical images, then, are uniquely specified within a scale factor by the Fourier transform phase alone.

Two approaches to reconstructing a sequence from its Fourier transform phase alone have been considered. The first approach leads to a closed-form solution and the second to an iterative procedure. In the first approach, $\tan \theta_x(\omega_1, \omega_2)$ is expressed as

$$\tan \theta_x(\omega_1, \omega_2) = \frac{X_I(\omega_1, \omega_2)}{X_R(\omega_1, \omega_2)} = -\frac{\displaystyle\sum_{(n_1,n_2) \in R_x} \sum x(n_1, n_2) \sin(\omega_1 n_1 + \omega_2 n_2)}{\displaystyle\sum_{(n_1,n_2) \in R_x} \sum x(n_1, n_2) \cos(\omega_1 n_1 + \omega_2 n_2)} \tag{1.43}$$

where R_x is the region of support of $x(n_1, n_2)$. Rewriting (1.43), we have

$$\sum_{(n_1,n_2) \in R_x} \sum x(n_1, n_2) \cos(\omega_1 n_1 + \omega_2 n_2) \tan \theta_x(\omega_1, \omega_2)$$

$$= -\sum_{(n_1,n_2) \in R_x} \sum x(n_1, n_2) \sin(\omega_1 n_1 + \omega_2 n_2). \tag{1.44}$$

Equation (1.44) is a linear equation for the unknown values in $x(n_1, n_2)$ for each frequency (ω_1, ω_2). If there are N^2 unknown values in $x(n_1, n_2)$, we can obtain a set of N^2 linear equations for $x(n_1, n_2)$ by sampling (ω_1, ω_2) at N^2 points. If the frequencies are sampled at distinctly different points, noting that $\theta_x(\omega_1, \omega_2)$ is an odd function and is periodic with a period of $2\pi \times 2\pi$, the solution to the set of N^2 linear equations can be shown to be $kx(n_1, n_2)$, where k is an arbitrary real scaling factor. An example of signal reconstruction from phase using (1.44) is shown in Figure 1.29. Figure 1.29(a) shows an image of 12×12 pixels, and Figure

Figure 1.28 Example of image synthesis from the Fourier transform phase of one image and the Fourier transform magnitude of another image. (a) Original image $x(n_1, n_2)$ of 128 × 128 pixels; (b) original image $y(n_1, n_2)$ of 128 × 128 pixels; (c) result of synthesis from $\theta_x(\omega_1, \omega_2)$ and $|Y(\omega_1, \omega_2)|$; (d) result of synthesis from $\theta_y(\omega_1, \omega_2)$ and $|X(\omega_1, \omega_2)|$.

1.29(b) shows the reconstruction. The scaling factor of the reconstructed sequence in the figure is chosen such that the reconstruction will match the original sequence.

The reconstruction algorithm discussed above is reasonable for a small size image, but is not practical for an image of typical size. For example, reconstructing an image of 512 × 512 pixels using (1.44) requires the solution of approximately

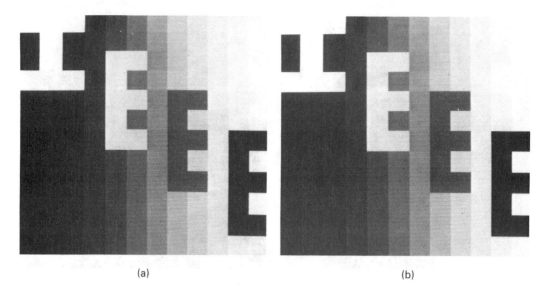

(a)   (b)

Figure 1.29 Example of phase-only reconstruction by a closed-form algorithm. (a) Original image of 12 × 12 pixels. One pixel in the image is a large square block; (b) phase-only reconstruction of the image in (a) by solving a set of linear equations in (1.44).

a quarter of a million linear equations. An alternate approach is to recognize that the solution to the phase-only reconstruction problem must satisfy constraints in both the spatial and frequency domains. Specifically, the solution must be real, must be zero outside the known region of support, and must have a nonfactorable Fourier transform. In addition, the phase of the Fourier transform of the solution must be the same as the $\theta_x(\omega_1, \omega_2)$ given. A useful approach to solving such a problem is an iterative procedure, in which we impose the spatial and frequency domain constraints separately in each domain. An iterative procedure for the phase-only reconstruction is shown in Figure 1.30. In the procedure, we begin with an initial estimate of the signal. This can be any real sequence with the same region of support as $x(n_1, n_2)$. We next compute its Fourier transform. We then replace the Fourier transform with the given $\theta_x(\omega_1, \omega_2)$. The Fourier transform magnitude is not affected. We then compute the inverse Fourier transform of the modified Fourier transform. Due to the modification in the Fourier transform domain, the sequence is no longer zero outside the known region of support of $x(n_1, n_2)$. We now impose the spatial domain constraint by setting the sequence to zero outside the known region of support. The resulting sequence is a new estimate of the solution. This completes one iteration in the iterative procedure. When the initial estimate of the sequence chosen is real, the constraint that the solution is real will automatically be satisfied. The above algorithm can be shown to converge to the desired solution [Tom, et al.]. An example of signal reconstruction from phase using the iterative procedure in Figure 1.30 is shown in Figure 1.31. Figure 1.31(a) shows an original image of 128 × 128 pixels. Figures 1.31(b), (c), and (d) show the results of the iterative procedure after one iteration [phase-

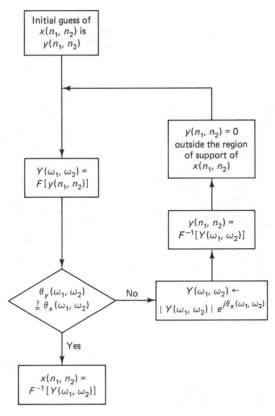

Figure 1.30 Iterative procedure for phase-only reconstruction of $x(n_1, n_2)$ from its phase $\theta_x(\omega_1, \omega_2)$.

only synthesis of (1.39)], 10 iterations, and 50 iterations. The initial estimate used is $\delta(n_1, n_2)$.

Although the magnitude-only synthesis of (1.40) does not capture the intelligibility of typical signals, almost all typical images are also uniquely specified by the Fourier transform magnitude. Specifically, if $x(n_1, n_2)$ is real, has finite extent, and has a nonfactorable Fourier transform, then $x(n_1, n_2)$ is uniquely specified by its Fourier transform magnitude $|X(\omega_1, \omega_2)|$ within a sign factor, translation, and rotation by 180 degrees [Bruck and Sodin, Hayes]. This raises the possibility of exactly reconstructing $x(n_1, n_2)$ from $|X(\omega_1, \omega_2)|$ within a sign factor, translation and rotation by 180 degrees. This is known in the literature as the *phase-retrieval* problem, and has many more potential applications than the phase-only reconstruction problem. Unfortunately, none of the algorithms developed to date are as straightforward or well-behaved as the algorithms developed for the phase-only reconstruction problem. It is possible to derive a closed-form algorithm or a set of linear equations that can be used in solving for $x(n_1, n_2)$ from $|X(\omega_1, \omega_2)|$, but their derivation is quite involved [Izraelevitz and Lim, Lane, et al.]. In addition, the closed-form solution is not practical for an image of reasonable size due to the large number of linear equations that must be solved. It is also possible to derive

(a)

(b)

(c)

(d)

Figure 1.31 Example of phase-only reconstruction by an iterative algorithm. (a) Original image of 128 × 128 pixels; (b) result of phase-only reconstruction of the image in (a) after one iteration of the iterative procedure in Figure 1.30. Since the initial estimate used is $\delta(n_1, n_2)$, this is the same as the phase-only synthesis of (1.39); (c) result after 10 iterations; (d) result after 50 iterations.

an iterative procedure similar to that in Figure 1.30, which was developed for the phase-only reconstruction. The only modification required is to replace the Fourier transform magnitude with the given $|X(\omega_1, \omega_2)|$ rather than to replace the Fourier transform phase with the given $\theta_x(\omega_1, \omega_2)$ when the frequency domain constraints are imposed. The algorithm has been observed to converge to the desired solution when the initial estimate used is quite accurate or the signal $x(n_1, n_2)$ has a special characteristic such as a triangular region of support. The magnitude-only reconstruction problem specifies $x(n_1, n_2)$ within a sign factor, translation, and rotation by 180°, and, therefore, more than one solution is possible. Imposing an initial estimate sufficiently close to a possible solution or imposing additional constraints such as a triangular region of support appear to prevent the iterative procedure from wandering around from one possible solution to another. In general, however, the algorithm does not converge to the desired solution. Figure 1.32 shows an example of signal reconstruction from the magnitude using a closed-form algorithm [Izraelevitz and Lim]. Figures 1.32(a) and (b) show the original and the reconstruction respectively. Developing a practical procedure that can be used to reconstruct $x(n_1, n_2)$ from $|X(\omega_1, \omega_2)|$ remains a problem for further research.

In addition to the phase-only and magnitude-only signal synthesis and reconstruction problems discussed above, a variety of results on the synthesis and reconstruction of a signal from other partial Fourier transform information—for instance, one bit of Fourier transform phase or signed Fourier transform magnitude—have been reported [Oppenheim, et al. (1983)].

1.4.2 The Fourier Transform of Typical Images

The Fourier transforms of typical images have been observed to have most of their energy concentrated in a small region in the frequency domain, near the origin

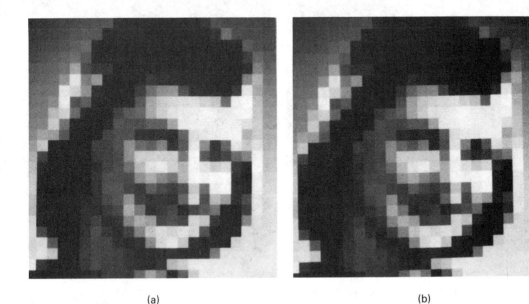

(a)   (b)

Figure 1.32 Example of magnitude-only reconstruction by a closed-form algorithm. (a) Original image of 24 × 24 pixels; (b) result of magnitude-only reconstruction of the image in (a) using a closed-form algorithm. After [Izraelevitz and Lim.]

and along the ω_1 and ω_2 axes. One reason for the energy concentration near the origin is that images typically have large regions where the intensities change slowly. Furthermore, sharp discontinuities such as edges contribute to low-frequency as well as high-frequency components. The energy concentration along the ω_1 and ω_2 axes is in part due to a rectangular window used to obtain a finite-extent image. The rectangular window creates artificial sharp discontinuities at the four boundaries. Discontinuities at the top and bottom of the image contribute energy along the ω_2 axis and discontinuities at the two sides contribute energy along the ω_1 axis. Figure 1.33 illustrates this property. Figure 1.33(a) shows an original image of 512×512 pixels, and Figure 1.33(b) shows $|X(\omega_1, \omega_2)|^{1/4}$ of the image in Figure 1.33(a). The operation $(\cdot)^{1/4}$ has the effect of compressing large amplitudes while expanding small amplitudes, and therefore shows $|X(\omega_1, \omega_2)|$ more clearly for higher-frequency regions. In this particular example, energy concentration along approximately diagonal directions is also visible. This is because of the many sharp discontinuities in the image along approximately diagonal directions. This example shows that most of the energy is concentrated in a small region in the frequency plane.

Since most of the signal energy is concentrated in a small frequency region, an image can be reconstructed without significant loss of quality and intelligibility from a small fraction of the transform coefficients. Figure 1.34 shows images that were obtained by inverse Fourier transforming the Fourier transform of the image in Figure 1.33(a) after setting most of the Fourier transform coefficients to zero. The percentages of the Fourier transform coefficients that have been preserved in

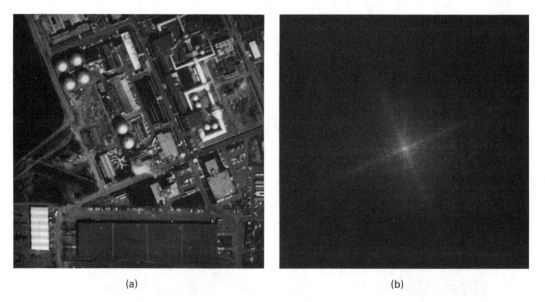

(a)   (b)

Figure 1.33 Example of the Fourier transform magnitude of an image. (a) Original image $x(n_1, n_2)$ of 512×512 pixels; (b) $|X(\omega_1, \omega_2)|^{1/4}$, scaled such that the smallest value maps to the darkest level and the largest value maps to the brightest level. The operation $(\cdot)^{1/4}$ has the effect of compressing large amplitudes while expanding small amplitudes, and therefore shows $|X(\omega_1, \omega_2)|$ more clearly for higher-frequency regions.

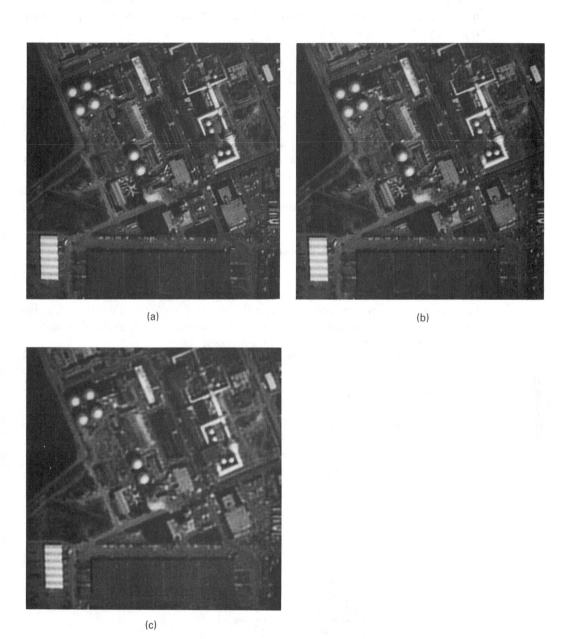

(a)

(b)

(c)

Figure 1.34 Illustration of energy concentration in the Fourier transform domain for a typical image. (a) Image obtained by preserving 12.4% of Fourier transform coefficients of the image in Figure 1.33(a). All other coefficients are set to 0. (b) Same as (a) with 10% of Fourier transform coefficients preserved; (c) same as (a) with 4.8% of Fourier transform coefficients preserved.

Figures 1.34(a), (b), and (c) are 12.4%, 10%, and 4.8%, respectively. The frequency region that was preserved in each of the three cases has the shape (shaded region) shown in Figure 1.35.

The notion that an image with good quality and intelligibility can be reconstructed from a small fraction of transform coefficients for some transforms, for instance the Fourier transform, is the basis of a class of image coding systems known collectively as *transform coding techniques*. One objective of image coding is to represent an image with as few bits as possible while preserving a certain level of image quality and intelligibility. Reduction of transmission channel or storage requirements is a typical application of image coding. In transform coding, the transform coefficients of an image rather than its intensities are coded. Since only a small fraction of the transform coefficients need to be coded in typical applications, the bit rate required in transform coding is often significantly lower than image coding techniques that attempt to code image intensities. The topic of image coding is discussed in Chapter 10.

1.4.3 The Projection-Slice Theorem

Another property of the Fourier transform is the projection-slice theorem, which is the mathematical basis of computed tomography (CT). Computed tomography has a number of applications, including the medical application of reconstructing cross sections of a human body from x-ray images. The impact of computed tomography on medicine requires no elaboration.

Consider a 2-D analog function $f_c(t_1, t_2)$ where t_1 and t_2 are continuous variables. The subscript c denotes that the signal is a function of a continuous variable or variables. The analog Fourier transform $F_c(\Omega_1, \Omega_2)$ is related to $f_c(t_1, t_2)$ by

$$F_c(\Omega_1, \Omega_2) = \int_{t_1 = -\infty}^{\infty} \int_{t_2 = -\infty}^{\infty} f_c(t_1, t_2) e^{-j\Omega_1 t_1} e^{-j\Omega_2 t_2} \, dt_1 \, dt_2 \tag{1.45a}$$

$$f_c(t_1, t_2) = \frac{1}{(2\pi)^2} \int_{\Omega_1 = -\infty}^{\infty} \int_{\Omega_2 = -\infty}^{\infty} F_c(\Omega_1, \Omega_2) e^{j\Omega_1 t_1} e^{j\Omega_2 t_2} \, d\Omega_1 \, d\Omega_2. \tag{1.45b}$$

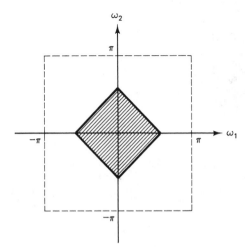

Figure 1.35 Shape of the frequency region where Fourier transform coefficients are preserved in obtaining the images in Figure 1.34.

Signals, Systems, and the Fourier Transform   Chap. 1

Let us integrate $f_c(t_1, t_2)$ along the parallel rays shown in Figure 1.36. The angle that the rays make with the t_2-axis is denoted by θ. The result of the integration at a given θ is a 1-D function, and we denote it by $p_\theta(t)$. In this figure, $p_\theta(0)$ is the result of integrating $f_c(t_1, t_2)$ along the ray passing through the origin. The function $p_\theta(t)$, which is called the *projection* of $f_c(t_1, t_2)$ at angle θ or *Radon transform* of $f_c(t_1, t_2)$, can be expressed in terms of $f_c(t_1, t_2)$ by

$$p_\theta(t) = \int_{u=-\infty}^{\infty} f_c(t_1, t_2) \big|_{t_1 = t\cos\theta - u\sin\theta, t_2 = t\sin\theta + u\cos\theta} \, du. \tag{1.46}$$

Equation (1.46) arises naturally from the analysis of an x-ray image. Consider a 2-D object (a slice of a 3-D object, for example) through which we radiate a monoenergetic x-ray beam, as shown in Figure 1.36. On the basis of the Lambert-Beer law, which describes the attenuation of the x-ray beam as it passes through an object, and of a model of a typical film used to record the output x-ray beam, the image recorded on film can be modeled by $p_\theta(t)$ in (1.46), where $f_c(t_1, t_2)$ is the attenuation coefficient of the 2-D object as a function of two spatial variables t_1 and t_2. The function $f_c(t_1, t_2)$ depends on the material that composes the 2-D object at the spatial position (t_1, t_2). To the extent that the attenuation coefficients of different types of material such as human tissue and bone differ, $f_c(t_1, t_2)$ can be used to determine the types of material. Reconstructing $f_c(t_1, t_2)$ from the recorded $p_\theta(t)$ is, therefore, of considerable interest.

Consider the 1-D analog Fourier transform of $p_\theta(t)$ with respect to the variable t and denote it by $P_\theta(\Omega)$, so that

$$P_\theta(\Omega) = \int_{t=-\infty}^{\infty} p_\theta(t) e^{-j\Omega t} \, dt. \tag{1.47}$$

It can be shown (see Problem 1.33) that there is a simple relationship between $P_\theta(\Omega)$ and $F_c(\Omega_1, \Omega_2)$, given by

$$P_\theta(\Omega) = F_c(\Omega_1, \Omega_2) \big|_{\Omega_1 = \Omega\cos\theta, \Omega_2 = \Omega\sin\theta} = F_c(\Omega\cos\theta, \Omega\sin\theta). \tag{1.48}$$

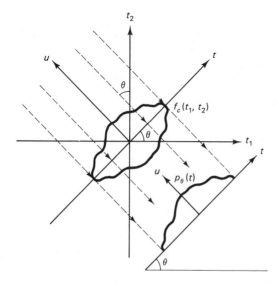

Figure 1.36 Projection of $f_c(t_1, t_2)$ at angle θ.

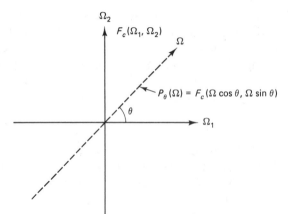

Figure 1.37 Projection slice theorem. $P_\theta(\Omega)$ is the 2-D Fourier transform $F_c(\Omega_1, \Omega_2)$ evaluated along the dotted line.

Expressed graphically, (1.48) states that the 1-D Fourier transform of the projection $p_\theta(t)$ is $F_c(\Omega_1, \Omega_2)$ evaluated along the slice that passes through the origin and makes an angle of θ with the Ω_1 axis, as shown in Figure 1.37. The relationship in (1.48) is called the *projection-slice theorem*.

The projection-slice theorem of (1.48) can be used in developing methods to reconstruct the 2-D function $f_c(t_1, t_2)$ from its projections $p_\theta(t)$. One method is to compute the inverse Fourier transform of $F_c(\Omega_1, \Omega_2)$ obtained from $p_\theta(t)$. Specifically, if we compute the 1-D Fourier transform of $p_\theta(t)$ with respect to t for all $0 \le \theta < \pi$, we will have complete information on $F_c(\Omega_1, \Omega_2)$. In practice, of course, $p_\theta(t)$ cannot be measured for all possible angles $0 \le \theta < \pi$, so $F_c(\Omega_1, \Omega_2)$ must be estimated by interpolating known slices of $F_c(\Omega_1, \Omega_2)$.

Another reconstruction method, known as the *filtered back-projection method*, is more popular in practice and can be derived from (1.45b) and (1.48). It can be shown [Kak] that

$$f_c(t_1, t_2) = \frac{1}{2\pi} \int_{\theta=0}^{\pi} q_\theta(t) \Big|_{t=t_1 \cos\theta + t_2 \sin\theta} d\theta = \frac{1}{2\pi} \int_{\theta=0}^{\pi} q_\theta(t_1 \cos\theta + t_2 \sin\theta)\, d\theta \qquad (1.49)$$

where $q_\theta(t)$ is related to $p_\theta(t)$ by

$$q_\theta(t) = p_\theta(t) * h(t) = \int_{\tau=-\infty}^{\infty} p_\theta(\tau) h(t - \tau)\, d\tau. \qquad (1.50)$$

The function $h(t)$, which can be viewed as the impulse response of a filter, is given by

$$h(t) = \frac{1}{2\pi} \int_{\Omega=-\Omega_c}^{\Omega_c} |\Omega| e^{j\Omega t}\, d\Omega \qquad (1.51)$$

where Ω_c is the frequency above which the energy in any projection $p_\theta(t)$ can be assumed to be zero. From (1.49) and (1.50), we can see that one method of reconstructing $f_c(t_1, t_2)$ from $p_\theta(t)$ is to first compute $q_\theta(t)$ by filtering (convolving) $p_\theta(t)$ with $h(t)$ and then to determine $f_c(t_1, t_2)$ from $q_\theta(t)$ by using (1.49). The

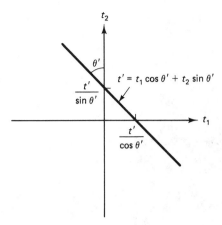

Figure 1.38 Values of (t_1, t_2) for which $f_c(t_1, t_2)$ is affected by $q_{\theta'}(t')$ in the filtered back-projection reconstruction method. They can be described by $t' = t_1 \cos \theta' + t_2 \sin \theta'$.

process of determining $f_c(t_1, t_2)$ from $q_\theta(t)$ using (1.49) can be viewed as a back-projection. Consider a particular θ and t, say θ' and t'. From (1.49), the values of (t_1, t_2) for which $f_c(t_1, t_2)$ is affected by $q_{\theta'}(t')$ are given by $t' = t_1 \cos \theta' + t_2 \sin \theta'$. These values are shown by the straight line in Figure 1.38. Furthermore, the contribution that $q_{\theta'}(t')$ makes to $f_c(t_1, t_2)$ is equal at all points along this line. In essence, $q_{\theta'}(t')$ is back-projected in the (t_1, t_2) domain. This back-projection takes place for all values of t' and is integrated over all values of θ'. Since $q_\theta(t)$ is a filtered version of $p_\theta(t)$, this technique is called the filtered back-projection method. In practice, $p_\theta(t)$ is not available for all values of θ. As a result, $q_\theta(t)$ must be interpolated from the known slices of $q_\theta(t)$.

In addition to the interpolation involved in both the direct Fourier transform method and the filtered back-projection method, a number of practical issues arise in reconstructing $f_c(t_1, t_2)$ from $p_\theta(t)$. For example, the Fourier transform, inverse Fourier transform, filtering, and integration require a discretization of the problem, which raises a variety of important issues, including sampling and aliasing. In practice, the measured function $p_\theta(t)$ may be only an approximate projection of $f_c(t_1, t_2)$. In addition, the measured data may not have been obtained from parallel-beam projection, but instead from fan-beam projection, in which case a different set of equations governs. More details on these and other theoretical and practical issues can be found in [Scudder, Kak]. We will close this section with an example in which a cross section of a human head was reconstructed from its x-ray projections. Figure 1.39 shows the reconstruction by the back-projection method.

1.5 DIGITAL PROCESSING OF ANALOG SIGNALS

Most signals that occur in practice are analog. In this section, we discuss digital processing of analog signals. Since the issues that arise in digital processing of analog signals are essentially the same in both the 1-D and 2-D cases, we will briefly summarize the 2-D results.

Consider an analog 2-D signal $x_c(t_1, t_2)$. We'll denote its analog Fourier

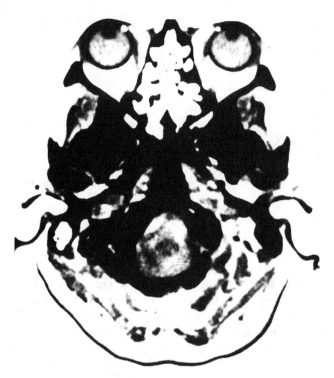

Figure 1.39 Cross section of a human head reconstructed from its projections by the filtered back-projection method. Courtesy of Tamas Sandor.

transform by $X_c(\Omega_1, \Omega_2)$. Suppose we obtain a discrete-space signal $x(n_1, n_2)$ by sampling the analog signal $x_c(t_1, t_2)$ with sampling period (T_1, T_2) as follows:

$$x(n_1, n_2) = x_c(t_1, t_2)\big|_{t_1 = n_1 T_1, t_2 = n_2 T_2}. \tag{1.52}$$

Equation (1.52) represents the input-output relationship of an ideal analog-to-digital (A/D) converter. The relationship between $X(\omega_1, \omega_2)$, the discrete-space Fourier transform of $x(n_1, n_2)$, and $X_c(\Omega_1, \Omega_2)$, the continuous-space Fourier transform of $x_c(t_1, t_2)$, is given by

$$X(\omega_1, \omega_2) = \frac{1}{T_1 T_2} \sum_{r_1 = -\infty}^{\infty} \sum_{r_2 = -\infty}^{\infty} X_c\left(\frac{\omega_1 - 2\pi r_1}{T_1}, \frac{\omega_2 - 2\pi r_2}{T_2}\right). \tag{1.53}$$

Two examples of $X_c(\Omega_1, \Omega_2)$ and $X(\omega_1, \omega_2)$ are shown in Figure 1.40. Figure 1.40(a) shows a case in which $1/T_1 > \Omega_c'/\pi$ and $1/T_2 > \Omega_c''/\pi$, where Ω_c' and Ω_c'' are the cutoff frequencies of $X_c(\Omega_1, \Omega_2)$, as shown in the figure. Figure 1.40(b) shows a case in which $1/T_1 < \Omega_c'/\pi$ and $1/T_2 < \Omega_c''/\pi$. From the figure, when $1/T_1 > \Omega_c'/\pi$ and $1/T_2 > \Omega_c''/\pi$, $x_c(t_1, t_2)$ can be recovered from $x(n_1, n_2)$. Otherwise, $x_c(t_1, t_2)$ cannot be exactly recovered from $x(n_1, n_2)$ without additional information on $x_c(t_1, t_2)$. This is the 2-D sampling theorem, and is a straightforward extension of the 1-D result.

An ideal digital-to-analog (D/A) converter recovers $x_c(t_1, t_2)$ from $x(n_1, n_2)$ when the sampling frequencies $1/T_1$ and $1/T_2$ are high enough to satisfy the require-

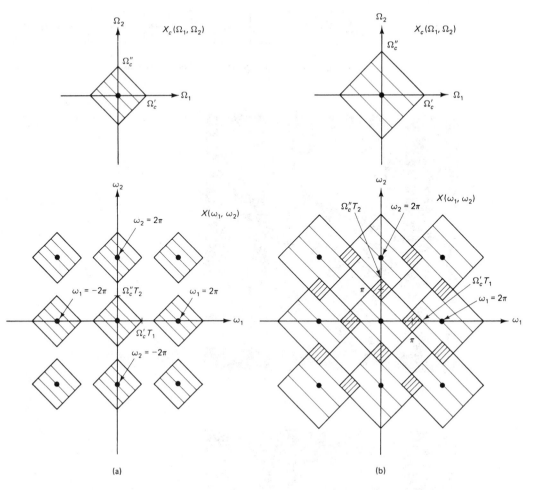

Figure 1.40 Example that illustrates the relationship between $X_c(\Omega_1, \Omega_2)$ and $X(\omega_1, \omega_2)$ given by (1.53). (a) No aliasing; (b) aliasing. Aliased regions are shown shaded.

ments of the sampling theorem. The output of the ideal D/A converter, $y_c(t_1, t_2)$, is given by

$$y_c(t_1, t_2) = \sum_{n_1=-\infty}^{\infty} \sum_{n_2=-\infty}^{\infty} x(n_1, n_2) \frac{\sin \frac{\pi}{T_1} (t_1 - n_1 T_1)}{\frac{\pi}{T_1} (t_1 - n_1 T_1)} \frac{\sin \frac{\pi}{T_2} (t_2 - n_2 T_2)}{\frac{\pi}{T_2} (t_2 - n_2 T_2)}.$$

(1.54)

The function $y_c(t_1, t_2)$ is identical to $x_c(t_1, t_2)$ when the sampling frequencies used in the ideal A/D converter are sufficiently high. Otherwise, $y_c(t_1, t_2)$ is an aliased version of $x_c(t_1, t_2)$. Equation (1.54) is a straightforward extension of the 1-D result.

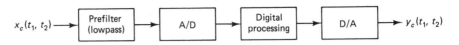

Figure 1.41 Digital processing of analog signals.

(a)

(b)

Figure 1.42 (a) Image of 128×128 pixels with little aliasing due to an effective antialiasing filter; (b) image of 128×128 pixels with noticeable aliasing.

An analog signal can often be processed by digital processing techniques using the A/D and D/A converters discussed above. The digital processing of analog signals can, in general, be represented by the system in Figure 1.41. The analog lowpass filter limits the bandwidth of the analog signal to reduce the effect of aliasing. In digital image processing, the analog prefiltering operation is often performed by a lens and the scanning aperture used in converting an optical image to an electrical signal. The importance of the antialiasing filter is illustrated in Figure 1.42. Figure 1.42(a) shows an image of 128×128 pixels with little aliasing due to an effective antialiasing filter used. Figure 1.42(b) shows an image of 128×128 pixels with noticeable aliasing.

The A/D converter of (1.52) is based on sampling on the Cartesian grid. The analog signal can also be sampled on a different type of grid. Sampling on a hexagonal grid is discussed in Problem 1.35.

REFERENCES

In this text, we have assumed that the reader is familiar with fundamentals of 1-D digital signal processing. For a comprehensive treatment of 1-D digital signal processing concepts, see [Oppenheim and Schafer (1975); Rabiner and Gold; Lim and Oppenheim; Oppenheim and Schafer (1989)].

For different viewpoints or more detailed treatment of some topics in 2-D digital signal processing, see [Huang; Huang; Dudgeon and Mersereau]. For collections of selected papers on 2-D digital signal processing, see [Mitra and Ekstrom; IEEE].

For a more detailed treatment of the Fourier transform theory, see [Papoulis]. For processing data obtained from sampling on any regular periodic lattice including the rectangular lattice and hexagonal lattice, see [Mersereau; Mersereau and Speake].

Y. M. Bruck and L. G. Sodin, On the ambiguity of the image reconstruction problem, *Opt. Commun.*, September 1979, pp. 304–308.

D. E. Dudgeon and R. M. Mersereau, *Multidimensional Digital Signal Processing*. Englewood Cliffs, NJ: Prentice-Hall, 1983.

M. H. Hayes, The reconstruction of a multidimensional sequence from the phase or magnitude of its Fourier transform, *IEEE Trans. on Acoust., Speech, and Sig. Proc.*, Vol. ASSP-30, April 1982, pp. 140–154.

T. S. Huang, ed., *Two-Dimensional Digital Signal Processing I*, in "Topics in Applied Physics," Vol. 42. Berlin: Springer-Verlag, 1981.

T. S. Huang, ed., *Two-Dimensional Digital Signal Processing II*, in "Topics in Applied Physics," Vol. 43. Berlin: Springer-Verlag, 1981.

IEEE, ASSP Society's MDSP Committee, editor, *Selected Papers in Multidimensional Digital Signal Processing*, IEEE Press, New York, 1986.

D. Izraelevitz and J. S. Lim, A new direct algorithm for image reconstruction from Fourier transform magnitude. *IEEE Trans. on Acoust., Speech, and Sig. Proc.*, Vol. ASSP-35, April 1987, pp. 511–519.

A. C. Kak, "Image Reconstruction from Projections," in *Digital Image Processing Techniques*, edited by M. Ekstrom. Orlando, FL: Academic Press, 1984, Chapter 4.

R. G. Lane, W. R. Fright, and R. H. T. Bates, Direct phase retrieval, *IEEE Trans. on Acoust., Speech, and Sig. Proc.*, Vol. ASSP-35, April 1987, pp. 520–525.

J. S. Lim and A. V. Oppenheim, ed., *Advanced Topics in Signal Processing*, Englewood Cliffs, NJ: Prentice-Hall, 1988.

R. M. Mersereau, "The processing of hexagonally sampled two-dimensional signals," *Proc. IEEE*, Vol. 67, May 1979, pp. 930–949.

R. M. Mersereau and T. C. Speake, The processing of periodically sampled multidimensional signals, *IEEE Trans. on Acoust., Speech, and Sig. Proc.*, Vol. ASSP-31, February 1983, pp. 188–194.

S. K. Mitra and M. P. Ekstrom, eds. *Two-Dimensional Digital Signal Processing*. Stroudsburg, PA: Dowden, Hutchinson and Ross, 1978.

A. V. Oppenheim, J. S. Lim, and S. R. Curtis, Signal synthesis and reconstruction from partial Fourier domain information, *J. Opt. Soc. Amer.*, Vol. 73, November 1983, pp. 1413–1420.

A. V. Oppenheim and R. W. Schafer, *Digital Signal Processing*. Englewood Cliffs, NJ: Prentice Hall, 1975.

A. V. Oppenheim and R. W. Schafer, *Discrete-Time Signal Processing*. Englewood Cliffs, NJ: Prentice Hall, 1989.

A. Papoulis, *The Fourier Integral and Its Applications*. New York: McGraw-Hill, 1962.

L. R. Rabiner and B. Gold, *Theory and Application of Digital Signal Processing*. Englewood Cliffs, NJ: Prentice Hall, 1975.

G. N. Ramachandran and R. Srinivasan, *Fourier Methods in Crystallography*. New York: Wiley-Interscience, 1978.

W. O. Saxton, *Computer Techniques for Image Processing in Electron Microscopy*. New York: Academic Press, 1970.

H. J. Scudder, Introduction to computer aided tomography, *Proc. IEEE*, Vol. 66, June 1978, pp. 628–637.

V. T. Tom, T. F. Quatieri, M. H. Hayes, and J. H. McClellan, Convergence of iterative nonexpansive signal reconstruction algorithms, *IEEE Trans. on Acoust., Speech, and Sig. Proc.*, Vol. ASSP-29, October 1981, pp. 1052–1058.

PROBLEMS

1.1. Sketch the following sequences:

    (a) $\delta(n_1 + 2, n_2 - 3) + 2\delta(n_1, -n_2 + 2)$

    (b) $\delta_T(n_1)u(n_1, n_2)$

    (c) $(\frac{1}{2})^{n_2} \delta_T(n_1 + n_2)u(-n_1 + 1, -n_2)$

1.2. Consider a sequence $x(n_1, n_2)$ sketched below:

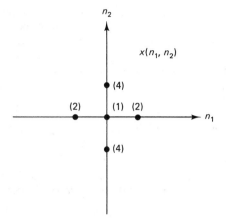

n_2

$x(n_1, n_2)$

(4)

(2)　　(1) (2)

n_1

(4)

Figure P1.2

Express $x(n_1, n_2)$ as a linear combination of $\delta(n_1, n_2)$ and its shifts.

1.3. We have defined a sequence $x(n_1, n_2)$ to be periodic with a period of $N_1 \times N_2$ if

$$x(n_1, n_2) = x(n_1 + N_1, n_2) = x(n_1, n_2 + N_2) \quad \text{for all } (n_1, n_2). \qquad (1)$$

More generally defined, the condition is

$$x(n_1, n_2) = x(n_1 + N_{11}, n_2 + N_{12}) = x(n_1 + N_{21}, n_2 + N_{22}) \text{ for all } (n_1, n_2) \qquad (2)$$

with the number of points in a period given by

$$|N_{11}N_{22} - N_{12}N_{21}|.$$

(a) Show that the condition in (2) reduces to the condition in (1) with a proper choice of N_{11}, N_{12}, N_{21}, and N_{22}.

(b) Consider the periodic sequence $x(n_1, n_2)$, which was shown in Figure 1.8. If we use (1), the minimum choices of N_1 and N_2 are 6 and 2, respectively, and the number of points in one period is 12. If we use (2), N_{11}, N_{12}, N_{21}, and N_{22} can be chosen such that $|N_{11}N_{22} - N_{12}N_{21}| = 6$. Determine one such set of N_{11}, N_{12}, N_{21}, and N_{22}.

(c) Show that any sequence that satisfies the condition in (2) will also satisfy the condition in (1) as long as N_1 and N_2 are chosen appropriately. This result shows that (1) can be used in representing any periodic sequence that can be represented by (2), although the number of points in one period may be much larger when (1) rather than (2) is used.

1.4. For each of the following systems, determine whether or not the system is (1) linear, (2) shift invariant, and (3) stable.

(a) $y(n_1, n_2) = T[x(n_1, n_2)] = e^{3x(n_1, n_2)}$

(b) $y(n_1, n_2) = T[x(n_1, n_2)] = \sum_{k_1 = n_1 - 3}^{n_1 + 6} x(k_1, n_2)$

(c) $y(n_1, n_2) = T[x(n_1, n_2)] = \sum_{k_2 = 0}^{n_2} x(n_1, k_2)$

1.5. The median filter is used in a number of signal processing applications, including image processing. When a median filter is applied to an image, a window slides along the image, and the median intensity value of the pixels (picture elements) within the window replaces the intensity of the pixel being processed. For example, when the pixel values within a window are 5, 6, 35, 10, and 5, and the pixel being processed has a value of 35, its value is changed to 6, the median of the five values. Answer each of the following questions. In your answer, use a 2-D median filter of size 3×3, with the center of the window corresponding to the pixel being processed.

(a) Is a median filter linear, shift invariant, and/or stable?

(b) Using an example, illustrate that a median filter tends to preserve sharp discontinuities, such as steps.

(c) Using an example, illustrate that a median filter is capable of eliminating impulsive values without seriously affecting the value of the pixel near those with the impulsive values. A pixel has an impulsive value when its value is significantly different from its neighborhood pixel values.

1.6. Consider a system T. When the input to the system is the unit step sequence $u(n_1, n_2)$, the response of the system is $s(n_1, n_2)$ as shown below.

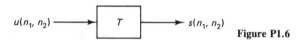

Figure P1.6

For each of the following three cases, determine the class of inputs for which we can determine the output of the system in terms of $s(n_1, n_2)$. For each input in the class, express the output in terms of $s(n_1, n_2)$.

(a) T is linear, but not shift invariant.

(b) T is shift invariant, but not linear.

(c) T is linear and shift invariant.

1.7. Compute $x(n_1, n_2) * h(n_1, n_2)$ for each of the following two problems.

(a)

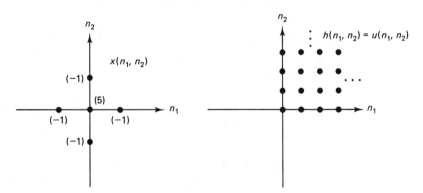

Figure P1.7

(b) $x(n_1, n_2) = (\frac{1}{2})^{n_1}(\frac{1}{3})^{n_2}u(n_1, n_2)$
$h(n_1, n_2) = u(n_1 - 1, n_2 - 2)$

1.8. Convolve the sequence $x(n_1, n_2) = a^{n_1}b^{n_2}u(n_1, n_2)$ with the sequence shown in the figure below. Assume $|a| < 1$. The filled-in circles represent samples with amplitude of 1. The vertical lines at $n_1 = -1, 0, 1$ extend to ∞, and the horizontal line at $n_2 = 0$ extends to $-\infty$ and $+\infty$.

Signals, Systems, and the Fourier Transform   Chap. 1

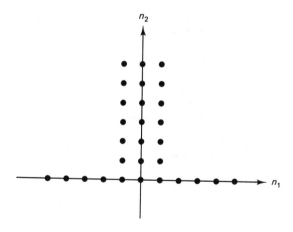

Figure P1.8

1.9. Consider a 2-D LSI system whose impulse response $h(n_1, n_2)$ has first-quadrant support. When the input $x(n_1, n_2)$ to the system is given by

$$x(n_1, n_2) = 2^{n_1 n_2} u(n_1, n_2),$$

some portion of the output $y(n_1, n_2)$ has been observed. Suppose the observed portion of $y(n_1, n_2)$ is as shown in the following figure.

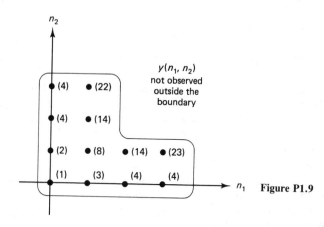

Figure P1.9

Determine $h(1, 1)$, the impulse response $h(n_1, n_2)$ evaluated at $n_1 = n_2 = 1$.

1.10. If the input $x(n_1, n_2)$ to an LSI system is periodic with a period of $N_1 \times N_2$, is the output $y(n_1, n_2)$ of the system periodic? If so, determine the periodicity of $y(n_1, n_2)$.

1.11. Consider the following system in which $x(n_1, n_2)$ represents an input sequence and $h_i(n_1, n_2)$ for $i = 1, 2, 3, 4, 5$ represents the impulse response of an LSI system.

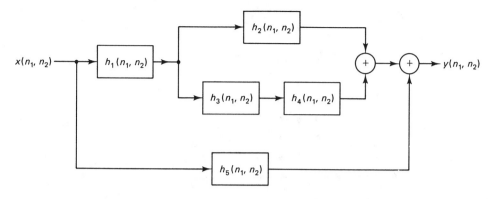

Figure P1.11

Without affecting the input-output relationship, the above system can be simplified as

$$x(n_1, n_2) \rightarrow \boxed{h(n_1, n_2)} \rightarrow y(n_1, n_2).$$

Express $h(n_1, n_2)$ in terms of $h_1(n_1, n_2)$, $h_2(n_1, n_2)$, $h_3(n_1, n_2)$, $h_4(n_1, n_2)$ and $h_5(n_1, n_2)$.

1.12. Let $h(n_1, n_2, n_3)$ be a separable, finite-extent, 3-D sequence of $M \times M \times M$ points which can be expressed as

$$h(n_1, n_2, n_3) = a(n_1) \, b(n_2) \, c(n_3).$$

(a) The Fourier transform of $h(n_1, n_2, n_3)$, $H(\omega_1, \omega_2, \omega_3)$, is defined by

$$H(\omega_1, \omega_2, \omega_3) = \sum_{n1 = -\infty}^{\infty} \sum_{n2 = -\infty}^{\infty} \sum_{n3 = -\infty}^{\infty} h(n_1, n_2, n_3) e^{-j\omega_1 n_1} e^{-j\omega_2 n_2} e^{-j\omega_3 n_3}.$$

Show that $H(\omega_1, \omega_2, \omega_3)$ is a separable function that can be expressed in the form of $A(\omega_1)B(\omega_2)C(\omega_3)$.

(b) We wish to filter an input sequence $x(n_1, n_2, n_3)$ of $N \times N \times N$ points using an LSI system with impulse response $h(n_1, n_2, n_3)$ as given above. Develop a computationally efficient way to compute the output $y(n_1, n_2, n_3)$.

(c) How does your method compare to direct evaluation of the convolution sum for each output point when $N = 512$ and $M = 10$?

1.13. An LSI system can be specified by its impulse response $h(n_1, n_2)$. An LSI system can also be specified by its unit step response $s(n_1, n_2)$, the response of the system when the input is the unit step sequence $u(n_1, n_2)$.

(a) Express $y(n_1, n_2)$, the output of an LSI system, in terms of the input $x(n_1, n_2)$ and the unit step response $s(n_1, n_2)$.

(b) In determining the output $y(n_1, n_2)$ of an LSI system, which of the two methods requires less computation: your result in (a), or convolving $x(n_1, n_2)$ with $h(n_1, n_2)$?

1.14. Show that an LSI system is stable in the BIBO (bounded-input-bounded-output) sense if and only if the impulse response of the system $h(n_1, n_2)$ is absolutely summable, that is,

$$\sum_{n1 = -\infty}^{\infty} \sum_{n2 = -\infty}^{\infty} |h(n_1, n_2)| < \infty.$$

1.15. For each of the following cases, determine the region of support of $y(n_1, n_2) = x(n_1, n_2) * h(n_1, n_2)$.

    (a) $x(n_1, n_2)$ and $h(n_1, n_2)$ are first-quadrant support sequences.

    (b) $x(n_1, n_2)$ and $h(n_1, n_2)$ are second-quadrant support sequences.

    (c) $x(n_1, n_2)$ is a first-quadrant support sequence and $h(n_1, n_2)$ is a fourth-quadrant support sequence.

    (d) $x(n_1, n_2)$ is a first-quadrant support sequence and $h(n_1, n_2)$ is a third-quadrant support sequence.

1.16. Let $x(n_1, n_2)$ and $y(n_1, n_2)$ denote an input and the corresponding output sequence of a 2-D system. We define a system to be *pseudo-causal* if $y(n_1, n_2)$ does not depend on $x(n_1 - k_1, n_2 - k_2)$ for $k_1 < 0$ or $k_2 < 0$. Show that a necessary and sufficient condition for an LSI system to be pseudo-causal is that its impulse response $h(n_1, n_2)$ must be a first-quadrant support sequence.

1.17. It is known that any wedge support sequence $x(n_1, n_2)$ can be mapped to a first-quadrant support sequence $y(n_1, n_2)$ without affecting its stability by linear mapping of variables:

$$y(n_1, n_2) = x(m_1, m_2) \big|_{m_1 = l_1 n_1 + l_2 n_2, m_2 = l_3 n_1 + l_4 n_2}$$

where $l_1, l_2, l_3,$ and l_4 are integers. Consider the following wedge support sequence $x(n_1, n_2)$.

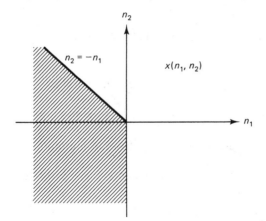

Figure P1.17

The shaded region in the figure is the region of (n_1, n_2) for which $x(n_1, n_2)$ is nonzero. Determine one specific linear mapping of variables that maps $x(n_1, n_2)$ to a first-quadrant support sequence $y(n_1, n_2)$ without affecting its stability.

1.18. Determine if each of the following sequences is an eigenfunction of a general LSI system.

    (a) $x(n_1, n_2) = e^{j\omega_2 n_2}$

    (b) $x(n_1, n_2) = a^{n_1} + b^{n_2}$

    (c) $x(n_1, n_2) = e^{j\omega_1 n_1} e^{j\omega_2 n_2}$

    (d) $x(n_1, n_2) = e^{j\omega_1 n_1} e^{j\omega_2 n_2} u(n_1, n_2)$

1.19. Determine the Fourier transform of each of the following sequences.

    (a) $\delta(n_1, n_2)$

    (b) $a^{2n_1 + n_2} u(n_1, n_2), |a| < 1$

    (c) $a^{n_1} b^{n_2} \delta_T(4n_1 - n_2) u_T(n_1), |a| < 1, |b| < 1$

    (d) $n_1 (\tfrac{1}{2})^{n_1} (\tfrac{1}{2})^{n_2} u(n_1, n_2)$

1.20. The Fourier transform $X(\omega_1, \omega_2)$ of the sequence $x(n_1, n_2)$ is given by

$$X(\omega_1, \omega_2) = 3 + 2 \cos \omega_1 + j4 \sin \omega_2 + 8e^{-j\omega_1}e^{-j\omega_2}.$$

Determine $x(n_1, n_2)$.

1.21. Consider the following sequence:

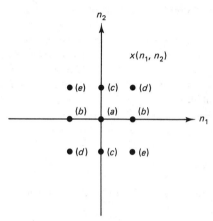

Figure P1.21

The coefficients a, b, c, d, and e are real.
(a) What can you say about $X(\omega_1, \omega_2)$ without explicitly computing $X(\omega_1, \omega_2)$?
(b) Determine $X(0, 0)$.
(c) Determine $X(\omega_1, \omega_2)$.

1.22. Using the Fourier transform pair of Equation (1.31), show that

$$\sum_{n_1 = -\infty}^{\infty} \sum_{n_2 = -\infty}^{\infty} |x(n_1, n_2)|^2 = \frac{1}{(2\pi)^2} \int_{\omega_1 = -\pi}^{\pi} \int_{\omega_2 = -\pi}^{\pi} |X(\omega_1, \omega_2)|^2 \, d\omega_1 \, d\omega_2.$$

1.23. We wish to design a highpass filter with impulse response $h(n_1, n_2)$ and frequency response $H(\omega_1, \omega_2)$.
(a) Determine one $h(n_1, n_2)$ which can be viewed as a highpass filter and which has the following property:

$$H(\omega_1, \omega_2) = \begin{cases} 0, & \omega_1 = \omega_2 = 0 \\ 1, & (\omega_1, \omega_2) = (\pi, 0), (\pi, \pi), (0, \pi), (-\pi, \pi), (-\pi, 0), \\ & (-\pi, -\pi), (0, -\pi) \text{ and } (\pi, -\pi). \end{cases}$$

(b) For your answer in (a), determine $H(\omega_1, \omega_2)$. Demonstrate that it can be viewed as a highpass filter by evaluating $H(\omega_1, \omega_2)$ at a reasonable number of values of (ω_1, ω_2).
(c) Let $x(n_1, n_2)$ represent the intensity of a digital image. The amplitude of $x(n_1, n_2)$ is real and nonnegative. We process $x(n_1, n_2)$ with the highpass filter designed in (a) and denote the resulting output by $y(n_1, n_2)$. Determine

$$\sum_{n_1 = -\infty}^{\infty} \sum_{n_2 = -\infty}^{\infty} y(n_1, n_2).$$

(d) From your answer to (c), discuss how $y(n_1, n_2)$ will appear on a display device that sets all negative amplitudes of $y(n_1, n_2)$ to 0 (the darkest level) before it displays $y(n_1, n_2)$.

1.24. The impulse response of a circularly symmetric ideal lowpass filter with cutoff frequency of ω_c is given by

$$h(n_1, n_2) = \frac{\omega_c}{2\pi\sqrt{n_1^2 + n_2^2}} J_1(\omega_c\sqrt{n_1^2 + n_2^2})$$

where $J_1(\cdot)$ is the Bessel function of the first kind and first order. In this problem, we derive this result.

(a) The frequency response of the filter is given by

$$H(\omega_1, \omega_2) = \begin{cases} 1, & \omega_1^2 + \omega_2^2 \leq \omega_c^2 \\ 0, & \text{otherwise.} \end{cases}$$

The sequence $h(n_1, n_2)$ is then given by

$$h(n_1, n_2) = \frac{1}{(2\pi)^2} \int_{(\omega_1,\omega_2) \in [\omega_1^2 + \omega_2^2 \leq \omega_c^2]} 1 e^{j\omega_1 n_1} e^{j\omega_2 n_2} \, d\omega_1 \, d\omega_2.$$

We now make the following change of variables:

$$r \cos \theta = \omega_1$$

$$r \sin \theta = \omega_2.$$

Show that $h(n_1, n_2)$ can be expressed as

$$h(n_1, n_2) = \frac{1}{(2\pi)^2} \int_{r=0}^{\omega_c} r \, dr \int_{\theta=a}^{a+2\pi} e^{jr(n_1 \cos \theta + n_2 \sin \theta)} \, d\theta \qquad (1)$$

for any real constant a.

(b) We next make the following change of variables:

$$n_1 = n \cos \phi$$

$$n_2 = n \sin \phi$$

Note that $n^2 = n_1^2 + n_2^2$. Show that $h(n_1, n_2)$ can be written as

$$h(n_1, n_2) = \frac{1}{(2\pi)^2} \int_{r=0}^{\omega_c} rf(r) \, dr \qquad (2)$$

where

$$f(r) = \int_{\theta=a}^{a+2\pi} e^{jrn \cos(\theta - \phi)} \, d\theta. \qquad (3)$$

(c) It is known that

$$J_0(x) = \frac{1}{2\pi} \int_{\theta=0}^{2\pi} \cos (x \sin \theta) \, d\theta = \frac{1}{2\pi} \int_{\theta=0}^{2\pi} \cos (x \cos \theta) \, d\theta \qquad (4)$$

where $J_0(x)$ is the Bessel function of the first kind, zeroth order. From (3) and (4) with $a = \phi$, show that

$$f(r) = 2\pi J_0(r\sqrt{n_1^2 + n_2^2}). \qquad (5)$$

(d) It is known that

$$xJ_1(x)\big|_{x=a}^{b} = \int_{x=a}^{b} xJ_0(x) \, dx \qquad (6)$$

where $J_1(x)$ is the Bessel function of the first kind, first order. From (2), (5), and (6), show that

$$h(n_1, n_2) = \frac{\omega_c}{2\pi\sqrt{n_1^2 + n_2^2}} J_1(\omega_c\sqrt{n_1^2 + n_2^2}).$$

This is the desired result.

1.25. Determine the impulse response of each of the following two filters. You may use the results of Problem 1.24.

(a) Circularly symmetric ideal highpass filter:

$$H(\omega_1, \omega_2) = \begin{cases} 0, & \omega_1^2 + \omega_2^2 \le \omega_c^2 \text{ (unshaded region)} \\ 1, & \text{otherwise (shaded region)} \end{cases}$$

(b) Circularly symmetric ideal bandpass filter:

$$H(\omega_1, \omega_2) = \begin{cases} 1, & R_1^2 \le \omega_1^2 + \omega_2^2 \le R_2^2 \text{ (shaded region)} \\ 0, & \text{otherwise (unshaded region)} \end{cases}$$

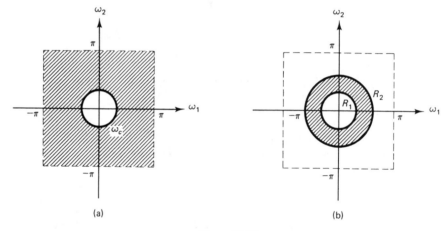

(a)   (b)

Figure P1.25

1.26. It is well known that circular symmetry of $X(\omega_1, \omega_2)$ implies circular symmetry of $x(n_1, n_2)$. However, circular symmetry of $x(n_1, n_2)$ does not imply circular symmetry of $X(\omega_1, \omega_2)$. To show the latter, determine a circularly symmetric sequence $x(n_1, n_2)$ that is a function of $n_1^2 + n_2^2$ with the property that $X(\omega_1, \omega_2)$ cannot be expressed as a function of $\omega_1^2 + \omega_2^2$ for $\omega_1^2 + \omega_2^2 \le \pi^2$.

1.27. Evaluate the following expression:

$$\sum_{n_1=-\infty}^{\infty} \sum_{n_2=-\infty}^{\infty} \frac{\left[J_1\left(\frac{1}{2}\sqrt{n_1^2 + n_2^2}\right)\right]^2}{n_1^2 + n_2^2}$$

where $J_1(\cdot)$ is the Bessel function of the first kind and first order.

1.28. Let $f(x, y)$ denote a 2-D analog function that is circularly symmetric and can therefore be expressed as

$$f(x, y) = g(r)\big|_{r=\sqrt{x^2+y^2}}.$$

Let $F(\Omega_x, \Omega_y)$ denote the 2-D analog Fourier transform of $f(x, y)$. For a circularly symmetric $f(x, y)$, $F(\Omega_x, \Omega_y)$ is also circularly symmetric and can therefore be expressed as

$$F(\Omega_x, \Omega_y) = G(\rho)\big|_{\rho = \sqrt{\Omega_x^2 + \Omega_y^2}}.$$

Note that the analog case is in sharp contrast with the discrete-space case, in which circular symmetry of a sequence $x(n_1, n_2)$ does not imply circular symmetry of its Fourier transform $X(\omega_1, \omega_2)$. The relationship between $g(r)$ and $G(\rho)$ is called the *zeroth-order Hankel transform* pair and is given by

$$G(\rho) = 2\pi \int_{r=0}^{\infty} rg(r)J_0(r\rho) \, dr$$

and

$$g(r) = \frac{1}{2\pi} \int_{\rho=0}^{\infty} \rho G(\rho)J_0(r\rho) \, d\rho,$$

where $J_0(\cdot)$ is the Bessel function of the first kind and zeroth order. Determine the Fourier transform of $f(x, y)$ when $f(x, y)$ is given by

$$f(x, y) = \begin{cases} 1, & \sqrt{x^2 + y^2} \leq 2 \\ 0, & \text{otherwise.} \end{cases}$$

Note that
$$xJ_1(x)\big|_{x=a}^{b} = \int_{x=a}^{b} xJ_0(x) \, dx$$

where $J_1(x)$ is the Bessel function of the first kind and first order.

1.29. Cosine transforms are used in many signal processing applications. Let $x(n_1, n_2)$ be a real, finite-extent sequence which is zero outside $0 \leq n_1 \leq N_1 - 1$, $0 \leq n_2 \leq N_2 - 1$. One of the possible definitions of the cosine transform $C_x(\omega_1, \omega_2)$ is

$$C_x(\omega_1, \omega_2) = \sum_{n_1=0}^{N_1-1} \sum_{n_2=0}^{N_2-1} x(n_1, n_2) \cos \omega_1 n_1 \cos \omega_2 n_2.$$

(a) Express $C_x(\omega_1, \omega_2)$ in terms of $X(\omega_1, \omega_2)$, the Fourier transform of $x(n_1, n_2)$.
(b) Derive the inverse cosine transform relationship; that is, express $x(n_1, n_2)$ in terms of $C_x(\omega_1, \omega_2)$.

1.30. In reconstructing an image from its Fourier transform phase, we have used an iterative algorithm, shown in Figure 1.30. The method of imposing constraints separately in each domain in an iterative manner in order to obtain a solution that satisfies all the required constraints is useful in a variety of applications. One such application is the band-limited extrapolation of a signal. As an example of a band-limited extrapolation problem, consider $x(n_1, n_2)$, which has been measured only for $0 \leq n_1 \leq N - 1$, $0 \leq n_2 \leq N - 1$. From prior information, however, we know that $x(n_1, n_2)$ is band-limited and that its Fourier transform $X(\omega_1, \omega_2)$ satisfies $X(\omega_1, \omega_2) = 0$ for $\sqrt{\omega_1^2 + \omega_2^2} \geq \omega_c$. Develop an iterative algorithm that may be used for determining $x(n_1, n_2)$ for all (n_1, n_2). You do not have to show that your algorithm converges to a desired solution. However, using $N = 1$, $x(0, 0) = 1$, and $\omega_c = \frac{\pi}{2}$, carry out a few iterations of your algorithm and illustrate that it behaves reasonably for at least this particular case.

1.31. Let $x(n_1, n_2)$ represent the intensity of a digital image. Noting that $|X(\omega_1, \omega_2)|$ decreases rapidly as the frequency increases, we assume that an accurate model of $|X(\omega_1, \omega_2)|$ is

$$|X(\omega_1, \omega_2)| = \begin{cases} Ae^{-2\sqrt{\omega_1^2 + \omega_2^2}}, & \sqrt{\omega_1^2 + \omega_2^2} \leq \pi \\ 0, & \text{otherwise.} \end{cases}$$

Suppose we reconstruct $y(n_1, n_2)$ by retaining only a fraction of the frequency components of $x(n_1, n_2)$. Specifically,

$$Y(\omega_1, \omega_2) = \begin{cases} X(\omega_1, \omega_2), & \sqrt{\omega_1^2 + \omega_2^2} \leq \dfrac{\pi}{10} \\ 0, & \text{otherwise.} \end{cases}$$

The fraction of the frequency components retained is $\dfrac{\pi(\pi/10)^2}{4\pi^2}$, or approximately 1%. By evaluating the quantity

$$\frac{\displaystyle\sum_{n1 = -\infty}^{\infty} \sum_{n2 = -\infty}^{\infty} (y(n_1, n_2) - x(n_1, n_2))^2}{\displaystyle\sum_{n1 = -\infty}^{\infty} \sum_{n2 = -\infty}^{\infty} x^2(n_1, n_2)},$$

discuss the amount of distortion in the signal caused by discarding 99% of the frequency components.

1.32. For a typical image, most of the energy has been observed to be concentrated in the low-frequency regions. Give an example of an image for which this observation may not be valid.

1.33. In this problem, we derive the projection-slice theorem, which is the basis for computed tomography. Let $f(t_1, t_2)$ denote an analog 2-D signal with Fourier transform $F(\Omega_1, \Omega_2)$.

(a) We integrate $f(t_1, t_2)$ along the t_2 variable and denote the result by $p_0(t_1)$; that is,

$$p_0(t_1) = \int_{t2 = -\infty}^{\infty} f(t_1, t_2)\, dt_2.$$

Express $P_0(\Omega)$ in terms of $F(\Omega_1, \Omega_2)$, where $P_0(\Omega)$ is the 1-D Fourier transform of $p_0(t_1)$ given by

$$P_0(\Omega) = \int_{t1 = -\infty}^{\infty} p_0(t_1)e^{-j\Omega t_1}\, dt_1.$$

(b) We integrate $f(t_1, t_2)$ along the t_1 variable and denote the result by $p_{\pi/2}(t_2)$; that is,

$$p_{\pi/2}(t_2) = \int_{t1 = -\infty}^{\infty} f(t_1, t_2)\, dt_1.$$

Express $P_{\pi/2}(\Omega)$ in terms of $F(\Omega_1, \Omega_2)$, where $P_{\pi/2}(\Omega)$ is the 1-D Fourier transform of $p_{\pi/2}(t_2)$ given by

$$P_{\pi/2}(\Omega) = \int_{t2 = -\infty}^{\infty} p_{\pi/2}(t_2)e^{-j\Omega t_2}\, dt_2.$$

(c) Suppose we obtain $a(t, u)$ from $f(t_1, t_2)$ by the coordinate rotation given by

$$a(t, u) = f(t_1, t_2)\big|_{t_1 = t\cos\theta - u\sin\theta, t_2 = t\sin\theta + u\cos\theta}$$

where θ is the angle shown in Figure P1.33(a). In addition, we obtain $B(\Omega'_1, \Omega'_2)$ from $F(\Omega_1, \Omega_2)$ by coordinate rotation given by

$$B(\Omega'_1, \Omega'_2) = F(\Omega_1, \Omega_2)\big|_{\Omega_1 = \Omega'_1\cos\theta - \Omega'_2\sin\theta, \Omega_2 = \Omega'_1\sin\theta + \Omega'_2\cos\theta}$$

where θ is the angle shown in Figure P1.33(b).

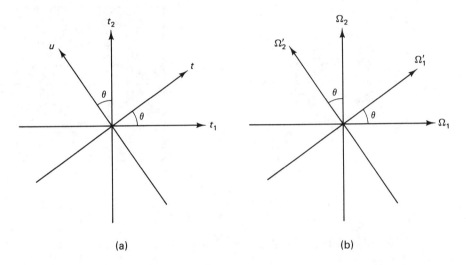

(a)   (b)

Figure P1.33

Show that $B(\Omega'_1, \Omega'_2) = A(\Omega'_1, \Omega'_2)$ where

$$A(\Omega'_1, \Omega'_2) = \int_{t=-\infty}^{\infty}\int_{u=-\infty}^{\infty} a(t, u)e^{-j\Omega'_1 t}e^{-j\Omega'_2 u}\, dt\, du.$$

The result states that when $f(t_1, t_2)$ is rotated by an angle θ with respect to the origin in the (t_1, t_2) plane, its Fourier transform $F(\Omega_1, \Omega_2)$ rotates by the same angle in the same direction with respect to the origin in the (Ω_1, Ω_2) plane. This is a property of the 2-D analog Fourier transform.

(d) Suppose we integrate $f(t_1, t_2)$ along the u variable where the u variable axis is shown in Figure P1.33(a). Let the result of integration be denoted by $p_\theta(t)$. The function $p_\theta(t)$ is called the projection of $f(t_1, t_2)$ at angle θ. Using the results of (a) and (c) or the results of (b) and (c), discuss how $P_\theta(\Omega)$ can be simply related to $F(\Omega_1, \Omega_2)$, where

$$P_\theta(\Omega) = \int_{t=-\infty}^{\infty} p_\theta(t)e^{-j\Omega t}\, dt.$$

The relationship between $P_\theta(\Omega)$ and $F(\Omega_1, \Omega_2)$ is the projection-slice theorem.

1.34. Consider an analog 2-D signal $s_c(t_1, t_2)$ degraded by additive noise $w_c(t_1, t_2)$. The degraded observation $y_c(t_1, t_2)$ is given by

$$y_c(t_1, t_2) = s_c(t_1, t_2) + w_c(t_1, t_2).$$

Suppose the spectra of $s_c(t_1, t_2)$ and $w_c(t_1, t_2)$ are nonzero only over the shaded regions shown in the following figure.

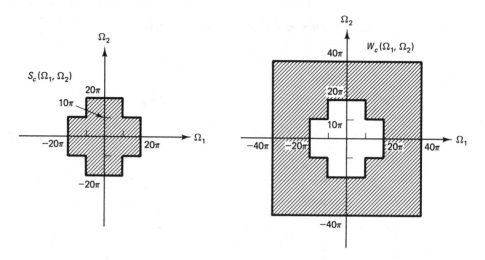

Figure P1.34

We wish to filter the additive noise $w_c(t_1, t_2)$ by digital filtering, using the following system:

$$y_c(t_1, t_2) \rightarrow \boxed{\text{Ideal A/D}} \xrightarrow{y(n_1,n_2)} \boxed{H(\omega_1, \omega_2)} \xrightarrow{\hat{s}(n_1,n_2)} \boxed{\text{Ideal D/A}} \xrightarrow{\hat{s}_c(t_1,t_2)}$$

$$y(n_1, n_2) = y_c(t_1, t_2)\big|_{t_1 = n_1 T_1, t_2 = n_2 T_2}$$

$$\hat{s}_c(t_1, t_2) = \sum_{n_1 = -\infty}^{\infty} \sum_{n_2 = -\infty}^{\infty} \hat{s}(n_1, n_2) \frac{\sin \dfrac{\pi}{T_1}(t_1 - n_1 T_1) \sin \dfrac{\pi}{T_2}(t_2 - n_2 T_2)}{\dfrac{\pi}{T_1}(t_1 - n_1 T_1) \dfrac{\pi}{T_2}(t_2 - n_2 T_2)}$$

Assuming that it is possible to have any desired $H(\omega_1, \omega_2)$, determine the maximum T_1 and T_2 for which $\hat{s}_c(t_1, t_2)$ can be made to equal $s_c(t_1, t_2)$.

1.35. In Section 1.5, we discussed the results for the ideal A/D and D/A converters when the analog signal is sampled on a rectangular grid. In this problem, we derive the corresponding results when the analog signal is sampled on a hexagonal grid. Let $x_c(t_1, t_2)$ and $X_c(\Omega_1, \Omega_2)$ denote an analog signal and its analog Fourier transform. Let $x(n_1, n_2)$ and $X(\omega_1, \omega_2)$ denote a sequence and its Fourier transform. An ideal A/D converter converts $x_c(t_1, t_2)$ to $x(n_1, n_2)$ by

$$x(n_1, n_2) = \begin{cases} x_c(t_1, t_2)\big|_{t_1 = n_1 T_1, t_2 = n_2 T_2}, & \text{if both } n_1 \text{ and } n_2 \text{ are even, or} \\ & \text{both } n_1 \text{ and } n_2 \text{ are odd} \\ 0, & \text{otherwise.} \end{cases} \quad (1)$$

The sampling periods T_1 and T_2 are related by $T_2 = \dfrac{\sqrt{3}}{3} T_1$. The sampling grid used in (1) is shown in the following figure.

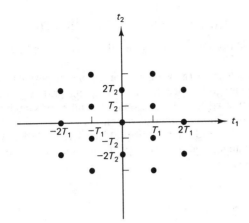

Figure P1.35a

We wish to derive the relationship between $X(\omega_1, \omega_2)$ and $X_c(\Omega_1, \Omega_2)$. It is convenient to represent (1) by the system shown below:

$$x_c(t_1, t_2) \longrightarrow \otimes \xrightarrow{\ x_s(t_1, t_2)\ } \boxed{G} \longrightarrow x(n_1, n_2)$$

$$\uparrow$$

$$p_c(t_1, t_2)$$

The function $p_c(t_1, t_2)$ is a periodic train of impulses given by

$$p_c(t_1, t_2) = \sum_{n1=-\infty}^{\infty} \sum_{n2=-\infty}^{\infty} \delta(t_1 - 2n_1 T_1, t_2 - 2n_2 T_2)$$
$$+ \sum_{n1=-\infty}^{\infty} \sum_{n2=-\infty}^{\infty} \delta(t_1 - 2n_1 T_1 - T_1, t_2 - 2n_2 T_2 - T_2)$$

(2)

where $\delta(t_1, t_2)$ is a dirac-delta function. The system G converts an analog signal $x_s(t_1, t_2)$ to a sequence $x(n_1, n_2)$ by measuring the area under each impulse and using it as the amplitude of the sequence $x(n_1, n_2)$.

(a) Sketch an example of $x_c(t_1, t_2)$, $x_s(t_1, t_2)$ and $x(n_1, n_2)$. Note, from (1), that $x(n_1, n_2)$ is zero for even n_1 and odd n_2 or odd n_1 and even n_2.

(b) Determine $P_c(\Omega_1, \Omega_2)$. Note that the Fourier transform of

$$\sum_{n1=-\infty}^{\infty} \sum_{n2=-\infty}^{\infty} \delta(t_1 - n_1 T, t_2 - n_2 T)$$

is given by

$$\left(\frac{2\pi}{T}\right)^2 \sum_{r1=-\infty}^{\infty} \sum_{r2=-\infty}^{\infty} \delta\left(\Omega_1 - \frac{2\pi r_1}{T}, \Omega_2 - \frac{2\pi r_2}{T}\right).$$

(c) Express $X_s(\Omega_1, \Omega_2)$ in terms of $X_c(\Omega_1, \Omega_2)$.
(d) Express $X(\omega_1, \omega_2)$ in terms of $X_s(\Omega_1, \Omega_2)$.

(e) From the results from (c) and (d), express $X(\omega_1, \omega_2)$ in terms of $X_c(\Omega_1, \Omega_2)$.

(f) Using the result from (e), sketch an example of $X_c(\Omega_1, \Omega_2)$ and of $X(\omega_1, \omega_2)$.

(g) Suppose $x_c(t_1, t_2)$ is band-limited to a circular region such that $X_c(\Omega_1, \Omega_2)$ has the following property:

$$X_c(\Omega_1, \Omega_2) = 0, \qquad \sqrt{\Omega_1^2 + \Omega_2^2} \geq \Omega_c.$$

Determine the conditions on T_1 and T_2 such that $x_c(t_1, t_2)$ can be exactly recovered from $x(n_1, n_2)$.

(h) Comparing the result from (g) with the corresponding result based on using a rectangular sampling grid, discuss which sampling grid is more efficient for exactly reconstructing $x_c(t_1, t_2)$ from a smaller number of samples of $x_c(t_1, t_2)$ if $X_c(\Omega_1, \Omega_2)$ is band-limited to a circular region.

(i) Show that the sampling grid used in this problem is very efficient compared with sampling on a rectangular grid when $x_c(t_1, t_2)$ is band-limited to the hexagonal region shown below.

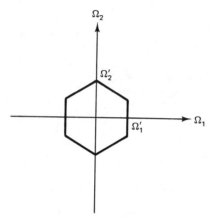

Figure P1.35b

We now derive the results for the ideal D/A converter for a hexagonal sampling grid. When $x_c(t_1, t_2)$ is sampled at a sufficiently high rate, the ideal D/A converter recovers $x_c(t_1, t_2)$ exactly from $x(n_1, n_2)$. It is convenient to represent the ideal D/A converter by the system shown below:

$$x(n_1, n_2) \longrightarrow \boxed{G^{-1}} \xrightarrow{\;x_s(t_1, t_2)\;} \boxed{H_c(\Omega_1, \Omega_2)} \xrightarrow{\;y_c(t_1, t_2)\;}$$

The system G^{-1} is the inverse of the system G discussed above.

(j) Express $x_s(t_1, t_2)$ in terms of $x(n_1, n_2)$.

(k) Determine the frequency response $H_c(\Omega_1, \Omega_2)$ such that $y_c(t_1, t_2) = x_c(t_1, t_2)$. Assume that $x(n_1, n_2)$ was obtained by sampling $x_c(t_1, t_2)$ at a sufficiently high rate on the hexagonal sampling grid.

(l) Using the results from (j) and (k), express $y_c(t_1, t_2)$ in terms of $x(n_1, n_2)$.

2

The z-Transform

2.0 INTRODUCTION

The Fourier transform discussed in Chapter 1 converges uniformly only for stable sequences. As a result, many important classes of sequences, for instance, the unit step sequence $u(n_1, n_2)$, cannot be represented by the Fourier transform. In this chapter, we study the z-transform signal representation and related topics. The z-transform converges for a much wider class of signals than does the Fourier transform. In addition, the z-transform representation of signals and systems is very useful in studying such topics as difference equations and system stability. In Section 2.1, we discuss the z-transform and its properties. Section 2.2 deals with the linear constant coefficient difference equation, which is important in studying digital filters. Section 2.3 treats topics related to system stability. In our discussion of the z-transform, the linear constant coefficient difference equation, and system stability, we will find fundamental differences between one-dimensional and two-dimensional results.

2.1 THE z-TRANSFORM

2.1.1 Definition of the z-Transform

The z-transform of a sequence $x(n_1, n_2)$ is denoted by $X(z_1, z_2)$ and is defined by*

$$X(z_1, z_2) = \sum_{n_1 = -\infty}^{\infty} \sum_{n_2 = -\infty}^{\infty} x(n_1, n_2) z_1^{-n_1} z_2^{-n_2} \qquad (2.1)$$

*The reader is cautioned that the notation used here creates an ambiguity between $X(\omega_1, \omega_2)$ and $X(z_1, z_2)$. For example, $X(1, 1)$ could mean either $X(z_1, z_2)|_{z_1=1, z_2=1}$ or $X(\omega_1, \omega_2)|_{\omega_1=1, \omega_2=1}$, which is different. In addition, we also denote the discrete Fourier transform discussed in Chapter 3 as $X(k_1, k_2)$. Although this is clearly an abuse of notation, we have adopted this notation for simplicity. In most cases, what is meant is clear from the context or the variables used in the arguments. For example, $X(z_1, z_2)$ refers to the z-transform. In cases where ambiguities arise, we will be explicit about what is meant, by using such notations as $X(z_1, z_2)|_{z_1=1, z_2=1}$ and $X(\omega_1, \omega_2)|_{\omega_1=1, \omega_2=1}$.

where z_1 and z_2 are complex variables. Since each of these two variables is complex-valued and therefore spans a 2-D space, the space spanned by (z_1, z_2) is 4-D. It is thus extremely difficult to visualize points or segments of $X(z_1, z_2)$.

From (2.1), $X(z_1, z_2)$ is related to $X(\omega_1, \omega_2)$ by

$$X(z_1, z_2)\big|_{z_1 = e^{j\omega_1}, z_2 = e^{j\omega_2}} = \sum_{n_1 = -\infty}^{\infty} \sum_{n_2 = -\infty}^{\infty} x(n_1, n_2) e^{-j\omega_1 n_1} e^{-j\omega_2 n_2} = X(\omega_1, \omega_2). \quad (2.2)$$

From (2.2), $X(\omega_1, \omega_2)$ is $X(z_1, z_2)$ evaluated at $z_1 = e^{j\omega_1}$ and $z_2 = e^{j\omega_2}$. This is one reason why the z-transform is considered a generalization of the Fourier transform. The 2-D space spanned by $(z_1 = e^{j\omega_1}, z_2 = e^{j\omega_2})$ is called the *unit surface*.

Suppose $X(z_1, z_2)$ in (2.1) is evaluated along $(z_1 = r_1 e^{j\omega_1}, z_2 = r_2 e^{j\omega_2})$, where r_1 and ω_1 are the radius and argument in the z_1 plane and r_2 and ω_2 are the radius and argument in the z_2 plane. The function $X(z_1, z_2)$ can be expressed as

$$X(z_1, z_2)\big|_{z_1 = r_1 e^{j\omega_1}, z_2 = r_2 e^{j\omega_2}} = \sum_{n_1 = -\infty}^{\infty} \sum_{n_2 = -\infty}^{\infty} x(n_1, n_2) r_1^{-n_1} r_2^{-n_2} e^{-j\omega_1 n_1} e^{-j\omega_2 n_2}$$

$$= F[x(n_1, n_2) r_1^{-n_1} r_2^{-n_2}] \quad (2.3)$$

where $F[x(n_1, n_2) r_1^{-n_1} r_2^{-n_2}]$ represents the Fourier transform of $x(n_1, n_2) r_1^{-n_1} r_2^{-n_2}$. Since uniform convergence of $F[x(n_1, n_2)]$ requires the absolute summability of $x(n_1, n_2)$, from (2.3), uniform convergence of the z-transform of $x(n_1, n_2)$ requires the absolute summability of $x(n_1, n_2) r_1^{-n_1} r_2^{-n_2}$:

$$\sum_{n_1 = -\infty}^{\infty} \sum_{n_2 = -\infty}^{\infty} |x(n_1, n_2)| r_1^{-n_1} r_2^{-n_2} < \infty \quad (2.4)$$

where $r_1 = |z_1|$ and $r_2 = |z_2|$.

From (2.4), the convergence of $X(z_1, z_2)$ will generally depend on the value of $r_1 = |z_1|$ and $r_2 = |z_2|$. For example, for the unit step sequence $u(n_1, n_2)$, $r_1^{-n_1} r_2^{-n_2} u(n_1, n_2)$ is absolutely summable only for $(|z_1| > 1, |z_2| > 1)$, so its z-transform converges only for $(|z_1| > 1, |z_2| > 1)$. The region in the (z_1, z_2) space where $X(z_1, z_2)$ uniformly converges is called the *region of convergence* (ROC).

For 1-D signals, the ROC is typically bounded by two circles in the z plane whose origins are at $|z| = 0$, as shown in Figure 2.1(a). For 2-D signals, (z_1, z_2) spans a 4-D space, so the ROC cannot be sketched analogously to Figure 2.1(a). Fortunately, however, the ROC depends only on $|z|$ for 1-D signals and $|z_1|$ and $|z_2|$ for 2-D signals. Therefore, an alternative way to sketch the ROC for 1-D signals would be to use the $|z|$ axis. The ROC in Figure 2.1(a) sketched using the $|z|$ axis is shown in Figure 2.1(b). In this sketch, each point on the $|z|$ axis corresponds to a 1-D contour in the z plane. For 2-D signals, we can use the $(|z_1|, |z_2|)$ plane to sketch the ROC; an example is shown in Figure 2.1(c). In this sketch, the point at $(|z_1| = 1, |z_2| = 1)$ corresponds to the 2-D unit surface, and each point in the $(|z_1|, |z_2|)$ plane corresponds to a 2-D subspace in the 4-D (z_1, z_2) space. The ROC plays an important role in the z-transform representation of a sequence, as we shall see shortly.

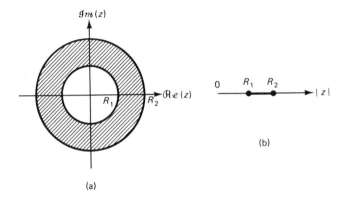

(a)

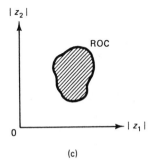

(b)

(c)

Figure 2.1 Representation of ROCs (regions of convergence) for 1-D and 2-D z-transforms.

2.1.2 Examples of the z-Transform

Example 1

We wish to determine the z-transform and its ROC for the following sequence:

$$x(n_1, n_2) = a^{n_1}b^{n_2}u(n_1, n_2).$$

The sequence is sketched in Figure 2.2(a). Using the z-transform definition in (2.1),

$$X(z_1, z_2) = \sum_{n_1 = -\infty}^{\infty} \sum_{n_2 = -\infty}^{\infty} a^{n_1}b^{n_2}u(n_1, n_2)z_1^{-n_1}z_2^{-n_2}$$

$$= \sum_{n_1 = 0}^{\infty} (az_1^{-1})^{n_1} \sum_{n_2 = 0}^{\infty} (bz_2^{-1})^{n_2}$$

$$= \frac{1}{1 - az_1^{-1}} \frac{1}{1 - bz_2^{-1}}, \qquad |z_1| > |a| \qquad \text{and} \qquad |z_2| > |b|.$$

The ROC is sketched in Figure 2.2(b).

For 1-D signals, poles of $X(z)$ are points in the z plane. For 2-D signals, poles of $X(z_1, z_2)$ are 2-D surfaces in the 4-D (z_1, z_2) space. In Example 1, for instance, the poles of $X(z_1, z_2)$ can be represented as $(z_1 = a,$ any $z_2)$ and (any $z_1,$

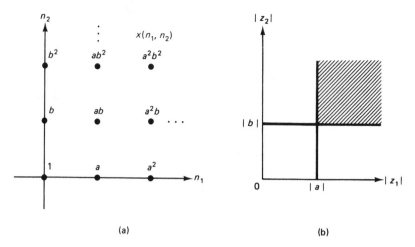

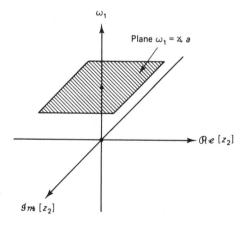

Figure 2.2 (a) Sequence $x(n_1, n_2) = a^{n_1}b^{n_2}u(n_1, n_2)$ and (b) ROC of its z-transform.

$z_2 = b$). Each of the two pole representations is a 2-D surface in the 4-D (z_1, z_2) space.

For the 1-D case, the ROC is bounded by poles. For the 2-D case, the ROC is bounded by pole surfaces. To illustrate this, consider the pole surfaces in Example 1. Taking the magnitude of z_1 and z_2 that corresponds to the pole surfaces, we have $|z_1| = |a|$ and $|z_2| = |b|$. These are the two solid lines that bound the ROC, as shown in Figure 2.2(b). Note that each of the two solid lines in Figure 2.2(b) is a 3-D space, since each point in the $(|z_1|, |z_2|)$ plane is a 2-D space. The pole surfaces, therefore, lie in the 2-D subspace within the 3-D spaces corresponding to the two solid lines in Figure 2.2(b). To illustrate this point more clearly, consider the 3-D space [the vertical solid line in Figure 2.2(b)] corresponding to $|z_1| = |a|$. This 3-D space can be represented as $(z_1 = |a|e^{j\omega_1}$, any $z_2)$, as shown in Figure 2.3. The pole surface corresponding to $(z_1 = a$, any $z_2)$ is sketched as the shaded region in Figure 2.3.

Figure 2.3 Pole surfaces within the 3-D space of $|z_1| = |a|$ for the sequence in Figure 2.2(a).

Example 2

We wish to determine the z-transform and its ROC for the following sequence:

$$x(n_1, n_2) = -a^{n_1}b^{n_2}u(-n_1 - 1, n_2).$$

The sequence is sketched in Figure 2.4(a). Using the z-transform definition in (2.1), and after a little algebra, we have

$$X(z_1, z_2) = \frac{1}{1 - az_1^{-1}}\frac{1}{1 - bz_2^{-1}},$$

ROC: $|z_1| < |a|, |z_2| > |b|$

Pole surfaces: $(z_1 = a, \text{ any } z_2), (\text{any } z_1, z_2 = b).$

The ROC is sketched in Figure 2.4(b).

Examples 1 and 2 show the importance of the ROC in the z-transform representation of a sequence. Even though the two sequences in Examples 1 and 2 are very different, their z-transforms are exactly the same. Given only the z-transform, therefore, the sequence cannot be uniquely determined. Uniquely determining the sequence requires not only the z-transform but also its ROC.

Example 3

We wish to determine the z-transform and its ROC for the following sequence:

$$x(n_1, n_2) = a^{n_1}u(n_1, n_2)u_T(n_1 - n_2).$$

The sequence is sketched in Figure 2.5(a). Using the z-transform definition in (2.1), we have

$$X(z_1, z_2) = \sum_{n1=-\infty}^{\infty} \sum_{n2=-\infty}^{\infty} a^{n_1}u(n_1, n_2)u_T(n_1 - n_2)z_1^{-n_1}z_2^{-n_2}$$

$$= \sum_{n1=0}^{\infty} \sum_{n2=0}^{n1} a^{n_1}z_1^{-n_1}z_2^{-n_2}$$

$$= \sum_{n1=0}^{\infty} \left(a^{n_1}z_1^{-n_1} \sum_{n2=0}^{n1} z_2^{-n_2}\right)$$

$$= \sum_{n1=0}^{\infty} a^{n_1}z_1^{-n_1} \frac{1 - (z_2^{-1})^{n_1+1}}{1 - z_2^{-1}}$$

$$= \frac{1}{1 - z_2^{-1}} \left(\sum_{n1=0}^{\infty} a^{n_1}z_1^{-n_1} - z_2^{-1}\sum_{n1=0}^{\infty} a^{n_1}z_1^{-n_1}z_2^{-n_1}\right).$$

Since $\sum_{n1=0}^{\infty} a^{n_1}z_1^{-n_1}$ converges to $1/(1 - az_1^{-1})$ for $|az_1^{-1}| < 1$ and $\sum_{n1=0}^{\infty} a^{n_1}z_1^{-n_1}z_2^{-n_1}$ converges to $1/(1 - az_1^{-1}z_2^{-1})$ for $|az_1^{-1}z_2^{-1}| < 1$, after a little algebra,

$$X(z_1, z_2) = \frac{1}{(1 - az_1^{-1})(1 - az_1^{-1}z_2^{-1})},$$

ROC: $|z_1| > |a|, |z_1z_2| > |a|$

Pole surfaces: $(z_1 = a, \text{ any } z_2), \left(\text{any } z_1 \neq 0, z_2 = \frac{a}{z_1}\right).$

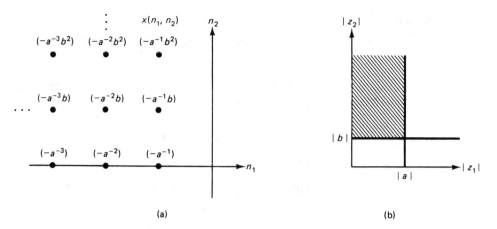

Figure 2.4 (a) Sequence $x(n_1, n_2) = -a^{n_1}b^{n_2}u(-n_1 - 1, n_2)$ and (b) ROC of its z-transform.

The ROC is sketched in Figure 2.5(b). The two pole surfaces lie in the 2-D subspaces within the two solid lines (3-D spaces) in Figure 2.5(b).

Example 4

We wish to determine the z-transform and its ROC for the following sequence:

$$x(n_1, n_2) = \frac{(n_1 + n_2)!}{n_1!n_2!} a^{n_1}b^{n_2}u(n_1, n_2).$$

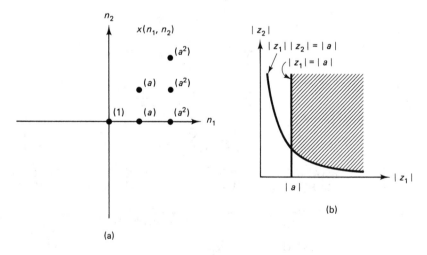

Figure 2.5 (a) Sequence $x(n_1, n_2) = a^{n_1}u(n_1, n_2)u_T(n_1 - n_2)$ and (b) ROC of its z-transform.

The sequence is sketched in Figure 2.6(a). Using the z-transform definition in (2.1),

$$X(z_1, z_2) = \sum_{n1=-\infty}^{\infty} \sum_{n2=-\infty}^{\infty} \frac{(n_1 + n_2)!}{n_1!n_2!} a^{n_1}b^{n_2}u(n_1, n_2)z_1^{-n_1}z_2^{-n_2}$$

$$= \sum_{n1=0}^{\infty} \sum_{n2=0}^{\infty} \frac{(n_1 + n_2)!}{n_1!n_2!} (az_1^{-1})^{n_1}(bz_2^{-1})^{n_2}.$$

Letting $m = n_1 + n_2$, $X(z_1, z_2)$ can be expressed as

$$X(z_1, z_2) = \sum_{n1=0}^{\infty} \sum_{m=n1}^{\infty} \frac{m!}{n_1!(m - n_1)!} (az_1^{-1})^{n_1}(bz_2^{-1})^{m-n_1}$$

$$= \sum_{m=0}^{\infty} \sum_{n1=0}^{m} \frac{m!}{n_1!(m - n_1)!} (az_1^{-1})^{n_1}(bz_2^{-1})^{m-n_1}.$$

Noting that

$$(x + y)^m = \sum_{n1=0}^{m} \frac{m!}{n_1!(m - n_1)!} x^{n_1}y^{m-n_1} \text{ for } m \geq 0,$$

$$X(z_1, z_2) = \sum_{m=0}^{\infty} (az_1^{-1} + bz_2^{-1})^m$$

$$= \frac{1}{1 - (az_1^{-1} + bz_2^{-1})}, \quad |az_1^{-1} + bz_2^{-1}| < 1.$$

The pole surface is given by (any z_1, $z_2 = bz_1/(z_1 - a)$) and is the shaded (dotted) region in Figure 2.6(b). The pole surface is a 2-D surface in the 4-D (z_1, z_2) space and it is visible in this example as a 2-D space in the $(|z_1|, |z_2|)$ plane. To determine the ROC, we note that the ROC depends on $(|z_1|, |z_2|)$. For $(|z_1'|, |z_2'|)$ to be in the ROC, $(z_1 = |z_1'|e^{j\omega_1}, z_2 = |z_2'|e^{j\omega_2})$ has to satisfy $|az_1^{-1} + bz_2^{-1}| < 1$ for all (ω_1, ω_2). After some algebra, the ROC is given by

$$\text{ROC: } |a| |z_1|^{-1} + |b||z_2|^{-1} < 1.$$

The ROC is the shaded (lined) region in Figure 2.6(b). The ROC can also be obtained using (2.4). Note that some values of (z_1, z_2) which satisfy $|az_1^{-1} + bz_2^{-1}| < 1$ are not in the ROC. For example, $(z_1 = a, z_2 = -b)$ satisfies $|az_1^{-1} + bz_2^{-1}| < 1$, but it is not in the ROC. This is because $(z_1 = |a|e^{j\omega_1}, z_2 = |b|e^{j\omega_2})$ does not satisfy $|az_1^{-1} + bz_2^{-1}| < 1$ for all (ω_1, ω_2).

Example 5

We wish to determine the z-transform and its ROC for the following sequence:

$$x(n_1, n_2) = a^{n_1}u_T(n_1).$$

The sequence is sketched in Figure 2.7. Using the z-transform definition in (2.1), we have

$$X(z_1, z_2) = \sum_{n1=0}^{\infty} \sum_{n2=-\infty}^{\infty} a^{n_1}z_1^{-n_1}z_2^{-n_2}.$$

The above expression does not converge for any (z_1, z_2), and therefore the z-transform does not exist.

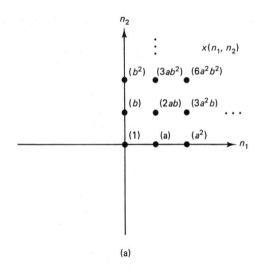

(a)

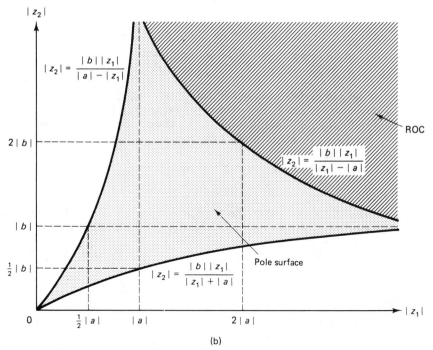

(b)

Figure 2.6 (a) Sequence $x(n_1, n_2) = ((n_1 + n_2)!/n_1! \, n_2!)a^{n_1}b^{n_2}u(n_1, n_2)$ and (b) ROC of its z-transform.

2.1.3 Properties of the Region of Convergence

Many useful properties of the ROC can be obtained from the z-transform definition in (2.1). Some of the more important properties are listed in Table 2.1.

Property 2 provides a necessary and sufficient condition for a sequence with

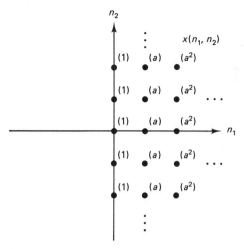

Figure 2.7 Sequence $x(n_1, n_2) = a^{n_1} u_T(n_1)$.

a valid z-transform to have first-quadrant support. The condition is that for any (z_1', z_2') in the ROC, all (z_1, z_2) in the shaded region in Figure 2.8 are also in the ROC. The sketch in Figure 2.8 is called a *constraint map*, since it shows the constraints that the ROC of any first-quadrant support sequence has to satisfy. This property can be demonstrated straightforwardly. To see that a first-quadrant support sequence implies the condition in the constraint map, note that the z-transform of any first-quadrant support sequence $x(n_1, n_2)$ can be expressed as

$$X(z_1, z_2) = \sum_{n_1=0}^{\infty} \sum_{n_2=0}^{\infty} x(n_1, n_2) z_1^{-n_1} z_2^{-n_2}. \tag{2.5}$$

Consider any (z_1', z_2') in the ROC. From the definition of the ROC, then,

$$\sum_{n_1=0}^{\infty} \sum_{n_2=0}^{\infty} |x(n_1, n_2)| |z_1'|^{-n_1} |z_2'|^{-n_2} < \infty. \tag{2.6}$$

Now consider any (z_1, z_2) such that $|z_1| \geq |z_1'|$ and $|z_2| \geq |z_2'|$. We have

$$\sum_{n_1=0}^{\infty} \sum_{n_2=0}^{\infty} |x(n_1, n_2)| |z_1|^{-n_1} |z_2|^{-n_2} \leq \sum_{n_1=0}^{\infty} \sum_{n_2=0}^{\infty} |x(n_1, n_2)| |z_1'|^{-n_1} |z_2'|^{-n_2} < \infty. \tag{2.7}$$

TABLE 2.1 PROPERTIES OF THE REGION OF CONVERGENCE

Property 1. An ROC is bounded by pole surfaces and is a connected region with no pole surfaces inside the ROC.

Property 2. First-quadrant support sequence $\Longleftrightarrow$ For any $(|z_1'|, |z_2'|)$ in the ROC, all $(|z_1| \geq |z_1'|, |z_2| \geq |z_2'|)$ are in the ROC.

Property 3. Finite-extent sequence $\Longleftrightarrow$ ROC is everywhere except possibly ($|z_1| = 0$ or ∞, $|z_2| = 0$ or ∞).

Property 4. Stable sequence $\Longleftrightarrow$ ROC includes the unit surface, ($|z_1| = 1$, $|z_2| = 1$).

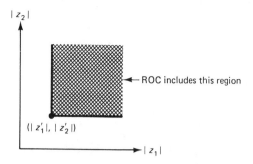

ROC includes this region

Figure 2.8 Constraint map for ROC of a first-quadrant support sequence. For any (z_1', z_2') in the ROC, all (z_1, z_2) in the shaded region are also in the ROC.

Equation (2.7) shows that if (z_1', z_2') is in the ROC, then all (z_1, z_2) such that $|z_1| \geq |z_1'|$ and $|z_2| \geq |z_2'|$ will also be in the ROC; this is the condition in the constraint map of Figure 2.8. To demonstrate that the condition in the constraint map implies that the sequence will be a first-quadrant support sequence, we will show that if the sequence is not a first-quadrant support sequence, then the condition in the constraint map cannot be true. We can do this because "B implies A" is logically equivalent to "Not A implies not B." If a sequence is not a first-quadrant support sequence, then the z-transform of such a sequence can be expressed as

$$X(z_1, z_2) = \sum_{n_1=0}^{\infty} \sum_{n_2=0}^{\infty} x(n_1, n_2) z_1^{-n_1} z_2^{-n_2} + \text{additional term(s)} \quad (2.8)$$

where the additional term(s) are in the form $z_1^{m_1} z_2^{m_2}$, where m_1 or m_2 is a positive integer. Because of the additional term(s), there is a pole surface at $|z_1| = \infty$, $|z_2| = \infty$, or both. Hence, it is not possible for the ROC to include all (z_1, z_2) such that $|z_1| > |z_1'|$ and $|z_2| > |z_2'|$ where (z_1', z_2') is in the ROC.

Examples of ROCs that satisfy the condition in the constraint map of Figure 2.8 were shown in Figures 2.2(b), 2.5(b), and 2.6(b). Three sequences whose z-transforms have the three ROCs are $x(n_1, n_2) = a^{n_1} b^{n_2} u(n_1, n_2)$, $x(n_1, n_2) = a^{n_1} u(n_1, n_2) u_T(n_1 - n_2)$, and $x(n_1, n_2) = ((n_1 + n_2)!/n_1! n_2!) a^{n_1} b^{n_2} u(n_1, n_2)$, as discussed in Section 2.1.2. All three are first-quadrant support sequences.

Constraint maps can also be obtained for other quadrant support sequences and other special support sequences. These are shown in Figure 2.9 along with the constraint map for first-quadrant support sequences.

2.1.4 Properties of the z-Transform

Many properties of the z-transform can be obtained from the z-transform definition in (2.1). Some of the more important properties are listed in Table 2.2. All the properties except Property 3 and Property 7 can be viewed as straightforward extensions of the 1-D case. Property 3 applies to separable sequences, and Property 7 can be used in determining the z-transform of a first-quadrant support sequence obtained by linear mapping of the variables of a wedge support sequence.

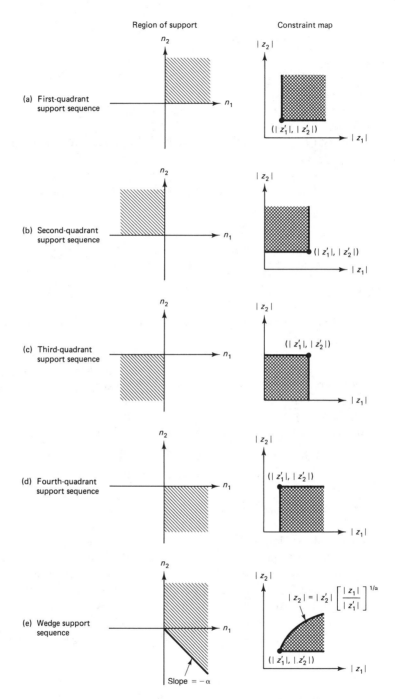

Region of support   Constraint map

(a) First-quadrant support sequence

(b) Second-quadrant support sequence

(c) Third-quadrant support sequence

(d) Fourth-quadrant support sequence

(e) Wedge support sequence

Figure 2.9   Constraint maps of ROCs of various special support sequences.

TABLE 2.2. PROPERTIES OF THE Z-TRANSFORM

$$x(n_1, n_2) \longleftrightarrow X(z_1, z_2), \text{ ROC: } R_x$$
$$y(n_1, n_2) \longleftrightarrow Y(z_1, z_2), \text{ ROC: } R_y$$

Property 1. Linearity
$$ax(n_1, n_2) + by(n_1, n_2) \longleftrightarrow aX(z_1, z_2) + bY(z_1, z_2) \qquad \text{ROC: at least } R_x \cap R_y$$

Property 2. Convolution
$$x(n_1, n_2) * y(n_1, n_2) \longleftrightarrow X(z_1, z_2)Y(z_1, z_2) \qquad \text{ROC: at least } R_x \cap R_y$$

Property 3. Separable Sequence
$$x_1(n_1)x_2(n_2) \longleftrightarrow X_1(z_1)X_2(z_2) \qquad \text{ROC: } |z_1| \in \text{ROC of } X_1(z_1) \text{ and } |z_2| \in \text{ROC of } X_2(z_2)$$

Property 4. Shift of a Sequence
$$x(n_1 - m_1, n_2 - m_2) \longleftrightarrow X(z_1, z_2)z_1^{-m_1}z_2^{-m_2}$$
$$\text{ROC: } R_x \text{ with possible exception of } |z_1| = 0, \infty, |z_2| = 0, \infty$$

Property 5. Differentiation

(a) $-n_1x(n_1, n_2) \longleftrightarrow \dfrac{\partial X(z_1, z_2)}{\partial z_1} z_1, \qquad \text{ROC: } R_x$

(b) $-n_2x(n_1, n_2) \longleftrightarrow \dfrac{\partial X(z_1, z_2)}{\partial z_2} z_2, \qquad \text{ROC: } R_x$

Property 6. Symmetry Properties
(a) $x^*(n_1, n_2) \longleftrightarrow X^*(z_1^*, z_2^*) \qquad \text{ROC: } R_x$
(b) $x(-n_1, -n_2) \longleftrightarrow X(z_1^{-1}, z_2^{-1}) \qquad \text{ROC: } |z_1^{-1}|, |z_2^{-1}| \text{ in } R_x$

Property 7. Linear Mapping of Variables
$$x(n_1, n_2) = y(m_1, m_2)\big|_{m_1 = l_1n_1 + l_2n_2, m_2 = l_3n_1 + l_4n_2} \longleftrightarrow Y(z_1, z_2) = X(z_1^{l_1}z_2^{l_3}, z_1^{l_2}z_2^{l_4})$$
$$\text{ROC: } (|z_1^{l_1}z_2^{l_3}|, |z_1^{l_2}z_2^{l_4}|) \text{ in } R_x$$
Note: For those points of $y(m_1, m_2)$ that do not correspond to any $x(n_1, n_2)$, $y(m_1, m_2)$ is taken to be zero.

Property 8. Useful Relations
(a) For a first-quadrant support sequence, $x(0, 0) = \lim\limits_{z_1 \to \infty} \lim\limits_{z_2 \to \infty} X(z_1, z_2)$
(b) For a third-quadrant support sequence, $x(0, 0) = \lim\limits_{z_1 \to 0} \lim\limits_{z_2 \to 0} X(z_1, z_2)$

2.1.5 The Inverse z-Transform

The z-transform definition in (2.1) can be used in determining the z-transform and ROC of a 2-D sequence. As in the 1-D case, using (2.1) and Cauchy's integral theorem, we can determine the inverse z-transform relation that expresses $x(n_1, n_2)$ as a function of $X(z_1, z_2)$ and its ROC. The inverse z-transform relation, which is a straightforward extension of the 1-D result, is given by

$$x(n_1, n_2) = \frac{1}{(2\pi j)^2} \oint_{C_1} \oint_{C_2} X(z_1, z_2)z_1^{n_1-1}z_2^{n_2-1} \, dz_1 \, dz_2 \qquad (2.9)$$

where C_1 and C_2 are both in the ROC of $X(z_1, z_2)$, C_1 is a closed contour that encircles in a counterclockwise direction the origin in the z_1 plane for any fixed z_2 on C_2, and C_2 is a closed contour that encircles in a counterclockwise direction the origin in the z_2 plane for any fixed z_1 on C_1.

Although the conditions that the contours C_1 and C_2 in (2.9) must satisfy appear complicated, there exists a simple way to determine the contours C_1 and C_2 when the ROC is given. Suppose $(|z_1'|, |z_2'|)$ lies in the ROC. One set of the contours C_1 and C_2 that satisfies the conditions in (2.9) is

$$C_1: z_1 = |z_1'|e^{j\omega_1}, \qquad \omega_1: 0 \text{ to } 2\pi \tag{2.10}$$
$$C_2: z_2 = |z_2'|e^{j\omega_2}, \qquad \omega_2: 0 \text{ to } 2\pi.$$

If the sequence is stable so that the ROC includes the unit surface, one possible set of contours is

$$C_1: z_1 = e^{j\omega_1}, \qquad \omega_1: 0 \text{ to } 2\pi \tag{2.11}$$
$$C_2: z_2 = e^{j\omega_2}, \qquad \omega_2: 0 \text{ to } 2\pi.$$

If this set of contours is chosen in (2.9), then (2.9) reduces to the inverse Fourier transform equation of (1.31b). This again shows that the Fourier transform representation is a special case of the z-transform representation.

From (2.9), the result of the contour integration differs depending on the contours. Therefore, a sequence $x(n_1, n_2)$ is not uniquely specified by $X(z_1, z_2)$ alone. Since the contours C_1 and C_2 can be determined from the ROC of $X(z_1, z_2)$, however, a sequence $x(n_1, n_2)$ is uniquely specified by both $X(z_1, z_2)$ and its ROC:

$$x(n_1, n_2) \leftrightarrow X(z_1, z_2), \text{ ROC}. \tag{2.12}$$

In theory, (2.9) can be used directly in determining $x(n_1, n_2)$ from $X(z_1, z_2)$ and its ROC. In practice, however, it is extremely tedious to perform the contour integration even for simple problems. The curious reader may attempt (see Problem 2.13) to determine a stable sequence by performing the contour integration in (2.9) when $X(z_1, z_2)$ is given by

$$X(z_1, z_2) = \frac{1}{1 - \frac{1}{2}z_1^{-1} - \frac{1}{4}z_1^{-1}z_2^{-1}}. \tag{2.13}$$

The sequence $x(n_1, n_2)$ in this problem is given by

$$x(n_1, n_2) = \begin{cases} \dfrac{n_1!}{(n_1 - n_2)!n_2!} (\tfrac{1}{2})^{n_1} (\tfrac{1}{2})^{n_2}, & n_1 \geq 0, n_2 \geq 0, n_1 \geq n_2 \\ 0 & , \quad \text{otherwise.} \end{cases} \tag{2.14}$$

For 1-D signals, approaches that have been useful in performing the inverse z-transform operation without explicit evaluation of a contour integral are series expansion, partial fraction expansion, and the inverse z-transform using the z-transform properties. The partial fraction expansion method, combined with the inverse z-transform using properties, is a general method that can always be used in performing the inverse z-transform operation for any 1-D rational z-transform. In the 1-D partial fraction expansion method, the z-transform $X(z)$ is first expressed as a sum of simpler z-transforms by factoring the denominator polynomial as a product of lower-order polynomials as follows:

$$X(z) = \frac{N(z)}{D(z)} = N_0(z) + \frac{N_1(z)}{D_1(z)D_2(z)D_3(z) \cdots D_M(z)}$$

$$= N_0(z) + \frac{N_2(z)}{D_1(z)} + \frac{N_3(z)}{D_2(z)} + \cdots + \frac{N_{M+1}(z)}{D_M(z)}.$$

(2.15)

The inverse z-transform is then performed for each of the simpler z-transforms, and the results are combined to obtain $x(n)$.

For 2-D signals, unfortunately, the partial fraction expansion method is not a general procedure. In the 1-D case, the factorization of any 1-D polynomial $D(z)$ as a product of lower-order polynomials is guaranteed by the fundamental theorem of algebra. A 2-D polynomial, however, cannot in general be factored as a product of lower-order polynomials. Therefore, it is not generally possible to use a procedure analogous to (2.15).

Partly because of the difficulty involved in the partial fraction expansion method, no known practical method exists for performing the inverse z-transform of a general 2-D rational z-transform. This limits to some extent the usefulness of the 2-D z-transform as compared to the 1-D z-transform. For example, a procedure involving an inverse z-transform operation may not be useful in 2-D signal processing.

2.2 LINEAR CONSTANT COEFFICIENT DIFFERENCE EQUATIONS

2.2.1 Introduction to Linear Constant Coefficient Difference Equations

Difference equations play a more important role for discrete-space systems than do differential equations for analog systems. In addition to representing a wide class of discrete-space systems, difference equations can also be used to recursively generate their solutions. This latter feature can be exploited in realizing digital filters with infinite-extent impulse responses.

In this section, we consider a class of linear constant coefficient difference equations of the following form:

$$\sum_{(k_1,k_2)\in R_a} \sum a(k_1, k_2)y(n_1 - k_1, n_2 - k_2) = \sum_{(k_1,k_2)\in R_b} \sum b(k_1, k_2)x(n_1 - k_1, n_2 - k_2)$$

(2.16)

where $a(k_1, k_2)$ and $b(k_1, k_2)$ are known real sequences with a finite number of nonzero amplitudes, R_a represents the region in (k_1, k_2) such that $a(k_1, k_2)$ is nonzero, and R_b is similarly defined. For convenience, we will refer to a linear constant coefficient difference equation simply as a *difference equation*. In addition, we assume that $a(k_1, k_2)$ is zero for $(k_1, k_2) \notin R_a$ and $b(k_1, k_2)$ is zero for $(k_1, k_2) \notin R_b$.

A difference equation alone does not specify a system, since there are many solutions of $y(n_1, n_2)$ in (2.16) for a given $x(n_1, n_2)$. For example, if $y_1(n_1, n_2)$ is a solution to $y(n_1, n_2) = (1/2)y(n_1 - 1, n_2 + 1) + (1/2)y(n_1 + 1, n_2 - 1) + x(n_1, n_2)$, then so is $y_1(n_1, n_2) + f(n_1 + n_2)$ for any function f. To uniquely specify a solution, a set of boundary conditions is needed. Since these boundary conditions must provide sufficient information for us to determine specific functional forms and the constants associated with the functions, they typically consist of an infinite number of values in the output $y(n_1, n_2)$. In this regard, the 2-D case differs fundamentally from the 1-D case. In the 1-D case, a difference equation specifies a solution within arbitrary constants. For example, an Nth-order 1-D difference equation specifies a solution within N constants, so N initial conditions (N values in the output $y(n)$) are generally sufficient to uniquely specify a solution. In the 2-D case, we typically need a set of boundary conditions that comprise an infinite number of points in the output $y(n_1, n_2)$.

2.2.2 Difference Equations with Boundary Conditions

The problem of solving a difference equation with boundary conditions can be stated as follows: Given $x(n_1, n_2)$, and $y(n_1, n_2)$ for $(n_1, n_2) \in R_{BC}$, find the solution to

$$\sum_{(k_1, k_2) \in R_a} a(k_1, k_2)y(n_1 - k_1, n_2 - k_2) = \sum_{(k_1, k_2) \in R_b} b(k_1, k_2)x(n_1 - k_1, n_2 - k_2),$$

(2.17)

where R_{BC} represents the region where the boundary conditions are given. To develop methods for solving (2.17), we will attempt to extend the methods used for solving the corresponding 1-D problem to the 2-D case.

For the 1-D case, the problem of solving a difference equation with initial conditions can be stated as follows: Given $x(n)$, and $y(n)$ for $n \in R_{IC}$, find the solution to

$$\sum_{k \in R_a} a(k)y(n - k) = \sum_{k \in R_b} b(k)x(n - k)$$

(2.18)

where R_{IC} represents the region where the initial conditions are given. One approach to solving (2.18) is to obtain the homogeneous solution and a particular solution, form a total solution as the sum of the homogeneous solution and a particular solution, and then impose initial conditions. The homogeneous solution represents all possible solutions to (2.18) with $x(n) = 0$ and without the initial conditions. It can always be obtained by assuming an exponential sequence ka^n to be a solution, substituting it into (2.18), and then solving the resulting characteristic equation for a. The homogeneous solution contains a set of unknown constants that are determined by the initial conditions. A particular solution is any one solution that satisfies (2.18) without the initial conditions. A particular solution can typically be determined by inspection or by z-transforming (2.18), solving for $Y(z)$, and then inverse z-transforming $Y(z)$ with any valid ROC. The last step, imposing the initial conditions to the total solution to determine the unknown constants in the homogeneous solution, requires solving a set of linear equations. This approach can be used in solving (2.18), and typically leads to a closed-form expression for $y(n)$.

The above approach cannot be used for solving a 2-D difference equation with boundary conditions. First, there is no general procedure for obtaining the homogeneous solution. The homogeneous solution consists of unknown functions, and the specific functional form of ka^n used in the 1-D case cannot be used in the 2-D case. Second, a particular solution cannot generally be obtained by either inspection or the z-transform method, since there is no practical procedure for performing the inverse z-transform for the 2-D case. Furthermore, determining the unknown functions in the homogeneous solution by imposing the boundary conditions (an infinite number of known values of $y(n_1, n_2)$) is not a simple linear problem.

Another approach to solving (2.18) is to compute $y(n)$ recursively. This is the obvious way to determine $y(n)$ by using a computer, and can be used in solving the 2-D difference equation with boundary conditions in (2.17). To illustrate this approach, consider the following 2-D difference equation with boundary conditions:

$$y(n_1, n_2) = y(n_1 - 1, n_2) + y(n_1, n_2 - 1) + y(n_1 - 1, n_2 - 1) + x(n_1, n_2)$$

$$x(n_1, n_2) = \delta(n_1, n_2) \tag{2.19}$$

Boundary conditions: $y(n_1, n_2) = 1$, for $n_1 = -1$ or $n_2 = -1$.

The output $y(n_1, n_2)$ for $n_1 \geq 0$ and $n_2 \geq 0$ can be obtained recursively as

$$\begin{aligned}
y(0, 0) &= y(-1, 0) + y(0, -1) + y(-1, -1) + x(0, 0) = 4 \\
y(1, 0) &= y(0, 0) + y(1, -1) + y(0, -1) + x(1, 0) = 6 \\
y(0, 1) &= y(-1, 1) + y(0, 0) + y(-1, 0) + x(0, 1) = 6 \\
y(2, 0) &= \cdots.
\end{aligned} \tag{2.20}$$

The output $y(n_1, n_2)$ for $n_1 \geq 0$ and $n_2 \geq 0$ is shown in Figure 2.10.

Although the approach of computing the output recursively appears quite simple, the choice of boundary conditions that ensures that the difference equation will have a unique solution is not straightforward. If we impose too many boundary conditions, no solution may satisfy both the difference equation and boundary conditions. If we impose too few boundary conditions, there may be an infinite number of possible solutions. In the 1-D case, N initial conditions are typically both necessary and sufficient for an Nth-order difference equation. In the 2-D case, the choice of boundary conditions such that the difference equation has a unique solution is not that obvious.

Suppose we have chosen the boundary conditions such that the difference equation has a unique solution. The difference equation with boundary conditions can then be considered a system. In both the 1-D and the 2-D cases, the system is in general neither linear nor shift invariant. Consider the example in (2.19). The output $y(n_1, n_2)$ for $n_1 \geq 0$ and $n_2 \geq 0$ which results when the input $x(n_1, n_2)$ in (2.19) is doubled so that $x(n_1, n_2) = 2\delta(n_1, n_2)$ is shown in Figure 2.11. Com-

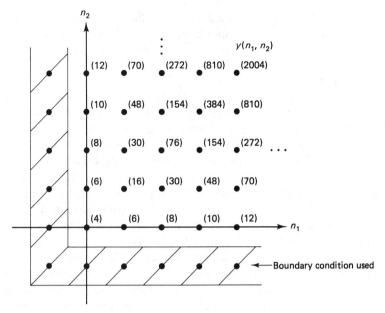

Figure 2.10 Output $y(n_1, n_2)$ for $n_1 \geq 0$ and $n_2 \geq 0$ of the system specified by (2.19) with $x(n_1, n_2) = \delta(n_1, n_2)$.

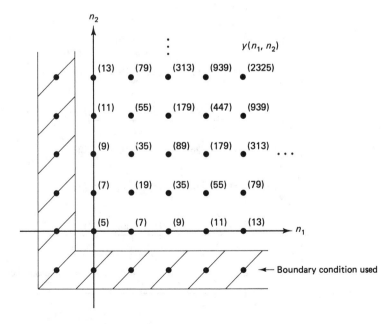

Figure 2.11 Output $y(n_1, n_2)$ for $n_1 \geq 0$ and $n_2 \geq 0$ of the system specified by (2.19) when $x(n_1, n_2) = 2\delta(n_1, n_2)$.

parison of Figures 2.10 and 2.11 shows that doubling the input does not double the output. When the input $x(n_1, n_2)$ in (2.19) is shifted so that $x(n_1, n_2) = \delta(n_1 - 1, n_2 - 1)$, the corresponding output $y(n_1, n_2)$ for $n_1 \geq 0$ and $n_2 \geq 0$ is shown in Figure 2.12. Comparison of Figures 2.10 and 2.12 shows that the system is not shift invariant.

Our main interest in the difference equation stems from the fact that the difference equation is the only practical way to realize an infinite impulse response (IIR) digital filter. Since a digital filter is a linear and shift-invariant (LSI) system, we need to force the difference equation to become an LSI system. This can be done by choosing a proper set of boundary conditions. Consider the difference equation in (2.19) and its solution (2.20). From (2.20), we can express each output value as a linear combination of the input and boundary conditions. Unless we choose the boundary conditions as a linear combination of the input values, it is impossible to have the output related linearly to only the input without setting the boundary conditions to zero. The linearity constraint, therefore, requires zero boundary conditions. To see a necessary condition imposed by the shift invariance property, note that the boundary conditions can be viewed as part of the output. If the output is to shift as the input is shifted, the boundary conditions must shift following the input. For the difference equation with boundary conditions to represent an LSI system, we require zero boundary conditions that shift as the input is shifted.

The above conditions are necessary but are not sufficient to guarantee that the difference equation with boundary conditions will be an LSI system. We have to ensure, for example, that the difference equation with boundary conditions will

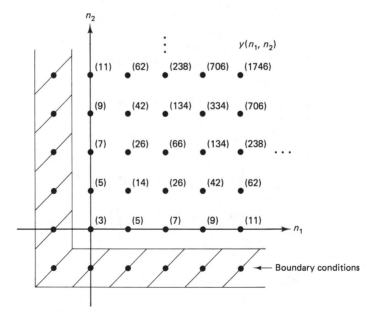

Figure 2.12 Output $y(n_1, n_2)$ for $n_1 \geq 0$ and $n_2 \geq 0$ of the system specified by (2.19) with $x(n_1, n_2) = \delta(n_1 - 1, n_2 - 1)$.

have a unique solution, so that it can be considered a system. We also have to ensure that the boundary conditions which are part of the output will not overlap with the result of convolving the input with the system's impulse response. If the overlap occurs, the part of the output specified by the boundary conditions is inconsistent with the output obtained from convolving the input with the system's impulse response, meaning that the system is not linear and shift invariant.

In the 1-D case, one way to force a difference equation to be an LSI system is by imposing an initial rest condition. An initial rest condition is defined as an initial condition obtained by requiring that the output $y(n)$ be zero for $n < n_0$ whenever the input $x(n)$ is zero for $n < n_0$. Clearly, the initial rest condition requires that zero initial conditions shift as the input shifts. The initial rest condition leads to a causal system. Since causality is a constraint for typical 1-D systems, imposing the initial rest condition is often the method of choice. In 2-D, causality is not typically a constraint, and boundary conditions that lead to LSI systems which are not "causal" are useful in practice. It is possible, though not straightforward, to generalize the basic idea behind the initial rest condition such that its extension to 2-D leads to 2-D LSI systems which are not restricted to be causal. Generalization of the basic idea behind the initial rest condition, its extension to the 2-D case, and the methods that follow are discussed in the next section.

2.2.3 Difference Equations as Linear Shift-Invariant Systems

One way to force a difference equation with boundary conditions to be an LSI system is to follow the three steps below in choosing the boundary conditions. As we discuss later, the initial rest condition can be viewed as a special case which results from following these three steps.

Step 1. Interpret the difference equation as a specific computational procedure.

Step 2. Determine R_{BC}, the region (n_1, n_2) for which the boundary conditions are applied, as follows:
(a) Determine R_h, the region of support of the impulse response $h(n_1, n_2)$.
(b) Determine R_y, the region of support of the output $y(n_1, n_2)$, from R_h and R_x, where R_x is similarly defined.
(c) R_{BC}: all $(n_1, n_2) \notin R_y$

Step 3. Boundary conditions: Set $y(n_1, n_2) = 0$ for $(n_1, n_2) \in R_{BC}$.

Step 1 ensures that we will choose one out of the many systems that can be derived from a given difference equation. Step 2 ensures that boundary conditions will be chosen such that they will shift as the input is shifted, and they will not overlap with the result of convolving the input with the impulse response of the system. Step 3 ensures that the boundary conditions we impose are zero boundary conditions. When we choose the boundary conditions following these three steps, we shall find that the resulting difference equation with boundary conditions will become an LSI system. We will now discuss each of the three steps in detail.

Step 1. In this step, we interpret a difference equation as a specific computational procedure. The best way to explain this step is with an example. Consider the following difference equation:

$$y(n_1, n_2) + 2y(n_1 - 1, n_2) + 3y(n_1, n_2 - 1)$$

$$+ 4y(n_1 - 1, n_2 - 1) = x(n_1, n_2). \quad (2.21)$$

Since there are four terms of the form $y(n_1 - k_1, n_2 - k_2)$, we obtain four equations by leaving only one of the four terms on the left-hand side of the equation:

$$y(n_1, n_2) = -2y(n_1 - 1, n_2) - 3y(n_1, n_2 - 1) \quad (2.22a)$$

$$- 4y(n_1 - 1, n_2 - 1) + x(n_1, n_2)$$

$$y(n_1 - 1, n_2) = -\tfrac{1}{2}y(n_1, n_2) - \tfrac{3}{2}y(n_1, n_2 - 1) \quad (2.22b)$$

$$- 2y(n_1 - 1, n_2 - 1) + \frac{1}{2}x(n_1, n_2)$$

$$y(n_1, n_2 - 1) = -\tfrac{1}{3}y(n_1, n_2) - \tfrac{2}{3}y(n_1 - 1, n_2) \quad (2.22c)$$

$$- \tfrac{4}{3}y(n_1 - 1, n_2 - 1) + \tfrac{1}{3}x(n_1, n_2)$$

$$y(n_1 - 1, n_2 - 1) = -\tfrac{1}{4}y(n_1, n_2) - \tfrac{1}{2}y(n_1 - 1, n_2) \quad (2.22d)$$

$$- \tfrac{3}{4}y(n_1, n_2 - 1) + \tfrac{1}{4}x(n_1, n_2).$$

By a simple change of variables, (2.22) can be rewritten so that the left-hand side of each equation has the form $y(n_1, n_2)$ as follows:

$$y(n_1, n_2) = -2y(n_1 - 1, n_2) - 3y(n_1, n_2 - 1) \quad (2.23a)$$

$$- 4y(n_1 - 1, n_2 - 1) + x(n_1, n_2)$$

$$y(n_1, n_2) = -\tfrac{1}{2}y(n_1 + 1, n_2) - \tfrac{3}{2}y(n_1 + 1, n_2 - 1) \quad (2.23b)$$

$$- 2y(n_1, n_2 - 1) + \tfrac{1}{2}x(n_1 + 1, n_2)$$

$$y(n_1, n_2) = -\tfrac{1}{3}y(n_1, n_2 + 1) - \tfrac{2}{3}y(n_1 - 1, n_2 + 1) \quad (2.23c)$$

$$- \tfrac{4}{3}y(n_1 - 1, n_2) + \tfrac{1}{3}x(n_1, n_2 + 1)$$

$$y(n_1, n_2) = -\tfrac{1}{4}y(n_1 + 1, n_2 + 1) - \tfrac{1}{2}y(n_1, n_2 + 1) \quad (2.23d)$$

$$- \tfrac{3}{4}y(n_1 + 1, n_2) + \tfrac{1}{4}x(n_1 + 1, n_2 + 1).$$

Even though all four equations in (2.23) are derived from the same difference equation, by proper interpretation they will correspond to four different specific computational procedures and therefore four different systems. In the interpretation we use, the left-hand side $y(n_1, n_2)$ is always computed from the right-hand side expression for all (n_1, n_2). When this interpretation is strictly followed, each of the four equations in (2.23) will correspond to a different computational procedure. This will be clarified when we discuss Step 2.

It is often convenient to represent a specific computational procedure pictorially. Suppose we wish to pictorially represent the computational procedure corresponding to (2.23a). To do so, we consider computing $y(0, 0)$. Since $y(n_1, n_2)$ on the left-hand side is always computed from the right-hand side, $y(0, 0)$ in this case is computed by

$$y(0, 0) \leftarrow -2y(-1, 0) - 3y(0, -1) - 4y(-1, -1) + x(0, 0). \quad (2.24)$$

We have used an arrow, ($\leftarrow$), to emphasize that $y(0, 0)$ is always computed from the right-hand side. Equation (2.24) is represented in Figure 2.13. The value $y(0, 0)$ that is computed is denoted by "x" in Figure 2.13(a). The values $y(-1, 0)$, $y(0, -1)$ and $y(-1, -1)$ that are used in obtaining $y(0, 0)$ are marked by a filled-in dot ($\bullet$) with the proper coefficient attached to the corresponding point. The value $x(0, 0)$ used in obtaining $y(0, 0)$ is marked by a filled-in dot in Figure 2.13(b). To compute $y(0, 0)$, therefore, we look at $y(n_1, n_2)$ and $x(n_1, n_2)$ at the points marked by filled-in dots, multiply each of these values by the corresponding scaling factor indicated, and sum all the terms. This is illustrated in Figure 2.13(c). Figure

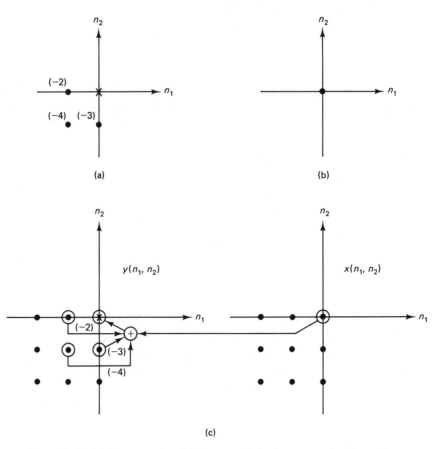

Figure 2.13 (a) Output mask and (b) input mask for the computational procedure in (2.24) and (c) graphical sketch of how $y(0, 0)$ is computed.

2.13(a) is called the *output mask* and Figure 2.13(b) is called the *input mask*, since they are masks that are applied to the output and input to compute $y(0, 0)$.

In the illustration, the output and input masks are sketched for the case when $y(0, 0)$ is computed. They are also very useful in visualizing how other points of $y(n_1, n_2)$ are computed. Suppose we wish to compute $y(3, 2)$ using the computational procedure in Figure 2.13. The points used in determining $y(3, 2)$ are marked by filled-in dots in Figure 2.14 with the proper scaling factors attached. Figure 2.14 is simply a shifted version of Figure 2.13.

As the above discussion suggests, a difference equation can lead to a number of different computational procedures. Which procedure is chosen from these possibilities depends on the context of the given problem. This point will be discussed after we discuss Steps 2 and 3.

Step 2. In this step, we determine R_y, the region of support of the system's output $y(n_1, n_2)$. To determine R_y, we first determine R_h, the region of support of the system's impulse response. To see how R_h is determined, consider the following computational procedure:

$$y(n_1, n_2) \leftarrow -2y(n_1 - 1, n_2) - 3y(n_1, n_2 - 1)$$
$$- 4y(n_1 - 1, n_2 - 1) + x(n_1, n_2). \qquad (2.25)$$

The output and input masks for this computational procedure are those shown in Figure 2.13. The region of support R_h is the region of (n_1, n_2) for which $y(n_1, n_2)$

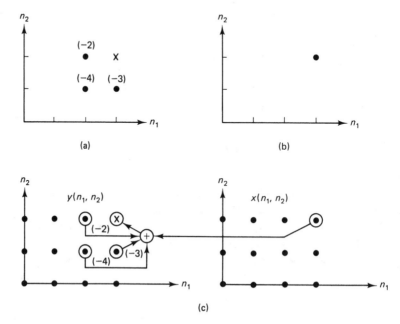

Figure 2.14 (a) Output mask and (b) input mask in Figure 2.13 shifted to illustrate (c) computation of $y(3, 2)$.

is influenced by the impulse* $\delta(n_1, n_2)$ when we set $x(n_1, n_2) = \delta(n_1, n_2)$. Consider $y(0, 0)$. From (2.25) or Figure 2.13, $y(0, 0)$ is influenced by $y(-1, 0)$, $y(0, -1)$, $y(-1, -1)$ and $\delta(0, 0)$. Clearly $y(0, 0)$ is influenced by the impulse $\delta(n_1, n_2)$. Let us now consider $y(1, 0)$, $y(0, 1)$, and $y(1, 1)$. From (2.25) or Figure 2.13,

$$y(1, 0) \leftarrow -2y(0, 0) - 3y(1, -1) - 4y(0, -1) + \delta(1, 0)$$

$$y(0, 1) \leftarrow -2y(-1, 1) - 3y(0, 0) - 4y(-1, 0) + \delta(0, 1) \qquad (2.26)$$

$$y(1, 1) \leftarrow -2y(0, 1) - 3y(1, 0) - 4y(0, 0) + \delta(1, 1).$$

Since $\delta(n_1, n_2)$ has already influenced $y(0, 0)$, and $y(0, 0)$ in turn influences $y(1, 0), y(0, 1),$ and $y(1, 1)$ from (2.26), $y(1, 0), y(0, 1),$ and $y(1, 1)$ will be influenced by the impulse $\delta(n_1, n_2)$. Now consider $y(-1, 0)$. From (2.25) or Figure 2.13, $y(-1, 0)$ is obtained from

$$y(-1, 0) \leftarrow -2y(-2, 0) - 3y(-1, -1) - 4y(-2, -1) + \delta(-1, 0). \qquad (2.27)$$

This is shown in Figure 2.15. The terms that influence $y(-1, 0)$ in (2.27) are obtained from

$$y(-2, 0) \leftarrow -2y(-3, 0) - 3y(-2, -1) - 4y(-3, -1) + \delta(-2, 0)$$

$$y(-1, -1) \leftarrow -2y(-2, -1) - 3y(-1, -2) - 4y(-2, -2) + \delta(-1, -1)$$

$$y(-2, -1) \leftarrow -2y(-3, -1) - 3y(-2, -2) - 4y(-3, -2) + \delta(-2, -1).$$

$$(2.28)$$

These are also shown in Figure 2.15. From Figure 2.15 and continuing the above procedure, we can see that the points that influence $y(-1, 0)$ are $y(n_1, n_2)$ in the shaded region in Figure 2.16. Since the impulse $x(n_1, n_2) = \delta(n_1, n_2)$ has its first impact on $y(0, 0)$, and $y(0, 0)$ does not in any way affect $y(-1, 0)$, $y(-1, 0)$ is not influenced by the impulse $\delta(n_1, n_2)$. If we consider all other points of $y(n_1, n_2)$ analogously, we can easily argue that the region of (n_1, n_2) for which $y(n_1, n_2)$ is influenced by $x(n_1, n_2) = \delta(n_1, n_2)$ is the shaded region shown in Figure 2.17. This is the region R_h. In essence, the impulse $\delta(n_1, n_2)$ has a direct effect only on $y(0, 0)$. Because of the specific computational procedure in (2.25), $y(0, 0)$ influences only the first-quadrant region. As a result, R_h is given by the shaded region in Figure 2.17.

Once R_h is determined, R_y can be obtained from R_h and R_x. Suppose $x(n_1, n_2)$ is given by the 3×3-point sequence shown in Figure 2.18(a). By convolving $x(n_1, n_2)$ and $h(n_1, n_2)$, R_y can be easily determined. Note that explicit convolution is not necessary to determine R_y. For the R_h considered above, R_y is given by the shaded region in Figure 2.18(b). R_{BC}, the region (n_1, n_2) for which the boundary conditions are applied, is determined to be all (n_1, n_2) outside R_y. In the current example, R_{BC} is given by the shaded region in Figure 2.18(c). Given

---

*If the right-hand side of (2.25) contains an additional term $x(n_1 - 1, n_2)$, then setting $x(n_1, n_2) = \delta(n_1, n_2)$ will cause two impulses $\delta(n_1, n_2)$ and $\delta(n_1 - 1, n_2)$ to be present in (2.25). In this case, R_h is the region of (n_1, n_2) for which $y(n_1, n_2)$ is influenced by either $\delta(n_1, n_2)$ and/or $\delta(n_1 - 1, n_2)$.

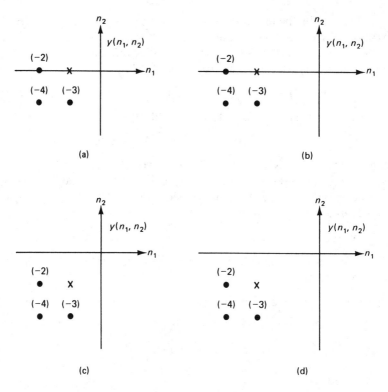

Figure 2.15 Output points that determine (a) $y(-1, 0)$, (b) $y(-2, 0)$, (c) $y(-1, -1)$, and (d) $y(-2, -1)$.

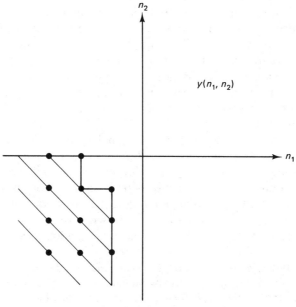

Figure 2.16 Output points that influence $y(-1, 0)$.

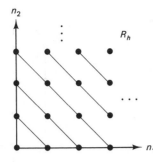

Figure 2.17 Region of support of the impulse response R_h for the computational procedure specified by (2.25).

a specific computational procedure, R_h is totally specified. As we shift the input, R_y is shifted by the same amount. This in turn shifts R_{BC} by the same amount that the input shifts, a necessary condition for the system to be shift invariant. In addition, boundary conditions do not overlap with R_y. In essence, we are determining the boundary conditions consistent with the resulting system being linear and shift invariant.

Step 3. In this step, we choose boundary conditions such that $y(n_1, n_2) = 0$ for all $(n_1, n_2) \in R_{BC}$. This is a necessary condition for the system to be linear. In the example considered in Step 2 above, the boundary conditions are chosen as

$$y(n_1, n_2) = 0 \quad \text{for} \quad n_1 < -1 \quad \text{or} \quad n_2 < -1. \tag{2.29}$$

Once the boundary conditions have been determined, we can compute the output $y(n_1, n_2)$ from the specific computational procedure, the input, and the boundary conditions. For the specific computational procedure of (2.25) and the input $x(n_1, n_2)$ in Figure 2.18(a), the output $y(n_1, n_2)$ is shown in Figure 2.19.

In the example above, we have assumed a specific computational procedure. As we mentioned before, many computational procedures are possible for a given difference equation. Which one is chosen depends on the problem context. The following example will show how this choice is made. Consider an infinite impulse response (IIR) filter whose system function $H(z_1, z_2)$ is given by

$$H(z_1, z_2) = \frac{1}{1 + 2z_1^{-1} + 3z_2^{-1} + 4z_1^{-1} z_2^{-1}}. \tag{2.30}$$

Suppose the IIR filter is designed so that the impulse response of the designed system is as close as possible in some sense to an ideal impulse response that is a first-quadrant support sequence. Then, we know that the filter is at least approximately a first-quadrant support system.* We can use that information to choose a specific computational procedure.

From (2.30), we can determine the corresponding difference equation by setting $H(z_1, z_2)$ to $Y(z_1, z_2)/X(z_1, z_2)$, cross-multiplying the resulting equation,

---

*In practice, we may even know the specific output and input masks from the design procedure. In this case, choosing a specific computational procedure follows directly from the known output and input masks. This is discussed further in Chapter 5.

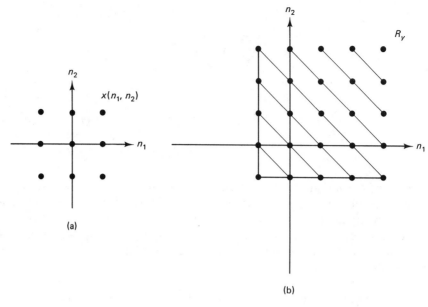

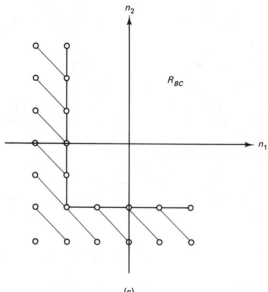

Figure 2.18 (a) $x(n_1, n_2)$; (b) R_y; (c) R_{BC}. R_y and R_{BC} are the region of support of $y(n_1, n_2)$ and boundary conditions, respectively, for the computational procedure specified by (2.25).

and then performing inverse z-transformation. The resulting difference equation is given by

$$y(n_1, n_2) + 2y(n_1 - 1, n_2) + 3y(n_1, n_2 - 1)$$

$$+ 4y(n_1 - 1, n_2 - 1) = x(n_1, n_2), \qquad (2.31)$$

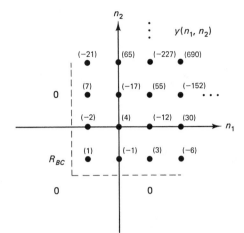

Figure 2.19 Output of the LSI system corresponding to (2.25), when the input used is shown in Figure 2.18(a).

which is identical to (2.21). From (2.31), we can derive four possible computational procedures, as shown in (2.23). We have already analyzed the first computational procedure given by (2.23a) in detail and have found that the corresponding $h(n_1, n_2)$ is a first-quadrant support sequence. Therefore, this computational procedure is consistent with the assumed information in this problem. If we analyze all three other computational procedures in detail, we can easily show that none corresponds to a first-quadrant support system. For example, consider the second computational procedure given by (2.23b). The output and input masks for this computational procedure are shown in Figures 2.20(a) and (b). The region of support R_h is the region of (n_1, n_2) for which $y(n_1, n_2)$ is influenced by the impulse $\delta(n_1, n_2)$ when we set $x(n_1, n_2) = \delta(n_1, n_2)$. The region R_h in this case is shown in Figure 2.20(c). This computational procedure is clearly inconsistent with the assumed information in the problem. In a similar manner, we can show that the two remaining computational procedures, given by (2.23c) and (2.23d), are also inconsistent with the assumption that the system is at least approximately a first-quadrant support system.

We have considered an example in which we began with $H(z_1, z_2)$, $x(n_1, n_2)$, and the approximate region of support of $h(n_1, n_2)$ and determined the output $y(n_1, n_2)$. This is how an IIR filter can be implemented. A summary of the steps involved is shown in Figure 2.21. From the system function $H(z_1, z_2)$, we can obtain the difference equation. From the difference equation and approximate region of support of $h(n_1, n_2)$, we can determine a specific computational procedure. From the specific computational procedure and the region of support of $x(n_1, n_2)$, we can determine the boundary conditions. From the specific computational procedure, boundary conditions, and $x(n_1, n_2)$, we can determine the output $y(n_1, n_2)$.

The procedure discussed above can be viewed as a generalization of the initial rest condition. We can illustrate this with a specific example. Consider a 1-D difference equation given by

$$y(n) - \tfrac{3}{4}y(n - 1) + \tfrac{1}{8}y(n - 2) = x(n). \tag{2.32}$$

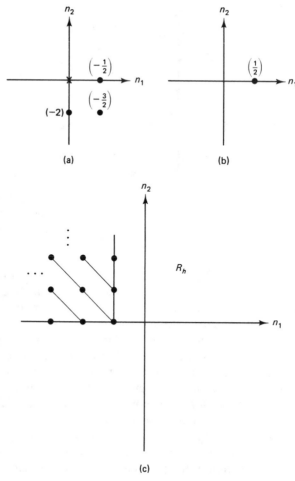

Figure 2.20 (a) Output mask, (b) input mask and (c) region of support of the impulse response for the computational procedure specified by (2.23b).

We assume that the input $x(n)$ to the system is given by

$$x(n) = \delta(n) + 2\delta(n - 1) + 4\delta(n - 2). \tag{2.33}$$

The initial rest condition in this example, from its definition, is

$$y(n) = 0, \quad n < 0. \tag{2.34}$$

Equation (2.34) can also be derived by using the procedure discussed in this section. From the difference equation in (2.32), we choose the specific computational procedure given by

$$y(n) \leftarrow \tfrac{3}{4}y(n - 1) - \tfrac{1}{8}y(n - 2) + x(n). \tag{2.35}$$

To determine R_h, the region of support of $h(n)$, we set $x(n) = \delta(n)$ and determine the values of n for which $y(n)$ is influenced by $\delta(n)$. From (2.35), R_h is given by $n \geq 0$. From R_x in (2.33) and R_h, R_y is given by $n \geq 0$. The initial conditions,

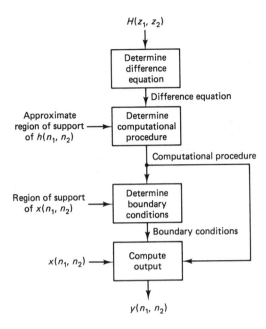

$H(z_1, z_2)$

Determine
difference
equation

Difference equation

Approximate
region of support → Determine
of $h(n_1, n_2)$ computational
procedure

Computational procedure

Region of support → Determine
of $x(n_1, n_2)$ boundary
conditions

Boundary conditions

$x(n_1, n_2)$ → Compute
output

$y(n_1, n_2)$

Figure 2.21 Steps involved in the implementation of an IIR filter.

then, are given by

$$y(n) = 0, \qquad n < 0. \tag{2.36}$$

This is identical to (2.34), showing that the initial rest condition in this example is a special case of the more general procedure to determine the boundary conditions discussed in this section.

2.2.4 Recursive Computability

The difference equation with boundary conditions plays a particularly important role in digital signal processing, since it is the only practical way to realize an IIR digital filter. As we discussed in previous sections, the difference equation can be used as a recursive procedure in computing the output. We define a system to be *recursively computable* when there exists a path we can follow in computing every output point recursively, one point at a time. The example in Section 2.2.3 corresponds to a recursively computable system, and all our discussions in that section were based on the implicit assumption that we were dealing with recursively computable systems.

From the definition of a recursively computable system, it can be easily shown that not all computational procedures resulting from difference equations are recursively computable. For example, consider a computational procedure whose output mask is shown in Figure 2.22. From the output mask, it is clear that computing $y(0, 0)$ requires $y(1, 0)$ and computing $y(1, 0)$ requires $y(0, 0)$. Therefore, we cannot compute $y(0, 0)$ and $y(1, 0)$ one at a time recursively, so the example in Figure 2.22 is not a recursively computable system. It can be shown that if a system is recursively computable, the output mask has wedge support.

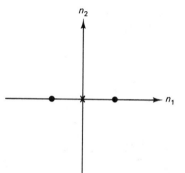

Figure 2.22 Example of an output mask corresponding to a system that is not recursively computable.

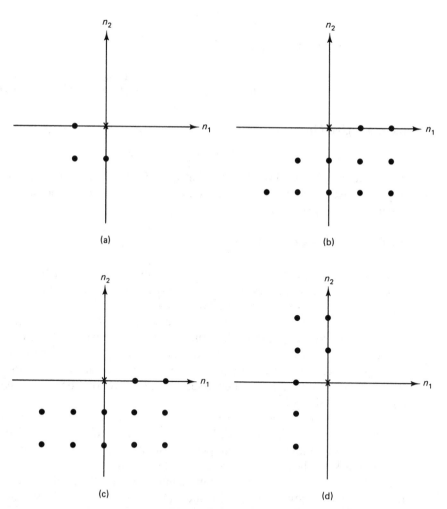

(a)

(b)

(c)

(d)

Figure 2.23 Examples of wedge support output masks. For finite-extent input, they correspond to recursively computable systems.

The z-Transform   Chap. 2

Examples of wedge support output masks are shown in Figure 2.23. The output mask in Figure 2.23(a) is said to have *quadrant support*. The output masks in Figures 2.23(c) and (d) are said to have *nonsymmetric half plane support*. Examples of output masks that do not have wedge support are shown in Figure 2.24. The output masks in Figures 2.24(a) and (b) are said to have *half plane support*. For a finite-extent input $x(n_1, n_2)$, the wedge support output mask is not only necessary, but also sufficient for the system to be recursively computable.

For a recursively computable system, there may be many different paths we can follow in computing all the output points needed. For example, consider a computational procedure whose output and input masks are shown in Figure 2.25. For the input $x(n_1, n_2)$ shown in Figure 2.26(a), the boundary conditions we need are shown in Figure 2.26(b). We can compute $y(n_1, n_2)$ in many different orders, using the graph shown in Figure 2.27. The figure shows which output points are needed to compute a given output point recursively. For example, $y(2, 0)$ and $y(1, 1)$ must be computed before we can compute $y(2, 1)$. Specific orders that can be derived from Figure 2.27 include

$$y(0, 0), y(1, 0), y(2, 0), y(3, 0), y(0, 1), y(1, 1), y(2, 1), \ldots$$

$$\text{or } y(0, 0), y(0, 1), y(0, 2), y(0, 3), y(1, 0), y(1, 1), y(1, 2), \ldots$$

$$\text{or } y(0, 0), y(1, 0), y(0, 1), y(2, 0), y(1, 1), y(0, 2), y(3, 0), \ldots .$$

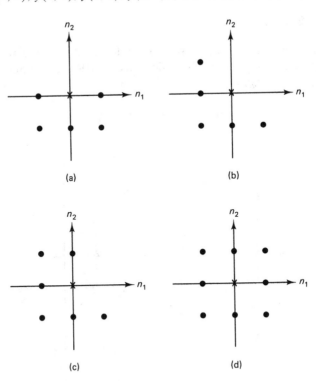

(a)

(b)

(c)

(d)

Figure 2.24 Examples of output masks of systems that are not recursively computable.

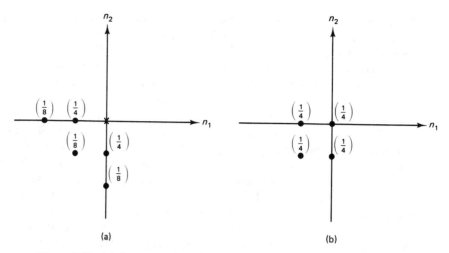

Figure 2.25 (a) Output mask and (b) input mask of a specific computational procedure.

These three different orders are illustrated in Figure 2.28. Although there may be many different orders in which $y(n_1, n_2)$ can be computed, the result does not depend on the specific order in which the output is computed.

2.2.5 Example

In this section, we will present one additional example to illustrate the steps in Figure 2.21. Suppose we have designed an IIR digital filter whose system function $H(z_1, z_2)$ is given by

$$H(z_1, z_2) = \frac{1 + 2z_1^{-1}}{1 - \frac{1}{2}z_1^{-1} + \frac{1}{4}z_2^{-1} + \frac{1}{8}z_2^{-2}}. \qquad (2.37)$$

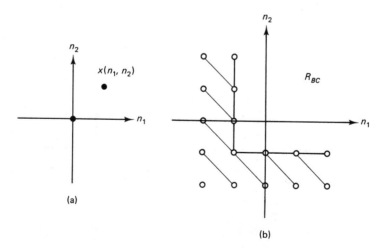

Figure 2.26 (a) Input $x(n_1, n_2)$ and (b) resulting boundary conditions for the system whose output and input masks are shown in Figure 2.25.

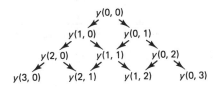

Figure 2.27 Illustration of the output points that are required to compute a given output point recursively, for the example considered in Figures 2.25 and 2.26. This graph can be used in determining an order in which all the output points needed are computed recursively.

The IIR filter was designed such that the impulse response of the designed system will be as close as possible in some sense to an ideal impulse response that is a first-quadrant support sequence. Therefore, we know that the filter is a first-quadrant support system. We wish to determine the output of the filter when the input $x(n_1, n_2)$ is given by

$$x(n_1, n_2) = \begin{cases} 1, & -1 \le n_1 \le 1, \ -1 \le n_2 \le 1 \\ 0, & \text{otherwise.} \end{cases} \tag{2.38}$$

Since the only practical way to implement an IIR filter is by using a difference equation, we first convert (2.37) to a difference equation by

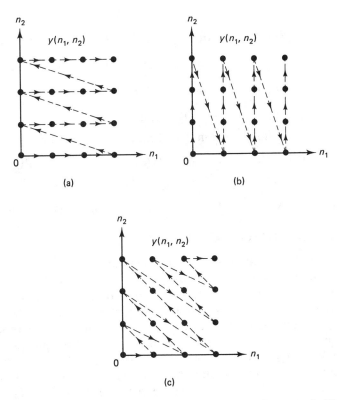

Figure 2.28 Possible ordering for computing the output of the system in Figure 2.25 with the input in Figure 2.26(a).

$$H(z_1, z_2) = \frac{Y(z_1, z_2)}{X(z_1, z_2)} = \frac{1 + 2z_1^{-1}}{1 - \frac{1}{2}z_1^{-1} + \frac{1}{4}z_2^{-1} + \frac{1}{8}z_2^{-2}},$$

$$Y(z_1, z_2) - \tfrac{1}{2}z_1^{-1}Y(z_1, z_2) + \tfrac{1}{4}z_2^{-1}Y(z_1, z_2) + \tfrac{1}{8}z_2^{-2}Y(z_1, z_2)$$

$$= X(z_1, z_2) + 2z_1^{-1}X(z_1, z_2),$$

$$y(n_1, n_2) - \tfrac{1}{2}y(n_1 - 1, n_2) + \tfrac{1}{4}y(n_1, n_2 - 1) + \tfrac{1}{8}y(n_1, n_2 - 2)$$

$$= x(n_1, n_2) + 2x(n_1 - 1, n_2). \qquad (2.39)$$

Since the IIR filter is an LSI system, we choose the proper set of boundary conditions such that the difference equation will become an LSI system. There are four specific computational procedures that correspond to the difference equation in (2.39). One computational procedure is

$$y(n_1, n_2) \leftarrow \tfrac{1}{2}y(n_1 - 1, n_2) - \tfrac{1}{4}y(n_1, n_2 - 1) - \tfrac{1}{8}y(n_1, n_2 - 2)$$

$$+ x(n_1, n_2) + 2x(n_1 - 1, n_2). \qquad (2.40)$$

The second computational procedure can be obtained by

$$-\tfrac{1}{2}y(n_1 - 1, n_2) = -y(n_1, n_2) - \tfrac{1}{4}y(n_1, n_2 - 1) - \tfrac{1}{8}y(n_1, n_2 - 2)$$

$$+ x(n_1, n_2) + 2x(n_1 - 1, n_2),$$

$$y(n_1, n_2) = 2y(n_1 + 1, n_2) + \tfrac{1}{2}y(n_1 + 1, n_2 - 1) + \tfrac{1}{4}y(n_1 + 1, n_2 - 2)$$

$$- 2x(n_1 + 1, n_2) - 4x(n_1, n_2),$$

$$y(n_1, n_2) \leftarrow 2y(n_1 + 1, n_2) + \tfrac{1}{2}y(n_1 + 1, n_2 - 1) + \tfrac{1}{4}y(n_1 + 1, n_2 - 2)$$

$$- 2x(n_1 + 1, n_2) - 4x(n_1, n_2). \qquad (2.41)$$

The two remaining computational procedures, obtained similarly, are

$$y(n_1, n_2) \leftarrow -8y(n_1, n_2 + 2) + 4y(n_1 - 1, n_2 + 2) - 2y(n_1, n_2 + 1)$$

$$+ 8x(n_1, n_2 + 2) + 16x(n_1 - 1, n_2 + 2) \qquad (2.42)$$

$$y(n_1, n_2) \leftarrow -4y(n_1, n_2 + 1) + 2y(n_1 - 1, n_2 + 1) - \tfrac{1}{2}y(n_1, n_2 - 1)$$

$$+ 4x(n_1, n_2 + 1) + 8x(n_1 - 1, n_2 + 1). \qquad (2.43)$$

The output and input masks corresponding to each of these four computational procedures are shown in Figure 2.29. From the output mask corresponding to (2.43), we recognize that (2.43) is not a recursively computable system, and we therefore eliminate (2.43) from further consideration. The region of support of $h(n_1, n_2)$ for each of the three remaining computational procedures is shown in Figure 2.30. Since we know that the filter is a first-quadrant support system, we choose the computational procedure given by (2.40). To determine the boundary conditions for the computational procedure chosen, we determine R_y, the region of support for the output sequence $y(n_1, n_2)$, from R_x in (2.38) and R_h in Figure 2.30(a). The region R_y is shown in Figure 2.31(a). The boundary conditions that we use are shown in Figure 2.31(b). With the boundary conditions shown in Figure

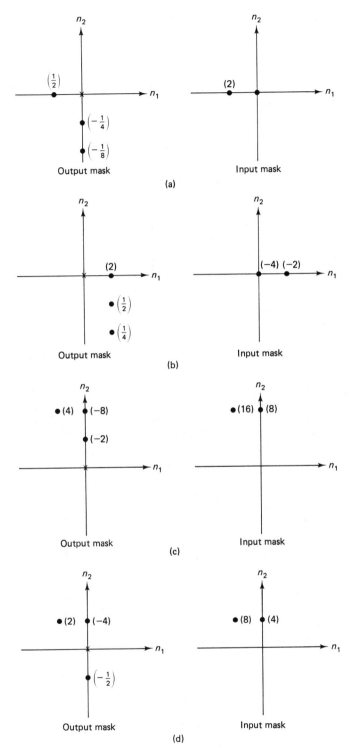

Figure 2.29 Output and input masks corresponding to the computational procedures of (a) (2.40), (b) (2.41), (c) (2.42), and (d) (2.43).

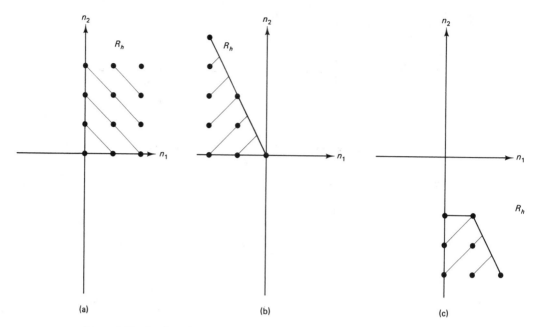

Figure 2.30 Region of support of $h(n_1, n_2)$ for systems specified by (a) (2.40), (b) (2.41), and (c) (2.42).

2.31(b), the output $y(n_1, n_2)$ can be computed recursively from (2.38) and (2.40). The result is shown in Figure 2.32. We can verify that the computational procedure is indeed an LSI system. If we double the input, the output also doubles. If we shift the input, then the output shifts by a corresponding amount.

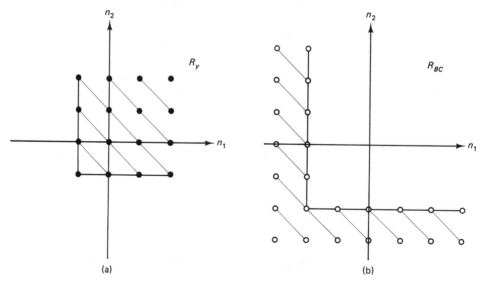

Figure 2.31 (a) R_y and (b) R_{BC} for the computational procedure given by (2.40) and $x(n_1, n_2)$ in (2.38).

The z-Transform   Chap. 2

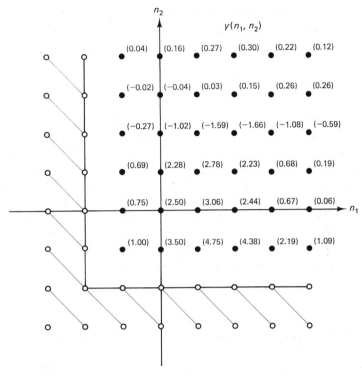

Figure 2.32 Output of the system corresponding to (2.40) when the input $x(n_1, n_2)$ is given by (2.38).

2.2.6 System Functions for Difference Equations

In previous sections, we showed that with a proper choice of boundary conditions, a difference equation with boundary conditions can be considered an LSI system. It is straightforward to determine the system function of the LSI system. Consider a difference equation of the form

$$\sum_{(k_1, k_2)\ \in R_a} \sum a(k_1, k_2)y(n_1 - k_1, n_2 - k_2)$$

$$= \sum_{(k_1, k_2)\ \in R_b} \sum b(k_1, k_2)x(n_1 - k_1, n_2 - k_2) \qquad (2.44)$$

where $a(k_1, k_2)$ and $b(k_1, k_2)$ are finite-extent sequences with nonzero amplitudes in R_a and R_b respectively. By z-transforming both sides of (2.44) and then solving for $H(z_1, z_2)$, we have

$$H(z_1, z_2) = \frac{Y(z_1, z_2)}{X(z_1, z_2)} = \frac{\displaystyle\sum_{(n_1, n_2)\ \in R_b} \sum b(n_1, n_2)z_1^{-n_1}z_2^{-n_2}}{\displaystyle\sum_{(n_1, n_2)\ \in R_a} \sum a(n_1, n_2)z_1^{-n_1}z_2^{-n_2}}. \qquad (2.45)$$

The denominator and numerator in (2.45) are z-transforms of $a(n_1, n_2)$ and

$b(n_1, n_2)$, and therefore can be expressed as $A(z_1, z_2)$ and $B(z_1, z_2)$ respectively. The system function $H(z_1, z_2)$ can then be expressed as

$$H(z_1, z_2) = \frac{B(z_1, z_2)}{A(z_1, z_2)} = \frac{\sum\sum\limits_{(n_1, n_2) \in R_b} b(n_1, n_2)z_1^{-n_1}z_2^{-n_2}}{\sum\sum\limits_{(n_1, n_2) \in R_a} a(n_1, n_2)z_1^{-n_1}z_2^{-n_2}}. \qquad (2.46)$$

The functions $A(z_1, z_2)$ and $B(z_1, z_2)$ are finite-order polynomials* in z_1 and z_2. The function $H(z_1, z_2)$, which is the ratio of two finite-order polynomials, is called a *rational z-transform*. It is clear from (2.46) that the system function derived from the difference equation of (2.44) is a rational z-transform. If $A(z_1, z_2)$ cannot be written as a product of two finite-order polynomials, it is said to be *irreducible*. If there is no common factor (except a constant or linear phase term) that divides both $A(z_1, z_2)$ and $B(z_1, z_2)$, $A(z_1, z_2)$ and $B(z_1, z_2)$ are said to be *co-prime*. A rational $H(z_1, z_2)$ is said to be *irreducible* if $A(z_1, z_2)$ and $B(z_1, z_2)$ are co-prime. An irreducible $H(z_1, z_2)$ does not imply, however, either an irreducible $A(z_1, z_2)$ or an irreducible $B(z_1, z_2)$.

2.3 STABILITY

2.3.1 The Stability Problem

In designing a discrete-space system such as a digital filter, an important consideration is the stability of the system. In this section, we consider the problem of testing the stability of a discrete-space system.

As was discussed in Section 1.2.3, a system is considered stable in the bounded-input–bounded-output (BIBO) sense if and only if a bounded input always leads to a bounded output. For an LSI system, a necessary and sufficient condition for the system to be stable is that its impulse response $h(n_1, n_2)$ be absolutely summable:

$$\sum_{n_1 = -\infty}^{\infty} \sum_{n_2 = -\infty}^{\infty} |h(n_1, n_2)| < \infty. \qquad (2.47)$$

Although this condition is a straightforward extension of 1-D results, the 2-D case differs significantly from the 1-D case with regard to stability.

In a typical stability problem, we are given the system function $H(z_1, z_2)$ and information about its ROC, and we wish to determine the system's stability. If the ROC is explicitly given, determining the system stability is straightforward. This is because a system is stable if and only if the ROC includes the unit surface ($|z_1| = 1$, $|z_2| = 1$). Unfortunately, however, the ROC is seldom given explicitly, and usually only implicit information about the ROC is available. When the system function is obtained from a digital filter, for example, the region of support of its impulse response $h(n_1, n_2)$ is typically known from the filter design step. Since

---

*The polynomial in z_1 and z_2 used here is a linear combination of 1, z_1, z_2, $z_1 z_2$, z_1^{-1}, z_2^{-1}, $z_1^{-1} z_2^{-1}$, and so on. "Finite order" means that there is a finite number of such terms.

our main interest is testing the stability of digital filters, we will consider the problem of testing the system stability given $H(z_1, z_2)$ and the region of support of $h(n_1, n_2)$.

When $H(z_1, z_2)$ is a system function of a digital filter, there are restrictions imposed on $H(z_1, z_2)$ and the region of support of $h(n_1, n_2)$. One restriction is that $H(z_1, z_2)$ must be a rational z-transform, which can be expressed as

$$H(z_1, z_2) = \frac{B(z_1, z_2)}{A(z_1, z_2)} \tag{2.48}$$

where $A(z_1, z_2)$ and $B(z_1, z_2)$ are finite-order polynomials in z_1 and z_2. We will assume that $A(z_1, z_2)$ and $B(z_1, z_2)$ are co-prime, so $H(z_1, z_2)$ is irreducible. Another restriction is that the system must be recursively computable. With these two restrictions, the system can be realized by a difference equation with boundary conditions and the output can be computed recursively, one value at a time.

In the 1-D case, when the system function $H(z)$ is expressed as $B(z)/A(z)$, where there are no common factors between $A(z)$ and $B(z)$, $B(z)$ does not affect the system stability. In the 2-D case, however, the presence of $B(z_1, z_2)$ in (2.48) can stabilize an otherwise unstable system, even when there are no common factors between $A(z_1, z_2)$ and $B(z_1, z_2)$. In this case, pole surfaces $(A(z_1, z_2) = 0)$ and zero surfaces $(B(z_1, z_2) = 0)$ intersect at the unit surface and the specific values of (z_1, z_2) at which they intersect are known as *nonessential singularities of the second kind*. An example of pole and zero surfaces where this situation occurs is shown in Figure 2.33. In the figure, zero surfaces (horizontal and vertical solid lines) intersect pole surfaces (shaded region) on the unit surface, but they do not cancel the entire pole surface. Since this case occurs rarely and there is no known procedure that can be used systematically to test the stability when it does occur, and since an unstable system stabilized by $B(z_1, z_2)$ is unstable for most practical purposes, we will assume that the numerator polynomial $B(z_1, z_2)$ does not affect the system stability. To make this explicit, we will assume that $B(z_1, z_2) = 1$. The system function can, then, be expressed as $1/A(z_1, z_2)$.

As we discussed in Section 2.2.4, recursive computability requires that the output mask have wedge support, which in turn requires the impulse response $h(n_1, n_2)$ corresponding to $1/A(z_1, z_2)$ to have wedge support with a finite shift* of $h(n_1, n_2)$. A finite shift of $h(n_1, n_2)$ does not affect the stability of $h(n_1, n_2)$, and we will assume that this shift was made so that $h(n_1, n_2)$ would be a wedge support sequence. As we saw in Section 1.2.3, it is always possible to find a linear mapping of variables that transforms a wedge support sequence to a first-quadrant support sequence without affecting the sequence's stability. Therefore, stability results that apply to first-quadrant support sequences can be used for all recursively computable systems. In our approach, a recursively computable system will first be transformed to a first-quadrant support system by a linear mapping of variables. This transformation changes the system function $H(z_1, z_2)$ to a new system function $H'(z_1, z_2)$. Stability results that apply to first-quadrant support systems will then be applied to $H'(z_1, z_2)$. This approach is sketched in Figure 2.34. As we have

---

*A finite shift of $h(n_1, n_2)$ accounts for an arbitrary linear phase term that may have been included as part of $A(z_1, z_2)$.

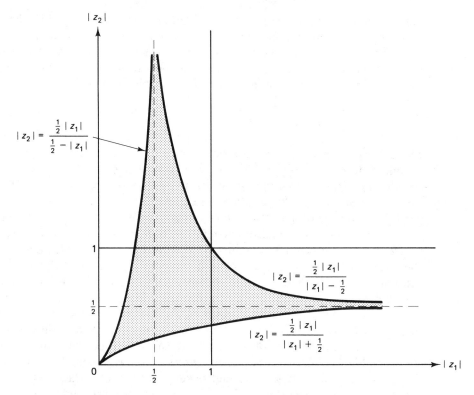

$|z_2| = \dfrac{\frac{1}{2}|z_1|}{\frac{1}{2}-|z_1|}$

$|z_2| = \dfrac{\frac{1}{2}|z_1|}{|z_1|-\frac{1}{2}}$

$|z_2| = \dfrac{\frac{1}{2}|z_1|}{|z_1|+\frac{1}{2}}$

Figure 2.33 Pole (shaded region) and zero (horizontal and vertical solid lines) surfaces of $H(z_1, z_2)$ with nonessential singularities of the second kind. The function $H(z_1, z_2)$ used is

$$H(z_1, z_2) = \frac{(1 - z_1^{-1})^8(1 - z_2^{-1})^8}{2 - z_1^{-1} - z_2^{-1}}.$$

The first-quadrant support system with the above system function $H(z_1, z_2)$ has been shown [Goodman] to be stable, even though

$$H(z_1, z_2) = \frac{1}{2 - z_1^{-1} - z_2^{-1}}$$

is clearly unstable. Note that the pole and zero surfaces are 2-D surfaces. The zero surfaces lie all within the horizontal and vertical solid lines. The pole surface, however, is visible as a 2-D plane in $(|z_1|, |z_2|)$ plane.

already discussed, the reason for first transforming a wedge support system to a first-quadrant support system is that developing stability results that apply to first-quadrant support systems is much easier notationally and conceptually than developing stability results that apply to wedge support systems.

Based on the above discussion, the stability problem we will consider can be stated as follows.

Stability Problem:
Given that $H(z_1, z_2) = 1/A(z_1, z_2)$ and $h(n_1, n_2)$ is a first-quadrant support sequence, determine the system stability.

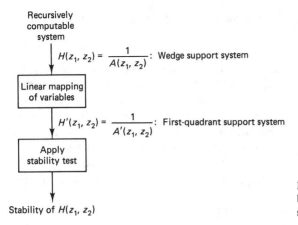

Figure 2.34 Approach to test the stability of a 2-D recursively computable system.

Because $H(z_1, z_2)$ is a rational system function of a first-quadrant support system, and since scaling and shifting $h(n_1, n_2)$ do not affect system stability, we can assume without loss of generality that $A(z_1, z_2)$ is in the form

$$A(z_1, z_2) = \sum_{n_1=0}^{M} \sum_{n_2=0}^{M} a(n_1, n_2) z_1^{-n_1} z_2^{-n_2} \quad \text{with} \quad a(0, 0) = 1 \qquad (2.49)$$

where M is sufficiently large. Unless specified otherwise, we assume that $A(z_1, z_2)$ is in the form of (2.49). Note that, with this assumption on the form of $A(z_1, z_2)$, the computational procedure is

$$y(n_1, n_2) \longleftarrow - \sum_{\substack{k_1=0 \\ (k_1,k_2) \neq (0,0)}}^{M} \sum_{k_2=0}^{M} a(k_1, k_2) y(n_1 - k_1, n_2 - k_2) + x(n_1, n_2). \qquad (2.50)$$

In the following sections, we discuss several stability theorems and how they can be used in testing the system stability.

2.3.2 Stability Theorems

In the 1-D case, the problem of testing the stability of a causal system whose system function is given by $H(z) = 1/A(z)$ is quite straightforward. Since a 1-D polynomial $A(z)$ can always be factored straightforwardly as a product of first-order polynomials, we can easily determine the poles of $H(z)$. The stability of the causal system is equivalent to having all the poles inside the unit circle.

The above approach cannot be used in testing the stability of a 2-D first-quadrant support system. That approach requires the specific location of all poles to be determined. Partly because a 2-D polynomial $A(z_1, z_2)$ cannot in general be factored as a product of lower-order polynomials, it is extremely difficult to determine all the pole surfaces of $H(z_1, z_2) = 1/A(z_1, z_2)$, and the approach based on explicit determination of all pole surfaces has not led to successful practical procedures for testing the system stability. In this section, we discuss representative stability theorems that can be used in developing practical procedures for testing the stability of a 2-D first-quadrant support system without explicit determination of all pole surfaces.

Theorem 1. The first theorem we will discuss is known as Shanks's theorem. This theorem does not in itself lead directly to practical stability-testing procedures. However, it was historically one of the first stability theorems developed, is conceptually very simple, and has led to other stability theorems to be discussed later. In addition, this theorem illustrates a case in which a 1-D result that is not useful for 1-D signals can be very useful when extended to the 2-D case.

In the 1-D case, the stability of a causal system with system function $H(z) = 1/A(z)$ is equivalent to requiring all the poles of $H(z)$ to be within the unit circle:

$$\text{Stability} \iff \text{all the poles (solutions to } A(z) = 0\text{) are inside } |z| = 1 \qquad (2.51)$$

A statement equivalent to (2.51) is

$$\text{Stability} \iff A(z) \neq 0 \quad \text{for} \quad |z| \geq 1. \qquad (2.52)$$

If all the poles are inside $|z| = 1$, then $A(z)$ cannot be zero for any $|z| \geq 1$. If $A(z) \neq 0$ for any $|z| \geq 1$, then all the poles must be inside $|z| = 1$. Therefore, (2.51) and (2.52) are equivalent.

Although (2.51) and (2.52) are equivalent statements, their implications for testing the system stability are quite different. The condition in (2.51) suggests a procedure in which we explicitly determine all pole locations and see if they are all inside $|z| = 1$. The condition in (2.52), however, suggests a procedure in which we evaluate $A(z)$ for each z such that $|z| \geq 1$ and see if $A(z)$ is zero for any $|z| \geq 1$. This requires a search in the 2-D plane. In the 1-D case, the procedure suggested by (2.51) is extremely simple because $A(z)$ can always be factored as a product of first-order polynomials, by using simple procedures. Therefore, the procedure suggested by (2.52), which requires a 2-D search, is not useful. In the 2-D case, however, the procedure suggested by (2.51) is extremely difficult, since $A(z_1, z_2)$ cannot in general be factored as a product of lower-order polynomials, and simple procedures do not exist for solving $A(z_1, z_2) = 0$. The extension of (2.52) to the 2-D case is Shanks's theorem, or Theorem 1, which can be stated as follows:

Theorem 1.

$$\text{Stability} \iff A(z_1, z_2) \neq 0 \quad \text{for any } |z_1| \geq 1, |z_2| \geq 1. \qquad (2.53)$$

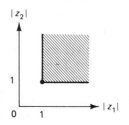

Theorem 1 in (2.53) is simple to show. To show that stability implies the condition in (2.53), we first note that stability implies that the unit surface ($|z_1| = 1$, $|z_2| = 1$) is in the ROC of $H(z_1, z_2)$. Because $h(n_1, n_2)$ is a first-quadrant support sequence, the ROC of $H(z_1, z_2)$ has to satisfy the conditions given by the constraint map in Figure 2.8. Therefore all (z_1, z_2) such that $|z_1| \geq 1$, $|z_2| \geq 1$ have to be in the ROC. Since the ROC cannot have any pole surfaces ($A(z_1, z_2) = 0$), $A(z_1, z_2) \neq 0$ for any $|z_1| \geq 1$, $|z_2| \geq 1$. To show that the condition in (2.53) implies system stability, we note that this condition implies that there are no pole surfaces for any $|z_1| \geq 1$, $|z_2| \geq 1$. Because $h(n_1, n_2)$ is a first-quadrant support sequence, the ROC of $H(z_1, z_2)$ has to satisfy the constraint map in Figure 2.8. This requirement, combined with the property that the ROC is bounded by pole surfaces, implies that the ROC includes the unit surface, which in turn implies the system stability.

The condition in (2.53) suggests a procedure in which we evaluate $A(z_1, z_2)$ in the 4-D space ($|z_1| \geq 1$, $|z_2| \geq 1$) shown in the shaded region in (2.53). The search in the 4-D space is, of course, a tremendous amount of work, and this procedure cannot itself be used in practice. We will discuss other theorems that considerably reduce the space in which we search. The proofs of these theorems are quite involved and do not themselves provide much insight into the theorems. Therefore, we will only sketch how these theorems can be proven, while attempting to provide some insights into their interpretation.

Theorem 2. Theorem 2 was also developed by Shanks and can be stated as follows:

Theorem 2.

$$\text{Stability} \iff \begin{array}{l} \text{(a) } A(z_1, z_2) \neq 0 \quad \text{for} \quad |z_1| = 1, |z_2| \geq 1 \\ \text{and (b) } A(z_1, z_2) \neq 0 \quad \text{for} \quad |z_1| \geq 1, |z_2| = 1. \end{array} \quad (2.54)$$

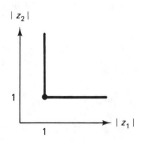

Let us first consider Condition (a). To satisfy Condition (a), we need to ensure that $A(z_1, z_2)$ is not zero for any (z_1, z_2) such that $|z_1| = 1$ and $|z_2| \geq 1$. This requires a 3-D search and is shown in the figure by the solid vertical line. To ensure Condition (b), we need to ensure that $A(z_1, z_2)$ is not zero for any (z_1, z_2) such that $|z_1| \geq 1$ and $|z_2| = 1$. This also requires a 3-D search and is shown in the figure by the solid horizontal line. From the search point of view, then, the conditions in Theorem 2 that require two 3-D searches are considerably simpler to satisfy than the condition in Theorem 1, which requires a 4-D search.

We will now illustrate how Theorem 2 can be derived from Theorem 1. This can be done by showing that the conditions in Theorem 2 are equivalent to the condition in Theorem 1, as shown in Figure 2.35. It is straightforward to see that the condition in Theorem 1 implies the two conditions in Theorem 2, since the two 3-D spaces in Theorem 2 are just a subset of the 4-D space in Theorem 1. To show that the conditions in Theorem 2 imply the condition in Theorem 1, we will perform an experiment. Suppose we determine all the pole surfaces (z_1, z_2) by finding all the roots of the following equation:

$$A(z_1, z_2) = 0. \tag{2.55}$$

If we consider a particular value of z_1 in (2.55), say z_1', then (2.55) becomes a 1-D polynomial equation with respect to z_2 given by

$$A(z_1', z_2) = 0. \tag{2.56}$$

Supposing (2.56) to be an Nth-order polynomial equation in z_2, there will be N values (which may or may not be distinct) of z_2 that satisfy (2.56). Suppose we solve this 1-D polynomial equation in (2.56) for all possible values of z_1. This will, in theory, determine all values of (z_1, z_2) that satisfy (2.55). An important property from a theorem of algebraic functions states that if we change z_1 in a continuous manner, then each of the N values of z_2 that satisfies (2.56) also must change in a continuous manner [Coolidge]. Let us begin with the roots of z_2 in (2.56) for $z_1' = 1$. Since Condition (a) in Theorem 2 states that $A(z_1, z_2)$ cannot be zero for $|z_2| \geq 1$ when $z_1' = 1$, the magnitude of each of the roots of z_2 in (2.56) when $z_1' = 1$ must be less than one. One possible set of $(z_1 = 1, z_2)$ that satisfies (2.56) in the $(|z_1|, |z_2|)$ plane is shown as "$\times$" in Figure 2.36. We now change z_1 in a continuous manner and cover all possible values of z_1. Since the roots of z_2 have to change in a continuous manner, and since the conditions in Theorem 2 state that the roots cannot cross the solid lines in Figure 2.36, all the pole surfaces (all the values of (z_1, z_2) that satisfy (2.55)) can lie only in the shaded region shown

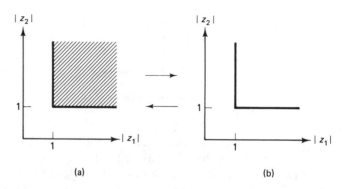

(a)   (b)

Figure 2.35   Derivation of Theorem 2 from Theorem 1. $A(z_1, z_2) \neq 0$ in the shaded region in (a) and the solid lines in (b). (a) Condition in Theorem 1; (b) two conditions in Theorem 2.

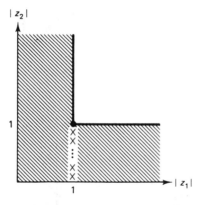

Figure 2.36 Illustration that the two conditions in Theorem 2 imply the condition in Theorem 1. Symbols "×" are one possible set of roots for $A(z_1 = 1, z_2) = 0$. Since the roots can only change continuously and cannot cross the solid lines, they remain in the shaded region.

in Figure 2.36. This implies that $A(z_1, z_2) \neq 0$ for any $(|z_1| \geq 1, |z_2| \geq 1)$, which is the condition in Theorem 1. The conditions in Theorem 2 are thus equivalent to the condition in Theorem 1, and Theorem 2 is demonstrated.

The 3-D search problem corresponding to Condition (a) can be substituted for by many 1-D stability tests. To satisfy Condition (a), we need to ensure that $A(z_1, z_2) \neq 0$ in the 3-D space corresponding to $(|z_1| = 1, |z_2| \geq 1)$. One approach to satisfying this condition is the following:

Step 1. Solve for all (z_1, z_2) such that $A(|z_1| = 1, z_2) = 0$. This is equivalent to solving all (ω_1, z_2) such that $A(e^{j\omega_1}, z_2) = 0$.

Step 2. Check if all $|z_2|$ obtained in Step 1 are less than 1.

In Step 1, we determine all $(|z_1| = 1, z_2)$ such that $A(|z_1| = 1, z_2) = 0$. The solutions obtained in this step will contain all solutions to $A(z_1, z_2) = 0$ in the 3-D space sketched in Figure 2.37. If none of these solutions has $|z_2|$ greater than or equal to 1, then $A(z_1, z_2) \neq 0$ for any $(|z_1| = 1, |z_2| \geq 1)$, which satisfies Condition (a).

Step 2 is clearly a simple operation. In Step 1, we must solve the following equation:

$$A(e^{j\omega_1}, z_2) = 0. \tag{2.57}$$

Consider a fixed value of ω_1, say ω_1'. Then $A(e^{j\omega_1}, z_2)$ is a 1-D polynomial in the variable z_2, and solving for all z_2 such that $A(e^{j\omega_1}, z_2) = 0$ is equivalent to a 1-D stability test with complex coefficients. If we vary ω_1 continuously from 0 to 2π, and we perform the 1-D stability test for each ω_1, then we will find all possible values of (ω_1, z_2) such that $A(e^{j\omega_1}, z_2) = 0$. In practice, we cannot change ω_1 continuously, and must consider discrete values of ω_1. We can obtain a table like Table 2.3 by performing many 1-D stability tests. If we choose a sufficiently small Δ, we can essentially determine all possible values of (ω_1, z_2) such that $A(e^{j\omega_1}, z_2) = 0$. By checking if all the values of $|z_2|$ in Table 2.3 are less than 1, we can satisfy Condition (a) without a 3-D search.

There are various ways to improve the efficiency of the 1-D stability tests in (2.57) for different values of ω_1. As we change ω_1 in (2.57) in a continuous manner,

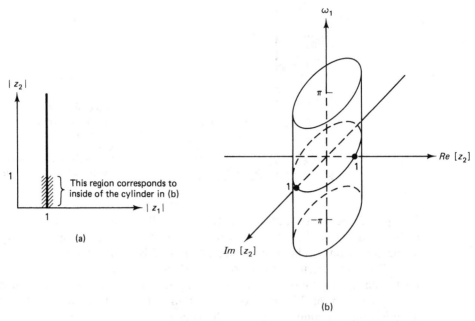

Figure 2.37 3-D space corresponding to $|z_1| = 1$. Condition (a) of Theorem 2 requires that all the roots to $A(|z_1| = 1, z_2) = 0$ lie in the inside of the cylinder shown in (b). (a) $(|z_1|, |z_2|)$ plane; (b) (ω_1, z_2) space.

the roots for z_2 also change in a continuous manner. Therefore, once the roots for $A(e^{j\omega_1}, z_2) = 0$ are known, they can be used as initial estimates for finding the roots of $A(e^{j(\omega_1 + \Delta)}, z_2) = 0$ for any small Δ. Furthermore, if any of the roots has magnitude greater than 1, the system is unstable and no additional tests are necessary. In addition, if we sample ω_1 at equally spaced points, then a fast Fourier transform (FFT) algorithm can be used to compute the coefficients of the 1-D polynomial in (2.57). Specifically, from (2.49) and (2.57),

$$A(e^{j\omega_1}, z_2) = \sum_{n_1=0}^{M} \sum_{n_2=0}^{M} a(n_1, n_2)e^{-j\omega_1 n_1}z_2^{-n_2}. \tag{2.58}$$

If we sample (2.58) at $\omega_1 = (2\pi/N)k$ for $0 \le k \le N - 1$ and rewrite (2.58),

$$A(e^{j(2\pi/N)k}, z_2) = \sum_{n_2=0}^{M} c(k, n_2)z_2^{-n_2} \tag{2.59a}$$

where

$$c(k, n_2) = \sum_{n_1=0}^{M} a(n_1, n_2)e^{-j(2\pi/N)kn_1}. \tag{2.59b}$$

For a sufficiently small M relative to N, computing $c(k, n_2)$ in (2.59b) directly is the most efficient approach. Otherwise, using an FFT algorithm may be more efficient. From (2.59b), $c(k, n_2)$ for a fixed n_2 is the 1-D discrete Fourier transform (DFT) of one row of $a(n_1, n_2)$. By computing $c(k, n_2)$ using a 1-D FFT algorithm for each n_2 and storing the results, all the coefficients of the 1-D polynomial in (2.57) for different values of ω_1 will have been obtained.

TABLE 2.3 SOLUTION TO $A(e^{j\omega_1}, z_2) = 0$

ω_1	z_2
0	$a_0, b_0, c_0, d_0, \cdots$
Δ	$a_1, b_1, c_1, d_1, \cdots$
2Δ	$a_2, b_2, c_2, d_2, \cdots$
.	.
.	.
.	.
.	
.	
2π	$a_0, b_0, c_0, d_0, \cdots$

Suppose we plot the values of z_2 such that $A(e^{j\omega_1}, z_2) = 0$ in the z_2 plane as we vary ω_1 continuously from 0 to 2π. Then the values of z_2 will vary continuously. The resulting sketch is called a *root map*. We will refer to it as Root Map 1. Two examples of Root Map 1 are shown in Figure 2.38 for $A(z_1, z_2)$ given by

$$A(z_1, z_2) = 1 - \tfrac{1}{4}z_1^{-1} + \tfrac{1}{4}z_2^{-2} - \tfrac{1}{4}z_1^{-1} z_2^{-1} \tag{2.60a}$$

$$A(z_1, z_2) = 1 - \tfrac{2}{3}z_1^{-1} - \tfrac{1}{2}z_2^{-1} - \tfrac{1}{3}z_1^{-2} z_2^{-2}. \tag{2.60b}$$

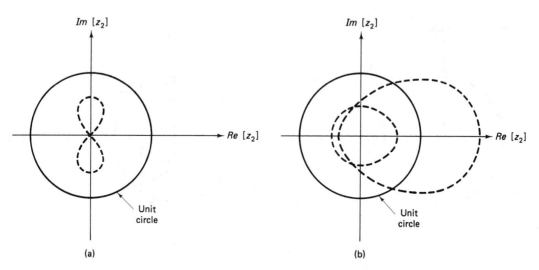

Figure 2.38 Root Map 1 for each of the system functions. Root Map 1 is a sketch of the roots of $A(e^{j\omega_1}, z_2)$, as we vary ω_1 from 0 to 2π.

$$\text{(a)} \quad H(z_1, z_2) = \frac{1}{A(z_1, z_2)} = \frac{1}{1 - \tfrac{1}{4}z_1^{-1} + \tfrac{1}{4}z_2^{-2} - \tfrac{1}{4}z_1^{-1}z_2^{-2}};$$

$$\text{(b)} \quad H(z_1, z_2) = \frac{1}{A(z_1, z_2)} = \frac{1}{1 - \tfrac{2}{3}z_1^{-1} - \tfrac{1}{2}z_2^{-1} - \tfrac{1}{3}z_1^{-2}z_2^{-2}}.$$

Clearly, the contours of Root Map 1 are continuous. Since the contours in this root map are the values of z_2 in Table 2.3, Condition (a) in Theorem 2 is equivalent to having all the contours in the root map lie inside the unit circle in the z_2 plane. All the contours in Figure 2.38(a) are inside the unit circle, and therefore $A(z_1, z_2)$ in (2.60a) has the potential to correspond to a stable system. Not all the contours in Figure 2.38(b) are inside the unit circle, so $A(z_1, z_2)$ in (2.60b) corresponds to an unstable system.

Condition (b) in Theorem 2 is equivalent to Condition (a) in Theorem 2, except that the roles of z_1 and z_2 are reversed. Therefore, our discussion of Condition (a) applies to Condition (b) with the roles of z_1 and z_2 interchanged. Thus, one approach to satisfying Condition (b) is to perform many 1-D stability tests as follows:

Step 1. Solve for all (z_1, z_2) such that $A(z_1, |z_2| = 1) = 0$. This is equivalent to solving all (z_1, ω_2) such that $A(z_1, e^{j\omega_2}) = 0$.

Step 2. Check if all $|z_1|$ obtained in Step 1 are less than 1.

If we plot the values of z_1 such that $A(z_1, e^{j\omega_2}) = 0$ in the z_1 plane as we vary ω_2 continuously from 0 to 2π, then the values of z_1 will vary continuously, and the resulting sketch will form another root map. We will refer to this as Root Map 2. Condition (b) in Theorem 2 is equivalent to having all the contours in Root Map 2 lie inside the unit circle in the z_1 plane. Two examples of Root Map 2 corresponding to (2.60a) and (2.60b) are shown in Figure 2.39. All the contours

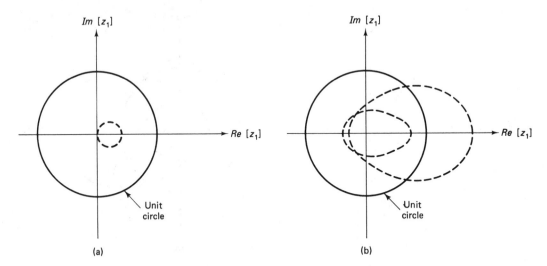

Figure 2.39 Root Map 2 for each of two system functions. Root Map 2 is a sketch of the roots of $A(z_1, e^{j\omega_2}) = 0$, as we vary ω_2 from 0 to 2π.

$$(a)\ H(z_1, z_2) = \frac{1}{A(z_1, z_2)} = \frac{1}{1 - \frac{1}{4}z_1^{-1} + \frac{1}{4}z_2^{-2} - \frac{1}{4}z_1^{-1}z_2^{-2}};$$

$$(b)\ H(z_1, z_2) = \frac{1}{A(z_1, z_2)} = \frac{1}{1 - \frac{2}{3}z_1^{-1} - \frac{1}{2}z_2^{-1} - \frac{1}{3}z_1^{-2}z_2^{-2}}.$$

in Figure 2.39(a) are inside the unit circle. This, combined with the previous result, shows that both Conditions (a) and (b) in Theorem 2 have been satisfied, and $A(z_1, z_2)$ in (2.60a) corresponds to a stable system. We have already determined that $A(z_1, z_2)$ in (2.60b) corresponds to an unstable system.

From the above discussion, it is clear that the conditions in Theorem 2 are equivalent to having all the contours in Root Maps 1 and 2 lie inside the unit circle in the respective z_2 and z_1 plane. Since constructing each root map is equivalent to performing many 1-D stability tests, Theorem 2 shows that a 2-D stability test can be performed by performing many 1-D stability tests twice.

Theorem 3. Theorem 3, which further simplifies the conditions in Theorem 2, will be referred to as Huang's theorem. This theorem* can be stated as follows:

Theorem 3.

$$\text{Stability} \iff \quad \begin{array}{l} (a)\ A(z_1, z_2) \neq 0 \quad \text{for} \quad |z_1| = 1, |z_2| \geq 1 \\ \text{and } (b)\ A(z_1, z_2) \neq 0 \quad \text{for} \quad |z_1| \geq 1, z_2 = 1. \end{array} \quad (2.61)$$

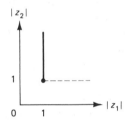

Condition (a) in this theorem is identical to Condition (a) in Theorem 2, and this 3-D search problem can be solved by solving many 1-D stability tests. To satisfy Condition (b), we must ensure that $A(z_1, z_2)$ is not zero for any (z_1, z_2) such that $|z_1| \geq 1$ and $z_2 = 1$. This corresponds to a 2-D search problem, where the space to be searched is shown by the dashed line in the figure above. The dashed line emphasizes that this is a 2-D search problem, in which the search is performed in the 2-D subspace $(|z_1| \geq 1, z_2 = 1)$ of the 3-D space $(|z_1| \geq 1, |z_2| = 1)$.

The 2-D search problem corresponding to Condition (b) can be substituted for by one 1-D stability test. To satisfy Condition (b), we must make sure that $A(z_1, z_2) \neq 0$ in the 2-D space corresponding to $(|z_1| \geq 1, z_2 = 1)$. The following two steps comprise one approach to satisfying this condition.

Step 1. Solve all (z_1, z_2) such that $A(z_1, z_2 = 1) = 0$.
Step 2. Check if all values of $|z_1|$ obtained in Step 1 are less than 1.

Step 1 determines all z_1 such that $A(z_1, 1) = 0$. If all values of $|z_1|$ are less than 1, then $A(z_1, 1)$ cannot be zero for $|z_1| \geq 1$, which satisfies Condition (b). Step

---

*The theorem originally developed by T. S. Huang is slightly different from Theorem 3. See Problem 2.31 for the theorem by T. S. Huang.

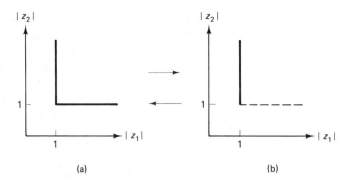

(a) (b)

Figure 2.40 Derivation of Theorem 3 from Theorem 2. $A(z_1, z_2) \neq 0$ in the solid lines in (a) and in the solid and dashed lines in (b). The dashed line emphasizes that it corresponds to a 2-D space. (a) Conditions in Theorem 2; (b) conditions in Theorem 3.

2 is a simple operation, and Step 1 is equivalent to a 1-D stability test, since $A(z_1, 1)$ is a 1-D polynomial in the variable z_1.

From the above discussion, it is clear that a 2-D stability test can be performed by many 1-D stability tests and one 1-D stability test. This can be used in developing a practical procedure to test the stability of a 2-D system.

We will now illustrate how Theorem 3 can be derived from Theorem 2. This can be done by showing that the conditions in Theorem 3 are equivalent to the conditions in Theorem 2, as shown in Figure 2.40. The conditions in Theorem 2 imply the conditions in Theorem 3 in a straightforward manner, since the 3-D and the 2-D spaces in Theorem 3 form a subset of the two 3-D spaces in Theorem 2. To see that the conditions in Theorem 3 imply the conditions in Theorem 2, consider the two root maps. Condition (a) of Theorem 3 is identical to Condition (a) of Theorem 2. As has been discussed, this is equivalent to having all the contours in Root Map 1 inside the unit circle. Now consider Root Map 2. All the contours in a root map are continuous. Since $A(z_1, z_2) \neq 0$ for $(|z_1| = 1, |z_2| = 1)$ from Condition (a), the root map contours cannot cross the unit circle. Therefore, each

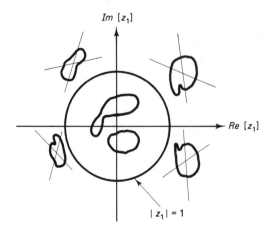

Figure 2.41 Root Map 2 used in deriving Theorem 3 from Theorem 2. The condition that $A(|z_1| = 1, |z_2| = 1) \neq 0$ implies that each contour in the root map is completely inside or outside $|z_1| = 1$.

114 The z-Transform Chap. 2

contour in the root map is either completely outside or completely inside the unit circle, as shown in Figure 2.41. Now we impose Condition (b) of Theorem 3. Consider all the values of z_1 such that $A(z_1, 1) = 0$. Condition (b) of Theorem 3 requires that these values of z_1 all have magnitude less than 1. Since the contours in Root Map 2 represent the values of z_1 such that $A(z_1, e^{j\omega_2}) = 0$ as we change ω_2 continuously from 0 (this corresponds to $z_2 = 1$) to 2π, one point of each contour in Root Map 2 must be inside the unit circle. This implies that all the contours in Root Map 2 are inside the unit circle, equivalent to Condition (b) of Theorem 2. This demonstrates the validity of Theorem 3.

There are many variations to Theorem 3. One is

Stability $\iff$ (a) $A(z_1, z_2) \neq 0$ for $|z_1| \geq 1$, $|z_2| = 1$

and (b) $A(z_1, z_2) \neq 0$ for $z_1 = 1$, $|z_2| \geq 1$.

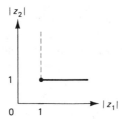

This variation is the same as Theorem 3 except that the roles of z_1 and z_2 have been interchanged.

Theorem 4. One additional theorem, DeCarlo-Strintzis's theorem, can be stated as follows:

Theorem 4.

Stability $\iff$ (a) $A(z_1, z_2) \neq 0$, $|z_1| = |z_2| = 1$

and (b) $A(z_1, 1) \neq 0$, $|z_1| \geq 1$ (2.62)

and (c) $A(1, z_2) \neq 0$, $|z_2| \geq 1$.

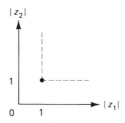

Each of the three conditions in this theorem corresponds to a 2-D search problem. Condition (a) requires $A(z_1, z_2)$ to be nonzero on the 2-D unit surface shown by the filled-in dot in the figure. Conditions (b) and (c) require $A(z_1, z_2)$ to be nonzero in the 2-D spaces shown by the dotted lines in the figure. From the search perspective, therefore, the conditions imposed by Theorem 4 are considerably simpler than those in Theorem 3, which require a 3-D search.

In practice, however, this theorem is not much simpler than Theorem 3. Condition (b) in Theorem 4 is the same as Condition (b) in Theorem 3, which can be checked by a 1-D stability test. In Theorem 4, Conditions (b) and (c) are the same, with the roles of z_1 and z_2 interchanged. Condition (c) can therefore be checked by one 1-D stability test. In the case of Condition (a), however, the 2-D search problem cannot be simplified by one 1-D stability test, and a full 2-D search on the unit surface is generally necessary. The amount of computation required for this 2-D search is often comparable to that required to run a set of 1-D stability tests to test Condition (a) of Theorem 3.

We will now derive Theorem 4 from Theorem 2. This can be done by showing that the conditions in Theorem 4 are equivalent to the conditions in Theorem 2, as shown in Figure 2.42. It is straightforward to show that the conditions in Theorem 2 imply the conditions in Theorem 4. We now show that the conditions in Theorem 4 imply the conditions in Theorem 2. Consider again the two root maps. From Condition (a) in Theorem 4, none of the contours in the two root maps can cross the unit circle. Each contour in the two root maps, then, must lie either completely inside or completely outside the unit circle, as shown in Figure 2.43. As we discussed in the derivation of Theorem 3, Condition (b) of Theorem 4, which is identical to Condition (b) of Theorem 3, requires that all the contours in Root Map 2 be inside the unit circle ($|z_1| = 1$). Since Conditions (b) and (c) of Theorem 4 are identical except that the roles of z_1 and z_2 are interchanged, Condition (c) requires that all the contours in Root Map 1 be inside the unit circle ($|z_2| = 1$). Since all the contours in both root maps are inside their respective unit circles, the conditions of Theorem 2 are satisfied. This demonstrates the validity of Theorem 4.

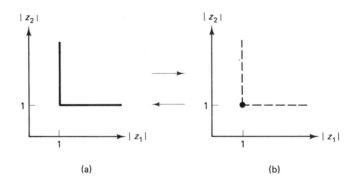

(a)   (b)

Figure 2.42 Derivation of Theorem 4 from Theorem 2. $A(z_1, z_2) \neq 0$ in the solid lines in (a), and the filled-in dot and dashed lines in (b). (a) Conditions in Theorem 2; (b) conditions in Theorem 4.

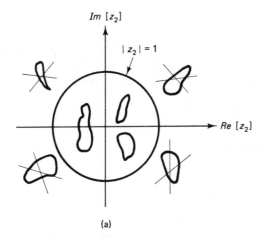

(a)

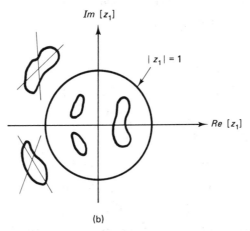

(b)

Figure 2.43 (a) Root Map 1 and (b) Root Map 2 used in deriving Theorem 4 from Theorem 2. Condition (a) in Theorem 4 implies that each contour in the root map is completely inside or outside the respective unit circle.

2.3.3 Algorithms for Stability Tests

From the theorems discussed in Section 2.3.2, we can develop many different methods useful for testing the stability of 2-D systems in practical applications. One such method is shown in Figure 2.44. Test 1 in the figure checks Condition (b) of Theorem 4. Test 2 checks Condition (c) of Theorem 4. Test 3 checks Condition (a) of Theorem 4. Test 3 can be replaced by testing whether all the contours in one of the two root maps are inside the unit circle. This checks Condition (a) of Theorem 3. If the system passes all three tests, it is stable. Otherwise, it is unstable.

To illustrate how the procedure in Figure 2.44 may be used in testing the stability of a 2-D system, we consider three examples.

Example 1

$$H(z_1, z_2) = \frac{1}{A(z_1, z_2)} = \frac{1}{1 - 0.6z_1^{-1} - 0.6z_2^{-1}}.$$

$h(n_1, n_2)$ is a first-quadrant support sequence.

Test 1 $A(z_1, 1) = 1 - 0.6z_1^{-1} - 0.6 = 0$

$$z_1 = \tfrac{3}{2} \geq 1$$

The system fails Test 1 and is therefore unstable.

Example 2

$$H(z_1, z_2) = \frac{1}{A(z_1, z_2)} = \frac{1}{1 - \tfrac{1}{2}z_1^{-1} - \tfrac{1}{4}z_1^{-1}z_2^{-1}}.$$

$h(n_1, n_2)$ is a first-quadrant support sequence.

Test 1 $A(z_1, 1) = 1 - \tfrac{1}{2}z_1^{-1} - \tfrac{1}{4}z_1^{-1} = 0$

$$z_1 = \tfrac{3}{4} < 1$$

Test 1 is passed.

Test 2 $A(1, z_2) = 1 - \tfrac{1}{2} - \tfrac{1}{4}z_2^{-1} = 0$

$$z_2 = \tfrac{1}{2} < 1$$

Test 2 is passed.

Test 3 $A(\omega_1, \omega_2) = A(z_1, z_2)|_{z_1 = e^{j\omega_1}, z_2 = e^{j\omega_2}} = 1 - \tfrac{1}{2}e^{-j\omega_1} - \tfrac{1}{4}e^{-j\omega_1}e^{-j\omega_2}$

$|A(\omega_1, \omega_2)| = |1 - \tfrac{1}{2}e^{-j\omega_1} - \tfrac{1}{4}e^{-j\omega_1}e^{-j\omega_2}| \geq 1 - |\tfrac{1}{2}e^{-j\omega_1}| - |\tfrac{1}{4}e^{-j\omega_1}e^{-j\omega_2}|$

$$= \tfrac{1}{4} > 0 \quad \text{for any } (\omega_1, \omega_2).$$

Therefore, $A(\omega_1, \omega_2) \neq 0$ for any (ω_1, ω_2).   Test 3 is passed.

The system passes all three tests and is therefore stable.

In this particular example, we could perform Test 3 by inspection.   In typical cases, however, Test 3 requires a considerable amount of computation.

Example 3

$$H(z_1, z_2) = \frac{1}{A(z_1, z_2)} = \frac{1}{1 + \tfrac{1}{4}z_1^{-1} - \tfrac{1}{2}z_2^{-1} - \tfrac{1}{3}z_1^{-1}z_2}.$$

The region of support for $h(n_1, n_2)$ is known to be the shaded region in Figure 2.45. Since $h(n_1, n_2)$ is not a first-quadrant support sequence, we will first map $h(n_1, n_2)$ to a first-quadrant support sequence $h'(n_1, n_2)$ without affecting the stability, by linear mapping of variables.   One such mapping is given by

$$h'(n_1, n_2) = h(m_1, m_2)|_{m_1 = n_1, m_2 = n_2 - n_1}.$$

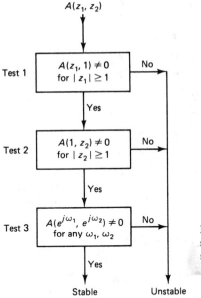

$A(z_1, z_2)$

Test 1 — $A(z_1, 1) \neq 0$ for $|z_1| \geq 1$ — No →

Yes ↓

Test 2 — $A(1, z_2) \neq 0$ for $|z_2| \geq 1$ — No →

Yes ↓

Test 3 — $A(e^{j\omega_1}, e^{j\omega_2}) \neq 0$ for any ω_1, ω_2 — No →

Yes ↓

Stable   Unstable

Figure 2.44   One approach to test the stability of a first-quadrant support system

$$H(z_1, z_2) = \frac{1}{A(z_1, z_2)}.$$

Using Property 7 of the z-transform properties from Table 2.2, we have

$$H(z_1, z_2) = H'(z_1 z_2^{-1}, z_2).$$

From this relationship and the given $H(z_1, z_2)$,

$$H'(z_1, z_2) = \frac{1}{1 + \frac{1}{4}z_1^{-1}z_2^{-1} - \frac{1}{2}z_2^{-1} - \frac{1}{3}z_1^{-1}} = \frac{1}{A'(z_1, z_2)}.$$

We now apply the stability test to $H'(z_1, z_2)$.

Test 1 $A'(z_1, 1) = \frac{1}{2} - \frac{1}{12}z_1^{-1} = 0$

$$z_1 = \frac{1}{6} < 1$$

Test 1 is passed.

Test 2 $A'(1, z_2) = \frac{2}{3} - \frac{1}{4}z_2^{-1} = 0$

$$z_2 = \frac{3}{8} < 1$$

Test 2 is passed.

Test 3.   To test whether $A'(\omega_1, \omega_2) = 0$ for any (ω_1, ω_2), we sketch Root Map 1, which can be obtained by solving $A'(e^{j\omega_1}, z_2) = 0$ as we vary ω_1 from 0 to 2π. Root Map 1 is shown in Figure 2.46. From Root Map 1, all the contours are inside $|z_2| = 1$. Therefore, Test 3 is passed.

The system passes all three tests and is therefore stable.

2.3.4 One-Dimensional Stability Tests

In Sections 2.3.2 and 2.3.3, we have described how a 2-D stability test can be performed by a set of many 1-D stability tests. From (2.49) and (2.57), the 1-D stability test problem that arises can be stated as follows:

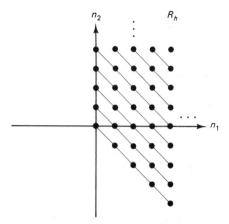

Figure 2.45 Region of support of $h(n_1, n_2)$ in Example 3.

Given a causal $H(z)$ of the form

$$H(z) = \frac{1}{A(z)} = \frac{1}{\sum\limits_{n=0}^{M} a(n)z^{-n}} \qquad (2.63)$$

where $a(n)$ is complex with $a(0) \neq 0$ and $a(M) \neq 0$, determine the system stability.

In this section, we discuss a few approaches to solving the above 1-D stability test problem. One method of solving (2.63) is to explicitly determine the roots of $A(z)$ by a root-finding method and check if all the roots are inside the unit circle. This method is conceptually very simple, and is computationally efficient for low values of M. For values of up to $M = 4$, a closed-form expression can be used to determine the roots. Even when closed-form expressions are not available, the roots determined in one test can be used as initial estimates for the roots in the next test. This is because the roots obtained in the 1-D stability tests typically do not vary much from one test to the next, as we discussed in Section 2.3.2. This approach has been reported [Shaw and Mersereau] to be quite useful in practice for M up to 8 or so. Another advantage of this approach is that explicit evaluation

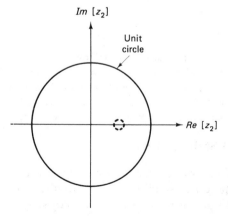

Figure 2.46 Root Map 1 for $A'(z_1, z_2)$
$= 1 + \frac{1}{4}z_1^{-1}z_2^{-1} - \frac{1}{2}z_2^{-1} - \frac{1}{3}z_1^{-1}$.

The z-Transform   Chap. 2

of the roots provides us considerable insight into exactly how stable a stable filter is. If none of the roots determined is very close to the unit circle, the filter is very stable.

Algebraic methods make up another class of methods for solving (2.63). One algebraic method that is quite efficient computationally is the modified Marden-Jury test [Marden, Jury (1964)]. This method requires the construction of a Marden-Jury table. To illustrate the table's construction, it is useful to rewrite $A(z)$ in (2.63) as

$$A(z) = \sum_{n=0}^{M} a(n)z^{-n}$$

$$= \frac{\sum_{n=0}^{M} a(n)z^{M-n}}{z^M} \qquad (2.64)$$

$$= \frac{\sum_{m=0}^{M} a(M-m)z^m}{z^M}.$$

Denoting $a(M-m)$ by a_m, we have

$$A(z) = \frac{\sum_{m=0}^{M} a_m z^m}{z^M}. \qquad (2.65)$$

From (2.65), the roots to $A(z) = 0$ are the same as the roots to $\sum_{m=0}^{M} a_m z^m = 0$. In the construction of the Marden-Jury table, we will consider the roots to

$$A'(z) = \sum_{m=0}^{M} a_m z^m = 0. \qquad (2.66)$$

Table 2.4 illustrates the construction of the Marden-Jury table. Rows 1 and 2 are obtained directly from the coefficients a_m in (2.66). The remaining rows are obtained iteratively, that is, rows 3 and 4 from rows 1 and 2, rows 5 and 6 from rows 3 and 4, and so on. The necessary and sufficient conditions for stability (all roots inside $|z| = 1$) are given by the following M conditions:

$$b_0 < 0, \ c_0 > 0, \ d_0 > 0, \ e_0 > 0, \ \cdots \ r_0 > 0, \ s_0 > 0. \qquad (2.67)$$

Since the coefficients $b_0, c_0 \cdots$ are obtained recursively one at a time, the procedure can terminate early when the system is unstable. This method is computationally very efficient, requiring approximately M^2 multiplications and M^2 additions. Another advantage, common to all algebraic methods, is that the stability is exactly determined in a finite number of steps, if infinite precision for the arithmetic is assumed. Other methods, such as explicit root determination, lack this property. This method has been reported to be computationally efficient and reliable for M up to 20 or so, which covers most infinite impulse response (IIR) filters considered in practice [O'Connor and Huang]. One disadvantage of this method is that it

TABLE 2.4 CONSTRUCTION OF MARDEN-JURY TABLE TO TEST STABILITY OF $\dfrac{1}{\displaystyle\sum_{m=0}^{M} a_m z^m}$ WITH $a_0 \neq 0$ AND $a_M \neq 0$

Row	1	z	z^2	$\cdots$	z^{M-1}	z^M
1	a_0	a_1	a_2	$\cdots$	a_{M-1}	a_M
2	a_M^*	a_{M-1}^*	a_{M-2}^*	$\cdots$	a_1^*	a_0^*
3	b_0	b_1	b_2	$\cdots$	b_{M-1}	
4	b_{M-1}^*	b_{M-2}^*	b_{M-3}^*		b_0^*	
5	c_0	c_1	c_2	$\cdots$	c_{M-2}	
6	c_{M-2}^*	c_{M-3}^*	c_{M-4}^*		c_0^*	
.	.					
.	.					
.	.					
$2M-1$	r_0	r_1				
$2M$	r_1^*	r_0^*				
$2M+1$	s_0					

$$\text{where } b_k = \begin{vmatrix} a_0 & a_{M-k} \\ a_M^* & a_k^* \end{vmatrix}, \quad c_k = \begin{vmatrix} b_0 & b_{M-1-k} \\ b_{M-1}^* & b_k^* \end{vmatrix}, \ldots$$

does not tell us just how stable a filter is. The method can be used in determining the number of roots inside the unit circle, but cannot be used in explicitly determining the root locations.

Another approach to solving (2.63) is to exploit the *Argument principle* [Marden]. Consider the net change in the argument of $A(z)$ in (2.63) as we follow the unit circle contour given by $z = e^{j\omega}$ from $\omega = 0$ to $\omega = 2\pi$ in a counterclockwise direction. Denoting the net argument change by $\Delta\theta_A(\omega{:}0, 2\pi)$, the Argument principle states

$$\Delta\theta_A(\omega{:}0, 2\pi) = 2\pi(N_z - M) \qquad (2.68)$$

where N_z is the number of zeros inside the unit circle. When all roots are inside the unit circle so that $N_z = M$,

$$\Delta\theta_A(\omega{:}0, 2\pi) = 0. \qquad (2.69)$$

From (2.69), one approach to testing the stability is to check if the net phase change is zero. The net phase change can be determined by *unwrapping* the phase [Oppenheim and Schafer]. The result of this procedure is a continuous phase, which is well defined except when any root is on the unit circle. A typical phase unwrapping procedure computes the principal values of the phase and/or the phase derivatives at many frequencies ω using an FFT algorithm. Starting from $\omega = 0$, the continuity assumption of the unwrapped phase is used to find a continuous phase function, which results in the unwrapped phase. If the unwrapped phase is continuous and the unwrapped phases at $\omega = 0$ and $\omega = 2\pi$ are identical, the system is stable. Otherwise, it is unstable. If there is a root on the unit circle, then a continuous unwrapped phase cannot be defined, and a phase discontinuity of π will occur. The system in this case is, of course, unstable. Phase unwrapping is further discussed in Chapter 5, in which we discuss IIR filters. This approach has been reported to be reliable and computationally efficient in comparison to other methods for M greater than approximately 20. For typical 2-D IIR filters, however, M is rarely greater than 20. Note also that phase unwrapping is known to be quite difficult when some roots are close to the unit circle. The unwrapped phase has some qualitative features that can be related to the degree of system stability. For a stable filter, rapid change in the unwrapped phase typically occurs when some zeros are close to the unit circle.

2.3.5 Additional Comments on Stability

The stability test procedures we discussed in Section 2.3 are based on the notion that a 2-D stability test can be performed by a set of many 1-D stability tests. Although we can be more certain about the stability of a 2-D system by increasing the number of 1-D stability tests, stability is not absolutely guaranteed with a finite number of 1-D stability tests, even with infinite precision in arithmetic.

There does exist a class of methods, known as *algebraic tests*, which can guarantee the stability of a system in a finite number of steps if infinite precision in arithmetic is assumed. All existing methods of this type test the conditions in Theorem 3 from Section 2.3.2 by modifying 1-D algebraic tests. In a 1-D algebraic stability test, such as the modified Marden-Jury test discussed in Section 2.3.4, a sequence of real numbers that contain the stability information is derived directly from the 1-D polynomial coefficients. In a 2-D algebraic test, $A(z_1, z_2)$ is viewed as a 1-D polynomial with respect to one variable, say z_2, and the complex coefficients of the polynomial are themselves 1-D real polynomials with respect to the other variable z_1. Applying 1-D algebraic tests to this case results in a sequence of real polynomials that contain the stability information. This sequence of real polynomials must be checked as a function of z_1 to determine the system stability. Although algorithms of this class can guarantee stability in a finite number of steps, they are very difficult to program and require a large number of computations and infinite precision in arithmetic to guarantee stability. Therefore, they do not appear to be very useful in practice.

Stability tests of another class are based on some properties of the complex cepstrum. These are called *cepstral methods*. The 2-D complex cepstrum is a straightforward extension of the 1-D complex cepstrum. One property of the complex cepstrum is that a recursively computable system is stable if and only if the sequence $a(n_1, n_2)$ corresponding to the denominator polynomial $A(z_1, z_2)$ and its complex cepstrum $\hat{a}(n_1, n_2)$ have the same wedge-shape support. By computing $\hat{a}(n_1, n_2)$ and checking its region of support, we can in principle test the stability. Unfortunately, $\hat{a}(n_1, n_2)$ is typically an infinite-extent sequence despite the fact that $a(n_1, n_2)$ is a finite-extent sequence, and computer computation of $\hat{a}(n_1, n_2)$ leads to an aliased version of the true $\hat{a}(n_1, n_2)$. In addition, computation of $\hat{a}(n_1, n_2)$

typically involves phase unwrapping. As we discussed in Section 2.3.4, the unwrapped phase can be used directly for stability testing and explicit computation of $\hat{a}(n_1, n_2)$ from the unwrapped phase has not proved to be useful. For these and other reasons, cepstral methods are not generally used in testing the stability of 2-D IIR systems. The complex cepstrum, however, is useful in other applications, for example spectral factorization and stabilization of an unstable filter. This is discussed in greater detail in Chapter 5, where we study 2-D IIR digital filters.

A 2-D rational system function can also be represented by using a state-space formulation [Roesser; Kung, et al.], in which the input of the system is related to the internal states of the system and the states are then related to the output of the system. In this formulation, stability conditions are stated in terms of the parameters associated with the state-space representation. Existing results on stability based on this approach are somewhat limited. The usefulness of this approach in testing the system stability in practice remains to be seen.

It is clear that testing the stability of a 2-D system is considerably more complex than testing the stability of a 1-D system. The complexity of testing a 2-D system's stability explains, in part, why 2-D finite impulse response (FIR) digital filters, which are always stable, are much preferred over 2-D IIR digital filters in practice. The preference for FIR filters over IIR filters is much more marked in 2-D than in 1-D signal processing applications.

REFERENCES

The 2-D z-transform is a polynomial in two variables. See [Blis; Mostowski and Stark; Walker] for many well-established results on polynomials in two variables. For further reading on recursive computability, see [Mutluay and Fahmy].

For overviews on stability, see [Jury (1978); O'Connor and Huang; Shaw and Mersereau]. Nonessential singularities of the second kind are discussed in [Goodman (1977)]. Stability theorems by Shanks are discussed in [Shanks et al.]. Stability theorems by Huang are discussed in [Huang; Davies; Goodman (1976)]. Stability theorems by DeCarlo and Strintzis are discussed in [DeCarlo et al.; Strintzis; Rajan and Reddy]. See [Huang; Maria and Fahmy; Anderson and Jury; Siljak] for various algebraic stability tests. See [Ekstrom and Woods] for cepstral methods for stability testing. For some results on stability based on state-space representation, see [Roesser; Lu and Lee; Anderson et al.].

B. D .O. Anderson, P. Agathoklis, E. I. Jury, and M. Mansour, Stability and the matrix Lyapunov equation for discrete 2-dimensional systems, *IEEE Trans. on Circuits and Systems*, Vol. CAS-33, March 1986, pp. 261–267.

B. D. O. Anderson and E. I. Jury, Stability test for two-dimensional recursive filters. *IEEE Trans. Audio Electroacoustics*, Vol. AU-21, August 1973, pp. 366–372.

G. A. Bliss, *Algebraic Functions*, New York: Math. Soc. 1933.

J. L. Coolidge, The continuity of the roots of an algebraic equation. *Ann. of Math.*, Vol. 9, 1908, pp. 116–118.

D. L. Davis, A correct proof of Huang's theorem on stability. *IEEE Trans. on Acoust., Speech, and Sig. Proc.*, Vol. ASSP-24, 1976, pp. 425–426.

R. A. DeCarlo, J. Murray, and R. Saeks, Multivariable Nyquist theory, *Int. J. Control*, Vol. 25, 1976, pp. 657–675.

M. P. Ekstrom and J. W. Woods, Two-dimensional spectral factorization with applications in recursive digital filtering, *IEEE Trans. Acoust., Speech, and Sig. Proc.*, Vol. ASSP-24, April 1976, pp. 115–128.

D. Goodman, An alternate proof of Huang's stability theorem, *IEEE Trans. Acoust., Speech, and Sig. Proc.*, Vol. ASSP-24, 1976, pp. 426–427.

D. Goodman, Some stability properties of two-dimensional linear shift-invariant digital filters, *IEEE Trans. on Circuits and Systems*, Vol. CAS-24, April 1977, pp. 201–208.

T. S. Huang, Stability of two-dimensional recursive filters, *IEEE Trans. Audio Electroacoustics*, Vol. AU-20, June 1972, pp. 158–163.

E. I. Jury, *Theory and Applications of the z-Transform Method.* New York: Wiley, 1964.

E. I. Jury, Stability of multidimensional scalar and matrix polynomials, *Proc. IEEE*, Vol. 66, Sept. 1978, pp. 1018–1047.

S. Y. Kung, B. C. Levy, M. Morf, and T. Kailath, New results in 2-D system theory, part II: 2-D state-space models realization and the notions of controllability, observability, and minimality, *Proc. IEEE*, Vol. 65, June 1977, pp. 945–961.

W. S. Lu and E. B. Lee, Stability analysis for two-dimensional systems via a Lyapunov approach, *IEEE Trans. on Circuits and Systems*, Vol. CAS-32, January 1985, pp. 61–68.

M. Marden, *The Geometry of the Zeros of a Polynomial in a Complex Variable.* New York: Amer. Math. Soc., 1949.

G. A. Maria and M. M. Fahmy, On the stability of two-dimensional digital filters, *IEEE Trans. Audio Electroacoustics*, Vol AU-21, 1973, pp. 470–472.

A. Mostowski and M. Stark, *Introduction to Higher Algebra.* New York: Macmillan Co., 1964.

H. E. Mutluay and M. M. Fahmy, Recursibility of N-dimensional IIR digital filters, *IEEE Trans. on Acoust., Speech and Sig. Proc.*, Vol. ASSP-32, April 1984, pp. 397–402.

B. T. O'Connor and T. S. Huang, Stability of general two-dimensional recursive digital filters, *IEEE Trans. on Acoust., Speech, and Sig. Proc.*, Vol. ASSP-26, December 1978, pp. 550–560.

A. V. Oppenheim and R. W. Schafer, *Discrete-Time Signal Processing.* Englewood Cliffs, NJ: Prentice Hall, 1989.

P. K. Rajan and H. C. Reddy, A simple deductive proof of a stability test for two-dimensional digital filters, *Proc. IEEE*, Vol. 72, September 1984, pp. 1221–1222.

R. P. Roesser, A discrete state-space model for linear image processing, *IEEE Trans. on Automatic Control*, Vol. AC-20, February 1975, pp. 1–10.

J. L. Shanks, S. Treitel, and J. H. Justice, Stability and synthesis of two-dimensional recursive filters. *IEEE Trans. Audio Electroacoustics*, Vol. AU-20, June 1972, pp. 115–128.

G. A. Shaw and R. M. Mersereau, Design, stability and performance of two-dimensional recursive digital filters, *Tech. Rep. E21-B05-1*, Georgia Tech, December 1979.

D. D. Siljak, Stability criteria for two-variable polynomials, *IEEE Trans. on Circuits and Systems*, Vol. CAS-22, March 1975, pp. 185–189.

M. G. Strintzis, Test of stability of multidimensional filters, *IEEE Trans. on Circuits and Systems*, Vol. CAS-24, August 1977, pp. 432–437.

R. J. Walker, *Algebraic Curves*, New York: Springer-Verlag, 1978.

PROBLEMS

2.1. Determine the z-transform and its region of convergence (ROC) for each of the following sequences.
(a) $\delta(n_1, n_2)$
(b) $(\frac{1}{2})^{n_1}(3)^{n_2}u(-n_1 - 1, -n_2 - 2)$
(c) $\delta_T(n_1 - 2n_2)u(-n_1, -n_2)$
(d) $\sin\frac{\pi}{12}n_1\,\delta(n_1 - 6, n_2) + e^{n_2}\,\delta(n_1, n_2 + 2)$
(e) $(\frac{1}{2})^{n_2}u(-n_1, n_2)u_T(n_1 + n_2)$
(f) $b^{n_1}u(-n_1, -n_2)u_T(n_2 - 2n_1)$

2.2. Let $X(z_1, z_2)$ and R_x denote the z-transform and its ROC for $x(n_1, n_2)$. For each of the following sequences, determine the z-transform and ROC in terms of $X(z_1, z_2)$ and R_x.
(a) $x(n_1 + 5, n_2 - 6)$
(b) $a^{n_1}x(n_1, n_2)$
(c) $n_1 n_2 x(n_1, n_2)$
(d) $x(n_1, n_2) * x(-n_1, -n_2)$

2.3. Let $X(z_1, z_2)$ and R_x denote the z-transform and its ROC for $x(n_1, n_2)$. Suppose we form a new sequence $y(n_1, n_2)$ by the following linear mapping of variables:

$$y(n_1, n_2) = x(m_1, m_2)|_{m_1 = -n_1, m_2 = -n_2 + 3n_1}.$$

Determine $Y(z_1, z_2)$ and R_y, the z-transform and its ROC for $y(n_1, n_2)$.

2.4. Consider a sequence $x(n_1, n_2)$ given by

$$x(n_1, n_2) = (\tfrac{1}{2})^{n_1} (\tfrac{1}{3})^{n_2}u(n_1, n_2).$$

The ROC of $X(z_1, z_2)$, the z-transform of $x(n_1, n_2)$, is sketched below:

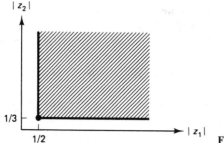

Figure P2.4

Consider the 2-D surface represented by $|z_1| = 1/2$ and $|z_2| = 1/3$, shown as a filled-in dot in the above figure. Determine and sketch the intersection of this 2-D surface with the pole surfaces of $X(z_1, z_2)$.

2.5. Consider a sequence $x(n_1, n_2)$ given by

$$x(n_1, n_2) = \delta_T(n_1 - n_2)u(n_1, n_2).$$

Determine and sketch the intersection of the pole surfaces of $X(z_1, z_2)$ with the unit surface.

2.6. Suppose $x(n_1, n_2)$ is a second-quadrant support sequence. Show that the constraint map in Figure 2.9(b) is a necessary and sufficient condition for the sequence to be a second-quadrant support sequence.

2.7. Consider a sequence $x(n_1, n_2)$ whose region of support is the shaded region of the (n_1, n_2) plane shown below. Let $X(z_1, z_2)$ denote the z-transform of $x(n_1, n_2)$.

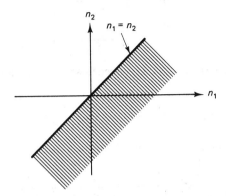

Figure P2.7

(a) What constraints does the ROC of $X(z_1, z_2)$ have to satisfy?

(b) Consider a sequence $y(n_1, n_2)$ given by

$$y(n_1, n_2) = u_T(n_1)\,\delta_T(n_2) + a^{n_2}\,\delta_T(n_1 - n_2)u(-n_1, -n_2).$$

Determine $Y(z_1, z_2)$ and its ROC. Is your answer consistent with your result in (a)?

2.8. Consider a separable system with system function $H(z_1, z_2) = H_1(z_1)H_2(z_2)$. How is the stability of the 2-D system $H(z_1, z_2)$ related to the stability of the two 1-D systems $H_1(z_1)$ and $H_2(z_2)$?

2.9. Consider a sequence $x(n_1, n_2)$ given by

$$x(n_1, n_2) = a^{n_1}b^{n_2}u(n_1, n_2).$$

(a) Determine the range of values of a and b for which $x(n_1, n_2)$ is stable by imposing the condition of absolute summability.

(b) Answer (a) by determining the ROC of $X(z_1, z_2)$ and using a property of the ROC.

2.10. Determine all sequences $x(n_1, n_2)$ with the same z-transform, which is given by

$$X(z_1, z_2) = \frac{1}{(1 - \frac{1}{2}z_1^{-1})(1 - 2z_2^{-1})}.$$

2.11. Consider a stable sequence $x(n_1, n_2)$ whose z-transform $X(z_1, z_2)$ is given by

$$X(z_1, z_2) = \frac{1}{1 - 2z_1^{-2}z_2^{-1} - \frac{1}{3}z_1^{-5}z_2^{-1} - \frac{1}{4}z_1^{-8}z_2^{-1}}.$$

Determine $x(-8, 1)$, the sequence $x(n_1, n_2)$ evaluated at $n_1 = -8$ and $n_2 = 1$. Explain your answer.

2.12. Determine the inverse z-transform of each of the following system functions. In all cases, assume that the ROC includes the unit surface.

(a)
$$H(z_1, z_2) = \frac{z_1^{-2}}{1 - \frac{1}{2}z_1^{-1}z_2^{-1}}$$

(b)
$$H(z_1, z_2) = \frac{z_1^{-2}}{1 - 2z_1^{-1}z_2^{-1}}$$

(c)
$$H(z_1, z_2) = \frac{1}{(1 - \frac{1}{2}z_1^{-1})(1 + 3z_2^{-1})}$$

2.13. In this problem, we determine a stable sequence whose z-transform $X(z_1, z_2)$ is given by

$$X(z_1, z_2) = \frac{1}{1 - \frac{1}{2}z_1^{-1} - \frac{1}{4}z_1^{-1}z_2^{-1}}.$$

(a) Using the inverse z-transform formula, we can express $x(n_1, n_2)$ as

$$x(n_1, n_2) = \frac{1}{2\pi j}\oint_{C_2} \left(\frac{1}{2\pi j}\oint_{C_1} \frac{z_1^{n_1-1}}{1 - \frac{1}{2}z_1^{-1} - \frac{1}{4}z_1^{-1}z_2^{-1}} \, dz_1\right) z_2^{n_2-1} \, dz_2.$$

For a fixed z_2, the expression inside the parenthesis can be viewed as the 1-D inverse z-transform of $G(z_1)$. Determine $g(n_1)$, the 1-D inverse z-transform of $G(z_1)$. The sequence $g(n_1)$ is a function of z_2.

(b) Using the result of (a), we obtain

$$x(n_1, n_2) = \frac{1}{2\pi j}\oint_{C_2} g(n_1)z_2^{n_2-1} \, dz_2.$$

The right-hand side expression of this equation can be viewed as the 1-D inverse z-transform of $F(z_2)$. Using the series expansion formula

$$(1 + x)^n = \sum_{k=0}^{n} \frac{n!}{(n-k)!k!} x^k$$

show that $x(n_1, n_2)$ is given by

$$x(n_1, n_2) = \begin{cases} (\frac{1}{2})^{n_1+n_2}\dfrac{n_1!}{(n_1 - n_2)!n_2!}, & n_1 \ge 0, n_2 \ge 0, n_1 \ge n_2 \\ 0, & \text{otherwise.} \end{cases}$$

(c) Sketch $x(n_1, n_2)$ in (b).

2.14. When the input $x(n_1, n_2)$ to an LSI system is the unit step sequence $u(n_1, n_2)$, the output $y(n_1, n_2)$ is

$$y(n_1, n_2) = (\tfrac{1}{2})^{n_1-1}u(n_1 + 1, n_2)$$

(a) Find the system function $H(z_1, z_2)$ and sketch its pole surfaces.
(b) Determine the impulse response $h(n_1, n_2)$.
(c) Is the system stable?

2.15. Consider an LSI system whose input and output are denoted by $x(n_1, n_2)$ and

$y(n_1, n_2)$, and whose system function $H(z_1, z_2)$ is given as shown in the following figure.

$$x(n_1, n_2) \longrightarrow \boxed{H(z_1, z_2) = \dfrac{1}{1 - \frac{1}{2}z_1^{-1} - \frac{1}{3}z_2^{-1}}} \longrightarrow y(n_1, n_2)$$

We wish to recover $x(n_1, n_2)$ from $y(n_1, n_2)$ with an LSI system with impulse response $f(n_1, n_2)$, as shown in the following figure.

$$y(n_1, n_2) \longrightarrow \boxed{f(n_1, n_2)} \longrightarrow x(n_1, n_2)$$

Determine $f(n_1, n_2)$. Does $f(n_1, n_2)$ depend on the ROC of $H(z_1, z_2)$?

2.16. In general, a system corresponding to a 1-D difference equation with initial conditions is neither linear nor shift invariant. As discussed in the text, one approach to forcing the difference equation with initial conditions to result in an LSI system is to impose the initial rest condition (IRC). Another approach is to impose the final rest condition (FRC). In this problem, we show that both the IRC and the FRC can be viewed as special cases of the following approach to obtaining the initial conditions.

Step 1. Interpret the difference equation as a specific computational procedure.

Step 2. Determine the initial conditions as follows.
   (a) Determine R_h, the region of support of $h(n)$.
   (b) Determine R_y, the region of support of $y(n)$.
   (c) R_{IC} is all $n \notin R_y$.

Step 3. Initial conditions are given by $y(n) = 0$ for $n \in R_{IC}$.

Consider the following difference equation:

$$y(n) + 2y(n - 1) + 3y(n - 2) = x(n),$$

where $x(n)$ is a finite-extent sequence.

   (a) One possible computational procedure obtained from the above difference equation is given by

$$y(n) \leftarrow -2y(n - 1) - 3y(n - 2) + x(n).$$

   Show that the initial conditions obtained from the above three-step approach are identical to the IRC. The IRC states that the output $y(n)$ is zero for $n < n_0$ whenever the input $x(n)$ is zero for $n < n_0$.

   (b) If you choose a different computational procedure obtained from the same difference equation, then the initial conditions obtained from the above three-step approach are identical to the FRC. The FRC states that the output $y(n)$ is zero for $n > n_0 + m_0$ for some fixed finite m_0 whenever the input $x(n)$ is zero for $n > n_0$. Determine the computational procedure and show that the resulting initial conditions are identical to the FRC.

   (c) Is it possible to derive one more computational procedure from the same difference equation? Is this computational procedure recursively computable?

2.17. Consider the following difference equation.

$$y(n_1, n_2) - 2y(n_1 - 1, n_2) - 3y(n_1, n_2 - 1) + 6y(n_1 - 1, n_2 - 1) = x(n_1, n_2).$$

Assume that $x(n_1, n_2)$ is a finite-extent sequence. Suppose the above difference equation corresponds to an LSI system with an ROC that satisfies the conditions in the following constraint map.

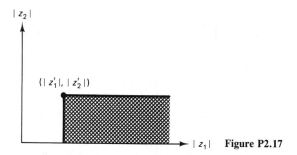

Figure P2.17

The constraint map shows that for any (z_1', z_2') in the ROC, all (z_1, z_2) such that $|z_1| \geq |z_1'|$ and $|z_2| \leq |z_2'|$ are also in the ROC.

(a) Determine the output mask and input mask for the system.

(b) Determine the impulse response of the above LSI system. Your answer should be a closed-form expression. (*Hint:* Consider doing an inverse z-transform.)

2.18. Consider the following difference equation:

$$y(n_1, n_2) + y(n_1 + 1, n_2) + y(n_1 + 1, n_2 + 1) + y(n_1, n_2 + 1)$$

$$+ y(n_1 - 1, n_2 + 1) + y(n_1 - 1, n_2) = x(n_1, n_2).$$

Assume that $x(n_1, n_2)$ is a finite-extent sequence.

(a) How many different recursively computable LSI systems can be obtained from the above difference equation? For each of the systems, sketch the output mask and input mask.

(b) For each of the systems in (a), sketch the region of support of $y(n_1, n_2)$ when $x(n_1, n_2) = \delta(n_1, n_2)$.

2.19. Consider a recursively computable LSI system with the output mask and input mask shown in the following figure. Assume that $x(n_1, n_2)$ is a first-quadrant support sequence.

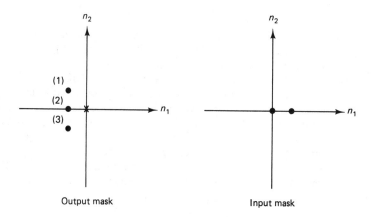

Figure P2.19

(a) Write the difference equation that has the above output and input masks.

(b) Find the boundary conditions that lead to a linear and shift-invariant system.

(c) Determine R_h, the region of support of the impulse response of the LSI system obtained in (b).

(d) Determine one recursion direction that can be used for the LSI system obtained in (b).

(e) Determine the system function.

2.20. Suppose we have designed an IIR filter whose system function is given by

$$H(z_1, z_2) = \frac{1 + z_1^{-1}}{1 + 2z_1^{-1} + 4z_2^{-1}}.$$

The IIR filter is an LSI system. Suppose the filter was designed by attempting to approximate a desired second-quadrant support sequence $h_d(n_1, n_2)$ with $h(n_1, n_2)$. We wish to filter the following 2×2-point input $x(n_1, n_2)$.

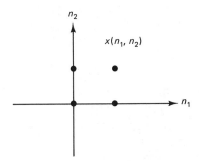

Figure P2.20

Determine the output $y(n_1, n_2)$.

2.21. Consider an LSI system with a third-quadrant support output mask and a third-quadrant support input mask. Show that the impulse response of the system has first-quadrant support. A digital filter with a first-quadrant support impulse response is sometimes referred to as a *causal* or *spatially causal* filter.

2.22. Consider a digital filter whose impulse response $h(n_1, n_2)$ has all its nonzero values in the shaded region in the following figure.

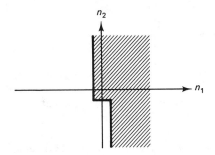

Figure P2.22a

A digital filter of this type is called a *nonsymmetric half plane* filter. Show that a filter whose output and input masks are shown in the following figure is a nonsymmetric half plane filter.

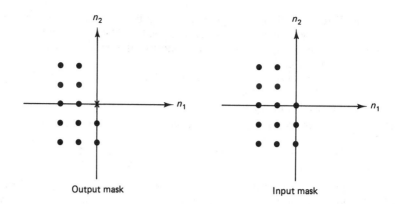

Output mask   Input mask

Figure P2.22b

2.23. Consider a 2-D LSI system with its system function denoted by $H(z_1, z_2)$. Consider the following four pieces of information.

(I1) $$H(z_1, z_2) = \frac{1}{1 + z_1^{-1} + z_2^{-1} + z_1^{-1}z_2 + z_1^{-1}z_2^{-1} + z_1 z_2^{-1}}.$$

(I2) The system is recursively computable.

(I3) The ROC of $H(z_1, z_2)$ satisfies the conditions of the following constraint map, which states that for any (z_1', z_2') in the ROC, all (z_1, z_2) such that $|z_1| \le |z_1'|$ and $|z_2| \le |z_2'|$ is also in the ROC.

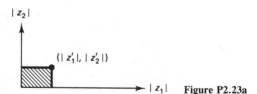

Figure P2.23a

(I4) The input to the system, $x(n_1, n_2)$, is a 3×3-point sequence given by

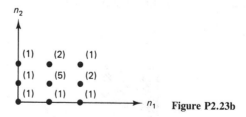

Figure P2.23b

(a) Suppose that only the information in (I1), (I2), and (I4) is available. How many different systems would satisfy the constraints imposed by the available information?

(b) Suppose all the above information in (I1), (I2), (I3), and (I4) is available. Sketch the output and input masks and determine $y(0, 0)$, the response of the system evaluated at $n_1 = n_2 = 0$.

2.24. When the input $x(n_1, n_2)$ is a finite-extent sequence, a necessary and sufficient condition for a computational procedure to be recursively computable is that the output mask have wedge support. When the input $x(n_1, n_2)$ is an infinite-extent sequence, the condition that the output mask have wedge support is necessary but not sufficient for the system to be recursively computable. To illustrate this, consider the following specific computational procedure:

$$y(n_1, n_2) \leftarrow y(n_1 - 1, n_2) + y(n_1, n_2 - 1) + y(n_1 - 1, n_2 - 1) + x(n_1, n_2).$$

Determine an infinite-extent quadrant support input $x(n_1, n_2)$ such that the boundary conditions obtained to force the computational procedure to be an LSI system do not make sense. In this case, the system is not recursively computable, even though the output mask has wedge support.

2.25. Determine whether or not the following filters are stable. Assume that they have first-quadrant support impulse responses.

(a)
$$H(z_1, z_2) = \frac{1}{(1 - \frac{1}{2}z_1^{-1})(1 - \frac{1}{3}z_2^{-1})(1 - 2z_2^{-1})}$$

(b)
$$H(z_1, z_2) = \frac{1}{1 + 5z_1^{-1} + 3z_2^{-1} + 2z_1^{-1}z_2^{-3} + 3z_2^{-6}}$$

(c)
$$H(z_1, z_2) = \frac{1}{10 - 2z_1^{-1} - 2z_2^{-1} - 3z_1^{-1}z_2^{-1}}$$

(d)
$$H(z_1, z_2) = \frac{1}{1 + 4z_1^{-1} + 3z_2^{-1} + z_1^{-1}z_2^{-2} + z_2^{-4}}$$

(e)
$$H(z_1, z_2) = \frac{1}{1 - 0.6z_2^{-1} - 0.81z_1^{-1}z_2^{-1}}$$

(f)
$$H(z_1, z_2) = \frac{1}{1 - 0.8z_1^{-1} - 0.7z_2^{-1} + 0.56z_1^{-1}z_2^{-1}}$$

2.26. Consider a first-quadrant support system with the system function $H(z_1, z_2)$ given by

$$H(z_1, z_2) = \frac{1}{1 - az_1^{-1} - bz_2^{-1}}.$$

Determine the conditions on a and b for which the system is stable.

2.27. Consider an LSI system whose output mask is sketched below.

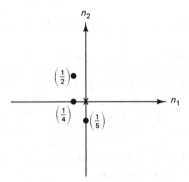

Figure P2.27

Determine whether or not the system is stable.

2.28. The stability theorems discussed in Section 2.3.2 apply to first-quadrant support systems.

(a) Modify Theorem 2 so that it applies to fourth-quadrant support systems.

(b) State your result in (a) in terms of root maps.

2.29. Let $h(n_1, n_2)$ denote the impulse response of a one-quadrant support system. The system function $H(z_1, z_2)$ is in the following form:

$$H(z_1, z_2) = \frac{1}{A(z_1, z_2)}$$

where $A(z_1, z_2)$ is a finite-order polynomial in the two variables. It is known that a necessary and sufficient condition for the stability of the system is given by

$$A(z_1, z_2) \neq 0 \text{ for (a)} \quad |z_1| = 1, |z_2| \geq 1$$
$$\text{(b)} \quad |z_1| \leq 1, |z_2| = 1.$$

(a) Show that the following set of conditions is equivalent to the above set of conditions.

   i. $A(z_1, z_2) \neq 0$ for $|z_1| = |z_2| = 1$

   ii. $A(z_1, -j) \neq 0$ for $|z_1| \leq 1$

   iii. $A(j, z_2) \neq 0$ for $|z_2| \geq 1$

(b) Suppose that for every ω_2 we find all values of z_1 such that $A(z_1, 3e^{j\omega_2}) = 0$. Suppose we sketch all the above values of z_1 in the z_1 plane. If the system is stable, where do the values of z_1 lie in the z_1 plane?

2.30. We have a one-quadrant support system whose system function $H(z_1, z_2)$ is given by

$$H(z_1, z_2) = \frac{1}{4z_1 + az_1^{-4}z_2^{-6}},$$

where a is a real constant. Determine the range of a for which the system is stable.

2.31. Consider a first-quadrant support LSI system whose system function is denoted by

$$H(z_1, z_2) = \frac{1}{A(z_1, z_2)}.$$

The stability theorem originally obtained by Huang is

$$\text{Stability} \Longleftrightarrow A(z_1, z_2) \neq 0 \text{ for (a)} \quad |z_1| = 1, |z_2| \geq 1$$
$$\text{(b)} \quad |z_1| \geq 1, z_2 = \infty.$$

Show the above theorem. You may use any of the stability theorems discussed in Section 2.3.2.

2.32. Theorem 2 from Section 2.3.2 suggests a procedure that involves performing many 1-D stability tests twice, while Theorem 3 from the same section suggests a procedure that involves performing many 1-D stability tests once. It would appear, therefore, that Theorem 3 reduces the problem complexity by half. This is not the case. Suppose we fix the total number of 1-D stability tests. We can use all of them to test Condition (a) of Theorem 3. Alternatively, we can use half of them to test Condition (a) of Theorem 2 and the remaining half to test Condition (b) of Theorem 2. Explain why one approach is not necessarily better than the other by relating each of the two approaches to Condition (a) of Theorem 4.

2.33. A 2-D stability problem can be solved by solving many 1-D stability problems. An efficient method of testing the stability of a 1-D system is by using the Marden-Jury table discussed in Section 2.3.4. Using the Marden-Jury table, determine the stability of the following 1-D causal system functions.

(a)
$$H(z) = \frac{1}{1 - 2z - 3z^2}$$

(b)
$$H(z) = \frac{1}{1 - \frac{1}{2}z - 2z^2 + 3z^3 - \frac{1}{4}z^4}$$

(c)
$$H(z) = \frac{1}{1 - 3z + 5z^2 + 2z^3 - z^4 + z^5 + z^6}$$

3

The Discrete Fourier Transform

3.0 INTRODUCTION

In many signal processing applications, such as image processing, we deal with finite-extent sequences. For such sequences, the Fourier transform and z-transform uniformly converge and are well defined. The Fourier transform and z-transform representations $X(\omega_1, \omega_2)$ and $X(z_1, z_2)$, however, are functions of continuous variables (ω_1, ω_2) and (z_1, z_2). For finite-extent sequences, which can be represented by a finite number of values, the Fourier transform and z-transform are not efficient frequency domain representations. The discrete Fourier transform represents a finite-extent sequence in the frequency domain with a finite number of values.

In this chapter, we study the discrete Fourier transform and algorithms to compute it efficiently. We also study the discrete cosine transform, which is closely related to the discrete Fourier transform. In Section 3.1, we discuss the discrete Fourier series representation of periodic sequences. The discrete Fourier series representation can be used in deriving the discrete Fourier transform representation. In Section 3.2, the discrete Fourier transform representation of finite-extent sequences is discussed. In Section 3.3, the discrete cosine transform representation of finite-extent sequences is discussed. In Section 3.4, we discuss algorithms that can be used in computing the discrete Fourier transform efficiently.

3.1 THE DISCRETE FOURIER SERIES

The *discrete Fourier series* (DFS) is a frequency domain representation of a periodic sequence. We will begin with a discussion of the DFS, since it naturally leads to the discrete Fourier transform (DFT), the main topic of this chapter.

A sequence $\tilde{x}(n_1, n_2)$ is said to be periodic with a period of $N_1 \times N_2$ when $\tilde{x}(n_1, n_2) = \tilde{x}(n_1 + N_1, n_2) = \tilde{x}(n_1, n_2 + N_2)$ for all (n_1, n_2). Since $\tilde{x}(n_1, n_2)r_1^{-n_1}r_2^{-n_2}$ is not absolutely summable for any r_1 and r_2, neither the Fourier transform nor the z-transform uniformly converges for a periodic sequence.

As in the 1-D case, a periodic sequence $\tilde{x}(n_1, n_2)$ with a period of $N_1 \times N_2$ can be obtained by appropriately combining complex exponentials of the form $\tilde{X}(k_1, k_2)e^{j(2\pi/N_1)k_1n_1}e^{j(2\pi/N_2)k_2n_2}$. The exponential sequence $\tilde{X}(k_1, k_2)e^{j(2\pi/N_1)k_1n_1}e^{j(2\pi/N_2)k_2n_2}$ for $0 \leq k_1 \leq N_1 - 1, 0 \leq k_2 \leq N_2 - 1$ represents all complex exponential sequences that are periodic with a period of $N_1 \times N_2$. The sequence $\tilde{X}(k_1, k_2)$, which represents the amplitude associated with the complex exponential, can be obtained from $\tilde{x}(n_1, n_2)$. The relationship between $\tilde{x}(n_1, n_2)$ and $\tilde{X}(k_1, k_2)$ is given by

---

Discrete Fourier Series Pair

$$\tilde{X}(k_1, k_2) = \sum_{n_1=0}^{N_1-1} \sum_{n_2=0}^{N_2-1} \tilde{x}(n_1, n_2)e^{-j(2\pi/N_1)k_1n_1}e^{-j(2\pi/N_2)k_2n_2} \qquad (3.1a)$$

$$\tilde{x}(n_1, n_2) = \frac{1}{N_1N_2} \sum_{k_1=0}^{N_1-1} \sum_{k_2=0}^{N_2-1} \tilde{X}(k_1, k_2)e^{j(2\pi/N_1)k_1n_1}e^{j(2\pi/N_2)k_2n_2} \qquad (3.1b)$$

Equation (3.1a) shows how the amplitude $\tilde{X}(k_1, k_2)$ associated with the exponential $e^{j(2\pi/N_1)k_1n_1}e^{j(2\pi/N_2)k_2n_2}$ can be determined from $\tilde{x}(n_1, n_2)$. The sequence $\tilde{X}(k_1, k_2)$ is called the DFS coefficients of $\tilde{x}(n_1, n_2)$. Equation (3.1b) shows how complex exponentials $\tilde{X}(k_1, k_2)e^{j(2\pi/N_1)k_1n_1}e^{j(2\pi/N_2)k_2n_2}$ are specifically combined to form $\tilde{x}(n_1, n_2)$. The sequence $\tilde{x}(n_1, n_2)$ is called the inverse DFS of $\tilde{X}(k_1, k_2)$. By combining (3.1a) and (3.1b), they can easily be shown to be consistent with each other.

From (3.1a) and (3.1b), $\tilde{x}(n_1, n_2)$ is represented by $\tilde{X}(k_1, k_2)$ for $0 \leq k_1 \leq N_1 - 1, 0 \leq k_2 \leq N_2 - 1$. The sequence $\tilde{X}(k_1, k_2)$ can, therefore, be defined arbitrarily for (k_1, k_2) outside $0 \leq k_1 \leq N_1 - 1, 0 \leq k_2 \leq N_2 - 1$. For convenience and by convention, $\tilde{X}(k_1, k_2)$ is defined to be periodic with a period of $N_1 \times N_2$, with one period of $\tilde{X}(k_1, k_2)$ given by $\tilde{X}(k_1, k_2)$ for $0 \leq k_1 \leq N_1 - 1, 0 \leq k_2 \leq N_2 - 1$. Also by convention, (3.1b) has a scaling factor of $1/(N_1N_2)$. This scaling factor can be distributed to (3.1a) and (3.1b) equally as $1/\sqrt{N_1N_2}$ if desired. Since $\tilde{x}(n_1, n_2)$ and $\tilde{X}(k_1, k_2)$ are periodic with a period of $N_1 \times N_2$ in the variables (n_1, n_2) and (k_1, k_2), respectively, and since $e^{\pm j(2\pi/N_1)k_1n_1}e^{\pm j(2\pi/N_2)k_2n_2}$ is periodic with a period of $N_1 \times N_2$ in both sets of variables (n_1, n_2) and (k_1, k_2), the limits of summation in (3.1a) and (3.1b) can be over any one period.

We can derive a number of useful properties from the DFS pair of (3.1a) and (3.1b). Some of the more important properties are listed in Table 3.1. These are essentially the same as in the 1-D case, except Property 4, which applies to separable 2-D sequences. Property 2 states that when two sequences $\tilde{x}(n_1, n_2)$ and $\tilde{y}(n_1, n_2)$ are periodically convolved, their DFS coefficients $\tilde{X}(k_1, k_2)$ and $\tilde{Y}(k_1, k_2)$ multiply. The periodic convolution, denoted by a circled asterisk (⊛), is very similar in form to the linear convolution. The difference lies in the limits of the summation. Specifically, $\tilde{x}(l_1, l_2)\tilde{y}(n_1 - l_1, n_2 - l_2)$ is summed over only one

TABLE 3.1   PROPERTIES OF THE DISCRETE FOURIER SERIES

$\tilde{x}(n_1, n_2), \tilde{y}(n_1, n_2)$: periodic with a period of $N_1 \times N_2$

$\tilde{x}(n_1, n_2) \longleftrightarrow \tilde{X}(k_1, k_2)$

$\tilde{y}(n_1, n_2) \longleftrightarrow \tilde{Y}(k_1, k_2)$

$N_1 \times N_2$-point DFS and IDFS are assumed

Property 1.   Linearity

$$a\tilde{x}(n_1, n_2) + b\tilde{y}(n_1, n_2) \longleftrightarrow a\tilde{X}(k_1, k_2) + b\tilde{Y}(k_1, k_2)$$

Property 2.   Periodic Convolution

$$\tilde{x}(n_1, n_2) \circledast \tilde{y}(n_1, n_2) \qquad\qquad \longleftrightarrow \tilde{X}(k_1, k_2)\tilde{Y}(k_1, k_2)$$

$$= \sum_{l_1=0}^{N_1-1} \sum_{l_2=0}^{N_2-1} \tilde{x}(l_1, l_2)\tilde{y}(n_1 - l_1, n_2 - l_2)$$

Property 3.   Multiplication

$$\tilde{x}(n_1, n_2)\tilde{y}(n_1, n_2) \longleftrightarrow \frac{1}{N_1 N_2} \tilde{X}(k_1, k_2) \circledast \tilde{Y}(k_1, k_2)$$

$$= \frac{1}{N_1 N_2} \sum_{l_1=0}^{N_1-1} \sum_{l_2=0}^{N_2-1} \tilde{X}(l_1, l_2)\tilde{Y}(k_1 - l_1, k_2 - l_2)$$

Property 4.   Separable Sequence

$$\tilde{x}(n_1, n_2) = \tilde{x}_1(n_1)\tilde{x}_2(n_2) \longleftrightarrow \tilde{X}(k_1, k_2) = \tilde{X}_1(k_1)\tilde{X}_2(k_2)$$

$$\tilde{X}_1(k_1): N_1\text{-point 1-D DFS}$$

$$\tilde{X}_2(k_2): N_2\text{-point 1-D DFS}$$

Property 5.   Shift of a Sequence

$$\tilde{x}(n_1 - m_1, n_2 - m_2) \longleftrightarrow \tilde{X}(k_1, k_2)e^{-j(2\pi/N_1)k_1 m_1}e^{-j(2\pi/N_2)k_2 m_2}$$

Property 6.   Initial Value and DC Value Theorem

$$\text{(a) } \tilde{x}(0, 0) = \frac{1}{N_1 N_2} \sum_{k_1=0}^{N_1-1} \sum_{k_2=0}^{N_2-1} \tilde{X}(k_1, k_2)$$

$$\text{(b) } \tilde{X}(0, 0) = \sum_{n_1=0}^{N_1-1} \sum_{n_2=0}^{N_2-1} \tilde{x}(n_1, n_2)$$

Property 7.   Parseval's Theorem

$$\text{(a) } \sum_{n_1=0}^{N_1-1} \sum_{n_2=0}^{N_2-1} \tilde{x}(n_1, n_2)\tilde{y}^*(n_1, n_2) = \frac{1}{N_1 N_2} \sum_{k_1=0}^{N_1-1} \sum_{k_2=0}^{N_2-1} \tilde{X}(k_1, k_2)\tilde{Y}^*(k_1, k_2)$$

$$\text{(b) } \sum_{n_1=0}^{N_1-1} \sum_{n_2=0}^{N_2-1} |\tilde{x}(n_1, n_2)|^2 = \frac{1}{N_1 N_2} \sum_{k_1=0}^{N_1-1} \sum_{k_2=0}^{N_2-1} |\tilde{X}(k_1, k_2)|^2$$

Property 8.   Symmetry Properties

(a) $\tilde{x}^*(n_1, n_2) \longleftrightarrow \tilde{X}^*(-k_1, -k_2)$

(b) real $\tilde{x}(n_1, n_2) \longleftrightarrow \tilde{X}(k_1, k_2) = \tilde{X}^*(-k_1, -k_2)$

$$\tilde{X}_R(k_1, k_2) = \tilde{X}_R(-k_1, -k_2)$$

$$\tilde{X}_I(k_1, k_2) = -\tilde{X}_I(-k_1, -k_2)$$

$$|\tilde{X}(k_1, k_2)| = |\tilde{X}(-k_1, -k_2)|$$

$$\theta_{\tilde{X}}(k_1, k_2) = -\theta_{\tilde{X}}(-k_1, -k_2)$$

period ($0 \le l_1 \le N_1 - 1, 0 \le l_2 \le N_2 - 1$) in the periodic convolution $\tilde{x}(n_1, n_2)$ $\circledast \tilde{y}(n_1, n_2)$, while $x(l_1, l_2)y(n_1 - l_1, n_2 - l_2)$ is summed over all values of l_1 and l_2 in the linear convolution $x(n_1, n_2) * y(n_1, n_2)$. An example of $\tilde{x}(n_1, n_2) \circledast$ $\tilde{y}(n_1, n_2)$ is shown in Figure 3.1. Figures 3.1(a) and (b) show one period of $\tilde{x}(n_1, n_2)$ and $\tilde{y}(n_1, n_2)$, each periodic with a period of 3×2. Figures 3.1(c) and (d) show one period of $\tilde{x}(l_1, l_2)$ and $\tilde{y}(n_1 - l_1, n_2 - l_2)|_{n_1 = n_2 = 0}$. The result of

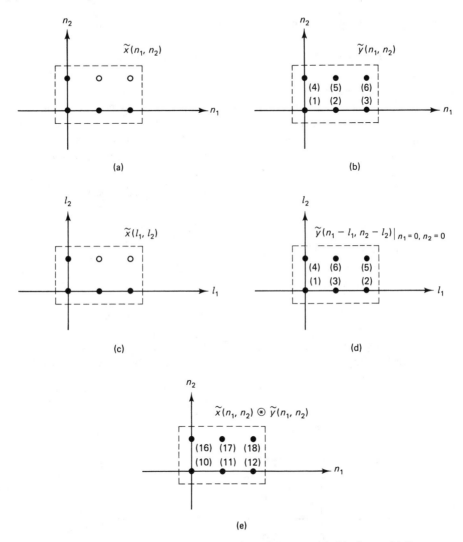

Figure 3.1 Example of periodic convolution. Only one period is shown. (a) Periodic sequence $\tilde{x}(n_1, n_2)$ with a period of 3×2; (b) periodic sequence $\tilde{y}(n_1, n_2)$ with a period of 3×2; (c) $\tilde{x}(l_1, l_2)$; (d) $\tilde{y}(n_1 - l_1, n_2 - l_2)|_{n_1 = n_2 = 0}$; (e) $\tilde{x}(n_1, n_2)$ $\circledast y(n_1, n_2)$ computed with an assumed period of 3×2. Note that $\tilde{x}(n_1, n_2) \circledast$ $y(n_1, n_2)$ at $n_1 = n_2 = 0$ is given by

$$\sum_{l_1 = 0}^{2} \sum_{l_2 = 0}^{1} \tilde{x}(l_1, l_2)\tilde{y}(n_1 - l_1, n_2 - l_2)|_{n_1 = n_2 = 0} = 1 + 3 + 2 + 4 = 10$$

$\tilde{x}(n_1, n_2) \circledast \tilde{y}(n_1, n_2)$ at $n_1 = n_2 = 0$ can be obtained by multiplying $\tilde{x}(l_1, l_2)$ and $\tilde{y}(n_1 - l_1, n_2 - l_2)|_{n_1 = n_2 = 0}$ and then summing over one period. One period of $\tilde{x}(n_1, n_2) \circledast \tilde{y}(n_1, n_2)$ is shown in Figure 3.1(e). In the figure, $\tilde{x}(n_1, n_2) \circledast \tilde{y}(n_1, n_2)$ is computed explicitly by using the periodic convolution sum. An alternative way to compute $\tilde{x}(n_1, n_2) \circledast \tilde{y}(n_1, n_2)$ is by performing the inverse DFS operation of $\tilde{X}(k_1, k_2)\tilde{Y}(k_1, k_2)$.

3.2 THE DISCRETE FOURIER TRANSFORM

3.2.1 The Discrete Fourier Transform Pair

The *discrete Fourier transform* (DFT) is a frequency domain representation of finite-extent sequences. The DFT representation can easily be derived from the DFS representation of a periodic sequence discussed in Section 3.1. Consider a sequence $\tilde{x}(n_1, n_2)$, which is periodic with a period of $N_1 \times N_2$. Suppose we form a finite-extent sequence $x(n_1, n_2)$ by preserving one period of $\tilde{x}(n_1, n_2)$ but setting all other values to zero. Specifically, we have

$$x(n_1, n_2) = \tilde{x}(n_1, n_2)R_{N_1 \times N_2}(n_1, n_2) \tag{3.2a}$$

where $\quad R_{N_1 \times N_2}(n_1, n_2) = \begin{cases} 1, & 0 \leq n_1 \leq N_1 - 1, 0 \leq n_2 \leq N_2 - 1 \\ 0, & \text{otherwise.} \end{cases} \tag{3.2b}$

An example of $\tilde{x}(n_1, n_2)$ and $x(n_1, n_2)$ when $N_1 = 3$ and $N_2 = 2$ is shown in Figure 3.2. Clearly, the operation given by (3.2) is invertible in that $\tilde{x}(n_1, n_2)$ can be determined from $x(n_1, n_2)$ by

$$\tilde{x}(n_1, n_2) = \sum_{r_1 = -\infty}^{\infty} \sum_{r_2 = -\infty}^{\infty} x(n_1 - r_1 N_1, n_2 - r_2 N_2). \tag{3.3}$$

Now consider $\tilde{X}(k_1, k_2)$, the DFS coefficients of $\tilde{x}(n_1, n_2)$. Suppose we form a finite-extent sequence $X(k_1, k_2)$ by preserving one period of $\tilde{X}(k_1, k_2)$ but setting all other values to zero as follows:

$$X(k_1, k_2) = \tilde{X}(k_1, k_2)R_{N_1 \times N_2}(k_1, k_2) \tag{3.4}$$

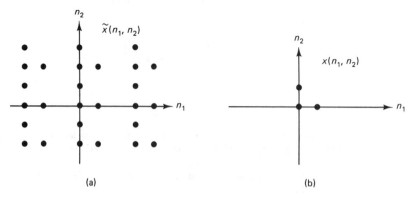

Figure 3.2 Finite-extent sequence obtained from one period of a periodic sequence. (a) $\tilde{x}(n_1, n_2)$; (b) $x(n_1, n_2)$.

where $R_{N_1 \times N_2}(k_1, k_2)$ is defined by (3.2b). This operation is also invertible, since $\tilde{X}(k_1, k_2)$ can be obtained from $X(k_1, k_2)$ by

$$\tilde{X}(k_1, k_2) = \sum_{r_1=-\infty}^{\infty} \sum_{r_2=-\infty}^{\infty} X(k_1 - r_1 N_1, k_2 - r_2 N_2). \tag{3.5}$$

From the above, the sequence $x(n_1, n_2)$ is related to $X(k_1, k_2)$ by

$$x(n_1, n_2) \leftrightarrow \tilde{x}(n_1, n_2) \leftrightarrow \tilde{X}(k_1, k_2) \leftrightarrow X(k_1, k_2) \tag{3.6}$$

where "$\leftrightarrow$" denotes an invertible operation. The relationship between $x(n_1, n_2)$ and $X(k_1, k_2)$ can be easily obtained from (3.6) and the DFS pair in (3.1). The DFT pair is given by

---

Discrete Fourier Transform Pair

$$X(k_1, k_2) = \begin{cases} \displaystyle\sum_{n_1=0}^{N_1-1} \sum_{n_2=0}^{N_2-1} x(n_1, n_2)e^{-j(2\pi/N_1)k_1 n_1} e^{-j(2\pi/N_2)k_2 n_2}, \\ \qquad 0 \leq k_1 \leq N_1 - 1, \, 0 \leq k_2 \leq N_2 - 1 \\ 0, \qquad \text{otherwise.} \end{cases} \tag{3.7a}$$

$$x(n_1, n_2) = \begin{cases} \displaystyle\frac{1}{N_1 N_2} \sum_{k_1=0}^{N_1-1} \sum_{k_2=0}^{N_2-1} X(k_1, k_2)e^{j(2\pi/N_1)k_1 n_1} e^{j(2\pi/N_2)k_2 n_2}, \\ \qquad 0 \leq n_1 \leq N_1 - 1, \, 0 \leq n_2 \leq N_2 - 1 \\ 0, \qquad \text{otherwise.} \end{cases} \tag{3.7b}$$

From (3.7), an $N_1 \times N_2$-point sequence $x(n_1, n_2)$ is represented in the frequency domain by an $N_1 \times N_2$-point sequence $X(k_1, k_2)$. The sequence $X(k_1, k_2)$ is called the DFT of $x(n_1, n_2)$, and $x(n_1, n_2)$ is called the inverse DFT (IDFT) of $X(k_1, k_2)$.

The DFT pair above is defined only for a finite-extent first-quadrant support sequence. This is not a serious restriction in practice, since a finite-extent sequence can always be shifted to have first-quadrant support, and this shift can easily be taken into account in many application problems.

For a finite-extent first-quadrant support sequence $x(n_1, n_2)$ that is zero outside $0 \leq n_1 \leq N_1 - 1$ and $0 \leq n_2 \leq N_2 - 1$, the DFT $X(k_1, k_2)$ is very simply related to the discrete-space Fourier transform $X(\omega_1, \omega_2)$. From (3.7a), it is clear that

$$X(k_1, k_2) = X(\omega_1, \omega_2)\big|_{\omega_1 = (2\pi/N_1)k_1, \omega_2 = (2\pi/N_2)k_2}, \tag{3.8}$$

$$0 \leq k_1 \leq N_1 - 1, \, 0 \leq k_2 \leq N_2 - 1.$$

Equation (3.8) states that the DFT coefficients of $x(n_1, n_2)$ are samples of $X(\omega_1, \omega_2)$ at equally spaced points on the Cartesian grid, beginning at $\omega_1 = \omega_2 = 0$. Since $X(k_1, k_2)$ completely specifies $x(n_1, n_2)$, which in turn completely specifies $X(\omega_1, \omega_2)$, $X(\omega_1, \omega_2)$ has considerable redundant information; that is, $N_1 \times N_2$ samples of $X(\omega_1, \omega_2)$ completely specify $X(\omega_1, \omega_2)$.

TABLE 3.2 PROPERTIES OF THE DISCRETE FOURIER TRANSFORM

$x(n_1, n_2), y(n_1, n_2) = 0$ outside $0 \le n_1 \le N_1 - 1, 0 \le n_2 \le N_2 - 1$
$x(n_1, n_2) \longleftrightarrow X(k_1, k_2)$
$y(n_1, n_2) \longleftrightarrow Y(k_1, k_2)$
$N_1 \times N_2$-point DFT and IDFT are assumed.

Property 1. *Linearity*
$$ax(n_1, n_2) + by(n_1, n_2) \longleftrightarrow aX(k_1, k_2) + bY(k_1, k_2)$$

Property 2. *Circular Convolution*
$$x(n_1, n_2) \circledast y(n_1, n_2) \longleftrightarrow X(k_1, k_2)Y(k_1, k_2)$$
$$= [\tilde{x}(n_1, n_2) \circledast \tilde{y}(n_1, n_2)]R_{N_1 \times N_2}(n_1, n_2)$$

Property 3. *Relation between Circular and Linear Convolution*
$$f(n_1, n_2) = 0 \text{ outside } 0 \le n_1 \le N_1' - 1, 0 \le n_2 \le N_2' - 1$$
$$g(n_1, n_2) = 0 \text{ outside } 0 \le n_1 \le N_1'' - 1, 0 \le n_2 \le N_2'' - 1$$
$$f(n_1, n_2) * g(n_1, n_2) = f(n_1, n_2) \circledast g(n_1, n_2) \text{ with periodicity } N_1 \ge N_1' + N_1'' - 1,$$
$$N_2 \ge N_2' + N_2'' - 1$$

Property 4. *Multiplication*
$$x(n_1, n_2)y(n_1, n_2) \longleftrightarrow \frac{1}{N_1 N_2} X(k_1, k_2) \circledast Y(k_1, k_2)$$
$$= \frac{1}{N_1 N_2} [\tilde{X}(k_1, k_2) \circledast \tilde{Y}(k_1, k_2)]R_{N_1 \times N_2}(k_1, k_2)$$

Property 5. *Separable Sequence*
$$x(n_1, n_2) = x_1(n_1)x_2(n_2) \longleftrightarrow X(k_1, k_2) = X_1(k_1)X_2(k_2)$$
$$X_1(k_1): N_1\text{-point 1-D DFT}$$
$$X_2(k_2): N_2\text{-point 1-D DFT}$$

Property 6. *Circular Shift of a Sequence*
$$\tilde{x}(n_1 - m_1, n_2 - m_2)R_{N_1 \times N_2}(n_1, n_2) \longleftrightarrow X(k_1, k_2)e^{-j(2\pi/N_1)k_1 m_1}e^{-j(2\pi/N_2)k_2 m_2}$$
$$= x((n_1 - m_1)_{N_1}, (n_2 - m_2)_{N_2})$$

Property 7. *Initial Value and DC Value Theorem*
$$\text{(a) } x(0, 0) = \frac{1}{N_1 N_2} \sum_{k_1=0}^{N_1-1} \sum_{k_2=0}^{N_2-1} X(k_1, k_2)$$
$$\text{(b) } X(0, 0) = \sum_{n_1=0}^{N_1-1} \sum_{n_2=0}^{N_2-1} x(n_1, n_2)$$

Property 8. *Parseval's Theorem*
$$\text{(a) } \sum_{n_1=0}^{N_1-1} \sum_{n_2=0}^{N_2-1} x(n_1, n_2)y^*(n_1, n_2) = \frac{1}{N_1 N_2} \sum_{k_1=0}^{N_1-1} \sum_{k_2=0}^{N_2-1} X(k_1, k_2)Y^*(k_1, k_2)$$
$$\text{(b) } \sum_{n_1=0}^{N_1-1} \sum_{n_2=0}^{N_2-1} |x(n_1, n_2)|^2 = \frac{1}{N_1 N_2} \sum_{k_1=0}^{N_1-1} \sum_{k_2=0}^{N_2-1} |X(k_1, k_2)|^2$$

TABLE 3.2 *Continued*

Property 9. Symmetry Properties

(a) $x^*(n_1, n_2) \longleftrightarrow \tilde{X}^*(-k_1, -k_2)R_{N_1 \times N_2}(k_1, k_2) = X^*((-k_1)_{N_1}, (-k_2)_{N_2})$

(b) real $x(n_1, n_2) \longleftrightarrow X(k_1, k_2) = \tilde{X}^*(-k_1, -k_2)R_{N_1 \times N_2}(k_1, k_2)$

$$X_R(k_1, k_2) = \tilde{X}_R(-k_1, -k_2)R_{N_1 \times N_2}(k_1, k_2)$$

$$X_I(k_1, k_2) = -\tilde{X}_I(-k_1, -k_2)R_{N_1 \times N_2}(k_1, k_2)$$

$$|X(k_1, k_2)| = |\tilde{X}(-k_1, -k_2)|R_{N_1 \times N_2}(k_1, k_2)$$

$$\theta_x(k_1, k_2) = -\tilde{\theta}_x(-k_1, -k_2)R_{N_1 \times N_2}(k_1, k_2)$$

3.2.2 Properties

We can derive a number of useful properties from the DFT pair in (3.7). On the other hand, the relationship in (3.6) and the DFS properties discussed in Section 3.1 are often more convenient and easier to use in deriving the DFT properties. Some of the more important properties are listed in Table 3.2. Most of these properties are straightforward extensions of 1-D results.

To illustrate how DFT properties can be obtained from (3.6) and the DFS properties discussed in Section 3.1, we will derive the circular convolution property. Consider the periodic convolution property in Table 3.1:

$$\tilde{x}(n_1, n_2) \circledast \tilde{y}(n_1, n_2) \leftrightarrow \tilde{X}(k_1, k_2)\tilde{Y}(k_1, k_2). \tag{3.9}$$

From (3.6) and (3.9),

$$[\tilde{x}(n_1, n_2) \circledast \tilde{y}(n_1, n_2)]R_{N_1 \times N_2}(n_1, n_2) \leftrightarrow \tilde{X}(k_1, k_2)\tilde{Y}(k_1, k_2)R_{N_1 \times N_2}(k_1, k_2). \tag{3.10}$$

Noting that $R_{N_1 \times N_2}(k_1, k_2) = R^2_{N_1 \times N_2}(k_1, k_2)$, $X(k_1, k_2) = \tilde{X}(k_1, k_2)R_{N_1 \times N_2}(k_1, k_2)$, and $Y(k_1, k_2) = \tilde{Y}(k_1, k_2)R_{N_1 \times N_2}(k_1, k_2)$, and from (3.10), we have

$$[\tilde{x}(n_1, n_2) \circledast \tilde{y}(n_1, n_2)]R_{N_1 \times N_2}(n_1, n_2) \leftrightarrow X(k_1, k_2)Y(k_1, k_2). \tag{3.11}$$

We define $x(n_1, n_2) \circledast y(n_1, n_2)$, the *circular convolution* of $x(n_1, n_2)$ and $y(n_1, n_2)$, by

$$x(n_1, n_2) \circledast y(n_1, n_2) = [\tilde{x}(n_1, n_2) \circledast \tilde{y}(n_1, n_2)]R_{N_1 \times N_2}(n_1, n_2). \tag{3.12}$$

From (3.11) and (3.12),

$$x(n_1, n_2) \circledast y(n_1, n_2) \leftrightarrow X(k_1, k_2)Y(k_1, k_2), \tag{3.13}$$

which is the desired result.

Under some conditions, the result of circular convolution is identical to the result of linear convolution. Suppose $f(n_1, n_2)$ is zero outside $0 \le n_1 \le N_1' - 1$ and $0 \le n_2 \le N_2' - 1$ and $g(n_1, n_2)$ is zero outside $0 \le n_1 \le N_1'' - 1$ and $0 \le n_2 \le N_2'' - 1$. Then Property 3 states that $f(n_1, n_2) \circledast g(n_1, n_2) = f(n_1, n_2) * g(n_1, n_2)$ if $f(n_1, n_2) \circledast g(n_1, n_2)$ is obtained with assumed periodicity of $N_1 \times N_2$ such that

$N_1 \geq N_1' + N_1'' - 1$ and $N_2 \geq N_2' + N_2'' - 1$. This is illustrated by the example in Figure 3.3, where $N_1' = N_1'' = N_2' = N_2'' = 2$. Figures 3.3(a) and (b) show two 2×2-point sequences $f(n_1, n_2)$ and $g(n_1, n_2)$ respectively. Figure 3.3(c) shows the result of the linear convolution $f(n_1, n_2) * g(n_1, n_2)$. Figures 3.3(d)–(i) show the results of the circular convolution $f(n_1, n_2) \circledast g(n_1, n_2)$ with an assumed periodicity of 2×2, 2×3, 3×2, 3×3, 3×4, and 4×3 respectively. In the examples given in Figure 3.3, $f(n_1, n_2) * g(n_1, n_2) = f(n_1, n_2) \circledast g(n_1, n_2)$ for $N_1 \geq N_1' + N_1'' - 1 = 3$ and $N_2 \geq N_2' + N_2'' - 1 = 3$.

Properties 2 and 3 offer an alternative way to perform linear convolution of two finite-extent sequences. To linearly convolve $x(n_1, n_2)$ and $y(n_1, n_2)$, we can assume the proper periodicity $N_1 \times N_2$, determine $X(k_1, k_2)$ and $Y(k_1, k_2)$, multiply $X(k_1, k_2)$ and $Y(k_1, k_2)$, and then perform the inverse DFT operation of

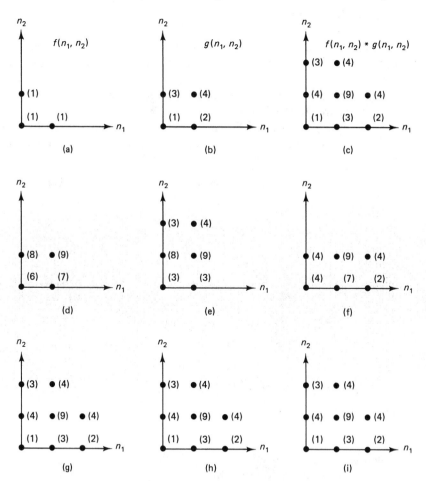

Figure 3.3 Comparison between circular convolution and linear convolution. (a) $f(n_1, n_2)$; (b) $g(n_1, n_2)$; (c) $f(n_1, n_2) * g(n_1, n_2)$; (d) $f(n_1, n_2) \circledast g(n_1, n_2)$ with assumed periodicity of $N_1 = 2$, $N_2 = 2$; (e) same as (d) with $N_1 = 2$, $N_2 = 3$; (f) same as (d) with $N_1 = 3$, $N_2 = 2$; (g) same as (d) with $N_1 = 3$, $N_2 = 3$; (h) same as (d) with $N_1 = 3$, $N_2 = 4$; (i) same as (d) with $N_1 = 4$, $N_2 = 3$.

$X(k_1, k_2)Y(k_1, k_2)$. Although this approach appears quite cumbersome, it sometimes reduces the amount of computation involved in performing linear convolution in practical applications. This is discussed in the next section.

3.2.3 The Overlap-Add and Overlap-Save Methods

Linear convolution of two sequences is desirable in a variety of contexts. In many cases, the region of support of the input data $x(n_1, n_2)$ is very large, while the region of support of the impulse response $h(n_1, n_2)$ is relatively small. In image processing, for example, an image may be 512×512 pixels (picture elements) in size, while the impulse response of a filter may be about 10×10 points. If the image signal $x(n_1, n_2)$ were filtered by using the convolution sum directly, we would need to perform about 100 multiplications and 100 additions per output sample.

An alternative to the direct convolution in the example above is to perform the inverse DFT of $X(k_1, k_2)H(k_1, k_2)$, where the DFT size is chosen to be greater than or equal to $(512 + 10 - 1) \times (512 + 10 - 1)$. As we will discuss in Section 3.4, there exist fast Fourier transform (FFT) algorithms that compute the 2-D DFT and inverse DFT very efficiently. If these methods are used in the DFT and inverse DFT computations, performing the inverse DFT of $X(k_1, k_2)H(k_1, k_2)$ requires less computation than performing the convolution directly. One problem with this approach, however, is the amount of on-line storage involved. Specifically, we need the entire $x(n_1, n_2)$ before we can compute $X(k_1, k_2)$, and $X(k_1, k_2)$ and $H(k_1, k_2)$ must be stored.

In the example above, the DFT size has to be greater than or equal to 521×521. As will be discussed in Section 3.4, 2-D FFT algorithms typically require the size to be a highly composite number. If, for example, we require the size to be $2^{M_1} \times 2^{M_2}$, then the DFT size required will be 1024×1024, four times the original size.

To reduce the amount of memory and computation required in this example, one ad-hoc method is to use DFT and inverse DFT sizes of 512×512. This will result in circular rather than linear convolution. The result of circular convolution is an aliased version of the result of linear convolution, and the regions in which circular and linear convolution give differing results are the image boundaries. When the effective region of support of $h(n_1, n_2)$ is small, the affected regions are small, and circular convolution often results in acceptable images. When the region of support of $h(n_1, n_2)$ is large, however, noticeable degradation sometimes results near the image boundaries. One such example is shown in Figure 3.4. Figure 3.4(a) shows an image of 512×512 pixels degraded by blurring and additive noise. Figure 3.4(b) shows the same image processed by Wiener filtering, which has been implemented by using DFTs and IDFTs of size 512×512. The effect of circular convolution is clearly visible near the image boundaries. Image restoration systems will be discussed in greater detail in Chapter 9.

Suppose the DFT size of 512×512 still requires too much memory. Suppose in addition that we really wish to perform the true linear convolution but still exploit the computational efficiency of FFT algorithms. Two methods that will satisfy all these requirements are the overlap-add and the overlap-save methods.

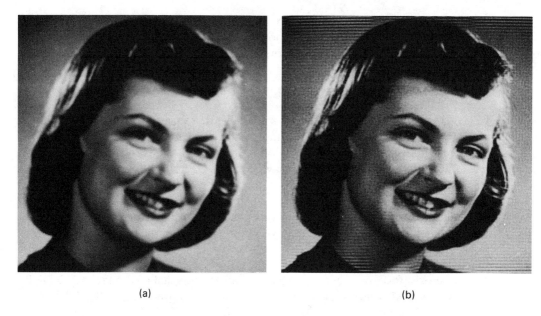

(a)   (b)

Figure 3.4 Illustration of boundary effect due to circular convolution. (a) Image degraded by blurring and additive noise; (b) image processed by Wiener filtering.

Since these methods are straightforward extensions of 1-D techniques, we will briefly summarize them.

In the overlap-add method, $x(n_1, n_2)$ is divided into many segments $x_{ij}(n_1, n_2)$, as shown in Figure 3.5. The sequence $x(n_1, n_2)$, then, can be represented by

$$x(n_1, n_2) = \sum_{i=1}^{L_1} \sum_{j=1}^{L_2} x_{ij}(n_1, n_2), \tag{3.14}$$

where $L_1 L_2$ is the number of segments in the image. Convolving $x(n_1, n_2)$ with $h(n_1, n_2)$ and using the distributive property of convolution, we obtain

$$\begin{aligned} x(n_1, n_2) * h(n_1, n_2) &= \left(\sum_{i=1}^{L_1} \sum_{j=1}^{L_2} x_{ij}(n_1, n_2) \right) * h(n_1, n_2) \\ &= \sum_{i=1}^{L_1} \sum_{j=1}^{L_2} (x_{ij}(n_1, n_2) * h(n_1, n_2)). \end{aligned} \tag{3.15}$$

Since $x_{ij}(n_1, n_2)$ is a sequence whose region of support is much smaller than $x(n_1, n_2)$, $x_{ij}(n_1, n_2) * h(n_1, n_2)$ can be computed by performing the inverse DFT operation of $X_{ij}(k_1, k_2)H(k_1, k_2)$ with a much smaller size DFT and IDFT. Since the DFT and IDFT used are much smaller in size, this computation requires much less memory. The overlap-add method is so named because the results of $x_{ij}(n_1, n_2) * h(n_1, n_2)$ are overlapped and added, as shown in Figure 3.5.

In the overlap-save method, we again consider a small segment of $x(n_1, n_2)$. Let $h(n_1, n_2)$ be an $M_1 \times M_2$-point sequence, zero outside $0 \leq n_1 \leq M_1 - 1$, $0 \leq n_2 \leq M_2 - 1$. Consider a segment of $x(n_1, n_2)$. The segment is zero outside

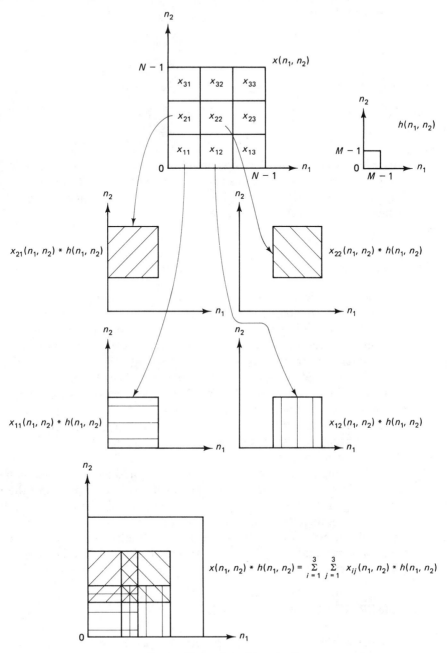

Figure 3.5 Overlap-add method.

$0 \leq n_1 \leq N_1' - 1, 0 \leq n_2 \leq N_2' - 1$. We will denote the segment by $x'(n_1, n_2)$. We choose N_1' and N_2' sufficiently large relative to M_1 and M_2, but sufficiently small relative to the image size:

$$M_1 \ll N_1' \ll N_1 \tag{3.16a}$$

$$M_2 \ll N_2' \ll N_2 \tag{3.16b}$$

where $N_1 \times N_2$ is the image size. It is straightforward to show that

$$x(n_1, n_2) * h(n_1, n_2) = x'(n_1, n_2) \circledast h(n_1, n_2) \tag{3.17}$$

$$\text{for } M_1 - 1 \leq n_1 \leq N_1' - 1, M_2 - 1 \leq n_2 \leq N_2' - 1$$

where the circular convolution is based on the assumed periodicity of $N_1' \times N_2'$.

The linear convolution can be performed by using (3.17), as is illustrated in Figure 3.6. We first segment the output $y(n_1, n_2)$ into small nonoverlapping segments $y_{ij}(n_1, n_2)$ of $N_1'' \times N_2''$ points so that

$$y(n_1, n_2) = \sum_{i=1}^{L_1} \sum_{j=1}^{L_2} y_{ij}(n_1, n_2). \tag{3.18}$$

Consider a particular segment $y_{ij}(n_1, n_2)$. To compute $y_{ij}(n_1, n_2)$, we look at $x(n_1, n_2)$ in the region identical to that occupied by $y_{ij}(n_1, n_2)$. We then extend this portion of the input by $M_1 - 1$ points in the negative n_1 direction and by $M_2 - 1$ points in the negative n_2 direction. Next, we perform circular convolution of the extended input segment and $h(n_1, n_2)$ with the assumed periodicity of $(N_1'' + M_1 - 1) \times (N_2'' + M_2 - 1)$. This circular convolution can be performed by computing DFTs and inverse DFTs. The upper right-hand corner of the result of the circular convolution is $y_{ij}(n_1, n_2)$. Since $(N_1'' + M_1 - 1) \times (N_2'' + M_2 - 1)$ is much smaller than the original image size, the computation requires less memory. Since the processed segments are overlapped and portions of them are saved, it is called the overlap-save method.

The computations involved in the overlap-add and overlap-save methods are roughly the same. In some image processing applications where the size of the support region of the impulse response is on the order of 10×10, both methods reduce the number of arithmetic operations (multiplications and additions) by a factor of five to ten relative to direct convolution. The overlap-add method is much easier to understand conceptually relative to the overlap-save method, and is more often used in practice.

3.3 THE DISCRETE COSINE TRANSFORM

In some applications, for example image coding, the discrete cosine transform (DCT) is extensively used. The DCT is the most widely used transform in a class of image coding systems known as transform coders: this is discussed further in Chapter 10. In this section, we discuss the DCT, which is closely related to the DFT.

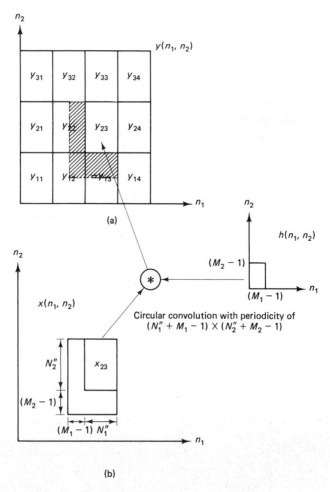

Figure 3.6 Overlap-save method.

3.3.1 The One-Dimensional Discrete Cosine Transform

We begin by discussing the 1-D DCT, which is notationally simpler and provides useful insights for understanding the DCT representation in general. The 1-D DCT can be extended straightforwardly to the 2-D DCT.

Let $x(n)$ denote an N-point sequence that is zero outside $0 \le n \le N - 1$. Among several variations, we will consider one variation known as *even symmetrical DCT*, which is most often used in signal coding applications. To derive the DCT relationship, it is convenient to relate the N-point sequence $x(n)$ to a new $2N$-point sequence $y(n)$, which is then related to its $2N$-point DFT $Y(k)$. We then relate $Y(k)$ to $C_x(k)$, the N-point DCT of $x(n)$. Specifically,

$$
\begin{array}{cccc}
N\text{-point} & 2N\text{-point DFT} & 2N\text{-point} & N\text{-point} \\
x(n) \longleftrightarrow & y(n) \longleftrightarrow & Y(k) \longleftrightarrow & C_x(k)
\end{array} \tag{3.19}
$$

The sequence $x(n)$ is related to $y(n)$ by

$$y(n) = x(n) + x(2N - 1 - n) \qquad (3.20)$$
$$= \begin{cases} x(n), & 0 \le n \le N - 1 \\ x(2N - 1 - n), & N \le n \le 2N - 1. \end{cases}$$

An example of $x(n)$ and $y(n)$ when $N = 4$ is shown in Figure 3.7. The sequence $y(n)$ is symmetric with respect to the half-sample point at $n = N - \frac{1}{2}$. When we form a periodic sequence $\tilde{x}(n)$ by repeating $x(n)$ every N points, $\tilde{x}(n)$ has artificial discontinuities, since the beginning and end part of $x(n)$ are joined in the repetition process. When we form a periodic sequence $\tilde{y}(n)$ by repeating $y(n)$ every $2N$ points, however, $\tilde{y}(n)$ no longer contains the artificial discontinuities. This is shown in Figure 3.8 for the $x(n)$ and $y(n)$ shown in Figure 3.7. As we discuss in Chapter 10, eliminating the artificial boundary discontinuities contributes to the energy compaction property that is exploited in transform image coding.

The $2N$-point DFT $Y(k)$ is related to $y(n)$ by

$$Y(k) = \sum_{n=0}^{2N-1} y(n) W_{2N}^{kn}, \quad 0 \le k \le 2N - 1 \qquad (3.21a)$$

where

$$W_{2N} = e^{-j(2\pi/2N)}. \qquad (3.21b)$$

From (3.20) and (3.21),

$$Y(k) = \sum_{n=0}^{N-1} x(n) W_{2N}^{kn} + \sum_{n=N}^{2N-1} x(2N - 1 - n) W_{2N}^{kn}, \quad 0 \le k \le 2N - 1. \qquad (3.22)$$

With a change of variables and after some algebra, (3.22) can be expressed as

$$Y(k) = W_{2N}^{-k/2} \sum_{n=0}^{N-1} 2x(n) \cos \frac{\pi}{2N} k(2n + 1), \quad 0 \le k \le 2N - 1. \qquad (3.23)$$

The N-point DCT of $x(n)$, $C_x(k)$, is obtained from $Y(k)$ by

$$C_x(k) = \begin{cases} W_{2N}^{k/2} Y(k), & 0 \le k \le N - 1 \\ 0, & \text{otherwise} \end{cases} \qquad (3.24)$$

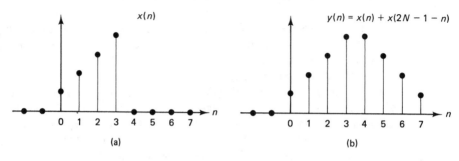

Figure 3.7 Example of (a) $x(n)$ and (b) $y(n) = x(n) + x(2N - 1 - n)$. The sequence $y(n)$ is used in the intermediate step in defining the discrete cosine transform of $x(n)$.

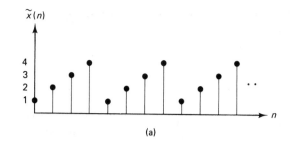

(a)

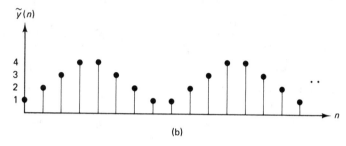

(b)

Figure 3.8 Periodic sequences $\tilde{x}(n)$ and $\tilde{y}(n)$ obtained from $x(n)$ and $y(n)$ in Figure 3.7.

From (3.23) and (3.24),

$$
C_x(k) = \begin{cases} \displaystyle\sum_{n=0}^{N-1} 2x(n) \cos \frac{\pi}{2N} k(2n + 1), & 0 \le k \le N - 1 \\ 0, & \text{otherwise.} \end{cases}
\tag{3.25}
$$

Equation (3.25) is the definition of the DCT of $x(n)$. From (3.25), $C_x(k)$ is an N-point sequence, and therefore N values of $x(n)$ are represented by N values of $C_x(k)$. If $x(n)$ is real, $C_x(k)$ is real. If $x(n)$ is complex, so is $C_x(k)$.

To derive the inverse DCT relation, we relate $C_x(k)$ to $Y(k)$, $Y(k)$ to $y(n)$, and then $y(n)$ to $x(n)$. We first consider determining $Y(k)$ from $C_x(k)$. Although $Y(k)$ is a $2N$-point sequence and $C_x(k)$ is an N-point sequence, redundancy in $Y(k)$ due to the symmetry of $y(n)$ allows us to determine $Y(k)$ from $C_x(k)$. Specifically, from (3.23),

$$
Y(k) = \begin{cases} W_{2N}^{-k} Y(2N - k), & 0 \le k \le 2N - 1 \\ 0, & k = N. \end{cases}
\tag{3.26}
$$

From (3.24) and (3.26),

$$
Y(k) = \begin{cases} W_{2N}^{-k/2} C_x(k), & 0 \le k \le N - 1 \\ 0, & k = N \\ -W_{2N}^{-k/2} C_x(2N - k), & N + 1 \le k \le 2N - 1. \end{cases}
\tag{3.27}
$$

Sec. 3.3   The Discrete Cosine Transform

The sequence $Y(k)$ is related to $y(n)$ through the $2N$-point inverse DFT relation given by

$$y(n) = \frac{1}{2N} \sum_{k=0}^{2N-1} Y(k) W_{2N}^{-kn}, \quad 0 \le n \le 2N - 1. \tag{3.28}$$

From (3.20), $x(n)$ can be recovered from $y(n)$ by

$$x(n) = \begin{cases} y(n), & 0 \le n \le N - 1 \\ 0, & \text{otherwise.} \end{cases} \tag{3.29}$$

From (3.27), (3.28), and (3.29), and after some algebra,

$$x(n) = \begin{cases} \dfrac{1}{N} \left[\dfrac{C_x(0)}{2} + \displaystyle\sum_{k=1}^{N-1} C_x(k) \cos \dfrac{\pi}{2N} k(2n + 1) \right], & 0 \le n \le N - 1 \\ 0, & \text{otherwise.} \end{cases} \tag{3.30}$$

Equation (3.30) can also be expressed as

$$x(n) = \begin{cases} \dfrac{1}{N} \displaystyle\sum_{k=0}^{N-1} w(k) C_x(k) \cos \dfrac{\pi}{2N} k(2n + 1), & 0 \le n \le N - 1 \\ 0, & \text{otherwise.} \end{cases} \tag{3.31a}$$

where

$$w(k) = \begin{cases} \dfrac{1}{2}, & k = 0 \\ 1, & 1 \le k \le N - 1. \end{cases} \tag{3.31b}$$

Equation (3.31) is the inverse DCT relation. From (3.25) and (3.31),

---

Discrete Cosine Transform Pair

$$C_x(k) = \begin{cases} \displaystyle\sum_{n=0}^{N-1} 2x(n) \cos \dfrac{\pi}{2N} k(2n + 1), & 0 \le k \le N - 1 \\ 0, & \text{otherwise.} \end{cases} \tag{3.32a}$$

$$x(n) = \begin{cases} \dfrac{1}{N} \displaystyle\sum_{k=0}^{N-1} w(k) C_x(k) \cos \dfrac{\pi}{2N} k(2n + 1), & 0 \le n \le N - 1 \\ 0, & \text{otherwise.} \end{cases} \tag{3.32b}$$

From the derivation of the DCT pair, the DCT and inverse DCT can be computed by

Computation of Discrete Cosine Transform

Step 1. $y(n) = x(n) + x(2N - 1 - n)$

Step 2. $Y(k) = \text{DFT} [y(n)]$ ($2N$-point DFT computation)

Step 3. $C_x(k) = \begin{cases} W_{2N}^{k/2} Y(k), & 0 \le k \le N - 1 \\ 0, & \text{otherwise} \end{cases}$

Step 1. $Y(k) = \begin{cases} W_{2N}^{-k/2}C_x(k), & 0 \le k \le N - 1 \\ 0, & k = N \\ -W_{2N}^{-k/2}C_x(2N - k), & N + 1 \le k \le 2N - 1 \end{cases}$

Step 2. $y(n) = \text{IDFT}\,[Y(k)]$ (2N-point inverse DFT computation)

Step 3. $x(n) = \begin{cases} y(n), & 0 \le n \le N - 1 \\ 0, & \text{otherwise} \end{cases}$

In computing the DCT and inverse DCT, Steps 1 and 3 are computationally quite simple. Most of the computations are in Step 2, where a 2N-point DFT is computed for the DCT and a 2N-point inverse DFT is computed for the inverse DCT. The DFT and inverse DFT can be computed by using fast Fourier transform (FFT) algorithms. In addition, because $y(n)$ has symmetry, the 2N-point DFT and inverse DFT can be computed (see Problem 3.20) by computing the N-point DFT and the N-point inverse DFT of an N-point sequence. Therefore, the computation involved in using the DCT is essentially the same as that involved in using the DFT.

In the derivation of the DCT pair, we have used an intermediate sequence $y(n)$ that has symmetry and whose length is even. The DCT we derived is thus called an even symmetrical DCT. It is also possible to derive the odd symmetrical DCT pair in the same manner. In the odd symmetrical DCT, the intermediate sequence $y(n)$ used has symmetry, but its length is odd. For the sequence $x(n)$ shown in Figure 3.9(a), the sequence $y(n)$ used is shown in Figure 3.9(b). The length of $y(n)$ is $2N - 1$, and $\bar{y}(n)$, obtained by repeating $y(n)$ every $2N - 1$ points, has no artificial discontinuities. The detailed derivation of the odd symmetrical DCT is considered in Problem 3.22. The even symmetrical DCT is more commonly used, since the odd symmetrical DCT involves computing an odd-length DFT, which is not very convenient when one is using FFT algorithms.

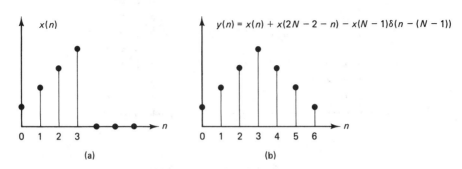

Figure 3.9 Example of (a) $x(n)$ and (b) $y(n) = x(n) + x(2N - 2 - n) - x(N - 1)\delta(n - (N - 1))$. The sequence $y(n)$ is used in the intermediate step in defining the odd symmetrical discrete cosine transform of $x(n)$.

3.3.2 The Two-Dimensional Discrete Cosine Transform

The 1-D DCT discussed in Section 3.3.1 can be extended straightforwardly to two dimensions. Let $x(n_1, n_2)$ denote a 2-D sequence of $N_1 \times N_2$ points that is zero outside $0 \leq n_1 \leq N_1 - 1$, $0 \leq n_2 \leq N_2 - 1$. We can derive the 2-D DCT pair by relating $x(n_1, n_2)$ to a new $2N_1 \times 2N_2$-point sequence $y(n_1, n_2)$, which is then related to its $2N_1 \times 2N_2$-point DFT $Y(k_1, k_2)$. We then relate $Y(k_1, k_2)$ to $C_x(k_1, k_2)$, the $N_1 \times N_2$-point DCT. Specifically,

$$
\begin{array}{ccccccc}
N_1 \times N_2\text{-point} & & 2N_1 \times 2N_2\text{-point} & \text{DFT} & 2N_1 \times 2N_2\text{-point} & & N_1 \times N_2\text{-point} \\
x(n_1, n_2) & \longleftrightarrow & y(n_1, n_2) & \longleftrightarrow & Y(k_1, k_2) & \longleftrightarrow & C_x(k_1, k_2).
\end{array}
$$

$$(3.33)$$

The sequence $x(n_1, n_2)$ is related to $y(n_1, n_2)$ by

$$
y(n_1, n_2) = x(n_1, n_2) + x(2N_1 - 1 - n_1, n_2) + x(n_1, 2N_2 - 1 - n_2) \\
+ x(2N_1 - 1 - n_1, 2N_2 - 1 - n_2). \tag{3.34}
$$

An example of $x(n_1, n_2)$ and $y(n_1, n_2)$ when $N_1 = 3$, $N_2 = 4$ is shown in Figures 3.10(a) and (b), respectively. A periodic sequence $\tilde{x}(n_1, n_2)$ with a period of $N_1 \times N_2$ obtained by repeating $x(n_1, n_2)$ is shown in Figure 3.11(a). A periodic sequence $\tilde{y}(n_1, n_2)$ with a period of $2N_1 \times 2N_2$ obtained by repeating $y(n_1, n_2)$ is shown in Figure 3.11(b). The artificial discontinuities present in $\tilde{x}(n_1, n_2)$ are not present in $\tilde{y}(n_1, n_2)$.

The sequence $y(n_1, n_2)$ is related to $Y(k_1, k_2)$ by

$$
Y(k_1, k_2) = \text{DFT} \left[y(n_1, n_2) \right]. \tag{3.35}
$$

The $N_1 \times N_2$-point DCT of $x(n_1, n_2)$, $C_x(k_1, k_2)$, is obtained from $Y(k_1, k_2)$ by

$$
C_x(k_1, k_2) = \begin{cases} W_{2N_1}^{k_1/2} W_{2N_2}^{k_2/2} Y(k_1, k_2), & 0 \leq k_1 \leq N_1 - 1, 0 \leq k_2 \leq N_2 - 1 \\ 0, & \text{otherwise.} \end{cases} \tag{3.36}
$$

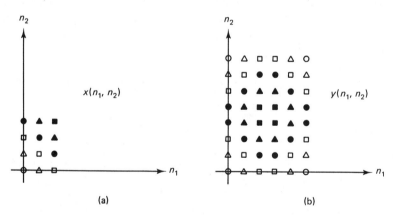

(a)   (b)

Figure 3.10 Example of (a) $x(n_1, n_2)$ and (b) $y(n_1, n_2) = x(n_1, n_2) + x(2N_1 - 1 - n_1, n_2) + x(n_1, 2N_2 - 1 - n_2) + x(2N_1 - 1 - n_1, 2N_2 - 1 - n_2)$. The sequence $y(n_1, n_2)$ is used in the intermediate step in defining the discrete cosine transform of $x(n_1, n_2)$.

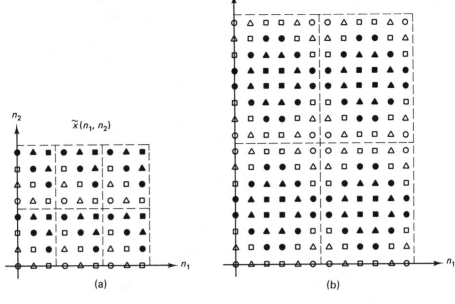

Figure 3.11 Periodic sequences $\tilde{x}(n_1, n_2)$ and $\tilde{y}(n_1, n_2)$ obtained from $x(n_1, n_2)$ and $y(n_1, n_2)$ in Figure 3.10.

From (3.34), (3.35), and (3.36), and after some algebra,

$$C_x(k_1, k_2) = \begin{cases} \sum_{n_1=0}^{N_1-1} \sum_{n_2=0}^{N_2-1} 4x(n_1, n_2) \cos \dfrac{\pi}{2N_1} k_1(2n_1 + 1) \cos \dfrac{\pi}{2N_2} k_2(2n_2 + 1), \\ \qquad \text{for } 0 \leq k_1 \leq N_1 - 1, 0 \leq k_2 \leq N_2 - 1 \\ 0, \qquad \text{otherwise.} \end{cases} \tag{3.37}$$

Equation (3.37) is the definition of the 2-D DCT.

The inverse DCT can be derived by relating $C_x(k_1, k_2)$ to $Y(k_1, k_2)$, exploiting the redundancy in $Y(k_1, k_2)$ due to the symmetry of $y(n_1, n_2)$, relating $Y(k_1, k_2)$ to $y(n_1, n_2)$ through the inverse DFT relationship, and then relating $y(n_1, n_2)$ to $x(n_1, n_2)$. The result is

$$x(n_1, n_2) = \begin{cases} \dfrac{1}{N_1 N_2} \sum_{k_1=0}^{N_1-1} \sum_{k_2=0}^{N_2-1} w_1(k_1)w_2(k_2)C_x(k_1, k_2) \cos \dfrac{\pi}{2N_1} k_1(2n_1 + 1) \cos \dfrac{\pi}{2N_2} k_2(2n_2 + 1), \\ \qquad \text{for } 0 \leq n_1 \leq N_1 - 1, 0 \leq n_2 \leq N_2 - 1 \\ 0, \qquad \text{otherwise.} \end{cases} \tag{3.38a}$$

where

$$w_1(k_1) = \begin{cases} \frac{1}{2}, & k_1 = 0 \\ 1, & 1 \leq k_1 \leq N_1 - 1 \end{cases} \tag{3.38b}$$

$$w_2(k_2) = \begin{cases} \frac{1}{2}, & k_2 = 0 \\ 1, & 1 \leq k_2 \leq N_2 - 1. \end{cases} \tag{3.38c}$$

From (3.37) and (3.38),

Two-Dimensional Discrete Cosine Transform Pair

$$C_x(k_1, k_2) = \begin{cases} \displaystyle\sum_{n_1=0}^{N_1-1} \sum_{n_2=0}^{N_2-1} 4x(n_1, n_2) \cos \frac{\pi}{2N_1} k_1(2n_1 + 1) \cos \frac{\pi}{2N_2} k_2(2n_2 + 1), \\ \qquad \text{for } 0 \le k_1 \le N_1 - 1,\, 0 \le k_2 \le N_2 - 1 \\ 0, \qquad \text{otherwise.} \end{cases} \quad (3.39a)$$

$$x(n_1, n_2) = \begin{cases} \displaystyle\frac{1}{N_1 N_2}\sum_{k_1=0}^{N_1-1} \sum_{k_2=0}^{N_2-1} w_1(k_1)w_2(k_2)C_x(k_1, k_2) \cos \frac{\pi}{2N_1} k_1(2n_1 + 1) \cos \frac{\pi}{2N_2} k_2(2n_2 + 1), \\ \qquad \text{for } 0 \le n_1 \le N_1 - 1,\, 0 \le n_2 \le N_2 - 1 \\ 0, \qquad \text{otherwise.} \end{cases} \quad (3.39b)$$

The DCT and inverse DCT can be computed by

Computation of the Discrete Cosine Transform

Step 1. $y(n_1, n_2) = x(n_1, n_2) + x(n_1, 2N_2 - 1 - n_2) + x(2N_1 - 1 - n_1, n_2)$
$\qquad\qquad + x(2N_1 - 1 - n_1, 2N_2 - 1 - n_2)$

Step 2. $Y(k_1, k_2) = \text{DFT}\,[y(n_1, n_2)]$ $(2N_1 \times 2N_2\text{-point DFT computation})$

Step 3. $C_x(k_1, k_2) = W_{2N_1}^{k_1/2}W_{2N_2}^{k_2/2}Y(k_1, k_2),\quad 0 \le k_1 \le N_1 - 1,\quad 0 \le k_2 \le N_2 - 1$

Computation of the Inverse Discrete Cosine Transform

Step 1.

$$Y(k_1, k_2) = \begin{cases} W_{2N_1}^{-k_1/2}W_{2N_2}^{-k_2/2}C_x(k_1, k_2),\quad 0 \le k_1 \le N_1 - 1,\, 0 \le k_2 \le N_2 - 1 \\ -W_{2N_1}^{-k_1/2}W_{2N_2}^{-k_2/2}C_x(2N_1 - k_1, k_2), \\ \qquad N_1 + 1 \le k_1 \le 2N_1 - 1,\, 0 \le k_2 \le N_2 - 1 \\ -W_{2N_1}^{-k_1/2}W_{2N_2}^{-k_2/2}C_x(k_1, 2N_2 - k_2), \\ \qquad 0 \le k_1 \le N_1 - 1,\, N_2 + 1 \le k_2 \le 2N_2 - 1 \\ W_{2N_1}^{-k_1/2}W_{2N_2}^{-k_2/2}C_x(2N_1 - k_1, 2N_2 - k_2), \\ \qquad N_1 + 1 \le k_1 \le 2N_1 - 1,\, N_2 + 1 \le k_2 \le 2N_2 - 1 \\ 0, \qquad k_1 = N_1 \ \text{or}\ k_2 = N_2. \end{cases}$$

Step 2. $y(n_1, n_2) = \text{IDFT}\,[Y(k_1, k_2)]$ $(2N_1 \times 2N_2\text{-point IDFT computation})$

Step 3. $x(n_1, n_2) = \begin{cases} y(n_1, n_2), & 0 \le n_1 \le N_1 - 1,\quad 0 \le n_2 \le N_2 - 1 \\ 0, & \text{otherwise.} \end{cases}$

The DFT and inverse DFT in Step 2 of the computation of the DCT and inverse DCT can be computed by using FFT algorithms. In addition, by exploiting the symmetry of $y(n_1, n_2)$, the $2N_1 \times 2N_2$-point DFT and inverse DFT can be computed by computing the $N_1 \times N_2$-point DFT and inverse DFT of an $N_1 \times N_2$-point sequence. The computations required for the DCT are essentially the same as those required for the DFT. The DCT discussed above is the 2-D even symmetrical discrete cosine transform. The 2-D odd symmetrical discrete cosine transform can also be derived.

3.3.3 Properties of the Discrete Cosine Transform

We can derive many useful DCT properties from the DCT pair in (3.39). However, the relationship in (3.33) and the DFT properties discussed in Section 3.2 are often more convenient and easier to use in deriving DCT properties. Some of the more important properties are listed in Table 3.3.

To illustrate how DCT properties can be obtained from (3.33) and the DFT properties, we will derive Property 3, which is analogous to Parseval's theorem for the DFT. From Parseval's theorem in Table 3.2,

$$\sum_{n_1=0}^{2N_1-1} \sum_{n_2=0}^{2N_2-1} |y(n_1, n_1)|^2 = \frac{1}{4N_1N_2} \sum_{k_1=0}^{2N_1-1} \sum_{k_2=0}^{2N_2-1} |Y(k_1, k_2)|^2. \tag{3.40}$$

From (3.33) and (3.34).

$$\sum_{n_1=0}^{2N_1-1} \sum_{n_2=0}^{2N_2-1} |y(n_1, n_2)|^2 = 4 \sum_{n_1=0}^{N_1-1} \sum_{n_2=0}^{N_2-1} |x(n_1, n_2)|^2. \tag{3.41}$$

From (3.33) and the redundancy in $Y(k_1, k_2)$,

$$\sum_{k_1=0}^{2N_1-1} \sum_{k_2=0}^{2N_2-1} |Y(k_1, k_2)|^2 = 4 \sum_{k_1=0}^{N_1-1} \sum_{k_2=0}^{N_2-1} w_1(k_1)w_2(k_2)|C_x(k_1, k_2)|^2 \tag{3.42}$$

where $w_1(k_1)$ and $w_2(k_2)$ are given by (3.38b) and (3.38c). From (3.40), (3.41), and (3.42),

$$\sum_{n_1=0}^{N_1-1} \sum_{n_2=0}^{N_2-1} |x(n_1, n_2)|^2 = \frac{1}{4N_1N_2} \sum_{k_1=0}^{N_1-1} \sum_{k_2=0}^{N_2-1} w_1(k_1)w_2(k_2)|C_x(k_1, k_2)|^2, \tag{3.43}$$

which is the desired result.

The DCT of a complex sequence is complex-valued, and a complex $N_1 \times N_2$-point sequence is represented by N_1N_2 complex DCT coefficients. For a real sequence, the symmetry property states that the DCT coefficients are real. A real $N_1 \times N_2$-point sequence is, therefore, represented by N_1N_2 real DCT coefficients. It is possible to derive other properties, such as the convolution property, but they are not as elegant as those of the DFT. For example, the DCT of the circular convolution of two sequences is not the product of two individual DCTs. More importantly, the DCT is used primarily in signal coding applications, and properties of this sort are not very useful in practice.

3.3.4 The Discrete-Space Cosine Transform

The DFT coefficients are samples of the discrete-space Fourier transform. In this section, we will discuss the discrete-space cosine transform, or *cosine transform* for short, whose samples can be related to the DCT coefficients. We again consider the even symmetrical cosine transform.

TABLE 3.3 PROPERTIES OF THE DISCRETE COSINE TRANSFORM

$x(n_1, n_2), y(n_1, n_2) = 0$ outside $0 \leq n_1 \leq N_1 - 1, 0 \leq n_2 \leq N_2 - 1$
$x(n_1, n_2) \longleftrightarrow C_x(k_1, k_2)$
$y(n_1, n_2) \longleftrightarrow C_y(k_1, k_2)$
$N_1 \times N_2$-point DCT and IDCT are assumed.

Property 1.   Linearity

$$ax(n_1, n_2) + by(n_1, n_2) \longleftrightarrow aC_x(k_1, k_2) + bC_y(k_1, k_2)$$

Property 2.   Separable Sequence

$$x(n_1, n_2) = x_1(n_1)x_2(n_2) \longleftrightarrow C_x(k_1, k_2) = C_{x1}(k_1)C_{x2}(k_2)$$

$$C_{x1}(k_1): N_1\text{-point 1-D DCT}$$

$$C_{x2}(k_2): N_2\text{-point 1-D DCT}$$

Property 3.   Energy Relationship

$$\sum_{n1=0}^{N_1-1} \sum_{n2=0}^{N_2-1} |x(n_1, n_2)|^2 = \frac{1}{4N_1 N_2} \sum_{k1=0}^{N_1-1} \sum_{k2=0}^{N_2-1} w_1(k_1)w_2(k_2)|C_x(k_1, k_2)|^2$$

$$w_1(k_1) = \begin{cases} \dfrac{1}{2}, & k_1 = 0 \\ 1, & 1 \leq k_1 \leq N_1 - 1 \end{cases}$$

$$w_2(k_2) = \begin{cases} \dfrac{1}{2}, & k_2 = 0 \\ 1, & 1 \leq k_2 \leq N_2 - 1 \end{cases}$$

Property 4.   Symmetry Properties
(a) $x^*(n_1, n_2) \longleftrightarrow C_x^*(k_1, k_2)$
(b) real $x(n_1, n_2) \longleftrightarrow$ real $C_x(k_1, k_2)$

The one-dimensional cosine transform.   Consider a causal 1-D sequence $x(n)$ that is zero for $n < 0$. To derive the cosine transform relation, we relate $x(n)$ to a new sequence $r(n)$, which is then related to its Fourier transform $R(\omega)$. We can then relate $R(\omega)$ to the cosine transform $C_x(\omega)$. Specifically,

$$x(n) \leftrightarrow r(n) \leftrightarrow R(\omega) \leftrightarrow C_x(\omega). \tag{3.44}$$

The sequence $x(n)$ is related to $r(n)$ by

$$r(n) = x(n) + x(-n - 1)$$
$$= \begin{cases} x(n), & n \geq 0 \\ x(-n - 1), & n < 0. \end{cases} \tag{3.45}$$

An example of $x(n)$ and $r(n)$ is shown in Figure 3.12. Note that $x(n)$ is causal and can be an infinite-extent sequence. For a causal, N-point sequence $x(n)$, $r(n)$ is not the same as the $y(n) = x(n) + x(2N - 1 - n)$ used in deriving the DCT

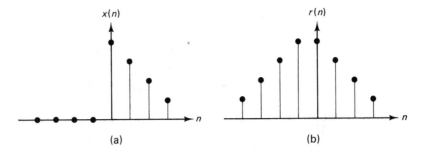

Figure 3.12 Example of (a) $x(n)$ and (b) $r(n) = x(n) + x(-n - 1)$. The sequence $r(n)$ is used in the intermediate step in defining the cosine transform of $x(n)$.

relation. However, a periodic sequence $\tilde{r}(n)$ formed by repeating $r(n)$ every $2N$ points is identical to $\tilde{y}(n)$ obtained by repeating $y(n)$ every $2N$ points.

The sequence $r(n)$ is related to its Fourier transform $R(\omega)$ by

$$R(\omega) = \sum_{n=-\infty}^{\infty} r(n)e^{-j\omega n}. \tag{3.46}$$

From (3.45) and (3.46),

$$R(\omega) = \sum_{n=0}^{\infty} x(n)e^{-j\omega n} + \sum_{n=-\infty}^{-1} x(-n - 1)e^{-j\omega n}. \tag{3.47}$$

Rewriting (3.47), we obtain

$$R(\omega) = e^{j\omega/2} \sum_{n=0}^{\infty} 2x(n) \cos (\omega(n + \tfrac{1}{2})). \tag{3.48}$$

The cosine transform $C_x(\omega)$ is related to $R(\omega)$ by

$$C_x(\omega) = e^{-j\omega/2}R(\omega) = \sum_{n=0}^{\infty} 2x(n) \cos \omega(n + \tfrac{1}{2}). \tag{3.49}$$

Equation (3.49) is the definition of the cosine transform of $x(n)$. Note that $C_x(\omega)$ is an even function of ω. From (3.32) and (3.49), for an N-point causal sequence $x(n)$,

$$C_x(k) = C_x(\omega) \big|_{\omega = (2\pi/2N)k}, \qquad 0 \le k \le N - 1. \tag{3.50}$$

This states that the DCT coefficients $C_x(k)$ are the cosine transform $C_x(\omega)$ sampled at equally spaced points. This result is analogous to the relation between the DFT and the Fourier transform.

To derive the inverse cosine transform relation, we relate $C_x(\omega)$ to $R(\omega)$, $R(\omega)$ to $r(n)$, and $r(n)$ to $x(n)$. From (3.49),

$$R(\omega) = e^{j\omega/2}C_x(\omega). \tag{3.51}$$

From (3.44) and (3.51),

$$r(n) = \frac{1}{2\pi} \int_{\omega=-\pi}^{\pi} R(\omega)e^{j\omega n} \, d\omega$$

$$= \frac{1}{2\pi} \int_{\omega=-\pi}^{\pi} C_x(\omega)e^{j\omega(n+1/2)} \, d\omega. \tag{3.52}$$

Noting that $C_x(\omega)$ is an even function of ω, from (3.45) and (3.52), we have

$$x(n) = r(n) = \begin{cases} \dfrac{1}{2\pi} \displaystyle\int_{\omega=-\pi}^{\pi} C_x(\omega) \cos \omega(n + \tfrac{1}{2}) \, d\omega, & n \geq 0, \\ 0, & \text{otherwise.} \end{cases} \tag{3.53}$$

Equation (3.53) is the inverse cosine transform relation. From (3.49) and (3.53),

---

Cosine Transform Pair

$$C_x(\omega) = \sum_{n=0}^{\infty} 2x(n) \cos \omega(n + \tfrac{1}{2}) \tag{3.54a}$$

$$x(n) = \begin{cases} \dfrac{1}{2\pi} \displaystyle\int_{\omega=-\pi}^{\pi} C_x(\omega) \cos \omega(n + \tfrac{1}{2}) \, d\omega, & n \geq 0 \\ 0, & \text{otherwise.} \end{cases} \tag{3.54b}$$

The two-dimensional cosine transform. The 1-D cosine transform discussed above can be extended straightforwardly to the 2-D case. Let $x(n_1, n_2)$ denote a first-quadrant support sequence. We can derive the 2-D cosine transform by relating $x(n_1, n_2)$ to a new sequence $r(n_1, n_2)$, which is then related to its Fourier transform $R(\omega_1, \omega_2)$. We can then relate $R(\omega_1, \omega_2)$ to the cosine transform $C_x(\omega_1, \omega_2)$. That is,

$$x(n_1, n_2) \leftrightarrow r(n_1, n_2) \leftrightarrow R(\omega_1, \omega_2) \leftrightarrow C_x(\omega_1, \omega_2). \tag{3.55}$$

The sequence $x(n_1, n_2)$ is related to $r(n_1, n_2)$ by

$$r(n_1, n_2) = x(n_1, n_2) + x(-n_1 - 1, n_2)$$
$$+ x(n_1, -n_2 - 1) + x(-n_1 - 1, -n_2 - 1). \tag{3.56}$$

An example of $x(n_1, n_2)$ and $r(n_1, n_2)$ is shown in Figure 3.13. The sequence $r(n_1, n_2)$ is symmetric with respect to the half-sample point at $n_1 = -\tfrac{1}{2}$ and $n_2 = -\tfrac{1}{2}$. By expressing $R(\omega_1, \omega_2)$ in terms of $x(n_1, n_2)$ using (3.56) and by defining the cosine transform $C_x(\omega_1, \omega_2)$ as

$$C_x(\omega_1, \omega_2) = e^{-j\omega_1/2}e^{-j\omega_2/2}R(\omega_1, \omega_2), \tag{3.57}$$

$$C_x(\omega_1, \omega_2) = \sum_{n_1=0}^{\infty} \sum_{n_2=0}^{\infty} 4x(n_1, n_2) \cos \omega_1(n_1 + \tfrac{1}{2}) \cos \omega_2(n_2 + \tfrac{1}{2}). \tag{3.58}$$

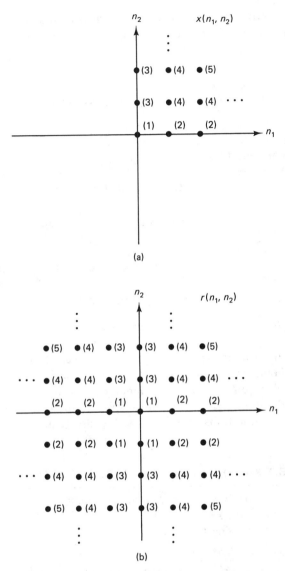

Figure 3.13 Example of (a) $x(n_1, n_2)$ and (b) $r(n_1, n_2) = x(n_1, n_2) + x(-n_1 - 1, n_2) + x(n_1, -n_2 - 1) + x(-n_1 - 1, -n_2 - 1)$. The sequence $r(n_1, n_2)$ is used in the intermediate step in defining the cosine transform of $x(n_1, n_2)$.

From (3.39) and (3.58), for an $N_1 \times N_2$-point first-quadrant support sequence,

$$C_x(k_1, k_2) = C_x(\omega_1, \omega_2) \Big|_{\omega_1 = (2\pi/2N_1)k_1, \omega_2 = (2\pi/2N_2)k_2}, \tag{3.59}$$

which states that the DCT coefficients $C_x(k_1, k_2)$ are samples of the cosine transform $C_x(\omega_1, \omega_2)$ at equally spaced points on the Cartesian grid.

To derive the inverse cosine transform relation, we relate $C_x(\omega_1, \omega_2)$ to $R(\omega_1, \omega_2)$ using (3.57), relate $R(\omega_1, \omega_2)$ to $r(n_1, n_2)$ using the inverse Fourier transform relation, and then relate $r(n_1, n_2)$ to $x(n_1, n_2)$ using (3.56). The result is

$$x(n_1, n_2) = \frac{1}{(2\pi)^2} \int_{\omega_1 = -\pi}^{\pi} \int_{\omega_2 = -\pi}^{\pi} C_x(\omega_1, \omega_2) \cos \omega_1(n_1 + \tfrac{1}{2}) \cos \omega_2(n_2 + \tfrac{1}{2}) \, d\omega_1 \, d\omega_2,$$

$$n_1 \geq 0, n_2 \geq 0. \tag{3.60}$$

From (3.58) and (3.60)

Cosine Transform Pair

$$C_x(\omega_1, \omega_2) = \sum_{n1=0}^{\infty} \sum_{n2=0}^{\infty} 4x(n_1, n_2) \cos \omega_1(n_1 + \tfrac{1}{2}) \cos \omega_2(n_2 + \tfrac{1}{2}). \tag{3.61a}$$

$$x(n_1, n_2) = \begin{cases} \dfrac{1}{(2\pi)^2} \displaystyle\int_{\omega_1 = -\pi}^{\pi} \int_{\omega_2 = -\pi}^{\pi} C_x(\omega_1, \omega_2) \cos \omega_1(n_1 + \tfrac{1}{2}) \cos \omega_2(n_2 + \tfrac{1}{2}) \, d\omega_1 \, d\omega_2 \\ \qquad\qquad n_1 \geq 0, n_2 \geq 0 \\[2mm] 0, \qquad\qquad\qquad \text{otherwise.} \end{cases} \tag{3.61b}$$

Many properties of the cosine transform can be derived from (3.61), or (3.55) and the Fourier transform properties. Some of the more important properties are listed in Table 3.4. From the symmetry properties, $C_x(\omega_1, \omega_2)$ is an even function and in addition is symmetric with respect to the ω_1 and ω_2 axes. When $x(n_1, n_2)$ is real, its cosine transform $C_x(\omega_1, \omega_2)$ is also real.

TABLE 3.4   PROPERTIES OF THE COSINE TRANSFORM

$x(n_1, n_2), y(n_1, n_2) = 0$ outside $n_1 \geq 0, n_2 \geq 0$

$x(n_1, n_2) \longleftrightarrow C_x(\omega_1, \omega_2)$

$y(n_1, n_2) \longleftrightarrow C_y(\omega_1, \omega_2)$

Property 1.   *Linearity*

$$ax(n_1, n_2) + by(n_1, n_2) \longleftrightarrow aC_x(\omega_1, \omega_2) + bC_y(\omega_1, \omega_2)$$

Property 2.   *Separable Sequence*

$$x(n_1, n_2) = x_1(n_1)x_2(n_2) \longleftrightarrow C_x(\omega_1, \omega_2) = C_{x1}(\omega_1)C_{x2}(\omega_2)$$

Property 3.   *Energy Relationship*

$$\sum_{n1=0}^{\infty} \sum_{n2=0}^{\infty} |x(n_1, n_2)|^2 = \frac{1}{4(2\pi)^2} \int_{\omega_1 = -\pi}^{\pi} \int_{\omega_2 = -\pi}^{\pi} |C_x(\omega_1, \omega_2)|^2 \, d\omega_1 \, d\omega_2$$

Property 4.   *Symmetry Properties*

(a) $C_x(\omega_1, \omega_2) = C_x(-\omega_1, \omega_2) = C_x(\omega_1, -\omega_2) = C_x(-\omega_1, -\omega_2)$

(b) $x^*(n_1, n_2) \longleftrightarrow C_x^*(\omega_1, \omega_2)$

(c) real $x(n_1, n_2) \longleftrightarrow$ real $C_x(\omega_1, \omega_2)$

3.4 THE FAST FOURIER TRANSFORM

3.4.1 Row-Column Decomposition

The DFT discussed in Section 3.2 is used in a variety of signal processing applications. Thus it is of considerable interest to efficiently compute the DFT and inverse DFT. As we discussed in Section 3.3, methods to compute the DFT and inverse DFT can also be used in computing the DCT and inverse DCT. In this section, we will discuss an efficient way to compute the 2-D DFT. We will refer to this method as the fast Fourier transform (FFT) by *row-column decomposition*. This method is simple, uses 1-D FFT algorithms, and offers considerable computational savings over direct computation. It is also the most popular 2-D FFT algorithm.

To appreciate the computational efficiency of the row-column decomposition method, we will first consider computing the DFT directly. We consider only the DFT computation, since the inverse DFT computation is essentially the same. Consider a complex $N_1 \times N_2$-point sequence $x(n_1, n_2)$ that is zero outside $0 \leq n_1 \leq N_1 - 1$, $0 \leq n_2 \leq N_2 - 1$. Since we are using complex data, we use the terms "multiplication" and "addition" in this section to refer to complex multiplications and complex additions, unless specified otherwise. The DFT of $x(n_1, n_2)$, $X(k_1, k_2)$, is related to $x(n_1, n_2)$ by

$$X(k_1, k_2) = \sum_{n_1=0}^{N_1-1} \sum_{n_2=0}^{N_2-1} x(n_1, n_2) e^{-j(2\pi/N_1)k_1 n_1} e^{-j(2\pi/N_2)k_2 n_2}. \tag{3.62}$$

From (3.62), directly computing $X(k_1, k_2)$ for each (k_1, k_2) requires $N_1 N_2 - 1$ additions and $N_1 N_2$ multiplications. Since there are $N_1 N_2$ different values of (k_1, k_2), the total number of arithmetic operations required in computing $X(k_1, k_2)$ from $x(n_1, n_2)$ is $N_1^2 N_2^2$ multiplications and $N_1 N_2(N_1 N_2 - 1)$ additions.

To develop the row-column decomposition method, we rewrite (3.62) as follows:

$$X(k_1, k_2) = \sum_{n_2=0}^{N_2-1} \underbrace{\sum_{n_1=0}^{N_1-1} x(n_1, n_2) e^{-j(2\pi/N_1)k_1 n_1}}_{f(k_1, n_2)} e^{-j(2\pi/N_2)k_2 n_2}. \tag{3.63}$$

We first compute $f(k_1, n_2)$ from $x(n_1, n_2)$ and then $X(k_1, k_2)$ from $f(k_1, n_2)$. Consider a fixed n_2, say $n_2 = 0$. Then $x(n_1, n_2)|_{n_2=0}$ represents a row of $x(n_1, n_2)$, and $f(k_1, n_2)|_{n_2=0}$ is the 1-D N_1-point DFT of $x(n_1, n_2)|_{n_2=0}$ with respect to the variable n_1. Therefore, $f(k_1, 0)$ can be computed from $x(n_1, n_2)$ by computing one 1-D N_1-point DFT. Since there are N_2 different values of n_2 in $f(k_1, n_2)$ that are of interest to us, $f(k_1, n_2)$ can be computed from $x(n_1, n_2)$ by computing N_2 1-D N_1-point DFTs. This is illustrated in Figure 3.14.

Once $f(k_1, n_2)$ is computed, from (3.63) we can compute $X(k_1, k_2)$ from $f(k_1, n_2)$ by

$$X(k_1, k_2) = \sum_{n_2=0}^{N_2-1} f(k_1, n_2) e^{-j(2\pi/N_2)k_2 n_2}. \tag{3.64}$$

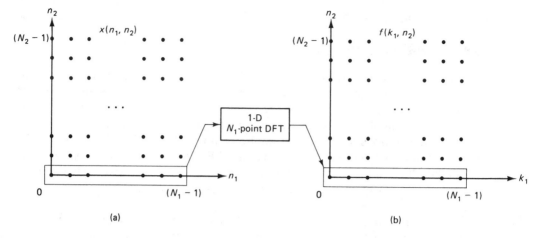

Figure 3.14 Computation of $f(k_1, n_2)$ from $x(n_1, n_2)$ by computing N_2 N_1-point 1-D DFTs.

To compute $X(k_1, k_2)$ from $f(k_1, n_2)$, consider a fixed k_1, say $k_1 = 0$. Then $f(k_1, n_2)|_{k_1=0}$ represents one column of $f(k_1, n_2)$, and $X(k_1, k_2)|_{k_1=0}$ in (3.64) is nothing but the 1-D N_2-point DFT of $f(k_1, n_2)|_{k_1=0}$ with respect to the variable n_2. Therefore, $X(0, k_2)$ can be computed from $f(k_1, n_2)$ by computing one 1-D N_2-point DFT. Since there are N_1 different values of k_1 in $X(k_1, k_2)$ that are of interest to us, $X(k_1, k_2)$ can be computed from $f(k_1, n_2)$ by computing N_1 1-D N_2-point DFTs. This is illustrated in Figure 3.15.

From the above discussion, $X(k_1, k_2)$ can be computed from $x(n_1, n_2)$ with a total of N_2 1-D N_1-point DFTs for the row operations and N_1 1-D N_2-point DFTs for the column operations. Suppose we compute the 1-D DFTs directly. Since direct computation of one 1-D N-point DFT requires N^2 multiplications and about

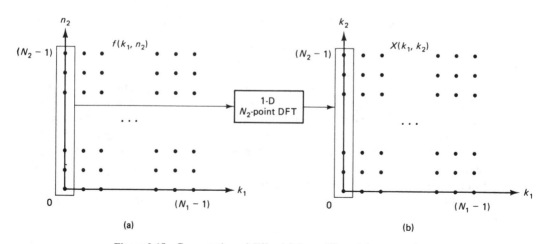

Figure 3.15 Computation of $X(k_1, k_2)$ from $f(k_1, n_2)$ by computing N_1 N_2-point 1-D DFTs.

The Discrete Fourier Transform   Chap. 3

N^2 additions, the total number of arithmetic operations involved in computing $X(k_1, k_2)$ is $N_1N_2(N_1 + N_2)$ multiplications and $N_1N_2(N_1 + N_2)$ additions. This is a significant computational saving over the $N_1^2N_2^2$ multiplications and $N_1^2N_2^2$ additions required for the direct computation of $X(k_1, k_2)$.

To further reduce the number of arithmetic operations, we can of course use any of a variety of 1-D FFT algorithms to compute the 1-D DFTs. When $N = 2^M$, a 1-D N-point FFT algorithm based on the Cooley-Tukey approach [Cooley and Tukey] requires $\dfrac{N}{2} \log_2 N$ multiplications and $N \log_2 N$ additions. To compute N_2 1-D N_1-point DFTs and N_1 1-D N_2-point DFTs using 1-D FFT algorithms when $N_1 = 2^{M_1}$ and $N_2 = 2^{M_2}$, we need a total of $\dfrac{N_1N_2}{2} \log_2 (N_1N_2)$ multiplications and $N_1N_2 \log_2 (N_1N_2)$ additions. This is a significant computational saving over the direct computation of the 1-D DFTs. If we represent the total number of points N_1N_2 as N, then the number of computations involved can be expressed as $\dfrac{N}{2} \log_2 N$ multiplications and $N \log_2 N$ additions. These are exactly the same expressions as those from a 1-D N-point DFT computation using an FFT algorithm such as a decimation-in-time algorithm.

To illustrate the computational savings involved, Table 3.5 shows the relative number of computations for the three methods considered. From the table, when $N_1 = N_2 = 512$, row-column decomposition alone reduces the number of multiplications and additions by a factor of approximately 250 as compared to the direct DFT computation. A reduction by an additional factor of approximately 110 for multiplications and 55 for additions is obtained by using 1-D FFT algorithms. The total reduction in the number of computations by row-column decomposition and 1-D FFT algorithms is a factor of approximately 28,000 for multiplications and 14,000 for additions for $N_1 = N_2 = 512$ as compared to the direct DFT computation.

In the derivation of the row-column decomposition approach, we expressed $X(k_1, k_2)$ in the form of (3.63). This led to a procedure in which row operations

TABLE 3.5 COMPUTATION OF REQUIRED NUMBER OF ARITHMETIC OPERATIONS FOR THREE METHODS OF DFT COMPUTATION

	Number of multiplications $(N_1 = N_2 = 512)$	Number of additions $(N_1 = N_2 = 512)$
Direct computation	$N_1^2N_2^2$ (100%)	$N_1^2N_2^2$ (100%)
Row-column decomposition with direct 1-D DFT computation	$N_1N_2(N_1 + N_2)$ (0.4%)	$N_1N_2(N_1 + N_2)$ (0.4%)
Row-column decomposition with 1-D FFT algorithm	$\dfrac{N_1N_2}{2} \log_2 (N_1N_2)$ (0.0035%)	$N_1N_2 \log_2 (N_1N_2)$ (0.007%)

were performed before the column operations. An alternative way to write (3.63) is

$$X(k_1, k_2) = \underbrace{\sum_{n_1=0}^{N_1-1} \sum_{n_2=0}^{N_2-1} x(n_1, n_2)e^{-j(2\pi/N_2)k_2n_2}}e^{-j(2\pi/N_1)k_1n_1}. \qquad (3.65)$$

$$g(n_1, k_2)$$

If (3.65) is used in computing $X(k_1, k_2)$, then the column operations are performed before the row operations. The number of computations involved in this case remains the same as in the previous case, where the row operations were performed first.

In the row-column decomposition approach, we compute a 2-D DFT by computing many 1-D DFTs. Other approaches to compute a 2-D DFT by computing many 1-D DFTs have also been developed. In one approach [Nussbaumer and Quandalle], a 2-D DFT is transformed to many 1-D DFTs using a transform called *polynomial transform*. For an $N \times N$-point DFT, this approach requires computation of $N + 1$ 1-D DFTs. This is approximately half of what is required in the row-column decomposition approach. Since the polynomial transform does not require any multiplications, the number of multiplications required in this approach is approximately half of that required in the row-column decomposition approach. The number of additions required remains approximately the same, however, due to the polynomial transform. The advantage of this approach compared to the row-column decomposition approach becomes more significant as we increase the number of dimensions. For an m-dimensional DFT of size N^m points, the approach requires $(N^m - 1)/(N - 1)$ 1-D DFTs. The row-column decomposition approach requires mN^{m-1} 1-D DFTs. The number of multiplications is reduced in this approach by a factor of approximately m, while the number of additions remains about the same. Other approaches [Nussbaumer (1981), Auslander, et al.] that require computation of $(N^m - 1)/(N - 1)$ 1-D DFTs to compute an N^m-point m-dimensional DFT have also been developed. In all these methods developed so far, the number of multiplications is reduced, but the number of additions remains approximately the same or significantly increases in comparison with the row-column decomposition method. In addition, they are more difficult to understand conceptually and more difficult to program. Detailed discussions of these approaches are found in [Nussbaumer (1977); Nussbaumer and Quandalle; Nussbaumer (1981); Auslander et al.].

3.4.2 Minicomputer Implementation of Row-Column Decomposition Method

The computation of a 2-D DFT of reasonable size requires a fair amount of memory. When we compute a 512×512-point DFT, we need about a quarter of a million memory units. As fast semiconductor memory becomes cheaper, the amount of memory space required for a DFT computation may become only a minor issue in practical applications. Where memory is relatively small, however, as in minicomputers, data may have to be stored on slow memory media such as disks. One I/O (input-output: read or write) operation on disk typically takes on the order of 10 to 100 milliseconds (msec), which is 10^4 to 10^5 times longer than one I/O operation on semiconductor memory. As a result, the required I/O time in a minicomputer environment is typically much greater than the time required for arithmetic operations in computing a DFT, and reducing the number of I/O operations is very important. In this section, we will consider the problem of reducing the required number of I/O operations in the row-column decomposition approach to the DFT computation.

Consider the problem of computing an $N \times N$-point DFT, where $N = 2^M$. Since the I/O operation using disk memory is slow, one block of data is read or written at a time. Suppose each block consists of N words, corresponding to one row or one column of data. We will assume that each row of $x(n_1, n_2)$ is stored as a block, as shown in Figure 3.16(a). In addition, we will assume that KN words of fast memory are available, as shown in Figure 3.16(b). In a typical implementation, K would be small, much less than N. For simplicity, we will assume that N/K is an integer.

We first consider a straightforward approach to performing the row and column operations. We then consider an approach that uses I/O operations much

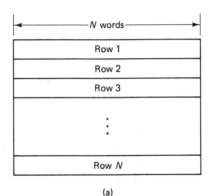

(a)

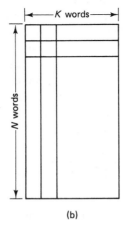

(b)

Figure 3.16 Memory arrangement assumed for efficient minicomputer implementation of row-column decomposition method. (a) One row of $x(n_1, n_2)$ is assumed to be stored in one block; (b) KN words of fast memory are assumed available.

more efficiently. In the straightforward approach, we first perform the row operations. Each row is read once, and it is written into fast memory, where the 1-D DFT is computed. The result is written back to the disk. If the original $x(n_1, n_2)$ does not need to be saved, the result can be stored in its original space. Processing each row in this manner requires two I/O operations: one read and one write. Since there are N rows, the row operations require a total of $2N$ I/O operations. Now consider the column operations. Since the data are stored row by row, we cannot read one column with one read operation. A simple approach is to read each row and save the first K words of each row in the fast memory. By reading all the rows, we would write K columns of data into the fast memory. We can now compute the 1-D DFT for each of the K columns. Once the K column operations are performed, they must be written on the disk. To do this, we read one row at a time, replace the first K words for each row with the data in the fast memory, and write the result back onto the disk. Before we write it onto the disk, however, we can read the next K words of data and store them in the fast memory, in anticipation of the next K column operations. The total number of I/O operations required so far is N reads, N reads, and N writes, in other words $3N$ I/O operations. This completes the first K column operations. For the next K column operations, the data are already in the fast memory, so the first N reads are not necessary and only $2N$ I/O operations are needed. Since we must perform K column operations N/K times, the total number of I/O operations needed for the column operations is $2N^2/K + N$. Combining this with the $2N$ I/O operations for the row operations, we have

$$\text{Total number of I/O operations} = 3N + \frac{2N^2}{K}. \tag{3.66}$$

For $N = 512$, $K = 2$, if it is assumed that an I/O operation takes an average of 20 msec, the amount of time required for the I/O operations alone in computing a 512×512-point DFT is approximately one and a half hours.

An alternative approach is to perform the row operations first and then transpose the data. The columns will now be rows. By again performing the row operations on the transposed data, we perform the column operations. The result can then be transposed to restore the rows and columns in their correct places. This is shown in Figure 3.17. In practice, the second transposition in Figure 3.17 is often not necessary. When a DFT is computed, it is modified in some way, and then the inverse DFT is computed. For the inverse DFT computation, we can perform the column operations first and then the row operations; this way, only one transposition will be needed for the DFT computation. In this discussion, we will assume that only one data transposition is necessary.

The row and column (row in the transposed data) operations in Figure 3.17 require $4N$ I/O operations. As we will discuss shortly, the data transposition when

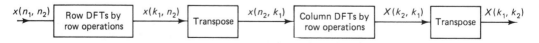

Figure 3.17 Implementation of row-column decomposition method.

efficiently performed requires $2N \log_2 N$ I/O operations for $K = 2$. In addition, the $2N$ I/O operations required for the row operations and $2N$ I/O operations required for the column operations can be eliminated by combining them with the data transposition. In total, we have

$$\text{Total number of I/O operations} = 2N \log_2 N. \tag{3.67}$$

For $N = 512$ and $K = 2$, $2N \log_2 N$ in (3.67) is approximately four percent of $3N + (2N^2/K)$ in (3.66). If we assume that an I/O operation takes an average of 20 msec, I/O operations require a few minutes. This is still considerably longer than the time required for the arithmetic operations, but significantly less than one and a half hours. With slow memory such as disk, the I/O operations often dominate the DFT computation time.

We will now discuss how the $N \times N$-point data can be transposed with $2N \log_2 N$ I/O operations using $2N$ words of fast memory ($K = 2$). This method was developed by [Eklundh], and is based on a divide-and-conquer strategy analogous to the development of the FFT algorithm. In this method, the $N \times N$-point data transposition is transformed to four $N/2 \times N/2$-point data transpositions. Each $N/2 \times N/2$-point data transposition is then transformed to four $N/4 \times N/4$-point data transpositions. This process can be continued until the size of the data to be transposed is 1×1.

We will next describe how an $N \times N$-point data transposition problem can be transformed to four $N/2 \times N/2$-point data transposition problems. Consider a matrix A of size $N \times N$, which can be represented by

$$A = \begin{bmatrix} A_{11} & A_{12} \\ A_{21} & A_{22} \end{bmatrix} \tag{3.68}$$

where A_{11}, A_{12}, A_{21}, and A_{22} have $N/2 \times N/2$ elements each. The transposition of A, A^T, can be expressed as

$$A^T = \begin{bmatrix} A_{11}^T & A_{21}^T \\ A_{12}^T & A_{22}^T \end{bmatrix}. \tag{3.69}$$

Equations (3.68) and (3.69) suggest a method of transforming an $N \times N$-point data transposition problem to four $N/2 \times N/2$-point data transposition problems. This is shown in Figure 3.18. The matrix A is first divided into four submatrices A_{11}, A_{12}, A_{21}, and A_{22}. The two submatrices A_{12} and A_{21} are interchanged. At this point, the task remaining is to transpose four submatrices. In the matrix representation, the element in the upper left-hand corner is the origin, and rows

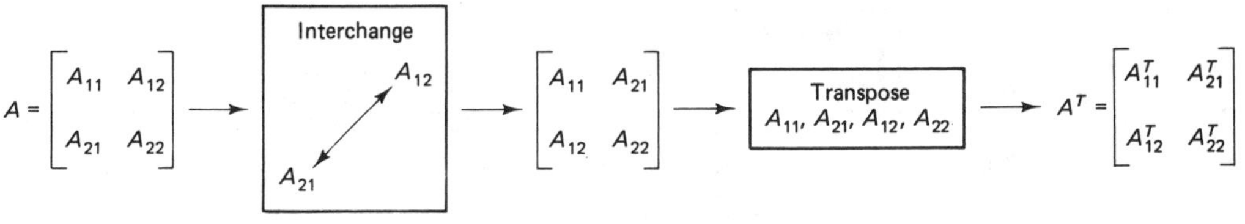

Figure 3.18 Transformation of one $N \times N$-point data transposition problem to four $N/2 \times N/2$-point data transposition problems.

and columns are counted relative to that point. In the signal representation, the origin is the element in the lower left-hand corner. The result of making this notational change to Figure 3.18 is shown in Figure 3.19. Each of the four $N/2 \times N/2$-point data transpositions can be carried out following a similar procedure.

In the above approach, the first stage transforms an $N \times N$-point problem to four $N/2 \times N/2$-point problems. The second stage transforms an $N/2 \times N/2$-point problem to four $N/4 \times N/4$-point problems. The number of stages required until we have 1×1-point problems, then, is given by

$$\text{Number of stages} = \log_2 N. \qquad (3.70)$$

We now will illustrate how each stage can be carried out with only $2N$ I/O operations when $K = 2$. We will use an 8×8-point data transposition problem as an example. Consider an 8×8-point sequence $x(n_1, n_2)$, as shown in Figure 3.20(a). In the first stage, the upper left-hand quarter of $x(n_1, n_2)$ is exchanged with the lower right-hand quarter. The result of the first stage is shown in Figure 3.20(b). To carry out this stage, we read the first row and the fifth $[(N/2) + 1]$ row into fast memory. In the fast memory, the first half of the first row is exchanged with the second half of the fifth row. The results are written back onto the disk. This requires four I/O operations (one read and one write for each row). We then read the second and sixth rows, and the process continues. In this way, the first stage is completed with $2N$ I/O operations. Figure 3.21(a) shows the input to the second stage, which is the result of the first stage in Figure 3.20. The operations required in the second stage are also shown in Figure 3.21(a). The result of this stage is shown in Figure 3.21(b). This time, we read the first row and the third $[(N/4) + 1]$ row. The first and third quarters of the first row are exchanged with the second and fourth quarters of the third row, respectively. After these exchanges are made in the fast memory, the results are written back to the disk. This requires four I/O operations. The next rows to be read are the second and fourth rows, and the process continues. The second stage, then, also requires a total of $2N$ I/O operations. The third stage, shown in Figure 3.22, can also be carried out by $2N$ I/O operations by first reading the first and second $[(N/8) + 1]$ rows, then reading the third and fourth rows, and so on. In this example, three $(\log_2 N = 3)$ stages are required. Since each stage requires $2N$ I/O operations, and since from (3.70) there is a total of $\log_2 N$ stages, the total number of I/O operations required in the data transposition is given by

$$\text{Total number of I/O operations for data transposition} = 2N \log_2 N, \qquad (3.71)$$

which is what we assumed in our previous analysis.

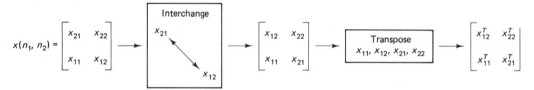

Figure 3.19 Transformation of one $N \times N$-point $x(n_1, n_2)$ transposition problem to four $N/2 \times N/2$-point sequence transposition problems.

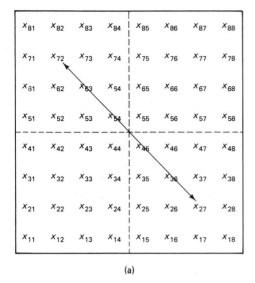

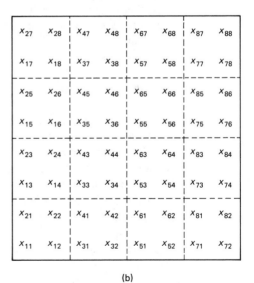

(a)

(b)

Figure 3.20 First stage of Eklundh's method. (a) Original sequence $x(n_1, n_2)$, showing exchanges; (b) output of the first stage.

Note that the row operations can be combined with the first stage of the data transposition. Specifically, the row operations can be performed right after two rows of data are read into the $2N$ words of fast memory in the first stage. This eliminates the $2N$ I/O operations which otherwise would have been necessary. Note also that the last stage of the data transposition can be combined with the row operations (which are really 1-D column DFT computations) that we perform

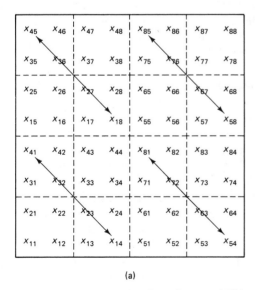

(a)

(b)

Figure 3.21 Second stage of Eklundh's method. (a) Input to the second stage, showing exchanges; (b) output of the second stage.

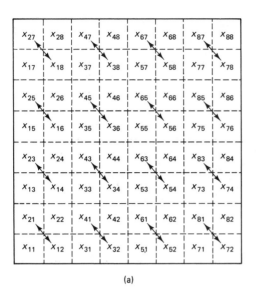

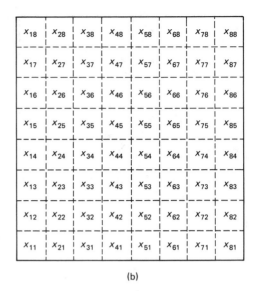

(a)   (b)

Figure 3.22 Third and final stage of Eklundh's method. (a) Input to the third stage, showing exchanges; (b) output $x^T(n_1, n_2) = x(n_2, n_1)$.

on the transposed data. Specifically, the results in the $2N$ words of fast memory, before they are written back to the disk in the last stage, are two rows [which are really columns of $x(n_1, n_2)$] of the transposed data. The row operations can be performed before the results are written onto the disk. This eliminates the $2N$ I/O operations that otherwise would have been necessary for the second set of row operations in Figure 3.17.

In this discussion, we have assumed that $N_1 = N_2 = N = 2^M$ and $K = 2$. This general idea can be extended to the case when $N_1 \neq N_2$, $N \neq 2^M$, and $K > 2$ as long as N_1 and N_2 are highly composite numbers. Examples of such extensions can be found in Problems 3.27 and 3.28.

3.4.3 The Vector Radix Fast Fourier Transform

In the section above, we discussed the row-column decomposition method, in which the 2-D DFT computation is transformed to many 1-D DFT computations and 1-D FFT algorithms are used to compute the 1-D DFTs. An alternative approach is to extend the idea behind the 1-D FFT algorithm development directly to the 2-D case. This leads to *vector radix FFT algorithms*.

Although there are many variations, most 1-D FFT algorithms are based on one simple principle: an N-point DFT can be computed by two $N/2$-point DFTs, or three $N/3$-point DFTs, and so on. This simple principle can be extended to the 2-D case in a straightforward manner. Specifically, an $N_1 \times N_2$-point DFT can be computed by four $N_1/2 \times N_2/2$-point DFTs, or six $N_1/2 \times N_2/3$-point DFTs, or nine $N_1/3 \times N_2/3$-point DFTs, and so on. In this manner, various 1-D FFT algorithms, such as the decimation-in-time and decimation-in-frequency algorithms, can be extended directly to the 2-D case. Since the extension is conceptually straightforward, we will only summarize the main results. Keeping track of the data indices, however, is more complicated in 2-D, so careful attention should be paid to them.

Consider an $N_1 \times N_2$-point sequence $x(n_1, n_2)$ that is complex and is zero outside $0 \le n_1 \le N_1 - 1, 0 \le n_2 \le N_2 - 1$. We will assume that $N_1 = N_2 = N = 2^M$ for convenience and consider the extension of the 1-D decimation-in-time algorithm. Rewriting $X(k_1, k_2)$, we have

$$X(k_1, k_2) = \sum_{n_1=0}^{N-1} \sum_{n_2=0}^{N-1} x(n_1, n_2)e^{-j(2\pi/N)k_1 n_1}e^{-j(2\pi/N)k_2 n_2}$$

$$= \sum_{n_1:\text{even}} \sum_{n_2:\text{even}} + \sum_{n_1:\text{odd}} \sum_{n_2:\text{even}} + \sum_{n_1:\text{even}} \sum_{n_2:\text{odd}} + \sum_{n_1:\text{odd}} \sum_{n_2:\text{odd}}$$

$$= G_{ee}(k_1, k_2) + G_{oe}(k_1, k_2)W_N^{k_1} + G_{eo}(k_1, k_2)W_N^{k_2} + G_{oo}(k_1, k_2)W_N^{k_1+k_2}$$

$$(3.72)$$

where $W_N = e^{-j(2\pi/N)}$, $G_{ee}(k_1, k_2)$ is the $N/2 \times N/2$-point DFT of $g_{ee}(n_1, n_2) = x(2n_1, 2n_2)$ and is assumed to be periodic with a period of $N/2 \times N/2$, and $G_{oe}(k_1, k_2)$, $G_{eo}(k_1, k_2)$, and $G_{oo}(k_1, k_2)$ are similarly defined. Equation (3.72) is illustrated in Figure 3.23 for the case $N = 4$. The symbol $\square$ in $x(n_1, n_2)$ shows the points that contribute to $g_{ee}(n_1, n_2)$. The symbols $\square$, $\bigcirc$, $\triangle$, and $\diamond$ in the figure indicate which points of $G_{ee}(k_1, k_2)$, $G_{oe}(k_1, k_2)$, $G_{eo}(k_1, k_2)$, and $G_{oo}(k_1, k_2)$ contribute to which points of $X(k_1, k_2)$. The figure clearly shows that the $N \times N$-point DFT computation has been transformed to four $N/2 \times N/2$-point DFT computations. In the next stage of the algorithm's development, each $N/2 \times N/2$-point DFT computation is transformed to four $N/4 \times N/4$-point DFT computations, and the process can continue. The part of the figure within the box in Figure 3.23 shows one stage of the resulting algorithm.

Figure 3.23 shows only one stage of the algorithm for a very small $N(N = 4)$, but with some imagination we can make various observations about the overall algorithm. From the figure, four elements of $X(k_1, k_2)$ are affected by only four elements, one each from $G_{ee}(k_1, k_2)$, $G_{oe}(k_1, k_2)$, $G_{eo}(k_1, k_2)$, and $G_{oo}(k_1, k_2)$. This is the basic computational element in all stages and is called a *vector radix* 2×2, or *radix* 2×2 for short, butterfly computation. To illustrate this more clearly, Figure 3.24 shows a general radix 2×2 butterfly computation where the input is arranged in the order of $G_{ee}(k_1, k_2)$, $G_{oe}(k_1, k_2)$, $G_{eo}(k_1, k_2)$, and $G_{oo}(k_1, k_2)$ and the output is arranged in the order of $X(k_1, k_2)$, $X(k_1 + N/2, k_2)$, $X(k_1, k_2 + N/2)$, and $X(k_1 + N/2, k_2 + N/2)$. Note that both the input and the output blocks are arranged with the bottom left coming first, then bottom right, next top left, and finally top right.

From Figure 3.24, a radix 2×2 butterfly computation requires three multiplications, one for each of the three nontrivial branch transmittances $W_N^{k_1}$, $W_N^{k_2}$, and $W_N^{k_1+k_2}$. Since four branches meet at each output node, a simple implementation requires three additions at each node, for a total of twelve additions per butterfly computation. To reduce the number of additions, we denote the values

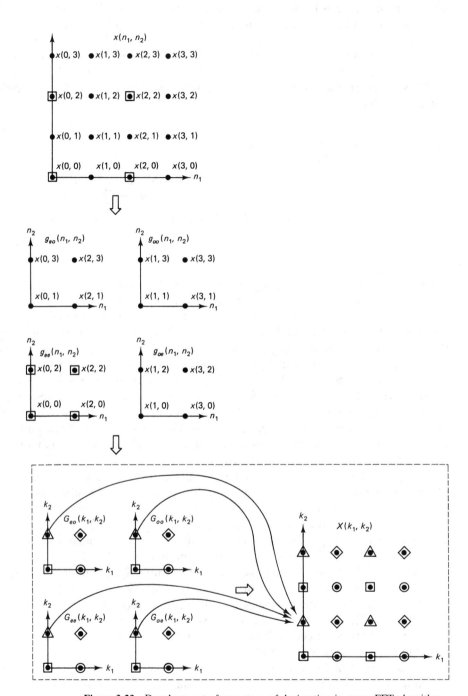

Figure 3.23 Development of one stage of decimation-in-space FFT algorithm.

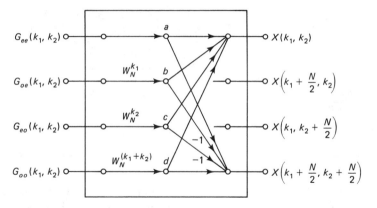

Figure 3.24 Vector radix 2 × 2 butterfly computation. For clarity, not all arrows are shown.

at the four nodes in the figure by a, b, c, and d. Then we can express $X(k_1, k_2)$, $X(k_1 + N/2, k_2)$, $X(k_1, k_2 + N/2)$, $X(k_1 + N/2, k_2 + N/2)$ by

$$X(k_1, k_2) = A + C \tag{3.73a}$$

$$X\left(k_1 + \frac{N}{2}, k_2\right) = B + D \tag{3.73b}$$

$$X\left(k_1, k_2 + \frac{N}{2}\right) = A - C \tag{3.73c}$$

$$X\left(k_1 + \frac{N}{2}, k_2 + \frac{N}{2}\right) = B - D \tag{3.73d}$$

where
$$A = a + b \tag{3.73e}$$

$$B = a - b \tag{3.73f}$$

$$C = c + d \tag{3.73g}$$

$$D = c - d. \tag{3.73h}$$

From (3.73), a radix 2 × 2 butterfly computation requires eight additions. From the above discussion, each radix 2 × 2 butterfly computation requires 3 multiplications and 8 additions. Since there are $\log_2 N$ stages and $N^2/4$ butterflies in each stage, the total number of arithmetic operations required in computing the $N \times N$-point DFT is $(3N^2/4) \log_2 N$ multiplications and $2N^2 \log_2 N$ additions. A comparison of the vector radix FFT and the row-column decomposition approach discussed in Section 3.4.1 is shown in Table 3.6. Compared to the row-column decomposition approach, the number of multiplications has been reduced by about twenty-five percent without affecting the number of additions. The same reduction can also be achieved in the row-column decomposition approach by using a radix 4 rather than a radix 2 FFT algorithm in the 1-D DFT computation. Note also that in-place computation is possible because the butterfly computation is the basic computational element.

TABLE 3.6 COMPARISON OF REQUIRED NUMBER OF ARITHMETIC
OPERATIONS FOR TWO DIFFERENT METHODS OF DFT COMPUTATION

	Number of multiplications	Number of additions
Row-column decomposition with 1-D FFT algorithm	$N^2 \log_2 N$	$2N^2 \log_2 N$
Vector radix FFT	$\dfrac{3N^2}{4} \log_2 N$	$2N^2 \log_2 N$

From Figure 3.23, we can also determine the number of I/O operations required when data are stored on slow memory media such as disk. We will assume,
as in Section 3.4.2, that each row is stored as one block on the disk, and that $2N$
words of fast memory which can store two rows of data are available. In the
vector radix FFT, we first need to perform bit reversal of input. From Figure
3.23, two rows of the bit-reversed data come from two rows of the input. We can
read these two rows of $x(n_1, n_2)$ into fast memory, bit-reverse the data within the
fast memory with respect to the n_1 dimension, and then write the data back to the
disk. Since each row is read and written once, the total number of I/O operations
for the complete bit reversal is $2N$. Now consider one stage of the FFT algorithm,
shown in the enclosed box in Figure 3.23. Again, two complete rows of the output
come from two complete rows of the input. Therefore, the two rows of input can
be read and stored in the fast memory. Within the fast memory, necessary linear
combinations are performed, and the results are written back to the disk. Since
each row is read from the disk once and written to the disk once, the total number
of I/O operations involved in this stage is $2N$. There are $\log_2 N$ stages, so we have
a total of $2N + 2N \log_2 N$ I/O operations, where the $2N$ comes from the bit-
reversal operation. This is approximately the same number of I/O operations
required in the row-column decomposition approach.

The complete vector radix FFT algorithm for the $N_1 = N_2 = 4$ case is shown
in Figure 3.25. This is obtained by carrying out one additional stage in Figure
3.23, where each $N/2 \times N/2$-point DFT computation is transformed to four
$N/4 \times N/4$-point DFT computations. The $N/4 \times N/4$-point DFT in this case is a
1×1-point DFT, which is an identity. The blocks of input in Figure 3.25 are
arranged in the following order: first, bottom left $[g_{ee}(n_1, n_2)]$, next, bottom right
$[g_{oe}(n_1, n_2)]$, then top left $[g_{eo}(n_1, n_2)]$, and last top right $[g_{oo}(n_1, n_2)]$. The output
blocks are arranged in the same manner. The four elements in each block are
also arranged in this order. From Figure 3.25, it is clear that data indexing using
the 1-D flowgraph representation is complicated. It is often easier to use the
2-D representation, for instance the one in Figure 3.23, in studying a vector radix
FFT algorithm.

Vector radix FFT algorithms are roughly equivalent in many respects to the
row-column decomposition approach discussed in Section 3.4.1, and do not offer
any significant advantages. The number of arithmetic operations required, the
number of I/O operations required when data are stored on disk, and the in-place

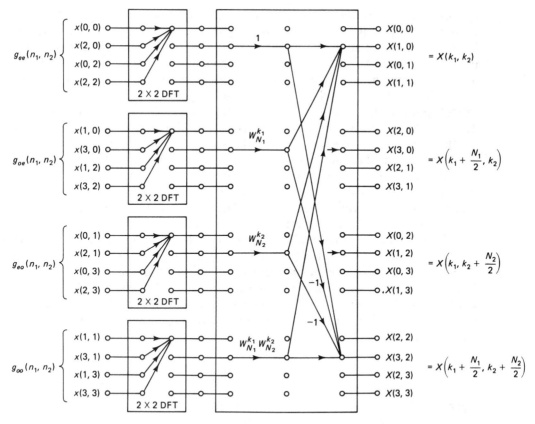

Figure 3.25 Decimation-in-space 2×2 vector radix FFT, for the case $N_1 = N_2 = 4$. For clarity, many values are omitted.

computation aspect are roughly the same. Since the row-column decomposition method is quite simple to implement by using 1-D FFT algorithms, it is used more often than the vector radix FFT algorithm.

3.4.4 Fast Algorithms for One-Dimensional Discrete Fourier Transform Computation

As we discussed in Section 3.4.1, efficient computation of a 2-D DFT is closely related to efficient computation of a 1-D DFT. In this section, we discuss two fast 1-D DFT computation algorithms that are significantly different from the Cooley-Tukey FFT algorithm.

Shortly after Cooley and Tukey published the FFT algorithm, there was a flurry of activity: identifying earlier related work,* extending the idea, and applying

---

*It has recently been discovered that Gauss used the Cooley-Tukey FFT algorithm in the early nineteenth century [Heideman et al.].

the result to practical problems. By the early 1970s, most of the interesting theoretical developments were thought to have been completed. In 1976, however, Winograd proposed an algorithm that significantly reduces the number of multiplications compared to the Cooley-Tukey approach. For some values of N, the number of multiplications required to compute an N-point DFT is proportional to N, not $N \log_2 N$. This led to another cycle of identifying earlier related work, extending the idea, and applying the result to practical algorithms. Now that the dust has settled somewhat, two algorithms appear to be worth considering. They are the *prime factor algorithm* (PFA), which was originally developed by [Good] and later developed further by [Kolba and Parks] and [Burrus and Eschenbacher], and the *Winograd Fourier transform algorithm* (WFTA). In this section, we briefly summarize the basic ideas behind these two algorithms and their practical significance. A detailed treatment of this topic requires a background in number theory, and the interested reader is referred to [McClellan and Rader].

Both the PFA and the WFTA are based on three key ideas. The first is the recognition that an N-point DFT computation for the case when N is a prime number or a power of a prime number can be expressed as a circular convolution. The result for the case when N is prime was originally described by [Rader], and was primarily a curiosity when the discovery was published in 1968. To illustrate this result, let us consider an example for the case $N = 5$. The 1-D DFT of a 5-point sequence $x(n)$ is given by

$$X(k) = \sum_{n=0}^{4} x(n) W_5^{kn} \tag{3.74}$$

where

$$W_5 = e^{-j(2\pi/5)}.$$

We first define $\overline{X}(k)$ by

$$\overline{X}(k) = \sum_{n=1}^{4} x(n) W_5^{kn}. \tag{3.75}$$

From (3.74) and (3.75), $X(k)$ can be obtained from $\overline{X}(k)$ by

$$X(0) = \sum_{n=0}^{4} x(n) \tag{3.76a}$$

$$X(k) = x(0) + \overline{X}(k), \quad 1 \le k \le 4. \tag{3.76b}$$

We will now illustrate that $\overline{X}(k)$ can be expressed as the result of a 4-point circular convolution ($N - 1$ being equal to four). We define three new sequences $u(n)$, $w(n)$, and $v(n)$. The sequence $u(n)$ is obtained from $x(n)$ by

$$u(0) = x(1) \tag{3.77a}$$

$$u(1) = x(3) \tag{3.77b}$$

$$u(2) = x(4) \tag{3.77c}$$

$$u(3) = x(2). \tag{3.77d}$$

The sequence $w(n)$ is obtained from W_5 by

$$w(0) = W_5^1 \qquad (3.78a)$$

$$w(1) = W_5^2 \qquad (3.78b)$$

$$w(2) = W_5^4 \qquad (3.78c)$$

$$w(3) = W_5^3. \qquad (3.78d)$$

The sequence $v(n)$ is obtained from $\overline{X}(k)$ by

$$v(0) = \overline{X}(1) \qquad (3.79a)$$

$$v(1) = \overline{X}(2) \qquad (3.79b)$$

$$v(2) = \overline{X}(4) \qquad (3.79c)$$

$$v(3) = \overline{X}(3). \qquad (3.79d)$$

From (3.77), (3.78), and (3.79), it can be shown (see Problem 3.33) after some algebra that

$$v(n) = u(n) \circledast w(n) \qquad (3.80)$$

with an assumed periodicity of $N - 1$ for the circular convolution. This method can be used in computing a DFT by performing a circular convolution. Specifically, from $x(n)$ and W_N, $u(n)$ and $w(n)$ are obtained. From (3.80), $v(n)$ is computed by a 4-point circular convolution of $u(n)$ and $w(n)$. Using (3.79), we obtain $\overline{X}(k)$ from $v(n)$. Using (3.76), we obtain $X(k)$ from $\overline{X}(k)$. This is illustrated in Figure 3.26. It is clear that a 5-point DFT can be computed from the result of a 4-point circular convolution. Since an N-point DFT for the case when N is a prime or a power of a prime can always be expressed in terms of an $N - 1$-point circular convolution, an efficient method for circular convolution potentially leads to an efficient method for DFT computation.

The second key idea due to Winograd is the result that the number of multiplications (assuming that only a certain type of multiplications is counted) required for a circular convolution has a theoretical minimum, and that this minimum can be achieved. As an example, Figure 3.27 shows the computation of a 3-point circular convolution of $x(n)$ and $h(n)$. In the figure, the coefficients associated with $h(n)$ are assumed to have been precomputed. A 3-point circular convolution,

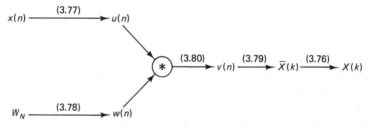

Figure 3.26 DFT computation by circular convolution.

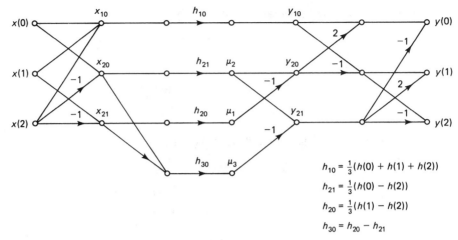

$$h_{10} = \tfrac{1}{3}(h(0) + h(1) + h(2))$$

$$h_{21} = \tfrac{1}{3}(h(0) - h(2))$$

$$h_{20} = \tfrac{1}{3}(h(1) - h(2))$$

$$h_{30} = h_{20} - h_{21}$$

Figure 3.27 Computation of 3-point circular convolution of two 3-point sequences $x(n)$ and $h(n)$ with four multiplications, the theoretically minimum number. After [McClellan and Rader].

when implemented by using the scheme in Figure 3.27, requires four multiplications. Winograd's findings indicate that this is the minimum possible for this case.

These two key ideas provide an approach to computing the DFT when its size N is a prime or a power of a prime. The result of applying these ideas to the problem of computing a 3-point DFT is given by [McClellan and Rader]:

$$a_1 = x(1) + x(2) \tag{3.81a}$$

$$a_2 = x(1) - x(2) \tag{3.81b}$$

$$a_3 = x(0) + a_1 \tag{3.81c}$$

$$m_1 = -\tfrac{1}{2}a_1 \tag{3.81d}$$

$$m_2 = -j\frac{\sqrt{3}}{2}a_2 \tag{3.81e}$$

$$c_1 = x(0) + m_1 \tag{3.81f}$$

$$X(0) = a_3 \tag{3.81g}$$

$$X(1) = c_1 + m_2 \tag{3.81h}$$

$$X(2) = c_1 - m_2. \tag{3.81i}$$

The above procedure requires one multiplication, 1 shift (multiplication by $\tfrac{1}{2}$), and six additions. Efficient methods of computing the DFT based on the two key ideas have been developed for many different DFT sizes including 2, 3, 2^2, 5, 7, 2^3, 3^3, 11, 13, 2^4, 17, 19, 5^2, and 2^5. Fewer multiplications are required for these cases than for a Cooley-Tukey FFT algorithm for a DFT computation of similar size. As N increases, however, the method becomes quite complicated, and the number of additions required goes up very rapidly. This problem is solved by the third key idea.

The third key idea is that when the DFT length N can be expressed as

$$N = P_1 P_2 \cdots \cdot P_M \tag{3.82}$$

where the set P_i is mutually prime, the 1-D DFT can be expressed as a multidimensional (m-D) DFT of size $P_1 \times P_2 \times \cdots \times P_M$. A set P_i $1 \le i \le M$ is called *mutually prime* if the only common divisor of all P_i is 1. This particular mapping of a 1-D problem to an m-D problem was first proposed by [Good]. The method is based on the Chinese remainder theorem, which requires that all the factors be mutually prime. The requirement that the factors be mutually prime eliminates the important case of $N = 2^M$. As an example, consider $N = P_1 \cdot P_2 = 2 \cdot 3$. Clearly 2 and 3 are mutually prime. From $x(n)$, we define a new $2 \times$ 3-point sequence $w(n_1, n_2)$ by

$$w(0, 0) = x(0) \tag{3.83a}$$

$$w(0, 1) = x(2) \tag{3.83b}$$

$$w(0, 2) = x(4) \tag{3.83c}$$

$$w(1, 0) = x(3) \tag{3.83d}$$

$$w(1, 1) = x(5) \tag{3.83e}$$

$$w(1, 2) = x(1) \tag{3.83f}$$

From $X(k)$, we define a new 2×3-point sequence $W(k_1, k_2)$ by

$$W(0, 0) = X(0) \tag{3.84a}$$

$$W(0, 1) = X(4) \tag{3.84b}$$

$$W(0, 2) = X(2) \tag{3.84c}$$

$$W(1, 0) = X(3) \tag{3.84d}$$

$$W(1, 1) = X(1) \tag{3.84e}$$

$$W(1, 2) = X(5). \tag{3.84f}$$

From (3.83), (3.84), and the 2-D DFT definition of (3.7), it can be shown that $W(k_1, k_2)$ is the 2×3-point DFT of $w(n_1, n_2)$, that is,

$$W(k_1, k_2) = \sum_{n_1=0}^{1} \sum_{n_2=0}^{2} w(n_1, n_2) W_2^{k_1 n_1} W_3^{k_2 n_2}. \tag{3.85}$$

From $x(n)$, a 2-D sequence $w(n_1, n_2)$ is obtained by using (3.83). The 2-D DFT $W(k_1, k_2)$ is next computed from $w(n_1, n_2)$. The 1-D DFT $X(k)$ is then obtained from $W(k_1, k_2)$ by using (3.84). This is illustrated in Figure 3.28.

$$x(n) \xrightarrow{\text{(3.83)}} w(n_1, n_2) \xrightarrow{\text{(3.85)}} W(k_1, k_2) \xrightarrow{\text{(3.84)}} X(k)$$

Figure 3.28 1-D DFT computation by m-D DFT computation.

The PFA differs from the WFTA in the way the m-D DFT is computed. In the PFA, the m-D DFT is computed by using the row-column decomposition, involving many 1-D DFT computations of size $P_1, P_2, \ldots P_M$, which are typically much shorter than N. In essence, we are computing a 1-D DFT by computing many shorter 1-D DFTs. The resulting shorter 1-D DFTs are computed by efficient circular convolution algorithms. In the WFTA, the special structure of short transforms is used to nest the multiplications. This decreases the number of multiplications, but increases the number of additions.

When we compare the various methods of computing the DFT, a number of issues must be considered, such as the number of computations, memory size, and regularity of structure. The WFTA has a clear advantage over the Cooley-Tukey FFT and the PFA in the number of multiplications required. However, it has a number of disadvantages. It requires significantly more additions than the other two methods, and the algorithm is quite complex and requires a large program. In addition, in-place computation does not appear to be possible, since the number of intermediate variables created in this method is much greater than the original data size. The number of coefficients that must be stored is also larger and the number of data transfers is greater. As a result, the WFTA appears to be useful only in a rather restricted environment where multiplications are much slower than additions and the cost of memory is very small.

The major advantages of the PFA over the Cooley-Tukey FFT are the significant reduction in the required number of multiplications without a significant increase in the number of additions, the absence of twiddle factors (the W_N term), the lack of need for bit-reversal data indexing, and the small number of stored constants. Because of these advantages, PFA benchmark programs run very quickly, and the PFA is a good choice for a floating-point software implementation in microcomputers or minicomputers. However, it is not as regular in structure as the Cooley-Tukey FFT. Each stage uses a different module or butterfly, and the structure of these modules is such that many additions are carried out separately from multiplications. As a result, the PFA cannot efficiently utilize many existing array processors and vector-oriented machines that are efficient for inner-product-type (multiply and add) operations. In addition, the basic concepts of the PFA are considerably more difficult to understand. Consequently, writing a program takes more effort. In contrast, the Cooley-Tukey FFT is very simple to understand. It has a clean, highly regular structure. The same butterfly or set of arithmetic instructions is used in each stage, and the structure is well suited to general-purpose computers, array processors with parallel or pipelined architecture, and VLSI machines.

REFERENCES

For books on the fast Fourier transform and related topics, see [Brigham; McClellan and Rader; Elliot and Rao; Blahut; Burrus and Parks].

For readings on the discrete cosine transform, see [Ahmed et al.; Chen et al.; Narasimha and Peterson; Makhoul; Clarke].

For methods in which we compute a 2-D DFT by computing many 1-D DFTs, see [Nussbaumer (1977); Nussbaumer and Quandalle; Nussbaumer (1981); Auslander et al.]. For a fast method for matrix transposition, see [Eklundh]. For vector radix FFT algorithms, see [Harris et al.; Rivard].

For a tutorial presentation of number theory and its application to fast Fourier transform algorithm development, see [McClellan and Rader]. For readings on Winograd Fourier transform algorithm, see [Winograd (1976, 1978); Silverman; Nawab and McClellan; Morris]. For prime factor algorithms, see [Kolba and Parks; Burrus and Eschenbacher; Johnson and Burrus].

For FFT algorithms for real-valued data, see [Vetterli and Nussbaumer; Sorensen et al.]. For in-place FFT algorithms, see [Pitas and Strintzis]. For FFT algorithms defined for arbitrary periodic sampling lattices such as hexagonal sampling, see [Mersereau and Speake; Guessoum and Mersereau].

N. Ahmed, T. Natarajan, and K. R. Rao, Discrete cosine transform, *IEEE Trans. Comput.*, Vol. C-23, January 1974, pp. 90–93.

L. Auslander, E. Feig, and S. Winograd, New algorithms for the multidimensional discrete Fourier transform, *IEEE Trans. on Acoust., Speech and Sig. Proc.*, Vol. ASSP-31, April 1983, pp. 388–403.

R. E. Blahut, *Fast Algorithms for Digital Signal Processing*, Reading, MA: Addison-Wesley, 1985.

E. O. Brigham, *The Fast Fourier Transform*. Englewood Cliffs, NJ: Prentice-Hall, 1974.

C. S. Burrus, Index mappings for multidimensional formulation of the DFT and convolution, *IEEE Trans. on Acoust., Speech and Sig. Proc.*, Vol. ASSP-25, June 1977, pp. 239–242.

C. S. Burrus and P. W. Eschenbacher, An in-place, in-order prime factor FFT algorithm, *IEEE Trans. on Acoust., Speech and Sig. Proc.*, Vol. ASSP-29, August 1981, pp. 806–817.

C. S. Burrus and T. W. Parks, *DFT/FFT and Convolution Algorithms*. New York: Wiley, 1985.

W. H. Chen, C. H. Smith, and S. C. Fralick, A fast computational algorithm for the discrete cosine transform, *IEEE Trans. Commun.*, Vol. COM-25, September 1977, pp. 1004–1009.

R. J. Clarke, *Transform Coding of Images*. London: Academic Press, 1985.

J. W. Cooley and J. W. Tukey, An algorithm for the machine calculation of complex Fourier series, *Math. Comput.*, Vol. 19, April 1965, pp. 297–301.

J. O. Eklundh, A fast computer method for matrix transposing, *IEEE Trans. Comput.*, Vol. C-21, July 1972, pp. 801–803.

D. F. Elliot and K. R. Rao, *Fast Transforms, Algorithms, Analyses, Applications*. New York: Academic Press, 1982.

I. J. Good, The relationship between two fast Fourier transforms, *IEEE Trans. on Computers*, Vol. C-20, March 1971, pp. 310–317.

A. Guessoum and R. M. Mersereau, Fast algorithms for the multidimensional discrete Fourier transform, *IEEE Trans. on Acoust., Speech and Sig. Proc.*, Vol. ASSP-34, August 1986, pp. 937–943.

D. B. Harris, J. H. McClellan, D. S. K. Chan, and H. W. Schuessler, Vector radix fast Fourier transform, *Proc. Int. Conf. Acoust., Speech and Sig. Proc.*, April 1977, pp. 548–551.

M. T. Heideman, D. H. Johnson, C. S. Burrus, Gauss and the history of the fast Fourier transform, *IEEE ASSP Magazine*, Vol. 1, October 1984, pp. 14–21.

H. W. Johnson and C. S. Burrus, The design of optimal DFT algorithms using dynamic programming, *IEEE Trans. on Acoust., Speech and Sig. Proc.*, Vol. ASSP-31, April 1983, pp. 378–387.

D. P. Kolba and T. W. Parks, A prime factor FFT algorithm using high-speed convolution, *IEEE Trans. on Acoust., Speech and Sig. Proc.*, Vol. ASSP-25, August 1977, pp. 281–294.

J. Makhoul, A fast cosine transform in one and two dimensions, *IEEE Trans. on Acoust., Speech and Sig. Proc.*, Vol. ASSP-28, February 1980, pp. 27–34.

J. H. McClellan and C. M. Rader, *Number Theory in Digital Signal Processing.* Englewood Cliffs, NJ: Prentice-Hall, 1979.

R. M. Mersereau and T. C. Speake, A unified treatment of Cooley-Tukey algorithms for the evaluation of the multidimensional DFT, *IEEE Trans. on Acoust., Speech and Sig. Proc.*, Vol. ASSP-29, October 1981, pp. 1011–1018.

L. R. Morris, A comparative study of time efficient FFT and WFTA programs for general purpose computers, *IEEE Trans. on Acoust., Speech and Sig. Proc.*, Vol. ASSP-26, April 1978, pp. 141–150.

M. J. Narasimha and A. M. Peterson, On the computation of the discrete cosine transform, *IEEE Trans. Commun.*, Vol. COM-26, June 1978, pp. 934–936.

H. Nawab and J. H. McClellan, Bounds on the minimum number of data transfers in WFTA and FFT programs, *IEEE Trans. on Acoust., Speech and Sig. Proc.*, Vol. ASSP-27, August 1979, pp. 394–398.

H. J. Nussbaumer, Digital filtering using polynomial transforms, *Electron. Lett.*, Vol. 13, June 1977, pp. 386–387.

H. J. Nussbaumer, New polynomial transform algorithms for multidimensional DFT's and convolutions, *IEEE Trans. on Acoust., Speech and Sig. Proc.*, Vol. ASSP-29, February 1981, pp. 74–83.

H. J. Nussbaumer and P. Quandalle, Fast computation of discrete Fourier transforms using polynomial transforms, *IEEE Trans. on Acoust., Speech and Sig. Proc.*, April 1979, pp. 169–181.

I. Pitas and M. G. Strintzis, "General in-place calculation of discrete Fourier transforms of multidimensional sequences, *IEEE Trans. on Acoust., Speech and Sig. Proc.*, Vol. ASSP-34, June 1986, pp. 565–572.

C. M. Rader, Discrete Fourier transforms when the number of data samples is prime, *Proc. IEEE*, Vol. 56, June 1968, pp. 1107–1108.

G. E. Rivard, Direct fast Fourier transform of bivariate functions, *IEEE Trans. on Acoust., Speech and Sig. Proc.*, Vol. ASSP-25, June 1977, pp. 250–252.

H. F. Silverman, An introduction to programming the Winograd Fourier transform algorithm (WFTA), *IEEE Trans. on Acoust., Speech and Sig. Proc.*, Vol. ASSP-25, April 1977, pp. 152–165.

H. V. Sorensen et al., Real-valued fast Fourier transform algorithms, *IEEE Trans. on Acoust., Speech and Sig. Proc.*, Vol. ASSP-35, June 1987, pp. 849–863.

M. Vetterli and H. J. Nussbaumer, Simple FFT and DCT algorithms with reduced number of operations, *Signal Processing*, Vol. 6, 1984, pp. 267–278.

S. Winograd, On computing the discrete Fourier transform, *Proc. National Academy of Sciences*, USA, Vol. 73, April 1976, pp. 1005–1006.

S. Winograd, On computing the discrete Fourier transform, *Math. Comput.*, Vol. 32, January 1978, pp. 175–199.

PROBLEMS

3.1. Suppose $\tilde{x}(n_1, n_2)$ is a periodic sequence with a period of $N_1 \times N_2$. The sequence $\tilde{x}(n_1, n_2)$, then, is also periodic with a period of $2N_1 \times 2N_2$. Let $\tilde{X}(k_1, k_2)$ denote the discrete fourier series (DFS) coefficients of $\tilde{x}(n_1, n_2)$, considered as a periodic sequence with a period of $N_1 \times N_2$ points, and $\tilde{X}'(k_1, k_2)$ denote the DFS coefficients of $\tilde{x}(n_1, n_2)$, considered as a periodic sequence with a period of $2N_1 \times 2N_2$ points.
(a) Express $\tilde{X}'(k_1, k_2)$ in terms of $\tilde{X}(k_1, k_2)$.
(b) Verify your result in (a) when $\tilde{x}(n_1, n_2)$ is as shown below.

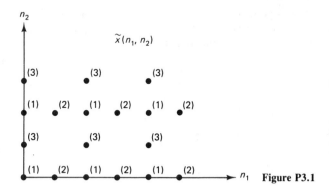

Figure P3.1

3.2. Prove Parseval's theorem, which states that

$$\sum_{n1=0}^{N_1-1} \sum_{n2=0}^{N_2-1} |\tilde{x}(n_1, n_2)|^2 = \frac{1}{N_1 N_2} \sum_{k1=0}^{N_1-1} \sum_{k2=0}^{N_2-1} |\tilde{X}(k_1, k_2)|^2,$$

where $\tilde{x}(n_1, n_2)$ is periodic with a period of $N_1 \times N_2$ and $\tilde{X}(k_1, k_2)$ is the $N_1 \times N_2$-point DFS of $\tilde{x}(n_1, n_2)$.

3.3. Let $\tilde{x}(n_1, n_2)$ denote a periodic sequence with a period of $N_1 \times N_2$ and $\tilde{X}(k_1, k_2)$ denote the $N_1 \times N_2$-point DFS of $\tilde{x}(n_1, n_2)$. Show each of the following properties.
(a) The DFS of $\tilde{x}^*(n_1, n_2)$ is $\tilde{X}^*(-k_1, -k_2)$.
(b) If $\tilde{x}(n_1, n_2)$ is real, $\tilde{X}(k_1, k_2) = \tilde{X}^*(-k_1, -k_2)$.
(c) If $\tilde{x}(n_1, n_2)$ is real, $|\tilde{X}(k_1, k_2)| = |\tilde{X}(-k_1, -k_2)|$.

3.4. The discrete Fourier transform (DFT) pair is given by Equation (3.7). Verify that the DFT pair is a consistent relationship.

3.5. The 2-D DFT pair that relates an $N_1 \times N_2$-point sequence $x(n_1, n_2)$ to its $N_1 \times N_2$-point DFT $X(k_1, k_2)$ can be expressed in a matrix form as

$$X = PxQ$$

$$x = P^{-1}XQ^{-1},$$

where X and x are $N_1 \times N_2$ matrices, P is an $N_1 \times N_1$ matrix, Q is an $N_2 \times N_2$ matrix, and P^{-1} and Q^{-1} are inverses of P and Q respectively. The element at the $(k_1 + 1)$th row and $(k_2 + 1)$th column of X is $X(k_1, k_2)$, and the element at the $(n_1 + 1)$th row and $(n_2 + 1)$th column of x is $x(n_1, n_2)$. Determine P and Q.

3.6. Let $x(n_1, n_2)$ denote an $N_1 \times N_2$-point sequence that is zero outside $0 \le n_1 \le N_1 - 1$, $0 \le n_2 \le N_2 - 1$. Let $X(k_1, k_2)$ denote the $N_1 \times N_2$-point DFT of $x(n_1, n_2)$. Determine the $N_1 \times N_2$-point DFT of $x(-n_1 + N_1 - 1, n_2)$.

3.7. Consider a 20×20-point finite-extent sequence $x(n_1, n_2)$ that is zero outside $0 \leq n_1 \leq 19$, $0 \leq n_2 \leq 19$, and let $X(\omega_1, \omega_2)$ represent the Fourier transform of $x(n_1, n_2)$.

(a) If we wish to evaluate $X(\omega_1, \omega_2)$ at $\omega_1 = (2\pi/3)k_1$ and $\omega_2 = (2\pi/3)k_2$ for all values of (k_1, k_2) by computing one $M \times M$-point DFT, what is the smallest value of M that can be used? Develop a method of computing the Fourier transform samples using the smallest possible M.

(b) If we wish to evaluate $X(\omega_1, \omega_2)$ at $\omega_1 = (10\pi/27)k_1$ and $\omega_2 = (10\pi/27)k_2$ for all values of (k_1, k_2) by computing one $L \times L$-point DFT, what is the smallest possible L we can use? Develop a method of computing the Fourier transform samples using the smallest L.

3.8. Consider a 4×4-point sequence $x(n_1, n_2)$ which is zero outside $0 \leq n_1 \leq 3$, $0 \leq n_2 \leq 3$. Develop a method of evaluating $X(\omega_1, \omega_2)$ at the following sixteen points in the (ω_1, ω_2) plane by computing only one 4×4-point DFT.

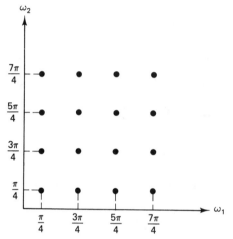

Figure P3.8

3.9. Consider the following two sequences, $x(n_1, n_2)$ and $h(n_1, n_2)$, sketched below:

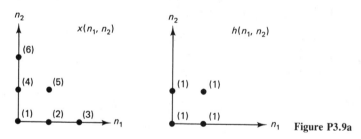

Figure P3.9a

(a) Determine $x(n_1, n_2) * h(n_1, n_2)$, the linear convolution of $x(n_1, n_2)$ and $h(n_1, n_2)$.

(b) Determine $x(n_1, n_2) \circledast h(n_1, n_2)$, with the assumed periodicity of 3×3 for the circular convolution.

(c) What is the minimum size of the assumed periodicity for the circular convolution for which $x(n_1, n_2) * h(n_1, n_2) = x(n_1, n_2) \circledast h(n_1, n_2)$?

(d) Determine $x(n_1, n_2) \circledast h(n_1, n_2)$ by computing

$$\text{IDFT}\left[\text{DFT}[x(n_1, n_2)]\text{DFT}[h(n_1, n_2)]\right]$$

with the assumed periodicity of the circular convolution given by your answer in (c). Is your answer the same as that in (a)?

(e) The DFT is defined only for first-quadrant support sequences. This is not a serious restriction in practice, however. To illustrate this, consider two new sequences, $x'(n_1, n_2)$ and $h'(n_1, n_2)$, shown below:

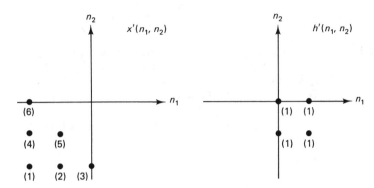

Figure P3.9b

Determine $x'(n_1, n_2) * h'(n_1, n_2)$, using your result in (d).

3.10. An FFT algorithm can be used in computing the result of linear convolution. Suppose we have a finite-extent sequence $x(n_1, n_2)$ that is zero outside $0 \leq n_1 \leq N_1 - 1, 0 \leq n_2 \leq N_2 - 1$. It is often useful to compute $R_x(n_1, n_2)$ defined by:

$$R_x(n_1, n_2) = x(n_1, n_2) * x(-n_1, -n_2).$$

Develop a procedure to compute $R_x(n_1, n_2)$ exploiting the computational efficiency of an FFT algorithm. Note that the DFT is not defined for $x(-n_1, -n_2)$.

3.11. Consider an image $x(n_1, n_2)$ of 512 × 512 pixels. We wish to filter $x(n_1, n_2)$ with a system whose impulse response is given by $h(n_1, n_2)$. The sequence $h(n_1, n_2)$ is an 11 × 11-point sequence that is zero outside $0 \leq n_1 \leq 10, 0 \leq n_2 \leq 10$. Let $y(n_1, n_2)$ and $w(n_1, n_2)$ be defined by

$$y(n_1, n_2) = x(n_1, n_2) * h(n_1, n_2)$$

and

$$w(n_1, n_2) = \text{IDFT}[\text{DFT}[x(n_1, n_2)]\text{DFT}[h(n_1, n_2)]],$$

where the DFT and IDFT are 512 × 512 points in size. For what values of (n_1, n_2) does $w(n_1, n_2)$ equal $y(n_1, n_2)$?

3.12. Consider an $N_1 \times N_2$-point image $x(n_1, n_2)$ which is zero outside $0 \leq n_1 \leq N_1 - 1$ and $0 \leq n_2 \leq N_2 - 1$. In transform coding, we discard the transform coefficients with small magnitudes and code only those with large magnitudes. Let $X(k_1, k_2)$ denote the $N_1 \times N_2$-point DFT of $x(n_1, n_2)$. Let $Y(k_1, k_2)$ denote $X(k_1, k_2)$ modified by

$$Y(k_1, k_2) = \begin{cases} X(k_1, k_2), & \text{when } |X(k_1, k_2)| \text{ is large} \\ 0, & \text{otherwise} \end{cases}$$

Let

$$\frac{\displaystyle\sum_{k_1=0}^{N_1-1} \sum_{k_2=0}^{N_2-1} |Y(k_1, k_2)|^2}{\displaystyle\sum_{k_1=0}^{N_1-1} \sum_{k_2=0}^{N_2-1} |X(k_1, k_2)|^2} = \frac{9}{10}.$$

We reconstruct an image $y(n_1, n_2)$ by computing the $N_1 \times N_2$-point inverse DFT of $Y(k_1, k_2)$. Express

$$\sum_{n1=0}^{N_1-1} \sum_{n2=0}^{N_2-1} (x(n_1, n_2) - y(n_1, n_2))^2$$

in terms of

$$\sum_{n1=0}^{N_1-1} \sum_{n2=0}^{N_2-1} x^2(n_1, n_2).$$

3.13 Let $x(n_1, n_2)$ be an $N_1 \times N_2$-point sequence that is zero outside $0 \le n_1 \le N_1 - 1$, $0 \le n_2 \le N_2 - 1$. We wish to express the $N_1 \times N_2$-point DFT of $x^*(n_1, n_2)$ in terms of $X(k_1, k_2)$, the $N_1 \times N_2$-point DFT of $x(n_1, n_2)$. Let $Y(k_1, k_2)$ denote the $N_1 \times N_2$-point DFT of $x^*(n_1, n_2)$. From the definition of the DFT,

$$Y(k_1, k_2) = \sum_{n1=0}^{N_1-1} \sum_{n2=0}^{N_2-1} x^*(n_1, n_2)e^{-j(2\pi/N_1)k_1n_1}e^{-j(2\pi/N_2)k_2n_2}. \tag{1}$$

Taking the complex conjugate of (1), we have

$$Y^*(k_1, k_2) = \sum_{n1=0}^{N_1-1} \sum_{n2=0}^{N_2-1} x(n_1, n_2)e^{j(2\pi/N_1)k_1n_1}e^{j(2\pi/N_2)k_2n_2}. \tag{2}$$

From the definition of the DFT,

$$X(k_1, k_2) = \sum_{n1=0}^{N_1-1} \sum_{n2=0}^{N_2-1} x(n_1, n_2)e^{-j(2\pi/N_1)k_1n_1}e^{-j(2\pi/N_2)k_2n_2}. \tag{3}$$

From (2) and (3), $\qquad Y^*(k_1, k_2) = X(-k_1, -k_2).$ \hfill (4)

From (4), $\qquad Y(k_1, k_2) = \mathrm{DFT}[x^*(n_1, n_2)] = X^*(-k_1, -k_2).$ \hfill (5)

(a) The result obtained in (5) is not correct. Which step in the above derivation is wrong?

(b) By relating the DFT to the DFS representation and using the properties of the DFS, express the DFT of $x^*(n_1, n_2)$ in terms of $X(k_1, k_2)$.

3.14. Determine a sequence $x(n_1, n_2)$ that satisfies all of the following conditions:

Condition 1. $X(\omega_1, \omega_2)$, the Fourier transform of $x(n_1, n_2)$, is in the form of

$$X(\omega_1, \omega_2) = A + B \cos(\omega_1 + \omega_2) + C \cos(\omega_1 - \omega_2)$$

where A, B, and C are all real numbers.

Condition 2. Let $Y(k_1, k_2)$ denote the 4×4-point DFT of $x(n_1 - 2, n_2 - 2)$. It is known that $Y(0, 0) = 13$.

Condition 3. Let $w(n_1, n_2)$ denote the following sequence:

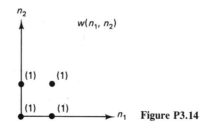

Figure P3.14

Let $v(n_1, n_2)$ denote $w(n_1, n_2) \circledast x(n_1 - 2, n_2 - 2)$ based on the assumed periodicity of 4×4. It is known that $v(0, 0) = 2$.

Condition 4. Let $s(n_1, n_2) = w(n_1, n_2) * x(n_1, n_2)$, where $w(n_1, n_2)$ is given in Condition 3. It is known that $s(1, 0) = 5$.

3.15. Consider the 2-D sequence sketched below.

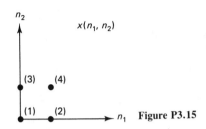

Figure P3.15

(a) Determine $C_x(k_1, k_2)$, the 2×2-point DCT of $x(n_1, n_2)$.
(b) For the sequence $x(n_1, n_2)$, verify the energy relationship given by Equation (3.43) with $N_1 = N_2 = 2$.

3.16. Let $x(n_1, n_2)$ be zero outside $0 \le n_1 \le N_1 - 1$, $0 \le n_2 \le N_2 - 1$. The sequence $x(n_1, n_2)$ is also a separable sequence so that

$$x(n_1, n_2) = x_1(n_1)x_2(n_2).$$

Show that $C_x(k_1, k_2) = C_{x_1}(k_1)C_{x_2}(k_2)$, where $C_x(k_1, k_2)$ is the 2-D $N_1 \times N_2$-point DCT of $x(n_1, n_2)$, $C_{x_1}(k_1)$ is the 1-D N_1-point DCT of $x_1(n_1)$, and $C_{x_2}(k_2)$ is the 1-D N_2-point DCT of $x_2(n_2)$.

3.17. Let $x(n_1, n_2)$ be an $N_1 \times N_2$-point sequence and $C_x(k_1, k_2)$ be an $N_1 \times N_2$-point DCT of $x(n_1, n_2)$.
(a) Show that the $N_1 \times N_2$-point DCT of $x^*(n_1, n_2)$ is $C_x^*(k_1, k_2)$.
(b) Show that if $x(n_1, n_2)$ is real, $C_x(k_1, k_2)$ will also be real.

3.18. Let $x(n_1, n_2)$ denote a first-quadrant support sequence, and let $C_x(\omega_1, \omega_2)$ denote the cosine transform of $x(n_1, n_2)$. Show each of the following symmetry properties.
(a) $C_x(\omega_1, \omega_2) = C_x(-\omega_1, \omega_2) = C_x(\omega_1, -\omega_2) = C_x(-\omega_1, -\omega_2)$.
(b) The cosine transform of $x^*(n_1, n_2)$ is given by $C_x^*(\omega_1, \omega_2)$.
(c) $C_x(\omega_1, \omega_2)$ of a real sequence $x(n_1, n_2)$ is real.

3.19. Let $x(n_1, n_2)$ denote a first-quadrant support sequence, and let $C_x(\omega_1, \omega_2)$ denote the cosine transform of $x(n_1, n_2)$. Show that

$$\sum_{n_1=0}^{\infty} \sum_{n_2=0}^{\infty} |x(n_1, n_2)|^2 = \frac{1}{4(2\pi)^2} \int_{\omega_1=-\pi}^{\pi} \int_{\omega_2=-\pi}^{\pi} |C_x(\omega_1, \omega_2)|^2 \, d\omega_1 \, d\omega_2.$$

3.20. Let $x(n)$ denote an N-point sequence that is zero outside $0 \le n \le N - 1$. In computing the N-point DCT of $x(n)$, the major computations involved are the $2N$-point DFT computation of $y(n)$, which is related to $x(n)$ by

$$y(n) = x(n) + x(2N - 1 - n).$$

In this problem, we show that the $2N$-point DFT $Y(k)$ can be computed by computing one N-point DFT.

(a) We first divide $y(n)$ into two N-point sequences $v(n)$ and $w(n)$ by

$$v(n) = y(2n), \qquad 0 \le n \le N - 1,$$
$$w(n) = y(2n + 1), \qquad 0 \le n \le N - 1.$$

By sketching an example of $y(n)$, $v(n)$, and $w(n)$, illustrate that

$$w(n) = v(N - 1 - n), \qquad 0 \le n \le N - 1.$$

(b) From the definition of $Y(k)$,

$$Y(k) = \sum_{n=0}^{2N-1} y(n) W_{2N}^{kn},$$

where

$$W_{2N}^{kn} = e^{-j(2\pi/2N)kn}.$$

Substituting $y(n)$ in the above expression with $v(n)$ and $w(n)$, show that

$$Y(k) = \sum_{n=0}^{N-1} v(n) W_N^{kn} + W_{2N}^{k} \sum_{n=0}^{N-1} w(n) W_N^{kn}, \qquad 0 \le k \le 2N - 1.$$

(c) Combining the results of (a) and (b), show that

$$Y(k) = \sum_{n=0}^{N-1} v(n) W_N^{kn} + W_{2N}^{-k} \sum_{n=0}^{N-1} v(n) W_N^{-kn}, \qquad 0 \le k \le 2N - 1.$$

(d) We define $V(k)$, the N-point DFT of $v(n)$, by

$$V(k) = \sum_{n=0}^{N-1} v(n) W_N^{kn}, \qquad 0 \le k \le N - 1.$$

From (c), express $Y(k)$ for $0 \le k \le N - 1$ in terms of $V(k)$ for $0 \le k \le N - 1$. The above result shows that the $2N$-point DFT $Y(k)$ for $0 \le k \le N - 1$ can be computed by computing one N-point DFT $V(k)$. We note that the N-point DCT $C_x(k)$ can be computed from $Y(k)$ for $0 \le k \le N - 1$, and $Y(k)$ for $N \le k \le 2N - 1$ does not have to be computed. This problem shows that the computations involved in computing the N-point DCT of an N-point sequence are similar to those of computing the N-point DFT of the N-point sequence.

(e) The major computation involved in computing the inverse DCT of $C_x(k)$ is the $2N$-point inverse DFT of $Y(k)$. To show that the $2N$-point inverse DFT of $Y(k)$ can be computed by computing one N-point inverse DFT, express $V(k)$ in terms of $Y(k)$. From $V(k)$, $v(n)$ can be obtained by computing the N-point inverse DFT of $V(k)$. From $v(n)$, $y(n)$ and therefore $x(n)$ can be obtained by using the result in (a). Alternatively, $x(n)$ can be obtained directly from $v(n)$.

3.21. In Problem 3.20, we showed that the computations involved in computing the N-point DCT of an N-point 1-D sequence are similar to those involved in computing the N-point DFT of the N-point sequence. In this problem, we show that this result can be extended to the 2-D case straightforwardly. Let $x(n_1, n_2)$ denote an $N_1 \times N_2$-

point sequence that is zero outside $0 \leq n_1 \leq N_1 - 1$ and $0 \leq n_2 \leq N_2 - 1$. We discussed in Section 3.3.2 that the major computations involved in computing the $N_1 \times N_2$-point DFT of $x(n_1, n_2)$ are the $2N_1 \times 2N_2$-point DFT computation of $y(n_1, n_2)$, which is related to $x(n_1, n_2)$ by

$$y(n_1, n_2) = x(n_1, n_2) + x(2N_1 - 1 - n_1, n_2) + x(n_1, 2N_2 - 1 - n_2)$$
$$+ x(2N_1 - 1 - n_1, 2N_2 - 1 - n_2).$$

To show that the $2N_1 \times 2N_2$-point DFT $Y(k_1, k_2)$ can be computed by computing one $N_1 \times N_2$-point DFT, we first divide $y(n_1, n_2)$ into four $N_1 \times N_2$-point sequences $v(n_1, n_2)$, $w_1(n_1, n_2)$, $w_2(n_1, n_2)$ and $w_3(n_1, n_2)$ by

$$\left. \begin{array}{l} v(n_1, n_2) = y(2n_1, 2n_2) \\ w_1(n_1, n_2) = y(2n_1 + 1, 2n_2) \\ w_2(n_1, n_2) = y(2n_1, 2n_2 + 1) \\ w_3(n_1, n_2) = y(2n_1 + 1, 2n_2 + 1) \end{array} \right\}, \quad 0 \leq n_1 \leq N_1 - 1, \quad 0 \leq n_2 \leq N_2 - 1.$$

(a) By following steps analogous to those taken in Problem 3.20, show that $Y(k_1, k_2)$ for $0 \leq k_1 \leq N_1 - 1$, $0 \leq k_2 \leq N_2 - 1$ can be expressed in terms of $V(k_1, k_2)$, the $N_1 \times N_2$-point DFT of $v(n_1, n_2)$. This problem shows that the computations involved in computing the $N_1 \times N_2$-point DCT of an $N_1 \times N_2$-point sequence are similar to those involved in computing the $N_1 \times N_2$-point DFT of the $N_1 \times N_2$-point sequence.

(b) The major computations involved in computing the inverse DCT of $C_x(k_1, k_2)$ are those in the $2N_1 \times 2N_2$-point inverse DFT of $Y(k_1, k_2)$. To show that the $2N_1 \times 2N_2$-point inverse DFT of $Y(k_1, k_2)$ can be computed by computing one $N_1 \times N_2$-point inverse DFT, express $V(k_1, k_2)$ in terms of $Y(k_1, k_2)$. From $v(n_1, n_2)$, $x(n_1, n_2)$ can be obtained directly.

3.22. Let $x(n)$ denote an N-point 1-D sequence that is zero outside $0 \leq n \leq N - 1$. In defining the DCT of $x(n)$, we first related $x(n)$ to $y(n)$ by

$$y(n) = \begin{cases} x(n), & 0 \leq n \leq N-1 \\ x(2N-1-n), & N \leq n \leq 2N-1. \end{cases}$$

We then related $y(n)$ to its DFT $Y(k)$, and then related $Y(k)$ to the N-point DCT $C_x(k)$. It is possible to define the DCT in a slightly different way. We first relate $x(n)$ to $r(n)$ by

$$r(n) = \begin{cases} x(n), & 0 \leq n \leq N-1 \\ x(2N-2-n), & N \leq n \leq 2N-2. \end{cases}$$

Note that $r(n)$ is a $2N-1$-point sequence, while $y(n)$ is a $2N$-point sequence. We then relate $r(n)$ to its $2N-1$-point DFT $R(k)$, which is then related to the N-point DCT of $x(n)$. Since $r(n)$ is an odd-length sequence, the resulting DCT $C_x'(k)$ is said to be an *odd symmetrical* DCT.

(a) Sketch an example of $x(n)$ and $r(n)$, and show that repeating $r(n)$ every $2N-1$-points does not result in artificial boundary discontinuities.

(b) Determine the relationship between $R(k)$ and $C_x'(k)$. Note that we want to ensure that $C_x'(k)$ will be real for a real sequence $x(n)$.

(c) Express the DCT $C_x'(k)$ in terms of $x(n)$.

(d) Express $x(n)$ in terms of $C_x'(k)$.

(e) Express $\sum_{n=0}^{N-1} |x(n)|^2$ in terms of $C_x'(k)$.

The results in this problem can be extended straightforwardly to define the odd symmetrical DCT for a 2-D sequence.

3.23. Let $x(n_1, n_2)$ denote an $N_1 \times N_2$-point sequence and let $X(k_1, k_2)$ denote the $N_1 \times N_2$-point DFT of $x(n_1, n_2)$. In Section 3.4.1, we showed that $X(k_1, k_2)$ can be computed by performing first row operations and then column operations. Show that $X(k_1, k_2)$ can be computed by column operations first and then row operations.

3.24. One approach to efficiently computing a 2-D DFT is the row-column approach, which involves many 1-D DFT computations. A 2-D DCT can also be computed efficiently with a row-column approach that involves many 1-D DCT computations.

(a) Show that the row-column approach can be used to compute the 2-D DCT.

(b) Show that the row-column decomposition alone significantly reduces computations by comparing the computational requirements for direct computation of an $N_1 \times N_2$-point DCT with that of the row-column approach, where 1-D DCTs are computed directly.

3.25. Consider a sequence $x(n_1, n_2)$ given by

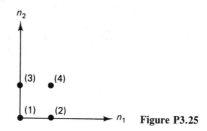

Figure P3.25

Let $X(k_1, k_2)$ denote the 2×2-point DFT of $x(n_1, n_2)$.

(a) Compute $X(k_1, k_2)$ directly from its definition.

(b) Compute $X(k_1, k_2)$ by the row-column decomposition method.

3.26. Consider a digital image of 512×512 pixels. We wish to process this image with a lowpass filter whose impulse response has a support of 16×16-points. Let N_1, N_2, and N_3 represent the following numbers:

N_1 = total number of complex multiplications when the convolution is performed directly

N_2 = total number of complex multiplications when the overlap-add method is used with a 64×64-point DFT and inverse DFT

N_3 = total number of complex multiplications when the overlap-save method is used with a 64×64-point DFT and inverse DFT

(a) Determine N_2/N_1.

(b) Determine N_3/N_1.

In your calculation, assume that both the image and the filter impulse response are complex sequences. Assume also that the DFT and inverse DFT are computed using the row-column decomposition method with a decimation-in-time 1-D FFT algorithm.

3.27. Let $x(n_1, n_2)$ denote a complex $N \times N$-point sequence with $N = 3^M$. We wish to compute $X(k_1, k_2)$, the $N \times N$-point DFT of $x(n_1, n_2)$, by using the row-column decomposition approach. Unfortunately, $x(n_1, n_2)$ is stored row by row in a slow disk memory and we have only $3N$ words of fast memory available to us. To reduce the number of I/O operations, we need an efficient way to transpose $x(n_1, n_2)$. Develop an efficient way to transpose $x(n_1, n_2)$ and determine the number of I/O operations required in your method. One I/O operation corresponds to reading or writing one row of data. Show concisely and clearly how you obtained the number of I/O operations. You may want to use a simple sketch.

3.28. We wish to filter a complex sequence $x(n_1, n_2)$ with an LSI system whose impulse response is given by the complex sequence $h(n_1, n_2)$. Answer the following questions.
(a) Suppose the size of $x(n_1, n_2)$ is 200×200 points and the size of $h(n_1, n_2)$ is 40×40 points. If the sequence is filtered by using DFT and inverse DFT operations to exploit the computational efficiency of the 1-D FFT through the row-column decomposition approach, how many complex multiplications are required? Assume that a radix-2 1-D FFT routine is used, and only one inverse DFT operation is allowed.
(b) Suppose we wish to compute an $N \times N$-point DFT with $N = 2^M$ for some integer M, using the row-column decomposition approach. If slow memory, for example disk memory, is used for the DFT computation, it is important to develop algorithms that will transpose the data in the slow memory with a small number of I/O operations. Suppose fast memory of $N \times K$ words in size is available for the DFT computation. If $K = 2$, Eklundh's method requires $2N \log_2 N$ I/O operations, where one I/O operation corresponds to reading or writing one row of data. If $K > 2$, then further reduction in the required I/O operations is possible. Determine the number of I/O operations required when $K = 2^L$ for an integer $L \le M$.

3.29. Let $x(n_1, n_2)$ denote a complex $N \times N$-point sequence with $N = 3^M$. We wish to use a decimation-in-time vector radix FFT to compute the $N \times N$-point DFT $X(k_1, k_2)$. We consider the first stage in the development of a decimation-in-time 9 $\times$ 9-point DFT computation. Sketch a figure analogous to Figure 3.23 that shows the first stage in the development.

3.30. Sketch a general vector radix 2×3 butterfly computation.

3.31. Suppose we have a complex $N_1 \times N_2$-point sequence $x(n_1, n_2)$ that is zero outside $0 \le n_1 \le N_1 - 1, 0 \le n_2 \le N_2 - 1$. Let $X(k_1, k_2)$ denote the $N_1 \times N_2$-point DFT of $x(n_1, n_2)$. Suppose $N_1 = 3^M$ and $N_2 = 2^M$. Determine the number of complex additions required to compue $X(k_1, k_2)$, using a decimation-in-time vector radix FFT algorithm.

3.32. Consider an $N \times N$-point sequence $x(n_1, n_2)$, where $N = 2^M$ for some integer M. In Section 3.4.3, we developed a vector radix FFT algorithm by extending the decimation-in-time FFT algorithm to the 2-D case. It is also possible to develop a vector-radix FFT algorithm by extending the decimation-in-frequency FFT algorithm to the 2-D case. Show the first stage of this development and derive a figure analogous to Figure 3.23.

3.33. One idea exploited in the development of the prime factor algorithm (PFA) and the Winograd Fourier transform algorithm (WFTA) is the concept that an N-point DFT computation for the case where N is a prime or a power of a prime number can be expressed as a circular convolution. Consider the 5-point sequence $x(n)$ sketched below.

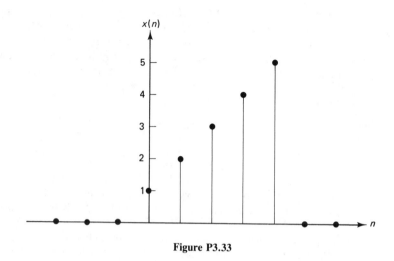

Figure P3.33

Using the results in Section 3.4.4, verify that the 5-point DFT $X(k)$ can be computed by a circular convolution.

4

Finite Impulse Response Filters

4.0 INTRODUCTION

Three steps are generally followed in using digital filters. In the first step, we specify the characteristics required of the filter. Filter specification depends, of course, on the intended application. For example, if we wish to restore a signal that has been degraded by background noise, the filter characteristics required depend on the spectral characteristics of the signal and the background noise. The second step is filter design. In this step, we determine $h(n_1, n_2)$, the impulse response of the filter, or its system function $H(z_1, z_2)$, that will meet the design specification. The third step is filter implementation, in which we realize a discrete system with the given $h(n_1, n_2)$ or $H(z_1, z_2)$.

These three steps are interrelated. It does not make much sense, for example, to specify a filter that cannot be designed. Nor is it worthwhile to design a filter that cannot be implemented. However, for convenience, we will discuss the three steps separately. From time to time, though, we will point out their interrelationships.

For practical reasons, we will restrict ourselves to a certain class of digital filters. One restriction is that $h(n_1, n_2)$ must be real. In practice, we will deal with real data. To ensure that the processed signal will be real, we will require $h(n_1, n_2)$ to be real. Another restriction is the stability of $h(n_1, n_2)$; that is, $\sum_{n_1=-\infty}^{\infty} \sum_{n_2=-\infty}^{\infty} |h(n_1, n_2)| < \infty$. In practice, an unbounded output can cause many difficulties, for example, system overload. We will restrict our discussion, then, to the class of digital filters whose impulse response $h(n_1, n_2)$ is real and stable.

Digital filters can be classified into two groups. In the first group, $h(n_1, n_2)$ is a finite-extent sequence, so the filters in this group are called *finite impulse response* (FIR) filters. In the second group, $h(n_1, n_2)$ is of infinite extent, so the filters in this group are called *infinite impulse response* (IIR) filters. We concentrate

195

on FIR filters in this chapter. IIR filters are discussed in Chapter 5. In 2-D, as in 1-D, FIR filters differ considerably in design and implementation from IIR filters. A major advantage of the FIR filter is that its stability is never an issue in design and implementation.

In Section 4.1, we discuss some FIR filter characteristics and properties. We refer to these properties throughout the chapter. In Section 4.2, we discuss the filter specification problem. In Sections 4.3, 4.4, and 4.5, we discuss different methods of FIR filter design. In Section 4.3, we discuss the window method and frequency sampling method, which are in many respects similar to their 1-D counterparts. Our discussions here are relatively brief. In Section 4.4, we discuss the frequency transformation method for filter design. Since this method is useful in practice and has no 1-D counterpart, we discuss it in some detail. In Section 4.5, we discuss the optimal (in the Chebyshev sense) filter design problem. We find that the 1-D results do not extend to 2-D, and that developing a practical, reliable method remains an area for further research. Section 4.6 treats the problem of realizing an FIR filter.

4.1 ZERO-PHASE FILTERS

A digital filter $h(n_1, n_2)$ is said to have *zero phase* when its frequency response $H(\omega_1, \omega_2)$ is a real function so that

$$H(\omega_1, \omega_2) = H^*(\omega_1, \omega_2). \tag{4.1}$$

Strictly speaking, not all filters with real frequency responses are necessarily zero-phase filters, since $H(\omega_1, \omega_2)$ can be negative. In practice, the frequency regions for which $H(\omega_1, \omega_2)$ is negative typically correspond to the stopband regions, and a phase of 180° in the stopband regions has little significance.

From the symmetry properties of the Fourier transform, (4.1) is equivalent in the spatial domain to the following:

$$h(n_1, n_2) = h^*(-n_1, -n_2). \tag{4.2}$$

Since we consider only real $h(n_1, n_2)$, (4.2) reduces to

$$h(n_1, n_2) = h(-n_1, -n_2). \tag{4.3}$$

Equation (4.3) states that the impulse response of a zero-phase filter is symmetric with respect to the origin.

One characteristic of a zero-phase filter is its tendency to preserve the shape of the signal component in the passband region of the filter. In applications such as speech processing, the zero-phase (or linear phase) characteristic of a filter is not very critical. The human auditory system responds to short time spectral magnitude characteristics, so the shape of a speech waveform can sometimes change drastically without the human listener's being able to distinguish it from the original. In image processing, the linear phase characteristic appears to be more important. Our visual world consists of lines, scratches, etc. A nonlinear phase distorts the proper registration of different frequency components that make up the lines and scratches. This distorts the signal shape in various ways, including blurring. Figure 4.1 illustrates the significance of the zero-phase characteristic in images. Figure 4.1(a) shows an original image of 512 × 512 pixels. Figure 4.1(b) shows the image processed by applying a zero-phase lowpass filter to the image in Figure 4.1(a). Figure 4.1(c) shows the image processed by applying a nonzero-phase lowpass filter

(a)

(b)

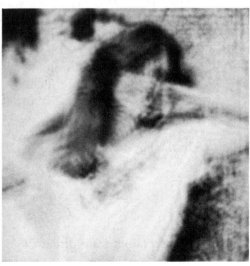

(c)

Figure 4.1 Illustration of the importance of zero-phase filters in image processing. (a) Original image of 512 × 512 pixels; (b) processed image by a zero-phase lowpass filter; (c) processed image by a nonzero-phase lowpass filter.

to the image in Figure 4.1(a). The magnitude responses of the two lowpass filters used in Figures 4.1(b) and (c) are approximately the same. The zero-phase characteristic is quite useful in applications such as image processing, and zero phase is very easy to achieve with FIR filters, due to (4.3). In addition, design and implementation are often simplified if we require zero phase. For these reasons, we will restrict our discussion of FIR filters to zero-phase filters.

Consider a zero-phase impulse response $h(n_1, n_2)$. From (4.3) and the Fourier transform definition of (1.31), the frequency response $H(\omega_1, \omega_2)$ can be expressed as

$$
\begin{aligned}
H(\omega_1, \omega_2) &= \sum_{(n_1, n_2) \in R_h} \sum h(n_1, n_2) e^{-j\omega_1 n_1} e^{-j\omega_2 n_2} \\
&= h(0, 0) + \sum_{(n_1, n_2) \in R'_h} \sum h(n_1, n_2) e^{-j\omega_1 n_1} e^{-j\omega_2 n_2} \qquad (4.4) \\
&+ \sum_{(n_1, n_2) \in R''_h} \sum h(n_1, n_2) e^{-j\omega_1 n_1} e^{-j\omega_2 n_2}
\end{aligned}
$$

where R_h is the region of support of $h(n_1, n_2)$ and consists of three mutually exclusive regions: $(0, 0)$, R'_h and R''_h. The region R''_h is R'_h flipped with respect to the origin. Combining the two terms in the right-hand side expression of (4.4) and noting that $h(n_1, n_2) = h(-n_1, -n_2)$, we can express (4.4) as

$$
\begin{aligned}
H(\omega_1, \omega_2) &= h(0, 0) + \sum_{(n_1, n_2) \in R'_h} \sum (h(n_1, n_2) e^{-j\omega_1 n_1} e^{-j\omega_2 n_2} + h(-n_1, -n_2) e^{j\omega_1 n_1} e^{j\omega_2 n_2}) \\
&= h(0, 0) + \sum_{(n_1, n_2) \in R'_h} \sum 2h(n_1, n_2) \cos (\omega_1 n_1 + \omega_2 n_2). \qquad (4.5)
\end{aligned}
$$

From (4.5), $H(\omega_1, \omega_2)$ for a zero-phase FIR filter can always be expressed as a linear combination of cosine terms of the form $\cos (\omega_1 n_1 + \omega_2 n_2)$.

The zero-phase filter $h(n_1, n_2)$ is symmetric with respect to the origin, so approximately half of the points in $h(n_1, n_2)$ are independent. As a result, the filter is said to have twofold symmetry. In some applications, such as the design of circularly symmetric filters, it may be useful to impose additional symmetries on $h(n_1, n_2)$. One such constraint is a fourfold symmetry given by

$$
h(n_1, n_2) = h(-n_1, n_2) = h(n_1, -n_2). \qquad (4.6)
$$

In the Fourier transform domain, (4.6) is equivalent to

$$
H(\omega_1, \omega_2) = H(-\omega_1, \omega_2) = H(\omega_1, -\omega_2). \qquad (4.7)
$$

Another such symmetry constraint is an eightfold symmetry given by

$$
h(n_1, n_2) = h(-n_1, n_2) = h(n_1, -n_2) = h(n_2, n_1). \qquad (4.8)
$$

In the Fourier transform domain, (4.8) is equivalent to

$$
H(\omega_1, \omega_2) = H(-\omega_1, \omega_2) = H(\omega_1, -\omega_2) = H(\omega_2, \omega_1). \qquad (4.9)
$$

The independent points of $h(n_1, n_2)$ in the twofold, fourfold, and eightfold symmetries are shown in Figure 4.2 for the case when the region of support of $h(n_1, n_2)$ is 11×11 points with square shape. For a given region of support

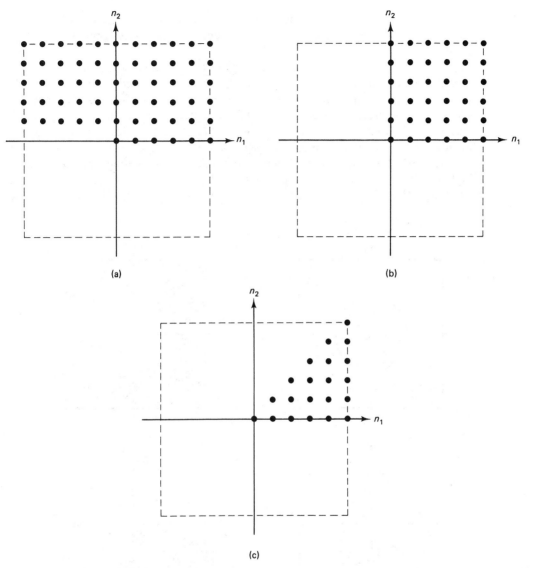

Figure 4.2 Independent points of an 11×11-point $h(n_1, n_2)$ with (a) twofold, (b) fourfold, and (c) eightfold symmetry.

of $h(n_1, n_2)$, imposing symmetry constraints reduces the number of independent parameters that must be estimated in the design and reduces the number of arithmetic operations required in the implementation.

4.2 FILTER SPECIFICATION

Like 1-D digital filters, 2-D digital filters are generally specified in the frequency domain. Since $H(\omega_1, \omega_2) = H(\omega_1 + 2\pi, \omega_2) = H(\omega_1, \omega_2 + 2\pi)$ for all (ω_1, ω_2), $H(\omega_1, \omega_2)$ for $-\pi \le \omega_1 \le \pi$, $-\pi \le \omega_2 \le \pi$ completely specifies $H(\omega_1, \omega_2)$. In addition, since $h(n_1, n_2)$ is assumed real, $H(\omega_1, \omega_2) = H^*(-\omega_1, -\omega_2)$. Specifying $H(\omega_1, \omega_2)$ for $-\pi \leq \omega_1 \leq \pi$, $0 \leq \omega_2 \leq \pi$ therefore completely specifies $H(\omega_1, \omega_2)$ for all (ω_1, ω_2).

A filter is said to have a circularly symmetric frequency response $H(\omega_1, \omega_2)$ if $H(\omega_1, \omega_2)$ is a function of $\omega_1^2 + \omega_2^2$ for $\sqrt{\omega_1^2 + \omega_2^2} \leq \pi$ and is constant outside the region within $-\pi \leq \omega_1 \leq \pi$, $-\pi \leq \omega_2 \leq \pi$. A filter is said to have a circularly symmetric impulse response $h(n_1, n_2)$ if $h(n_1, n_2)$ is a function of $n_1^2 + n_2^2$. As we discussed in Section 1.3.3, circular symmetry of $H(\omega_1, \omega_2)$ implies circular symmetry of $h(n_1, n_2)$. Circular symmetry of $h(n_1, n_2)$ does not, however, imply circular symmetry of $H(\omega_1, \omega_2)$. Frequency responses of circularly symmetric ideal lowpass, highpass, bandpass, and bandstop filters are shown in Figures 4.3(a), (b), (c), and (d), respectively. The shaded regions in the figures have amplitude 1, and the

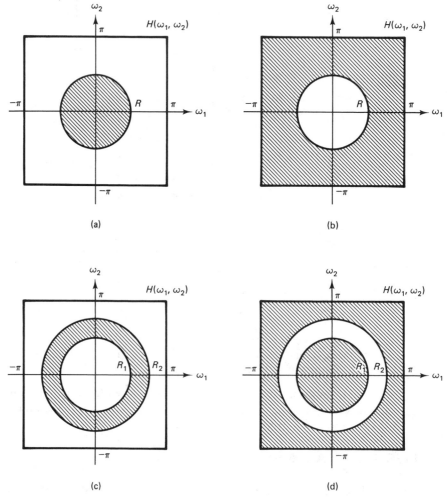

Figure 4.3 Frequency responses of circularly symmetric ideal filters. (a) Lowpass filter; (b) highpass filter; (c) bandpass filter; (d) bandstop filter.

unshaded regions have amplitude 0. Their corresponding impulse responses are given by

$$h_{lp}(n_1, n_2) = \frac{R}{2\pi\sqrt{n_1^2 + n_2^2}} J_1(R\sqrt{n_1^2 + n_2^2}) \qquad (4.10a)$$

$$h_{hp}(n_1, n_2) = \delta(n_1, n_2) - h_{lp}(n_1, n_2) \qquad (4.10b)$$

$$h_{bp}(n_1, n_2) = \frac{R_2}{2\pi\sqrt{n_1^2 + n_2^2}} J_1(R_2\sqrt{n_1^2 + n_2^2}) \qquad (4.10c)$$

$$- \frac{R_1}{2\pi\sqrt{n_1^2 + n_2^2}} J_1(R_1\sqrt{n_1^2 + n_2^2})$$

and $$\qquad h_{bs}(n_1, n_2) = \delta(n_1, n_2) - h_{bp}(n_1, n_2) \qquad (4.10d)$$

where $h_{lp}(n_1, n_2)$, $h_{hp}(n_1, n_2)$, $h_{bp}(n_1, n_2)$ and $h_{bs}(n_1, n_2)$ represent the lowpass, highpass, bandpass, and bandstop filters, respectively, and $J_1(x)$ is the Bessel function of the first kind and the first order. Equation (4.10) follows directly from (1.36). In a sense, a circularly symmetric filter does not give preferential treatment to any particular direction in the frequency domain. When we refer to such filters as ideal lowpass filters, circular symmetry is generally assumed.

Since $H(\omega_1, \omega_2)$ is in general a complex function of (ω_1, ω_2), we must specify both the magnitude and the phase of $H(\omega_1, \omega_2)$. For FIR filters, we require zero phase, so we need only to specify the magnitude response. The method most commonly used for magnitude specification is called the *tolerance scheme*. To illustrate this scheme, let us consider the specification of a lowpass filter. Ideally, a lowpass filter has only a *passband region* and a *stopband region*. In practice, a sharp transition between the two regions cannot be achieved; the passband region corresponds to $(\omega_1, \omega_2) \in R_p$ and the stopband region corresponds to $(\omega_1, \omega_2) \in R_s$, as shown in Figure 4.4. The frequency region R_t between the passband and the

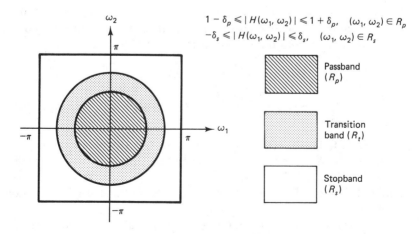

$$1 - \delta_p \leqslant |H(\omega_1, \omega_2)| \leqslant 1 + \delta_p, \quad (\omega_1, \omega_2) \in R_p$$
$$-\delta_s \leqslant |H(\omega_1, \omega_2)| \leqslant \delta_s, \quad (\omega_1, \omega_2) \in R_s$$

Passband (R_p)

Transition band (R_t)

Stopband (R_s)

Figure 4.4 Example of a 2-D lowpass filter specification using a tolerance scheme.

stopband regions is called the *transition band*. Ideally, the magnitude response $|H(\omega_1, \omega_2)|$ is unity in the passband region and zero in the stopband region. In practice, we require $1 - \delta_p < |H(\omega_1, \omega_2)| < 1 + \delta_p$ for $(\omega_1, \omega_2) \in R_p$ and $|H(\omega_1, \omega_2)| < \delta_s$ for $(\omega_1, \omega_2) \in R_s$. The variables δ_p and δ_s are called *passband tolerance* and *stopband tolerance*, respectively. A lowpass filter, then, is completely specified by δ_p, δ_s, R_p, and R_s. Other filters can also be specified in an analogous manner. The choice of the filter specification parameters depends on the intended application.

4.3 FILTER DESIGN BY THE WINDOW METHOD AND THE FREQUENCY SAMPLING METHOD

The problem of designing a filter is basically one of determining $h(n_1, n_2)$ or $H(z_1, z_2)$ that meets the design specification. The four standard approaches to designing FIR filters are the window method, the frequency sampling method, the frequency transformation method, and optimal filter design. The window and the frequency sampling methods are, in many respects, straightforward extensions of 1-D results. They are the topics of this section. Our discussion of these two methods is relatively brief. Other design methods are discussed in later sections.

4.3.1 The Window Method

In the window method, the desired frequency response $H_d(\omega_1, \omega_2)$ is assumed to be known. By inverse Fourier transforming $H_d(\omega_1, \omega_2)$, we can determine the desired impulse response of the filter, $h_d(n_1, n_2)$. In general, $h_d(n_1, n_2)$ is an infinite-extent sequence. In the window method, an FIR filter is obtained by multiplying $h_d(n_1, n_2)$ with a window $w(n_1, n_2)$; that is,

$$h(n_1, n_2) = h_d(n_1, n_2)w(n_1, n_2). \tag{4.11}$$

If $h_d(n_1, n_2)$ and $w(n_1, n_2)$ are both symmetric with respect to the origin, $h_d(n_1, n_2)w(n_1, n_2)$ will be also, so the resulting filter will be zero phase.

From (4.11) and the Fourier transform properties,

$$H(\omega_1, \omega_2) = H_d(\omega_1, \omega_2) \circledast W(\omega_1, \omega_2) \tag{4.12}$$

$$= \frac{1}{(2\pi)^2} \int_{\theta_1 = -\pi}^{\pi} \int_{\theta_2 = -\pi}^{\pi} H_d(\theta_1, \theta_2)W(\omega_1 - \theta_1, \omega_2 - \theta_2)\, d\theta_1\, d\theta_2.$$

From (4.12), the effect of the window in the frequency domain is to smooth $H_d(\omega_1, \omega_2)$. We like to have the mainlobe width of $W(\omega_1, \omega_2)$ small so that the transition width of $H(\omega_1, \omega_2)$ is small. We also want to have small sidelobe amplitude to ensure that the ripples in the passband and stopband regions will have small amplitudes.

A 2-D window used in filter design is typically obtained from a 1-D window.

We will discuss two specific methods. The first method is to obtain a 2-D window $w(n_1, n_2)$ by

$$w(n_1, n_2) = w_1(n_1)w_2(n_2) = w_c(t_1, t_2)|_{t_1 = n_1, t_2 = n_2} \qquad (4.13a)$$

$$w_c(t_1, t_2) = w_a(t_1)w_b(t_2). \qquad (4.13b)$$

The functions $w_a(t_1)$ and $w_b(t_2)$ in (4.13b) are 1-D analog windows. We can think of $w_1(n_1)$ and $w_2(n_2)$ as 1-D window sequences, or we can think of them as samples of 1-D analog window functions. The resulting sequence $w(n_1, n_2)$ is separable, and its Fourier transform $W(\omega_1, \omega_2)$ is very simply related to $W_1(\omega_1)$ and $W_2(\omega_2)$ by

$$W(\omega_1, \omega_2) = W_1(\omega_1)W_2(\omega_2) \qquad (4.14)$$

where $W_1(\omega_1)$ and $W_2(\omega_2)$ are 1-D Fourier transforms of $w_1(n_1)$ and $w_2(n_2)$, respectively. Because of (4.14), it is relatively straightforward to relate the sidelobe behavior and mainlobe behavior of $w(n_1, n_2)$ to those of $w_1(n_1)$ and $w_2(n_2)$. If the frequency response of the desired filter is a separable function, one simple approach to design a 2-D filter $h(n_1, n_2)$ is to design two 1-D filters $h_1(n_1)$ and $h_2(n_2)$ and multiply them. If each of the two 1-D filters is designed using the window method, the approach is equivalent to using a separable 2-D window in (4.13) in the 2-D filter design.

The second popular method [Huang] is to obtain a 2-D window $w(n_1, n_2)$ by

$$w(n_1, n_2) = w_c(t_1, t_2)|_{t_1 = n_1, t_2 = n_2} \qquad (4.15a)$$

where

$$w_c(t_1, t_2) = w_a(t)|_{t = \sqrt{t_1^2 + t_2^2}}. \qquad (4.15b)$$

The function $w_a(t)$ in (4.15b) is a 1-D analog window. In this method, a 2-D analog window $w_c(t_1, t_2)$ is obtained by rotating a 1-D analog window $w_a(t)$. Note that $W_c(\Omega_1, \Omega_2)$, the 2-D analog Fourier transform of $w_c(t_1, t_2)$, is a circularly symmetric function. It is not, however, a rotated version of $W_a(\Omega)$, the 1-D analog Fourier transform of $w_a(t)$. Specifically, $W_c(\Omega_1, \Omega_2)$ is related to $w_a(t)$ by

$$W_c(\Omega_1, \Omega_2) = G(\rho)|_{\rho = \sqrt{\Omega_1^2 + \Omega_2^2}} = 2\pi \int_{t=0}^{\infty} t w_a(t) J_0(t\rho) \, dt|_{\rho = \sqrt{\Omega_1^2 + \Omega_2^2}} \qquad (4.16)$$

where $J_0(\cdot)$ is the Bessel function of the first kind, zeroth order. The function $G(\rho)$ in (4.16) is the Hankel transform (see Problem 1.28 in Chapter 1) of $w_a(t)$. To obtain a 2-D window $w(n_1, n_2)$, the rotated 2-D analog window $w_c(t_1, t_2)$ is sampled. The resulting sequence $w(n_1, n_2)$ is a circularly symmetric window. From (4.15a) and (1.53),

$$W(\omega_1, \omega_2) = \sum_{r_1 = -\infty}^{\infty} \sum_{r_2 = -\infty}^{\infty} W_c(\Omega_1, \Omega_2)|_{\Omega_1 = \omega_1 - 2\pi r_1, \Omega_2 = \omega_2 - 2\pi r_2}. \qquad (4.17)$$

Due to the aliasing effect in (4.17), $W(\omega_1, \omega_2)$ is no longer circularly symmetric. As we discussed in Chapter 1 (Section 1.3.3), circular symmetry of $w(n_1, n_2)$ does not guarantee circular symmetry of its Fourier transform $W(\omega_1, \omega_2)$. The function $W(\omega_1, \omega_2)$ can deviate from circular symmetry considerably for (ω_1, ω_2) away from the origin. Near the origin, however, the aliasing effect is less and $W(\omega_1, \omega_2)$ tends to be close to being circularly symmetric. If the desired filter has a circularly symmetric frequency response, the circularly symmetric window of (4.15) tends to have better performance for a fixed window size than the separable window of (4.14).

Examples of 1-D analog windows that can be used in obtaining a 2-D window are the rectangular, the Hamming, and the Kaiser windows. Their functional forms are given below.

The Rectangular Window

$$w_a(t) = \begin{cases} 1, & |t| < \tau \\ 0, & \text{otherwise.} \end{cases} \tag{4.18}$$

The Hamming Window

$$w_a(t) = \begin{cases} 0.54 + 0.46 \cos(\pi t/\tau), & |t| < \tau \\ 0, & \text{otherwise.} \end{cases} \tag{4.19}$$

The Kaiser Window

$$w_a(t) = \begin{cases} \dfrac{I_0\left(\alpha \sqrt{1 - \left(\dfrac{t}{\tau}\right)^2}\right)}{I_0(\alpha)}, & |t| < \tau \\ 0, & \text{otherwise.} \end{cases} \tag{4.20}$$

where I_0 is the modified Bessel function of the first kind, order zero, and α is a parameter. Note that $w_a(t)$ is zero at $|t| = \tau$. If τ is an integer, therefore, the 1-D window $w(n)$ obtained by $w(n) = w_a(t)|_{t=n}$ will be a $2\tau - 1$-point sequence. The rectangular window has a very good mainlobe behavior (small mainlobe width), but a poor sidelobe behavior (large sidelobe peak). It is very simple to use and is optimal, based on the minimum mean square error criterion. Specifically, $h(n_1, n_2)$ designed by using the rectangular window minimizes

$$\text{Error} = \sum_{n_1 = -\infty}^{\infty} \sum_{n_2 = -\infty}^{\infty} |h(n_1, n_2) - h_d(n_1, n_2)|^2 \tag{4.21}$$

$$= \frac{1}{(2\pi)^2} \int_{\omega_1 = -\pi}^{\pi} \int_{\omega_2 = -\pi}^{\pi} |H(\omega_1, \omega_2) - H_d(\omega_1, \omega_2)|^2 \, d\omega_1 \, d\omega_2$$

for a fixed region of support of $h(n_1, n_2)$. The Hamming window does not have a good mainlobe behavior, but does have a good sidelobe behavior. It is also very simple to use. The Kaiser window is a family of windows. The Kaiser window's mainlobe and sidelobe behavior can be controlled by the parameter α. Improving the mainlobe behavior will generally result in a poorer sidelobe behavior and vice versa. The Kaiser window involves a Bessel function; evaluating the function is somewhat more complicated than using the rectangular or Hamming windows.

Figure 4.5 shows the Fourier transform of a separable window and a circularly symmetric window obtained from the analog rectangular window with $\tau = 8$. Figure 4.5(a) corresponds to the $(2\tau - 1) \times (2\tau - 1)$-point separable window and Figure 4.5(b) corresponds to the circularly symmetric window. The regions of support of the separable window (225 nonzero points) and the circularly symmetric window (193 nonzero points) used in Figure 4.5 are shown in Figures 4.6(a) and (b), respectively.

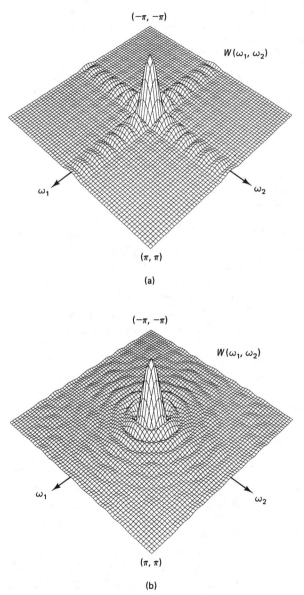

Figure 4.5 Fourier transform of (a) separable window and (b) circularly symmetric window obtained from the analog rectangular window with $\tau = 8$.

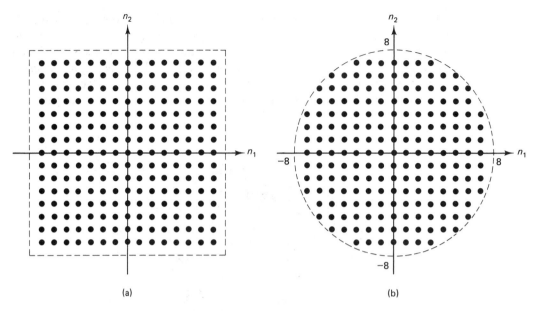

Figure 4.6 Support region of $w(n_1, n_2)$ for $\tau = 8$. (a) Separable window; (b) circularly symmetric window.

As is expected from (4.14), the Fourier transform of a separable window has large amplitudes near the origin and along the ω_1 and ω_2 axes. The Fourier transform of a circularly symmetric window is circularly symmetric approximately with a very good approximation near the origin in the (ω_1, ω_2) plane. Figure 4.7 shows the Fourier transforms of circularly symmetric windows obtained from the Hamming window in (4.19) and Kaiser window in (4.20). Figure 4.7(a) corresponds to the Hamming window and Figures 4.7(b) and (c) correspond to the Kaiser window with $\alpha = 1$ and 3, respectively. In all cases, the value of τ is 8.

The shape of the window affects both the mainlobe and sidelobe behavior. The size of the window, however, primarily affects the mainlobe behavior. Since the sidelobe behavior (and therefore passband and stopband tolerance) is affected by only the shape of the window, the window shape is chosen first on the basis of the passband and stopband tolerance requirements. The window size is then determined on the basis of the transition requirements. Two examples of digital filters designed by the window method are shown in Figures 4.8 and 4.9. Figure 4.8 shows the result of a lowpass filter design using the Kaiser window with $\tau = 8$ and $\alpha = 2$. The desired impulse response $h_d(n_1, n_2)$ used is the circularly symmetric ideal lowpass filter of (4.10a) with cutoff frequency $R = 0.4\pi$. Figure 4.8(a) is based on the separable window design, and Figure 4.8(b) is based on the circularly symmetric window design. In each case, both the perspective plot and contour plot are shown. Note that the number of nonzero values of $h(n_1, n_2)$ is not the same in the two cases (see Figure 4.6). It is 225 for Figure 4.8(a) and 193 for Figure 4.8(b). Figure 4.9 is the same as Figure 4.8 except that the desired impulse response used is the circularly symmetric ideal bandpass filter of (4.10c) with cutoff frequencies $R_1 = 0.3\pi$ and $R_2 = 0.7\pi$.

In Figure 4.9(a), the deviation from the desired frequency response in the stop-

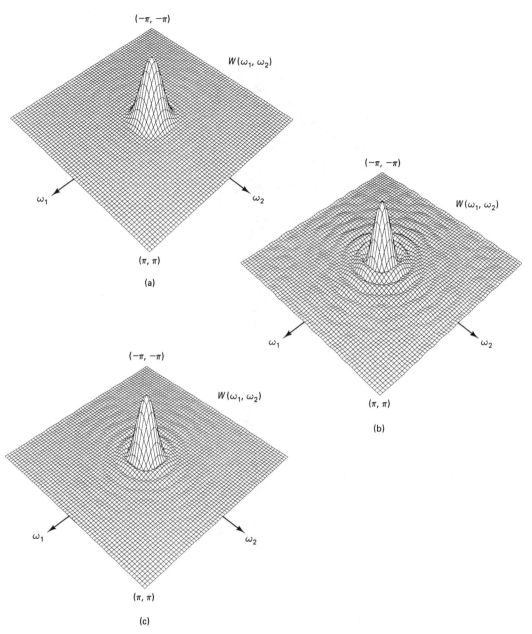

Figure 4.7 Fourier transform of circularly symmetric windows with $\tau = 8$. (a) Hamming window; (b) Kaiser window with $\alpha = 1$; (c) Kaiser window with $\alpha = 3$.

band is larger along the ω_1 and ω_2 axes than along the diagonal directions ($\omega_1 = \omega_2$, $\omega_1 = -\omega_2$). This is typically the case for a lowpass or bandpass filter design using a separable window. As was illustrated in Figure 4.5(a), the Fourier transform of a separable window has large amplitudes near the origin and along the ω_1 and ω_2 axes. The convolution operation in (4.12) of the Fourier transform of a separable window

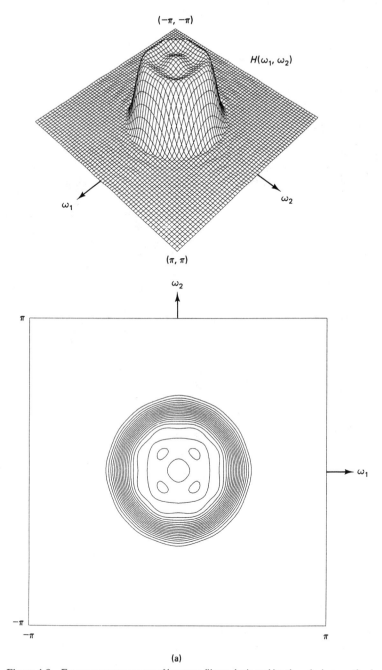

$(-\pi, -\pi)$

$H(\omega_1, \omega_2)$

ω_1

ω_2

(π, π)

ω_2

π

ω_1

$-\pi$

$-\pi$

π

(a)

Figure 4.8 Frequency responses of lowpass filters designed by the window method. The desired impulse response was obtained by using (4.10a) with $R = 0.4\pi$. The 1-D Kaiser window was used. The support regions of the windows are those shown in Figure 4.6. Both perspective and contour plots are shown. (a) Separable window design; (b) rotated circularly symmetric window design.

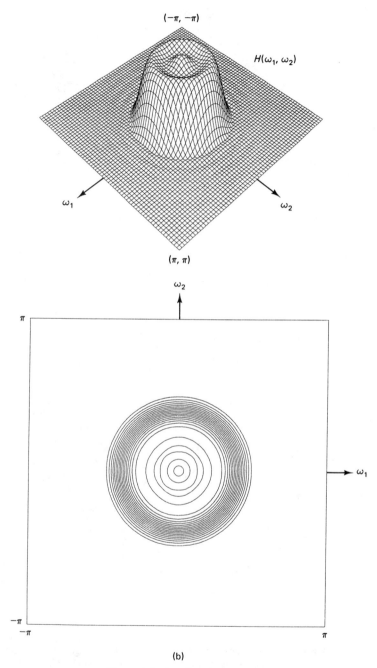

(b)

Figure 4.8 (continued)

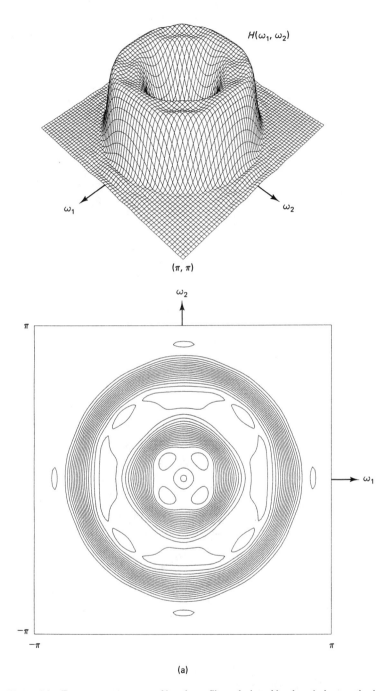

$H(\omega_1, \omega_2)$

(π, π)

(a)

Figure 4.9 Frequency responses of bandpass filters designed by the window method. The desired impulse response was obtained by using (4.10c) with $R_1 = 0.3\pi$ and $R_2 = 0.7\pi$. The 1-D Kaiser window was used. The support regions of the windows are those shown in Figure 4.6. Both perspective and contour plots are shown. (a) Separable window design; (b) rotated circularly symmetric window design.

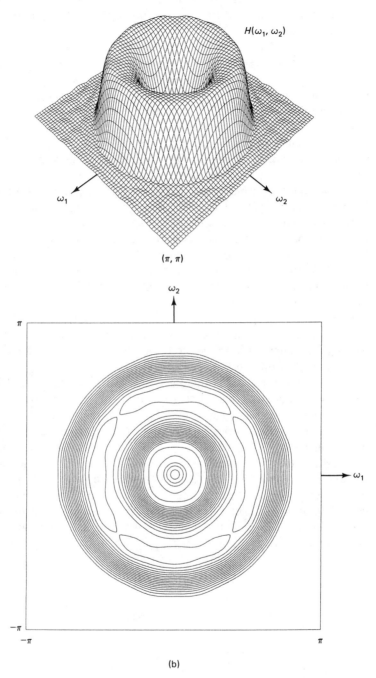

$H(\omega_1, \omega_2)$

(π, π)

(b)

Figure 4.9   (continued)

with a circularly symmetric ideal lowpass filter results in a larger deviation in the stopband from the desired frequency response along the ω_1 and ω_2 axes than along the diagonal directions.

It has been empirically observed that the passband and stopband tolerances of a typical 2-D filter designed by the window method depend not only on the sidelobe amplitudes of the window but also on their distributions in the frequency plane. In addition, for a typical lowpass filter designed (see Figure 4.8), the passband tolerance has been observed to be larger than the stopband tolerance by an average of approximately 25% for the separable window design and 50% for the circularly symmetric window design. This is in sharp contrast with the 1-D case. The passband and stopband tolerances of a typical 1-D filter depend primarily on the maximum sidelobe amplitude of the window and are approximately equal.

The separable window and circularly symmetric window are designed from a 1-D window in the spatial domain and therefore preserve spatial domain characteristics of the 1-D window. A 2-D window can also be designed from a 1-D window in the frequency domain to preserve frequency domain characteristics of the 1-D window. In one such method [Kato and Furukawa], the Fourier transform of a 1-D window is circularly rotated to obtain the Fourier transform of a 2-D window. In another method [Yu and Mitra], a 2-D window is obtained from a 1-D window by the frequency transformation method discussed in Section 4.4. Even though these methods can preserve such frequency domain characteristics as maximum sidelobe amplitude and can design a 2-D window whose Fourier transform is more circularly symmetric than the methods discussed in this section, the filters designed by them do not appear to have better performance.

The window method is quite general. We have discussed it in the context of designing a zero-phase filter, but it can be used for nonzero-phase filter design as well. When $H_d(\omega_1, \omega_2)$ has a nonzero-phase component, we can still determine $h_d(n_1, n_2)$ by inverse Fourier transforming $H_d(\omega_1, \omega_2)$, and a window such as a zero-phase window can be applied to $h_d(n_1, n_2)$. The resulting filter, of course, will not have exactly the same phase as $H_d(\omega_1, \omega_2)$, but it can be reasonably close. The window method can also be used to design a filter with linear phase due to a half-sample delay. From (4.5), the number of nonzero values of $h(n_1, n_2)$ must be odd when $h(0, 0) \neq 0$. Therefore, $h(n_1, n_2)$ with rectangular regions of support of such sizes as 2×2, 2×3, 3×4, and 4×4 cannot have zero phase. By allowing a linear phase term $\theta_{h_d}(\omega_1, \omega_2)$ in $H_d(\omega_1, \omega_2)$ where $\theta_{h_d}(\omega_1, \omega_2)$ is of the form

$$\theta_{h_d} = -\frac{\omega_1}{2}, \tag{4.22a}$$

$$= -\frac{\omega_2}{2}, \text{ or} \tag{4.22b}$$

$$= -\frac{\omega_1 + \omega_2}{2}, \tag{4.22c}$$

a filter with an even number of nonzero values and with essentially the zero-phase characteristic (which preserves the shape of the waveform in the passband region) can

Finite Impulse Response Filters   Chap. 4

be designed. To do this, we first require that $H_d(\omega_1, \omega_2)$ have the form given by

$$H_d(\omega_1, \omega_2) = H_{zp}(\omega_1, \omega_2)e^{j\theta_{hd}(\omega_1,\omega_2)} \quad\quad (4.23)$$

where $H_{zp}(\omega_1, \omega_2)$ is the desired zero-phase filter and $\theta_{hd}(\omega_1, \omega_2)$ is of the form given by (4.22). Equations (4.22a), (4.22b), and (4.22c) correspond to a half-sample shift along the n_1 dimension, the n_2 dimension, and both n_1 and n_2 dimensions, respectively. We then determine $w(n_1, n_2)$ by

$$w(n_1, n_2) = w_c(t_1, t_2)|_{t_1 = n_1 - 1/2, t_2 = n_2} \quad \text{for (4.22a)} \quad\quad (4.24a)$$

$$= w_c(t_1, t_2)|_{t_1 = n_1, t_2 = n_2 - 1/2} \quad \text{for (4.22b)} \quad\quad (4.24b)$$

$$= w_c(t_1, t_2)|_{t_1 = n_1 - 1/2, t_2 = n_2 - 1/2} \quad \text{for (4.22c)} \quad\quad (4.24c)$$

where $w_c(t_1, t_2)$ is the 2-D analog window function used in (4.13) and (4.15). The resulting filter $h(n_1, n_2)$ is symmetric with respect to the origin shifted by a half sample along the n_1 dimension, the n_2 dimension, or both the n_1 and n_2 dimensions, and tends to preserve the shape of the waveform in the passband regions. The region of support of the resulting filter consists of an even number of points and can accommodate such sizes as 2×2, 2×3, 3×4, and 4×4.

The window method is not optimal. There exists in general a filter that meets the given design specification and whose region of support is smaller than the filter designed by the window method. For an arbitrary $H_d(\omega_1, \omega_2)$, determining $h_d(n_1, n_2)$ from $H_d(\omega_1, \omega_2)$ may require a large inverse discrete Fourier transform computation. In addition, because of a lack of control over the frequency domain specification parameters, it is sometimes necessary to design several filters to meet a given design specification. Despite these disadvantages, the window method is often used in practice because of its conceptual and computational simplicity.

4.3.2 The Frequency Sampling Method

In the frequency sampling method, the desired frequency response $H_d(\omega_1, \omega_2)$ is sampled at equally spaced points on the Cartesian grid, and the result is inverse discrete Fourier transformed. Specifically, let $H'(k_1, k_2)$ be obtained by

$$H'(k_1, k_2) = H_d(\omega_1, \omega_2)e^{-j\omega_1(N_1-1)/2}e^{-j\omega_2(N_2-1)/2}|_{\omega_1 = (2\pi/N_1)k_1, \omega_2 = (2\pi/N_2)k_2},$$

$$0 \leq k_1 \leq N_1 - 1, 0 \leq k_2 \leq N_2 - 1 \quad\quad (4.25)$$

where $H_d(\omega_1, \omega_2)$ is the desired zero-phase frequency response, and N_1 and N_2 are odd. The corresponding sequence $h'(n_1, n_2)$ is obtained from (4.25) by

$$h'(n_1, n_2) = \text{IDFT}[H'(k_1, k_2)]. \quad\quad (4.26)$$

Finally, the zero-phase filter $h(n_1, n_2)$ designed is given by

$$h(n_1, n_2) = h'\left(n_1 + \frac{N_1 - 1}{2}, n_2 + \frac{N_2 - 1}{2}\right). \qu\quad (4.27)$$

The linear phase term in (4.25) and the resulting shift of the sequence in (4.27) are due to the fact that the DFT is defined only for a first-quadrant support sequence, which cannot be zero phase. The inverse DFT in (4.26) can be computed by using an FFT algorithm for highly composite numbers N_1 and N_2. From the definition of the Fourier transform and the DFT, the frequency response of the designed filter $H(\omega_1, \omega_2)$ can be shown to be related to $H'(k_1, k_2)$ and therefore to $H_d(\omega_1, \omega_2)$ by

$$H(\omega_1, \omega_2) = \frac{e^{j\omega_1(N_1-1)/2}e^{j\omega_2(N_2-1)/2}(1 - e^{-j\omega_1 N_1})(1 - e^{-j\omega_2 N_2})}{N_1 N_2}$$

$$\cdot \sum_{k_1=0}^{N_1-1} \sum_{k_2=0}^{N_2-1} \frac{H'(k_1, k_2)}{(1 - e^{j(2\pi/N_1)k_1}e^{-j\omega_1})(1 - e^{j(2\pi/N_2)k_2}e^{-j\omega_2})}. \quad (4.28)$$

When $H_d(\omega_1, \omega_2)$ is sampled by using (4.25) strictly to design piecewise constant filters such as lowpass filters, $H'(k_1, k_2)$ changes sharply from 1 to 0 near transition regions. As a result, both the stopband and passband behaviors of the filter $H(\omega_1, \omega_2)$ in (4.28) are rather poor. They can be improved considerably if some transition samples are taken in the frequency region where $H_d(\omega_1, \omega_2)$ has a sharp transition. Determining the values of the transition samples optimally is not simple and requires linear programming methods. However, a reasonable choice, such as linear interpolation of samples, will work well. Two examples of digital filters designed by the frequency sampling method are shown in Figures 4.10 and 4.11. Figure 4.10 shows the result of a lowpass filter design with $N_1 = 15$ and $N_2 = 15$. The region of the transition samples used is shown in Figure 4.10(a), and the transition samples used are samples of a circularly symmetric linear interpolation between 0 and 1. The perspective plot and the contour plot of the frequency response of the resulting filter are shown in Figures 4.10(b) and (c), respectively. Figure 4.11 is the same as Figure 4.10, except that a bandpass filter is designed. The values of the transition samples are again obtained from samples of a circularly symmetric linear interpolation between 0 and 1.

Like the window method, the frequency sampling method is quite general. When $H_d(\omega_1, \omega_2)$ has a nonzero-phase component, we can still use (4.25), (4.26), and (4.27). The method can also be used to design a filter with linear phase resulting from a half-sample delay. The modification made to $H_d(\omega_1, \omega_2)$ to incorporate the half-sample delay is identical to that in the window design case, and the modified $H_d(\omega_1, \omega_2)$ can be used in (4.25) and (4.26). The half-sample delay can be incorporated in the right-hand side expression of (4.27) by subtracting $\frac{1}{2}$ from n_1, n_2, or both n_1 and n_2. The frequency sampling method can also be used when the desired frequency samples are obtained on a non-Cartesian grid. The basic objective behind the frequency sampling method is to ensure that $H(\omega_1, \omega_2)$ is identical to some given $H_d(\omega_1, \omega_2)$ at a set of frequencies (ω_1', ω_2'), that is,

$$H(\omega_1, \omega_2)\big|_{\omega_1 = \omega_1', \omega_2 = \omega_2'} = H_d(\omega_1, \omega_2)\big|_{\omega_1 = \omega_1', \omega_2 = \omega_2'} \quad (4.29)$$

$$= \sum_{(n_1, n_2) \in R_h} \sum h(n_1, n_2)e^{-j\omega_1 n_1}e^{-j\omega_2 n_2}\big|_{\omega_1 = \omega_1', \omega_2 = \omega_2'}.$$

At a fixed frequency (ω_1', ω_2'), (4.29) is a linear equation for the unknown coefficients $h(n_1, n_2)$. By considering many different frequencies of (ω_1', ω_2'), not necessarily on the Cartesian grid, $h(n_1, n_2)$ can be determined by solving a set of

Finite Impulse Response Filters   Chap. 4

linear equations. If we impose a zero-phase constraint, from (4.5),

$$H_d(\omega_1, \omega_2)\big|_{\omega_1 = \omega_1', \omega_2 = \omega_2'} = h(0, 0) + \sum_{(n_1, n_2) \in R_h} \sum 2h(n_1, n_2) \cos(\omega_1' n_1 + \omega_2' n_2). \qquad (4.30)$$

The set of linear equations in (4.30) have coefficients that are all real, and the resulting filter is zero phase. For M independent values in $h(n_1, n_2)$, M linear equations in (4.30) almost always have a unique solution for $h(n_1, n_2)$ in practice.

As with the window method, the filter designed by the frequency sampling method is not optimal. That is, there exists in general a filter that meets the same design specification and whose region of support is smaller than the filter designed by the frequency sampling method. In addition, because of a lack of control over the frequency domain parameters, we may have to design several filters to meet a given specification. Despite those disadvantages, the frequency sampling method is sometimes used in practice because of its conceptual and computational simplicity. Determining specific values and the region of the transition samples is cumbersome compared to using the window method, but an inverse transform of $H_d(\omega_1, \omega_2)$ is not needed in the frequency sampling method. The performance of the two methods appears comparable as measured by the size of the region of support for the filter designed to meet a given design specification.

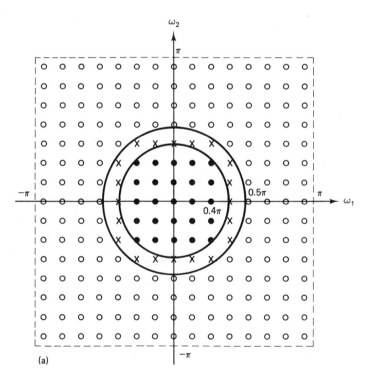

(a)

Figure 4.10 Example of a 15 × 15-point lowpass filter designed by the frequency sampling method. (a) Passband (filled-in dots), transition band (marked by "x"), and stopband (open dots) samples used in the design; (b) perspective plot of the filter frequency response; (c) contour plot of the filter frequency response.

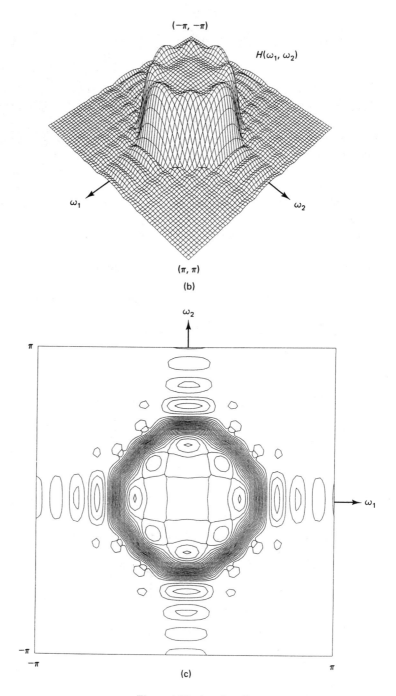

Figure 4.10 (continued)

Finite Impulse Response Filters   Chap. 4

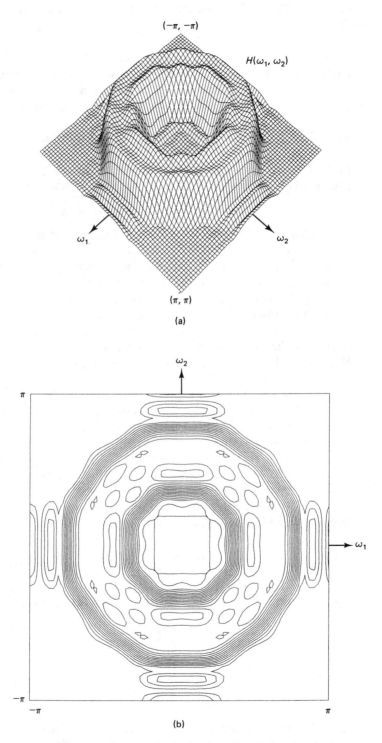

Figure 4.11 Example of a 15 × 15-point bandpass filter designed by the frequency sampling method. Transition regions used in the design are from 0.3π to 0.4π and from 0.7π to 0.8π. (a) Perspective plot of the filter frequency response; (b) contour plot of the filter frequency response.

4.4 FILTER DESIGN BY THE FREQUENCY TRANSFORMATION METHOD

Filter design by frequency transformation does not have a 1-D counterpart. This method leads to a practical means of designing a 2-D FIR filter with good performance. Under a restricted set of conditions, the method has been shown to be optimal. For these reasons, we discuss this method in some detail.

4.4.1 The Basic Idea of Frequency Transformation

In the frequency transformation method, a 2-D zero-phase FIR filter is designed from a 1-D zero-phase FIR filter. Consider the transformation given by

$$H(\omega_1, \omega_2) = H(\omega)|_{\omega = G(\omega_1, \omega_2)} \tag{4.31}$$

where $H(\omega)$ is a 1-D digital filter frequency response and $H(\omega_1, \omega_2)$ is the frequency response of the resulting 2-D digital filter. Suppose $H(\omega)$ is a bandpass filter, as shown in Figure 4.12. Consider one particular frequency $\omega = \omega_0'$. Suppose the function $\omega_0' = G(\omega_1, \omega_2)$ represents a contour in the (ω_1, ω_2) plane, shown in Figure 4.12. Then, according to (4.31), $H(\omega_1, \omega_2)$ evaluated on the contour equals $H(\omega_0')$. If we now consider other frequencies $\omega_1', \omega_2', \ldots$, and if their corresponding contours are as shown in Figure 4.12, then the resulting 2-D filter will be a bandpass filter. Furthermore, the amplitude characteristics of $H(\omega)$ will be preserved in $H(\omega_1, \omega_2)$. For example, if $H(\omega)$ has equiripple characteristics in the passband region, so will $H(\omega_1, \omega_2)$.

Two important issues must be considered in this method. One is whether or not the resulting 2-D filter is a zero-phase FIR filter. The second is whether or not a transformation function $G(\omega_1, \omega_2)$ exists such that there will be a nice mapping between ω and $\omega = G(\omega_1, \omega_2)$, as in Figure 4.12. Both of these issues are resolved by using a 1-D zero-phase filter and the appropriate transformation function. We will refer to this method as the *frequency transformation method*, or the *transformation method* for short.

Consider a 1-D zero-phase FIR filter $h(n)$ with length $2N + 1$. The frequency response $H(\omega)$ can be expressed as

$$
\begin{aligned}
H(\omega) &= \sum_{n=-N}^{N} h(n)e^{-j\omega n} \\[2mm]
&= h(0) + \sum_{n=1}^{N} 2h(n) \cos \omega n \\[2mm]
&= \sum_{n=0}^{N} a(n) \cos \omega n \\[2mm]
&= \sum_{n=0}^{N} b(n) (\cos \omega)^n.
\end{aligned}
\tag{4.32}
$$

In (4.32), the sequence $b(n)$ is not the same as $h(n)$, but can be obtained (see

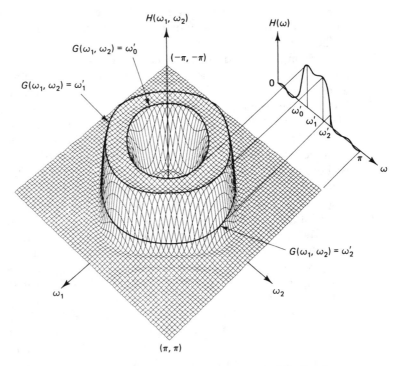

Figure 4.12 Illustration of the principle behind the design of a 2-D filter from a 1-D filter by frequency transformation.

Problem 4.16) simply from $h(n)$. The 2-D frequency response $H(\omega_1, \omega_2)$ is obtained by

$$H(\omega_1, \omega_2) = H(\omega)|_{\cos\omega = T(\omega_1, \omega_2)} = \sum_{n=0}^{N} b(n)(T(\omega_1, \omega_2))^n \qquad (4.33)$$

where $T(\omega_1, \omega_2)$ is the Fourier transform of a finite-extent zero-phase sequence, so that $T(\omega_1, \omega_2)$ can be expressed as

$$\begin{aligned}
T(\omega_1, \omega_2) &= \sum_{(n_1,n_2)\in R_t} \sum t(n_1, n_2)e^{-j\omega_1 n_1}e^{-j\omega_2 n_2} \\
&= \sum_{(n_1,n_2)\in R_c} \sum c(n_1, n_2)\cos(\omega_1 n_1 + \omega_2 n_2)
\end{aligned} \qquad (4.34)$$

where R_t and R_c represent the region of support of $t(n_1, n_2)$ and $c(n_1, n_2)$, respectively. The sequence $c(n_1, n_2)$ is simply related to $t(n_1, n_2)$ and can be easily obtained from $t(n_1, n_2)$. An example of $T(\omega_1, \omega_2)$ often used in practice is given by

$$\begin{aligned}
T(\omega_1, \omega_2) &= -\tfrac{1}{2} + \tfrac{1}{2}\cos\omega_1 + \tfrac{1}{2}\cos\omega_2 + \tfrac{1}{4}\cos(\omega_1 + \omega_2) + \tfrac{1}{4}\cos(\omega_1 - \omega_2) \\
&= -\tfrac{1}{2} + \tfrac{1}{2}\cos\omega_1 + \tfrac{1}{2}\cos\omega_2 + \tfrac{1}{2}\cos\omega_1\cos\omega_2.
\end{aligned} \qquad (4.35)$$

The sequences $t(n_1, n_2)$ and $c(n_1, n_2)$ that correspond to $T(\omega_1, \omega_2)$ in (4.35) are shown in Figure 4.13. Equation (4.35) originally proposed by [McClellan] is known as the McClellan transformation.

From (4.33) and (4.34), $H(\omega_1, \omega_2)$ is always real, so the resulting 2-D filter has zero phase. In addition, it is an FIR filter. For example, when $t(n_1, n_2)$ has a region of support of size $(2M_1 + 1) \times (2M_2 + 1)$ and the length of $h(n)$ is $2N + 1$, the resulting 2-D filter $H(\omega_1, \omega_2)$ will be a finite-extent sequence of size $(2M_1N + 1) \times (2M_2N + 1)$. If $N = 10$ and $M_1 = M_2 = 1$, $t(n_1, n_2)$ will be a 3×3-point sequence and $h(n_1, n_2)$ will be a 21×21-point sequence. In this example, the 2-D filter obtained has a large region of support, even though $h(n)$ is short and $t(n_1, n_2)$ has a small region of support. This is typically the case. Since $h(n_1, n_2)$ is completely specified by $h(n)$ and $t(n_1, n_2)$, there is considerable structure in $h(n_1, n_2)$. As discussed in Section 4.6, this structure can be exploited in reducing the number of arithmetic operations when the filter is implemented.

By choosing $T(\omega_1, \omega_2)$ in (4.34) properly, we can obtain many different sets of contours that can be used for 2-D filter design. For the transformation function $T(\omega_1, \omega_2)$ in (4.35), the set of contours obtained by $\cos \omega = T(\omega_1, \omega_2)$ for $\omega = 0, \pi/10, 2\pi/10, \ldots, \pi$ are shown in Figure 4.14. This can be used to design many different 2-D FIR filters. From a 1-D lowpass filter of 21 points in length whose $H(\omega)$ is shown in Figure 4.15(a), we can obtain a 2-D filter whose frequency response is shown in Figure 4.15(b). Note that the amplitude characteristics of $H(\omega)$ are preserved in $H(\omega_1, \omega_2)$. For example, the equiripple characteristics of $H(\omega)$ in the passband and stopband regions are preserved in $H(\omega_1, \omega_2)$. If we begin with a 1-D highpass or bandpass filter, the resulting 2-D filter based on (4.35) would be a highpass or bandpass filter. Due to the approximate circular symmetry of the contours corresponding to $\cos \omega = T(\omega_1, \omega_2)$ with $T(\omega_1, \omega_2)$ in (4.35), $H(\omega_1, \omega_2)$ obtained has approximately circular symmetry particularly in the low-frequency regions.

The basic idea behind the transformation method discussed in this section can be used to develop specific methods of designing 2-D zero-phase FIR filters. The next two sections treat two specific design methods.

4.4.2 Design Method 1

In the transformation method, $h(n_1, n_2)$ is completely specified by the transformation sequence $t(n_1, n_2)$ and the 1-D filter $h(n)$. Therefore, designing $h(n_1, n_2)$ involves designing $t(n_1, n_2)$ and $h(n)$. Designing $t(n_1, n_2)$ and $h(n)$ jointly is a highly nonlinear problem, and they are designed separately in practice. Even with this suboptimal approach, the transformation method often results in filters with very good performance.

One method of designing $h(n_1, n_2)$ is

Step 1. Choose $t(n_1, n_2)$.

Step 2. Translate the given specification of $H(\omega_1, \omega_2)$ to the specification of $H(\omega)$.

Step 3. Design $h(n)$.

Step 4. Determine $h(n_1, n_2)$ from $t(n_1, n_2)$ and $h(n)$.

Finite Impulse Response Filters   Chap. 4

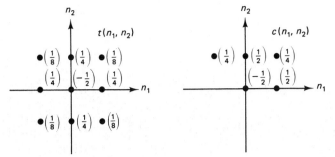

Figure 4.13 Transformation sequence $t(n_1, n_2)$ and the corresponding sequence $c(n_1, n_2)$ in (4.34) often used in the transformation method.

In Step 1, we choose $t(n_1, n_2)$ from the results that are already available in the open literature [McClellan; Mersereau et al.]. For a circularly symmetric filter design, for example, the McClellan transformation of (4.35) may be used. In Step 2, we translate the filter specification of $H(\omega_1, \omega_2)$ to the specification of $H(\omega)$, using the transformation function selected in Step 1. This translation should be made such that when $h(n)$ is designed meeting the 1-D filter specification, the resulting $h(n_1, n_2)$ will meet the 2-D filter specification. In Step 3, we design the 1-D filter $h(n)$ that meets the specification determined in Step 2. The optimal 1-D filter design technique developed by [Parks and McClellan] can be used here. In Step 4, we determine $h(n_1, n_2)$ from $t(n_1, n_2)$ and $h(n)$ using (4.32) and (4.33). To illustrate these four steps, let us consider a specific design example.

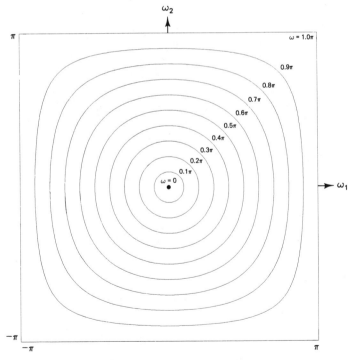

Figure 4.14 Contours obtained by $\cos \omega = T(\omega_1, \omega_2)$ for $\omega = 0, 0.1\pi, 0.2\pi, \ldots,$ 1.0π. The transformation function $T(\omega_1, \omega_2)$ used is given by (4.35).

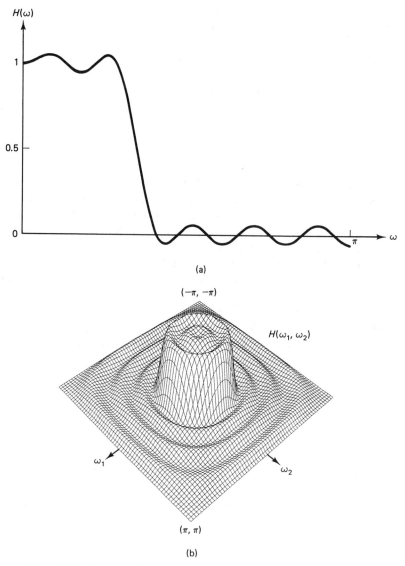

Figure 4.15 Example of a lowpass filter designed by the transformation method. (a) Frequency response of the 21-point 1-D filter used in the design; (b) frequency response of the 21 × 21-point 2-D filter designed.

Design Example

We wish to design a 2-D lowpass filter that meets the following circularly symmetric filter specification:

δ_p (passband tolerance) $= 0.05$, δ_s (stopband tolerance) $= 0.025$

ω_p (passband cutoff frequency) $= 0.4\pi$, ω_s(stopband cutoff frequency) $= 0.5\pi$

$$(4.36)$$

Step 1. Since the filter specification given has circular symmetry, we use the McClellan transformation of (4.35), repeated here.

$$T(\omega_1, \omega_2) = -\tfrac{1}{2} + \tfrac{1}{2} \cos \omega_1 + \tfrac{1}{2} \cos \omega_2 + \tfrac{1}{2} \cos \omega_1 \cos \omega_2. \qquad (4.37)$$

The contours of $\cos \omega = T(\omega_1, \omega_2)$ for different values of ω for this case are shown in Figure 4.16(a). In the same figure, we also show the passband and stopband regions given by (4.36).

Step 2. We now translate the 2-D filter specification to the 1-D filter specification. For the transformation of (4.37), a 1-D lowpass filter will result in a 2-D lowpass filter. Thus, we need to determine δ_p', δ_s', ω_p', ω_s' for the 1-D lowpass filter specification. We first choose $\delta_p' = \delta_p = 0.05$ and $\delta_s' = \delta_s = 0.025$, since the amplitude characteristics of $H(\omega)$ remain in $H(\omega_1, \omega_2)$ and therefore the passband and stopband ripples of the 1-D and 2-D filters are identical in this case. We now choose the smallest ω_p' such that $\cos \omega_p' = T(\omega_1, \omega_2)$ completely encircles the entire passband region of the 2-D filter specification. This guarantees that a 1-D filter which has the passband frequency given by ω_p' will lead to a 2-D filter that meets the specification in the passband region. Similarly, we choose the smallest ω_s' such that the stopband region is completely outside $\cos \omega_s' = T(\omega_1, \omega_2)$. The 1-D filter specification is then given by

$$\delta_p' = 0.05, \qquad \delta_s' = 0.025, \qquad \omega_p' = 0.4\pi, \qquad \omega_s' = 0.486\pi. \qquad (4.38)$$

Note that ω_s' is slightly different from 0.5π. This difference shows that the contours corresponding to $\cos \omega = T(\omega_1, \omega_2)$ are not exactly circularly symmetric. This deviation from circular symmetry is more clearly visible in high-frequency regions.

Step 3. Using the Parks-McClellan algorithm, we design a 1-D Type I filter ($h(n) = h(-n)$) that meets the design specification in (4.38). The frequency response of the 1-D filter designed is shown in Figure 4.16(b). The length of the filter $2N + 1$ in this case is 31.

Step 4. From $t(n_1, n_2)$ and $h(n)$ obtained above, a 2-D filter $h(n_1, n_2)$ is obtained by using (4.32) and (4.33). The frequency response $H(\omega_1, \omega_2)$ is shown as a perspective plot in Figure 4.16(c) and as a contour plot in Figure 4.16(d). This filter is guaranteed to meet the specification given by (4.36). The 2-D filter $h(n_1, n_2)$ obtained is a 31 × 31-point sequence.

    Figure 4.17 shows another example of a filter designed by using the method discussed in this section. Figures 4.17(a) and (b) show the perspective plot and the contour plot, respectively. The filter is a bandpass filter designed on the basis of a filter specification with circular symmetry. The four cutoff frequencies R_1, R_2, R_3, R_4 used are 0.3π, 0.4π, 0.6π, and 0.7π, respectively with R_1 and R_4 representing the stopband frequencies and R_2 and R_3 representing the passband frequencies. The passband and stopband tolerances δ_p and δ_s used are 0.054 and 0.027. The size of the filter is 41 × 41.

    In this method, we have simply selected a transformation function from existing ones. This, of course, can be quite restrictive. If there is not a good match between the contours obtained from the transformation function and the passband and stopband region contours in the 2-D filter specification, then the resulting filter may significantly exceed the given specification. As a result, the region of support

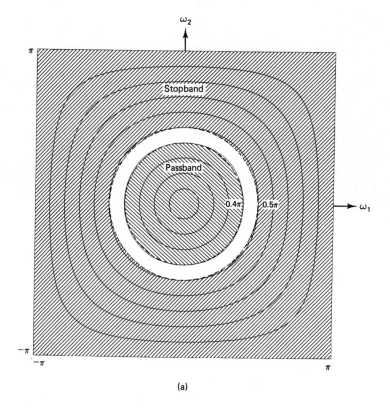

(a)

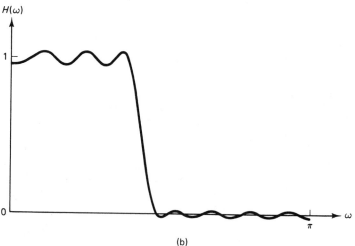

(b)

Figure 4.16 Illustration of a lowpass filter design using Design Method 1. (a) Passband (▨), transition band (☐), and stopband (▨) regions in the filter specification with circular symmetry. Note that boundaries of different regions in the filter specification do not coincide with contours given by $\cos \omega = T(\omega_1, \omega_2)$. This illustrates that contours given by $\cos \omega = T(\omega_1, \omega_2)$ are not circularly symmetric. (b) Frequency response of the 31-point 1-D filter used in the design; (c) perspective plot of the frequency response of the 2-D filter designed; (d) contour plot of the frequency response of the 2-D filter designed.

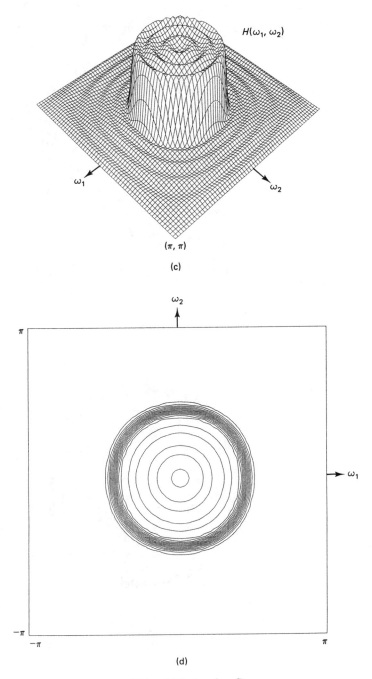

$H(\omega_1, \omega_2)$

ω_1

ω_2

(π, π)

(c)

ω_2

π

$-\pi$

$-\pi$ π

ω_1

(d)

Figure 4.16 (continued)

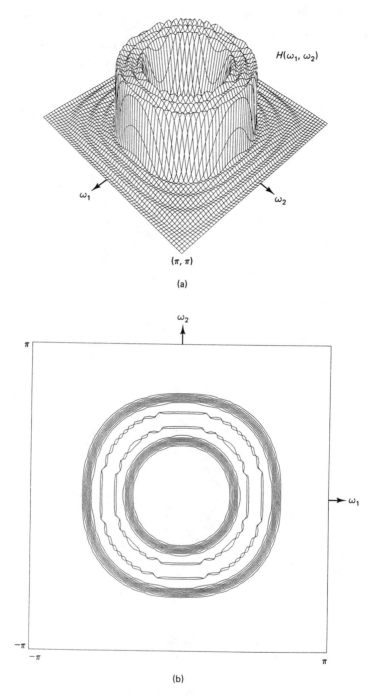

Figure 4.17 Frequency response of a bandpass filter designed by using Design Method 1. (a) Perspective plot; (b) contour plot.

of the designed filter may be considerably larger than necessary. In the next section, we discuss a method in which we design $t(n_1, n_2)$ rather than selecting one from existing ones.

4.4.3 Design Method 2

In this method, we first design the transformation sequence $t(n_1, n_2)$. Once $t(n_1, n_2)$ is designed, the remaining steps are identical to those of Design Method 1. The four steps involved in Method 2 can be stated as

Step 1. Design $t(n_1, n_2)$.
Step 2. Translate the given specification of $H(\omega_1, \omega_2)$ to the specification of $H(\omega)$.
Step 3. Design $h(n)$.
Step 4. Determine $h(n_1, n_2)$ from $t(n_1, n_2)$ and $h(n)$.

We will discuss Step 1 by considering a specific example. Consider the problem of designing a 2-D lowpass filter whose specification is shown in Figure 4.18(a). The passband contour C_p and the stopband contour C_s are ellipses that can be represented by

$$\text{Passband contour } C_p: \quad \left(\frac{\omega_1}{0.5\pi}\right)^2 + \left(\frac{\omega_2}{0.6\pi}\right)^2 = 1 \qquad (4.39a)$$

$$\text{Stopband contour } C_s: \quad \left(\frac{\omega_1}{0.6\pi}\right)^2 + \left(\frac{\omega_2}{0.7\pi}\right)^2 = 1. \qquad (4.39b)$$

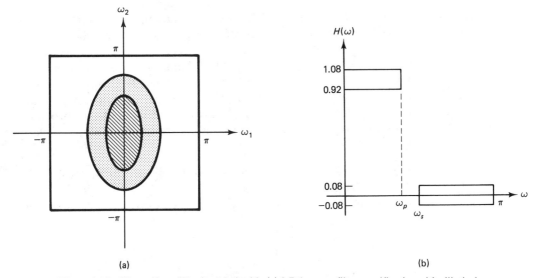

(a)   (b)

Figure 4.18 Illustration of Design Method 2. (a) 2-D lowpass filter specification with elliptical passband and stopband contours. (b) 1-D lowpass filter specification. The 1-D filter that meets this specification is used in designing the 2-D filter specified in (a).

The passband tolerance δ_p and the stopband tolerance δ_s are given by

$$\delta_p = 0.08, \qquad \delta_s = 0.08. \tag{4.39c}$$

Consider a 1-D lowpass filter whose specification is shown in Figure 4.18(b) and is given by

$$\delta_p = 0.08, \quad \delta_s = 0.08, \omega_p, \omega_s. \tag{4.40}$$

The 1-D filter specification in (4.40) comes from Step 2, but to show how Step 1 is done, we will assume that (4.40) is given. Despite this assumption, we will see that Step 1 does not require the result of Step 2. The passband and stopband tolerances are the same in both the 1-D and 2-D filter specifications, since we map the passband and stopband regions of the 1-D filter to the passband and stopband regions, respectively, of the 2-D filter.

One approach to designing $t(n_1, n_2)$ is to satisfy the following conditions:

$$\cos \omega_p = T(\omega_1, \omega_2), \qquad (\omega_1, \omega_2) \in C_p \tag{4.41a}$$

$$\cos \omega_s = T(\omega_1, \omega_2), \qquad (\omega_1, \omega_2) \in C_s \tag{4.41b}$$

$$\cos \omega|_{0 \le \omega \le \omega_p} = T(\omega_1, \omega_2)|_{(\omega_1, \omega_2) \in R_p} \tag{4.41c}$$

where R_p includes the passband region

$$\cos \omega|_{\omega_s \le \omega \le \pi} = T(\omega_1, \omega_2)|_{(\omega_1, \omega_2) \in R_s} \tag{4.41d}$$

where R_s includes the stopband region
and the region of support of $t(n_1, n_2)$ is as small as possible. $\tag{4.41e}$

Equations (4.41a) and (4.41b) are the constraints to map $\omega = \omega_p$ to the passband contour C_p and $\omega = \omega_s$ to the stopband contour C_s in (4.39). If this can be done, we can reduce the amount by which the designed filter exceeds the specification. Equations (4.41c) and (4.41d) are the constraints to map the 1-D filter's passband and stopband regions to the 2-D filter's passband and stopband regions. The constraint in (4.41e) is intended to minimize the region of support of the resulting 2-D filter $h(n_1, n_2)$. Since the conditions in (4.41a) and (4.41b) apply to an infinite number of points on the C_p and C_s contours, they cannot in general be satisfied exactly with a finite number of terms in $t(n_1, n_2)$. In addition, attempting to satisfy all the conditions in (4.41) simultaneously becomes a highly nonlinear problem.

One suboptimal method to solve (4.41), and thus to complete Step 1 of Design Method 2, is given by

Step 1.1. Assume a shape and size for the region of support of $t(n_1, n_2)$.

Step 1.2. Impose a structure on $t(n_1, n_2)$.

Step 1.3. Assume an ω_p and ω_s, and design $t(n_1, n_2)$ so that (4.41a) and (4.41b) are approximately satisfied:

$$\cos \omega_p \approx T(\omega_1, \omega_2), \qquad (\omega_1, \omega_2) \in C_p$$

$$\cos \omega_s \approx T(\omega_1, \omega_2), \qquad (\omega_1, \omega_2) \in C_s.$$

Step 1.4. Design a new $t'(n_1, n_2)$ from $t(n_1, n_2)$ in Step 1.3 to ensure that (4.41c) and (4.41d) are satisfied.

Because these steps are somewhat involved and interrelated, it is important to study each of the steps carefully. After understanding each of the steps involved, the reader will be able to develop an overall perspective of the design procedure. We now discuss each of the steps.

Step 1.1. To design $t(n_1, n_2)$, we first assume a size and shape of the region of support of $t(n_1, n_2)$. If the region of support of $t(n_1, n_2)$ is too large, it will unnecessarily increase the region of support of the resulting filter $h(n_1, n_2)$. If it is too small, there may not be enough degrees of freedom in $t(n_1, n_2)$, and the set of contours given by $\cos \omega = T(\omega_1, \omega_2)$ may not be flexible enough to accommodate the filter specification. For relatively simple filters, such as lowpass filters, a reasonable shape and size is a square of 3×3 points and is given by

$$t(n_1, n_2) = 0 \quad \text{outside} \quad -1 \leq n_1 \leq 1, \; -1 \leq n_2 \leq 1. \tag{4.42}$$

Step 1.2. We now impose some structure on $t(n_1, n_2)$. We first impose the zero-phase constraint on $t(n_1, n_2)$ so that

$$t(n_1, n_2) = t(-n_1, -n_2). \tag{4.43}$$

With this constraint, there are five independent parameters in $t(n_1, n_2)$. We then consider the filter specification. Since the specification in our example has fourfold symmetry, it is reasonable to impose the same fourfold symmetry on $t(n_1, n_2)$. The result is given by

$$t(1, 1) = t(-1, 1). \tag{4.44}$$

Imposing the constraints of (4.43) and (4.44) on $t(n_1, n_2)$, we can express $T(\omega_1, \omega_2)$ in the form of

$$T(\omega_1, \omega_2) = A + B \cos \omega_1 + C \cos \omega_2 + D \cos \omega_1 \cos \omega_2 \tag{4.45}$$

for some real constants A, B, C, and D. The McClellan transformation of (4.35) is a special case of (4.45). Since a 1-D lowpass filter is transformed to a 2-D lowpass filter, we may impose the constraint that $\omega = 0$ be mapped to $\omega_1 = \omega_2 = 0$ and therefore

$$\cos \omega|_{\omega=0} = T(\omega_1, \omega_2)|_{\omega_1=0, \omega_2=0}. \tag{4.46}$$

From (4.45) and (4.46),

$$A + B + C + D = 1. \tag{4.47}$$

From (4.45) and (4.47), designing $t(n_1, n_2)$ is equivalent to determining three independent parameters. It is possible to impose further constraints, such as $\cos \omega|_{\omega=\pi} = T(\omega_1, \omega_2)|_{\omega_1=\pi, \omega_2=\pi}$. However, imposing too many constraints reduces the number of degrees of freedom that may be necessary to satisfy other constraints, for instance, the passband and stopband constraints. Increasing the region of support size of $t(n_1, n_2)$ is possible, but it will increase the region of support size of the resulting 2-D filter $h(n_1, n_2)$.

Step 1.3. In this step, we impose the passband and stopband contour constraints. We first choose initial values of ω_p and ω_s. These frequencies will be modified later, and the performance of the resulting filter is not very sensitive to the specific initial choices of ω_p and ω_s. Since there are an infinite number of (ω_1, ω_2) on the contours C_p and C_s, it is not in general possible to satisfy the passband and stopband constraints of (4.41a) and (4.41b) exactly with a finite number of parameters in $t(n_1, n_2)$. Therefore, we define an error criterion such as

$$\text{Error} = \int\int_{(\omega_1, \omega_2) \in C_p} (\cos \omega_p - T(\omega_1, \omega_2))^2 \, d\omega_1 \, d\omega_2$$

$$+ \int\int_{(\omega_1, \omega_2) \in C_s} (\cos \omega_s - T(\omega_1, \omega_2))^2 \, d\omega_1 \, d\omega_2. \qquad (4.48)$$

Since $T(\omega_1, \omega_2)$ is a linear combination of the unknown parameters $t(n_1, n_2)$ and the Error in (4.48) is a quadratic form of the unknown parameters, minimization of the error expression with respect to $t(n_1, n_2)$ is a linear problem. In our particular design problem, we minimize the Error in (4.48) with respect to A, B, C, and D under the constraint of (4.47). The solution to this problem with the choice of $\omega_p = 0.5\pi$ and $\omega_s = 0.6\pi$ is given by

$$T(\omega_1, \omega_2) = -0.47 + 0.67 \cos \omega_1 + 0.47 \cos \omega_2 + 0.33 \cos \omega_1 \cos \omega_2. \qquad (4.49)$$

Step 1.4. The transformation function in (4.49) ensures that $T(\omega_1, \omega_2)$ will have fourfold symmetry, $\omega = 0$ will map exactly to $(\omega_1, \omega_2) = (0, 0)$, and $\omega = \omega_p$ and $\omega = \omega_s$ will map approximately to $(\omega_1, \omega_2) \in C_p$ and $(\omega_1, \omega_2) \in C_s$. It does not ensure, however, that $T(\omega_1, \omega_2)$ will be between -1 and $+1$ for all (ω_1, ω_2). Since the frequency transformation used is $\cos \omega = T(\omega_1, \omega_2)$, and $-1 \leq \cos \omega \leq 1$ for all real ω, $T(\omega_1, \omega_2)$ less than -1 or greater than $+1$ leads to an unacceptable result. To understand this problem more clearly, we consider (4.32) rewritten as

$$H'(x) = H(\omega)\Big|_{\cos\omega = x} = \sum_{n=0}^{N} b(n) (\cos \omega)^n \Big|_{\cos\omega = x} = \sum_{n=0}^{N} b(n) x^n. \qquad (4.50)$$

When a 1-D filter $H(\omega)$ is designed, $b(n)$ is estimated such that $H(\omega)$ approximates the desired frequency response over $0 \leq \omega \leq \pi$. This is equivalent to estimating $b(n)$ such that $H'(x)$ approximates some desired function over the interval $-1 \leq x \leq 1$. Examples of $H(\omega)$ and $H'(x)$ are shown in Figure 4.19. Figure 4.19(a) shows the 1-D lowpass filter designed by the Parks-McClellan algorithm for $\delta_p = 0.08$, $\delta_s = 0.08$, $\omega_p = 0.5\pi$, $\omega_s = 0.6\pi$. The corresponding function $H'(x)$ is shown in Figure 4.19(b). It is clear that $H'(x)$ is a well-controlled function for $-1 \leq x \leq 1$ but does not behave in a controlled manner outside that range. In 2-D filter design by the transformation method, $H(\omega_1, \omega_2)$ is obtained by

$$H(\omega_1, \omega_2) = H(\omega)\Big|_{\cos\omega = T(\omega_1, \omega_2)} = H'(x)\Big|_{x = T(\omega_1, \omega_2)}. \qquad (4.51)$$

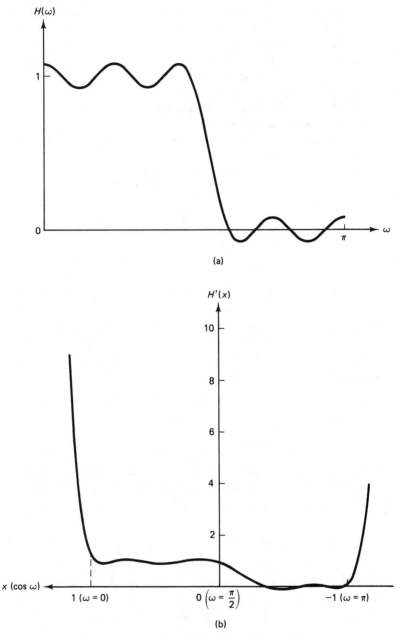

(a)

(b)

Figure 4.19 Example of (a) $H(\omega)$ and (b) $H'(x)$. $H(\omega)$ is the frequency response of a 1-D optimal filter. $H'(x)$ is obtained from $H(\omega)$ by

$$H'(x) = H(\omega)|_{\cos\omega = x}.$$

Note that the x-axis in $H'(x)$ is reversed in direction.

This equation shows that when $T(\omega_1, \omega_2)$ lies outside the range from -1 and $+1$ for some frequency region of (ω_1, ω_2), $H(\omega_1, \omega_2)$ is obtained over that frequency region from $H'(x)$ for x outside the range $-1 \leq x \leq +1$. Since this region of $H'(x)$ is not controlled, the resulting $H(\omega_1, \omega_2)$ will not behave properly. For $T(\omega_1, \omega_2)$ in (4.49), the frequency region controlled by $H'(x)$ for x outside $-1 \leq x \leq 1$ is shown by the shaded region in Figure 4.20(a), along with contours given by $\cos \omega = T(\omega_1, \omega_2)$ for $0 \leq \omega \leq \pi$. The 2-D filter that results from $T(\omega_1, \omega_2)$ in (4.49) and the 1-D filter shown in Figure 4.19 is shown in Figure 4.20(b). Clearly, it is not a good lowpass filter.

To avoid this problem, we obtain a new transformation function $T'(\omega_1, \omega_2)$ from $T(\omega_1, \omega_2)$ by

$$T'(\omega_1, \omega_2) = k_1 T(\omega_1, \omega_2) + k_2 \tag{4.52a}$$

$$k_1 = \frac{2}{T_{\text{MAX}} - T_{\text{MIN}}} \tag{4.52b}$$

$$k_2 = -\frac{T_{\text{MAX}} + T_{\text{MIN}}}{T_{\text{MAX}} - T_{\text{MIN}}} \tag{4.52c}$$

where T_{MAX} and T_{MIN} are the maximum and minimum values of $T(\omega_1, \omega_2)$, respectively over $-\pi \leq \omega_1 \leq \pi$, $-\pi \leq \omega_2 \leq \pi$. In essence, the constants k_1 and k_2 are chosen such that the minimum and maximum of $T(\omega_1, \omega_2)$ become -1 and $+1$ in $T'(\omega_1, \omega_2)$, respectively, and all in-between values are mapped monotonically between -1 and $+1$. Equation (4.52) ensures that $T'(\omega_1, \omega_2)$ will satisfy the condition that

$$-1 \leq T'(\omega_1, \omega_2) \leq 1 \quad \text{for all } (\omega_1, \omega_2). \tag{4.53}$$

From (4.52a),

$$t'(n_1, n_2) = k_1 t(n_1, n_2) + k_2 \delta(n_1, n_2). \tag{4.54}$$

The sequence $t'(n_1, n_2)$ in (4.54) is the transformation sequence designed. If $t(n_1, n_2)$ has fourfold symmetry, $t'(n_1, n_2)$ will also have fourfold symmetry. When (4.52) is applied to (4.49), the resulting $T'(\omega_1, \omega_2)$ for our specific design problem is

$$T'(\omega_1, \omega_2) = -0.28 + 0.58 \cos \omega_1 + 0.41 \cos \omega_2 + 0.29 \cos \omega_1 \cos \omega_2. \tag{4.55}$$

This completes the design of the transformation sequence.

We will now consider the effect of the modification in (4.52) on the constraints that we previously imposed on $t(n_1, n_2)$. We first consider the general nature of the modification. For the transformation function $T(\omega_1, \omega_2)$, the mapping used is

$$\cos \omega = T(\omega_1, \omega_2). \tag{4.56}$$

For the new transformation function $T'(\omega_1, \omega_2)$, we have

$$\cos \omega = T'(\omega_1, \omega_2) = k_1 T(\omega_1, \omega_2) + k_2. \tag{4.57}$$

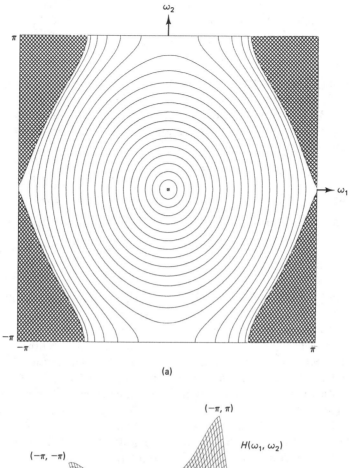

(a)

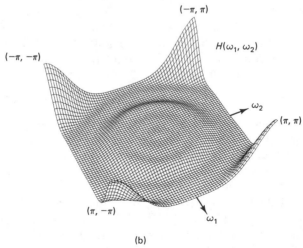

(b)

Figure 4.20 (a) Contours given by $\cos \omega = T(\omega_1, \omega_2)$ with $T(\omega_1, \omega_2)$ in (4.49). Regions of (ω_1, ω_2) where $1 < |T(\omega_1, \omega_2)|$ are cross-hatched. (b) Frequency response of the filter designed by the transformation method using the 1-D filter in Figure 4.19(a) and the transformation function $T(\omega_1, \omega_2)$ in (4.49). Note that the filter has serious problems in the frequency regions cross-hatched in (a).

Rewriting (4.57), we obtain

$$\frac{\cos \omega - k_2}{k_1} = T(\omega_1, \omega_2). \qquad (4.58)$$

From (4.56) and (4.58), the modification does not affect the shape of the contours. It only affects the 1-D frequency ω associated with each contour.

We now consider the effect of the modification on specific constraints imposed on $t(n_1, n_2)$. We first consider the passband and stopband contour constraints. From Step 1.3,

$$T(\omega_1, \omega_2)\big|_{(\omega_1, \omega_2) \in C_p} \approx \cos \omega_p \qquad (4.59a)$$

and
$$T(\omega_1, \omega_2)\big|_{(\omega_1, \omega_2) \in C_s} \approx \cos \omega_s. \qquad (4.59b)$$

From (4.52) and (4.59),

$$T'(\omega_1, \omega_2)\big|_{(\omega_1, \omega_2) \in C_p} \approx k_1 \cos \omega_p + k_2 = \cos \omega_p' \qquad (4.60a)$$

$$T'(\omega_1, \omega_2)\big|_{(\omega_1, \omega_2) \in C_s} \approx k_1 \cos \omega_s + k_2 = \cos \omega_s'. \qquad (4.60b)$$

From (4.60), it is clear that the C_p and C_s contours will map approximately to new frequencies ω_p' and ω_s'. Therefore, the 1-D filter specification we assumed in Step 1.1 for the purpose of designing $t'(n_1, n_2)$ must be modified. The result is consistent with the observation that the modification does not affect the shape of the contours, but affects the 1-D frequency associated with each contour. The modification of the 1-D filter specification is automatically done in Step 2, where the 2-D filter specification is translated to the 1-D filter specification based on the $t'(n_1, n_2)$ designed in Step 1. Step 2, together with a small number of degrees of freedom for $T(\omega_1, \omega_2)$, will ensure that the passband regions of the 1-D filter specification will map to at least the passband regions of the 2-D filter specification. This satisfies the condition in (4.41c). Similarly, (4.41d) that applies to the stopband region is also satisfied. From (4.49) and (4.55),

$$T(0, 0) = T'(0, 0) = T_{\max} = 1 = \cos 0. \qquad (4.61)$$

Therefore, the constraint in (4.46) that $\omega = 0$ must map to $\omega_1 = \omega_2 = 0$ is also satisfied in this example. This type of constraint, however, may not be satisfied in general after the original transformation function is modified. Even though (4.46) was a reasonable constraint to impose, it was not necessary.

Steps 2, 3, and 4 were discussed in detail in Section 4.4.2. This completes Design Method 2. Figure 4.21(a) shows the contours given by $\cos \omega = T'(\omega_1, \omega_2)$ with $T'(\omega_1, \omega_2)$ in (4.55). Figures 4.21(b) and (c) show the perspective plot and the contour plot of the 2-D filter $H(\omega_1, \omega_2)$ designed by following Steps 2, 3, and 4 with $T'(\omega_1, \omega_2)$ in (4.55). The size of the filter's region of support is 27×27. One additional example of a filter designed by Design Method 2 is shown in Figure 4.22. Figure 4.22 shows the design of another lowpass filter. In designing the filter, the passband and stopband contours in the specification had square shape, the passband and stopband frequencies used were 0.5π and 0.75π, respectively, on the ω_1 axis, and the passband and stopband tolerances used were 0.05 and 0.025, respectively. The size of the filter's region of support is 25×25. Both the contour and perspective plots are shown in the figure. Due to the square-shaped passband and stopband contours used in the filter specification, the resulting figure has approximately square shape near the transition regions. Note that this example is intended primarily to illustrate the flexibility of the transformation method. For truly square-shaped passband and stopband contours, designing a separable 2-D filter by designing two 1-D filters and then multiplying their impulse responses would be preferable to the transformation method.

Although the transformation method is somewhat more complex conceptually than either the window method or the frequency sampling method, its performance appears to be better than that of the other two methods. The filter designed by the transformation method has been shown to be optimal in a certain restricted set of cases, to be discussed in the next section. Since reliable, practical procedures for designing optimal filters have not yet been developed, the frequency transformation method should be considered in sophisticated applications requiring high performance.

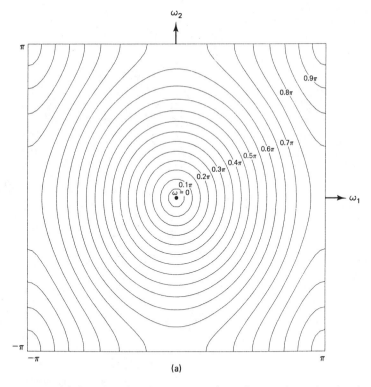

(a)

Figure 4.21 (a) Contours given by $\cos \omega = T(\omega_1, \omega_2)$ with $T(\omega_1, \omega_2)$ in (4.55); (b) frequency response of the 2-D filter designed by Design Method 2. Perspective plot; (c) contour plot.

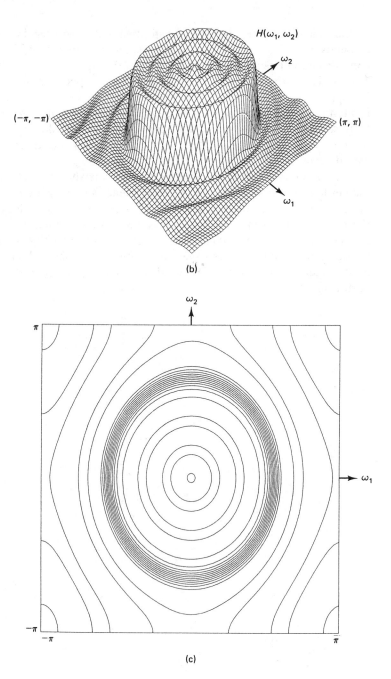

(b)

(c)

Figure 4.21 (continued)

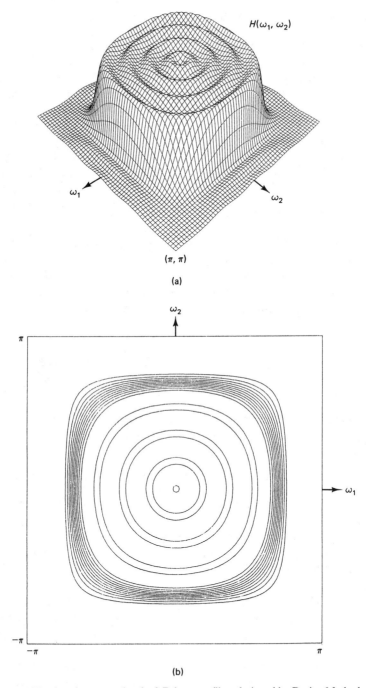

Figure 4.22 Another example of a 2-D lowpass filter designed by Design Method 2. (a) Perspective plot; (b) contour plot.

4.5 OPTIMAL FILTER DESIGN

In the previous two sections, we discussed three methods for designing FIR filters. These methods are relatively simple to understand and use, but they are not optimal. In this section, we discuss the problem of designing optimal FIR digital filters.

In contrast to the 1-D case, development of a practical procedure for reliably designing a 2-D optimal FIR filter remains as an area of research. A detailed discussion of the 2-D optimal FIR filter design problem requires a background in functional approximation theory. Here, we will confine our discussion to some of the major differences between the 1-D and 2-D cases. These contrasts will suggest the complexity of the 2-D case relative to the 1-D case.

4.5.1 Summary of 1-D Optimal Filter Design

We will first briefly review the 1-D optimal filter design problem. For simplicity, we will concentrate on the design of a zero-phase lowpass filter with design specification parameters δ_p (passband tolerance), δ_s (stopband tolerance), ω_p (passband frequency), and ω_s (stopband frequency). From (4.32), the frequency response of a $2N + 1$-point zero-phase 1-D filter can be expressed as

$$H(\omega) = \sum_{n=0}^{N} a(n) \cos \omega n$$

$$= \sum_{i=0}^{N} a_i \phi_i(\omega)$$

(4.62)

where $a(n)$ is simply related to the impulse response $h(n)$, $a_i = a(i)$ and $\phi_i(\omega) = \cos \omega i$. From (4.62), $H(\omega)$ can be viewed as a linear combination of $N + 1$ known basis functions $\phi_i(\omega)$. The problem of designing an optimal filter can be stated as

Given ω_p, ω_s, $k = \delta_p/\delta_s$, N, determine a_i so that δ_s is minimized. (4.63)

It can be shown that a solution to the problem in (4.63) can be used (see Problem 4.17) to solve other formulations of the optimal design problem, such as the problem of determining a_i that minimizes N.

The problem stated above can be shown to be a special case of a weighted Chebyshev approximation problem that can be stated as

Determine a_i that minimizes the maximum error magnitude $|E(\omega)|$ over a closed (compact) subset K of $0 \leq \omega \leq \pi$, where $E(\omega)$ is given by

$$E(\omega) = W(\omega)(D(\omega) - H(\omega)).$$

(4.64)

The function $W(\omega)$ in (4.64) is a positive weighting function, and $D(\omega)$ is the desired function that we wish to approximate with $H(\omega)$. The error criterion used in (4.64) is the minimization of the maximum error between the desired frequency response and the resulting filter frequency response, and (4.64) is sometimes referred to as a *min-max problem* as well as a *weighted Chebyshev approximation problem*. By choosing K, the interval of approximation, such that it consists of the passband region R_p ($0 \leq \omega \leq \omega_p$) and the stopband region R_s ($\omega_s \leq \omega \leq \pi$), by choosing the weighting function $W(\omega)$ to be 1 in R_p and k in R_s, and by choosing the desired frequency response $D(\omega)$ to be 1 in R_p and 0 in R_s, it can be shown that (4.64) reduces to (4.63).

The weighted Chebyshev approximation problem in (4.64) has been studied extensively in mathematics. Suppose we define an $(N + 1) \times 1$-column vector $\mathbf{V}(\omega)$ at a particular frequency ω by

$$\mathbf{V}(\omega) = [\phi_0(\omega), \phi_1(\omega), \phi_2(\omega), \ldots, \phi_N(\omega)]^T. \tag{4.65}$$

If $\mathbf{V}(\omega_i)$ for $0 \leq i \leq N$ are independent vectors for any choice of distinctly different $\omega_i \in K$, then the basis functions $\phi_i(\omega)$ are said to form a *Chebyshev set* and to satisfy the *Haar condition*. The set of basis functions $\phi_i(\omega) = \cos \omega i$ can be shown to satisfy the Haar condition for $0 \leq \omega \leq \pi$. When $\phi_i(\omega)$ satisfies the Haar condition, a powerful theorem known as the *alternation theorem* applies to the weighted Chebyshev approximation problem. The alternation theorem states that there is a unique solution to (4.64). In addition, the theorem states that a necessary and sufficient condition for the unique solution is that the error function $E(\omega)$ exhibit at least $N + 2$ alternations on K. The frequencies at which alternations occur are called *alternation frequencies*. At an alternation frequency, $|E(\omega)|$ reaches its maximum over $\omega \in K$, the domain of approximation. When the theorem is applied to the lowpass filter design problem of (4.63), the alternation frequencies of the resulting lowpass filter will include ω_p, ω_s, and all local extrema (minimum or maximum) of $H(\omega)$ with the possible exception of $\omega = 0$ or π. An example of an optimal filter $H(\omega)$ is shown in Figure 4.23 for the case $N = 9$, $\omega_p = 0.5\pi$, $\omega_s = 0.6\pi$, and $k = \delta_p/\delta_s = 1$. The value δ_s in this case is 0.08. The function

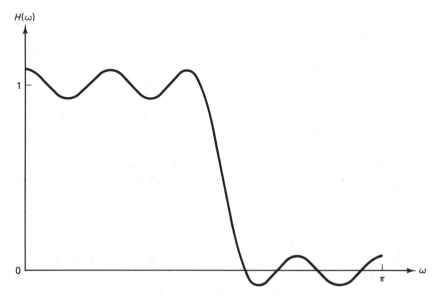

Figure 4.23 Frequency response of a 19-point 1-D optimal lowpass filter.

$H(\omega)$ in Figure 4.23 shows equiripple behavior in the passband and stopband regions. This is typical of optimal filters, and optimal filters are thus sometimes called *equiripple filters*.

The Remez multiple exchange algorithm exploits the necessary and sufficient condition given by the alternation theorem to solve the weighted Chebyshev approximation problem. This exchange algorithm was first used to solve the optimal filter design problem by [Parks and McClellan], so it is referred to as the Parks-McClellan algorithm. The optimal filter design method based on the Remez exchange algorithm is an iterative procedure in which the filter is improved in each iteration. Each iteration consists of two steps. One step is determining candidate coefficients a_i from candidate alternation frequencies. In this step, we attempt to impose the condition that the magnitude of the error function $E(\omega)$ must reach its maximum at the alternation frequencies. The step involves solving $N + 2$ linear equations for δ_s and $N + 1$ coefficients of a_i of the form

$$\vdots$$

$$H(\omega_{i-1}) = 1 - k\delta_s$$

$$H(\omega_i) = 1 + k\delta_s$$

$$H(\omega_p) = 1 - k\delta_s$$

$$H(\omega_s) = \delta_s \tag{4.66}$$

$$H(\omega_j) = -\delta_s$$

$$H(\omega_{j+1}) = \delta_s$$

$$\vdots$$

where ω_i denotes the candidate alternation frequencies and $H(\omega)$ is a linear combination of a_i given by (4.62). Note that (4.66) can be obtained straightforwardly once the alternation frequencies are known. The other step is the determination of candidate alternation frequencies from the candidate coefficients a_i. In this step, we attempt to impose the condition that the local extrema of the error function $E(\omega)$, with the possible exception of $\omega = 0$ or π, constitute most of the alternation frequencies. The step involves evaluating $H(\omega)$ on a dense grid on ω and looking for local extrema of $H(\omega)$. Once the local extrema are found, new candidate alternation frequencies can be determined straightforwardly from the local extrema and bandedge frequencies (ω_p and ω_s). The iterative algorithm is guaranteed to converge and is widely used to design 1-D optimal FIR filters.

4.5.2 Comparison of 1-D and 2-D Optimal Filter Design

We will now contrast the 1-D optimal filter design problem, discussed above, with the 2-D case. Consider the problem of designing a 2-D zero-phase lowpass filter with design specification parameters δ_p, δ_s, R_p (passband region), and R_s (stopband region). From (4.5), the frequency response of a 2-D zero-phase filter can be expressed as

$$H(\omega) = H(\omega_1, \omega_2) = \sum_{(n_1,n_2) \in R_a} \sum a(n_1, n_2) \cos (\omega_1 n_1 + \omega_2 n_2)$$

$$= \sum_{i=0}^{N} a_i \phi_i(\omega_1, \omega_2) \qquad (4.67)$$

$$= \sum_{i=0}^{N} a_i \phi_i(\omega)$$

where $N + 1$ is the number of parameters in $a(n_1, n_2)$, a_i is one parameter in $a(n_1, n_2)$ for a particular (n_1^i, n_2^i), and $\phi_i(\omega_1, \omega_2) = \cos (\omega_1 n_1^i + \omega_2 n_2^i)$. If there is additional symmetry to $h(n_1, n_2)$, for instance, fourfold or eightfold symmetry, $H(\omega)$ can still be expressed in a form similar to (4.67). From (4.67), $H(\omega)$ can be viewed as a linear combination of $N + 1$ known basis functions $\phi_i(\omega)$. The problem of designing an optimal filter can be stated as

Given R_p, R_s, $k = \delta_p/\delta_s$, and N, determine a_i so that δ_s is minimized. (4.68)

As in the 1-D case, it can be shown that a solution to (4.68) can be used to solve other formulations of the optimal filter design problem such as the problem of determining a_i that minimizes N.

The problem stated in (4.68) can be shown to be a special case of a more general problem, which can be stated as

Determine a_i that minimizes the maximum error magnitude

$|E(\omega)|$ over a compact subset K of $(-\pi \leq \omega_1 \leq \pi, 0 \leq \omega_2 \leq \pi)$ (4.69)

where $E(\omega)$ is given by $E(\omega) = W(\omega)(D(\omega) - H(\omega))$.

In (4.69), $W(\omega)$ is a positive weighting function and $D(\omega)$ is the desired frequency response that we wish to approximate with $H(\omega)$. By proper choice of K, $W(\omega)$, $D(\omega)$, and $H(\omega)$, the problem in (4.69) can be shown to reduce to the problem in (4.68). So far, the extensions are straightforward.

In contrast to the 1-D case, the 2-D basis functions $\phi_i(\omega)$ do not satisfy the Haar condition. In other words, the $N + 1$ column vectors $V(\omega_i)$ are not always independent for any $N + 1$ distinct sets of ω_i in K. In fact, no set of nontrivial 2-D functions satisfies the Haar condition, and the powerful alternation theorem does not apply to the problem in (4.69). The theoretical results that apply to (4.69), although less powerful than the alternation theorem, do serve as a basis for an algorithm development for solving (4.69). One theoretical result states that the functional approximation problem in (4.69) does not have a unique solution. This is not too much of an issue, since we need to find only one of many possible solutions. One theorem states that the necessary and sufficient condition for an optimal solution is that the solution have a *critical point set* that consists of *p critical points* where $p \leq N + 2$. The critical points are similar to the alternation frequencies in that $|E(\omega)|$ reaches the maximum at the critical points. But they are different from the alternation frequencies in that only a small subset of the extremal points where $|E(\omega)|$ reaches maximum forms a critical point set and they have to satisfy some additional conditions.*

An iterative procedure analogous to the Remez exchange algorithm can be developed by exploiting the necessary and sufficient conditions for a solution. In the algorithm, we need to determine p. In the 1-D case, the alternation theorem states that the unique solution must have at least $N + 2$ alternations, so the choice of p in the algorithm is $p = N + 2$. In the 2-D case, however, all we know is that $p \leq N + 2$. What is typically done is to choose $p = N + 2$. The case when $p < N + 2$ is called the *degeneracy case*. When it happens, the iterative procedure does not converge. This does not appear to happen very often, however. As in the 1-D case, the iterative procedure attempts to improve the filter in each iteration. Each iteration consists of two steps. One step is the determination of candidate coefficients a_i from candidate critical point frequencies. This step involves solving a set of $N + 2$ linear equations for δ_s and $N + 1$ parameters in a_i of the form

$$H(\omega_i) = 1 + \sigma(\omega_i)k\delta_s \quad \text{for } \omega_i \text{ in } R_p$$
$$H(\omega_j) = \sigma(\omega_j)\delta_s \quad \text{for } \omega_j \text{ in } R_s$$

(4.70)

where ω_i and ω_j are the candidate critical point frequencies, $H(\omega)$ is a linear combination of a_i, and $\sigma(\omega_i)$ and $\sigma(\omega_j)$ are either 1 or -1 and are determined when candidate critical point frequencies are chosen.

The other step in the iteration is determining candidate critical point frequencies from the candidate coefficients a_i. This step involves evaluating $H(\omega_1, \omega_2)$ on a dense grid of (ω_1, ω_2) and looking for some local extrema of $H(\omega_1, \omega_2)$. Evaluating $H(\omega_1, \omega_2)$ on a dense grid typically requires several orders of magnitude more computations than does evaluating $H(\omega)$ on a dense grid. For example, for a 1-D filter design, the rule of thumb is to evaluate $H(\omega)$ on a grid about 20 times the length of the filter. When $h(n)$ has 51 points, the number of grid points used

---

*A critical point set is a set of extremal points of minimum size such that the zero vector lies in the convex hull of the associated signed characteristic vectors $\sigma(\omega_i)V_i(\omega_i)$

$$\text{where } \sigma(\omega_i) = \begin{cases} -1, & E(\omega_i) > 0 \\ 1, & E(\omega_i) < 0. \end{cases}$$

By minimum size, we mean that if any point were removed from the set, the zero vector would no longer be within the convex hull of the reduced set.

If a set of vectors v_i is finite, then a vector v is said to lie in the convex hull of the set v_i if and only if there exist constants α_i that satisfy all the following three conditions:

$$\text{(a) } v = \sum_i \alpha_i v_i \qquad \text{(b) } \alpha_i \geq 0 \quad \text{for all } i \qquad \text{(c) } \sum_i \alpha_i = 1.$$

Qualitatively, the convex hull of a set of vectors is the space bounded by straight lines that join one vector to another vector in the set, where the straight lines are chosen so that the enclosed space includes all straight lines that join any vector in the set to any other vector in the set.

Finite Impulse Response Filters   Chap. 4

is around 1000. In designing a 2-D filter whose sizes range from 5×5 to 9×9, the number of grid points used is typically 256×256 on the Cartesian grid. In addition, $H(\omega_1, \omega_2)$ is also evaluated on a dense grid on the bandedge (such as the passband and stopband) contours, since a critical point set often includes critical points on the bandedge contours. Once $H(\omega_1, \omega_2)$ is evaluated on a dense grid, local extrema are located. Finding local extrema is much more involved in the 2-D than in the 1-D case. The 2-D function $H(\omega_1, \omega_2)$ can have ridges, or nearly flat edges along which the error function may not vary much. As a result, the 2-D function must be carefully searched along many directions. In the 1-D case, finding local extrema involves searching $H(\omega)$ in one direction. Once the local extrema are found, new candidate critical point frequencies are determined from local extrema and current candidate critical point frequencies. This is straightforward in the 1-D case, since all local extrema, with the possible exception of $\omega = 0$ or π are alternation frequencies. Furthermore, all the alternation frequencies can be replaced with new ones in each iteration. In the 2-D case, choosing a set of frequencies that form a new improved critical point set from local extrema and current critical points is quite involved, partly because the critical point set has to satisfy a certain set of conditions.* In addition, not all the critical points are replaced in each iteration, and this tends to increase the number of iterations required relative to the 1-D case.

Partly due to the difficulties cited above, iterative algorithms of the Remez exchange type developed so far are very expensive computationally, and have not been demonstrated to reliably converge to a correct solution when used to design optimal filters. The reason that existing algorithms do not always converge to a correct solution is not well understood. When the roundoff error in computation is reduced in one existing algorithm, for example, the convergence characteristic becomes worse. A comparison of several existing algorithms which are based on the Remez exchange type of iterative procedure discussed above and which differ in their implementation details can be found in [Harris and Mersereau]. Developing a computationally efficient algorithm to design 2-D optimal filters remains an area for research.

Figure 4.24 shows four examples of optimal lowpass filters designed by using the algorithm developed by [Harris and Mersereau]. In all cases, circular symmetry was assumed in the design specification, and eightfold symmetry was assumed for $h(n_1, n_2)$. This assumption significantly reduces the number of coefficients a_i for a given size of $h(n_1, n_2)$. In all cases, the passband and stopband frequencies were 0.4π and 0.6π, respectively, and $k = \delta_p/\delta_s = 1$. The sizes of the filters are 5×5, 7×7, 9×9, and 11×11 in Figures 4.24(a), (b), (c), and (d), respectively.

In this section, we have confined our discussion to some of the major differences between the 1-D and 2-D cases to suggest the complexity of the 2-D case relative to the 1-D case. A detailed discussion of the 2-D optimal FIR filter design problem requires a background in functional approximation theory. An interested reader is referred to [Cheney; Rice; Kamp and Thiran] for a more detailed discussion on the topic.

*See the footnote, p. 242.

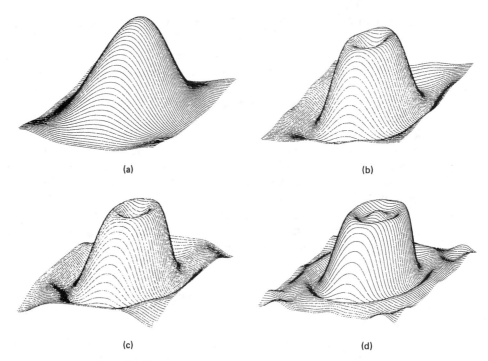

(a)   (b)

(c)   (d)

Figure 4.24 Frequency responses of 2-D optimal lowpass filters. In all cases, circularly symmetric filter specification with $\omega_p = 0.4\pi$, $\omega_s = 0.6\pi$, and $k = \delta_p/\delta_s = 1$ was used. After [Harris and Mersereau]. (a) 5×5-point filter with $\delta_p = 0.2670$; (b) 7×7-point filter with $\delta_p = 0.1269$; (c) 9×9-point filter with $\delta_p = 0.1141$; (d) 11×11-point filter with $\delta_p = 0.0569$.

4.5.3 Optimal Filter Design by Frequency Transformation

A filter designed by the transformation method discussed in Section 4.4 is not in general an optimal filter. To see this qualitatively, note that the filter $h(n_1, n_2)$ designed by using the transformation method has a great deal of structure built into it. As part of this structure, the contours given by $\cos \omega = T(\omega_1, \omega_2)$ are constrained by the region of support size of $t(n_1, n_2)$ and the functional form of $T(\omega_1, \omega_2)$. Since the boundaries of passband and stopband regions in the filter specification do not generally coincide with contours given by $\cos \omega = T(\omega_1, \omega_2)$, the filter designed by the transformation method has to exceed the filter specification. By eliminating the structure imposed on $h(n_1, n_2)$ by the transformation method, we can in general design a filter that exceeds the given filter specification by less than what is possible with the transformation method.

In a very restricted set of cases, however, it is possible to show that an optimal filter can be designed by using the transformation method discussed in Section 4.4. Suppose the transformation sequence $t(n_1, n_2)$ used in the transformation method is first-order (3×3 size) with fourfold symmetry along the n_1 and n_2 axes, and the contours (such as the passband and stopband contours in the lowpass filter design) in the filter specification exactly match the contours given by $\cos \omega =$

$T(\omega_1, \omega_2)$. In addition, suppose the 1-D filter used is optimal in the Chebyshev sense. Under the above assumptions, it has been shown [Harris (1976)] that if the transformation $\cos \omega = T(\omega_1, \omega_2)$ maps an edge (side) of the square $[0, \pi] \times [0, \pi]$ in the (ω_1, ω_2) plane one to one onto $[0, \pi]$ in the ω axis, then the resulting 2-D filter by the transformation method is optimal in the Chebyshev sense. The square $[0, \pi] \times [0, \pi]$ has four possible edges. According to Harris's analysis, a 2-D lowpass filter designed from a 1-D optimal lowpass filter using the McClellan transformation of (4.35) is optimal in the Chebyshev sense if the passband and stopband contours in the filter specification are identical to the contours generated by $\cos \omega = T(\omega_1, \omega_2)$. To see that $\cos \omega = T(\omega_1, \omega_2)$ maps an edge of the square $[0, \pi] \times [0, \pi]$ in the (ω_1, ω_2) plane one to one onto $[0, \pi]$ in the ω axis in this case, consider the edge given by $0 \leq \omega_1 \leq \pi$, $\omega_2 = 0$. Then $\cos \omega = T(\omega_1, \omega_2)$ in the McClellan transformation of (4.35) becomes $\cos \omega = T(\omega_1, 0) = \cos \omega_1$. Clearly the edge maps one to one onto $[0, \pi]$ in the ω axis. An optimal 2-D filter designed using the transformation method with first-order $T(\omega_1, \omega_2)$ chosen such that the contours given by $\cos \omega = T(\omega_1, \omega_2)$ have elliptical shapes is shown in Figure 4.25. This example shows that equiripple points in the passband and stopband regions lie on continuous contours and do not have to be isolated points. This is often the case in 2-D optimal filters.

4.6 IMPLEMENTATION OF FIR FILTERS

In the previous sections, we discussed the problem of specifying and designing an FIR filter. Once the filter is designed, the remaining task is implementation. In this section, we discuss the implementation of an FIR filter.

4.6.1 Implementation of General FIR Filters

The filter implementation problem is to realize a discrete system with $h(n_1, n_2)$, the impulse response of the designed filter. The simplest method of implementing an FIR filter is to use the convolution sum. Let $x(n_1, n_2)$ and $y(n_1, n_2)$ denote the input and output of the filter. Then $y(n_1, n_2)$ is related to $x(n_1, n_2)$ by

$$y(n_1, n_2) = \sum_{(k_1, k_2) \in R_h} \sum h(k_1, k_2) x(n_1 - k_1, n_2 - k_2) \qquad (4.71)$$

where R_h is the region of support of $h(n_1, n_2)$. From (4.71), the number of arithmetic operations required for each output point is about N multiplications and N additions, where N is the number of nonzero coefficients of $h(n_1, n_2)$. As in the 1-D case, the realization can be improved by exploiting the symmetry of $h(n_1, n_2)$. Since $h(n_1, n_2) = h(-n_1, -n_2)$, by rewriting (4.71) and combining the two terms that have the same value for $h(k_1, k_2)$, the number of multiplications can be reduced by about fifty percent without affecting the number of additions.

Any FIR filter can also be implemented by using an FFT algorithm. As shown in Section 3.2.3, the overlap-add method and the overlap-save method can be used to perform the filtering operation. In some cases, this method reduces

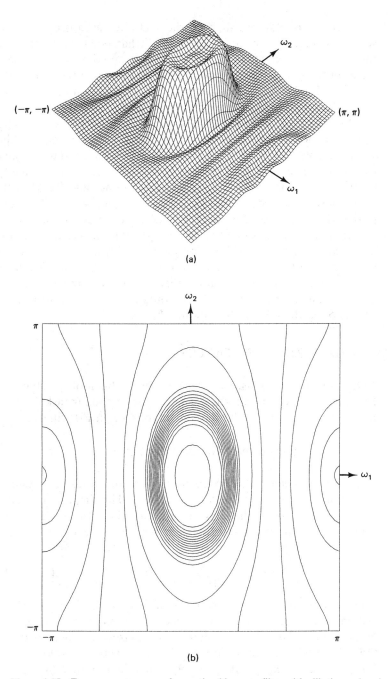

(a)

(b)

Figure 4.25 Frequency response of an optimal lowpass filter with elliptic passband and stopband contours, designed by the transformation method. In a certain restricted set of cases, filters designed by the transformation method are optimal. (a) Perspective plot; (b) contour plot.

the number of arithmetic operations significantly, relative to realization by direct convolution.

Using the convolution sum or an FFT algorithm to implement an FIR filter is very general, in that it applies to any FIR filter independent of how it is designed. If there is some structure in $h(n_1, n_2)$, however, it can be used to reduce the number of arithmetic operations required. For example, the number of multiplications required in the direct convolution case can be reduced by about fifty percent by exploiting the symmetry of $h(n_1, n_2)$. An implementation that exploits a special structure of $h(n_1, n_2)$ is generally applicable to only that class of $h(n_1, n_2)$. The next two sections treat the implementation of FIR filters whose impulse responses have special structures in addition to the structure resulting from $h(n_1, n_2) = h(-n_1, -n_2)$.

4.6.2 Implementation of FIR Filters Designed by the Transformation Method

If a filter is designed by the transformation method discussed in Section 4.4, the number of arithmetic operations required in its implementation can be reduced significantly. Suppose a 2-D filter has been derived from a 1-D filter of length $2N + 1$ and a transformation sequence $t(n_1, n_2)$ in (4.34) of size $(2M_1 + 1) \times (2M_2 + 1)$. The resulting filter $h(n_1, n_2)$ is of size $(2M_1N + 1) \times (2M_2N + 1)$. If the filter is implemented by direct convolution exploiting the property that $h(n_1, n_2) = h(-n_1, -n_2)$, then the number of arithmetic operations per output sample will be around $(2M_1N + 1)(2M_2N + 1)/2$ multiplications and $(2M_1N + 1) \times (2M_2N + 1)$ additions, which is proportional to N^2. The filter designed by the transformation method, however, has a great deal of structure in addition to the structure resulting from $h(n_1, n_2) = h(-n_1, -n_2)$. To see this, suppose $N = 10$ and $M_1 = M_2 = 1$. The resulting filter $h(n_1, n_2)$ is of size 21×21. The sequence $h(n_1, n_2)$, which consists of 441 parameters, is totally specified by a 21-point $h(n)$ that consists of 11 independent parameters and a 3×3-point $t(n_1, n_2)$ that consists of 5 independent parameters.

To exploit the structure present in $h(n_1, n_2)$, we exploit the fact [Equations (4.32) and (4.33)] that $H(\omega_1, \omega_2)$ designed by the transformation method has the following form:

$$H(\omega_1, \omega_2) = \sum_{n=0}^{N} a(n) \cos \omega n \big|_{\cos\omega = T(\omega_1, \omega_2)} \tag{4.72a}$$

or
$$H(\omega_1, \omega_2) = \sum_{n=0}^{N} b(n) (\cos \omega)^n \big|_{\cos\omega = T(\omega_1, \omega_2)}. \tag{4.72b}$$

The most straightforward implementation results when (4.72b) is used directly. From (4.72b), $H(\omega_1, \omega_2)$ can be viewed as a parallel combination of $b(0)$, $b(1)T(\omega_1, \omega_2)$, $b(2)T^2(\omega_1, \omega_2)$, . . . , and $b(N)T^N(\omega_1, \omega_2)$. Since $T^i(\omega_1, \omega_2)$ can be obtained from cascading $T^{i-1}(\omega_1, \omega_2)$ with $T(\omega_1, \omega_2)$, a straightforward implementation of (4.72b) results in the structure shown in Figure 4.26. Since $T(\omega_1, \omega_2)$ corresponds to a finite-extent sequence of $(2M_1 + 1) \times (2M_2 + 1)$ points, the number of arithmetic operations per output point is about $(2M_1 + 1)(2M_2 + 1)N$ multiplications and $(2M_1 + 1)(2M_2 + 1)N$ which is proportional to N. For large values of N, this represents considerable computational savings. When $N = 20$ and $M_1 = M_2 = 1$, direct convolution with only symmetry exploitation requires approximately 850 multiplications and 1700 additions per output point, while the realization in Figure 4.26 involves about 180 multiplications and 180 additions. If the symmetry of $t(n_1, n_2)$ is also exploited, then the number of multiplications can be further reduced by approximately half.

From Figure 4.26, keeping two intermediate results would be sufficient in the implementation. For example, once $f_1(n_1, n_2)$ and $g_1(n_1, n_2)$ are computed, $x(n_1, n_2)$ is not needed. Once $f_2(n_1, n_2)$ and $g_2(n_1, n_2)$ are computed, $f_1(n_1, n_2)$ and $g_1(n_1, n_2)$ can be eliminated. At any given time, therefore, we need to store only two intermediate results, and the amount of storage required in this implementation will be approximately twice as much as in the direct convolution case.

It has been observed that the implementation in Figure 4.26 tends to be somewhat sensitive to finite precision arithmetic. This characteristic can be improved by using (4.72a) and expressing $\cos \omega n$ in terms of $\cos \omega(n - 1)$ and $\cos \omega(n - 2)$. See Problem 4.22 for further discussion.

4.6.3 Cascade Implementation

The cascade structure is a general method of implementing a 1-D filter. In the 1-D cascade form, the system function $H(z)$ is expressed as

$$H(z) = A \prod_k H_k(z). \tag{4.73}$$

Since a 1-D polynomial can always be factored as a product of lower-order polynomials, $H(z)$ can always be expressed in the form of (4.73). In contrast, the cascade form is not a general structure that can be used in implementing 2-D filters.

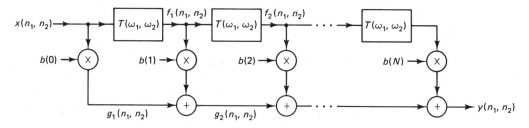

Figure 4.26 Computationally efficient method for implementation of a 2-D filter designed by the transformation method. In the figure,

$$f_i(n_1, n_2) = f_{i-1}(n_1, n_2) * t(n_1, n_2)$$

$$\text{and } g_i(n_1, n_2) = g_{i-1}(n_1, n_2) + b(i - 1)f_{i-1}(n_1, n_2).$$

Finite Impulse Response Filters   Chap. 4

A 2-D polynomial cannot, in general, be factored as a product of lower-order polynomials, and $H(z_1, z_2)$ cannot generally be written in the form of

$$H(z_1, z_2) = A \prod_k H_k(z_1, z_2). \qquad (4.74)$$

To use a cascade form in the realization of a 2-D FIR filter, therefore, the form in (4.74) should be used explicitly in the design step.

If $H(z_1, z_2)$ can be expressed in the form given by (4.74), that information can be exploited in reducing the number of arithmetic operations. To see this, suppose $H(z_1, z_2)$ can be expressed as

$$H(z_1, z_2) = H_1(z_1, z_2)H_2(z_1, z_2). \qquad (4.75)$$

Suppose further that $H_1(z_1, z_2)$ corresponds to an $M \times M$-point FIR filter and that $H_2(z_1, z_2)$ corresponds to an $N \times N$-point FIR filter. The resulting $H(z_1, z_2)$ will correspond to an FIR filter of size $(M + N - 1) \times (M + N - 1)$. If $H(z_1, z_2)$ is directly implemented by using convolution without exploiting the symmetry of $h(n_1, n_2)$, computation of each output sample will require approximately $(M + N - 1)^2$ multiplications and $(M + N - 1)^2$ additions. If $H(z_1, z_2)$ is implemented by exploiting the structure in (4.75), but without exploiting the symmetry of $h_1(n_1, n_2)$ or $h_2(n_1, n_2)$, computation of each output sample will require approximately $M^2 + N^2$ multiplications and $M^2 + N^2$ additions. The M^2 multiplications and M^2 additions result from $H_1(z_1, z_2)$ and the N^2 multiplications and N^2 additions result from $H_2(z_1, z_2)$. When $M = N$, exploiting the structure in (4.75) reduces the number of arithmetic operations by fifty percent. If $H(z_1, z_2)$ can be expressed as a product of L terms, where each term has the same size, the number of arithmetic operations can be reduced by a factor of L by exploiting the cascade form.

The notion that the cascade form in (4.75) imposes a structural constraint on $H(z_1, z_2)$ can be easily demonstrated. In the above example, since $H(z_1, z_2)$ is totally specified by $H_1(z_1, z_2)$ and $H_2(z_1, z_2)$, $(M + N - 1)^2$ values of $H(z_1, z_2)$ are totally specified by M^2 values of $H_1(z_1, z_2)$ and N^2 values of $H_2(z_1, z_2)$. As a result, $H(z_1, z_2)$ in the form of (4.75) has a smaller number of parameters or degrees of freedom (by about fifty percent when $N = M$) relative to a general $H(z_1, z_2)$ of the same size. A general $H(z_1, z_2)$, therefore, cannot always be expressed in the form of (4.75).

In the 1-D case, the cascade form does not impose any structural constraint on $H(z)$. Suppose $H(z)$ is expressed as

$$H(z) = H_1(z)H_2(z). \qquad (4.76)$$

If $H_1(z)$ and $H_2(z)$ in (4.76) correspond to an M-point and an N-point sequence, respectively, $H(z)$ corresponds to an $M + N - 1$-point sequence. In contrast to the 2-D case, the total number of parameters in $H_1(z)$ and $H_2(z)$ is not less than that in $H(z)$. The cascade form in (4.76) does not impose any structural constraint on $H(z)$ and cannot be exploited to reduce the number of arithmetic operations.

Imposing the structural constraint on $H(z_1, z_2)$ in the form of (4.75) typically makes the design problem considerably more complicated, since $H(\omega_1, \omega_2)$ is no

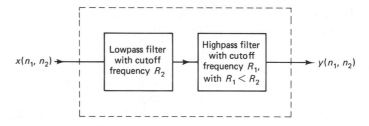

Figure 4.27 Realization of a bandpass filter by cascading a highpass filter with a lowpass filter.

longer a linear combination of the parameters that need to be determined in the design. For example, suppose $H_1(z_1, z_2)$ and $H_2(z_1, z_2)$ have forms given by

$$H_1(z_1, z_2) = a + bz_1^{-1} + bz_1 + cz_2^{-1} + cz_2 \tag{4.77}$$

and

$$H_2(z_1, z_2) = d + ez_1^{-1} + ez_1.$$

The function $H(z_1, z_2)$ is given by

$$
\begin{aligned}
H(z_1, z_2) = {} & ad + 2be + (ae + bd)z_1^{-1} + (ae + bd)z_1 + cdz_2^{-1} + cdz_2 \\
& + bez_1^{-2} + bez_1^2 + cez_1^{-1}z_2 + cez_1z_2^{-1} + cez_1^{-1}z_2^{-1} + cez_1z_2.
\end{aligned}
\tag{4.78}
$$

Clearly, the coefficients in $H(z_1, z_2)$ that need to be determined in the design step are nonlinear functions of a, b, c, d, and e.

In some cases, it is possible to design a complicated filter by cascading simpler filters. For example, by cascading a highpass filter with a lowpass filter, it is possible to design a bandpass filter. This is shown in Figure 4.27. If we have efficient ways to design a lowpass filter and a highpass filter with impulse response of a certain size, then a bandpass filter with a larger region of support for the impulse response may be obtained by cascading the two simpler filters. In this case, the frequency responses of the filters designed are no longer nonlinear functions of the parameters determined in the design step.

REFERENCES

For a review of 1-D FIR filter design methods, see [Rabiner; Rabiner et al.]. For the effect of symmetry on 2-D FIR filter design, see [Aly and Fahmy; Rajan et al.]. For the importance of phase in digital filters applied to image processing problems, see [Huang et al.].

For a survey of 1-D windows, see [Harris (1978)]. For 2-D window design methods, see [Huang; Kato and Furukawa; Yu and Mitra]. For a discussion on the initial choice of window parameters in the window design method, see [Speake and Mersereau].

For readings on the frequency transformation method for FIR filter design, see [McClellan; Mersereau et al.; Mersereau]. For design of circularly symmetric filters using the transformation method, see [Rajan and Swamy; Hazra and Reddy]. For implementation of filters designed by the transformation method, see [Mecklenbrauker and Mersereau; McClellan and Chan].

For books on functional approximation theory, see [Cheney; Rice]. For the Parks-McClellan algorithm for 1-D optimal filter design, see [Parks and McClellan; McClellan et al.]. For one of the earlier attempts to design 2-D optimal FIR filters based on a computationally very expensive linear programming approach, see [Hu and Rabiner]. For the Remez exchange type approach to the 2-D optimal FIR filter design problem, see [Kamp and Thiran; Harris and Mersereau]. For a design based on the l_p norm, which can be used to design an optimal filter approximately, see [Lodge and Fahmy]. For an optimal filter design approach based on general functional optimization methods, see [Charalambous].

For an approach to the design of FIR filters with a special structure such as the cascade of simple filters, see [Abramatic and Faugeras].

J. F. Abramatic and O. D. Faugeras, Sequential convolution techniques for image filtering, *IEEE Trans. Acoust., Speech and Sig. Proc.*, Vol. ASSP-30, February 1982, pp. 1–10.

S. A. H. Aly and M. M. Fahmy, Symmetry in two-dimensional rectangularly sampled digital filters, *IEEE Trans. Acoust., Speech and Sig. Proc.*, Vol. ASSP-29, August 1981, pp. 794–805.

C. Charalambous, The performance of an algorithm for minimax design of two-dimensional linear phase FIR digital filters, *IEEE Trans. Circuits Syst.*, Vol. CAS-32, October 1985, pp. 1016–1028.

E. W. Cheney, *Introduction to Approximation Theory*. New York: McGraw-Hill, 1966.

D. B. Harris, Iterative procedures for optimal Chebyshev design of FIR digital filters, S.M. thesis, MIT EECS Dept., February 1976.

D. B. Harris and R. M. Mersereau, A comparison of algorithms for minimax design of two-dimensional linear phase FIR digital filters, *IEEE Trans. Acoust., Speech and Sig. Proc.*, Vol. ASSP-25, December 1977, pp. 492–500.

F. J. Harris, On the use of windows for harmonic analysis with the discrete Fourier transform, *Proc. IEEE*, Vol. 66, January 1978, pp. 51–83.

S. N. Hazra and M. S. Reddy, Design of circularly symmetric low-pass two-dimensional FIR digital filters using transformation, *IEEE Trans. Circuits Syst.*, Vol. CAS-33, October 1986, pp. 1022–1026.

J. V. Hu and L. R. Rabiner, Design techniques for two-dimensional digital filters, *IEEE Trans. Audio Electroacoust.*, Vol. AU-20, October 1972, pp. 249–257.

T. S. Huang, Two-dimensional windows, *IEEE Trans. Audio Electroacoust.*, Vol. AU-20, March 1972, pp. 88–90.

T. S. Huang, J. W. Burnett, and A. G. Deczky, The importance of phase in image processing filters, *IEEE Trans. Acoust., Speech and Sig. Proc.*, Vol. ASSP-23, December 1975, pp. 529–542.

Y. Kamp and J. P. Thiran, Chebyshev approximation for two-dimensional nonrecursive digital filters, *IEEE Trans. Circuits Syst.*, Vol. CAS-22, March 1975, pp. 208–218.

H. Kato and T. Furukawa, Two-dimensional type-preserving circular windows, *IEEE Trans. Acoust., Speech and Sig. Proc.*, Vol. ASSP-29, August 1981, pp. 926–928.

J. H. Lodge and M. F. Fahmy, An efficient l_p optimization technique for the design of two-dimensional linear phase FIR digital filters, *IEEE Trans. Acoust., Speech and Sig. Proc.*, Vol. ASSP-28, June 1980, pp. 308–313.

J. H. McClellan, The design of two-dimensional digital filters by transformation, *Proc. 7th Annu. Princeton Conf. Inform. Sci. Syst.*, 1973, pp. 247–251.

J. H. McClellan and D. S. K. Chan, A 2-D FIR filter structure derived from the Chebyshev recursion, *IEEE Trans. Circuits Syst.*, Vol. CAS-24, July 1977, pp. 372–378.

J. H. McClellan, T. W. Parks, and L. R. Rabiner, A computer program for designing optimum FIR linear phase digital filters, *IEEE Trans. Audio Electroacoust.*, Vol. AU-21, December 1973, pp. 506–526.

W. F. G. Mecklenbrauker and R. M. Mersereau, McClellan transformations for two-dimensional digital filtering: II. Implementation, *IEEE Trans. Circuits Syst.*, Vol. CAS-23, July 1976, pp. 414–422.

R. M. Mersereau, The design of arbitrary 2-D zero-phase FIR filters using transformations, *IEEE Trans. Circuits Syst.*, Vol. CAS-27, February 1980, pp. 142–144.

R. M. Mersereau, W. F. G. Mecklenbrauker and T. F. Quatieri, Jr., McClellan transformations for two-dimensional digital filtering: I. Design, *IEEE Trans. Circuits Syst.*, Vol. CAS-23, July 1976, pp. 405–414.

T. W. Parks and J. H. McClellan, Chebyshev approximation for nonrecursive digital filters with linear phase, *IEEE Trans. Circuit Theory*, Vol. CT-19, March 1972, pp. 189–194.

L. R. Rabiner, Techniques for designing finite-duration impulse-response digital filters, *IEEE Trans. Commun. Technol.*, Vol. COM-19, April 1971, pp. 188–195.

L. R. Rabiner, J. H. McClellan, and T. W. Parks, FIR digital filter design techniques using weighted Chebyshev approximation, *Proc. IEEE*, Vol. 63, April 1975, pp. 595–610.

P. K. Rajan, H. C. Reddy, and M. N. S. Swamy, Fourfold rational symmetry in two-dimensional functions, *IEEE Trans. Acoust., Speech and Sig. Proc.*, Vol. ASSP-30, June 1982, pp. 488–499.

P. K. Rajan and M. N. S. Swamy, Design of circularly symmetric two-dimensional FIR digital filters employing transformations with variable parameters, *IEEE Trans. Acoust., Speech and Sig. Proc.*, Vol. ASSP-31, June 1983, pp. 637–642.

J. R. Rice, *The Approximation of Functions, Vol. 2, Nonlinear and Multivariate Theory*, Reading, MA: Addison-Wesley, 1969.

T. S. Speake and R. M. Mersereau, A note on the use of windows for two-dimensional FIR filter design, *IEEE Trans. Acoust., Speech and Sig. Proc.*, Vol. ASSP-29, February 1981, pp. 125–127.

T. H. Yu and S. K. Mitra, A new two-dimensional window, *IEEE Trans. Acoust., Speech and Sig. Proc.*, Vol. ASSP-33, August 1985, pp. 1058–1061.

PROBLEMS

4.1. Let $H(\omega_1, \omega_2)$ and $h(n_1, n_2)$ denote the frequency response and impulse response of an FIR filter. For each of the following constraints that $H(\omega_1, \omega_2)$ satisfies, give an example of a 3 × 3-point $h(n_1, n_2)$.

(a) $H(\omega_1, \omega_2) = H(\omega_1, -\omega_2)$

(b) $H(\omega_1, \omega_2) = H(\omega_2, -\omega_1)$

4.2. Let $h(n_1, n_2)$ and $w(n_1, n_2)$ denote two zero-phase sequences. Show that $h(n_1, n_2)w(n_1, n_2)$ is also a zero-phase sequence.

4.3. Consider a band-limited analog signal $x_a(t_1, t_2)$ whose Fourier transform $X_a(\Omega_1, \Omega_2)$ has the following property:

$$X_a(\Omega_1, \Omega_2) = 0 \quad \text{outside} \quad |\Omega_1| \le 2\pi \cdot 10{,}000, \; |\Omega_2| \le 2\pi \cdot 10{,}000.$$

We wish to filter $x_a(t)$ with an analog filter whose magnitude response $|H_a(\Omega_1, \Omega_2)|$ satisfies a certain set of specifications. Suppose the analog filter above is implemented by digital processing as follows:

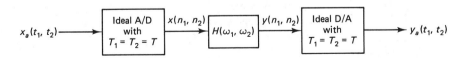

Figure P4.3

The sampling frequency $f_s = 1/T$ in both the A/D and D/A converters is 30 kHz. Suppose the digital filter $H(\omega_1, \omega_2)$ is a highpass filter that meets the following design specification:

$$1 - \delta_p \le |H(\omega_1, \omega_2)| \le 1 + \delta_p, \quad \text{for } \omega_p \le \sqrt{\omega_1^2 + \omega_2^2}$$

$$|H(\omega_1, \omega_2)| \le \delta_s, \quad \text{for } 0 \le \sqrt{\omega_1^2 + \omega_2^2} \le \omega_s.$$

Suppose $\omega_p = \pi/2$, $\omega_s = \pi/4$, $\delta_p = 0.1$, $\delta_s = 0.05$. What filter specifications will the analog filter $|H_a(\Omega_1, \Omega_2)|$ satisfy?

4.4. Let $h_l(n)$ denote the impulse response of a 1-D FIR lowpass filter. One simple way to design a 1-D FIR highpass filter $h(n)$ is by

$$h(n) = (-1)^n h_l(n).$$

(a) Show that $h(n)$ is the impulse response of a highpass filter.

(b) Let $h_l(n_1, n_2)$ denote the impulse response of a 2-D FIR lowpass filter. A straightforward extension of the above method is to design a 2-D FIR highpass filter $h(n_1, n_2)$ by

$$h(n_1, n_2) = (-1)^{n_1}(-1)^{n_2} h_l(n_1, n_2).$$

Is this a good method of designing a highpass filter?

4.5. In this problem, we use the window method to design a 2-D FIR filter.

(a) Suppose we wish to design a zero-phase bandstop filter with the following design specifications:

$$0.95 < H(\omega_1, \omega_2) < 1.05 \quad \text{for} \quad (\omega_1, \omega_2) \in \{\omega_1^2 + \omega_2^2 \le (0.3\pi)^2\}$$

$$-0.04 < H(\omega_1, \omega_2) < 0.04 \quad \text{for} \quad (\omega_1, \omega_2) \in \{(0.4\pi)^2 \le \omega_1^2 + \omega_2^2 \le (0.6\pi)^2\}$$

$$0.95 < H(\omega_1, \omega_2) < 1.05 \quad \text{for} \quad (\omega_1, \omega_2) \in \{(0.7\pi)^2 \le \omega_1^2 + \omega_2^2\}.$$

Determine $h_d(n_1, n_2)$, the impulse response of the ideal bandstop filter, to be used in the window method.

(b) A flowchart of the process of designing a lowpass filter with circularly symmetric filter specification is shown below.

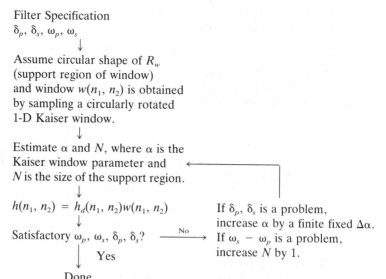

Filter Specification
δ_p, δ_s, ω_p, ω_s
↓

Assume circular shape of R_w
(support region of window)
and window $w(n_1, n_2)$ is obtained
by sampling a circularly rotated
1-D Kaiser window.
↓

Estimate α and N, where α is the
Kaiser window parameter and
N is the size of the support region.
↓

$h(n_1, n_2) = h_d(n_1, n_2)w(n_1, n_2)$
↓

Satisfactory ω_p, ω_s, δ_p, δ_s? $\xrightarrow{\text{No}}$

If δ_p, δ_s is a problem,
increase α by a finite fixed $\Delta\alpha$.
If $\omega_s - \omega_p$ is a problem,
increase N by 1.

↓ Yes

Done

Suppose a filter that meets the design specification is obtained by using the above procedure. Let the values of α and N of the designed filter be denoted by α' and N'. Give two reasons why a different choice of α and N ($\alpha = \alpha''$, $N = N''$) might lead to a more optimal filter than the filter designed (more optimal in the sense that $N'' < N'$ still meets the required design specification).

4.6. We wish to design a zero-phase bandpass filter whose ideal magnitude response $|H_d(\omega_1, \omega_2)|$ is given by

$$|H_d(\omega_1, \omega_2)| = \begin{cases} 1, & 0.2\pi \leq \sqrt{\omega_1^2 + \omega_2^2} \leq 0.4\pi \\ 0, & \text{otherwise.} \end{cases}$$

We wish to design an FIR filter whose impulse response $h(n_1, n_2)$ is zero outside $-5 \leq n_1 \leq 5$ and $-5 \leq n_2 \leq 5$ by minimizing

$$\text{Error} = \frac{1}{(2\pi)^2} \int_{\omega_1 = -\pi}^{\pi} \int_{\omega_2 = -\pi}^{\pi} |H_d(\omega_1, \omega_2) - H(\omega_1, \omega_2)|^2 \, d\omega_1 \, d\omega_2$$

where $H(\omega_1, \omega_2)$ is the frequency response of the resulting digital filter. Determine $h(n_1, n_2)$ and show that it minimizes the above error expression.

4.7. In the frequency sampling method, we ensure that $H(\omega_1, \omega_2)$, the 2-D filter designed, will be identical to some given $H_d(\omega_1, \omega_2)$ at a set of frequencies. The frequencies in the chosen set do not have to be on the Cartesian grid, and the zero-phase filter $h(n_1, n_2)$ can be determined by solving a set of linear equations of the form

$$H_d(\omega_1, \omega_2)\big|_{\omega_1 = \omega_1^i, \omega_2 = \omega_2^i} = h(0, 0) + \sum\sum_{(n_1, n_2) \in R_h} 2h(n_1, n_2) \cos(\omega_1^i n_1 + \omega_2^i n_2).$$

For M independent values in $h(n_1, n_2)$, M linear equations almost always have a unique solution for $h(n_1, n_2)$. However, it is possible to select a set of M distinct frequencies

(ω_1', ω_2') such that the resulting set of equations does not have a unique solution. Consider a 3×3-point $h(n_1, n_2)$. In this case, $M = 5$. Determine one set of five distinctly different frequencies (ω_1', ω_2') for which the five linear equations will not be consistent (will have no solution).

4.8. In the filter design by the frequency transformation method, a 2-D FIR filter $H(\omega_1, \omega_2)$ is designed from a 1-D FIR filter $H(\omega)$ by

$$H(\omega_1, \omega_2) = H(\omega)\big|_{\cos\omega = T(\omega_1, \omega_2)}.$$

We assume that $H(\omega)$ is real so that the 1-D filter used will have zero phase.

(a) Suppose the transformation sequence $t(n_1, n_2)$ has a fourfold symmetry given by

$$t(n_1, n_2) = t(-n_1, n_2) = t(n_1, -n_2).$$

What type of symmetry does $h(n_1, n_2)$ have?

(b) Suppose $t(n_1, n_2)$ has an eightfold symmetry given by

$$t(n_1, n_2) = t(-n_1, n_2) = t(n_1, -n_2) = t(n_2, n_1).$$

What type of symmetry does $h(n_1, n_2)$ have?

4.9. Suppose we design a 2-D FIR filter $H(\omega_1, \omega_2)$ from a 1-D FIR filter $H(\omega)$ by

$$H(\omega_1, \omega_2) = H(\omega)\big|_{\cos\omega = -1/2 + (\cos\omega_1)/2 + (\cos\omega_2)/2 + (\cos\omega_1\cos\omega_2)/2}.$$

Suppose $h(n)$, the impulse response of the 1-D filter, is given by

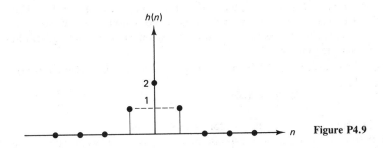

Figure P4.9

Determine $h(n_1, n_2)$, the impulse response of the 2-D filter.

4.10. Suppose we design a 2-D zero-phase lowpass filter $H(\omega_1, \omega_2)$ from a 1-D zero-phase lowpass filter $H(\omega)$ by

$$H(\omega_1, \omega_2) = H(\omega)\big|_{\cos\omega = A + B\cos\omega_1 + C\cos\omega_2}.$$

We wish the resulting filter $H(\omega_1, \omega_2)$ to be symmetric with respect to the ω_1 axis, the ω_2 axis, the $\omega_2 = \omega_1$ axis, and the $\omega_2 = -\omega_1$ axis.

(a) Determine one set of A, B, and C that could be used. Explain your answer. You may assume that a reasonable 1-D lowpass filter $H(\omega)$ is available to you.

(b) Suppose $h(n)$, the impulse response of the 1-D filter used, is nonzero for $-2 \leq n \leq 2$ and zero outside $-2 \leq n \leq 2$. Determine the number of nonzero coefficients in $h(n_1, n_2)$.

4.11. We wish to design a 2-D FIR filter from a 1-D FIR filter by the transformation method. Suppose the transformation sequence $t(n_1, n_2)$ is zero except for the five points shown in the following figure.

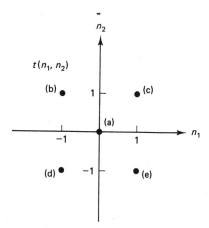

Figure P4.11

(a) What are the minimum constraints on the coefficients a, b, c, d, and e for the resulting 2-D filter to be zero phase [i.e., for the filter frequency response $H(\omega_1, \omega_2)$ to be real]? Assume the 1-D filter used in the transformation method is zero phase.

(b) Suppose $a = b = e = 0$ and $c = d = 1/2$. Suppose also that the 1-D zero-phase filter used is a lowpass filter with a passband region given by $-\pi/2 \le \omega \le \pi/2$. Sketch the complete passband region (or regions) of the resulting 2-D filter. Label the axes in your sketch.

4.12. Suppose we design a 2-D FIR filter $H(\omega_1, \omega_2)$ from a 1-D FIR filter $H(\omega)$ by

$$H(\omega_1, \omega_2) = H(\omega)\big|_{\cos \omega = T(\omega 1, \omega 2) = \cos(\omega 1 - \omega 2)}.$$

(a) Determine $t(n_1, n_2)$, the transformation sequence.

(b) If $H(\omega)$ is the frequency response of a highpass filter, what type of filter is $H(\omega_1, \omega_2)$?

4.13. In FIR filter design by the frequency transformation method, the 2-D filter frequency response $H(\omega_1, \omega_2)$ is obtained from the 1-D filter frequency response $H(\omega)$ by

$$H(\omega_1, \omega_2) = H(\omega)\big|_{\cos \omega = T(\omega 1, \omega 2)}.$$

The filter designed by using the above transformation is typically not circularly symmetric. One clever student suggested the following approach:

$$H(\omega_1, \omega_2) = H(\omega)\big|_{\omega = \sqrt{\omega_1^2 + \omega_2^2}}.$$

Is this a good approach? If so, explain why. If not, discuss the problems of this approach.

4.14. Consider a 3×3-point transformation sequence $t(n_1, n_2)$. We design a 2-D zero-phase filter $H(\omega_1, \omega_2)$ from a 1-D zero-phase filter $H(\omega)$ by

$$H(\omega_1, \omega_2) = H(\omega)|_{\cos\omega = T(\omega_1, \omega_2)}$$

where $T(\omega_1, \omega_2)$ is the Fourier transform of $t(n_1, n_2)$.

(a) If we impose the zero-phase constraint of $t(n_1, n_2)$ such that $t(n_1, n_2) = t(-n_1, -n_2)$, show that $T(\omega_1, \omega_2)$ can be expressed in the form

$$T(\omega_1, \omega_2) = A + B \cos \omega_1 + C \cos \omega_2 + D \cos \omega_1 \cos \omega_2 + E \sin \omega_1 \sin \omega_2.$$

(b) We wish to design a filter whose ideal frequency response $H_d(\omega_1, \omega_2)$ is shown in the following figure.

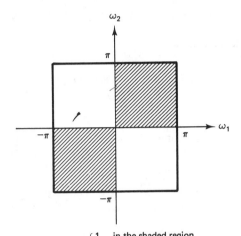

$$H_d(\omega_1, \omega_2) = \begin{cases} 1, & \text{in the shaded region} \\ 0, & \text{in the unshaded region} \end{cases}$$

Figure P4.14

Determine one set of A, B, C, D, and E that could be used in the design of a 2-D filter.

(c) With the choice of parameters in (b), what type of 1-D filter $H(\omega)$ do we need to design the filter in (b)?

4.15. Suppose we design a 2-D zero-phase lowpass filter $H(\omega_1, \omega_2)$ from a 1-D zero-phase filter $H(\omega)$ by

$$H(\omega_1, \omega_2) = H(\omega)|_{\cos\omega = -1/2 + (\cos\omega_1)/2 + (\cos\omega_2)/2 + (\cos\omega_1\cos\omega_2)/2}.$$

The 2-D filter $H(\omega_1, \omega_2)$ has to satisfy the following specification:

$$0.95 \le H(\omega_1, \omega_2) \le 1.05, \quad (\omega_1, \omega_2) \in \text{passband region (see figure)}$$

$$-0.02 \le H(\omega_1, \omega_2) \le 0.02, \quad (\omega_1, \omega_2) \in \text{stopband region (see figure)}$$

Determine the filter specification that the 1-D filter $H(\omega)$ has to satisfy so that the resulting 2-D filter $H(\omega_1, \omega_2)$ will be guaranteed to meet the above filter specification.

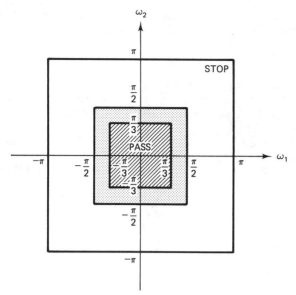

Figure P4.15

4.16. Consider a 1-D zero-phase FIR filter $h(n)$ with length $2N + 1$. The frequency response $H(\omega)$ can be expressed as

$$H(\omega) = \sum_{n=-N}^{N} h(n)e^{-j\omega n} \tag{1}$$

$$= \sum_{n=0}^{N} a(n)\cos \omega n \tag{2}$$

where $a(0) = h(0)$ and $a(n) = 2h(n)$ for $n \geq 1$. In this problem, we show that $H(\omega)$ can also be expressed in the form of

$$H(\omega) = \sum_{n=0}^{N} b(n)(\cos \omega)^n \tag{3}$$

where $b(n)$ can be simply related to $a(n)$ or $h(n)$. Note that

$$\cos (A + B) = \cos A \cos B - \sin A \sin B$$

and $\sin (A + B) = \sin A \cos B + \cos A \sin B$.

(a) Show that $\cos 2\omega = 2 \cos \omega \cos \omega - 1$.
(b) Show that $\cos 3\omega = 2 \cos \omega \cos 2\omega - \cos \omega$.
(c) More generally, show that for $n \geq 2$,

$$\cos \omega n = 2 \cos \omega \cos \omega(n - 1) - \cos \omega(n - 2).$$

(d) Let cos ωn be denoted by $P_n[\cos \omega]$. From the results in (a), (b), and (c), show that

$$P_0[\cos \omega] = 1$$

$$P_1[\cos \omega] = \cos \omega$$

$$P_2[\cos \omega] = 2 \cos \omega P_1[\cos \omega] - P_0[\cos \omega] \tag{4}$$

$$\vdots$$

$$P_n[\cos \omega] = 2 \cos \omega P_{n-1}[\cos \omega] - P_{n-2}[\cos \omega].$$

The polynomial $P_n[\cdot]$ is known as the *Chebyshev polynomial*.

(e) From the result of (d), show that cos ωn can be expressed as a linear combination of 1, cos ω, $(\cos \omega)^2$, . . . , $(\cos \omega)^n$. In other words, show that

$$\cos \omega n = \sum_{k=0}^{n} q(k) (\cos \omega)^k \quad \text{for some } q(k). \tag{5}$$

(f) Using (4), determine $q(k)$ in (5) for cos 3ω.

(g) From (2) and (5), show that $H(\omega)$ can be expressed in the form of (3).

(h) Suppose $h(n)$ is given by

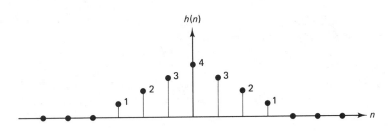

Figure P4.16

Determine $a(n)$ in (2) and $b(n)$ in (3).

4.17. We wish to design a 2-D zero-phase lowpass filter with design specification parameters δ_p (passband tolerance), δ_s (stopband tolerance), R_p (passband region), and R_s (stopband region). We assume that the impulse response of the filter designed has a region of support of $(2N + 1) \times (2N + 1)$ points. Suppose we have developed an algorithm that solves the following problem:

Given R_p, R_s, $k = \delta_p/\delta_s$, and N, determine $h(n_1, n_2)$ so that δ_s is minimized.

We'll refer to the algorithm as Algorithm A.

(a) Using Algorithm A, develop a method that solves the following problem:

Given R_p, R_s, δ_p, and δ_s, determine $h(n_1, n_2)$ so that N is minimized.

(b) Suppose R_p is given by

$$\sqrt{\omega_1^2 + \omega_2^2} \leq \omega_p$$

and R_s is given by

$$\sqrt{\omega_1^2 + \omega_2^2} \geq \omega_s.$$

Using Algorithm A, develop a method that solves the following problem:

Given ω_p, δ_p, δ_s, and N, determine $h(n_1, n_2)$ so that ω_s is minimized.

4.18. Consider an FIR filter with a region of support of $(2N + 1) \times (2N + 1)$ points. If we implement the filter by direct convolution, the number of multiplications required per output point will be $(2N + 1)^2$. If $h(n_1, n_2)$ has some symmetry, this symmetry can be exploited to reduce the number of multiplications. Suppose $h(n_1, n_2)$ has a fourfold symmetry given by

$$h(n_1, n_2) = h(-n_1, n_2) = h(n_1, -n_2).$$

Discuss how this can be used to reduce the number of multiplications. Approximately how many multiplications can be reduced in this case? Is it possible to reduce the number of additions required by exploiting the symmetry?

4.19. Suppose we define an $(N + 1) \times 1$-column vector $\mathbf{V}(\boldsymbol{\omega}) = \mathbf{V}(\omega_1, \omega_2)$ at a particular frequency $\boldsymbol{\omega}$ by

$$\mathbf{V}(\boldsymbol{\omega}) = [\phi_0(\boldsymbol{\omega}), \phi_1(\boldsymbol{\omega}), \phi_2(\boldsymbol{\omega}), \ldots , \phi_N(\boldsymbol{\omega})]^T.$$

If $\mathbf{V}(\boldsymbol{\omega}_i)$ for $0 \leq i \leq N$ are independent vectors for any choice of distinctly different $\boldsymbol{\omega}_i \in K$ where K is some known region, then the functions $\phi_i(\boldsymbol{\omega})$ are said to form a Chebyshev set or satisfy the Haar condition. Show that $\phi_i(\boldsymbol{\omega}) = \cos(\omega_1 n_1^i + \omega_2 n_2^i)$ for integers n_1^i and n_2^i do not satisfy the Haar condition for K given by $0 \leq \omega_1 \leq \pi, 0 \leq \omega_2 \leq \pi$.

4.20. In some geographical applications, it is useful to design a *fan filter* whose ideal frequency response $H_d(\omega_1, \omega_2)$ is shown in the following figure.

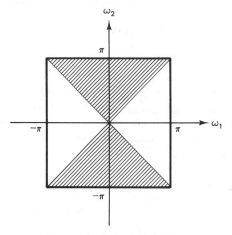

$$H_d(\omega_1, \omega_2) = \begin{cases} 1, & \text{in the shaded region} \\ 0, & \text{in the unshaded region} \end{cases}$$

Figure P4.20

One approach suggested is to design a 2-D zero-phase fan filter $H(\omega_1, \omega_2)$ from a 1-D zero-phase filter $H(\omega)$ by

$$H(\omega_1, \omega_2) = H(\omega)\big|_{\cos\omega = T(\omega1,\omega2) = (\cos\omega1)/2 - (\cos\omega2)/2}.$$

(a) What type of 1-D filter do we need?

(b) Suppose the boundaries of the passband and stopband regions in the fan filter specification coincide with the contours given by $\cos\omega = T(\omega_1, \omega_2)$. Also suppose that the 1-D filter is designed by using the Parks-McClellan algorithm. Is the 2-D filter $H(\omega_1, \omega_2)$ designed optimal in the Chebyshev sense?

4.21. Suppose an FIR filter that meets its design specification has been obtained. The region of support of the digital filter $h(n_1, n_2)$ is 25×25 points in size. We consider realizing this filter using the overlap-add method to exploit the computational efficiency of an FFT algorithm. The size of the DFT used is $N \times N$, where N can be expressed as a power of 2. What is the approximate number of multiplications per output point in terms of N? Assume that the region of support of the input $x(n_1, n_2)$ is very large and that the row-column decomposition method with a decimation-in-time FFT algorithm for the 1-D DFT computation is used.

4.22. The frequency response $H(\omega_1, \omega_2)$ of an FIR filter designed by the transformation method can be expressed as

$$H(\omega_1, \omega_2) = \sum_{n=0}^{N} a(n) \cos \omega n \big|_{\cos\omega = T(\omega1,\omega2)} \tag{1}$$

or

$$H(\omega_1, \omega_2) = \sum_{n=0}^{N} b(n) (\cos \omega)^n \big|_{\cos\omega = T(\omega1,\omega2)}. \tag{2}$$

In Section 4.6.2, we developed an implementation method based directly on (2). An alternate implementation method can be developed by using (1) and expressing $\cos \omega n$ in terms of $\cos \omega(n - 1)$ and $\cos \omega(n - 2)$. Specifically, let $\cos \omega n$ be denoted by $P_n[\cos \omega]$. From (1),

$$H(\omega_1, \omega_2) = \sum_{n=0}^{N} a(n) P_n[T(\omega_1, \omega_2)]. \tag{3}$$

From Problem 4.16, $P_n[\cdot]$ is the nth-order Chebyshev polynomial and is given by

$$P_0[x] = 1$$

$$P_1[x] = x$$

$$P_2[x] = 2x P_1[x] - P_0[x] \tag{4}$$

$$\vdots$$

$$P_n[x] = 2x P_{n-1}[x] - P_{n-2}[x].$$

(a) Show that $P_n[T(\omega_1, \omega_2)]$ for $n \geq 2$ can be obtained from $P_{n-1}[T(\omega_1, \omega_2)]$ and $P_{n-2}[T(\omega_1, \omega_2)]$ by

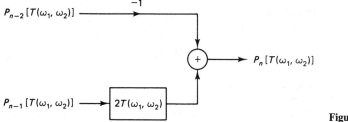

$P_{n-2}[T(\omega_1, \omega_2)]$ -1

$+$

$P_n[T(\omega_1, \omega_2)]$

$P_{n-1}[T(\omega_1, \omega_2)]$ $2T(\omega_1, \omega_2)$

Figure P4.22(a)

(b) From (3) and the result in (a), show that $H(\omega_1, \omega_2)$ for $N = 6$ can be implemented by Figure P4.22(b).

(c) Compare the computational requirements (arithmetic operations) using the structure in (b) with direct convolution of $h(n_1, n_2)$ with the input.

4.23. Let $H(z_1, z_2, z_3)$ represent a 3-D FIR filter. Suppose $H(z_1, z_2, z_3)$ can be expressed as

$$H(z_1, z_2, z_3) = H_1(z_1, z_2, z_3)H_2(z_1, z_2, z_3)$$

where $H_1(z_1, z_2, z_3)$ and $H_2(z_1, z_2, z_3)$ each has a region of support of $N \times N \times N$ points.

(a) Compare the number of arithmetic operations required for convolving the input with $h(n_1, n_2, n_3)$ to the number required for convolving the input with $h_1(n_1, n_2, n_3)$ first and then convolving the result with $h_2(n_1, n_2, n_3)$.

(b) Considering the number of degrees of freedom, argue that $H(z_1, z_2, z_3)$ in general cannot be expressed as $H_1(z_1, z_2, z_3)H_2(z_1, z_2, z_3)$. A cascade structure, therefore, is not a general structure that can be used in the implementation of a general FIR filter $H(z_1, z_2, z_3)$.

(c) Repeat (a) when $H(z_1, z_2, z_3)$ can be expressed as

$$H(z_1, z_2, z_3) = \prod_{i=1}^{K} H_i(z_1, z_2, z_3),$$

where each component of $H_i(z_1, z_2, z_3)$ has a region of support of $N \times N \times N$ points.

Finite Impulse Response Filters   Chap. 4

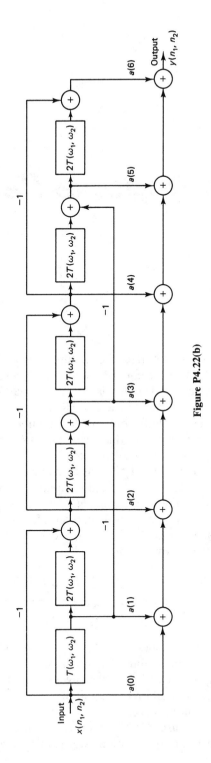

Figure P4.22(b)

5

Infinite Impulse Response Filters

5.0 INTRODUCTION

An infinite impulse response (IIR) digital filter has an impulse response that is infinite in extent. As a result, an IIR filter differs in some major respects from an FIR filter. For example, stability is never an issue for FIR filters, while it is a very important and problematic issue for IIR filters. On the other hand, an IIR filter typically requires a significantly smaller number of coefficients to meet a particular magnitude specification than does an FIR filter. In this chapter, we study a number of issues related to IIR filters.

An IIR filter with an arbitrary impulse response $h(n_1, n_2)$ cannot be realized, since computing each output sample requires a large number of arithmetic operations. As a result, in addition to requiring $h(n_1, n_2)$ to be real and stable, we require $h(n_1, n_2)$ to have a rational z-transform corresponding to a recursively computable system. Specifically, we require $H(z_1, z_2)$, the z-transform of $h(n_1, n_2)$, to be a rational function of the following form:

$$H(z_1, z_2) = \frac{\displaystyle\sum_{(k_1,k_2) \in R_b} \sum b(k_1, k_2)z_1^{-k_1}z_2^{-k_2}}{\displaystyle\sum_{(k_1,k_2) \in R_a} \sum a(k_1, k_2)z_1^{-k_1}z_2^{-k_2}} \tag{5.1}$$

where R_a is the region of support of $a(k_1, k_2)$ and R_b is the region of support of $b(k_1, k_2)$. In addition, we require the system to have a wedge support output mask. As discussed in Section 2.2, a recursively computable system derived from a linear constant coefficient difference equation requires a rational system function with a wedge support output mask.

As with FIR filters, using IIR filters involves filter specification, design, and implementation. In filter specification, we need to specify both the magnitude and phase of the frequency response $H(\omega_1, \omega_2)$. For an FIR filter, linear phase is very easy to achieve, and we have required all FIR filters to have linear phase in our discussion. With an IIR filter, however, controlling the phase characteristic is very difficult. As a result, only the magnitude response is specified when an IIR filter is designed. The phase characteristic of the resulting filter is then regarded as acceptable. If zero phase is necessary, two or more IIR filters are combined to obtain zero phase for the overall filter. This is discussed further in Sections 5.2 and 5.5. The magnitude specification that can be used is the tolerance scheme discussed in Section 4.2.

The IIR filter design problem is treated in Section 5.1. In Sections 5.2, 5.3, 5.4 and 5.5, we discuss two different approaches to designing IIR filters. In Section 5.2, the spatial domain design approach is discussed. In Section 5.3, we discuss the complex cepstrum representation of signals, which is useful in developing a method for stabilizing an unstable filter and in understanding the frequency domain design. In Section 5.4, the problem of stabilizing an unstable filter is discussed. In Section 5.5, the frequency domain design approach is discussed. In Section 5.6, we discuss the problem of realizing IIR filters. In Section 5.7, we discuss the advantages and disadvantages of FIR and IIR filters.

5.1 THE DESIGN PROBLEM

The problem of designing an IIR filter is to determine a rational and stable $H(z_1, z_2)$ with a wedge support output mask that meets a given design specification. In other words, we wish to determine a stable computational procedure that is recursively computable and meets a design specification.

In discussing the filter design problem, it is helpful to use a system function and a computational procedure interchangeably. However, as we discussed in Section 2.2.3, a given rational system function $H(z_1, z_2)$ in (5.1) can lead to many different computational procedures. To make the relationship unique, we will adopt a convention in expressing $H(z_1, z_2)$. Specifically, we will assume that $a(0, 0)$ in (5.1) is always 1, so $H(z_1, z_2)$ will then be in the form

$$H(z_1, z_2) = \frac{\displaystyle\sum_{(k_1,k_2)\ \in R_b} b(k_1, k_2)z_1^{-k_1}z_2^{-k_2}}{1 + \displaystyle\sum_{(k_1,k_2)\ \in R_a-(0,0)} a(k_1, k_2)z_1^{-k_1}z_2^{-k_2}} \tag{5.2}$$

where $R_a - (0, 0)$ represents the region of support of $a(k_1, k_2)$ except the origin $(0, 0)$. In addition, we will assume that the term $a(0, 0) = 1$ contributes $y(n_1, n_2)$ in the expression of a computational procedure. The unique computational procedure corresponding to (5.2) is then given by

$$y(n_1, n_2) \leftarrow -\sum_{(k_1,k_2)\ \in R_a-(0,0)} a(k_1, k_2)y(n_1 - k_1, n_2 - k_2)$$

$$+ \sum_{(k_1,k_2)\ \in R_b} b(k_1, k_2)x(n_1 - k_1, n_2 - k_2). \tag{5.3}$$

To obtain $H(z_1, z_2)$ from (5.3), we compute the z-transform of both sides of (5.3), without allowing any multiplication of terms (for instance, by a constant or $z_1^{M_1} z_2^{M_2}$), and solve for $H(z_1, z_2) = Y(z_1, z_2)/X(z_1, z_2)$. This will result in (5.2). If we follow the convention stated above for interpreting $H(z_1, z_2)$, then there will be a unique relationship between the system function in (5.2) and the computational procedure in (5.3). Just as we earlier obtained a specific computational procedure from a difference equation by adopting a notational convention, we have here obtained a specific computational procedure from $H(z_1, z_2)$ by adopting a notational convention. Unless otherwise specified, the convention of expressing $H(z_1, z_2)$ in the form of (5.2) and equating it to the computational procedure in (5.3) will be followed throughout this chapter.

The computational procedure in (5.3) can be represented by the output and input masks shown in Figure 5.1 for a specific choice of R_a and R_b. The system function in (5.2) can be represented by the two sequences $a(n_1, n_2)$ and $b(n_1, n_2)$ shown in Figure 5.2. From Figures 5.1 and 5.2, the output and input masks are very simply related to $a(n_1, n_2)$ and $b(n_1, n_2)$. The regions of support of the output and input masks, for example, are reflections of the regions of support of $a(n_1, n_2)$ and $b(n_1, n_2)$ with respect to the origin. Since the output and input masks are uniquely related to $a(n_1, n_2)$ and $b(n_1, n_2)$, they can be used interchangeably. The sequences $a(n_1, n_2)$ and $b(n_1, n_2)$ will be referred to as the filter coefficients of an IIR filter.

The first step in the design of an IIR filter is usually an initial determination of R_a and R_b, the regions of support of $a(n_1, n_2)$ and $b(n_1, n_2)$. The regions of support of $a(n_1, n_2)$ and $b(n_1, n_2)$ are determined by several considerations. One is that the resulting system must be recursively computable; this requires $a(n_1, n_2)$ to be a wedge support sequence. Another consideration is the approximate region of support that we wish the resulting filter $h(n_1, n_2)$ to have. If we determine the filter coefficients by attempting to approximate some desired impulse response

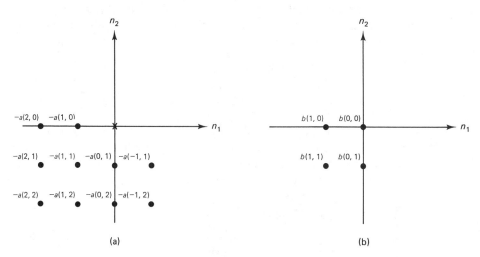

Figure 5.1 (a) Output mask and (b) input mask corresponding to the computational procedure in (5.3) for a specific choice of R_a and R_b.

Infinite Impulse Response Filters   Chap. 5

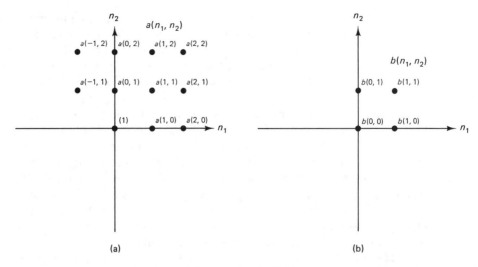

Figure 5.2 Sequences (a) $a(n_1, n_2)$ and (b) $b(n_1, n_2)$ corresponding to the computational procedure with output and input masks shown in Figure 5.1.

$h_d(n_1, n_2)$ in the spatial domain, we will want to choose R_a and R_b such that $h(n_1, n_2)$ will have at least approximately the same region of support as $h_d(n_1, n_2)$. If R_a and R_b have the same wedge support, then it can be shown (Problem 5.3) that $h(n_1, n_2)$ also has exactly the same wedge support. If we wish to have a first-quadrant support $h(n_1, n_2)$, therefore, a reasonable choice would be first-quadrant support $a(n_1, n_2)$ and $b(n_1, n_2)$. Another consideration relates to the filter specification parameters. In lowpass filter design, for example, a small δ_p, δ_s, and transition region will generally require a larger number of filter coefficients. It is often difficult to determine the number of filter coefficients required to meet a given filter specification for a particular design algorithm, and an iterative procedure may become necessary.

One major difference between IIR and FIR filters is on issues related to stability. An FIR filter is always stable as long as $h(n_1, n_2)$ is bounded (finite) for all (n_1, n_2), so stability is never an issue. With an IIR filter, however, ensuring stability is a major task. One approach to designing a stable IIR filter is to impose a special structure on $H(z_1, z_2)$ such that testing the stability and stabilizing an unstable filter become relatively easy tasks. Such an approach, however, tends to impose a severe constraint on the design algorithm or to highly restrict the class of filters that can be designed. For example, if $H(z_1, z_2)$ has a separable denominator polynomial of the form $A_1(z_1)A_2(z_2)$, testing the stability and stabilizing an unstable $H(z_1, z_2)$ without affecting the magnitude response is a 1-D problem, and is consequently quite straightforward. However, the class of filters that can be designed with a separable denominator polynomial without a significant increase in the number of coefficients in the numerator polynomial of $H(z_1, z_2)$ is restricted. In addition, the filter coefficients $a(n_1, n_2)$ are nonlinear functions of the unknown parameters to be determined, and this imposes a severe restriction on some design approaches. An alternative is to design a filter without considering the stability issue, and then test the stability of the resulting filter and attempt to stabilize it if it proves unstable. Although testing stability and stabilizing an unstable filter are not easy problems, this approach does not impose severe constraints on design algorithms and does not unduly restrict the class of filters that can be designed. For these reasons, we will stress this approach to designing stable IIR filters in this chapter.

In the 1-D case, there are two standard approaches to designing IIR filters. One is to design the filter from an analog system function, and the other is to design directly in the discrete domain. The first approach is typically much simpler and more useful than the second. Using an elliptic analog filter's system function and bilinear transformation, for example, we can design optimal IIR lowpass, highpass, bandpass, and bandstop filters by following a few simple steps. Unfortunately, this approach is not useful in the 2-D case. In the 1-D case, this approach exploits the availability of many simple methods to design 1-D analog filters. Simple methods do not exist for the design of 2-D analog filters.

The second approach, designing an IIR filter directly in the discrete domain, can be classified into two categories. The first is the spatial domain design approach, where filters are designed by using an error criterion in the spatial domain. The second is the frequency domain design approach, where filters are designed by using an error criterion in the frequency domain. The magnitude response of an IIR filter is often specified by using the tolerance scheme discussed in Section 4.2, and the performances of different design algorithms are often compared on the basis of the tolerance scheme. Therefore, the weighted Chebyshev error criterion, also known as the min-max error criterion, is a natural choice for designing IIR filters. An error criterion of this type, however, leads to a highly nonlinear problem. As a result, IIR filters are often designed in the spatial or frequency domain based on some reasonable error criterion that leads to simpler design algorithms. This is analogous to FIR filter design by the window, frequency sampling, and transformation methods, which are not based on the weighted Chebyshev error criterion. In Section 5.2, we discuss spatial domain IIR filter design. In Section 5.5, frequency domain IIR filter design is discussed.

5.2 SPATIAL DOMAIN DESIGN

In spatial domain IIR filter design, some desired or ideal spatial domain response of a system to a known input is assumed given. The filter coefficients are estimated such that the response of the designed filter to the known input is as close as possible in some sense to the desired response. This does not, of course, minimize the Chebyshev error norm, but it is a natural approach if we wish the filter to preserve some desirable spatial domain properties. The input often used in IIR filter design is $\delta(n_1, n_2)$, and the desired impulse response that is assumed given is denoted by $h_d(n_1, n_2)$.

Spatial domain design can be viewed as a system identification problem. The system identification problem occurs in a number of different contexts, so it has received considerable attention in the literature. Suppose we have an unknown system that we wish to model with a rational system function $H(z_1, z_2)$. Suppose further that we use $\delta(n_1, n_2)$ as an input to the system and observe $h_d(n_1, n_2)$, the output of the unknown system. One approach to estimating the system model parameters [filter coefficients $a(n_1, n_2)$ and $b(n_1, n_2)$ in our case] is to require the impulse response of the designed system to be as close as possible in some sense to $h_d(n_1, n_2)$.

In the system identification literature, modeling a system by $H(z_1, z_2)$ in (5.2) is called *auto-regressive moving-average* (ARMA) modeling. When all $b(k_1, k_2)$ in (5.2) are zero except $b(0, 0)$, so that the numerator is a constant, it is called *auto-regressive* (AR) or *all-pole* modeling. When all the values of $a(k_1, k_2)$ in (5.2) are zero except $a(0, 0)$, so that the denominator is a constant, it is called *moving-average* (MA) or *all-zero* modeling. Designing an IIR filter can be viewed as ARMA modeling, and designing an FIR filter can be viewed as MA modeling.

Among many possibilities, one error criterion often used in the filter design is

$$\text{Error} = \sum_{(n_1, n_2) \in R_e} \sum e^2(n_1, n_2) \tag{5.4a}$$

where

$$e(n_1, n_2) = h_d(n_1, n_2) - h(n_1, n_2) \tag{5.4b}$$

and R_e is the region of support of the error sequence. Ideally, R_e is all values of (n_1, n_2). In practice, however, it extends only over a finite region of (n_1, n_2), where $h_d(n_1, n_2)$ has significant energy. The mean square error in (5.4) is chosen because it is used widely in a number of system identification problems and some variation of it serves as the basis for a number of simple methods developed to estimate system parameters.

Minimizing the Error in (5.4) with respect to $a(n_1, n_2)$ and $b(n_1, n_2)$ is a nonlinear problem. To illustrate this, let us consider a simple example where $H(z_1, z_2)$ is given by

$$H(z_1, z_2) = \frac{b}{1 + a_1 z_1^{-1} + a_2 z_2^{-1}} \tag{5.5}$$

where a_1, a_2, and b are the filter coefficients to be estimated. The computational procedure corresponding to (5.5) is given by

$$y(n_1, n_2) \leftarrow -a_1 y(n_1 - 1, n_2) - a_2 y(n_1, n_2 - 1) + bx(n_1, n_2). \tag{5.6}$$

When $x(n_1, n_2) = \delta(n_1, n_2)$, $y(n_1, n_2)$ is $h(n_1, n_2)$ which is given by

$$h(0, 0) = b \tag{5.7a}$$

$$h(1, 0) = -a_1 b \tag{5.7b}$$

$$h(0, 1) = -a_2 b \tag{5.7c}$$

$$h(1, 1) = 2a_1 a_2 b \tag{5.7d}$$

$$\vdots$$

From (5.4) and (5.7), the error is given by

$$\text{Error} = (h_d(0, 0) - b)^2 + (h_d(1, 0) + a_1 b)^2 + (h_d(0, 1) + a_2 b)^2 \quad (5.8)$$
$$+ (h_d(1, 1) - 2a_1 a_2 b)^2 + \cdots.$$

Clearly, minimizing the Error in (5.8) with respect to the filter coefficients a_1, a_2, and b is a nonlinear problem.

One approach to minimizing the Error in (5.4) is to use well-known standard descent or hill-searching algorithms, which apply to a wide class of system identification problems, or to use somewhat ad hoc iterative procedures that are specific to the IIR filter design problem. An alternate approach is to slightly modify the Error in (5.4), so that the resulting algorithm leads to closed form solutions that require solving only sets of linear equations. We will begin with these suboptimal, but much simpler algorithms.

5.2.1 Linear Closed-Form Algorithms

The algorithms we discuss in this section lead to closed-form solutions that require solving only sets of linear equations. They are considerably simpler computationally than the iterative procedures discussed in the next two sections. However, none of them minimizes the Error in (5.4).

The methods discussed in this section are based on the observation that a reasonable modification of the error expression in (5.4) transforms a nonlinear problem to a linear one. Consider a computational procedure given by

$$y(n_1, n_2) \leftarrow - \sum_{(k_1,k_2) \in R_a - (0,0)} \sum a(k_1, k_2)y(n_1 - k_1, n_2 - k_2)$$

$$+ \sum_{(k_1,k_2) \in R_b} \sum b(k_1, k_2)x(n_1 - k_1, n_2 - k_2). \quad (5.9)$$

We will assume that there are p unknown values of $a(n_1, n_2)$ and $q + 1$ unknown values of $b(n_1, n_2)$, and thus a total of $N = p + q + 1$ filter coefficients to be determined. Replacing* $x(n_1, n_2)$ with $\delta(n_1, n_2)$ and $y(n_1, n_2)$ with $h(n_1, n_2)$ in (5.9) and noting that $\sum_{(k_1,k_2) \in R_b} \sum b(k_1, k_2) \delta(n_1 - k_1, n_2 - k_2)$ is $b(n_1, n_2)$, we have

$$h(n_1, n_2) \leftarrow - \sum_{(k_1,k_2) \in R_a - (0,0)} \sum a(k_1, k_2)h(n_1 - k_1, n_2 - k_2) + b(n_1, n_2). \quad (5.10)$$

If we replace $h(n_1, n_2)$ in both sides of (5.10) with the given $h_d(n_1, n_2)$, then the equality in (5.10) will not hold:

$$h_d(n_1, n_2) \not\leftarrow - \sum_{(k_1,k_2) \in R_a - (0,0)} \sum a(k_1, k_2)h_d(n_1 - k_1, n_2 - k_2) + b(n_1, n_2). \quad (5.11)$$

Since we wish to approximate $h_d(n_1, n_2)$ as well as we can with $h(n_1, n_2)$, it is

---

*The general philosophy behind the methods we develop does not restrict us to this choice of $x(n_1, n_2)$.

reasonable to define an error sequence $e_M(n_1, n_2)$ as the difference between the left-hand and right-hand side expressions of (5.11):

$$e_M(n_1, n_2) = h_d(n_1, n_2) + \sum_{(k_1,k_2) \in R_a - (0,0)} \sum a(k_1, k_2)h_d(n_1 - k_1, n_2 - k_2) - b(n_1, n_2).$$
(5.12)

It is clear that $e_M(n_1, n_2)$ in (5.12) is not the same as $e(n_1, n_2)$ in (5.4b). The subscript M in $e_M(n_1, n_2)$ is used to emphasize that $e_M(n_1, n_2)$ is a modification of $e(n_1, n_2)$. However, it is linear in the unknown filter coefficients $a(n_1, n_2)$ and $b(n_1, n_2)$.

There are many different interpretations of (5.12). One useful interpretation can be obtained by rewriting (5.12). Expressing the right-hand side of (5.12) using the convolution operation, we find that (5.12) leads to

$$e_M(n_1, n_2) = a(n_1, n_2) * h_d(n_1, n_2) - b(n_1, n_2). \qquad (5.13)$$

This equation also shows that $e_M(n_1, n_2)$ is linear in $a(n_1, n_2)$ and $b(n_1, n_2)$. Since $H(z_1, z_2) = B(z_1, z_2)/A(z_1, z_2)$ and therefore $b(n_1, n_2) = h(n_1, n_2) * a(n_1, n_2)$, (5.13) can be rewritten as

$$e_M(n_1, n_2) = a(n_1, n_2) * h_d(n_1, n_2) - a(n_1, n_2) * h(n_1, n_2). \qquad (5.14)$$

From (5.14), it is clear that $e_M(n_1, n_2)$ in (5.12) is not defined in the domain of $h_d(n_1, n_2)$ and $h(n_1, n_2)$. Instead, it is defined in a new domain where both $h_d(n_1, n_2)$ and $h(n_1, n_2)$ are prefiltered with $a(n_1, n_2)$. For this reason, methods based on (5.12) or equivalently (5.14) are called *indirect signal modeling methods*. Figure 5.3 shows the difference between $e(n_1, n_2)$ in (5.4b) and $e_M(n_1, n_2)$ in (5.12). Figure 5.3(a) shows $e(n_1, n_2)$, which is defined as the difference between $h_d(n_1, n_2)$ and $h(n_1, n_2)$, corresponding to (5.4b). Figure 5.3(b) shows $e_M(n_1, n_2)$ defined in the domain where both $h_d(n_1, n_2)$ and $h(n_1, n_2)$ are prefiltered, corresponding to (5.14). A simplification of Figure 5.3(b) is shown in Figure 5.3(c), which corresponds to (5.13).

The observation that $e_M(n_1, n_2)$ defined in (5.12) or equivalently in (5.14) is linear in the filter coefficients $a(n_1, n_2)$ and $b(n_1, n_2)$ can be used in a variety of ways to develop closed-form algorithms for estimating $a(n_1, n_2)$ and $b(n_1, n_2)$. We will now discuss a few representative methods.

Pade matching. In Pade matching [Pade], more often known as "Pade approximation," $e_M(n_1, n_2)$ in (5.12) is set to zero for a finite region of (n_1, n_2) that includes $N = p + q + 1$ points. This results in

$$e_M(n_1, n_2) = 0, \qquad (n_1, n_2) \in R_{\text{Pade}} \qquad (5.15)$$

where R_{Pade} consists of $N = p + q + 1$ points. Since $e_M(n_1, n_2)$ is linear in $a(n_1, n_2)$ and $b(n_1, n_2)$, (5.15) is a set of N linear equations in N unknowns. The region R_{Pade} includes R_b and is usually chosen such that $h_d(n_1, n_2)$ has large amplitudes for $(n_1, n_2) \in R_{\text{Pade}}$. If $h_d(n_1, n_2)$ has wedge support, R_{Pade} is typically chosen near the origin. If R_{Pade} is properly chosen, $h(n_1, n_2)$ will match $h_d(n_1, n_2)$ exactly for N points of (n_1, n_2). The set of N linear equations for N unknown

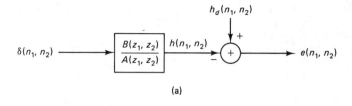

(a)

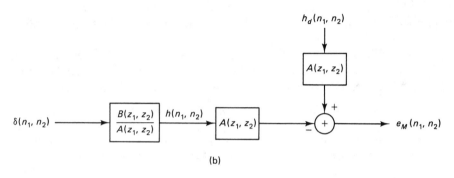

(b)

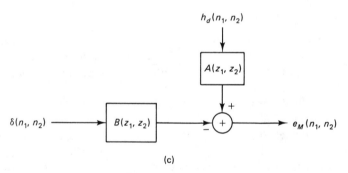

(c)

Figure 5.3 Mean square and modified mean square errors. (a) $e(n_1, n_2)$ used in the mean square error criterion; (b) $e_M(n_1, n_2)$ used in the modified mean square error criterion; (c) simplification of (b).

parameters in (5.15) typically has a unique solution. If there are multiple solutions, the number of unknown parameters N can be reduced or R_{Pade} can be expanded. If there are no solutions, one approach is to choose a different region R_{Pade}.

One problem with Pade matching is that the resulting filter may not be stable. This is typically the case with other spatial domain design techniques, which we will discuss, but the problem is more severe with Pade matching. As we will discuss later, other methods attempt to reduce $e_M(n_1, n_2)$ for all regions of (n_1, n_2). As long as $h_d(n_1, n_2)$ is stable, it is likely that $h(n_1, n_2)$ will be stable. Another, much more serious, problem with Pade matching is that $e_M(n_1, n_2)$ is minimized only over $N = p + q + 1$ points of (n_1, n_2), an area that is typically much smaller than the effective extent of $h_d(n_1, n_2)$. If we wish to reduce $e_M(n_1, n_2)$ over a larger region of (n_1, n_2), then we have to increase the model order of the filter. The only exception is when the desired $h_d(n_1, n_2)$ can be exactly represented by a low-order rational model; in practice, this is not likely to be the case. If we wish to reduce $e_M(n_1, n_2)$ over a larger region of (n_1, n_2) without increasing the model order, we must employ a different method. Prony's method, discussed next, accomplishes this objective.

Prony's method. In Prony's method, the error expression minimized is

$$\text{Error} = \sum_{n_1 = -\infty}^{\infty} \sum_{n_2 = -\infty}^{\infty} e_M^2(n_1, n_2) \tag{5.16}$$

where $e_M(n_1, n_2)$ is given by (5.12).

The Error in (5.16) is a quadratic form of the unknown parameters $a(n_1, n_2)$ and $b(n_1, n_2)$, and minimizing the error expression with respect to the unknown parameters requires solving a set of linear equations. The $p + q + 1$ linear equations are given by

$$\frac{\partial \text{ Error}}{\partial a(l_1, l_2)} = 0, \qquad (l_1, l_2) \in R_a - (0, 0) \tag{5.17a}$$

and

$$\frac{\partial \text{ Error}}{\partial b(l_1, l_2)} = 0, \qquad (l_1, l_2) \in R_b. \tag{5.17b}$$

Careful observation of the Error in (5.16) shows that (5.17) can be solved by first solving p linear equations for $a(n_1, n_2)$ and then solving $q + 1$ linear equations for $b(n_1, n_2)$. To see this, it is useful to rewrite (5.16) as

$$\text{Error} = E_1 + E_2 \tag{5.18a}$$

where

$$E_1 = \sum_{(n_1,n_2) \in R_b} \sum e_M^2(n_1, n_2) \tag{5.18b}$$

and

$$E_2 = \sum_{(n_1,n_2) \notin R_b} \sum e_M^2(n_1, n_2). \tag{5.18c}$$

The expression E_1 in (5.18b) consists of $q + 1$ terms, and E_2 in (5.18c) consists of a large number of terms. Consider E_2 first. From (5.12) and (5.18c), E_2 can be expressed as

$$E_2 = \sum_{(n_1,n_2) \notin R_b} \sum \left(h_d(n_1, n_2) + \sum_{(k_1,k_2) \in R_a - (0,0)} \sum a(k_1, k_2) h_d(n_1 - k_1, n_2 - k_2) \right.$$
$$\left. - b(n_1, n_2) \right)^2. \tag{5.19}$$

For $(n_1, n_2) \notin R_b$, $b(n_1, n_2) = 0$, and therefore (5.19) simplifies to

$$E_2 = \sum_{(n_1,n_2) \notin R_b} \sum \left(h_d(n_1,n_2) + \sum_{(k_1,k_2) \in R_a - (0,0)} \sum a(k_1, k_2) h_d(n_1 - k_1, n_2 - k_2) \right)^2. \tag{5.20}$$

We observe that E_2 does not depend on $b(n_1, n_2)$. Now consider E_1. Since E_1 consists of $q + 1$ terms which are quadratic in $a(n_1, n_2)$ and $b(n_1, n_2)$ and there are $q + 1$ terms of $b(n_1, n_2)$, we can choose $b(n_1, n_2)$ such that E_1 is zero for any set of $a(n_1, n_2)$. This means that minimizing the Error in (5.18a) with respect to $a(n_1, n_2)$ is equivalent to minimizing E_2 with respect to $a(n_1, n_2)$. This observation can also be verified by explicitly writing down $p + q + 1$ linear equations given by (5.17).

Minimizing E_2 in (5.18) with respect to $a(n_1, n_2)$ results in p linear equations for p unknowns given by

$$\sum_{(k_1,k_2) \in R_a - (0,0)} \sum a(k_1, k_2) r(k_1, k_2; l_1, l_2) = -r(0, 0; l_1, l_2), \quad (l_1, l_2) \in R_a - (0,0) \quad (5.21a)$$

where $r(k_1, k_2; l_1, l_2) = \sum_{(n_1,n_2) \in R_b} \sum h_d(n_1 - k_1, n_2 - k_2) h_d(n_1 - l_1, n_2 - l_2)$. $\quad (5.21b)$

Equation (5.21) is called a set of *normal equations*, and is a straightforward extension of 1-D normal equations.

Once $a(n_1, n_2)$ is determined, we can minimize the Error in (5.18a) with respect to $b(n_1, n_2)$. Since E_2 does not depend on $b(n_1, n_2)$, minimizing the Error in (5.18a) with respect to $b(n_1, n_2)$ is equivalent to minimizing E_1 in (5.18b) with respect to $b(n_1, n_2)$. The error E_1 can be minimized by differentiating E_1 with respect to $b(l_1, l_2)$ and setting the result to zero, or alternatively by observing that E_1 can be reduced to zero by

$$e_M(n_1, n_2) = 0 \quad \text{for} \quad (n_1, n_2) \in R_b. \quad (5.22)$$

In either case, the resulting equations for $b(n_1, n_2)$ are given by

$$b(n_1, n_2) = h_d(n_1, n_2) + \sum_{(k_1,k_2) \in R_a - (0,0)} \sum a(k_1, k_2) h_d(n_1 - k_1, n_2 - k_2), (n_1, n_2) \in R_b.$$

$$(5.23)$$

We can see from (5.23) that $b(n_1, n_2)$ does not require solving $q + 1$ simultaneous linear equations, and the coefficients of $b(n_1, n_2)$ can be solved one at a time.

Noting that $a(0, 0) = 1$, we can also express (5.23) as

$$b(n_1, n_2) = h_d(n_1, n_2) * a(n_1, n_2), \quad (n_1, n_2) \in R_b. \quad (5.24)$$

Equation (5.24) can also be obtained from (5.13) and (5.22). The method discussed above can be viewed as a straightforward application of a system identification method developed by [Prony] to the problem of designing a 2-D IIR filter. Since this method was first applied to the 2-D IIR filter design problem by [Shanks et al.], it is sometimes referred to as Shanks's method.

The advantages of Prony's method over Pade matching are clear. Like Pade matching, Prony's method will estimate the model parameters exactly if $h_d(n_1, n_2)$ can be represented exactly by a low-order rational model. In addition, the error $e_M(n_1, n_2)$ is reduced over a much larger region of (n_1, n_2). Although there is no guarantee that Prony's method will result in a stable filter, it is more likely than Pade matching to do so. Qualitatively, an unstable filter is likely to result in $h(n_1, n_2)$ with large amplitude, and $e(n_1, n_2)$ in (5.4) or $e_M(n_1, n_2)$ in (5.12) will tend to be large for some region of (n_1, n_2) for a stable $h_d(n_1, n_2)$. Since Prony's method attempts to reduce the total square error, the resulting filter is likely to be stable.

There are a number of variations to the method described above. For example, both $a(n_1, n_2)$ and $b(n_1, n_2)$ have been estimated by minimizing the Error in (5.16). Because of the specific choice of $e_M(n_1, n_2)$, however, $b(n_1, n_2)$ was determined essentially from only $q + 1$ terms of $e_M(n_1, n_2)$. One approach to improving the estimate of $b(n_1, n_2)$ is to use the original error expression in (5.4), which is repeated:

$$\text{Error} = \sum_{(n_1,n_2) \in R_e} \sum e^2(n_1, n_2)$$

$$= \sum_{(n_1,n_2) \in R_e} \sum (h_d(n_1, n_2) - h(n_1, n_2))^2. \tag{5.25}$$

Even though minimization of the Error in (5.25) with respect to both $a(n_1, n_2)$ and $b(n_1, n_2)$ is a nonlinear problem, it becomes linear if we assume that $a(n_1, n_2)$ is given. This is illustrated in Figure 5.4. We first estimate $a(n_1, n_2)$ using Prony's method. Then, assuming that $a(n_1, n_2)$ is given, we compute $v(n_1, n_2)$, the output of $1/A(z_1, z_2)$ with the input of $\delta(n_1, n_2)$. If $1/A(z_1, z_2)$ is unstable, $v(n_1, n_2)$ is likely to grow in amplitude, and we will not be able to continue. In this case, it is not useful to estimate $b(n_1, n_2)$. Instead, we should estimate $a(n_1, n_2)$ better. Assuming that we can proceed, we can rewrite (5.25) in terms of $v(n_1, n_2)$ as

$$\text{Error} = \sum_{(n_1,n_2) \in R_e} \sum (h_d(n_1, n_2) - v(n_1, n_2) * b(n_1, n_2))^2$$

$$= \sum_{(n_1,n_2) \in R_e} \sum \left(h_d(n_1, n_2) - \sum_{(k_1,k_2) \in R_b} \sum b(k_1, k_2) v(n_1 - k_1, n_2 - k_2) \right)^2. \tag{5.26}$$

Since $h_d(n_1, n_2)$ and $v(n_1, n_2)$ are known in (5.26), the Error in (5.26) is a quadratic form of the unknown parameters $b(n_1, n_2)$ and therefore can be estimated by solving a set of linear equations. Differentiating the Error in (5.26) with respect to $b(l_1, l_2)$ for $(l_1, l_2) \in R_b$ and setting it to zero yields

$$\sum_{(k_1,k_2) \in R_b} \sum b(k_1, k_2) \left(\sum_{(n_1,n_2) \in R_e} \sum v(n_1 - k_1, n_2 - k_2) v(n_1 - l_1, n_2 - l_2) \right)$$

$$= \sum_{(n_1,n_2) \in R_e} \sum h_d(n_1, n_2) v(n_1 - l_1, n_2 - l_2), \qquad (l_1, l_2) \in R_b. \tag{5.27}$$

Equation (5.27) is a set $q + 1$ linear equations for $q + 1$ unknowns of $b(n_1, n_2)$.

One major advantage of all the methods discussed in this section is that they lead to closed-form solutions that require solving sets of linear equations. This has been accomplished by modifying the original error function in (5.4). Although such methods as Prony's produce reasonable results in terms of the Error in (5.4),

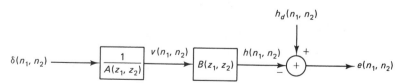

Figure 5.4 Error signal $e(n_1, n_2)$ used in estimating $b(n_1, n_2)$ in the modified Prony's method.

none of these methods minimizes it. Some examples of filters designed by the methods discussed in this section will be shown in Section 5.2.5 after we discuss other spatial domain design techniques. In the next two sections, we will consider iterative algorithms that attempt to reduce the Error in (5.4) better than do the closed-form methods discussed in this section.

5.2.2 Iterative Algorithms

The methods discussed in the previous section suggest an ad hoc iterative procedure that may improve algorithm performance in reducing the Error in (5.4). Suppose we estimate $a(n_1, n_2)$ by solving (5.21). We can then estimate $b(n_1, n_2)$ by solving (5.27). If we assume we know $b(n_1, n_2)$, we will be able to reestimate $a(n_1, n_2)$ by minimizing the Error in (5.18). When we estimate both $a(n_1, n_2)$ and $b(n_1, n_2)$, E_1 in (5.18b) is used in estimating $b(n_1, n_2)$ and the equations in (5.21) are derived by minimizing only E_2 in (5.18c). Since $b(n_1, n_2)$ is now assumed known, both E_1 and E_2 can be used in deriving a new set of equations for $a(n_1, n_2)$. With a new estimate of $a(n_1, n_2)$, we can reestimate $b(n_1, n_2)$ by solving (5.27), and the iterative procedure continues.

Another iterative procedure that has somewhat better properties than that above is an extension of a 1-D system identification method [Steiglitz and McBride]. From (5.14), $e(n_1, n_2) = h_d(n_1, n_2) - h(n_1, n_2)$ is related to $e_M(n_1, n_2)$ by

$$e_M(n_1, n_2) = a(n_1, n_2) * e(n_1, n_2). \tag{5.28}$$

Equation (5.28) can be rewritten as

$$e(n_1, n_2) = v(n_1, n_2) * e_M(n_1, n_2) \tag{5.29a}$$

where

$$V(z_1, z_2) = \frac{1}{A(z_1, z_2)}. \tag{5.29b}$$

The sequence $v(n_1, n_2)$ is the inverse of $a(n_1, n_2)$.
From (5.13) and (5.29a),

$$
\begin{aligned}
e(n_1, n_2) &= v(n_1, n_2) * e_M(n_1, n_2) \\
&= v(n_1, n_2) * (a(n_1, n_2) * h_d(n_1, n_2) - b(n_1, n_2)).
\end{aligned}
\tag{5.30}
$$

Equation (5.30) can also be derived from

$$
\begin{aligned}
E(z_1, z_2) &= H_d(z_1, z_2) - H(z_1, z_2) \\
&= H_d(z_1, z_2) - \frac{B(z_1, z_2)}{A(z_1, z_2)} \\
&= \frac{1}{A(z_1, z_2)} (A(z_1, z_2)H_d(z_1, z_2) - B(z_1, z_2)) \\
&= V(z_1, z_2)(A(z_1, z_2)H_d(z_1, z_2) - B(z_1, z_2)).
\end{aligned}
\tag{5.31}
$$

From (5.30), if $v(n_1, n_2)$ is somehow given, then $e(n_1, n_2)$ is linear in both $a(n_1, n_2)$ and $b(n_1, n_2)$, so minimization of $\Sigma_{n_1} \Sigma_{n_2} e^2(n_1, n_2)$ with respect to $a(n_1, n_2)$ and $b(n_1, n_2)$ is a linear problem. Of course, if $v(n_1, n_2)$ is given, then $a(n_1, n_2)$ will automatically be given by (5.29b). However, the above discussion suggests an iterative procedure, where we begin with some initial estimate of $a(n_1, n_2)$, obtained using a method such as Prony's, obtain $v(n_1, n_2)$ from $a(n_1, n_2)$, and then minimize $\Sigma_{n_1} \Sigma_{n_2} e^2(n_1, n_2)$ with respect to both $a(n_1, n_2)$ and $b(n_1, n_2)$ by solving a set of linear equations. We now have a new estimate of $a(n_1, n_2)$, and the process continues. Since $v(n_1, n_2)$ is obtained from the previous estimate of $a(n_1, n_2)$ and it "prefilters" $e_M(n_1, n_2)$ to obtain $e(n_1, n_2)$, this procedure is called the *iterative prefiltering method*. The method is sketched in Figure 5.5.

In neither of the two iterative procedures discussed in this section, is the algorithm guaranteed to converge. In addition, the conditions under which either algorithm converges are not known. Both methods require computation of $v(n_1, n_2)$, the inverse of $a(n_1, n_2)$, in each iteration. If $1/A(z_1, z_2)$ is unstable at any point, the iterative procedure cannot proceed any further. Despite these problems which have been encountered in practice, the iterative prefiltering method has been successfully used in the design of some IIR filters [Shaw and Mersereau]. The algorithm is usually terminated before convergence. The filter designed, when the method is successful, appears to be "good" within a few iterations, and the error based on $e(n_1, n_2)$ in (5.30) decreases very slowly after a few iterations. If the iterative prefiltering method does converge, then the converging solution can

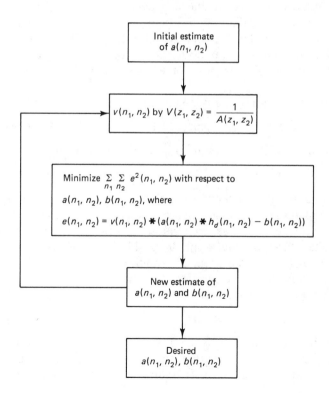

Figure 5.5 Iterative prefiltering method.

be shown to be the solution that minimizes $\Sigma_{n_1} \Sigma_{n_2} e^2(n_1, n_2)$. Examples of filters designed using the iterative prefiltering method along with filters designed using the linear closed-form methods described in the previous section are shown in Section 5.2.5.

The iterative procedures developed in this section are ad hoc in the sense that they attempt to minimize $\Sigma_{n_1} \Sigma_{n_2} (h_d(n_1, n_2) - h(n_1,n_2))^2$ indirectly by some "reasonable" exploitation of the specific structure present in the problem. An alternative to this is to solve the nonlinear problem of minimizing $\Sigma_{n_1} \Sigma_{n_2} (h_d(n_1, n_2) - h(n_1, n_2))^2$ directly by using well-known functional minimization methods. This is the topic of the next section.

5.2.3 Descent Algorithms

The problem of estimating unknown variables by minimizing (or maximizing) a scalar function arises in many different contexts. A number of approaches to this problem have been developed. In this section, we briefly summarize a few representative methods.

In theory, the methods discussed in this section are applicable not only to spatial domain IIR filter design, but also to FIR filter design and optimal IIR filter design. However, these methods are not usually considered for FIR filter design, since there exist computationally simple methods that produce reasonable results. In addition, if the error is defined in the spatial domain as $\Sigma_{n_1} \Sigma_{n_2} (h_d(n_1, n_2) - h(n_1, n_2))^2$, the window method with a rectangular window minimizes it. Nor are these methods usually considered in optimal filter design, since they require evaluation of either the first-order or both the first- and second-order partial derivatives with respect to the unknown parameters. As a result, they become computationally very expensive when applied to the minimization of the weighted Chebyshev error criterion.

Suppose we wish to solve the following minimization problem:

$$\text{Minimize } f(\boldsymbol{\theta}) \text{ with respect to } \boldsymbol{\theta}. \tag{5.32}$$

In (5.32), $\boldsymbol{\theta}$ is an $N \times 1$-column vector consisting of N parameters given by

$$\boldsymbol{\theta} = [\theta_1, \theta_2, \theta_3, \ldots, \theta_N]^T \tag{5.33}$$

and $f(\boldsymbol{\theta})$ is a scalar function of the parameters $\boldsymbol{\theta}$ we wish to estimate. In the filter design problem, $\boldsymbol{\theta}$ would consist of p parameters in $a(n_1, n_2)$ and $q + 1$ parameters in $b(n_1, n_2)$. We use $\boldsymbol{\theta}$ to represent the unknown parameters for notational simplicity and to emphasize that these methods apply to a wide range of problems. A standard approach to solving (5.32) is a descent algorithm, where the parameters denoted by $\boldsymbol{\theta}$ are perturbed in each iteration to decrease $f(\boldsymbol{\theta})$.

Any iterative descent algorithm for estimating $\boldsymbol{\theta}$ can be expressed as

$$\boldsymbol{\theta}^{i+1} = \boldsymbol{\theta}^i + \alpha^i \mathbf{p}^i \tag{5.34}$$

where $\boldsymbol{\theta}^{i+1}$ is the result of the ith iteration, $\mathbf{p}^i$ is the direction of search used in the ith iteration, and α^i is a positive scaling factor that can be used to adjust the step size. For the class of descent algorithms known as gradient methods, the direction of $\mathbf{p}^i$ used in (5.34) can be expressed in terms of the gradient of $f(\boldsymbol{\theta})$ at $\boldsymbol{\theta} = \boldsymbol{\theta}^i$ as

$$\mathbf{p}^i = -R^i \nabla f(\boldsymbol{\theta}^i) \tag{5.35}$$

where R^i is an $N \times N$ matrix and $\nabla f(\boldsymbol{\theta}^i)$ is the gradient of $f(\boldsymbol{\theta})$ at $\boldsymbol{\theta} = \boldsymbol{\theta}^i$ given by

$$\nabla f(\boldsymbol{\theta}^i) = \left[\frac{\partial f(\boldsymbol{\theta})}{\partial \theta_1}, \frac{\partial f(\boldsymbol{\theta})}{\partial \theta_2}, \ldots, \frac{\partial f(\boldsymbol{\theta})}{\partial \theta_N} \right]^T_{\boldsymbol{\theta} = \boldsymbol{\theta}^i} \tag{5.36}$$

From (5.34) and (5.35),

$$\boldsymbol{\theta}^{i+1} = \boldsymbol{\theta}^i - \alpha^i R^i \nabla f(\boldsymbol{\theta}^i). \tag{5.37}$$

One desirable characteristic for R^i is that it be positive definite. For a positive definite R^i, if $\nabla f(\boldsymbol{\theta}^i) \neq 0$, it is guaranteed that

$$f(\boldsymbol{\theta}^{i+1}) < f(\boldsymbol{\theta}^i) \tag{5.38}$$

for a sufficiently small positive α^i. In addition, it is reasonable to choose α^i such that (5.38) holds. We will briefly discuss three representative and well-known descent algorithms: the steepest descent method, the Newton-Raphson (NR) method, and the Davidon-Fletcher-Powell (DFP) method. All three methods can be expressed in the form of (5.37), but they differ in how α^i and R^i are specified. These methods have many variations, so we will discuss only the basic ideas behind each method.

The steepest descent method. The steepest descent method is based on the well-known fact that a function decreases locally fastest in the direction of the negative gradient. Because of its simplicity, this is the most widely used method in system identification. In this method, we have

$$\boldsymbol{\theta}^{i+1} = \boldsymbol{\theta}^i - \alpha^i \nabla f(\boldsymbol{\theta}^i). \tag{5.39}$$

For the special case when $\boldsymbol{\theta}$ consists of one element $(\theta_1 = a)$, (5.39) reduces to

$$a^{i+1} = a^i - \alpha^i \frac{df(a)}{da} \Big|_{a = a^i}. \tag{5.40}$$

Figure 5.6 illustrates this. In the figure, $f(a)$ has a positive gradient (slope), so the search will be in the negative direction of a. Figure 5.6 illustrates why the

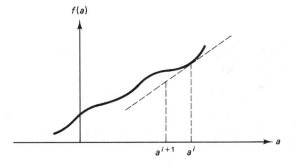

$f(a)$

a^{i+1} a^i a

Figure 5.6 Method of steepest descent, with a^i and a^{i+1} denoting the estimate of a after $i - i$th and ith iterations.

gradient is a reasonable indicator of the search direction. Comparison of (5.37) and (5.39) shows that R^i chosen is the identity matrix I. The step size α^i should be chosen carefully. If α^i is too large, θ^{i+1} may overshoot and go into an oscillation around the true solution. If α^i is too small, the method will approach the solution very slowly. Many reasonable methods of choosing α^i exist. One approach is to choose α^i such that $f(\theta^{i+1}) = f(\theta^i - \alpha^i \nabla f(\theta^i))$ is minimum. This does introduce a separate optimization problem, but in many cases it reduces the number of iterations needed. It is thus useful when computing $\nabla f(\theta^i)$ is difficult.

The steepest descent method relies only on the first-order derivative information, so it is very simple computationally compared to the NR and DFP methods discussed later. It is also very effective in decreasing the function in the initial stage of the algorithm, when the iterative solution is far from the true solution. However, the algorithm is known to converge only linearly, and the method can be very slow as the iterative solution approaches the true solution. In addition, the steepest descent method always searches in the direction of negative gradient, which can cause a problem in the presence of valleys or ridges in $f(\theta)$. As a result, this method is often used only in the initial stage, with a different method which has better convergence characteristics being substituted as the method nears convergence.

The Newton-Raphson method. The Newton-Raphson (NR) method exploits both the first- and second-order derivative information in determining the search direction. This significantly improves the convergence characteristics as the iterative solution approaches the true solution. The NR method is based on the Taylor series expansion of $f(\theta)$ at $\theta = \theta^i$. Specifically, $f(\theta)$ near $\theta = \theta^i$ can be expressed as

$$f(\theta) = f(\theta^i) + \nabla f(\theta^i)(\theta - \theta^i) + \tfrac{1}{2}(\theta - \theta^i)^T H_f(\theta^i)(\theta - \theta^i) \qquad (5.41)$$
$$+ \text{ higher-order terms}$$

where $H_f(\theta)$ is the Hessian matrix given by

$$H_f(\theta) = \begin{bmatrix} \dfrac{\partial^2 f(\theta)}{\partial \theta_1 \partial \theta_1}, & \cdots, & \dfrac{\partial^2 f(\theta)}{\partial \theta_1 \partial \theta_N} \\ \vdots & & \vdots \\ \dfrac{\partial^2 f(\theta)}{\partial \theta_N \partial \theta_1}, & \cdots, & \dfrac{\partial^2 f(\theta)}{\partial \theta_N \partial \theta_N} \end{bmatrix} \qquad (5.42)$$

When θ is sufficiently close to θ^i, the higher-order terms in (5.41) are negligible. In the NR method, it is assumed that

$$f(\theta) = f(\theta^i) + \nabla f(\theta^i)(\theta - \theta^i) + \tfrac{1}{2}(\theta - \theta^i)^T H_f(\theta^i)(\theta - \theta^i). \qquad (5.43)$$

Since (5.43) is a quadratic form in θ, minimization of $f(\theta)$ in (5.43) with respect to θ can be performed straightforwardly. Choosing θ^{i+1} as the solution to this minimization problem, we have

$$\boldsymbol{\theta}^{i+1} = \boldsymbol{\theta}^i - H_f^{-1}(\boldsymbol{\theta}^i)\nabla f(\boldsymbol{\theta}^i). \tag{5.44}$$

When $\boldsymbol{\theta}$ consists of one element $(\theta_1 = a)$, (5.43) and (5.44) reduce to

$$f(a) = f(a^i) + f'(a^i)(a - a^i) + \tfrac{1}{2}f''(a^i)(a - a^i)^2 \tag{5.45}$$

and

$$a^{i+1} = a^i - \frac{1}{f''(a^i)} f'(a^i) \tag{5.46}$$

where $f'(a)$ is $df(a)/da$ and $f''(a)$ is $d^2f(a)/da^2$. Comparison of (5.37) and (5.44) shows that $\alpha^i R^i$ is chosen to be $H_f^{-1}(\boldsymbol{\theta}^i)$. The step size α^i is automatically chosen in the NR method.

The choice of R^i in the NR method may not have the positive definite property desirable for convergence. In fact, the NR method may not converge if the iterative solution is not close to the true solution. The NR method is considerably more complex than the steepest descent method. It requires the computation of $N(N + 1)/2$ second-order partial derivatives and the inverse of an $N \times N$ matrix $H_f(\boldsymbol{\theta})$ in each iteration. When the iterative solution is sufficiently close to the true solution, the method has the highly desirable property of quadratic convergence.

The Davidon-Fletcher-Powell method. The Davidon-Fletcher-Powell (DFP) method, which was originated by Davidon and later modified by Fletcher and Powell, is in a sense a combination of the steepest descent and the NR methods. In the earlier stage in the iterative algorithm, it behaves like the steepest descent method. In the later stage, it behaves like the NR method. For this reason, it is sometimes referred to as the *steepest descent with accelerated convergence method*.

In the DFP method, $\boldsymbol{\theta}^{i+1}$ is again updated from $\boldsymbol{\theta}^i$ by

$$\boldsymbol{\theta}^{i+1} = \boldsymbol{\theta}^i - \alpha^i R^i \nabla f(\boldsymbol{\theta}^i) \tag{5.47}$$

where R^i is also updated recursively. The key idea is to recursively build R^i into an approximation to the inverse Hessian matrix while using the first-order derivatives of $f(\boldsymbol{\theta})$. The matrix R^0 in the initial iteration is usually chosen to be the identity matrix I. Therefore, this method starts out as a steepest descent method. The matrix R^i in (5.47) is then updated recursively from R^{i-1}, $\boldsymbol{\theta}^i - \boldsymbol{\theta}^{i-1}$, and $\nabla f(\boldsymbol{\theta}^i)$ as follows:

$$R^i = R^{i-1} + \frac{\mathbf{r}^i(\mathbf{r}^i)^T}{(\mathbf{r}^i)^T \mathbf{s}^i} - \frac{(R^{i-1}\mathbf{s}^i)(R^{i-1}\mathbf{s}^i)^T}{(\mathbf{s}^i)^T R^{i-1}\mathbf{s}^i}, \qquad i = 1, 2, 3, \ldots \tag{5.48a}$$

where

$$\mathbf{r}^i = \boldsymbol{\theta}^i - \boldsymbol{\theta}^{i-1} \tag{5.48b}$$

and

$$\mathbf{s}^i = \nabla f(\boldsymbol{\theta}^i) - \nabla f(\boldsymbol{\theta}^{i-1}). \tag{5.48c}$$

When $\boldsymbol{\theta}$ consists of one element $(\theta_1 = a)$, (5.48) reduces to

$$R^i = \frac{a^i - a^{i-1}}{f'(a^i) - f'(a^{i-1})}. \tag{5.49}$$

Equation (5.47) in this case reduces to

$$a^{i+1} = a^i - \alpha^i \frac{a^i - a^{i-1}}{f'(a^i) - f'(a^{i-1})} f'(a^i). \tag{5.50}$$

It is not obvious why the update equation for R is a good choice. However, this choice has some very desirable properties. Fletcher and Powell showed that if R^i is updated by (5.48) and α^i is chosen as the value α that minimizes $f(\theta^i - \alpha R \nabla f(\theta^i))$, then R^i will always be a positive definite matrix as long as the initial choice R^0 is positive definite. $R^0 = I$ is, of course, positive definite. The property that R^i is positive definite is desirable for algorithm convergence. In addition, under some conditions R^i becomes identical to $H_f^{-1}(\theta^i)$, which is the same R^i used for the NR method. The DFP method can be observed to behave at least approximately like the NR method by comparing the two methods for the case when θ contains only one element ($\theta_1 = a$). Comparing (5.46) corresponding to the NR method with (5.50) corresponding to the DFP method, we see that

$$R^i = \frac{a^i - a^{i-1}}{f'(a^i) - f'(a^{i-1})} \approx \frac{1}{f''(a^i)}. \tag{5.51}$$

From (5.51), the DFP matrix R^i can be viewed as the inverse of a finite difference approximation to the second-order derivative. Equation (5.51) is the approximate result from the Taylor series expansion. More generally, R^i in the DFP method can be shown to approximate the inverse Hessian matrix $H_f^{-1}(\theta^i)$ by the Taylor series expansion and requires calculation of only $\nabla f(\theta)$. Since no second-order derivatives are calculated, the DFP method requires far less computation in each iteration than the NR method.

The DFP method starts out like the steepest descent method, which is known to be very effective in reducing $f(\theta)$ in the earlier stage. The method then adapts itself to become more like an NR method, which is known to have the desirable behavior of quadratic convergence when the iterative solution is sufficiently close to the true solution. Unlike the NR method, the DFP method has the property that R^i can be forced to remain positive definite, which is desirable for algorithm convergence. The method has proven very successful in a wide variety of parameter and state estimation problems.

The methods discussed in this section require first-order or both first- and second-order derivative information of the form $\partial f(\theta)/\partial \theta_i$ and $\partial^2 f(\theta)/\partial \theta_i \partial \theta_j$. Since the error to be minimized in the spatial domain IIR filter design is of the form $\Sigma_{n_1} \Sigma_{n_2} (h_d(n_1, n_2) - h(n_1, n_2))^2$, we need to compute $\partial h(n_1, n_2)/\partial \theta_i$ and $\partial^2 h(n_1, n_2)/\partial \theta_i \partial \theta_j$, where the parameters are the filter coefficients $a(n_1, n_2)$ and $b(n_1, n_2)$. Exploiting the form of the computational procedure in (5.10) that governs $h(n_1, n_2)$, we can develop [Cadzow] relatively simple procedures for determining $\partial h(n_1, n_2)/\partial \theta_i$ and $\partial^2 h(n_1, n_2)/\partial \theta_i \partial \theta_j$.

These methods have been applied [Cadzow; Shaw and Mersereau] to the design of IIR filters by minimizing $\Sigma_{n_1} \Sigma_{n_2} (h_d(n_1, n_2) - h(n_1, n_2))^2$. The steepest descent method has not been very successful when used alone, due to its slow linear convergence. Nor has the NR method been very successful when used alone, since the inverse Hessian matrix $H_f^{-1}(\boldsymbol{\theta}^i)$ is not always positive definite, particularly when the iterative solution is not close to the true solution. The NR algorithm has been observed to diverge in many cases. The DFP method, which does not suffer from the limitations of the steepest-descent method or the NR method, has been successfully applied to the design of IIR filters. Some examples of filters designed by using the DFP method are shown in Section 5.2.5 after we discuss the problem of designing zero-phase IIR filters.

The techniques discussed in this section are well-known standard descent methods. They apply to a wide variety of problems in addition to the spatial domain IIR filter design problem treated here. Their convergence characteristics and the conditions under which they will converge are well known. The methods, however, are quite general and are not designed to exploit the specific characteristics of a particular problem. In contrast, other iterative methods, such as those discussed in Section 5.2.2, are ad hoc and less general, but cleverly exploit detailed structural information present in specific filter design problems. Therefore, if such algorithms do work and converge to a reasonable solution, they tend to do so faster than standard descent methods.

5.2.4 Zero-Phase Filter Design

In some applications, zero-phase design may be necessary. As we discussed in Chapter 4, it is simple to design zero-phase FIR filters. It is impossible, however, for a single recursively computable IIR filter to have zero phase. To have zero phase, $h(n_1, n_2)$ must be equal to $h(-n_1, -n_2)$. An IIR filter requires an infinite extent $h(n_1, n_2)$. Recursive computability requires the output mask to have wedge support. These requirements cannot all be satisfied at the same time. It is possible, however, to achieve zero phase by using more than one IIR filter. A method particularly well suited to spatial domain design is to divide $h_d(n_1, n_2)$ into different regions, design an IIR filter to approximate $h_d(n_1, n_2)$ in each region, and then combine the filters by using a parallel structure.

Suppose we have a desired $h_d(n_1, n_2)$. Since zero phase is desired, we assume that

$$h_d(n_1, n_2) = h_d(-n_1, -n_2). \tag{5.52}$$

We can divide $h_d(n_1, n_2)$ into an even number of regions: two, four, six, eight, or more. Suppose we divide $h_d(n_1, n_2)$ into four regions by

$$h_d^I(n_1, n_2) = h_d(n_1, n_2)w(n_1, n_2) \tag{5.53a}$$

$$h_d^{II}(n_1, n_2) = h_d(n_1, n_2)w(-n_1, n_2) \tag{5.53b}$$

$$h_d^{III}(n_1, n_2) = h_d(n_1, n_2)w(-n_1, -n_2) \tag{5.53c}$$

and $$h_d^{IV}(n_1, n_2) = h_d(n_1, n_2)w(n_1, -n_2) \tag{5.53d}$$

where $w(n_1, n_2)$ is a first-quadrant support sequence given by

$$w(n_1, n_2) = \begin{cases} 1, & n_1 \geq 1, n_2 \geq 1 \\ \frac{1}{2}, & n_1 \geq 1, n_2 = 0 \\ \frac{1}{2}, & n_1 = 0, n_2 \geq 1 \\ \frac{1}{4}, & n_1 = n_2 = 0. \end{cases} \tag{5.54}$$

The window sequence $w(n_1, n_2)$ is shown in Figure 5.7. It is chosen such that there will be maximum symmetry and the values in all four windows will add up to 1. From (5.53) and (5.54), it is clear that

$$h_d(n_1, n_2) = h_d^I(n_1, n_2) + h_d^{II}(n_1, n_2) + h_d^{III}(n_1, n_2) + h_d^{IV}(n_1, n_2). \tag{5.55}$$

Since $h_d^I(n_1, n_2)$, $h_d^{II}(n_1, n_2)$, $h_d^{III}(n_1, n_2)$, and $h_d^{IV}(n_1, n_2)$ are quadrant support sequences, they can be implemented by means of recursively computable systems. Suppose we use one of the spatial IIR filter design techniques discussed earlier to design $H^I(z_1, z_2)$ that approximates $h_d^I(n_1, n_2)$. Similarly, suppose we have designed $H^{II}(z_1, z_2)$ that approximates $h_d^{II}(n_1, n_2)$. From (5.52) and (5.53),

$$h_d^{III}(n_1, n_2) = h_d^I(-n_1, -n_2). \tag{5.56}$$

Therefore, $H^{III}(z_1, z_2)$ that approximates $h_d^{III}(n_1, n_2)$ can be obtained from $H^I(z_1, z_2)$ by

$$H^{III}(z_1, z_2) = H^I(z_1^{-1}, z_2^{-1}). \tag{5.57}$$

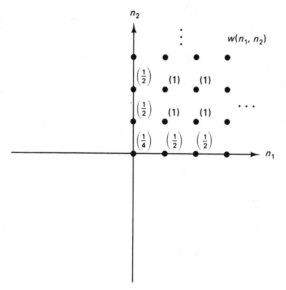

Figure 5.7 Window sequence $w(n_1, n_2)$ in (5.54).

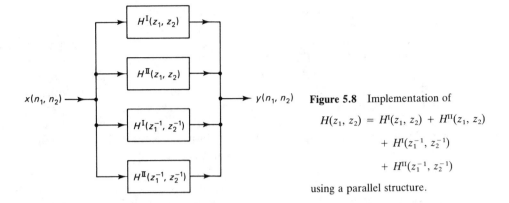

Figure 5.8 Implementation of

$$H(z_1, z_2) = H^I(z_1, z_2) + H^{II}(z_1, z_2)$$
$$+ H^I(z_1^{-1}, z_2^{-1})$$
$$+ H^{II}(z_1^{-1}, z_2^{-1})$$

using a parallel structure.

Similarly, $H^{IV}(z_1, z_2)$ can be obtained from $H^{II}(z_1, z_2)$ by

$$H^{IV}(z_1, z_2) = H^{II}(z_1^{-1}, z_2^{-1}). \tag{5.58}$$

Since $H^I(z_1, z_2)$, $H^{II}(z_1, z_2)$, $H^{III}(z_1, z_2)$, and $H^{IV}(z_1, z_2)$ approximate $h_d^I(n_1, n_2)$, $h_d^{II}(n_1, n_2)$, $h_d^{III}(n_1, n_2)$, and $h_d^{IV}(n_1, n_2)$, respectively, from (5.55), $h_d(n_1, n_2)$ will be approximated by $H(z_1, z_2)$ given by

$$H(z_1, z_2) = H^I(z_1, z_2) + H^{II}(z_1, z_2) + H^{III}(z_1, z_2) + H^{IV}(z_1, z_2) \tag{5.59}$$
$$= H^I(z_1, z_2) + H^{II}(z_1, z_2) + H^I(z_1^{-1}, z_2^{-1}) + H^{II}(z_1^{-1}, z_2^{-1}).$$

Each of the four filters approximates a one-quadrant support sequence and is recursively computable. In addition, $H(z_1, z_2)$ has zero phase since $H(z_1, z_2) = H(z_1^{-1}, z_2^{-1})$. The system in (5.59) can be implemented by using a parallel structure as shown in Figure 5.8. The input is filtered by each of the four recursively computable systems, and the results are combined to produce the output. If $h_d(n_1, n_2)$ has fourfold symmetry,

$$h_d^{II}(n_1, n_2) = h_d^I(-n_1, n_2) \tag{5.60}$$

and therefore $H^{II}(z_1, z_2)$ can be determined from $H^I(z_1, z_2)$ by

$$H^{II}(z_1, z_2) = H^I(z_1^{-1}, z_2). \tag{5.61}$$

In this case, $H(z_1, z_2)$ in (5.59) is given by

$$H(z_1, z_2) = H^I(z_1, z_2) + H^I(z_1^{-1}, z_2) + H^I(z_1^{-1}, z_2^{-1}) + H^I(z_1, z_2^{-1}). \tag{5.62}$$

From (5.62), only one filter needs to be designed in this case.

5.2.5 Examples

In previous sections, we discussed various methods of designing IIR filters in the spatial domain. In this section, we show examples of IIR filters designed by some of these methods. The examples in this section are for illustrative purposes only and are not intended to support a detailed comparison of different design methods' performances.

Figure 5.9 shows the frequency response of a zero-phase lowpass filter designed by using the modified Prony's method. Figures 5.9(a) and (b) show the perspective plot and contour plot, respectively. In the modified Prony's method, the denominator coefficients $a(n_1, n_2)$ are estimated first by using Prony's method and the numerator coefficients $b(n_1, n_2)$ are then estimated by solving (5.27). The desired impulse response used is the circularly symmetric ideal lowpass filter with cutoff frequency of 0.4π, to which a circularly symmetric Hamming window with radius of 11 points was applied. The window forces the desired impulse response used in the design to become a more stable sequence, and the filter designed is more likely to be stable. The zero-phase filter was designed by using the method discussed in Section 5.2.4. The first-quadrant filter with 3×3-point $a(n_1, n_2)$ and 3×3-point $b(n_1, n_2)$ was first designed and the overall system function was obtained by using (5.62).

Figure 5.10 shows another example of a filter designed by the modified Prony's method. The desired impulse response used is the circularly symmetric ideal bandpass filter with cutoff frequencies of 0.3π and 0.7π, to which a circularly symmetric Hamming window with radius of 11 points was applied. The zero-phase filter was designed by using the same method used to design the filter in Figure 5.9 except that a 4×4-point $a(n_1, n_2)$ and 4×4-point $b(n_1, n_2)$ were used. Even though the modified Prony's method does not guarantee the stability of the filter designed, the filters designed are often stable. If the filter designed is not stable, the stabilization methods discussed in Section 5.4 may be considered to stabilize the filter.

Figures 5.11 and 5.12 show the frequency responses of the zero-phase lowpass and bandpass filters designed by using the iterative prefiltering method discussed in Section 5.2.2. The desired impulse response and the zero-phase filter design method used are identical to those used to generate the examples in Figures 5.9 and 5.10. The initial estimate of the denominator coefficients $a(n_1, n_2)$ was obtained by using Prony's method. The iterative prefiltering method often has a stability problem within a few iterations. When the method does not encounter a stability problem, the method has been observed to decrease the error very slowly after a few iterations. The method is, therefore, terminated within a few iterations. The performance of the method, even when successful, does not appear to be significantly better than that of the modified Prony's method.

Figure 5.13 shows the frequency response of a zero-phase lowpass filter designed by using the Davidon-Fletcher-Powell (DFP) method to minimize the error expression in (5.4). The desired impulse response used is the circularly symmetric ideal lowpass filter with cutoff frequency of 0.5π, to which a circularly symmetric Kaiser window with radius of 11 points was applied. The zero-phase filter was designed by parallel combination of four one-quadrant filters. The first-quadrant filter with 4×4-point $a(n_1, n_2)$ and 4×4-point $b(n_1, n_2)$ was first designed and the overall system function was obtained by using (5.62).

Figure 5.14 shows another example of a filter designed by the DFP method.

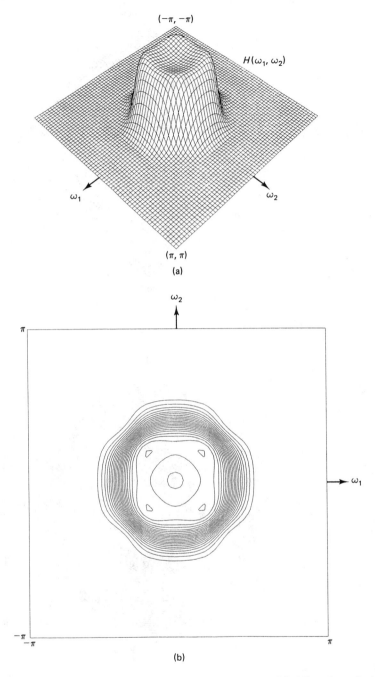

$(-\pi, -\pi)$

$H(\omega_1, \omega_2)$

ω_1

ω_2

(π, π)

(a)

ω_2

π

ω_1

$-\pi$
$-\pi$ π

(b)

Figure 5.9 Zero-phase lowpass filter designed by the modified Prony's method. (a) Perspective plot; (b) contour plot.

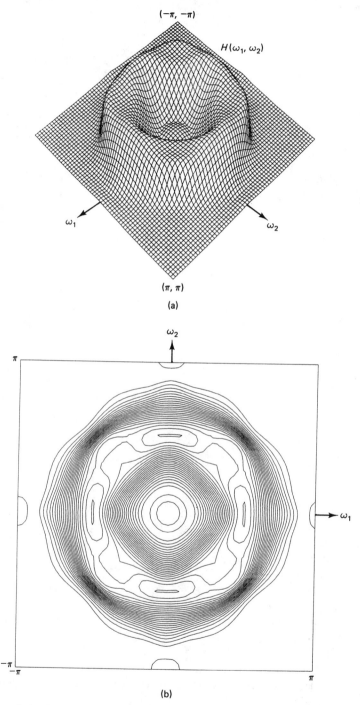

Figure 5.10 Zero-phase bandpass filter designed by the modified Prony's method. (a) Perspective plot; (b) contour plot.

Infinite Impulse Response Filters   Chap. 5

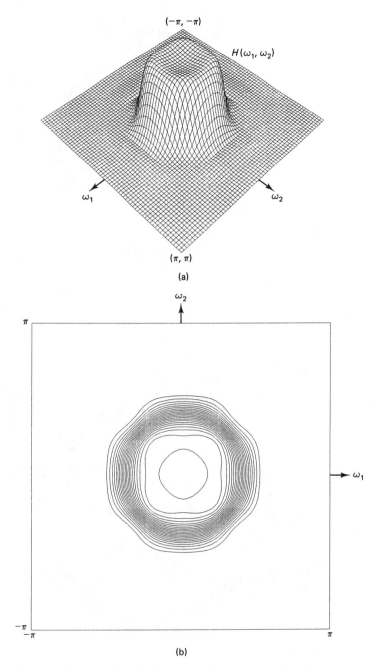

Figure 5.11 Zero-phase lowpass filter designed by the iterative prefiltering method. (a) Perspective plot; (b) contour plot.

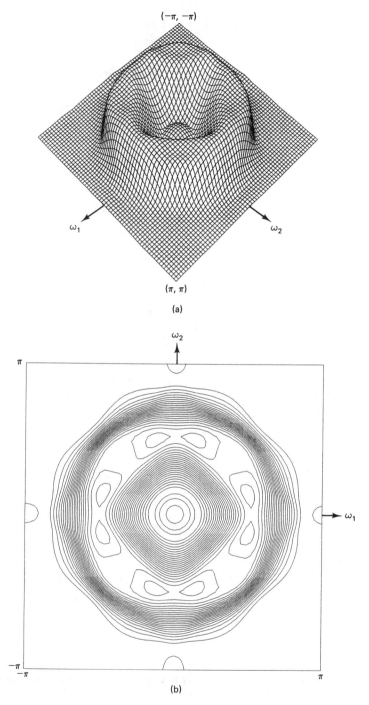

$(-\pi, -\pi)$

ω_1

ω_2

(π, π)

(a)

ω_2

π

ω_1

$-\pi$

$-\pi$

π

(b)

Figure 5.12 Zero-phase bandpass filter designed by the iterative prefiltering method. (a) Perspective plot; (b) contour plot.

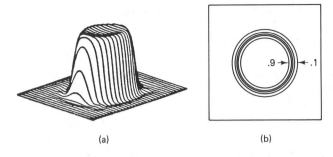

(a)   (b)

Figure 5.13 Zero-phase lowpass filter designed by the Davidon-Fletcher-Powell (DFP) method. After [Shaw and Mersereau]. (a) Perspective plot; (b) contour plot.

The desired impulse response used is the circularly symmetric ideal bandpass filter with cutoff frequencies of 0.3π and 0.7π, to which a circularly symmetric Kaiser window with radius of 16 points was applied. The zero-phase filter was again designed by parallel combination of four one-quadrant filters. The first-quadrant filter with 5×5-point $a(n_1, n_2)$ and 5×5-point $b(n_1, n_2)$ was first designed, and the overall system function was then obtained from the first-quadrant filter designed. The method typically requires a large number (on the order of 100) of iterations and is very expensive computationally. However, the method can be used with any error criterion. For a given error criterion, the method performs better than other methods we discussed in this section.

In this section, we discussed spatial domain design methods. As we discussed, simple design methods that require solving a set of linear equations are available. In addition, it is a natural approach if we wish the filter to preserve some desirable spatial domain properties such as the shape of the impulse response. However, with spatial domain design methods, we do not have much control over frequency domain design parameters. In addition, they are not optimal in the Chebyshev error sense. An alternative to the spatial domain design is the frequency domain design. We will first discuss the complex cepstrum representation of signals, which is useful in our discussion of the frequency domain design.

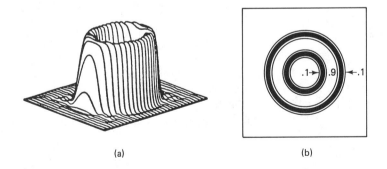

(a)   (b)

Figure 5.14 Zero-phase bandpass filter designed by the Davidon-Fletcher-Powell (DFP) method. After [Shaw and Mersereau]. (a) Perspective plot; (b) contour plot.

5.3 THE COMPLEX CEPSTRUM REPRESENTATION OF SIGNALS

5.3.1 The One-Dimensional Complex Cepstrum

The complex cepstrum representation of a 2-D sequence is in many respects a straightforward extension of the 1-D case. However, the problems to which they have been applied differ markedly. The 1-D complex cepstrum is primarily used for deconvolution problems; the 2-D complex cepstrum, for problems related to IIR filter design. We will first summarize results on the 1-D complex cepstrum representation and then discuss how they can be applied to solving problems related to 1-D IIR filter design. The complex cepstrum is not very useful in 1-D filter design applications, due to the existence of simpler methods, but the general ideas can be extended to the corresponding 2-D problems.

The complex cepstrum of a 1-D sequence $x(n)$ is defined by the system shown in Figure 5.15. The complex cepstrum is denoted by $\hat{x}(n)$, and its z-transform $\hat{X}(z)$ is related to $X(z)$ by

$$\hat{X}(z) = \log X(z). \tag{5.63}$$

Although it is not necessary from a theoretical point of view, we will assume that the region of convergence of $X(z)$ and $\hat{X}(z)$ includes the unit circle. Replacing the z-transform with the Fourier transform in Figure 5.15, we have the system shown in Figure 5.16. The relation between $X(\omega)$ and $\hat{X}(\omega)$ is given by

$$\hat{X}(\omega) = \log X(\omega). \tag{5.64}$$

Rewriting $X(\omega)$ in (5.64) in terms of its magnitude $|X(\omega)|$ and phase $\theta_x(\omega)$, we have

$$\begin{aligned}
\hat{X}(\omega) &= \log X(\omega) \\
&= \log \left[|X(\omega)| e^{j\theta_x(\omega)} \right] \\
&= \log |X(\omega)| + j\theta_x(\omega).
\end{aligned} \tag{5.65}$$

Denoting the real and imaginary parts of $\hat{X}(\omega)$ by $\hat{X}_R(\omega)$ and $\hat{X}_I(\omega)$, respectively, from (5.65), we obtain

$$\hat{X}_R(\omega) = \log |X(\omega)| \tag{5.66a}$$

$$\hat{X}_I(\omega) = \theta_x(\omega). \tag{5.66b}$$

The term *complex* is used because $X(\omega)$ is in general complex and the logarithmic operation used in (5.64) is a complex logarithm. The term *cepstrum* comes from reversing the letters in the first syllable of "spectrum." The complex cepstrum $\hat{x}(n)$ is the inverse Fourier transform of a spectrum, but in a sense it is not in the

Figure 5.15 Definition of the complex cepstrum of a sequence using the z-transform.

Figure 5.16 Definition of the complex cepstrum of a sequence using the Fourier transform.

same domain as $x(n)$ because $\hat{X}(\omega)$ is obtained from nonlinear modification of $X(\omega)$.

In (5.65) and (5.66), $\hat{X}_R(\omega)$ is very well defined. However, $\hat{X}_I(\omega) = \theta_x(\omega)$ is a multivalued function and can be expressed as

$$\hat{X}_I(\omega) = \theta_x(\omega) = \tilde{\theta}_x(\omega) + 2\pi K(\omega) \tag{5.67}$$

where $\tilde{\theta}_x(\omega)$ is the principal value of the phase of $X(\omega)$ and $K(\omega)$ is any integer function. To uniquely define $\theta_x(\omega)$, it is expressed in terms of its derivative $\theta_x'(u) = d\theta_x(u)/du$ by

$$\hat{X}_I(\omega) = \theta_x(\omega) = \theta_x(0) + \int_{u=0}^{\omega} \theta_x'(u)\, du \tag{5.68a}$$

where

$$\theta_x(0) = \tilde{\theta}_x(0) = 0 \tag{5.68b}$$

and

$$\hat{X}_I'(\omega) = \theta_x'(\omega) = \frac{X_R(\omega)X_I'(\omega) - X_I(\omega)X_R'(\omega)}{|X_R(\omega)|^2 + |X_I(\omega)|^2}. \tag{5.68c}$$

The phase function $\theta_x(\omega)$ defined in this way is called the *unwrapped phase function*. For the class of sequences that we will be concerned with, $\theta_x(\omega)$ is an odd and continuous function that is periodic with a period of 2π. Methods [Tribolet] that compute $\theta_x(\omega)$ from $\tilde{\theta}_x(\omega)$ typically exploit the continuity constraint of $\theta_x(\omega)$. Examples of the principal value phase $\tilde{\theta}_x(\omega)$ and the unwrapped phase $\theta_x(\omega)$ are shown in Figure 5.17 for $X(z)$ given by

$$X(z) = (1 - \tfrac{4}{5}z^{-1})(1 - \tfrac{5}{6}z^{-1}). \tag{5.69}$$

From $\hat{x}(n)$, $x(n)$ can be obtained by

$$X(\omega) = e^{\hat{X}(\omega)}. \tag{5.70}$$

A system that recovers $x(n)$ from $\hat{x}(n)$ is shown in Figure 5.18. The exponential operation is very well defined and has no ambiguity.

The complex cepstrum $\hat{x}(n)$ defined above exists only for a certain restricted class of sequences. For example, $\hat{x}(n)$ cannot be defined for $x(n) = 0$. Of the many possibilities, we will restrict ourselves, for both theoretical and practical reasons, to a class of sequences $x(n)$ for which $\hat{x}(n)$ defined in the above manner is real and stable. A sequence $x(n)$ which has a real and stable $\hat{x}(n)$ is said to have a *valid* complex cepstrum. Although $\hat{x}(n)$ is called a complex cepstrum, a valid $\hat{x}(n)$ is always real.

For any valid $\hat{x}(n)$, there exists a unique corresponding sequence $x(n)$ which is real and stable and whose z-transform $X(z)$ has no poles or zeros on the unit circle. To see this, note that $\hat{X}(\omega)$ is a well-defined Fourier transform for a stable

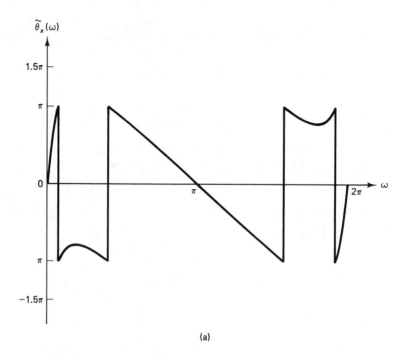

(a)

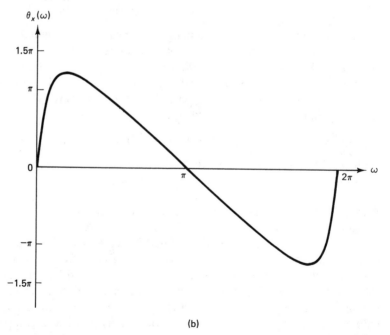

(b)

Figure 5.17 Example of an unwrapped phase function. (a) Principal value of the phase function of $X(\omega) = (1 - \frac{4}{5}e^{-j\omega})(1 - \frac{5}{8}e^{-j\omega})$; (b) unwrapped phase function of $X(\omega)$ in (a).

Figure 5.18 System for recovering $x(n)$ from its complex cepstrum $\hat{x}(n)$.

$\hat{x}(n)$. Since $X(\omega)$ is related to $\hat{X}(\omega)$ by (5.70), $X(\omega)$ is also a well-defined Fourier transform. Therefore, $x(n)$ is stable. Since $\hat{x}(n)$ is real, $\hat{X}_R(\omega)$ is even and $\hat{X}_I(\omega)$ is odd. From (5.70),

$$X(\omega) = e^{\hat{X}_R(\omega) + j\hat{X}_I(\omega)}$$

$$= |X(\omega)|\, e^{j\theta_x(\omega)}. \tag{5.71}$$

From (5.71), it is clear that $|X(\omega)|$ cannot be zero or infinite, and therefore $X(z)$ has no poles or zeros on the unit circle. In addition, $|X(\omega)|$ is even and $\theta_x(\omega)$ is odd. Therefore, $x(n)$ is real and stable. Furthermore, since $\hat{X}_I(\omega)$ is the imaginary part of $\hat{X}(\omega)$, $\hat{X}_I(\omega)$ is odd, analytic, and periodic with a period of 2π. Since $\hat{X}_I(\omega) = \theta_x(\omega)$, $\theta_x(\omega)$ is also odd, analytic, and periodic with a period of 2π.

The condition that $x(n)$ must be real and stable, with no zeros or poles on the unit circle, is necessary but not sufficient for $x(n)$ to have a valid $\hat{x}(n)$. For example, $x(n) = \delta(n) - 2\delta(n-1)$ satisfies all the above conditions, but does not have a valid $\hat{x}(n)$. If a sequence satisfies all the necessary conditions above and has a z-transform of the form

$$X(z) = A\, \frac{\prod_k (1 - c_k z^{-1}) \prod_k (1 - d_k z)}{\prod_k (1 - a_k z^{-1}) \prod_k (1 - b_k z)} \tag{5.72}$$

where A is positive and $|a_k|, |b_k|, |c_k|, |d_k| < 1$, then $x(n)$ has a valid $\hat{x}(n)$. This can be seen by considering each individual term in (5.72) and showing that the unwrapped phase of each individual term is odd, analytic, and periodic with a period of 2π.

The condition in (5.72) is highly restrictive, and a typical $x(n)$ does not have the form in (5.72). However, any real and stable $x(n)$ with a rational z-transform with no poles or zeros on the unit circle can be very simply modified such that the modified sequence $x'(n)$ will have a valid complex cepstrum. The modification involves multiplying $x(n)$ with -1 and/or shifting $x(n)$ by K points for some fixed integer K. This modification is shown in Figure 5.19. One way to determine if -1 should be multiplied to $x(n)$ and what value of K should be used in Figure 5.19 is

$$\text{Multiply by } -1, \quad \text{if} \quad X(\omega)|_{\omega=0} < 0 \tag{5.73a}$$

and

$$K = \frac{\theta_x(2\pi) - \theta_x(0)}{2\pi} \tag{5.73b}$$

where $\theta_x(\omega)$ is defined in (5.68). The integer K is also given, from the Argument

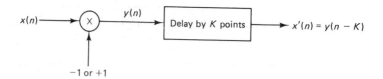

Figure 5.19 Modification of $x(n)$ such that the modified sequence has a valid complex cepstrum.

Principle, by

$$K = \text{number of zeros of } X(z) \text{ inside the unit circle}$$
$$- \text{ number of poles of } X(z) \text{ inside the unit circle.} \quad (5.74)$$

In some applications, the modification made to $x(n)$ can be very easily taken into account. If the modification cannot be taken into account in a simple manner, we are constrained to a highly restrictive set of sequences if we wish to use the complex cepstrum representation.

5.3.2 Properties of the One-Dimensional Complex Cepstrum

A number of useful properties can be derived from the definition of the complex cepstrum. They are listed in Table 5.1.

All the properties in Table 5.1 can be shown from the properties of the Fourier transform and from the definition of the complex cepstrum. Property 1 states that

---

TABLE 5.1 PROPERTIES OF THE ONE-DIMENSIONAL COMPLEX CEPSTRUM

Property 1. Suppose $x(n)$ and $w(n)$ have valid complex cepstra $\hat{x}(n)$ and $\hat{w}(n)$. We then have

$$y(n) = x(n) * w(n) \qquad \hat{y}(n) = \hat{x}(n) + \hat{w}(n)$$
$$Y(\omega) = X(\omega)W(\omega) \qquad \longleftrightarrow \qquad \hat{Y}(\omega) = \hat{X}(\omega) + \hat{W}(\omega)$$

Property 2. Suppose $x(n)$ has a valid complex cepstrum $\hat{x}(n)$. We define a new sequence $v(n)$ by

$$V(\omega) = \frac{1}{X(\omega)}.$$

The sequence $v(n)$ also has a valid complex cepstrum $\hat{v}(n) = -\hat{x}(n)$. We call $v(n)$ the *stable inverse* of $x(n)$.

$$x(n), v(n): \text{causal} \longleftrightarrow \hat{x}(n): \quad \text{causal}.$$

Property 3. Suppose $x(n)$ is real and stable, and has a rational z-transform with no poles or zeros on the unit circle. The sequence $x(n)$ may or may not have a valid complex cepstrum. Suppose we define $r(n)$ by

$$r(n) = x(n) * x(-n).$$

Then $r(n)$ is even and has a valid complex cepstrum $\hat{r}(n)$ which is also even.

Property 4. The complex cepstrum $\hat{x}(n)$ is typically an infinite-extent sequence, even though $x(n)$ may be a finite-extent sequence.

Infinite Impulse Response Filters   Chap. 5

if two sequences are convolved, then their complex cepstra add. This is the basis behind its application to deconvolution problems. Alternatively, if the complex cepstra add, then the corresponding spectra multiply. This is the basis behind its application to spectral factorization problems.

Property 2 defines the stable inverse $v(n)$ for a sequence $x(n)$ that has a valid complex cepstrum. The stable inverse $v(n)$ is related to $x(n)$ by

$$V(\omega) = \frac{1}{X(\omega)}. \tag{5.75}$$

It is important to note that $v(n)$ in (5.75) is guaranteed to be stable, but its region of support can be different from that of $x(n)$. From Property 2, a causal $\hat{x}(n)$ implies a causal $x(n)$ and $v(n)$. A sequence is called a *minimum phase sequence* if $\hat{x}(n)$ is causal. If a minimum phase sequence $x(n)$ has a rational z-transform, all its poles and zeros will be inside the unit circle. Since $v(n)$ is guaranteed to be stable, $v(n)$ is a causal and stable sequence for a minimum phase sequence $x(n)$. This property, together with Properties 3 and 4, can be used in stabilizing an unstable filter, as will be discussed in the next section.

Property 3 states that even if $x(n)$ does not have a valid complex cepstrum, $r(n) = x(n) * x(-n)$ can have a valid complex cepstrum. In this case, both $r(n)$ and $\hat{r}(n)$ are even sequences.

Property 4 states that $\hat{x}(n)$ is typically an infinite-extent sequence, even though $x(n)$ may be a finite-extent sequence. Suppose $x(n) = \delta(n) - \frac{1}{2}\delta(n - 1)$. Its complex cepstrum $\hat{x}(n)$ is given by

$$\hat{x}(n) = -\frac{1}{n}\left(\frac{1}{2}\right)^n u(n - 1). \tag{5.76}$$

This property shows that computing the complex cepstrum can be a problem in practice. If we replace the Fourier transform and inverse Fourier transform operations in Figure 5.18 by the N-point DFT and inverse DFT operations, then $\hat{x}_c(n)$, the computed complex cepstrum, will be given by

$$\hat{x}_c(n) = \sum_{r=-\infty}^{\infty} \hat{x}(n - rN). \tag{5.77}$$

Fortunately, $\hat{x}(n)$ typically decays fast. Equation (5.77) is based on the assumption that the samples of the unwrapped phase function are accurate. If N is not large, there may be an error in the unwrapped phase, and the computed $\hat{x}_c(n)$ will be degraded further.

5.3.3 Applications of the One-Dimensional Complex Cepstrum

The 1-D complex cepstrum representation is most often used in solving deconvolution problems. Suppose two sequences $x(n)$ and $w(n)$ are combined by convolution

$$y(n) = x(n) * w(n) \tag{5.78}$$

and we wish to separate $x(n)$ from $w(n)$. In general, $y(n)$ does not have a valid complex cepstrum. In typical applications, however, recovering $\pm x(n - K)$ for some integer K is sufficient. In such a case, assuming that $x(n)$ and $w(n)$ are real and stable and have rational z-transforms with no poles or zeros on the unit circle, we can modify $y(n)$, using (5.73), and compute its complex cepstrum. Modifying $y(n)$ is equivalent to modifying $x(n)$ and $w(n)$ individually and then combining them by using (5.78). Since the delay and multiplication by -1 are assumed to be unimportant, let us suppose that $x(n)$ and $w(n)$ and therefore $y(n)$ in (5.78) all have valid complex cepstra. Then from Property 1 in Table 5.1,

$$\hat{y}(n) = \hat{x}(n) + \hat{w}(n). \tag{5.79}$$

Suppose $\hat{x}(n)$ is separable from $\hat{w}(n)$ by a linear operation on $\hat{y}(n)$. For example, if $\hat{x}(n)$ consists of low-frequency components while $\hat{w}(n)$ consists of high-frequency components, then lowpass filtering $\hat{y}(n)$ will result in $\hat{x}(n)$. Then $\hat{x}(n)$ can be recovered from linearly operating on $\hat{y}(n)$. From $\hat{x}(n)$, $x(n)$ can be recovered by using (5.70). This is the basis of the homomorphic system for convolution. One application of the above idea is the development of a homomorphic vocoder. On a short time basis, human voiced speech may be approximately modeled by (5.78), where $x(n)$ is a vocal tract impulse response and $w(n)$ is a train of impulses with equal spacing. By exploiting the result that $\hat{x}(n)$ decays very fast while $\hat{w}(n)$ is another train of pulses with the same spacing as $w(n)$, low-time gating of $\hat{y}(n)$ can be used to approximately separate $\hat{x}(n)$ from $\hat{w}(n)$. This is the rationale for separating the vocal tract impulse response from the pitch information in the development of a homomorphic vocoder.

Another application of the 1-D complex cepstrum, one which is not very useful in 1-D signal processing but is significant in 2-D signal processing, is the stabilization of an unstable filter. Consider a system function $H(z) = B(z)/A(z)$ where $A(z)$ and $B(z)$ do not have a common factor. Since $B(z)$ does not affect system stability, we will assume that $B(z) = 1$. Suppose $H(z) = 1/A(z)$ has been designed so that it is a causal system. Then $A(z)$ is in the form of

$$A(z) = 1 + a(1)z^{-1} + a(2)z^{-2} + \cdots + a(p)z^{-p} \tag{5.80a}$$

and the corresponding computational procedure is given by

$$y(n) \leftarrow -a(1)y(n - 1) - a(2)y(n - 2) - \cdots - a(p)y(n - p) + x(n). \tag{5.80b}$$

Unfortunately, we find that $H(z)$ is unstable. We wish to find an $H_s(z)$ which is causal and at the same time stable. In addition, we want $|H_s(\omega)|$ to be given by

$$|H_s(\omega)| = |H(\omega)|. \tag{5.81}$$

We first assume that $A(z)$ has no zeros on the unit circle. If $A(z)$ has zeros on the unit circle, the problem is impossible to solve. It is also very important that $H_s(z)$ remain a causal system. If we are willing to sacrifice causality, then the stabilization problem could be solved by simply choosing a different region of convergence for $H(z)$ that includes the unit circle.

One simple approach to solving this stabilization problem is to compute the roots of $A(z) = 0$, reflect each root outside the unit circle to a root inside the unit circle at a conjugate reciprocal location, and scale the result appropriately. For example, suppose the causal system function $H(z)$ is given by

$$H(z) = \frac{1}{A(z)} = \frac{1}{1 - 2z^{-1}}. \tag{5.82}$$

Clearly, the system is unstable due to a pole at $z = 2$. We replace the pole at $z = 2$ with a pole at $1/2^* = 1/2$. The resulting system function $H_s(z)$ is given by

$$H_s(z) = \frac{k}{1 - \frac{1}{2}z^{-1}}. \tag{5.83}$$

The scaling factor k can be computed by requiring $H(\omega)|_{\omega=0}$ to have the same amplitude as $H_s(\omega)|_{\omega=0}$. This leads to one choice of $k = \frac{1}{2}$, and (5.83) becomes

$$H_s(z) = \frac{\frac{1}{2}}{1 - \frac{1}{2}z^{-1}}. \tag{5.84}$$

It is easy to verify that $H_s(z)$ is a causal and stable system with $|H_s(\omega)| = |H(\omega)|$.

An alternate approach is to use the complex cepstrum representation. This is shown in Figure 5.20. Consider the sequence $r(n)$ given by

$$r(n) = a(n) * a(-n). \tag{5.85}$$

If $a(n)$ is real and stable, and has a rational z-transform with no poles or zeros on the unit circle, then $r(n)$ has a valid complex cepstrum $\hat{r}(n)$, according to Property 3 in Table 5.1. From (5.85),

$$R(\omega) = |A(\omega)|^2. \tag{5.86}$$

We can therefore compute $R(\omega)$ directly from $a(n)$, as shown in Figure 5.20. Since $R(\omega)$ is always real and positive, $\theta_r(\omega)$ is zero, and computation of $\hat{r}(n)$ does not

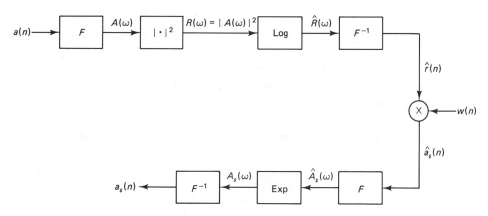

Figure 5.20 System for stabilizing an unstable system.

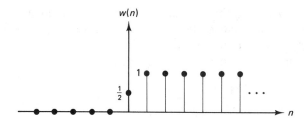

Figure 5.21 Window sequence $w(n)$ used in Figure 5.20.

require a phase-unwrapping operation. We compute a causal $\hat{a}_s(n)$ from $\hat{r}(n)$ by

$$\hat{a}_s(n) = \hat{r}(n)w(n) \tag{5.87}$$

where

$$w(n) = \begin{cases} 0, & n < 0 \\ \frac{1}{2}, & n = 0 \\ 1, & n > 0. \end{cases} \tag{5.88}$$

The sequence $w(n)$ is sketched in Figure 5.21. With this choice of $w(n)$,

$$w(n) + w(-n) = 1. \tag{5.89}$$

The complex cepstrum $\hat{a}_s(n)$ is clearly real and stable, so the corresponding sequence $a_s(n)$ can be computed from $\hat{a}_s(n)$. The stabilized system is

$$H_s(z) = \frac{1}{A_s(z)} \tag{5.90}$$

where $H_s(z)$ is the stable inverse of $A_s(z)$. To see that this is a solution of the stabilization problem, note that $\hat{a}_s(n)$ is causal and therefore, from Property 2 in Table 5.1, its stable inverse $H_s(z)$ is also causal and stable. All that must be shown, then, is that $|H_s(\omega)| = |H(\omega)|$. From (5.87) and (5.89), and noting from Property 3 in Table 5.1 that $\hat{r}(n)$ is even, we obtain

$$\hat{a}_s(n) + \hat{a}_s(-n) = \hat{r}(n)w(n) + \hat{r}(-n)w(-n)$$

$$= \hat{r}(n)[w(n) + w(-n)] \tag{5.91}$$

$$= \hat{r}(n).$$

From (5.70), (5.86), and (5.91),

$$R(\omega) = |A(\omega)|^2 = |A_s(\omega)|^2. \tag{5.92}$$

This shows that (5.90) solves the stabilization problem. In essence, we have factored $R(z)$ by dividing the even sequence $\hat{r}(n)$ into the causal part $\hat{a}_s(n)$ and the anticausal part $\hat{a}_s(-n)$. By exploiting the property that a causal $\hat{a}_s(n)$ has a stable and causal inverse and the fact that $R(\omega) = |A_s(\omega)|^2$, the stabilization problem is solved.

To illustrate this approach with an example, let us consider the causal but unstable system given by (5.82). The functions $A(\omega)$, $R(\omega)$, $\hat{R}(\omega)$, $\hat{r}(n)$, $\hat{a}_s(n)$, $\hat{A}_s(\omega)$, $A_s(\omega)$, and $H_s(z)$ are given by

$$A(\omega) = 1 - 2e^{-j\omega} \tag{5.93a}$$

Infinite Impulse Response Filters   Chap. 5

$$R(\omega) = A(\omega)A^*(\omega) = 4(1 - \tfrac{1}{2}e^{j\omega})(1 - \tfrac{1}{2}e^{-j\omega}) \tag{5.93b}$$

$$\hat{R}(\omega) = \log R(\omega) = \log 4 - \sum_{k=1}^{\infty} \frac{[(1/2)e^{j\omega}]^k}{k} - \sum_{k=1}^{\infty} \frac{[(1/2)e^{-j\omega}]^k}{k} \tag{5.93c}$$

$$\hat{r}(n) = 2 \log 2\, \delta(n) - \sum_{k=1}^{\infty} \frac{(1/2)^k}{k} \delta(n + k) - \sum_{k=1}^{\infty} \frac{(1/2)^k}{k} \delta(n - k) \tag{5.93d}$$

$$\hat{a}_s(n) = \hat{r}(n)w(n) = \log 2\, \delta(n) - \sum_{k=1}^{\infty} \frac{(1/2)^k}{k} \delta(n - k) \tag{5.93e}$$

$$\hat{A}_s(\omega) = \log 2 + \log (1 - \tfrac{1}{2}e^{-j\omega}) \tag{5.93f}$$

$$A_s(\omega) = 2(1 - \tfrac{1}{2}e^{-j\omega}) \tag{5.93g}$$

$$H_s(z) = \frac{1}{A_s(z)} = \frac{\tfrac{1}{2}}{1 - \tfrac{1}{2}z^{-1}}. \tag{5.93h}$$

In (5.93b), we have expressed $R(\omega)$ such that each of its factors corresponds to a sequence that has a valid complex cepstrum. In this way, we can use Property 1 in Table 5.1 to obtain $\hat{R}(\omega)$ in (5.93c) and $\hat{r}(n)$ in (5.93d). The result in (5.93h) is identical to the result in (5.84). Using the complex cepstrum is, of course, considerably more complicated than the approach that led to (5.84). However, the approach of flipping the poles does not extend readily to the 2-D case, while the approach based on the complex cepstrum extends to the 2-D case in a straightforward manner.

5.3.4 The Two-Dimensional Complex Cepstrum

The results developed in the previous three sections for the 1-D complex cepstrum representation can be extended to the 2-D case in a straightforward manner. In this section, we summarize the results of Sections 5.3.1 and 5.3.2 extended to the 2-D case.

The complex cepstrum of a 2-D sequence $x(n_1, n_2)$ is defined by the system shown in Figure 5.22. The complex cepstrum denoted by $\hat{x}(n_1, n_2)$ is related to $x(n_1, n_2)$ by

$$\hat{X}(\omega_1, \omega_2) = \log X(\omega_1, \omega_2). \tag{5.94}$$

Rewriting $\hat{X}(\omega_1, \omega_2)$ in terms of its real part $\hat{X}_R(\omega_1, \omega_2)$ and imaginary part $\hat{X}_I(\omega_1, \omega_2)$, we have

$$\hat{X}_R(\omega_1, \omega_2) = \log |X(\omega_1, \omega_2)| \tag{5.95a}$$

Figure 5.22 Definition of the complex cepstrum of a two-dimensional sequence.

and
$$\hat{X}_I(\omega_1, \omega_2) = \theta_x(\omega_1, \omega_2) = \tilde{\theta}_x(\omega_1, \omega_2) + 2\pi K(\omega_1, \omega_2) \qquad (5.95b)$$

where $\theta_x(\omega_1, \omega_2)$ is the unwrapped phase function, $\tilde{\theta}_x(\omega_1, \omega_2)$ is the principal value of the phase function, and $K(\omega_1, \omega_2)$ is an integer function. The unwrapped phase function $\theta_x(\omega_1, \omega_2)$ is defined in terms of its derivative by

$$\hat{X}_I(\omega_1, \omega_2) = \theta_x(\omega_1, \omega_2)$$

$$= \theta_x(0, 0) + \int_{u_2=0}^{\omega_2} \frac{\partial\theta_x(0, u_2)}{\partial u_2} du_2 + \int_{u_1=0}^{\omega_1} \frac{\partial\theta_x(u_1, \omega_2)}{\partial u_1} du_1 \qquad (5.96a)$$

where $\qquad \theta_x(0, 0) = \tilde{\theta}_x(0, 0) = 0.$ $\qquad\qquad\qquad (5.96b)$

The phase derivatives $\partial\theta_x(0, u_2)/\partial u_2$ and $\partial\theta_x(u_1, \omega_2)/\partial u_1$ are given by

$$\frac{\partial\theta_x(0, u_2)}{\partial u_2} = \frac{X_R(0, u_2) \dfrac{\partial X_I(0, u_2)}{\partial u_2} - X_I(0, u_2) \dfrac{\partial X_R(0, u_2)}{\partial u_2}}{|X_R(0, u_2)|^2 + |X_I(0, u_2)|^2} \qquad (5.96c)$$

and

$$\frac{\partial\theta_x(u_1, \omega_2)}{\partial u_1} = \frac{X_R(u_1, \omega_2) \dfrac{\partial X_I(u_1, \omega_2)}{\partial u_1} - X_I(u_1, \omega_2) \dfrac{\partial X_R(u_1, \omega_2)}{\partial u_1}}{|X_R(u_1, \omega_2)|^2 + |X_I(u_1, \omega_2)|^2}. \qquad (5.96d)$$

In (5.96a), the phase derivative is integrated along the path shown in Figure 5.23. For the class of sequences $x(n_1, n_2)$ we will consider, any continuous path from $(0, 0)$ to (ω_1, ω_2) can be followed without affecting the result. For sequences of this class, $\theta_x(\omega_1, \omega_2)$ is an odd and continuous function, which is periodic with a period of 2π in both ω_1 and ω_2 dimensions:

$$\theta_x(\omega_1, \omega_2) = -\theta_x(-\omega_1, -\omega_2) = \theta_x(\omega_1 + 2\pi, \omega_2) = \theta_x(\omega_1, \omega_2 + 2\pi). \qquad (5.97)$$

The 2-D unwrapped phase function $\theta_x(\omega_1, \omega_2)$ can be computed from $\tilde{\theta}_x(\omega_1, \omega_2)$ by imposing the continuity property of $\theta_x(\omega_1, \omega_2)$ in 2-D or by repeatedly using a 1-D phase-unwrapping algorithm [Dudgeon (1977); Tribolet] along the ω_1 dimension or along the ω_2 dimension, which exploits the phase continuity along the horizontal or vertical direction in the 2-D frequency plane.

From $\hat{x}(n_1, n_2)$, $x(n_1, n_2)$ can be obtained by

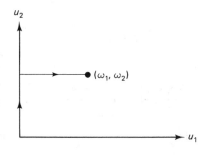

Figure 5.23 Integration path of the phase derivative used in defining the unwrapped phase in (5.96).

Figure 5.24 System for recovering $x(n_1, n_2)$ from its complex cepstrum $\hat{x}(n_1, n_2)$.

$$X(\omega_1, \omega_2) = e^{\hat{X}(\omega_1, \omega_2)}. \tag{5.98}$$

A system that recovers $x(n_1, n_2)$ from $\hat{x}(n_1, n_2)$ is shown in Figure 5.24.

The complex cepstrum $\hat{x}(n_1, n_2)$ defined above exists only for a certain restricted class of sequences $x(n_1, n_2)$. Of the many possibilities, we will again restrict ourselves to a class of sequences $x(n_1, n_2)$ for which $\hat{x}(n_1, n_2)$ defined in the above manner is real and stable. A sequence $x(n_1, n_2)$ that has a real and stable $\hat{x}(n_1, n_2)$ will be said to have a valid complex cepstrum.

For any valid $\hat{x}(n_1, n_2)$, there exists a unique corresponding $x(n_1, n_2)$ which is real and stable, and whose z-transform has no pole or zero surfaces on the unit surface. In addition, any real and stable $x(n_1, n_2)$ with a rational z-transform with no pole or zero surfaces on the unit surface can be very simply modified so that the modified sequence $x'(n_1, n_2)$ has a valid complex cepstrum. The modification involves multiplying $x(n_1, n_2)$ by -1 and/or shifting $x(n_1, n_2)$ by K_1 points along the n_1 dimension and K_2 points along the n_2 dimension for some fixed integers K_1 and K_2. This modification is shown in Figure 5.25. One way to determine if $x(n_1, n_2)$ should be multiplied by -1 and what values of K_1 and K_2 should be used is

$$\text{Multiply by } -1 \text{ if } X(\omega_1, \omega_2)|_{\omega_1 = 0, \omega_2 = 0} < 0, \tag{5.99a}$$

$$K_1 = \frac{\theta_x(2\pi, 0) - \theta_x(0, 0)}{2\pi} \tag{5.99b}$$

and

$$K_2 = \frac{\theta_x(0, 2\pi) - \theta_x(0, 0)}{2\pi} \tag{5.99c}$$

where $\theta_x(\omega_1, \omega_2)$ is obtained by (5.96). In some applications, the modification made to $x(n_1, n_2)$ can be easily taken into account. Otherwise, we are limited to a highly restrictive set of sequences if we are to use the complex cepstrum representation.

Many useful properties can be obtained from the definition of the complex cepstrum. They are listed in Table 5.2.

One potential application of the 2-D complex cepstrum representation is in deconvolution. If two sequences $x(n_1, n_2)$ and $w(n_1, n_2)$ that have valid complex

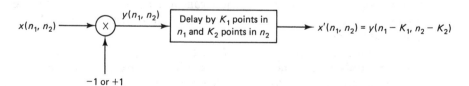

Figure 5.25 Modification of $x(n_1, n_2)$ such that the modified sequence has a valid complex cepstrum.

Sec. 5.3   The Complex Cepstrum Representation of Signals   **303**

TABLE 5.2 PROPERTIES OF THE TWO-DIMENSIONAL COMPLEX CEPSTRUM

Property 1. Suppose $x(n_1, n_2)$ and $w(n_1, n_2)$ have valid complex cepstra $\hat{x}(n_1, n_2)$ and $\hat{w}(n_1, n_2)$. We then have

$$y(n_1, n_2) = x(n_1, n_2) * w(n_1, n_2) \qquad \hat{y}(n_1, n_2) = \hat{x}(n_1, n_2) + \hat{w}(n_1, n_2)$$
$$Y(\omega_1, \omega_2) = X(\omega_1, \omega_2)W(\omega_1, \omega_2) \overset{\longleftrightarrow}{} \hat{Y}(\omega_1, \omega_2) = \hat{X}(\omega_1, \omega_2) + \hat{W}(\omega_1, \omega_2).$$

Property 2. Suppose $x(n_1, n_2)$ has a valid complex cepstrum $\hat{x}(n_1, n_2)$. We define a new sequence $v(n_1, n_2)$ by

$$V(\omega_1, \omega_2) = \frac{1}{X(\omega_1, \omega_2)}.$$

The sequence $v(n_1, n_2)$ also has a valid complex cepstrum $\hat{v}(n_1, n_2) = -\hat{x}(n_1, n_2)$. We call $v(n_1, n_2)$ the *stable inverse* of $x(n_1, n_2)$.

$x(n_1, n_2)$, $v(n_1, n_2)$: wedge support $\hat{x}(n_1, n_2)$: wedge support sequence with
sequences with identical wedge shape. $\overset{\longleftrightarrow}{}$ the same wedge shape as $x(n_1, n_2)$ and
                                                     $v(n_1, n_2)$.

Property 3. Suppose $x(n_1, n_2)$ is real and stable and has a rational z-transform with no pole or zero surfaces on the unit surface. The sequence $x(n_1, n_2)$ may or may not have a valid complex cepstrum. Suppose we define $r(n_1, n_2)$ by

$$r(n_1, n_2) = x(n_1, n_2) * x(-n_1, -n_2)$$

Then $r(n_1, n_2)$ is even and has a valid complex cepstrum $\hat{r}(n_1, n_2)$ that is also even.

Property 4. The complex cepstrum $\hat{x}(n_1, n_2)$ is typically an infinite-extent sequence, even though $x(n_1, n_2)$ may be a finite-extent sequence.

---

cepstra are combined by convolution, then their complex cepstra $\hat{x}(n_1, n_2)$ and $\hat{w}(n_1, n_2)$ add. If $\hat{x}(n_1, n_2)$ is linearly separable from $\hat{w}(n_1, n_2)$, linear operation on the complex cepstrum can lead to separation of $x(n_1, n_2)$ from $w(n_1, n_2)$. Unfortunately, problems where $\hat{x}(n_1, n_2)$ is linearly separable, even approximately, from $\hat{w}(n_1, n_2)$ are not common in practice. An important application of the 2-D complex cepstrum representation is the stabilization of an unstable IIR filter. This is discussed in the next section.

5.4 STABILIZATION OF AN UNSTABLE FILTER

The filters designed by the spatial domain design techniques discussed in Section 5.2 can be unstable. The filters designed by the frequency domain design techniques discussed in the next section can also be unstable. In this section, we discuss the problem of stabilizing an unstable filter without seriously affecting the magnitude response of the filter.

Consider a system function $H(z_1, z_2) = B(z_1, z_2)/A(z_1, z_2)$, where $A(z_1, z_2)$ and $B(z_1, z_2)$ are co-prime. As discussed in Section 2.3.1, $B(z_1, z_2)$ can affect the stability of $H(z_1, z_2)$. However, such occurrences are rare, and we will assume $B(z_1, z_2) = 1$. Suppose $H(z_1, z_2) = 1/A(z_1, z_2)$ has been designed to be a recursively computable system. Then $A(z_1, z_2)$ is in the form of

$$A(z_1, z_2) = 1 + \sum_{(n_1, n_2) \in Ra - (0,0)} \sum a(n_1, n_2) z_1^{-n_1} z_2^{-n_2} \tag{5.100a}$$

and the corresponding computational procedure is given by

$$y(n_1, n_2) \leftarrow - \sum_{(k_1, k_2) \in Ra - (0,0)} \sum a(k_1, k_2) y(n_1 - k_1, n_2 - k_2) + x(n_1, n_2) \tag{5.100b}$$

where $a(k_1, k_2)$ with $a(0, 0) = 1$ is a wedge support sequence. We wish to find an $H_s(z_1, z_2)$ that is recursively computable, with $h_s(n_1, n_2)$ having the same or similar region of support as $h(n_1, n_2)$, and at the same time stable, with $|H_s(\omega_1, \omega_2)|$ given by

$$|H_s(\omega_1, \omega_2)| = |H(\omega_1, \omega_2)|. \tag{5.101}$$

We assume* that $A(z_1, z_2)$ has no zeros on the unit surface. With zeros of $A(z_1, z_2)$ on the unit surface, the problem would be impossible to solve.

One simple approach to the 1-D stabilization problem discussed in Section 5.3.3 is to factor $A(z)$ and flip the poles that lie outside the unit circle. If we look at this procedure a little more carefully, it can be viewed as first determining $R(z)$ given by

$$R(z) = A(z)A(z^{-1}), \tag{5.102}$$

next factoring $R(z)$, and then collecting the terms whose poles are inside the unit circle. Since $R(z) = R(z^{-1})$, each pole inside the unit circle has a corresponding pole at a conjugate reciprocal location outside the unit circle. This procedure for determining $A_s(z)$ ensures that $H_s(z) = 1/A_s(z)$ will have all its poles inside the unit circle and $A_s(z)$ will satisfy

$$R(z) = A_s(z)A_s(z^{-1}). \tag{5.103}$$

From (5.102) and (5.103),

$$R(z) = A(z)A(z^{-1}) = A_s(z)A_s(z^{-1}). \tag{5.104}$$

From (5.104), it is clear that the method can be used only when $A(z)$ can be factored as a product of other polynomials.

An extension of (5.104) to the 2-D problem is

$$R(z_1, z_2) = A(z_1, z_2)A(z_1^{-1}, z_2^{-1}) = A_s(z_1, z_2)A_s(z_1^{-1}, z_2^{-1}). \tag{5.105}$$

Since $A(z_1, z_2)$ cannot in general be factored as a product of other finite-order polynomials, a simple extension of (5.104) cannot be used in solving the 2-D stabilization problem. Careful observation of (5.105) shows that the problem is more serious. The constraint that $|H_s(\omega_1, \omega_2)| = |H(\omega_1, \omega_2)|$ in (5.101) requires that $|A_s(\omega_1, \omega_2)| = |A(\omega_1, \omega_2)|$. A solution that satisfies $|A_s(\omega_1, \omega_2)| = |A(\omega_1, \omega_2)|$

---

*A 1-D IIR filter which is unstable seldom has poles on the unit circle. A 2-D IIR filter which is unstable has pole surfaces that cross the unit surface more frequently than the 1-D case. This assumption is, therefore, more restrictive in 2-D than in 1-D.

has to satisfy (5.105). However, $A(z_1, z_2)$ cannot in general be factored and (5.105) cannot in general be satisfied. This means that the 2-D stabilization problem does not have a solution. Various methods to solve the stabilization problem approximately have been developed. In this section, we discuss two methods. One is based on the complex cepstrum representation of signals. The other is based on the least squares inverse solution of a polynomial.

5.4.1 Complex Cepstrum Method

One approach to approximately solving the stabilization problem is to allow $a_s(n_1, n_2)$ to become an infinite-extent sequence, solve the stabilization problem, and then truncate $a_s(n_1, n_2)$, hoping that the truncation will not affect the solution too much. Once we allow $a_s(n_1, n_2)$ to be an infinite-extent sequence, (5.105) can be satisfied. A 2-D finite-order polynomial cannot in general be factored as a product of other finite-order polynomials, but can always be factored as a product of many different infinite-order polynomials. An approach that solves the stabilization problem with an infinite-extent $a_s(n_1, n_2)$ is a straightforward extension of the 1-D approach discussed in Section 5.3.3.

The 2-D stabilization system based on the complex cepstrum representation is shown in Figure 5.26. We consider $r(n_1, n_2)$ given by

$$r(n_1, n_2) = a(n_1, n_2) * a(-n_1, -n_2). \qquad (5.106)$$

Even though $a(n_1, n_2)$ may not have a valid complex cepstrum, $r(n_1, n_2)$ has a valid $\hat{r}(n_1, n_2)$, according to Property 3 in Table 5.2. Since $R(\omega_1, \omega_2)$ is always positive, the unwrapped phase is zero, and no phase-unwrapping operation is necessary. We compute a wedge support sequence $a_s(n_1, n_2)$ by

$$\hat{a}_s(n_1, n_2) = \hat{r}(n_1, n_2)w(n_1, n_2) \qquad (5.107)$$

where the window sequence $w(n_1, n_2)$ is chosen such that $w(n_1, n_2)$ has a region of support that includes the region of support of $a(n_1, n_2)$ and

$$w(n_1, n_2) + w(-n_1, -n_2) = 1. \qquad (5.108)$$

The window $w(n_1, n_2)$ that can be used when $a(n_1, n_2)$ corresponds to a nonsymmetric

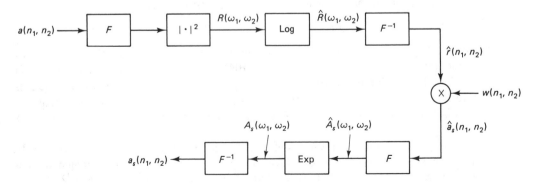

Figure 5.26 System for stabilizing an unstable system.

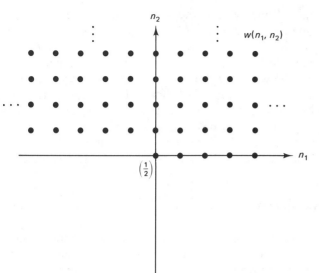

Figure 5.27 Window sequence $w(n_1, n_2)$ used in Figure 5.26.

half-plane filter whose region of support lies in the first and second quadrants, except the line corresponding to $n_1 < 0$, $n_2 = 0$, is shown in Figure 5.27. The complex cepstrum $\hat{a}_s(n_1, n_2)$ is clearly real and stable, and the corresponding sequence $a_s(n_1, n_2)$ can be computed from $\hat{a}_s(n_1, n_2)$. The stabilized system $H_s(z_1, z_2)$ is

$$H_s(z_1, z_2) = \frac{1}{A_s(z_1, z_2)} \qquad (5.109)$$

where $H_s(z_1, z_2)$ is the stable inverse of $A_s(z_1, z_2)$. From Property 2 in Table 5.2, $H_s(z_1, z_2)$ is stable and $h_s(n_1, n_2)$ has a wedge-shaped region of support that includes the region of support of $a(n_1, n_2)$. In addition, from (5.107) and (5.108),

$$\hat{r}(n_1, n_2) = \hat{a}_s(n_1, n_2) + \hat{a}_s(-n_1, -n_2). \qquad (5.110)$$

From (5.110),

$$R(\omega_1, \omega_2) = |A_s(\omega_1, \omega_2)|^2. \qquad (5.111)$$

This shows that (5.109) solves the stabilization problem. In essence, $R(z_1, z_2)$ in (5.105) is factored by dividing $\hat{r}(n_1, n_2)$ into two pieces. We choose, among many ways of dividing $\hat{r}(n_1, n_2)$ into two pieces, the way that will factor $R(z_1, z_2)$ such that the resulting factor $A_s(z_1, z_2)$ when used in (5.109) will solve the stabilization problem.

The sequence $a_s(n_1, n_2)$ obtained in this manner is typically an infinite-extent sequence. One approach to obtaining a finite-extent sequence from $a_s(n_1, n_2)$ is to window $a_s(n_1, n_2)$. This is, in a sense, similar to the window method for FIR filter design. In FIR filter design, the windowed sequence constitutes the coefficients of the numerator polynomial, while $a_s(n_1, n_2)$ denotes the coefficients of the denominator polynomial. Windowing $a_s(n_1, n_2)$ may affect the stability of $H_s(z_1, z_2)$, and the magnitude constraint in (5.101) is no longer exactly satisfied. In addition, spatial aliasing results when the Fourier transform and inverse Fourier transform operations are replaced by DFTs and inverse DFTs. In typical cases, $a_s(n_1, n_2)$ decays rapidly, and the windowing operation tends to preserve stability without significantly affecting the magnitude. The choice of sufficiently large DFT and inverse DFT sizes, combined with rapid decay of both $a_s(n_1, n_2)$ and $\hat{r}(n_1, n_2)$, reduces the spatial aliasing problem. In essence, solving the 2-D stabilization problem exactly is not theoretically possible, and some constraint must be relaxed. The system in Figure 5.26 provides one approach to approximately solving the 2-D stabilization problem.

5.4.2 Planar Least Squares Inverse Method

In the complex cepstrum method, we allowed $A_s(z_1, z_2)$ to be an infinite-order polynomial, solved the stabilization problem exactly, and then truncated the resulting $A_s(z_1, z_2)$. In the method discussed in this section, we require $A_s(z_1, z_2)$ to be a finite-order polynomial but allow $|A_s(\omega_1, \omega_2)|$ to be only approximately equal to $|A(\omega_1, \omega_2)|$.

Let $a(n_1, n_2)$ be an $N_1 \times N_2$-point first-quadrant support sequence. Let $i(n_1, n_2)$ denote an inverse of $a(n_1, n_2)$, so that

$$a(n_1, n_2) * i(n_1, n_2) = \delta(n_1, n_2) \qquad (5.112a)$$

and
$$A(z_1, z_2)I(z_1, z_2) = 1. \qquad (5.112b)$$

The sequence $i(n_1, n_2)$ is an infinite-extent sequence. We wish to approximate $i(n_1, n_2)$ with an $M_1 \times M_2$-point first-quadrant support sequence $c(n_1, n_2)$. If we replace $i(n_1, n_2)$ in (5.112a) with $c(n_1, n_2)$, then (5.112a) will not hold exactly. We can define an error sequence $e(n_1, n_2)$ by

$$e(n_1, n_2) = \delta(n_1, n_2) - a(n_1, n_2) * c(n_1, n_2). \qquad (5.113)$$

Given $a(n_1, n_2)$, we can estimate $c(n_1, n_2)$ by minimizing

$$\text{Error} = \sum_{n_1 = -\infty}^{\infty} \sum_{n_2 = -\infty}^{\infty} e^2(n_1, n_2)$$

$$= \sum_{n_1 = -\infty}^{\infty} \sum_{n_2 = -\infty}^{\infty} [\delta(n_1, n_2) - a(n_1, n_2) * c(n_1, n_2)]^2. \qquad (5.114)$$

The Error in (5.114) can also be expressed in the frequency domain as

$$\text{Error} = \frac{1}{(2\pi)^2} \int_{\omega_1 = -\pi}^{\pi} \int_{\omega_2 = -\pi}^{\pi} |1 - A(\omega_1, \omega_2)C(\omega_1, \omega_2)|^2 \, d\omega_1 \, d\omega_2. \qquad (5.115)$$

From (5.114), the error expression is a quadratic form of the unknown parameters $c(n_1, n_2)$, and therefore its minimization with respect to $c(n_1, n_2)$ is a linear problem. The solution $c(n_1, n_2)$ to the above minimization problem is called the least squares inverse of $a(n_1, n_2)$. The least squares inverse of a 2-D sequence is often referred to as the planar least squares inverse (PLSI). Since $c(n_1, n_2)$ is an approximate inverse of $a(n_1, n_2)$, we expect that

$$C(\omega_1, \omega_2) \approx \frac{1}{A(\omega_1, \omega_2)}. \tag{5.116}$$

As we increase M_1 and M_2 [the size of $c(n_1, n_2)$], $C(\omega_1, \omega_2)$ will approximate $1/A(\omega_1, \omega_2)$ better, but determining $c(n_1, n_2)$ will require more computations.

Shanks conjectured [Shanks et al.] that a first-quadrant support system with system function $1/C(z_1, z_2)$, where $C(z_1, z_2)$ is the z-transform of the PLSI $c(n_1, n_2)$ of any first-quadrant support sequence $a(n_1, n_2)$, is stable. This is known as *Shanks's conjecture*. Suppose Shanks's conjecture is true. We can determine an $N_1 \times N_2$-point first-quadrant support sequence $d(n_1, n_2)$ by computing the PLSI of $c(n_1, n_2)$. Note that the region of support size of $d(n_1, n_2)$ is chosen to be the same as that of $a(n_1, n_2)$. The sequence $d(n_1, n_2)$ is called a double PLSI of $a(n_1, n_2)$. Clearly, the first-quadrant support system with system function $1/D(z_1, z_2)$ would be stable. In addition,

$$D(\omega_1, \omega_2) \approx \frac{1}{C(\omega_1, \omega_2)}. \tag{5.117}$$

From (5.116) and (5.117),

$$D(\omega_1, \omega_2) \approx A(\omega_1, \omega_2). \tag{5.118}$$

An approximate solution to the stabilization problem would then be given by

$$H_s(z_1, z_2) = \frac{1}{D(z_1, z_2)} \tag{5.119}$$

where $d(n_1, n_2)$ is a double PLSI of $a(n_1, n_2)$.

Shanks's conjecture has been verified for many examples. However, for some examples with $A(\omega_1, \omega_2) = 0$ for some (ω_1, ω_2), it has been shown [Genin and Kamp] that Shanks's conjecture does not hold. A counterexample that disproves the modified Shanks's conjecture, which is Shanks's conjecture with the constraint that $A(\omega_1, \omega_2) \neq 0$ for any (ω_1, ω_2), has not yet been found.

The resulting filter obtained from the PLSI method has the magnitude response which is only approximately equal to the given magnitude response. In addition, it is not yet known if the resulting filter is guaranteed to be stable. However, the method is simple and can be easily generalized [Chang and Aggarwal (1978)] to solve stabilization problems that involve nonsymmetric half-plane filters.

5.5 FREQUENCY DOMAIN DESIGN

5.5.1 Design Approaches

In frequency domain IIR filter design, some desired or ideal frequency domain response to a known input is assumed given. The filter coefficients are estimated such that the response of the designed filter in the frequency domain will be as close as possible in some sense to the desired response. As with the spatial domain IIR filter design, an input often used is $\delta(n_1, n_2)$, and the desired frequency response that is assumed given is $H_d(\omega_1, \omega_2)$, which either includes both the magnitude and phase responses or only includes the magnitude response $|H_d(\omega_1, \omega_2)|$.

If we define a filter to be optimal when it minimizes the weighted Chebyshev error norm, an obvious error criterion will be the minimization of the maximum weighted error over the domain of approximation. This error criterion, however, makes the minimization problem highly nonlinear. Even though the standard descent algorithms discussed in Section 5.2.3 can in theory be used, computing the first- and second-order partial derivative information typically required in such algorithms is quite involved. As a result, this approach has been considered only for the design of very low-order filters.

Another frequency domain error criterion that has been considered is

$$\text{Error} = \frac{1}{(2\pi)^2} \int_{\omega_1 = -\pi}^{\pi} \int_{\omega_2 = -\pi}^{\pi} |W(\omega_1, \omega_2)|^2 |E(\omega_1, \omega_2)|^2 \, d\omega_1 \, d\omega_2 \qquad (5.120a)$$

where

$$E(\omega_1, \omega_2) = H_d(\omega_1, \omega_2) - H(\omega_1, \omega_2) \qquad (5.120b)$$

and where $W(\omega_1, \omega_2)$ is a positive weighting function that can take into account the relative importance of different frequency components. The error criterion in (5.120) can be very simply related to a spatial domain error criterion. Using Parseval's theorem, we find that (5.120) is equivalent to

$$\text{Error} = \sum_{n_1 = -\infty}^{\infty} \sum_{n_2 = -\infty}^{\infty} (w(n_1, n_2) * e(n_1, n_2))^2 \qquad (5.121a)$$

where

$$e(n_1, n_2) = h_d(n_1, n_2) - h(n_1, n_2). \qquad (5.121b)$$

When $W(\omega_1, \omega_2) = 1$ so that equal weight is given to all frequency components, $w(n_1, n_2) = \delta(n_1, n_2)$, and the Error in (5.120) reduces to

$$\text{Error} = \sum_{n_1 = -\infty}^{\infty} \sum_{n_2 = -\infty}^{\infty} (h_d(n_1, n_2) - h(n_1, n_2))^2. \qquad (5.122)$$

This is precisely the same error criterion that we considered in Section 5.2, and all the methods discussed in Section 5.2 apply to the minimization of the Error in (5.120) with $W(\omega_1, \omega_2) = 1$. Furthermore, the spatial domain design techniques discussed in Section 5.2 can easily be extended to incorporate $w(n_1, n_2)$ in (5.121). Since $w(n_1, n_2)$ is a known sequence, the presence of $w(n_1, n_2)$ does not affect the linearity or nonlinearity of the methods we discussed. In other words, solving a set of linear equations with $w(n_1, n_2) = \delta(n_1, n_2)$ will remain the problem of solving a set of linear equations with a general $w(n_1, n_2)$.

Although (5.120) and (5.121) are equivalent error criteria, there are some differences between the two in practice. In spatial domain design based on (5.121), a specific R_h (the region of support of $h(n_1, n_2)$) is assumed and $h(n_1, n_2)$ is compared to $h_d(n_1, n_2)$. Therefore $h(n_1, n_2)$ designed will always have R_h as its region of support, will approximate $h_d(n_1, n_2)$ for $(n_1, n_2) \in R_h$, and will tend to be stable for a stable $h_d(n_1, n_2)$. In frequency domain design based on (5.120), $H(\omega_1, \omega_2)$ is obtained by evaluating $H(z_1, z_2)$ on the unit surface:

$$H(\omega_1, \omega_2) = H(z_1, z_2)|_{z_1 = e^{j\omega_1}, z_2 = e^{j\omega_2}} \tag{5.123a}$$

where

$$H(z_1, z_2) = \frac{\displaystyle\sum_{(k_1, k_2) \in R_b} \sum b(k_1, k_2) z_1^{-k_1} z_2^{-k_2}}{1 + \displaystyle\sum_{(k_1, k_2) \in R_a - (0,0)} \sum a(k_1, k_2) z_1^{-k_1} z_2^{-k_2}}. \tag{5.123b}$$

The region of support of $h(n_1, n_2)$ obtained by minimizing (5.120) with $H(\omega_1, \omega_2)$ in (5.123a) now depends on the specific choice of the filter coefficients $a(n_1, n_2)$ and $b(n_1, n_2)$. If R_{h_d}, the region of support of $h_d(n_1, n_2)$, is approximately the same as R_h, $a(n_1, n_2)$, $b(n_1, n_2)$ and $h(n_1, n_2)$ obtained from both spatial and frequency domain designs are likely to be the same. If R_{h_d} is significantly different from the assumed R_h, $a(n_1, n_2)$ and $b(n_1, n_2)$ obtained can be different. In this case, the region of support of the filter designed may no longer be the same as the assumed R_h. We can require the filter in (5.123b) to have the same region of support as the assumed R_h, but we will then be affecting the stability of the filter. Therefore, if we require the resulting filter to have the assumed region of support R_h to ensure that the filter is recursively computable, the filter obtained from frequency domain design is likely to be unstable more often than the filter obtained from spatial domain design. In essence, we have assumed that the filter designed has a particular region of support R_h in spatial domain design, and we have assumed that the filter designed is stable in frequency domain design. These two assumptions are not always the same and can lead to two different filters. An example that illustrates this point can be found in Problem 5.22.

In addition to the difference between (5.120) and (5.121) discussed above, there is another difference in practice. In the spatial domain, $h_d(n_1, n_2)$ is typically truncated after a certain point, and $h(n_1, n_2)$ approximates a truncated version of $h_d(n_1, n_2)$. In the frequency domain, $H_d(\omega_1, \omega_2)$ and $H(\omega_1, \omega_2)$ are typically evaluated on a Cartesian grid, and samples of $H(\omega_1, \omega_2)$ approximate samples of $H_d(\omega_1, \omega_2)$. When this is interpreted in the spatial domain, an aliased version of $h(n_1, n_2)$ approximates an aliased version of $h_d(n_1, n_2)$. When a sufficiently large region of $h_d(n_1, n_2)$ is used in the spatial domain and a sufficiently large number of samples are used in the frequency domain, the difference should be minor. Experience has shown that this is the case. This difference in practice between (5.120) and (5.121) suggests a way of checking how significant truncating the tail of $h_d(n_1, n_2)$ is in spatial domain design. Once $h(n_1, n_2)$ is estimated, the error can be computed by both (5.120) and (5.121). Since the tail parts of $h_d(n_1, n_2)$ and $h(n_1, n_2)$ affect (5.120) and (5.121) differently, a small difference in the two computed errors is a reasonable indication that truncation of $h_d(n_1, n_2)$ is not very significant.

Another error criterion that has been considered is

$$E_p = \frac{1}{(2\pi)^2} \int_{\omega_1 = -\pi}^{\pi} \int_{\omega_2 = -\pi}^{\pi} W(\omega_1, \omega_2) |E(\omega_1, \omega_2)|^p \, d\omega_1 \, d\omega_2 \tag{5.124a}$$

where
$$E(\omega_1, \omega_2) = |H_d(\omega_1, \omega_2)| - |H(\omega_1, \omega_2)|. \qquad (5.124b)$$

If we let p in the pth-order error norm E_p approach ∞, E_p will become the weighted Chebyshev error, since the maximum error in the region of approximation will dominate E_p. The problem of minimizing E_p in (5.124) is, of course, nonlinear. In theory, functional minimization algorithms such as the standard descent algorithms discussed in Section 5.2.3 can be used to solve the minimization problem. Evaluating the first- and second-order derivative information is more involved here than in the spatial domain design case due to the magnitude operation on $H(\omega_1, \omega_2)$ in (5.124). When the filter is designed by minimizing E_p in (5.124), $|H(\omega_1, \omega_2)|$ is obtained from (5.123). The magnitude response $|H(\omega_1, \omega_2)|$ will approximate $|H_d(\omega_1, \omega_2)|$, but the region of support of $h(n_1, n_2)$ may not be the same as the assumed region of support R_h. We can require the filter in (5.123b) to have the assumed region of support R_h, but we will then affect the stability of the filter.

We have qualitatively argued that filters designed with spatial domain design techniques tend to be stable. We have also argued that filters designed with frequency domain design techniques tend to be unstable more often than filters designed with spatial domain design techniques. One approach to reducing the instability problem is to define an error criterion that combines a frequency domain error such as E_p in (5.124) with some error norm that penalizes an unstable filter. Minimizing the resulting error function will remain a nonlinear problem, and standard descent algorithms may be used to solve the minimization problem.

One error norm that penalizes an unstable filter is based on the complex cepstrum representation of signals. Specifically, the method discussed in Section 5.4.1, which stabilizes an unstable system, determines a stable denominator coefficient sequence $a_s(n_1, n_2)$ from an unstable denominator coefficient sequence. If $a(n_1, n_2)$ has wedge support and is a stable denominator coefficient sequence with $a(0, 0) = 1$, then the only solution to this stabilization problem is $a(n_1, n_2)$. Therefore, $a_s(n_1, n_2)$ obtained by the stabilization system will be $a(n_1, n_2)$:

$$a_s(n_1, n_2) = a(n_1, n_2). \qquad (5.125)$$

The notion that $a_s(n_1, n_2)$ is identical to $a(n_1, n_2)$ when the system is stable, but is different from $a(n_1, n_2)$ when the system is unstable, suggests a stability error criterion E_s defined by

$$E_s = \sum_{n_1 = -\infty}^{\infty} \sum_{n_2 = -\infty}^{\infty} (a_s(n_1, n_2) - a(n_1, n_2))^2$$
$$= \frac{1}{(2\pi)^2} \int_{\omega_1 = -\pi}^{\pi} \int_{\omega_2 = -\pi}^{\pi} |A_s(\omega_1, \omega_2) - A(\omega_1, \omega_2)|^2 \, d\omega_1 d\omega_2. \qquad (5.126)$$

If we combine E_s in (5.126) with the frequency domain error criterion E_p in (5.124) with $p = 2$,

$$\text{Error} = \frac{1}{(2\pi)^2} \int_{\omega_1 = -\pi}^{\pi} \int_{\omega_2 = -\pi}^{\pi} W(\omega_1, \omega_2)|E(\omega_1, \omega_2)|^2 \, d\omega_1 \, d\omega_2$$
$$+ \alpha \frac{1}{(2\pi)^2} \int_{\omega_1 = -\pi}^{\pi} \int_{\omega_2 = -\pi}^{\pi} |E_s(\omega_1, \omega_2)|^2 \, d\omega_1 \, d\omega_2 \qquad (5.127a)$$

where
$$E(\omega_1, \omega_2) = |H_d(\omega_1, \omega_2)| - |H(\omega_1, \omega_2)| \qquad (5.127b)$$

and
$$E_s(\omega_1, \omega_2) = A_s(\omega_1, \omega_2) - A(\omega_1, \omega_2). \qquad (5.127c)$$

The parameter α in (5.127a) controls the relative importance of the magnitude approximation error and the stability error. Since E_s penalizes an unstable filter, an IIR filter designed by using the error criterion in (5.127) is more likely to result in a stable filter than is a filter designed without incorporating the stability error. Minimization of the Error in (5.127) is, of course, a nonlinear problem, and descent algorithms such as those discussed in Section 5.2.3 must be employed. An example of a zero-phase IIR filter designed by minimizing the Error in (5.127) will be shown after we discuss the zero-phase design problem in the next section.

From (5.125), the denominator polynomial coefficients denoted by $a(n_1, n_2)$ corresponding to a stable and recursively computable system have a valid complex cepstrum. Then the unwrapped phase $\theta_a(\omega_1, \omega_2)$ is odd, continuous, and periodic with a period of 2π along both the ω_1 and ω_2 dimensions. In fact, the above condition on the unwrapped phase can be shown [O'Connor and Huang] to be necessary and sufficient for $a(n_1, n_2)$ with a positive $a(0, 0)$ to be a denominator polynomial coefficient sequence of a recursively computable and stable system:

$$\text{Stability} \iff \theta_a(\omega_1, \omega_2) = -\theta_a(-\omega_1, -\omega_2)$$

$$= \theta_a(\omega_1 + 2\pi, \omega_2) = \theta_a(\omega_1, \omega_2 + 2\pi) \qquad (5.128)$$

and $\theta_a(\omega_1, \omega_2)$ is continuous.

One advantage of (5.128) over the stability tests discussed in Section 2.3 is that $a(n_1, n_2)$ does not have to be a first-quadrant support sequence when (5.128) is used. Practical algorithms to test the stability of a 2-D filter can be developed based on (5.128). The computations involved in such algorithms, however, are essentially equivalent to performing many 1-D stability tests where each 1-D stability test is performed by a 1-D phase-unwrapping operation. Thus, such algorithms do not offer any computational advantages over the approaches discussed in Section 2.3.

5.5.2 Zero-Phase Design

As we discussed in Section 5.2.4, it is not possible to design a recursively computable IIR filter that has zero phase. For spatial domain design techniques, a natural approach for designing a zero-phase filter is to design filters that approximate different regions of the desired impulse response and then design the overall system by parallel combination. For frequency domain design techniques, however, this approach is not very useful. A parallel combination of two filters $h(n_1, n_2)$ and $h(-n_1, -n_2)$, for example, has an overall frequency response $H_T(\omega_1, \omega_2)$ given by

$$H_T(\omega_1, \omega_2) = H(\omega_1, \omega_2) + H(-\omega_1, -\omega_2)$$

$$= H(\omega_1, \omega_2) + H^*(\omega_1, \omega_2) \qquad (5.129)$$

$$= 2Re[H(\omega_1, \omega_2)] = 2H_R(\omega_1, \omega_2).$$

To approximate a particular desired magnitude response, we must design $h(n_1, n_2)$ such that $|H_R(\omega_1, \omega_2)|$ approximates the desired magnitude response. The function $|H_R(\omega_1, \omega_2)|$ is not simply related to the filter coefficients $a(n_1, n_2)$ and $b(n_1, n_2)$.

An alternative approach is to design an overall system by cascade combination. A cascade of two filters $h(n_1, n_2)$ and $h(-n_1, -n_2)$ has an overall frequency response $H_T(\omega_1, \omega_2)$ given by

$$H_T(\omega_1, \omega_2) = H(\omega_1, \omega_2)H(-\omega_1, -\omega_2)$$

$$= |H(\omega_1, \omega_2)|^2. \tag{5.130}$$

Therefore, if we design a filter whose magnitude response approximates the square root of the desired magnitude response $|H_d(\omega_1, \omega_2)|^{1/2}$, the resulting overall filter will approximate $|H_d(\omega_1, \omega_2)|$. Cascading two filters in the spatial domain is equivalent to combining in parallel the filters in the complex cepstrum domain, and is similar in this sense to zero-phase spatial domain design.

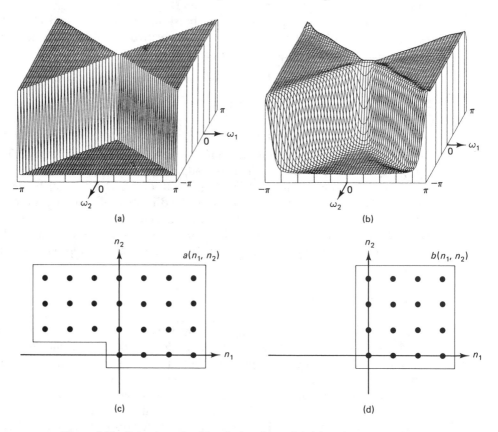

Figure 5.28 Zero-phase fan filter designed by minimizing the Error in (5.127), which is a sum of a frequency domain error and a stability error. After [Ekstrom et al.]. (a) Ideal frequency response of a fan filter; (b) frequency response of the filter designed; (c) region of support of $a(n_1, n_2)$ used in the filter design; (d) region of support of $b(n_1, n_2)$ used in the filter design.

Cascading two IIR filters, $h(n_1, n_2)$ and $h(-n_1, -n_2)$, has one major complication. In such applications as image processing, we are dealing with a finite-extent input and we wish to generate an output of the same size. In parallel combination, each filter is required only to generate the output in the desired region or in a region slightly larger than the desired region, and the results are simply added. In cascade combination, however, the two filters recurse in different directions; the second filter has to wait until the first filter generates an infinite amount of data, and then it has to generate an infinite amount of data to get to the output region of interest to us. In practice, of course, the first filter stops after the result has essentially died out, and the second filter operates on the result. This will affect the magnitude response of the filter, and both the first and second filter will still have to generate some output we will not use.

Figure 5.28 shows [Ekstrom et al.] an example of a zero-phase fan filter designed by minimizing the Error in (5.127), which is a sum of a frequency domain error and a stability error. The desired frequency response of a fan filter used is shown in Figure 5.28(a). Fan filters can discriminate certain directional information in 2-D signals and therefore are useful in some geophysical signal processing applications [Embree et al.]. The frequency response of the filter designed is shown in Figure 5.28(b). The zero-phase characteristic was obtained by cascading a nonsymmetric half-plane (NSHP) filter $h(n_1, n_2)$ with $h(-n_1, -n_2)$. The regions of support of $a(n_1, n_2)$ and $b(n_1, n_2)$ used in the NSHP filter design are shown in Figures 5.28(c) and (d), respectively.

5.6 IMPLEMENTATION

5.6.1 The Signal Flowgraph Representation

To implement a filter is to realize a discrete system with a given $h(n_1, n_2)$ or $H(z_1, z_2)$. An IIR filter has an infinite-extent impulse response, and it cannot be realized by direct convolution. Therefore, we have required the IIR filter to have a rational z-transform of the form in (5.2) so that the output can be computed recursively by the computational procedure in (5.3).

The signal flowgraph representation is useful in discussing different structures for IIR filter implementation. The basic elements of a signal flowgraph are shown in Figure 5.29. Despite the notational similarity, there are some major differences between 1-D and 2-D signal flowgraph representations. Consider a 1-D computational procedure given by

$$y(n) \leftarrow \tfrac{1}{2}y(n - 1) + x(n). \tag{5.131}$$

The signal flowgraph representing (5.131) is shown in Figure 5.30. By counting the number of delay elements in Figure 5.30, we can determine the number of memory units required. In addition, the order in which the output points are evaluated is completely determined by (5.131) or Figure 5.30.

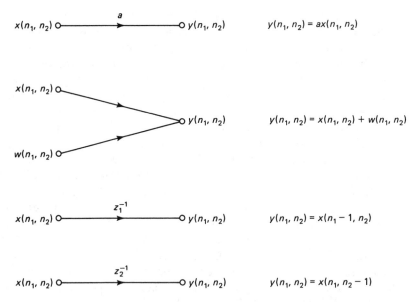

Figure 5.29 Basic elements used in the signal flowgraph representation of a digital filter.

Consider a 2-D computational procedure given by

$$y(n_1, n_2) \leftarrow \tfrac{1}{2}y(n_1 - 1, n_2) + x(n_1, n_2). \tag{5.132}$$

The signal flowgraph representing (5.132) is shown in Figure 5.31. Neither (5.132) nor its signal flowgraph specifies in what order the output elements should be computed. The output can be computed in many different orders. In addition, the number of memory units required by the delay element z_1^{-1} varies depending on the order in which the output values are computed. Suppose we have an $N \times N$-point input sequence $x(n_1, n_2)$ that is zero outside $0 \leq n_1 \leq N - 1, 0 \leq n_2 \leq N - 1$, and we wish to compute the output $y(n_1, n_2)$ in the same region, that is, $0 \leq n_1 \leq N - 1, 0 \leq n_2 \leq N - 1$. The output and input masks for (5.132) are shown in Figure 5.32. As was shown in Section 2.2.3, the boundary conditions can be computed from the output and input masks and $x(n_1, n_2)$. The region of support of $x(n_1, n_2)$ and the boundary conditions are shown in Figure 5.33. We will consider two of the many different orders in which the output can be computed,

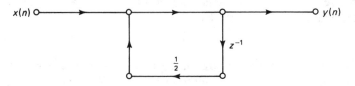

Figure 5.30 Signal flowgraph that represents the 1-D computational procedure given by

$$y(n) \leftarrow \tfrac{1}{2}y(n - 1) + x(n).$$

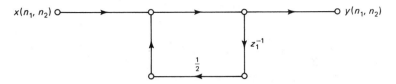

Figure 5.31 Signal flowgraph that represents the 2-D computational procedure given by

$$y(n_1, n_2) \leftarrow \tfrac{1}{2}y(n_1 - 1, n_2) + x(n_1, n_2).$$

row by row and column by column. In row-by-row computation, z_1^{-1} represents one memory unit. To compute $y(0, 0)$, we must have $y(-1, 0)$ stored. Once $y(0, 0)$ is computed, $y(-1, 0)$ is no longer necessary and we need to store only $y(0, 0)$. When $y(N - 1, 0)$ is computed, we do not need to compute further output points. The value $y(N - 1, 0)$ is not needed, and we store the boundary condition $y(-1, 1)$. In this manner, when $y(n_1', n_2')$ is computed, we need to store only $y(n_1' - 1, n_2')$ and all future output values can be computed. Now consider column-by-column computation. To compute $y(0, 0)$, we need $y(-1, 0)$. Even after $y(0, 0)$ has been computed and is made available to an output device, it must still be saved, since it is needed for computing $y(1, 0)$ later. To compute $y(0, 1)$, we need $y(-1, 1)$. Again, $y(0, 1)$ also has to be stored since it is needed for computing $y(1, 1)$. When we are computing $y(n_1', n_2')$, we should have N elements stored. The elements that must be stored when $y(n_1', n_2')$ is being computed are shown in Figure 5.34 for the row-by-row and column-by-column cases. From Figure 5.34,

$$z_1^{-1}: \quad \begin{cases} 1 \text{ memory unit for row-by-row computation} \\ N \text{ memory units for column-by-column computation.} \end{cases} \tag{5.133}$$

Similarly,

$$z_2^{-1}: \quad \begin{cases} N \text{ memory units for row-by-row computation} \\ 1 \text{ memory unit for column-by-column computation.} \end{cases} \tag{5.134}$$

The problem of counting the number of memory units corresponding to z_1^{-1} or z_2^{-1} is further complicated by the fact that the region of support of the

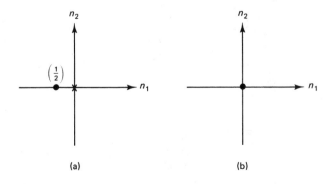

(a)   (b)

Figure 5.32 (a) Output mask and (b) input mask corresponding to the computational procedure in (5.132).

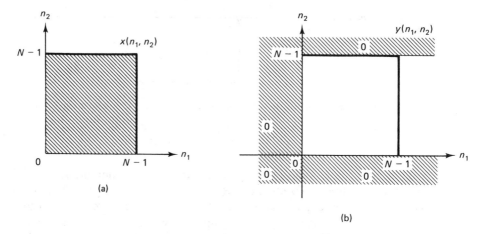

Figure 5.33 Boundary conditions for the computational procedure in (5.132). (a) Region of support of input $x(n_1, n_2)$; (b) boundary conditions.

boundary conditions depends on the output and input masks. Consider a computational procedure given by

$$y(n_1, n_2) \leftarrow \tfrac{1}{4}y(n_1 - 1, n_2) + \tfrac{1}{4}y(n_1 - 1, n_2 - 1) + \tfrac{1}{8}y(n_1, n_2 - 1)$$
$$+ \tfrac{1}{8}y(n_1 + 1, n_2 - 1) + x(n_1, n_2). \tag{5.135}$$

For the same input $x(n_1, n_2)$ above, the boundary conditions in this case are shown in Figure 5.35. The delay z_2^{-1} now corresponds to $2N$ memory units for row-by-row computation. Even though the output values in the shaded region (cross-hatched region) in Figure 5.35 are not themselves needed, we still have to compute them to obtain the needed output. This is one of the disadvantages of a nonsymmetric half-plane filter as compared to a quadrant filter.

As is clear from the above discussion, the number of memory units required and the amount of computation required in a given implementation depend on a number of factors, including the input and output masks, the region of support of $x(n_1, n_2)$, the desired region of the output, and the order in which the output is

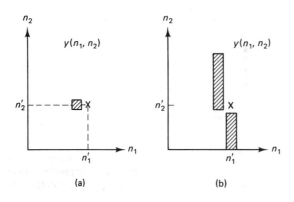

Figure 5.34 Required storage elements corresponding to z_1^{-1} in (a) row-by-row computation and (b) column-by-column computation.

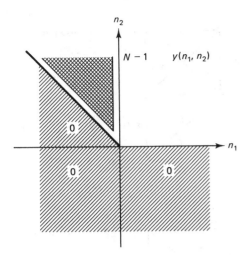

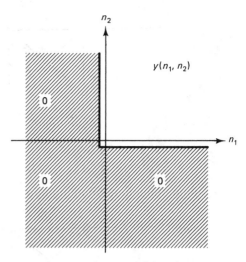

Figure 5.35 Boundary conditions for the computational procedure in (5.135) with the region of support of the input $x(n_1, n_2)$ in Figure 5.33(a). The output $y(n_1, n_2)$ in the cross-hatched region has to be computed to compute $y(n_1, n_2)$ for $0 \leq n_1 \leq N - 1$ and $0 \leq n_2 \leq N - 1$.

Figure 5.36 Boundary conditions for a computational procedure with first-quadrant support $a(n_1, n_2)$ and $b(n_1, n_2)$.

computed. There is an inherent flexibility in the implementation of a 2-D computational procedure, which is reflected in the 2-D signal flowgraph representation. To simplify some of our discussions in this section, we will assume, unless it is specified otherwise, that $x(n_1, n_2)$ is an $N \times N$-point sequence which is zero outside $0 \leq n_1 \leq N - 1$, $0 \leq n_2 \leq N - 1$, that the output $y(n_1, n_2)$ desired is for $0 \leq n_1 \leq N - 1$, $0 \leq n_2 \leq N - 1$, and that the filter coefficients $a(n_1, n_2)$ and $b(n_1, n_2)$ are first-quadrant support sequences. In this case, the region of support of the boundary conditions is all of (n_1, n_2) except the first-quadrant region, as shown in Figure 5.36.

5.6.2 General Structures

The standard methods of implementing 1-D IIR filters are direct, cascade, and parallel forms. These structures can be used for the realization of any causal IIR filter with a rational z-transform. In the 2-D case, only the direct form can be used to implement any recursively computable IIR filter with a rational z-transform. As we will discuss in Section 5.6.3, the cascade and parallel forms can be used to realize only a small subclass of recursively computable IIR filters.

To illustrate the direct form implementation, we will use a specific example. Consider a system function $H(z_1, z_2)$ given by

$$H(z_1, z_2) = \frac{B(z_1, z_2)}{A(z_1, z_2)} = \frac{6 + 7z_1^{-1} + 8z_2^{-1}}{1 - 2z_1^{-1} - 3z_2^{-1} - 4z_1^{-1}z_2^{-1} - 5z_1^{-2}z_2^{-1}}. \quad (5.136)$$

The filter coefficients in (5.136) were arbitrarily chosen to be simple numbers.

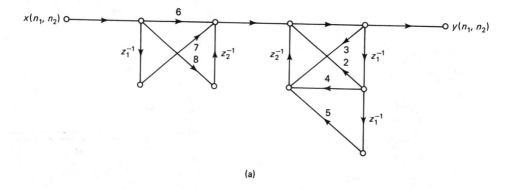

(a)

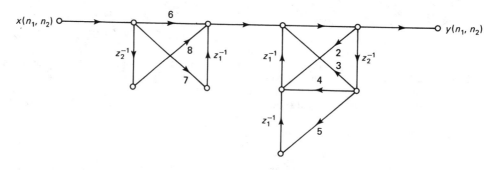

(b)

Figure 5.37 Signal flowgraphs corresponding to the computational procedure in (5.137).

The computational procedure corresponding to (5.136) is given by

$$y(n_1, n_2) \leftarrow 2y(n_1 - 1, n_2) + 3y(n_1, n_2 - 1) + 4y(n_1 - 1, n_2 - 1) + 5y(n_1 - 2, n_2 - 1)$$
$$+ 6x(n_1, n_2) + 7x(n_1 - 1, n_2) + 8x(n_1, n_2 - 1). \tag{5.137}$$

A structure which can be obtained directly from (5.137) is shown in Figure 5.37(a). If we reverse the locations of z_1^{-1} and z_2^{-1} in Figure 5.37(a) and make corresponding changes, we will also obtain the structure in Figure 5.37(b). Since delaying the signals first and then adding the results is equivalent to adding the signals first and then delaying the result, we can eliminate one z_2^{-1} element in Figure 5.37(a) and one z_1^{-1} element in Figure 5.37(b). The resulting structures are shown in Figure 5.38. Note that the signal flowgraph representation makes it quite easy to see which delay elements are redundant and thus can be eliminated. The advantage of the structures in Figure 5.38 over those in Figure 5.37 is the reduction in the number of memory units required. The structure in Figure 5.38(a) requires $3 + N$ memory units for row-by-row computation and $3N + 1$ units for column-by-column computation. The structure in Figure 5.38(b) requires $2 + 2N$ memory units for both row-by-row and column-by-column computations. From the perspective of memory requirements, then, the structure in Figure 5.38(a) is preferable for row-

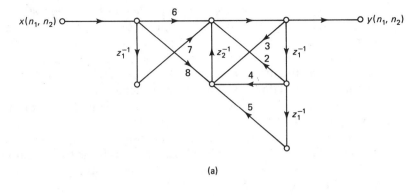

(a)

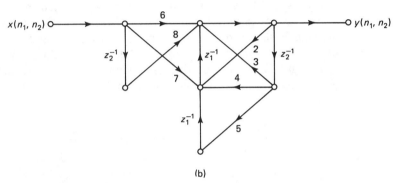

(b)

Figure 5.38 Signal flowgraphs obtained from the signal flowgraphs in Figure 5.37 by eliminating common delay elements.

by-row computation, while the structure in Figure 5.38(b) is preferable for column-by-column computation.

The signal flowgraphs in Figure 5.38 were obtained from those in Figure 5.37 by eliminating either redundant z_1^{-1} elements or redundant z_2^{-1} elements. It is natural to ask if we can eliminate both redundant z_1^{-1} elements and redundant z_2^{-1} elements. A structure which requires the theoretically minimum number of z_1^{-1} and z_2^{-1} elements is called a *minimal realization*. For a first-quadrant support $M_1 \times M_2$-point sequence $a(n_1, n_2)$ with $a(M_1 - 1, M_2 - 1) \neq 0$ and $N_1 \times N_2$-point sequence $b(n_1, n_2)$ with $b(N_1 - 1, N_2 - 1) \neq 0$, it is not possible to obtain a structure which requires fewer than $\max[M_1 - 1, N_1 - 1]$ z_1^{-1} elements and $\max[M_2 - 1, N_2 - 1]$ z_2^{-1} elements, where $\max[\cdot, \cdot]$ is the larger of the two arguments. The total number of delay elements cannot therefore be less than $\max[M_1 - 1, N_1 - 1] + \max[M_2 - 1, N_2 - 1]$. Methods [Fornasini and Marchesini; Kung et al.; Chan; Fornasini] have been developed to obtain minimal realizations. Unfortunately, however, branch transmittances in minimal realizations are obtained by solving nonlinear equations and are complex-valued. The advantages due to a smaller number of delay elements in minimal realizations often disappear due to the need to perform complex arithmetic and store complex-valued elements.

The structures in Figures 5.37 and 5.38 can be viewed as cascades of two systems whose system functions are given by $B(z_1, z_2)$ and $1/A(z_1, z_2)$ in (5.136). Since changing the order of $B(z_1, z_2)$ and $1/A(z_1, z_2)$ does not affect the overall system function, the structures in Figure 5.39 are equivalent to those in Figure 5.37. The two structures in Figure 5.40 can be derived from those in Figure 5.39 by eliminating the redundancy in the delay elements.

As in the 1-D case, the transposition of a signal flowgraph does not affect the overall system function in the 2-D case. The transposition of a signal flowgraph involves changing the roles of the input and output and reversing all the signal flow directions. The signal flowgraph in Figure 5.40(a) is the transposition of the signal flowgraph in Figure 5.38(b). The signal flowgraph in Figure 5.40(b) is the transposition of that in Figure 5.38(a).

From the signal flowgraph representation, we can determine the number of arithmetic operations required, the number of memory units required, which intermediate signals should be stored in the memory units, and in which order the intermediate values must be computed. To illustrate this, consider the signal flowgraph in Figure 5.38(a), which is redrawn in Figure 5.41 with each node marked with a specific signal value. We will assume the row-by-row computation, and the

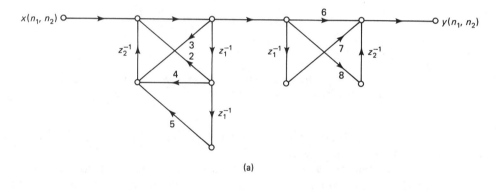

(a)

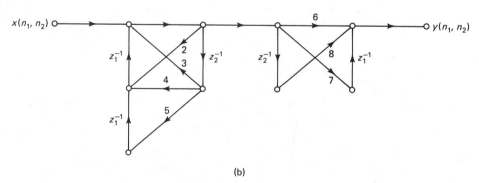

(b)

Figure 5.39 Signal flowgraphs obtained from the signal flowgraphs in Figure 5.37 by implementing $\dfrac{1}{A(z_1, z_2)}$ first and then cascading it with $B(z_1, z_2)$.

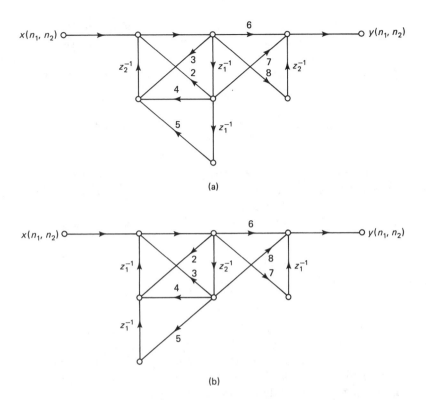

Figure 5.40 Signal flowgraphs obtained from the signal flowgraphs in Figure 5.39 by eliminating common delay elements.

structure requires $3 + N$ memory units. By counting the number of nontrivial branch transmittances, we see that each output requires seven multiplications. By counting the number of additions required at each node, we can see that $v_2(n_1, n_2)$ requires three additions and $v_3(n_1, n_2)$ requires three additions. The total number of additions required per output point is then six. The $3 + N$ memory units should be used for three z_1^{-1} elements and one z_2^{-1} element. Looking at the nodes from which z_1^{-1} or z_2^{-1} originates, we see that previously computed values of $x(n_1, n_2)$, $y(n_1, n_2)$, $v_4(n_1, n_2)$, and $v_3(n_1, n_2)$ have to be stored. Since z_1^{-1} originated from the three nodes marked by $x(n_1, n_2)$, $y(n_1, n_2)$, and $v_4(n_1, n_2)$, one memory unit should be allocated for each of these three signals. The other N memory units should be allocated for previously computed values of $v_3(n_1, n_2)$, since z_2^{-1} originates from the node with signal value $v_3(n_1, n_2)$. Immediately before the computation for $y(n_1', n_2')$ begins, the $3 + N$ memory units should contain the following set of values (from Figure 5.41):

$$x(n_1' - 1, n_2'), y(n_1' - 1, n_2'), v_4(n_1' - 1, n_2') \tag{5.138}$$

and
$$v_3(n_1, n_2) \quad \text{for } 0 \le n_1 \le n_1' - 1, n_2 = n_2'$$

$$\text{and } n_1' \le n_1 \le N - 1, n_2 = n_2' - 1.$$

The order in which intermediate signal values can be computed can be determined by noting that all the signals enter nodes $v_1(n_1, n_2)$, $v_4(n_1, n_2)$, and $v_5(n_1, n_2)$ through delay elements. Therefore, $v_1(n_1, n_2)$, $v_4(n_1, n_2)$, and $v_5(n_1, n_2)$ can be computed from previously computed and stored values by

$$v_1(n_1, n_2) = x(n_1 - 1, n_2) \tag{5.139a}$$

$$v_4(n_1, n_2) = y(n_1 - 1, n_2) \tag{5.139b}$$

$$v_5(n_1, n_2) = v_4(n_1 - 1, n_2). \tag{5.139c}$$

For this particular structure, this observation allows us to compute all the signal values in the leftmost and rightmost regions shown in the signal flowgraph except $y(n_1, n_2)$. We can then compute $v_2(n_1, n_2)$, since all signals entering $v_2(n_1, n_2)$ are either through the delay element or through the leftmost and rightmost portions. Therefore,

$$v_2(n_1, n_2) = 6x(n_1, n_2) + 7v_1(n_1, n_2) + v_3(n_1, n_2 - 1) + 2v_4(n_1, n_2). \tag{5.140}$$

The signal $y(n_1, n_2)$ is the same as $v_2(n_1, n_2)$,

$$y(n_1, n_2) = v_2(n_1, n_2). \tag{5.141}$$

Now we have computed all the signals in the leftmost and rightmost regions, including $y(n_1, n_2)$. Since all the signals in the middle region come from a delay element or the signals in the left-hand and right-hand regions, they can be all computed. In the signal flowgraph in Figure 5.41, we have only one signal $v_3(n_1, n_2)$ remaining in the middle region. It is given by

$$v_3(n_1, n_2) = 8x(n_1, n_2) + 5v_5(n_1, n_2) + 4v_4(n_1, n_2) + 3y(n_1, n_2). \tag{5.142}$$

Even though $y(n_1, n_2)$ has already been computed, $v_3(n_1, n_2)$ still must be computed and stored, since it will be needed later on when (5.140) is used to compute $v_2(n_1, n_2 + 1)$. From (5.139), (5.140), (5.141), and (5.142),

$$v_1(n_1, n_2) = x(n_1 - 1, n_2) \tag{5.143a}$$

$$v_4(n_1, n_2) = y(n_1 - 1, n_2) \tag{5.143b}$$

$$v_5(n_1, n_2) = v_4(n_1 - 1, n_2) \tag{5.143c}$$

$$v_2(n_1, n_2) = 6x(n_1, n_2) + 7v_1(n_1, n_2) + v_3(n_1, n_2 - 1) + 2v_4(n_1, n_2) \tag{5.143d}$$

$$y(n_1, n_2) = v_2(n_1, n_2) \tag{5.143e}$$

$$v_3(n_1, n_2) = 8x(n_1, n_2) + 5v_5(n_1, n_2) + 4v_4(n_1, n_2) + 3y(n_1, n_2). \tag{5.143f}$$

The above approach can also be used for other structures to determine the order in which the intermediate signals can be recursively computed.

The structures discussed in this section are called *direct form*, since they were obtained directly from the computational procedure. Direct form structures can be used to implement any recursively computable IIR filter. They can also be used to implement subsystems in an implementation with a special structure. This will be discussed in a later section.

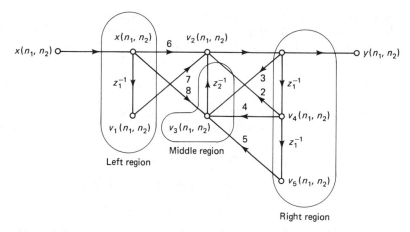

Figure 5.41   Signal flowgraph in Figure 5.38(a), redrawn with each node marked with a specific signal value.

5.6.3 State-Space Representation

In addition to the signal flowgraph representation, a 1-D computational procedure (linear constant coefficient difference equation) can also be represented by a set of first-order linear constant coefficient difference equations called *state equations*. As an example, consider a 1-D computational procedure given by

$$y(n) \leftarrow \tfrac{1}{2}y(n-1) + \tfrac{1}{4}y(n-2) + x(n) + 2x(n-1). \qquad (5.144)$$

A signal flowgraph representation of (5.144) is shown in Figure 5.42.   Denoting the output of the two z^{-1} elements in the signal flowgraph as $s_1(n)$ and $s_2(n)$, from the signal flowgraph we can express $s_1(n+1)$ and $s_2(n+1)$, the input to the two z^{-1} elements, as a linear combination of $s_1(n)$, $s_2(n)$, and $x(n)$:

$$\begin{bmatrix} s_1(n+1) \\ s_2(n+1) \end{bmatrix} = \begin{bmatrix} \tfrac{1}{2} & \tfrac{1}{4} \\ 1 & 0 \end{bmatrix} \begin{bmatrix} s_1(n) \\ s_2(n) \end{bmatrix} + \begin{bmatrix} 1 \\ 0 \end{bmatrix} x(n). \qquad (5.145)$$

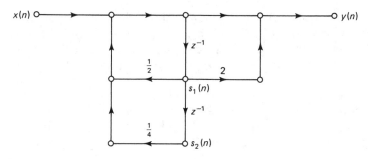

Figure 5.42   1-D signal flowgraph corresponding to the 1-D computational procedure in (5.144).

The two variables $s_1(n)$ and $s_2(n)$ are state variables, and the two equations in (5.145) are state equations. We can also express from the signal flowgraph the output $y(n)$ as a linear combination of $s_1(n)$, $s_2(n)$, and $x(n)$ as

$$y(n) = [2\tfrac{1}{2} \quad \tfrac{1}{4}] \begin{bmatrix} s_1(n) \\ s_2(n) \end{bmatrix} + x(n). \tag{5.146}$$

Equation (5.146) is called the *observation equation*. Equations (5.145) and (5.146) are a state-space representation of the computational procedure in (5.144).

A general state-space representation for any 1-D causal computational procedure with one input and one output is in the form of

$$s(n + 1) = As(n) + Bx(n) \tag{5.147a}$$

$$y(n) = Cs(n) + Dx(n) \tag{5.147b}$$

where $s(n)$ is a vector of state variables. In the state-space representation, we relate the input to state variables and relate the state variables to the output. The states summarize the past history of the system. The state and observation equations in (5.147) can be used in generating future state and output values.

As in 1-D, a 2-D computational procedure with first-quadrant support $a(n_1, n_2)$ and $b(n_1, n_2)$ can be expressed in a state-space form. Among several variations [Attasi; Roesser; Fornasini and Marchesini] proposed, the form proposed by [Roesser] is the most general, and we will consider this model. As an example of a 2-D state-space representation, consider the computational procedure in (5.137). A signal flowgraph corresponding to (5.137) was previously shown in Figure 5.41. Denoting the output of the three z_1^{-1} elements in the signal flowgraph as $s_1^h(n_1, n_2)$, $s_2^h(n_1, n_2)$, and $s_3^h(n_1, n_2)$,

$$s_1^h(n_1, n_2) = v_1(n_1, n_2) \tag{5.148a}$$

$$s_2^h(n_1, n_2) = v_4(n_1, n_2) \tag{5.148b}$$

$$s_3^h(n_1, n_2) = v_5(n_1, n_2). \tag{5.148c}$$

Denoting the output of the z_2^{-1} element in the signal flowgraph as $s_1^v(n_1, n_2)$, we have

$$s_1^v(n_1, n_2) = v_3(n_1, n_2 - 1). \tag{5.149}$$

Note that $s_1^v(n_1, n_2)$ is not $v_2(n_1, n_2)$. From the signal flowgraph and after some algebra, we can express $s_1^h(n_1 + 1, n_2)$, $s_2^h(n_1 + 1, n_2)$, $s_3^h(n_1 + 1, n_2)$ and $s_1^v(n_1, n_2 + 1)$, the input to the three z_1^{-1} elements and one z_2^{-1} element, as a linear combination of $s_1^h(n_1, n_2)$, $s_2^h(n_1, n_2)$, $s_3^h(n_1, n_2)$, $s_1^v(n_1, n_2)$, and $x(n_1, n_2)$:

$$\begin{bmatrix} s_1^h(n_1 + 1, n_2) \\ s_2^h(n_1 + 1, n_2) \\ s_3^h(n_1 + 1, n_2) \\ s_1^v(n_1, n_2 + 1) \end{bmatrix} = \begin{bmatrix} 0 & 0 & 0 & 0 \\ 7 & 2 & 0 & 1 \\ 0 & 1 & 0 & 0 \\ 21 & 10 & 5 & 3 \end{bmatrix} \begin{bmatrix} s_1^h(n_1, n_2) \\ s_2^h(n_1, n_2) \\ s_3^h(n_1, n_2) \\ s_1^v(n_1, n_2) \end{bmatrix} + \begin{bmatrix} 1 \\ 6 \\ 0 \\ 26 \end{bmatrix} x(n_1, n_2). \tag{5.150}$$

Clearly, $s_1^h(n_1, n_2)$, $s_2^h(n_1, n_2)$, and $s_3^h(n_1, n_2)$ propagate in the horizontal direction. They are called *horizontal state variables* and have the superscript h. The state variable $s_1^v(n_1, n_2)$, which has the superscript v, propagates in the vertical direction and is called a *vertical state variable*. Both the horizontal and vertical state variables are state variables in the state-space representation, and the four equations in (5.150) are state equations. We can express from the signal flowgraph the output $y(n_1, n_2)$ as a linear combination of the four state variables and $x(n_1, n_2)$ as

$$y(n_1, n_2) = \begin{bmatrix} 7 & 2 & 0 & 1 \end{bmatrix} \begin{bmatrix} s_1^h(n_1, n_2) \\ s_2^h(n_1, n_2) \\ s_3^h(n_1, n_2) \\ s_1^v(n_1, n_2) \end{bmatrix} + 6\, x(n_1, n_2). \tag{5.151}$$

Equation (5.151) is the observation equation. Equations (5.150) and (5.151) are a state-space representation of the signal flowgraph in Figure 5.41. The equations in (5.150) and (5.151) can be used to compute the output as an alternative to the equations in (5.143).

A general state-space representation for any 2-D computational procedure with first-quadrant support $a(n_1, n_2)$ and $b(n_1, n_2)$ for one input and one output is in the form of

$$\begin{bmatrix} \mathbf{s}^h(n_1 + 1, n_2) \\ \mathbf{s}^v(n_1, n_2 + 1) \end{bmatrix} = \begin{bmatrix} A_1 & A_2 \\ A_3 & A_4 \end{bmatrix} \begin{bmatrix} \mathbf{s}^h(n_1, n_2) \\ \mathbf{s}^v(n_1, n_2) \end{bmatrix} + \begin{bmatrix} B_1 \\ B_2 \end{bmatrix} x(n_1, n_2) \tag{5.152a}$$

$$y(n_1, n_2) = \begin{bmatrix} C_1 & C_2 \end{bmatrix} \begin{bmatrix} \mathbf{s}^h(n_1, n_2) \\ \mathbf{s}^v(n_1, n_2) \end{bmatrix} + Dx(n_1, n_2) \tag{5.152b}$$

where $\mathbf{s}^h(n_1, n_2)$ is a vector of horizontal state variables that propagate in the horizontal direction and $\mathbf{s}^v(n_1, n_2)$ is a vector of vertical state variables that propagate in the vertical direction. By taking the z-transform of (5.152), we can compute the system function $H(z_1, z_2) = Y(z_1, z_2)/X(z_1, z_2)$ given by

$$H(z_1, z_2) = \frac{Y(z_1, z_2)}{X(z_1, z_2)}$$

$$= \begin{bmatrix} C_1 & C_2 \end{bmatrix} \left(\begin{bmatrix} z_1 I & 0 \\ 0 & z_2 I \end{bmatrix} - \begin{bmatrix} A_1 & A_2 \\ A_3 & A_4 \end{bmatrix} \right)^{-1} \begin{bmatrix} B_1 \\ B_2 \end{bmatrix} + D. \tag{5.153}$$

Although the form of the 2-D state-space representation is a straightforward extension of its 1-D counterpart, there are major differences between the 1-D and 2-D cases. For example, for the 1-D state-space representation, current states contain sufficient information to generate all the future output. In the 2-D case, however, we need current states and the boundary conditions to generate all the future output. For this reason, the states in 2-D are sometimes referred to as *local states*, and the boundary conditions, together with the local states, are referred to as *global states*. This difference between 1-D and 2-D arises from fundamental differences between 1-D and 2-D difference equations. As we discussed in Section 2.2, to have a unique solution, an Nth-order 1-D difference equation requires N values of the output as initial conditions, while a finite-order 2-D difference equation requires an infinite number of output values.

The state-space representation is an alternative to the signal flowgraph representation and has some advantages over the signal flowgraph representation. For example, the state-space representation is more explicit in specifying how the output can be computed recursively. In addition, the significance of minimal realization in a parallel processor environment is more explicit in the state-space representation. If we assign a processor to evaluate one state variable, we need N processors to update N state variables. Since the number of states is the same as the number of delay elements z_1^{-1} and z_2^{-1} in the signal flowgraph representation and determines the number of processors required, minimal realization will minimize the number of processors required in this case. The state-space representation is also very useful in studying such concepts as controllability, observability, and minimal realization. For a summary of the 2-D state-space representation and related concepts, see [Kung et al.].

5.6.4 Special Structures

The cascade form is probably the most common method for realizing 1-D IIR filters because of its relative insensitivity to coefficient quantization. As we have discussed in Section 4.6.3, the cascade form in 2-D is not a general structure and can be used for only a very restricted class of IIR filters. Specifically, a 2-D polynomial cannot, in general, be factored as a product of lower-order polynomials, and $H(z_1, z_2)$ cannot generally be written in the form of

$$H(z_1, z_2) = A \prod_k H_k(z_1, z_2). \tag{5.154}$$

To use a cascade form in the realization of a 2-D IIR filter, therefore, the form in (5.154) should be explicitly used in the design step. Methods for designing IIR filters with the cascade form in (5.154) typically involve nonlinear functional minimization and can be found in [Costa and Venetsanopoulos; Maria and Fahmy; Goodman; Iijima et al.].

As discussed in Section 4.6.3, one disadvantage of the cascade structure is that the numerator filter coefficients of the overall filter are nonlinearly related to the numerator filter coefficients of individual sections. This also applies to the denominator filter coefficients. Another disadvantage is that the cascade structure is restrictive in approximating a given magnitude response.

The cascade structure has certain advantages, however. In addition to its low sensitivity to coefficient quantization, the cascade implementation requires a smaller number of arithmetic operations for a given filter size. As described in Section 4.6.3, there is a fair amount of structure in the cascade implementation, and this structure reduces the required number of arithmetic operations. In addition, issues related to stability are easier to deal with for filters with a smaller number of denominator coefficients.

One variation of the cascade structure is to require only the denominator polynomial of the system function to be a product of polynomials. The system function $H(z_1, z_2)$ in this case is of the form

$$H(z_1, z_2) = \frac{B(z_1, z_2)}{A(z_1, z_2)} = \frac{B(z_1, z_2)}{\prod_k A_k(z_1, z_2)}. \qquad (5.155)$$

This structure is designed to exploit the stability advantages of the cascade structure without being too restrictive in approximating the desired frequency response. Again, in this case, the form of (5.155) must be explicitly incorporated in the design algorithm.

A separable denominator system function is a special case of (5.155) and is of the form

$$H(z_1, z_2) = \frac{B(z_1, z_2)}{A_1(z_1)A_2(z_2)}. \qquad (5.156)$$

A separable denominator system function is more restrictive in approximating a desired frequency response than the form in (5.155). Limited preliminary studies [Shaw and Mersereau] based on zero-phase lowpass filters designed with circularly symmetric filter specification indicate that the number of independent filter coefficients required to meet a given specification is significantly greater with a separable denominator filter than with a general IIR filter. However, it has a major advantage in testing the system stability and stabilizing an unstable filter. Assuming that the numerator polynomial $B(z_1, z_2)$ does not affect the system stability, the stability of $H(z_1, z_2)$ can be determined by testing the stability of two 1-D systems with system functions $1/A_1(z_1)$ and $1/A_2(z_2)$. An unstable filter can be stabilized by stabilizing 1-D systems with system functions $1/A_1(z_1)$ and/or $1/A_2(z_2)$. Specific methods for designing IIR filters with separable denominator system functions can be found in [Twogood and Mitra; Abramatic et al.; Lashgari et al.; Hinamoto].

As discussed in Section 5.5.2, the cascade structure naturally arises in frequency domain zero-phase IIR filter design. In addition, as described in Section 4.6.3, one approach to designing a complicated filter is by cascading simpler filters. For example, a bandpass filter can be designed by cascading a highpass filter with a lowpass filter.

In the 1-D parallel form realization, the system function $H(z)$ is expressed as

$$H(z) = \sum_k H_k(z). \qquad (5.157)$$

The parallel form also requires the factorization of the denominator polynomial, so it cannot be used as a general procedure for realizing a 2-D IIR filter. Like the cascade form, the parallel form can be used when the form in (5.157) is used explicitly in the design step.

The disadvantage of the parallel form is that it is restrictive in approximating the desired frequency response. In addition, the filter coefficients of the overall filter are nonlinearly related to the filter coefficients of smaller individual filters. The parallel form's advantage is that issues related to stability are easier to deal with, since each individual filter has a smaller number of denominator coefficients than the overall filter. In addition, parallel form implementation requires fewer arithmetic operations than does the method of combining all the individual filters and then implementing the overall filter.

As discussed in Section 5.2.4, the parallel form naturally arises in spatial domain zero-phase IIR filter design. In addition, a complicated filter can also be designed by parallel combination of simpler filters. For example, the parallel combination of a lowpass filter and a highpass filter can result in a bandstop filter.

5.7 COMPARISON OF FIR AND IIR FILTERS

FIR filters have many advantages over IIR filters. Stability is not an issue in FIR filter design or implementation. For IIR filters, however, testing filter stability and stabilizing an unstable filter without significantly affecting the magnitude response are very big tasks. Zero phase is simple to achieve for FIR filters. Designing zero-phase IIR filters is possible, but is more involved than designing zero-phase FIR filters. In addition, design methods are simpler for FIR than for IIR filters.

The main advantage of an IIR filter over an FIR filter is the smaller number of arithmetic operations required for implementation in direct form. To meet the same magnitude specification, an FIR filter requires significantly more arithmetic operations (typically by a factor of 5 to 30) per output sample than an IIR filter. If an FIR filter is implemented by exploiting the computational efficiency of FFT algorithms, however, the IIR filter's advantage is likely to disappear. IIR filters are useful, therefore, mainly in those applications where low-cost implementation is very important. Because of their overwhelming general advantage over IIR filters, FIR filters are much more common in practice. However, design and implementation of 2-D FIR and IIR filters are active areas of research. The impact that this research will have on the practical application of 2-D FIR and IIR filters remains to be seen.

Significant differences between the 1-D and 2-D cases also exist in the design and implementation of digital filters. In 1-D, there are practical methods to design optimal FIR and IIR filters. In 2-D, practical methods that can be reliably used to design optimal FIR or IIR filters have not yet been developed. In 1-D, checking the stability of an IIR filter and stabilizing an unstable filter without affecting the magnitude response are quite simple. In contrast, checking the stability and stabilizing an unstable filter without significantly affecting the magnitude response is a big task in 2-D. In 1-D, the cascade and parallel forms are general implementation methods for rational system functions. In 2-D, the cascade and parallel forms cannot generally be used for the realization of rational system functions. Design and implementation are more complex for 2-D than for 1-D digital filters.

REFERENCES

For a tutorial overview of 2-D digital filtering, see [Mersereau and Dudgeon]. For constraints imposed by symmetry on filter design and implementation, see [Aly and Fahmy (1981); Pitas and Venetsanopoulos].

For one of the earliest works on stability theory, IIR filter design and sta-bilization of an unstable filter by the planar least squares inverse method, see [Shanks et al.]. For readings on some spatial domain design methods, see [Steiglitz and McBride; Bednar; Cadzow; Shaw and Mersereau; Aly and Fahmy (1980); Chaparro and Jury].

For the existence and computation of 2-D complex cepstrum, see [Dudgeon (1975); Dudgeon (1977)]. For stabilization of an unstable filter using complex cepstrum representation, see [Ekstrom and Woods]. For stabilization methods based on the planar least squares inverse, see [Shanks et al.; Genin and Kamp; Reddy et al.; Swamy et al.; Raghuramireddy et al.]. For readings on some fre-quency domain design methods, see [Ekstrom et al.; Mutluay and Fahmy].

Designing 1-D IIR filters by frequency transformation of 1-D analog prototype filters or 1-D digital filters is quite useful. In 2-D, however, design by frequency transformation of 1-D or 2-D analog prototype filters or 1-D or 2-D digital prototype filters is quite restrictive and has not been discussed in this chapter. For readings on this topic, see [Pendergrass et al.; Chakrabarti and Mitra; Goodman; Rama-moorthy and Bruton; Bernstein].

For realizations based on signal flowgraph representation, see [Mitra et al.; Mitra and Chakrabarti]. For minimal realizations that require the theoretically minimum number of delay elements and state-space realization, see [Roesser; Fornasini and Marchesini; Kung et al.; Chan; Fornasini]. For realization with high degrees of parallelism suitable to VLSI implementation, see [Nikias et al.; Lampropoulos and Fahmy].

For design methods based on cascade implementation, see [Maria and Fahmy; Costa and Venetsanopoulos; Chang and Aggarwal (1977); Goodman; Iijima et al.]. For design methods for separable denomination filters see [Twogood and Mitra; Abramatic et al.; Lashgari et al.; Hinamoto].

For readings on design methods based on both the magnitude and phase error criteria, see [Aly and Fahmy (1978); Woods et al.; Shimizu and Hirata; Hinamoto and Maekawa].

J. F. Abramatic, F. Germain, and E. Rosencher, Design of two-dimensional separable denominator recursive filters, *IEEE Trans. Acoust. Speech Sig. Proc.*, Vol. ASSP-27, October 1979, pp. 445–453.

S. A. H. Aly and M. M. Fahmy, Design of two-dimensional recursive digital filters with specified magnitude and group delay characteristics, *IEEE Trans. Circuits and Systems*, Vol. CAS-25, November 1978, pp. 908–916.

S. A. H. Aly and M. M. Fahmy, Spatial-domain design of two-dimensional recursive digital filters, *IEEE Trans. Circuits and Systems*, Vol. CAS-27, October 1980, pp. 892–901.

S. A. H. Aly and M. M. Fahmy, Symmetry exploitation in the design and implementation of recursive 2-D rectangularly sampled digital filters, *IEEE Trans. Acoust. Speech Sig. Proc.*, Vol. ASSP-29, October 1981, pp. 973–982.

A. Attasi, Systemes lineaires homogenes a deux indices, *Rapport Laboria*, No. 31, September 1973.

J. B. Bednar, Spatial recursive filter design via rational Chebyshev approximation, *IEEE Trans. Circuits and Systems*, Vol. CAS-22, June 1975, pp. 572–574.

R. Bernstein, Design of inherently stable two-dimensional recursive filters from one-dimensional filters, *IEEE Trans. Acoust. Speech Sig. Proc.*, Vol. ASSP-32, February 1984, pp. 164–168.

J. A. Cadzow, Recursive digital filter synthesis via gradient based algorithms, *IEEE Trans. Acoust. Speech Sig. Proc.*, Vol. ASSP-24, October 1976, pp. 349–355.

S. Chakrabarti and S. K. Mitra, Design of two-dimensional digital filters via spectral transformations, *Proc. IEEE*, Vol. 65, June 1977, pp. 905–914.

D. S. K. Chan, A simple derivation of minimal and non-minimal realizations of 2-D transfer functions, *Proc. IEEE*, Vol. 66, April 1978, pp. 515–516.

H. Chang and J. K. Aggarwal, Design of two-dimensional recursive filters by interpolation, *IEEE Trans. Circuits and Systems*, Vol. CAS-24, June 1977, pp. 281–291.

H. Chang and J. K. Aggarwal, Design of two-dimensional semicausal recursive filters, *IEEE Trans. Circuits and Systems*, Vol. CAS-25, December 1978, pp. 1051–1059.

L. F. Chaparro and E. I. Jury, Rational approximation of 2-D linear discrete systems, *IEEE Trans. Acoust. Speech Sig. Proc.*, Vol. ASSP-30, October 1982, pp. 780–787.

J. M. Costa and A. N. Venetsanopoulos, Design of circularly symmetric two-dimensional recursive filters, *IEEE Trans. Acoust. Speech Sig. Proc.*, Vol. ASSP-22, December 1974, pp. 432–442.

D. E. Dudgeon, The computation of two-dimensional cepstra, *IEEE Trans. Acoust. Speech Sig. Proc.*, Vol. ASSP-25, December 1977, pp. 476–484.

D. E. Dudgeon, The existence of cepstra for two-dimensional rational polynomials, *IEEE Trans. Acoust. Speech Sig. Proc.*, Vol. ASSP-23, April 1975, pp. 242–243.

M. P. Ekstrom, R. E. Twogood, and J. W. Woods, Two-dimensional recursive filter design—a spectral factorization approach, *IEEE Trans. Acoust. Speech Sig. Proc.*, Vol. ASSP-28, February 1980, pp. 16–26.

M. P. Ekstrom and J. W. Woods, Two-dimensional spectral factorization with applications in recursive digital filtering, *IEEE Trans. Acoust. Speech Sig. Proc.*, Vol. ASSP-24, April 1976, pp. 115–128.

P. Embree, J. P. Burg, and M. M. Backus, Wide-band velocity filtering—the pie-slice process, *Geophysics*, Vol. 28, 1963, pp. 948–974.

R. Fletcher and M. J. D. Powell, A rapidly convergent descent method for minimization, *Computer J.*, Vol. 6, July 1963, pp. 163–168.

E. Fornasini, On the relevance of noncommutative power series in spatial filters realization, *IEEE Trans. Circuits and Systems*, Vol. CAS-25, May 1978, pp. 290–299.

E. Fornasini and G. Marchesini, State-space realization theory of two-dimensional filters, *IEEE Trans. Automat. Contr.*, Vol. AC-21, August 1976, pp. 484–492.

Y. Genin and Y. Kamp, Counterexample in the least-square inverse stabilization of 2-D recursive filters, *Electron. Lett.*, Vol. 11, July 1975, pp. 330–331.

D. M. Goodman, A design technique for circularly symmetric low-pass filters, *IEEE Trans. Acoust. Speech Sig. Proc.*, August 1978, pp. 290–304.

T. Hinamoto, Design of 2-D separable-denominator recursive digital filters, *IEEE Trans. Circuits and Systems*, Vol. CAS-31, November 1984, pp. 925–932.

T. Hinamoto and S. Maekawa, Design of two-dimensional recursive digital filters using mirror-image polynomials, *IEEE Trans. Circuits and Systems*, Vol. CAS-33, August 1986, pp. 750–758.

R. Iijima, N. Haratani, and S. I. Takahashi, Design method for 2-D circularly symmetric recursive filters, *IEEE Trans. Acoust. Speech Sig. Proc.*, October 1983, pp. 1298–1299.

S. Y. Kung, B. Levy, M. Morf, and T. Kailath, New results in 2-D systems theory, Part II: 2-D state-space models-realization and the notions of controllability, observability and minimality, *Proc. IEEE*, Vol. 65, June 1977, pp. 945–961.

G. A. Lampropoulos and M. M. Fahmy, A new realization for 2-D digital filters, *IEEE Trans. Acoust. Speech Sig. Proc.*, Vol. ASSP-35, April 1987, pp. 533–542.

B. Lashgari, L. M. Silverman, and J. F. Abramatic, Approximation of 2-D separable in denominator filters, *IEEE Trans. Circuits and Systems*, Vol. CAS-30, February 1983, pp. 107–121.

G. A. Maria and M. M. Fahmy, An l_p design technique for two-dimensional digital recursive filters, *IEEE Trans. Acoust. Speech Sig. Proc.*, Vol. ASSP-22, February 1974, pp. 15–21.

R. M. Mersereau and D. E. Dudgeon, Two-dimensional digital filtering, *IEEE Proc.*, Vol. 63, April 1975, pp. 610–623.

S. K. Mitra and S. Chakrabarti, A new realization method for 2-D digital transfer functions, *IEEE Trans. Acoust. Speech Sig. Proc.*, Vol. ASSP-26, December 1978, pp. 544–550.

S. K. Mitra, A. D. Sagar, and N. A. Pendergrass, Realization of two-dimensional recursive digital filters, *IEEE Trans. Circuits and Systems*, Vol. CAS-22, March 1975, pp. 177–184.

H. E. Mutluay and M. M. Fahmy, Frequency-domain design of N-D digital filters, *IEEE Trans. Circuits and Systems*, Vol. CAS-32, December 1985, pp. 1226–1233.

C. L. Nikias, A. P. Chrysafis, and A. N. Venetsanopoulos, The LU decomposition theorem and its implications to the realization of two-dimensional digital filters, *IEEE Trans. Acoust. Speech Sig. Proc.*, Vol. ASSP-33, June 1985, pp. 694–711.

B. O'Connor and T. S. Huang, An efficient algorithm for bilinear transformation of multivariable polynomials, *IEEE Trans. Acoust. Speech Sig. Proc.*, Vol. ASSP-26, August 1978, pp. 380–381.

B. T. O'Connor and T. S. Huang, Stability of general two-dimensional recursive digital filters, *IEEE Trans. Acoust. Speech Sig. Proc.*, Vol. ASSP-26, December 1978, pp. 550–560.

H. H. Pade, Sur la representation approchee d'une fonction par des fractions rationelles, *Annales Scientifique de L'Ecole Normale Superieure*, Vol. 9, 1892, pp. 1–93.

N. A. Pendergrass, S. K. Mitra, and E. I. Jury, Spectral transformations for two-dimensional digital filters, *IEEE Trans. Circuits and Systems*, Vol. CAS-23, January 1976, pp. 26–35.

J. K. Pitas and A. N. Venetsanopoulos, The use of symmetries in the design of multidimensional digital filters, *IEEE Trans. Circuits and Systems*, Vol. CAS-33, September 1986, pp. 863–873.

G. R. B. Prony, Essai experimentale et analytique sur les lois de la dilatabilite de fluides elastiques et sur celles de la force expansion de la vapeur de l'alcohol, a differentes temperatures, *Journal de l'Ecole Polytechnique (Paris)*, Vol. 1, 1795, pp. 24–76.

D. Raghuramireddy, G. Schmeisser, and R. Unbehauen, PLSI polynomials: stabilization of 2-D recursive filters, *IEEE Trans. Circuits and Systems*, Vol. CAS-33, October 1986, pp. 1015–1018.

P. A. Ramamoorthy and L. T. Bruton, Design of stable two-dimensional analogue and digital filters with applications in image processing, *Int. J. Circuit Theory Appl.*, Vol. 7, April 1979, pp. 229–245.

P. S. Reddy, D. R. R. Reddy, and M. N. S. Swamy, Proof of a modified form of Shanks' conjecture on the stability of 2-D planar least square inverse polynomials and its implications, *IEEE Trans. Circuits and Systems*, Vol. CAS-31, December 1984, pp. 1009–1015.

R. P. Roesser, A discrete state-space model for linear image processing, *IEEE Trans. Automat. Contr.*, Vol. AC-20, February 1975, pp. 1–10.

J. L. Shanks, S. Treitel, and J. H. Justice, Stability and synthesis of two-dimensional recursive filters, *IEEE Trans. Audio Electro-acoust.*, Vol. AU-20, June 1972, pp. 115–128.

G. A. Shaw and R. M. Mersereau, Design stability and performance of two-dimensional recursive digital filters, *Tech. Rep. E21-B05-1*, Georgia Tech., December 1979.

K. Shimizu and T. Hirata, Optimal design using min-max criteria for two-dimensional recursive digital filters, *IEEE Trans. Circuits and Systems*, Vol. CAS-33, May 1986, pp. 491–501.

K. Steiglitz and L. E. McBride, A technique for the identification of linear systems, *IEEE Trans. Automatic Control*, Vol. AC-10, October 1965, pp. 461–464.

M. N. S. Swamy, L. M. Roytman, and E. I. Plotkin, Planar least squares inverse polynomials and practical-BIBO stabilization of *n*-dimensional linear shift-invariant filters, *IEEE Trans. Circuits and Systems*, Vol. CAS-32, December 1985, pp. 1255–1259.

J. M. Tribolet, A new phase unwrapping alogrithm, *IEEE Trans. Acoust. Speech Sig. Proc.*, Vol. ASSP-25, April 1977, pp. 170–177.

R. E. Twogood and S. K. Mitra, Computer-aided design of separable two-dimensional digital filters, *IEEE Trans. Acoust. Speech Sig. Proc.*, Vol. ASSP-25, April 1977, pp. 165–169.

J. W. Woods, J. H. Lee, and I. Paul, Two-dimensional IIR filter design with magnitude and phase error criteria, *IEEE Trans. Acoust. Speech Sig. Proc.*, Vol. ASSP-31, August 1983, pp. 886–894.

PROBLEMS

5.1. Consider a system function $H(z_1, z_2)$ given by

$$H(z_1, z_2) = \frac{1 + z_1^{-1}}{1 + 2z_1^{-1} + 3z_2^{-1}}.$$

(a) If we use the notational convention discussed in Section 5.1, $H(z_1, z_2)$ above will correspond to one particular computational procedure. Determine the computational procedure.

(b) Sketch the output and input masks of the computational procedure in (a).

(c) If we do not assume the notational convention discussed in Section 5.1, more than one computational procedure will have the above system function $H(z_1, z_2)$. Determine all computational procedures that have the above system function.

5.2. Consider a system function $H(z_1, z_2)$, given by

$$H(z_1, z_2) = \frac{\displaystyle\sum_{(k_1,k_2)\in Rb} \sum b(k_1, k_2)z_1^{-k_1}z_2^{-k_2}}{\displaystyle\sum_{(k_1,k_2)\in Ra} \sum a(k_1, k_2)z_1^{-k_1}z_2^{-k_2}}$$

$$= \frac{1 + 2z_1^{-1} + 3z_2^{-1}}{1 + 2z_1^{-1} + 3z_2^{-1} + 4z_1^{-1}z_2^{-1}}.$$

(a) Sketch the output and input masks of the system.

(b) Sketch $a(n_1, n_2)$ and $b(n_1, n_2)$.

(c) Discuss the relationship between the output mask in (a) and $a(n_1, n_2)$ in (b).

(d) Discuss the relationship between the input mask in (a) and $b(n_1, n_2)$ in (b).

5.3. Consider a system function $H(z_1, z_2)$, given by

$$H(z_1, z_2) = \frac{\displaystyle\sum_{(k_1, k_2) \in R_b} b(k_1, k_2) z_1^{-k_1} z_2^{-k_2}}{\displaystyle\sum_{(k_1, k_2) \in R_a} a(k_1, k_2) z_1^{-k_1} z_2^{-k_2}}$$

$$= \frac{1 + z_1^{-1} + z_2^{-1}}{1 + 2z_1^{-1} + 3z_2^{-1} + 4z_1^{-1}z_2^{-1}}.$$

(a) Show that the impulse response of the system $h(n_1, n_2)$ is a first-quadrant support sequence.

(b) The result in (a) is a special case of a more general result that states that the impulse response of a system with first-quadrant support sequences $a(n_1, n_2)$ and $b(n_1, n_2)$ is a first-quadrant support sequence. Show this more general result.

(c) The result in (b) is a special case of a more general result that states that a system with wedge support sequences $a(n_1, n_2)$ and $b(n_1, n_2)$ with the same wedge shape is also a wedge support sequence with the same wedge shape. Show this more general result.

5.4. Let $h_d(n_1, n_2)$ denote the desired impulse response of a system. The region of support of $h_d(n_1, n_2)$ is shown below.

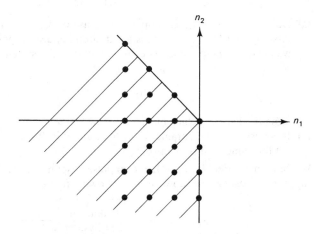

Figure P5.4

We wish to design $H(z_1, z_2)$, given by

$$H(z_1, z_2) = \frac{\displaystyle\sum_{(n_1, n_2) \in R_b} b(n_1, n_2) z_1^{-n_1} z_2^{-n_2}}{\displaystyle\sum_{(n_1, n_2) \in R_a} a(n_1, n_2) z_1^{-n_1} z_2^{-n_2}}$$

that best approximates $h_d(n_1, n_2)$. Determine one reasonable region of support of $a(n_1, n_2)$ and $b(n_1, n_2)$ that could be used.

5.5. Consider a system function $H(z_1, z_2) = B(z_1, z_2)/A(z_1, z_2)$, where $a(n_1, n_2)$ is an $M \times M$-point first-quadrant support sequence and $b(n_1, n_2)$ is an $N \times N$-point first-quadrant support sequence. We wish to design a zero-phase filter with a circularly symmetric filter specification by a parallel combination of $H(z_1, z_2)$, $H(z_1^{-1}, z_2)$, $H(z_1, z_2^{-1})$, and $H(z_1^{-1}, z_2^{-1})$. For a fixed M and N, imposing symmetry on $H(z_1, z_2)$ reduces the number of independent parameters in $H(z_1, z_2)$ and therefore may reduce the complexity of the filter design problem.
(a) What type of symmetry is reasonable to impose on $H(z_1, z_2)$?
(b) To achieve the symmetry in (a), what type of symmetry is reasonable to impose on $A(z_1, z_2)$ and $B(z_1, z_2)$?
(c) For the symmetry imposed on $A(z_1, z_2)$ and $B(z_1, z_2)$ in (b), how many independent parameters do we have in $H(z_1, z_2)$?

5.6. Suppose $H(z_1, z_2)$, the system function of an IIR filter, is given by

$$H(z_1, z_2) = \frac{d}{1 - az_1^{-1} - bz_2^{-1} - cz_1^{-1}z_2^{-1}}.$$

Assuming that the desired impulse response $h_d(n_1, n_2)$ is a first-quadrant support sequence, use the Pade matching method to estimate a, b, c, and d. Express your answer in terms of $h_d(0, 0)$, $h_d(0, 1)$, $h_d(1, 0)$, and $h_d(1, 1)$.

5.7. Consider a system function $H(z_1, z_2)$, given by

$$H(z_1, z_2) = \frac{1 + z_1^{-1}}{1 - \frac{1}{2}z_1^{-1} - \frac{1}{4}z_2^{-1}}.$$

(a) Determine $h(n_1, n_2)$, the impulse response of the system.
(b) Assume that $h(n_1, n_2)$ in (a) is the desired impulse response $h_d(n_1, n_2)$. Suppose we estimate the parameters a, b, c, and d in the system function $F(z_1, z_2)$ given by

$$F(z_1, z_2) = \frac{c + dz_1^{-1}}{1 + az_1^{-1} + bz_2^{-1}}$$

using the Pade matching method. Show that $a = -\frac{1}{2}$, $b = -\frac{1}{4}$, $c = 1$, and $d = 1$.
(c) If we estimate the parameters a, b, c, and d in (b) using Prony's method, do we get the same results as in (b)?

5.8. We have measured $h_d(n_1, n_2)$, the impulse response of an unknown first-quadrant support LSI system, as shown in the following figure.

$$\delta(n_1, n_2) \longrightarrow \boxed{\begin{array}{c} \text{Unknown} \\ \text{LSI System} \end{array}} \longrightarrow h_d(n_1, n_2)$$

We wish to model the unknown system by a rational system function, as shown in the following figure.

$$\delta(n_1, n_2) \longrightarrow \boxed{\dfrac{a + bz_1^{-1} + cz_2^{-1}}{1 - \frac{1}{2}z_1^{-1} - \frac{1}{4}z_2^{-1}}} \longrightarrow h(n_1, n_2)$$

We wish to determine the parameters a, b, and c by minimizing the following error function:

$$\text{Error} = \sum_{n1=0}^{\infty} \sum_{n2=0}^{\infty} (h_d(n_1, n_2) - h(n_1, n_2))^2.$$

Is the resulting set of equations linear or nonlinear? If the equations are linear, explicitly determine a, b, and c. If the equations are nonlinear, explain why.

5.9. Suppose we are given a desired zero-phase infinite impulse response $h_d(n_1, n_2)$. We wish to design a zero-phase IIR filter whose impulse response is as close as possible in some sense to $h_d(n_1, n_2)$. One approach, which we discussed in Section 5.2.4, is to divide $h_d(n_1, n_2)$ into four quadrants, design four one-quadrant IIR filters, and then implement them in an appropriate way. We will refer to this as Approach 1. Another approach, which we will call Approach 2, is to divide $h_d(n_1, n_2)$ into two segments, design two IIR filters, and then implement them in an appropriate way.

(a) To design the IIR filter only once in Approach 2, how should we segment $h_d(n_1, n_2)$? Specify the window to be used in the segmentation.

(b) After one filter is designed in (a), how should that filter be used for the second filter?

(c) If we wish to design an IIR filter whose impulse response approximates a desired impulse response that is nonzero everywhere, which approach would you recommend? Explain why one of the two approaches is inappropriate.

5.10. Suppose we wish to design a 2-D IIR filter whose impulse response $h(n_1, n_2)$ approximates the desired impulse response $h_d(n_1, n_2)$. The sequence $h_d(n_1, n_2)$ is circularly symmetric; that is, $h_d(n_1, n_2) = f(n_1^2 + n_2^2)$. One such example is

$$h_d(n_1, n_2) = \frac{\omega_c}{2\pi\sqrt{n_1^2 + n_2^2}} J_1(\omega_c\sqrt{n_1^2 + n_2^2}).$$

The 2-D IIR filter $h(n_1, n_2)$ should have the following three characteristics:

(1) The sequence $h(n_1, n_2)$ should have a maximum amount of symmetry and should be symmetric with respect to as many lines as possible that pass through the origin. One such line is the n_2 axis, in which case $h(n_1, n_2) = h(-n_1, n_2)$.

(2) The sequence $h(n_1, n_2)$ should be designed as a combination of recursively computable IIR subfilters $h_i(n_1, n_2)$. Only one subfilter should be designed. Let $H_1(z_1, z_2)$ denote the system function of this subfilter. All others should be derivable from $H_1(z_1, z_2)$ by inspection, exploiting only symmetry considerations.

(3) The desired impulse response that $H_1(z_1, z_2)$ approximates should have as small a region of support as possible. This simplifies the design of $H_1(z_1, z_2)$.

Develop an approach that will achieve the above objectives. Show that your approach will have the desired characteristics. Clearly state the system functions of all the subfilters in terms of $H_1(z_1, z_2)$, the system function of the subfilter you designed.

5.11. Consider a sequence $x(n_1, n_2)$ which has a valid complex cepstrum. We can express $X(\omega_1, \omega_2)$, the Fourier transform of $x(n_1, n_2)$, in terms of its real and imaginary part as

$$X(\omega_1, \omega_2) = X_R(\omega_1, \omega_2) + jX_I(\omega_1, \omega_2).$$

We can also express $\hat{X}(\omega_1, \omega_2)$, the Fourier transform of $\hat{x}(n_1, n_2)$, in terms of its real and imaginary part as

$$\hat{X}(\omega_1, \omega_2) = \log X(\omega_1, \omega_2) = \log |X(\omega_1, \omega_2)| + j\theta_x(\omega_1, \omega_2).$$

(a) Express $\partial\hat{X}(\omega_1, \omega_2)/\partial\omega_1$ in terms of $X_R(\omega_1, \omega_2)$ and $X_I(\omega_1, \omega_2)$.

(b) By considering the imaginary part of $\partial\hat{X}(\omega_1, \omega_2)/\partial\omega_1$ obtained in (a), express $\partial\theta_x(\omega_1, \omega_2)/\partial\omega_1$ in terms of $X_R(\omega_1, \omega_2)$ and $X_I(\omega_1, \omega_2)$. Compare your result with (5.96).

5.12. For any valid (real and stable) complex cepstrum $\hat{x}(n_1, n_2)$, show that there is a unique corresponding sequence $x(n_1, n_2)$, which is real and stable, and whose z-transform has no pole or zero surfaces on the unit surface.

5.13. Determine the complex cepstrum of each of the following sequences. If modification by shifting the sequence and/or multiplying the sequence by -1 is needed, determine the complex cepstrum of the modified sequence.

(a) $\delta(n_1, n_2)$
(b) $-\delta(n_1, n_2)$
(c) $2\delta(n_1 - 1, n_2)$
(d) $\delta(n_1, n_2) - \frac{1}{2}\delta(n_1 - 1, n_2 - 2)$
(e) $\delta(n_1, n_2) - 2\delta(n_1 - 1, n_2 - 2)$

5.14. Consider a sequence $x(n_1, n_2)$ given by

$$x(n_1, n_2) = \delta(n_1, n_2) - 2\delta(n_1 - 1, n_2).$$

(a) Show that $x(n_1, n_2)$ without any modification does not have a valid complex cepstrum.

(b) We define $r(n_1, n_2)$ by

$$r(n_1, n_2) = x(n_1, n_2) * x(-n_1, -n_2).$$

Show that $r(n_1, n_2)$ has a valid complex cepstrum and determine $\hat{r}(n_1, n_2)$, the complex cepstrum of $r(n_1, n_2)$.

5.15. Let $\hat{\theta}_x(\omega)$ represent the principal value of the phase of $X(\omega)$, the Fourier transform of $x(n)$, and let $\theta_x(\omega)$ represent the odd and continuous phase function of $X(\omega)$. Suppose $\tilde{\theta}_x(\omega)$ has been sampled at $\omega = (2\pi/N)k$ and that $\tilde{\theta}_x(k) = \tilde{\theta}_x(\omega)|_{\omega = 2\pi k/N}$ is shown below.

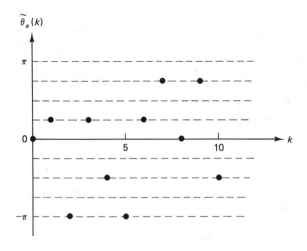

Figure P5.15

(a) Assuming that $|\theta_x(k) - \theta_x(k - 1)| < \pi$ for all k, determine $\theta_x(k)$ for $0 \le k \le 10$.

(b) If we cannot assume that $|\theta_x(k) - \theta_x(k - 1)| < \pi$ for all k, discuss why it may not be possible to determine $\theta_x(k)$ from $\hat{\theta}_x(k)$.

This problem illustrates the difficulty associated with the phase unwrapping operation. Increasing N reduces the possibility of the phase unwrapping errors, but increases the number of computations. In addition, there is no simple systematic way to determine N for a given $x(n)$ such that the phase-unwrapping errors can be avoided.

5.16. The real cepstrum of $x(n_1, n_2)$ is defined by

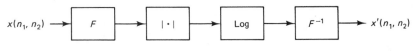

Figure P5.16

The sequence $x'(n_1, n_2)$ is called the real cepstrum of $x(n_1, n_2)$, since the logarithmic operation is applied to $|X(\omega_1, \omega_2)|$ and therefore does not have to be complex. Relate $x'(n_1, n_2)$ and $\hat{x}(n_1, n_2)$, the complex cepstrum of $x(n_1, n_2)$. Assume that $x(n_1, n_2)$ has a valid complex cepstrum.

5.17. Consider a 1-D causal system whose system function $H(z)$ is given by

$$H(z) = \frac{1}{A(z)} = \frac{1}{1 - \frac{1}{3}z^{-1}} \frac{1}{1 - 5z^{-1}} \frac{1}{1 - 2z^{-1}}.$$

Determine the system function of a system that is causal and stable without changing $|H(\omega)|$, using each of the following methods:

(a) By reflecting each root outside the unit circle to a root inside the unit circle at a conjugate reciprocal location.

(b) By using the complex cepstrum.

5.18. Let $x(n_1, n_2)$ denote a real and stable sequence that has a rational z-transform $X(z_1, z_2)$ with no pole or zero surfaces crossing the unit surface. Suppose we wish to factor $X(z_1, z_2)$ as a product of four z-transforms as follows:

$$X(z_1, z_2) = X_1(z_1, z_2) X_2(z_1, z_2) X_3(z_1, z_2) X_4(z_1, z_2).$$

The function $X_i(z_1, z_2)$ represents the z-transform of a sequence that has the ith-quadrant support region. For example, $X_2(z_1, z_2)$ represents the z-transform of a second-quadrant support sequence. Develop a method to perform the above spectral factorization. Note that $X_i(z_1, z_2)$ may not be a rational z-transform.

5.19. Let $a(n_1, n_2)$ denote a real, finite-extent, and stable sequence with its Fourier transform denoted by $A(\omega_1, \omega_2)$. The sequence $a(n_1, n_2)$, in addition, satisfies the following properties:

(1) $A(\omega_1, \omega_2) \ne 0$ at any (ω_1, ω_2).

(2) $a(n_1, n_2) = a(-n_1, n_2) = a(n_1, -n_2) = a(-n_1, -n_2)$.

(3) $\hat{a}(n_1, n_2)$, the complex cepstrum of $a(n_1, n_2)$, is well defined without any modification of $a(n_1, n_2)$.

Suppose we obtain a sequence $b(n_1, n_2)$ by the system in Figure P5.19. Note that $l(n_1, n_2)$ is a third-quadrant support sequence.

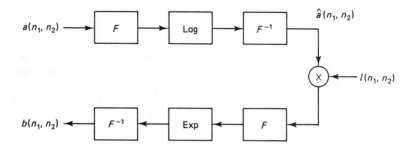

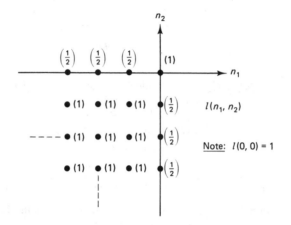

Figure P5.19

(a) Let a sequence $c(n_1, n_2)$ denote the inverse of $b(n_1, n_2)$; that is,

$$C(\omega_1, \omega_2) = \frac{1}{B(\omega_1, \omega_2)}.$$

Is $c(n_1, n_2)$ a well-defined sequence; that is, is $C(\omega_1, \omega_2)$ a valid Fourier transform? Explain your answer.

(b) Can you recover $a(n_1, n_2)$ from $b(n_1, n_2)$? If so, express $a(n_1, n_2)$ in terms of $b(n_1, n_2)$. If not, explain why $a(n_1, n_2)$ cannot be recovered from $b(n_1, n_2)$.

5.20. Suppose an image $x(n_1, n_2)$ is degraded by blurring in such a way that the blurred image $y(n_1, n_2)$ can be represented by

$$y(n_1, n_2) = x(n_1, n_2) * b(n_1, n_2)$$

where $b(n_1, n_2)$ represents the impulse response of the blurring system. Using the complex cepstrum, develop a method to reduce the effect of $b(n_1, n_2)$ by linear filtering in the complex cepstrum domain. This type of system is referred to as a homomorphic system for convolution. Determine the conditions under which the effect of $b(n_1, n_2)$ can be completely eliminated by linear filtering in the complex cepstrum domain.

5.21. Consider a 2 × 2-point first-quadrant support sequence $a(n_1, n_2)$ sketched in the following figure.

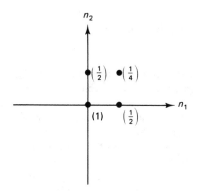

Figure P5.21

Let $b(n_1, n_2)$ be a 2×1-point sequence that is zero outside $0 \le n_1 \le 1$, $n_2 = 0$. Determine $b(n_1, n_2)$, the planar least squares inverse of $a(n_1, n_2)$.

5.22. In this problem, we study one significant difference between spatial and frequency domain methods. For simplicity, we consider the 1-D case. The result extends straightforwardly to the 2-D case. Let the desired impulse response $h_d(n)$ be given by

$$h_d(n) = (\tfrac{1}{2})^n u(n) + 8(4)^n u(-n-1). \tag{1}$$

Let the desired frequency response $H_d(\omega)$ be the Fourier transform of $h_d(n)$. We wish to design an IIR filter with the system function given by

$$H(z) = \frac{c}{1 + az^{-1} + bz^{-2}}. \tag{2}$$

Let $h(n)$ denote the impulse response of the filter.

(a) In a spatial domain design method, we assume that the computational procedure is given by

$$y(n) \leftarrow -ay(n-1) - by(n-2) + cx(n). \tag{3}$$

We estimate the filter coefficients a, b, and c by minimizing

$$\text{Error} = \sum_{n=-\infty}^{\infty} (h_d(n) - h(n))^2. \tag{4}$$

Show that $a = -\tfrac{1}{2}$, $b = 0$, $c = 1$, and $h(n)$ is given by $h(n) = (\tfrac{1}{2})^n u(n)$.

(b) Is the system in (a) stable?

(c) In a frequency domain design method, we assume that

$$H(\omega) = H(z)|_{z=e^{j\omega}} = \frac{c}{1 + ae^{-j\omega} + be^{-j2\omega}}. \tag{5}$$

We estimate a, b, and c by minimizing

$$\text{Error} = \frac{1}{2\pi} \int_{\omega=-\pi}^{\pi} |H_d(\omega) - H(\omega)|^2 \, d\omega. \tag{6}$$

Show that $a = -9/2$, $b = 2$, $c = -7$, and $h(n)$ obtained by inverse Fourier transforming $H(\omega)$ is given by

$$h(n) = (\tfrac{1}{2})^n u(n) + 8(4)^n u(-n - 1). \tag{7}$$

Note that the filter is stable, but it does not have the same region of support as $h(n)$ in (a), and it is not even recursively computable.

(d) Suppose we use the a, b, and c obtained in (c) but use the computational procedure in (3). Determine $h(n)$ and show that the system is unstable.

(e) Suppose $h_d(n) = (\tfrac{1}{2})^n u(n)$. Show that the impulse responses obtained from the spatial domain design method in (a) and the frequency domain design method in (c) are the same.

Even though Parseval's theorem states that the two error criteria in (4) and (6) are the same, they can lead to different results in practice, depending on how they are used.

5.23. One error criterion that penalizes an unstable filter is given by (5.127). If we let the parameter α in (5.127) approach ∞, is the filter designed guaranteed to be stable? Is minimizing the Error in (5.127) while letting α approach ∞ a good way to design an IIR filter?

5.24. In one design method by spectral transformation, a 2-D IIR filter is designed from a 1-D IIR filter. Let $H(z)$ represent a 1-D causal and stable IIR filter. The filter is a lowpass filter with cutoff frequency of $\pi/2$. We design two 2-D filters $H_1(z_1, z_2)$ and $H_2(z_1, z_2)$ by

$$H_1(z_1, z_2) = H(z_1)H(z_2)$$

$$H_2(z_1, z_2) = H(z_1 z_2)H(z_1 z_2^{-1}).$$

(a) What is the approximate magnitude response of $H_1(z_1, z_2)$?
(b) Is $H_1(z_1, z_2)$ stable and recursively computable?
(c) What is the approximate magnitude response of $H_2(z_1, z_2)$?
(d) Is $H_2(z_1, z_2)$ stable and recursively computable?
(e) We design another 2-D filter $H_T(z_1, z_2)$ by

$$H_T(z_1, z_2) = H_1(z_1, z_2)H_2(z_1, z_2).$$

What is the approximate magnitude response of $H_T(z_1, z_2)$?

5.25. Let $H(z_1, z_2)$ denote a system function given by

$$H(z_1, z_2) = \frac{1 - z_1^{-1}}{(1 - \tfrac{1}{2}z_1^{-1})(1 - \tfrac{1}{4}z_2^{-1})}.$$

(a) Sketch a signal flowgraph that realizes the above system function using a direct form.
(b) Sketch a signal flowgraph that realizes the above system function using a cascade form.

5.26. Consider the following computational procedure:

$$y(n_1, n_2) \leftarrow 2y(n_1 - 1, n_2) + 3y(n_1, n_2 - 1) + 4y(n_1 - 3, n_2) + x(n_1, n_2)$$

$$+ \tfrac{1}{2}x(n_1 - 2, n_2) + \tfrac{1}{3}x(n_1, n_2 - 4).$$

Draw a signal flowgraph that requires the smallest (within a few storage elements from the minimum possible) number of storage units when the input is available row by row. You may assume that the input $x(n_1, n_2)$ is zero outside $0 \le n_1 \le N - 1$, $0 \le n_2 \le N - 1$ where $N \gg 1$, and that we wish to compute $y(n_1, n_2)$ for $0 \le n_1 \le N - 1, 0 \le n_2 \le N - 1$.

5.27.

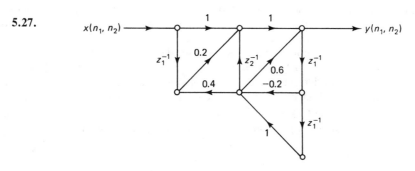

Figure P5.27

(a) Write a series of difference equations that will realize the above signal flowgraph.

(b) Determine the system function of the above signal flowgraph.

5.28. Consider a first-quadrant support IIR filter whose system function is given by

$$H(z_1, z_2) = \frac{1 + z_1^{-1} + z_2^{-1}}{1 + az_1^{-1} + bz_2^{-1} + cz_1^{-1}z_2^{-1}}.$$

The filter may or may not be stable, depending on the coefficients, a, b, and c.

(a) If $a = 2$, $b = 4$, and $c = 8$, the filter is unstable. Determine $H'(z_1, z_2)$ which is stable and is a first-quadrant support system, and which has the same magnitude response as $H(z_1, z_2)$. Note that the denominator polynomial is factorable for this choice of a, b, and c.

(b) We wish to implement $H(z_1, z_2)$, even though it may be unstable. Sketch a signal flowgraph of $H(z_1, z_2)$, which minimizes (within a few storage elements from the minimum possible) the number of storage elements needed when the input is available column by column. Find the total number of storage elements needed when the input is zero outside $0 \le n_1 \le N - 1, 0 \le n_2 \le N - 1$ and we wish to compute the output for $0 \le n_1 \le N - 1, 0 \le n_2 \le N - 1$.

5.29. Consider the following signal flowgraph, which implements an IIR filter:

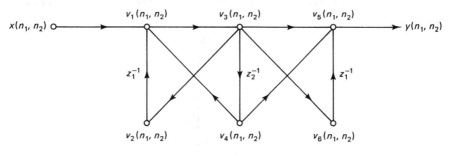

Figure P5.29

(a) Write a difference equation that relates $x(n_1, n_2)$ to $y(n_1, n_2)$.

(b) Determine the system function of the system represented by the above signal flowgraph.

(c) Determine an order in which $v_1(n_1, n_2)$, $v_2(n_1, n_2)$, $v_3(n_1, n_2)$, $v_4(n_1, n_2)$, $v_5(n_1, n_2)$, and $v_6(n_1, n_2)$ can be computed from $x(n_1, n_2)$. Assume that all previously computed values of $v_1, v_2, v_3, v_4, v_5, v_6$, and y are available. For example, you may assume that $y(n_1 - 2, n_2 - 3)$ is given.

(d) Suppose the input $x(n_1, n_2)$ is zero outside $0 \le n_1 \le 9, 0 \le n_2 \le 9$. We wish to compute $y(n_1, n_2)$ for $0 \le n_1 \le 9, 0 \le n_2 \le 9$ row by row. From the signal flowgraph, determine the number of storage elements required and what values they should have stored when we compute $y(n_1, n_2)$ at $n_1 = n_2 = 1$. You can ignore temporary storage elements.

5.30. Consider a system whose output and input masks are shown below.

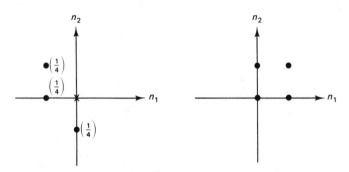

Figure P5.30a

(a) Determine the computational procedure.

(b) Determine the system function.

(c) Suppose the input $x(n_1, n_2)$ is zero outside $0 \le n_1 \le 511, 0 \le n_2 \le 511$. We wish to compute the output $y(n_1, n_2)$ for $0 \le n_1 \le 511, 0 \le n_2 \le 511$. Is it possible to compute $y(n_1, n_2)$ in only the desired region without computing any other output points? If not, what is the region for which the output is not desired but has to be computed?

(d) If the output is computed column by column, how many storage elements are required to realize z_1^{-1} and z_2^{-1} elements, respectively? Assume that the input is available in any order desired.

(e) Can the output be computed row by row?

(f) Answer (d) if the output is computed in the direction shown below.

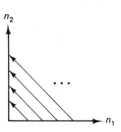

Figure P5.30b

5.31. For the signal flowgraph shown in Problem 5.27, give a state-space representation.

5.32. Consider the following state-space representation.

$$\begin{bmatrix} s^h(n_1 + 1, n_2) \\ s^v(n_1, n_2 + 1) \end{bmatrix} = \begin{bmatrix} 1 & 2 \\ 3 & 4 \end{bmatrix} \begin{bmatrix} s^h(n_1, n_2) \\ s^v(n_1, n_2) \end{bmatrix} + \begin{bmatrix} 0 \\ 1 \end{bmatrix} x(n_1, n_2)$$

$$y(n_1, n_2) = 5s^h(n_1, n_2) + 6s^v(n_1, n_2) + 2x(n_1, n_2).$$

Sketch a signal flowgraph that has the same input-output relationship as the state-space representation.

5.33. Consider a separable denominator system function $H(z_1, z_2)$ given by

$$H(z_1, z_2) = \frac{1}{A_1(z_1)A_2(z_2)}$$

where

$$A_1(z_1) = \sum_{n1=0}^{N} a_1(n_1)z_1^{-n1}$$

and

$$A_2(z_2) = \sum_{n2=0}^{N} a_2(n_2)z_2^{-n2}.$$

To implement the system, we consider two approaches. In the first approach (Approach 1), we use a computational procedure that relates $x(n_1, n_2)$ to $y(n_1, n_2)$ directly. In the second approach (Approach 2), we implement $1/A_1(z_1)$ first with a computational procedure that relates $x(n_1, n_2)$ to $r(n_1, n_2)$, and then implement $1/A_2(z_2)$ with a computational procedure that relates $r(n_1, n_2)$ to $y(n_1, n_2)$.

(a) How many multiplications and additions do we need to evaluate one output point using Approach 1?

(b) Determine the computational procedure that relates $x(n_1, n_2)$ to $r(n_1, n_2)$.

(c) Determine the computational procedure that relates $r(n_1, n_2)$ to $y(n_1, n_2)$.

(d) How many multiplications and additions do we need to evaluate one output point using Approach 2?

5.34. One approach to designing a zero-phase IIR filter is by cascading more than one IIR filter. Let $H(z_1, z_2)$ denote a first-quadrant support IIR filter. We now cascade $H(z_1, z_2)$, $H(z_1^{-1}, z_2)$, $H(z_1, z_2^{-1})$, and $H(z_1^{-1}, z_2^{-1})$, as shown below.

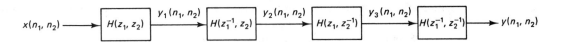

Figure P5.34

(a) Show that $F(z_1, z_2)$ given by

$$F(z_1, z_2) = H(z_1, z_2)H(z_1^{-1}, z_2)H(z_1, z_2^{-1})H(z_1^{-1}, z_2^{-1})$$

is a zero-phase filter.

(b) Let the input $x(n_1, n_2)$ denote an $N \times N$-point finite-extent sequence that is zero outside $0 \le n_1 \le N - 1$, $0 \le n_2 \le N - 1$. What is the region of support of $y_1(n_1, n_2)$, the output of the system $H(z_1, z_2)$?

(c) If we do not truncate the infinite-extent sequence $y_1(n_1, n_2)$, is it possible to implement $H(z_1^{-1}, z_2)$ with a recursively computable system?

6

Spectral Estimation

6.0 INTRODUCTION

Estimating the power spectrum associated with a random process is desirable in many applications. For example, Wiener filtering is one approach to solving image restoration problems in which a signal is degraded by additive random noise. The frequency response of a noncausal Wiener filter requires knowledge of the spectral contents of the signal and background noise. In practice, they often are not known and must be estimated from segments of data. As another example, spectral contents of signals received from an array of sensors contain information on the signal source, such as the direction of low-flying aircraft in the case of an acoustic sensor array application.

This chapter treats the 2-D spectral estimation problem. At a conceptual level, 2-D spectral estimation methods are straightforward extensions of 1-D methods, or can be derived straightforwardly from them. However, some 2-D spectral estimation methods differ considerably from the corresponding 1-D methods in such details as computational complexity and properties. We examine both similarities and differences, with more emphasis on topics where there are significant differences between the 1-D and 2-D cases.

In Section 6.1, we briefly summarize some fundamental results related to discrete-space random processes. Readers familiar with the material may wish to skim this section, but should give some attention to the notation used. In Section 6.2, we discuss various issues related to spectral estimation and different approaches to spectral estimation. In Section 6.3, the performance of several different spectral estimation methods is compared. Section 6.4 contains some additional comments on the spectral estimation problem. In Section 6.5, we illustrate the application of a spectral estimation technique to data from an array of acoustic microphones.

Parts of this chapter appeared in "Multi-dimensional Spectral Estimation," by Jae S. Lim, Chapter 6 of *Advances in Computer Vision and Image Processing: Image Enhancement and Restoration*, edited by Thomas S. Huang. Copyright © 1986 by JAI Press. Reprinted by permission of the publisher.

6.1 RANDOM PROCESSES

This section summarizes the fundamentals of random processes, which are useful in this chapter for studying spectral estimation and in later chapters for studying image processing. Since the reader is assumed to have some familiarity with random processes, many results are simply stated. Results with which some readers may not be familiar are derived. Despite the derivations, this section is intended to serve primarily for review and establishment of notation. For a detailed and complete presentation of topics related to random processes, the reader is referred to [Papoulis; Van Trees].

6.1.1 Random Variables

A real random variable x is a variable that takes on real values at random, for instance, from the outcome of flipping a coin. It is completely characterized by its probability density function $p_x(x_0)$. The subscript x in $p_x(x_0)$ denotes the random variable x, and x_0 is a dummy variable that denotes a specific value of x. The probability that x will lie between a and b is given by

$$\text{Prob } [a \leq x \leq b] = \int_{x_0=a}^{b} p_x(x_0) \, dx_0. \tag{6.1}$$

Since an event that is certain to occur is assumed to have a probability of 1,

$$\text{Prob } [-\infty \leq x \leq \infty] = \int_{x_0=-\infty}^{\infty} p_x(x_0) \, dx_0 = 1. \tag{6.2}$$

The expectation of a function of a random variable x, $E[f(x)]$, is defined by

$$E[f(x)] = \int_{x_0=-\infty}^{\infty} f(x_0)p_x(x_0) \, dx_0. \tag{6.3}$$

The expectation defined above is a linear operator and satisfies

$$E[f(x) + g(x)] = E[f(x)] + E[g(x)] \tag{6.4}$$

and
$$E[cf(x)] = cE[f(x)] \tag{6.5}$$

where c is any scalar constant.

The nth moment of a random variable x, $E[x^n]$, is defined by

$$E[x^n] = \int_{x_0=-\infty}^{\infty} x_0^n p_x(x_0) \, dx_0. \tag{6.6}$$

The first moment of x is called the mean or average of x. From (6.6),

$$E[x] = \int_{x_0=-\infty}^{\infty} x_0 p_x(x_0) \, dx_0. \tag{6.7}$$

The variance of x, $\text{Var}[x]$, is defined by

$$\text{Var}[x] = E[(x - E[x])^2] = E[x^2 - 2xE[x] + E^2[x]] = E[x^2] - E^2[x]. \tag{6.8}$$

The standard deviation of x, s.d. $[x]$, is defined by

$$\text{s.d.}[x] = (\text{Var}[x])^{1/2}. \tag{6.9}$$

Two random variables, x and y, are completely characterized by their joint probability density function $p_{x,y}(x_0, y_0)$. They are said to be *statistically independent* if they satisfy

$$p_{x,y}(x_0, y_0) = p_x(x_0)p_y(y_0) \quad \text{for all} \quad (x_0, y_0). \tag{6.10}$$

The expectation of a function of two random variables, $E[f(x, y)]$, is defined by

$$E[f(x, y)] = \int_{x_0 = -\infty}^{\infty} \int_{y_0 = -\infty}^{\infty} f(x_0, y_0)p_{x,y}(x_0, y_0) \, dx_0 \, dy_0. \tag{6.11}$$

Two random variables x and y are said to be *linearly independent* if

$$E[xy] = E[x]E[y]. \tag{6.12}$$

Statistical independence implies linear independence, but linear independence does not imply statistical independence.

The probability density function of a random variable x given (conditioned on) another random variable y is denoted by $p_{x|y}(x_0|y_0)$ and is defined by

$$p_{x|y}(x_0|y_0) = p_{x,y}(x_0, y_0)/p_y(y_0). \tag{6.13}$$

If x and y are statistically independent, knowing y does not tell us anything about x, and $p_{x|y}(x_0|y_0)$ reduces to $p_x(x_0)$. The expectation of a function of x conditioned on y, $E[f(x)|y]$, is defined by

$$E[f(x)|y] = \int_{x_0 = -\infty}^{\infty} f(x_0)p_{x|y}(x_0|y_0) \, dx_0. \tag{6.14}$$

A complex random variable w is defined by

$$w = x + jy \tag{6.15}$$

where x and y are the two random variables defined above. The expectation of a function of w, $E[f(w)]$, is defined by

$$E[f(w)] = E[f(x + jy)] \tag{6.16}$$

$$= \int_{x_0 = -\infty}^{\infty} \int_{y_0 = -\infty}^{\infty} f(x_0 + jy_0)p_{x,y}(x_0, y_0) \, dx_0 \, dy_0.$$

The mean of w is defined by

$$E[w] = E[x + jy] = E[x] + jE[y]. \tag{6.17}$$

The variance of w is defined by

$$\begin{aligned}
\text{Var}[w] &= E[(w - E[w])(w - E[w])^*] \\
&= E[ww^*] - E[w]E^*[w] \\
&= E[x^2 + y^2] - (E^2[x] + E^2[y]) \\
&= \text{Var}[x] + \text{Var}[y].
\end{aligned} \tag{6.18}$$

Note that the variance of w is real and nonnegative, even though w may be complex.

Many random variables x, y, z, ... are completely characterized by their joint probability density function $p_{x,y,z,...}(x_0, y_0, z_0, ...)$. The definitions of the expectation operation and linear and statistical independence in this case are similar to the definitions in the two random variable case.

6.1.2 Random Processes

A collection of an infinite number of random variables is called a random process. If the random variables are real, the collection is called a *real random process*. If the random variables are complex, the collection is called a *complex random process*. Unless stated otherwise, all the results in this section will be stated for complex random processes. The results for a real random process are special cases of the results for a complex random process.

Let us denote an infinite number of complex random variables by $x(n_1, n_2)$, where $x(n_1, n_2)$ for a particular (n_1, n_2) is a complex random variable. The random process $x(n_1, n_2)$ is completely characterized by the joint probability density function of all the random variables. If we obtain one sample, or realization, of the random process $x(n_1, n_2)$, the result will be a 2-D sequence. We will refer to this 2-D sequence as a random signal, and will denote it also by $x(n_1, n_2)$. Whether $x(n_1, n_2)$ refers to a random process or one realization of a random process will usually be clear from the context. Otherwise, what is meant will be specifically stated. The collection of all possible realizations is called the *ensemble* of the random process $x(n_1, n_2)$.

The *auto-correlation function*, or for short, the *correlation function* of the random process $x(n_1, n_2)$, $R_x(n_1, n_2; k_1, k_2)$, is defined by

$$R_x(n_1, n_2; k_1, k_2) = E[x(n_1, n_2)x^*(k_1, k_2)]. \tag{6.19}$$

The correlation is the expectation of the product of two complex random variables, $x(n_1, n_2)$ and $x^*(k_1, k_2)$. The *auto-covariance function*, or the *covariance function*, for short, of $x(n_1, n_2)$, $\gamma_x(n_1, n_2; k_1, k_2)$, is defined by

$$\gamma_x(n_1, n_2; k_1, k_2) = E[(x(n_1, n_2) - E[x(n_1, n_2)])(x(k_1, k_2) - E[x(k_1, k_2)])^*]$$

$$= E[x(n_1, n_2)x^*(k_1, k_2)] - E[x(n_1, n_2)]E^*[x(k_1, k_2)] \tag{6.20}$$

$$= R_x(n_1, n_2; k_1, k_2) - E[x(n_1, n_2)]E^*[x(k_1, k_2)].$$

A random process $x(n_1, n_2)$ is called a zero-mean process if

$$E[x(n_1, n_2)] = 0 \quad \text{for all } (n_1, n_2). \tag{6.21}$$

For a zero-mean random process,

$$R_x(n_1, n_2; k_1, k_2) = \gamma_x(n_1, n_2; k_1, k_2). \tag{6.22}$$

A random process $x(n_1, n_2)$ with nonzero mean can always be transformed to a zero-mean random process by subtracting $E[x(n_1, n_2)]$ from $x(n_1, n_2)$. Unless specified otherwise, we will assume that $x(n_1, n_2)$ is a zero-mean process and (6.22) is valid.

A random process $x(n_1, n_2)$ is said to be *stationary* or *homogeneous in the strict sense* if the joint probability density function does not depend on the origin of the index (n_1, n_2):

$$p_{x(n_1,n_2),x(n_1',n_2'),\ldots}(x_1, x_2, \ldots) = p_{x(n_1'+k_1,n_2'+k_2),x(n_1'+k_1,n_2'+k_2),\ldots}(x_1, x_2, \ldots) \qquad (6.23)$$

for any fixed k_1 and k_2. For a stationary random process $x(n_1, n_2)$, $E[x(n_1, n_2)]$ is a constant independent of n_1 and n_2, and $R_x(n_1, n_2; k_1, k_2)$ is a function of only $n_1 - k_1$ and $n_2 - k_2$:

$$E[x(n_1, n_2)] = m_x \quad \text{for all } (n_1, n_2) \qquad (6.24)$$

$$\begin{aligned} R_x(n_1 - k_1, n_2 - k_2) &= R_x(n_1 - k_1, n_2 - k_2; 0, 0) \\ &= E[x(n_1 - k_1, n_2 - k_2)x^*(0, 0)]. \end{aligned} \qquad (6.25)$$

Rewriting (6.25), we obtain

$$R_x(n_1, n_2) = E[x(k_1, k_2)x^*(k_1 - n_1, k_2 - n_2)] \quad \text{for all } (k_1, k_2). \qquad (6.26)$$

Note that the arguments n_1 and n_2 in $R_x(n_1, n_2)$ in (6.26) are $k_1 - n_1$ subtracted from k_1 and $k_2 - n_2$ subtracted from k_2. A random process that satisfies (6.24) and (6.26), but does not necessarily satisfy (6.23), is said to be *stationary in the wide sense*. From (6.26),

$$R_x(n_1, n_2) = R_x^*(-n_1, -n_2). \qquad (6.27)$$

From (6.27), it is clear that the correlation sequence has complex conjugate symmetry.

A stationary random process $x(n_1, n_2)$ is said to be *ergodic* if the time (or space) average equals the ensemble average. Suppose we wish to estimate $m_x = E[x(n_1, n_2)]$ from realizations, or samples, of a stationary $x(n_1, n_2)$. Since m_x represents an ensemble average, we need an ensemble (an entire collection of all possible outcomes) of $x(n_1, n_2)$ for any particular (n_1, n_2). If the random process is ergodic, then m_x can be computed from one realization of $x(n_1, n_2)$ by

$$m_x = E[x(n_1, n_2)] = \lim_{N \to \infty} \frac{1}{(2N + 1)^2} \sum_{n_1 = -N}^{N} \sum_{n_2 = -N}^{N} x(n_1, n_2). \qquad (6.28)$$

Similarly, for an ergodic process,

$$R_x(n_1, n_2) = E[x(k_1, k_2)x^*(k_1 - n_1, k_2 - n_2)]$$

$$= \lim_{N \to \infty} \frac{1}{(2N + 1)^2} \sum_{k_1 = -N}^{N} \sum_{k_2 = -N}^{N} x(k_1, k_2)x^*(k_1 - n_1, k_2 - n_2). \qquad (6.29)$$

Equations (6.28) and (6.29) allow us to determine m_x or $R_x(n_1, n_2)$ from one realization of $x(n_1, n_2)$. Note that ergodicity implies stationarity (in the wide sense), but stationarity does not imply ergodicity.

The *power spectrum* of a stationary random process $x(n_1, n_2)$, $P_x(\omega_1, \omega_2)$, is defined by

$$P_x(\omega_1, \omega_2) = F[R_x(n_1, n_2)] = \sum_{n_1 = -\infty}^{\infty} \sum_{n_2 = -\infty}^{\infty} R_x(n_1, n_2)e^{-j\omega_1 n_1}e^{-j\omega_2 n_2}. \qquad (6.30)$$

From (6.27) and (6.30),

$$P_x(\omega_1, \omega_2) = P_x^*(\omega_1, \omega_2). \tag{6.31}$$

From (6.31), $P_x(\omega_1, \omega_2)$ is always real. Furthermore, it can be shown that $P_x(\omega_1, \omega_2)$ is always nonnegative; that is,

$$P_x(\omega_1, \omega_2) \geq 0 \quad \text{for all } (\omega_1, \omega_2). \tag{6.32}$$

From (6.26) and (6.30),

$$R_x(0, 0) = \sigma_x^2 = E[x(n_1, n_2)x^*(n_1, n_2)] \tag{6.33}$$

$$= \frac{1}{(2\pi)^2} \int_{\omega_1 = -\pi}^{\pi} \int_{\omega_2 = -\pi}^{\pi} P_x(\omega_1, \omega_2) \, d\omega_1 \, d\omega_2.$$

It can also be shown that

$$R_x(0, 0) \geq |R_x(n_1, n_2)| \quad \text{for all } (n_1, n_2). \tag{6.34}$$

The value σ_x^2 is called the average power of the random process $x(n_1, n_2)$.

A random process is called a *white noise process* if

$$R_x(n_1, n_2; k_1, k_2) = E[x(n_1, n_2)x^*(k_1, k_2)] \tag{6.35}$$

$$= \begin{cases} \sigma_x^2(n_1, n_2), & n_1 = k_1, n_2 = k_2 \\ 0, & \text{otherwise.} \end{cases}$$

For a stationary white noise process, then,

$$R_x(n_1, n_2) = E[x(k_1, k_2)x^*(k_1 - n_1, k_2 - n_2)] \tag{6.36}$$

$$= \sigma_x^2 \, \delta(n_1, n_2).$$

From (6.30) and (6.36), the power spectrum of a stationary white noise process is given by

$$P_x(\omega_1, \omega_2) = \sigma_x^2 \quad \text{for all } (\omega_1, \omega_2). \tag{6.37}$$

The power spectrum is constant for all frequencies; hence the term "white."

For a real random process $x(n_1, n_2)$, (6.19), (6.26), (6.27), and (6.29) reduce to

$$R_x(n_1, n_2; k_1, k_2) = E[x(n_1, n_2)x(k_1, k_2)] \tag{6.38}$$

$$R_x(n_1, n_2) = E[x(k_1, k_2)x(k_1 - n_1, k_2 - n_2)] \tag{6.39}$$

for all (k_1, k_2) for a stationary process

$$R_x(n_1, n_2) = R_x(-n_1, -n_2) \tag{6.40}$$

and $R_x(n_1, n_2) = \lim_{N \to \infty} \frac{1}{(2N+1)^2} \sum_{k_1 = -N}^{N} \sum_{k_2 = -N}^{N} x(k_1, k_2)x(k_1 - n_1, k_2 - n_2)$ (6.41)

for an ergodic process.

From (6.40),

$$P_x(\omega_1, \omega_2) = P_x(-\omega_1, -\omega_2). \tag{6.42}$$

Since $P_x(\omega_1, \omega_2)$ is periodic with a period of 2π in both the variables ω_1 and ω_2, from (6.42) $P_x(\omega_1, \omega_2)$ for a real random process is completely specified by $P_x(\omega_1, \omega_2)$ for $-\pi \leq \omega_1 \leq \pi, 0 \leq \omega_2 \leq \pi$. For a real random process, therefore, $P_x(\omega_1, \omega_2)$ is often displayed only for $-\pi \leq \omega_1 \leq \pi, 0 \leq \omega_2 \leq \pi$.

Two complex random processes $x(n_1, n_2)$ and $y(n_1, n_2)$ are completely characterized by the joint probability density function of the random variables in $x(n_1, n_2)$ and $y(n_1, n_2)$. The *cross-correlation function* of $x(n_1, n_2)$ and $y(n_1, n_2)$, $R_{xy}(n_1, n_2; k_1, k_2)$ is defined by

$$R_{xy}(n_1, n_2; k_1, k_2) = E[x(n_1, n_2)y^*(k_1, k_2)]. \tag{6.43}$$

The *cross-covariance function* of $x(n_1, n_2)$ and $y(n_1, n_2)$, $\gamma_{xy}(n_1, n_2; k_1, k_2)$, is defined by

$$\gamma_{xy}(n_1, n_2; k_1, k_2) = E[(x(n_1, n_2) - E[x(n_1, n_2)])(y^*(k_1, k_2) - E^*[y(k_1, k_2)])].$$
$$\tag{6.44}$$

From (6.43) and (6.44), for zero-mean processes $x(n_1, n_2)$ and $y(n_1, n_2)$,

$$R_{xy}(n_1, n_2; k_1, k_2) = \gamma_{xy}(n_1, n_2; k_1, k_2). \tag{6.45}$$

For stationary processes $x(n_1, n_2)$ and $y(n_1, n_2)$,

$$R_{xy}(n_1, n_2) = E[x(k_1, k_2)y^*(k_1 - n_1, k_2 - n_2)] \quad \text{independent of } (k_1, k_2). \tag{6.46}$$

For ergodic processes $x(n_1, n_2)$ and $y(n_1, n_2)$,

$$R_{xy}(n_1, n_2) = E[x(k_1, k_2)y^*(k_1 - n_1, k_2 - n_2)]$$
$$= \lim_{N \to \infty} \frac{1}{(2N + 1)^2} \sum_{k_1 = -N}^{N} \sum_{k_2 = -N}^{N} x(k_1, k_2)y^*(k_1 - n_1, k_2 - n_2). \tag{6.47}$$

The *cross-power spectrum* of two jointly stationary processes $x(n_1, n_2)$ and $y(n_1, n_2)$, $P_{xy}(\omega_1, \omega_2)$, is defined by

$$P_{xy}(\omega_1, \omega_2) = F[R_{xy}(n_1, n_2)]$$
$$= \sum_{n_1 = -\infty}^{\infty} \sum_{n_2 = -\infty}^{\infty} R_{xy}(n_1, n_2)e^{-j\omega_1 n_1}e^{-j\omega_2 n_2}. \tag{6.48}$$

6.1.3 Random Signals as Inputs to Linear Systems

Consider a stationary complex random process $x(n_1, n_2)$ with mean of m_x and correlation of $R_x(n_1, n_2)$. Suppose we obtain a new random process $y(n_1, n_2)$ by passing $x(n_1, n_2)$ through an LSI system with the impulse response $h(n_1, n_2)$, as

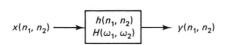

Figure 6.1 Deterministic linear shift-invariant system with an input random signal $x(n_1, n_2)$ and the corresponding output random signal $y(n_1, n_2)$.

shown in Figure 6.1. Clearly, $y(n_1, n_2)$ is related to $x(n_1, n_2)$ by

$$y(n_1, n_2) = h(n_1, n_2) * x(n_1, n_2)$$

$$= \sum_{k_1 = -\infty}^{\infty} \sum_{k_2 = -\infty}^{\infty} h(k_1, k_2) x(n_1 - k_1, n_2 - k_2).$$

(6.49)

Note that $x(n_1, n_2)$ is a random process, while $h(n_1, n_2)$ is a deterministic signal. In practice, $h(n_1, n_2)$ would typically be a real signal, but we will derive here a more general result that applies to a complex $h(n_1, n_2)$.

We wish to determine $E[y(n_1, n_2)]$, $R_y(n_1, n_2; k_1, k_2)$, $R_{xy}(n_1, n_2; k_1, k_2)$, and $R_{yx}(n_1, n_2; k_1, k_2)$. We shall find that $y(n_1, n_2)$ is a stationary random process. To determine $E[y(n_1, n_2)]$, the expectation operator is applied to (6.49), recognizing that $h(n_1, n_2)$ is a deterministic signal:

$$E[y(n_1, n_2)] = \sum_{k_1 = -\infty}^{\infty} \sum_{k_2 = -\infty}^{\infty} E[x(k_1, k_2)] h(n_1 - k_1, n_2 - k_2)$$

$$= m_x H(0, 0) \quad \text{for all } (n_1, n_2)$$

(6.50)

$$= m_y.$$

For a zero-mean $x(n_1, n_2)$, $y(n_1, n_2)$ is also zero mean.

To obtain $R_y(n_1, n_2; k_1, k_2)$ from (6.49),

$$R_y(n_1, n_2; k_1, k_2) = E[y(n_1, n_2) y^*(k_1, k_2)]$$

$$= \sum_{l_1 = -\infty}^{\infty} \sum_{l_2 = -\infty}^{\infty} \sum_{m_1 = -\infty}^{\infty} \sum_{m_2 = -\infty}^{\infty} h(l_1, l_2) h^*(m_1, m_2)$$

$$\cdot E[x(n_1 - l_1, n_2 - l_2) x^*(k_1 - m_1, k_2 - m_2)]$$

(6.51)

$$= \sum_{l_1 = -\infty}^{\infty} \sum_{l_2 = -\infty}^{\infty} \sum_{m_1 = -\infty}^{\infty} \sum_{m_2 = -\infty}^{\infty} h(l_1, l_2) h^*(m_1, m_2)$$

$$\cdot R_x(n_1 - k_1 - l_1 + m_1, n_2 - k_2 - l_2 + m_2).$$

From (6.51), $R_y(n_1, n_2; k_1, k_2)$ is a function of $n_1 - k_1$ and $n_2 - k_2$, and so, denoting $R_y(n_1, n_2; 0, 0)$ by $R_y(n_1, n_2)$, we can rewrite (6.51) as

$$R_y(n_1, n_2) = \sum_{l_1 = -\infty}^{\infty} \sum_{l_2 = -\infty}^{\infty} \sum_{m_1 = -\infty}^{\infty} \sum_{m_2 = -\infty}^{\infty} h(l_1, l_2) h^*(m_1, m_2)$$

$$\cdot R_x(n_1 - l_1 + m_1, n_2 - l_2 + m_2)$$

$$= R_x(n_1, n_2) * h(n_1, n_2) * h^*(-n_1, -n_2).$$

(6.52)

Similarly,

$$R_{xy}(n_1, n_2) = R_x(n_1, n_2) * h^*(-n_1, -n_2) \tag{6.53}$$

$$R_{yx}(n_1, n_2) = R_x(n_1, n_2) * h(n_1, n_2). \tag{6.54}$$

From (6.52), (6.53), and (6.54),

$$P_y(\omega_1, \omega_2) = P_x(\omega_1, \omega_2)|H(\omega_1, \omega_2)|^2 \tag{6.55}$$

$$P_{xy}(\omega_1, \omega_2) = P_x(\omega_1, \omega_2)H^*(\omega_1, \omega_2) \tag{6.56}$$

and
$$P_{yx}(\omega_1, \omega_2) = P_x(\omega_1, \omega_2)H(\omega_1, \omega_2). \tag{6.57}$$

Equation (6.55) is particularly important and is used extensively in many applications. It can also be used in showing that $P_x(\omega_1, \omega_2)$ and $P_y(\omega_1, \omega_2)$ are nonnegative and that it is reasonable to call them *power spectra* or *power spectral densities*.

6.1.4 The Noncausal Wiener Filter

Suppose we have a signal $s(n_1, n_2)$ and a noise $w(n_1, n_2)$ which are samples of zero-mean stationary random processes $s(n_1, n_2)$ and $w(n_1, n_2)$. The noisy observation $x(n_1, n_2)$ is given by

$$x(n_1, n_2) = s(n_1, n_2) + w(n_1, n_2). \tag{6.58}$$

We wish to determine $s(n_1, n_2)$ from $x(n_1, n_2)$ using a linear estimator given by

$$\hat{s}(n_1, n_2) = x(n_1, n_2) * h(n_1, n_2). \tag{6.59}$$

The linear estimator is an LSI system, since we are dealing with stationary random processes. The error criterion used is

$$\text{Error} = E[|e(n_1, n_2)|^2] \tag{6.60a}$$

$$e(n_1, n_2) = s(n_1, n_2) - \hat{s}(n_1, n_2) \tag{6.60b}$$

where $\hat{s}(n_1, n_2)$ denotes the estimate of $s(n_1, n_2)$. This is called a linear minimum mean square error estimation problem, since we are using a linear estimator and minimizing the mean squared error between $s(n_1, n_2)$ and $\hat{s}(n_1, n_2)$.

This signal estimation problem can be solved by using the orthogonality principle. This principle states that the Error in (6.60) is minimized by requiring that $e(n_1, n_2)$ be uncorrelated with any random variable in $x^*(n_1, n_2)$. From the orthogonality principle,

$$E[e(n_1, n_2)x^*(m_1, m_2)] = 0 \quad \text{for all } (n_1, n_2) \text{ and } (m_1, m_2). \tag{6.61}$$

From (6.59), (6.60b), and (6.61),

$$E[s(n_1, n_2)x^*(m_1, m_2)] = E[\hat{s}(n_1, n_2)x^*(m_1, m_2)]$$

$$= E[(h(n_1, n_2) * x(n_1, n_2))x^*(m_1, m_2)]$$

$$= \sum_{k_1 = -\infty}^{\infty} \sum_{k_2 = -\infty}^{\infty} h(k_1, k_2)E[x(n_1 - k_1, n_2 - k_2)x^*(m_1, m_2)].$$

$$(6.62)$$

Rewriting (6.62), we have

$$R_{sx}(n_1 - m_1, n_2 - m_2)$$

$$= \sum_{k_1 = -\infty}^{\infty} \sum_{k_2 = -\infty}^{\infty} h(k_1, k_2)R_x(n_1 - k_1 - m_1, n_2 - k_2 - m_2). \qquad (6.63)$$

From (6.63),

$$R_{sx}(n_1, n_2) = h(n_1, n_2) * R_x(n_1, n_2) \qquad (6.64)$$

and therefore

$$H(\omega_1, \omega_2) = \frac{P_{sx}(\omega_1, \omega_2)}{P_x(\omega_1, \omega_2)}. \qquad (6.65)$$

The filter $H(\omega_1, \omega_2)$ in (6.65) is called the *noncausal Wiener filter*.

Suppose $s(n_1, n_2)$ is uncorrelated with $w(n_1, n_2)$. Then we have

$$E[s(n_1, n_2)w^*(m_1, m_2)] = E[s(n_1, n_2)]E[w^*(m_1, m_2)]. \qquad (6.66)$$

From (6.66) and noting that $s(n_1, n_2)$ and $w(n_1, n_2)$ are zero-mean processes, we obtain

$$R_{sx}(n_1, n_2) = R_s(n_1, n_2) \qquad (6.67a)$$

$$R_x(n_1, n_2) = R_s(n_1, n_2) + R_w(n_1, n_2). \qquad (6.67b)$$

From (6.67),

$$P_{sx}(\omega_1, \omega_2) = P_s(\omega_1, \omega_2) \qquad (6.68a)$$

$$P_x(\omega_1, \omega_2) = P_s(\omega_1, \omega_2) + P_w(\omega_1, \omega_2). \qquad (6.68b)$$

From (6.65) and (6.68), the noncausal Wiener filter $H(\omega_1, \omega_2)$ is given by

$$H(\omega_1, \omega_2) = \frac{P_s(\omega_1, \omega_2)}{P_s(\omega_1, \omega_2) + P_w(\omega_1, \omega_2)}. \qquad (6.69)$$

Application of the noncausal Wiener filter in (6.69) to the signal estimation problem is shown in Figure 6.2. The noncausal Wiener filter has been the basis of many systems developed in such applications as speech enhancement and image restoration.

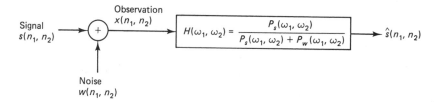

Figure 6.2   Application of the noncausal Wiener filter for linear minimum mean square error signal estimation.

These results can also be applied to solving the linear mean square error estimation problem when m_s and m_w, the means of the signal and noise processes, are nonzero.   The solution in this case involves subtracting $E[x(n_1, n_2)] = m_s + m_w$ from the noisy observation $x(n_1, n_2)$, applying the noncausal Wiener filter developed for zero-mean processes, and adding m_s to the resulting signal.   This is illustrated in Figure 6.3.

6.1.5 Statistical Parameter Estimation

Suppose $\boldsymbol{\theta}$ denotes a column vector of the parameters we wish to estimate and $\mathbf{x}$ denotes a column vector of the observations.   We assume that there is a probabilistic mapping between $\boldsymbol{\theta}$ and $\mathbf{x}$ with a point $\boldsymbol{\theta}$ in the parameter space mapped to a point $\mathbf{x}$ in the observation space.   The parameter estimation problem is to estimate $\boldsymbol{\theta}$ from the observation $\mathbf{x}$, using some estimation rule.   This is fundamentally different from the problem of estimating the filter coefficients discussed in Section 5.2.   In the filter design problem, the impulse response $h(n_1, n_2)$ is a deterministic function of the estimated parameters, and there is no probabilistic mapping between the parameters $\boldsymbol{\theta}$ and the observation $\mathbf{x}$.

The three estimation rules, known as maximum likelihood (ML), maximum a posteriori (MAP), and minimum mean square error (MMSE) estimation, have many desirable properties and have been extensively studied.   For nonrandom parameters, the ML estimation rule is often used.   In ML estimation, the parameter value is chosen such that it most likely results in the observation $\mathbf{x}$.   Thus, the value of $\boldsymbol{\theta}$ is chosen such that $p_{\mathbf{x}|\boldsymbol{\theta}}(\mathbf{x}_0|\boldsymbol{\theta}_0)$ is maximized for the observed $\mathbf{x}$ by the

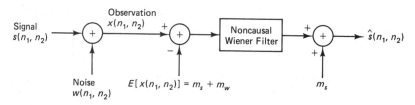

Figure 6.3   Linear minimum mean square error estimation of a signal when the signal and noise processes have nonzero means.

chosen value of θ. One important property of ML estimation is that the resulting estimate has the minimum variance under a certain set of conditions.

The MAP and MMSE estimation rules are commonly used for parameters that can be considered random variables whose a priori density functions are known. In the MAP estimation rule, the parameter value is chosen such that the a posteriori density $p_{\theta|x}(\theta_0|x_0)$ is maximized. ML and MAP estimation rules lead to identical estimates of the parameter value when the a priori density of the parameters in the MAP estimation rule is assumed to be flat over the parameter space. For this reason, the ML estimation rule is often viewed as a special case of the MAP estimation rule. In the MMSE estimation rule, the estimate $\hat{\theta}$ is obtained by minimizing the mean square error $E[(\hat{\theta} - \theta)^T(\hat{\theta} - \theta)]$. The MMSE estimate of θ is given by $E[\theta|x]$, the a posteriori mean of θ given x. Therefore, when the maximum of the a posteriori density function $p_{\theta|x}(\theta_0|x_0)$ coincides with its mean, the MAP estimation and MMSE estimation rules lead to identical estimates.

To illustrate these three estimation rules, let us consider a simple example. Suppose we have a biased coin that has probability θ of coming up heads and probability $1 - \theta$ of coming up tails in any given toss. We toss the coin N times, and we wish to estimate the parameter θ from the experimental outcome. Let a random variable x denote the experimental outcome as follows:

$$x = \begin{cases} 1, & \text{when the result is a head} \\ 0, & \text{when the result is a tail.} \end{cases} \tag{6.70}$$

The results of N tosses are denoted by $x_1, x_2, \ldots, x_N$. We will first assume that θ is nonrandom, and will estimate θ using the ML estimation rule. From (6.70), $p_{x|\theta}(x_0|\theta_0)$ is given by

$$p_{x|\theta}(x_0|\theta_0) = \theta_0^M (1 - \theta_0)^{N-M} \tag{6.71}$$

where

$$M = \sum_{i=1}^{N} x_i \tag{6.72}$$

and M represents the number of heads in N trials. To maximize $p_{x|\theta}(x_0|\theta_0)$ in (6.71) with respect to θ_0, we differentiate $p_{x|\theta}(x_0|\theta_0)$ with respect to θ_0, set the result to zero, and solve for θ_0. The result is

$$\hat{\theta}_{ML} = \frac{M}{N} = \frac{\sum\limits_{i=1}^{N} x_i}{N} \tag{6.73}$$

where $\hat{\theta}_{ML}$ is the ML estimate of θ. The result in (6.73) is quite reasonable. We count the number of heads and divide it by the total number of trials.

We will next assume that θ is random and that the a priori probability density function $p_\theta(\theta_0)$ is given by

$$p_\theta(\theta_0) = \begin{cases} 2\theta_0, & 0 \le \theta_0 \le 1 \\ 0, & \text{otherwise.} \end{cases} \tag{6.74}$$

To use the MAP estimation rule, we determine $p_{\theta|x}(\theta_0|x_0)$. From (6.13), we obtain

$$p_{\theta|x}(\theta_0|x_0) = \frac{p_{x|\theta}(x_0|\theta_0)p_\theta(\theta_0)}{p_x(x_0)}. \tag{6.75}$$

Since $p_x(x_0)$ in (6.75) does not depend on θ, maximizing $p_{\theta|x}(\theta_0|x_0)$ with respect to θ_0 is equivalent to maximizing $p_{x|\theta}(x_0|\theta_0)p_\theta(\theta_0)$ with respect to θ_0. From (6.71), (6.74), and (6.75), maximizing $p_{\theta|x}(\theta_0|x_0)$ with respect to θ_0 leads to

$$\hat{\theta}_{\text{MAP}} = \frac{M+1}{N+1} = \frac{\left[\sum_{i=1}^{N} x_i\right] + 1}{N+1}. \tag{6.76}$$

When $M = 1$ and $N = 2$ (one head out of two trials), $\hat{\theta}_{\text{ML}}$ is $\frac{1}{2}$, while $\hat{\theta}_{\text{MAP}}$ is $\frac{2}{3}$. The reason for the higher value for $\hat{\theta}_{\text{MAP}}$ is that $p_\theta(\theta_0)$ in (6.74) assumes higher probability for larger values of θ. If $p_\theta(\theta_0)$ is constant over $0 \le \theta_0 \le 1$, then $\hat{\theta}_{\text{MAP}}$ is identical to $\hat{\theta}_{\text{ML}}$.

The minimum mean square error estimate of θ, $\hat{\theta}_{\text{MMSE}}$, is given by

$$\hat{\theta}_{\text{MMSE}} = E[\theta|x] = \int_{\theta_0=0}^{1} \theta_0 p_{\theta|x}(\theta_0|x_0)\, d\theta_0 \tag{6.77}$$

where $p_{\theta|x}(\theta_0|x_0)$ is given by (6.75). Evaluating (6.77) for a general M and N is possible but somewhat involved, so we will consider the specific case of $M = 1$ and $N = 2$. Determining $\hat{\theta}_{\text{MMSE}}$ is typically more involved than determining $\hat{\theta}_{\text{ML}}$ or $\hat{\theta}_{\text{MAP}}$. From (6.71), (6.74), and (6.75),

$$p_{\theta|x}(\theta_0|x_0) = \frac{\theta_0(1-\theta_0)2\theta_0}{p_x(x_0)}. \tag{6.78}$$

To determine $p_x(x_0)$, note that $p_{\theta,x}(\theta_0, x_0)$ is given by $p_{x|\theta}(x_0|\theta_0)p_\theta(\theta_0)$ from (6.13). We have, then,

$$\begin{aligned} p_x(x_0) &= \int_{\theta_0=0}^{1} p_{\theta,x}(\theta_0, x_0)\, d\theta_0 = \int_{\theta_0=0}^{1} p_{x|\theta}(x_0|\theta_0)p_\theta(\theta_0)\, d\theta_0 \\ &= \int_{\theta_0=0}^{1} \theta_0(1-\theta_0)2\theta_0\, d\theta_0 = \frac{1}{6}. \end{aligned} \tag{6.79}$$

From (6.78) and (6.79),

$$p_{\theta|x}(\theta_0|x_0) = 12\theta_0^2(1-\theta_0). \tag{6.80}$$

From (6.77) and (6.80),

$$\hat{\theta}_{\text{MMSE}} = \int_{\theta_0=0}^{1} 12\theta_0^3(1-\theta_0)\, d\theta_0 = \frac{3}{5}. \tag{6.81}$$

The estimate $\hat{\theta}_{\text{MMSE}}$ for this case is greater than $\hat{\theta}_{\text{ML}}$, but is less than $\hat{\theta}_{\text{MAP}}$.

6.2 SPECTRAL ESTIMATION METHODS

Let $x(n_1, n_2)$ denote a stationary random process. The correlation sequence and power spectrum of the random process x are denoted by $R_x(n_1, n_2)$ and $P_x(\omega_1, \omega_2)$, respectively. The spectral estimation problem is to estimate $P_x(\omega_1, \omega_2)$ in (6.30) from a segment of a sample of the random process x. Specifically, with $x(n_1, n_2)$ denoting a sample of the random process x, the problem can be stated as

Problem Statement 1. Given $x(n_1, n_2)$ for $(n_1, n_2) \in D$, estimate $P_x(\omega_1, \omega_2)$.

The notation D represents the region in which the data are assumed available. In the spectral estimation problem, we hope to estimate $P_x(\omega_1, \omega_2)$ from a sample of the random process x. For this reason, the random process is typically assumed to be ergodic. For an ergodic process x, the ensemble average is the same as the time (or space) average [see (6.28) and (6.29)], and $R_x(n_1, n_2)$ can be determined from $x(n_1, n_2)$, a sample of the random process.

For the purpose of developing some spectral estimation techniques, it is useful to phrase the spectral estimation problem as

Problem Statement 2. Given $R_x(n_1, n_2)$ for $(n_1, n_2) \in C$, estimate $P_x(\omega_1, \omega_2)$.

The notation C represents the region in which the correlation is available. Problem Statement 2 is clearly related to the original problem statement, since $R_x(n_1, n_2)$ can be estimated more accurately and for a larger region C as the amount of $x(n_1, n_2)$ available increases. In practice, a segment of $R_x(n_1, n_2)$ is not available. To use spectral estimation techniques based on Problem Statement 2, therefore, $R_x(n_1, n_2)$ must first be estimated from $x(n_1, n_2)$. Issues related to estimating $R_x(n_1, n_2)$ from $x(n_1, n_2)$ are discussed in Section 6.4.1.

Many approaches to spectral estimation exist. We discuss some of them in this section. We treat Fourier-transform-based conventional techniques, dimension-dependent processing, the maximum likelihood method (MLM), autoregressive (AR) signal modeling, data or correlation extension, and the maximum entropy method (MEM).

Conceptually, 2-D spectral estimation techniques are straightforward extensions of 1-D techniques, or can be derived straightforwardly from them. For some methods, however, 2-D spectral estimation differs considerably in computational complexity or properties from 1-D spectral estimation. We give special attention to the methods for which the extension is not straightforward.

To illustrate the types of spectral estimates that can be expected from the various techniques, we show examples of spectral estimates. These examples, however, are only for illustrative purposes and are not intended to provide a comparison of the performance of different methods. A quantitative comparison of the resolution capabilities of several different methods is given in Section 6.3.

The data used in the examples in this section and those following are derived from sinusoids buried in white Gaussian noise. Signals of this class are used because they occur in some array processing applications, and they are often used in the literature to illustrate the performance of a spectral estimation technique. In an application where a signal of this type occurs, what we hope to determine from the power spectrum is the number of sinusoids, the frequencies of sinusoids, and, possibly, the amplitudes of the sinusoids. The complex data $x(n_1, n_2)$ used is given by

$$x(n_1, n_2) = \sum_{i=1}^{M} a_i e^{j(\omega_{i1}n_1 + \omega_{i2}n_2 + \phi_i)} + w(n_1, n_2) \tag{6.82}$$

where a_i^2, ω_{i1} and ω_{i2} denote the power and frequency location of the ith sinusoid, M is the number of sinusoids, ϕ_i is the phase term associated with the ith sinusoid, and $w(n_1, n_2)$ denotes zero-mean white noise of power σ^2. For the case of real data, $x(n_1, n_2)$ used is given by

$$x(n_1, n_2) = \sum_{i=1}^{M} \sqrt{2} \, a_i \cos(\omega_{i1}n_1 + \omega_{i2}n_2 + \phi_i) + w(n_1, n_2). \tag{6.83}$$

For both complex and real data, the data are assumed to be given over the region $(n_1, n_2) \in D$. The region D, unless otherwise noted, is a square, symmetric about the origin.

Since many spectral estimation techniques are based on the assumption that a segment of the correlation sequence is available, it is often useful to assume that the exact correlation sequence is available over a given region, which we will denote by C. For the case of M complex sinusoids in noise, the exact correlation sequence is given by

$$R_x(n_1, n_2) = \sum_{i=1}^{M} a_i^2 \, e^{j(\omega_{i1}n_1 + \omega_{i2}n_2)} + \sigma^2 \, \delta(n_1, n_2). \tag{6.84}$$

For the case of real sinusoids, the exact correlation sequence is given by

$$R_x(n_1, n_2) = \sum_{i=1}^{M} a_i^2 \cos(\omega_{i1}n_1 + \omega_{i2}n_2) + \sigma^2 \, \delta(n_1, n_2). \tag{6.85}$$

In both (6.84) and (6.85), $R_x(n_1, n_2)$ is assumed to be known for $(n_1, n_2) \in C$. The region C, unless otherwise noted, is again a square, symmetric about the origin.

Strictly speaking, sinusoids buried in noise cannot be considered as a stationary random process. In (6.82), for example, $E[x(n_1, n_2)]$ is not independent of n_1 and n_2, and $E[x(k_1, k_2)x^*(k_1 - n_1, k_2 - n_2)]$ is not independent of k_1 and k_2, for deterministic a_i and ϕ_i. Therefore, we cannot express the correlation sequence of such a random process with $R_x(n_1, n_2)$ as in (6.84), and cannot define the power spectrum $P_x(\omega_1, \omega_2)$ as the Fourier transform of $R_x(n_1, n_2)$. If the process is not stationary, it is of course not ergodic. One way to get around this problem is to define a sequence $\theta_x(n_1, n_2)$ by

$$\theta_x(n_1, n_2) = \lim_{N \to \infty} \frac{1}{(2N+1)^2} \sum_{k_1=-N}^{N} \sum_{k_2=-N}^{N} x(k_1, k_2)x^*(k_1 - n_1, k_2 - n_2). \tag{6.86}$$

If $x(n_1, n_2)$ were an ergodic random process, from (6.29), $\theta_x(n_1, n_2)$ would be identical to the correlation sequence $R_x(n_1, n_2)$. If we blindly apply (6.86) to (6.82) and (6.83), then the resulting $\theta_x(n_1, n_2)$ will be independent of k_1 and k_2, and will be given by (see Problem 6.16) the right-hand side expression of (6.84) and (6.85), respectively. Since (6.86) is identical to (6.29), and the spectral contents of $\theta_x(n_1, n_2)$ are very simply related to the spectral contents of the signal (sinusoids) in the data, we simply denote $\theta_x(n_1, n_2)$ as $R_x(n_1, n_2)$ and call it the correlation sequence of $x(n_1, n_2)$. If the data consist of signal and noise where both the signal and noise are samples of ergodic processes, this problem does not arise and the special interpretation of $R_x(n_1, n_2)$ above is not needed.

It is important to note that the methods discussed in this section apply to other types of data as well and they do not exploit the specific structure available in the data. If we know in advance that the correlation sequence is in the form of (6.84) or (6.85), and know enough exact values of $R_x(n_1, n_2)$, then we can exactly determine the frequencies and amplitudes of the sinusoids. We emphasize that the data of sinusoids in noise are used primarily to illustrate some characteristics of spectral estimation techniques.

The signal-to-noise ratio (SNR) is defined as the sum of the powers of each sinusoid divided by the noise power. Specifically, for the case of M sinusoids with a_i^2 representing the power of the ith sinusoid, the SNR is given by

$$\text{SNR} = \frac{\sum_{i=1}^{M} a_i^2}{\sigma^2} \tag{6.87}$$

where σ^2 is the noise power.

All the spectral estimates in the examples are displayed in the form of contour plots, with the highest contour level normalized to zero dB. Whenever the length of the contour permits, it is labeled with the contour level in dB below the maximum (0 dB). The contours are always equally spaced, and the increment between contours (CINC) in dB is always noted. In all the examples, the true peak location is noted with "x." For real data, the spectral estimates are symmetric with respect to the origin, and thus only half the 2-D frequency plane is displayed. The full 2-D plane is displayed for spectral estimates obtained from complex data.

6.2.1 Conventional Methods

Methods based on the Fourier transform were the first to be used for spectral estimation, and are known collectively as *conventional methods*. Since the Fourier transform can be computed by exploiting the computational efficiency of FFT (fast Fourier transform) algorithms in both 1-D and 2-D, its computational simplicity extends straightforwardly, except that the amount of data is generally greater in 2-D than in 1-D. The general behavior of the method has also been observed to extend in a straightforward manner from 1-D to 2-D.

The simplest of the conventional methods is the periodogram. Suppose $x(n_1, n_2)$ is given for $(n_1, n_2) \in D$. In the periodogram, $x(n_1, n_2)$ is assumed to be non-zero for $(n_1, n_2) \in D$ and is zero otherwise. The spectral estimate $\hat{P}_x(\omega_1, \omega_2)$ is then obtained by

$$\hat{P}_x(\omega_1, \omega_2) = \frac{1}{N} |X(\omega_1, \omega_2)|^2 \tag{6.88}$$

where N is the number of points in the region D. The periodogram $\hat{P}_x(\omega_1, \omega_2)$ in (6.88) is the Fourier transform of $\hat{R}_x(n_1, n_2)$ given by

$$\hat{R}_x(n_1, n_2) = \frac{1}{N} \sum_{k_1 = -\infty}^{\infty} \sum_{k_2 = -\infty}^{\infty} x(k_1, k_2) x^*(k_1 - n_1, k_2 - n_2). \tag{6.89}$$

Because of the windowing, the minimum frequency separation at which samples of $\hat{P}_x(\omega_1, \omega_2)$ are independent for a data size of $N_1 \times N_2$ points is on the order of $2\pi/N_1$ in ω_1 and $2\pi/N_2$ in ω_2. For small N_1 or N_2, therefore, the resolution of $P_x(\omega_1, \omega_2)$ can be rather poor. This is a problem in all conventional methods. In addition, the variance of $\hat{P}_x(\omega_1, \omega_2)$ is on the order of $P_x(\omega_1, \omega_2)$ independent of the size of D, which limits the usefulness of the periodogram as a spectral estimation method.

There are various ways of improving the variance of $\hat{P}_x(\omega_1, \omega_2)$ by sacrificing some resolution. One method is periodogram averaging. In this method, the region D is divided into smaller segments. The periodogram is obtained for each of the smaller segments, and the spectral estimate $\hat{P}_x(\omega_1, \omega_2)$ is taken as the average of the periodograms. If the data are divided into $K_1 \times K_2$ sections, the variance is reduced by a factor of $K_1 K_2$ and the resolution (inversely related to the minimum frequency separation at which $\hat{P}_x(\omega_1, \omega_2)$ is independent) is reduced by a factor of K_1 in ω_1 and K_2 in ω_2.

Another way to trade the variance and resolution of $\hat{P}_x(\omega_1, \omega_2)$ is by directly smoothing the periodogram $\hat{P}_x(\omega_1, \omega_2)$ in the frequency domain with some suitably chosen smoothing operator. An alternative method is correlation windowing. Since the correlation sequence and power spectrum are related by the Fourier transform, windowing the correlation corresponds to convolving, or smoothing, the power spectrum with the Fourier transform of the window. To use this method, a 2-D window must be designed. A 2-D window is typically designed from 1-D windows, as shown in Section 4.3.1. One way to apply this method to periodogram smoothing is to perform inverse Fourier transform operation of the periodogram $\hat{P}_x(\omega_1, \omega_2)$ to obtain an estimate of $R_x(n_1, n_2)$, window the estimated $\hat{R}_x(n_1, n_2)$, and then Fourier transform the resulting $\hat{R}_x(n_1, n_2)$. The trade-off between variance and resolution depends on the smoothing operator used. The performance of the correlation windowing method is similar to that of periodogram averaging.

A number of variations of the conventional methods exist. In all cases, they make trade-offs between the variance and resolution of the spectral estimate, and the product of the variance and resolution of the spectral estimate is about the same to a first-order approximation.

Figure 6.4(a) shows the spectral estimate obtained by a correlation windowing method. In obtaining the spectral estimate, real data derived from two sinusoids buried in white Gaussian noise are used. Specifically, the data $x(n_1, n_2)$ were generated for $0 \leq n_1 \leq 5$, $0 \leq n_2 \leq 5$ (Region D) by using (6.83) with $M = 2$, $a_1 = 1$, $a_2 = 1$, $(\omega_{11}/2\pi, \omega_{12}/2\pi) = (-0.2, 0.2)$, $(\omega_{21}/2\pi, \omega_{22}/2\pi) = (0.3, 0.3)$, $\phi_i = 0$, $\sigma^2 = 1$ and SNR = 3 dB. From $x(n_1, n_2)$ generated above, first a periodogram is obtained and then the correlation is estimated by inverse Fourier transforming the periodogram. The resulting estimated correlation is windowed by a 2-D separable triangular window of size 9×9 (Region C). The windowed correlation is then Fourier transformed to obtain the spectral estimate, which is displayed in Figure 6.4(a). Figure 6.4(b) shows the spectral estimate obtained by applying a 2-D separable triangular window directly to the exact correlation given by (6.85). All the parameters used, including the window size for Figure 6.4(b), are identical to those used for Figure 6.4(a). If we increase the data size without changing the shape and size of the window, the spectral estimate in Figure 6.4(a) will approach that in Figure 6.4(b). This is because the correlation estimated from the data will approach the exact correlation as we increase the data size. The effect of data size on the resulting spectral estimate is further discussed in Section 6.4.1.

When a large amount of data is available, so that many correlation points can be estimated reliably, conventional methods perform very well in both variance and resolution. In addition, such methods are quite simple computationally, since the major computations involved are Fourier transforms, which can be computed using FFT algorithms. In 2-D, however, the data are often limited in size. For example, in array processing, the amount of data available in the spatial dimension is limited by the number of sensors. Techniques that give higher resolution than do conventional techniques are called *modern* or *high resolution* spectral estimation techniques. These are discussed in the next sections.

6.2.2 Dimension-Dependent Processing

In some spectral estimation problems, there may be much more data along one dimension than along others. In such cases, it is reasonable to apply different spectral estimation techniques along different dimensions. For example, in array processing, there are typically much more data along the time dimension than along the spatial dimension. We may wish to apply Fourier-transform-based conventional techniques along the time dimension and high resolution spectral estimation techniques, such as the maximum entropy method (MEM), along the spatial dimensions.

In applying different spectral estimation techniques along different dimensions, it is important to preserve the phase information until the data have been processed for all dimensions. The necessity of preserving the phase information can be shown by rewriting the relationship between the correlation sequence $R_x(n_1, n_2)$ and the power spectrum $P_x(\omega_1, \omega_2)$ as

$$P_x(\omega_1, \omega_2) = \sum_{n_1 = -\infty}^{\infty} e^{-j\omega_1 n_1} \phi_x(n_1, \omega_2) \qquad (6.90)$$

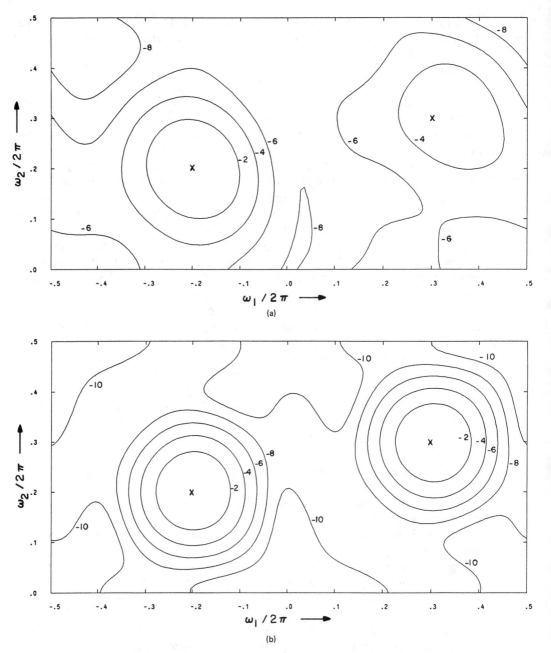

Figure 6.4 Conventional spectral estimates by correlation windowing. Parameters used for data generation: M (number of sinusoids) = 2; a_1 (amplitude of the first sinusoid) = 1; $a_2 = 1$; $(\omega_{11}/2\pi, \omega_{12}/2\pi) = (-0.2, 0.2)$; $(\omega_{21}/2\pi, \omega_{22}/2\pi) = (0.3, 0.3)$; ϕ_i (phase of the ith sinusoid) = 0; σ^2 (noise power) = 1; SNR = 3 dB; correlation size (region C), 9 × 9; CINC (increments in decibels between contours) = 2 dB. (a) Estimated correlation from data size of 6 × 6 was used; (b) exact correlation was used.

where $\phi_x(n_1, \omega_2)$ is given by

$$\phi_x(n_1, \omega_2) = \sum_{n_2 = -\infty}^{\infty} R_x(n_1, n_2)e^{-j\omega_2 n_2}. \qquad (6.91)$$

The function $\phi_x(n_1, \omega_2)$ can be determined by applying 1-D Fourier transform operations to $R_x(n_1, n_2)$ along the n_2 dimension, and $P_x(\omega_1, \omega_2)$ can be determined by applying 1-D Fourier transform operations to $\phi_x(n_1, \omega_2)$ along the n_1 dimension. Although $P_x(\omega_1, \omega_2)$ is real and nonnegative, the intermediate function $\phi_x(n_1, \omega_2)$ is in general complex. If $\phi_x(n_1, \omega_2)$ is forced to be real and nonnegative in (6.91), the expression on the right-hand side of (6.90) will not be equal to $P_x(\omega_1, \omega_2)$.

For dimension-dependent processing, it is not necessary to apply a 1-D spectral estimation method to each dimension. For 3-D data, for example, a 1-D spectral estimation method may be applied along one dimension and a 2-D spectral estimation method along the remaining two dimensions. In a planar array environment, where the data are 3-D (one time dimension and two spatial dimensions), for example, first a conventional 1-D spectral estimation method can be applied along the time dimension, and then a 2-D high-resolution spectral estimation technique, such as the maximum likelihood method (MLM), can be applied along the two spatial dimensions.

6.2.3 Maximum Likelihood Method

The maximum likelihood method (MLM) was originally developed [Capon] for nonuniformly sampled data. "Maximum likelihood" is a misnomer, in that this method is not an ML estimator as discussed in Section 6.1. The computation and performance aspects of this method are essentially the same in both 1-D and 2-D except for the amount of data involved.

Consider a specific frequency (ω_1', ω_2') at which we wish to estimate the power spectrum $P_x(\omega_1, \omega_2)$. To develop the MLM, we first obtain $y(n_1, n_2)$ by linearly combining (filtering) the random signal $x(n_1, n_2)$ by

$$y(n_1, n_2) = \sum_{(k_1, k_2) \in B} \sum b(k_1, k_2)x(n_1 - k_1, n_2 - k_2) \qquad (6.92)$$

where B is a specified region, to be discussed later. The coefficients $b(n_1, n_2)$ in (6.92) are determined by minimizing $E[|y(n_1, n_2)|^2]$, the average power in the output, subject to the constraint that $b(n_1, n_2)$ pass the frequency component of $x(n_1, n_2)$ at (ω_1', ω_2') without attenuation, that is,

$$\begin{align} &\text{Minimize} \quad E[|y(n_1, n_2)|^2] \\ &b(k_1, k_2) \end{align} \qquad (6.93)$$

$$\text{Subject to } B(\omega_1', \omega_2') = \sum_{(n_1, n_2) \in B} \sum b(n_1, n_2)e^{-j\omega_1' n_1}e^{-j\omega_2' n_2} = 1. \qquad (6.94)$$

The average power $E[|y(n_1, n_2)|^2]$ with this choice of $b(n_1, n_2)$ is then considered the spectral estimate $\hat{P}_x(\omega_1', \omega_2')$. In essence, the MLM estimates $P_x(\omega_1, \omega_2)$ by designing a separate finite impulse response (FIR) filter at each frequency (ω_1', ω_2') such that the filter will pass the frequency component (ω_1', ω_2') unaffected and will reject as much as possible of the power resulting from other frequency components. In this respect, the technique is fundamentally different from the conventional spectral estimation techniques discussed in the previous section. In the periodogram, for example, the window applied to the data is fixed. Applying a fixed window to the data and computing its Fourier transform can be interpreted as filtering the data with a set of bandpass filters which are identical except for the passband center frequencies. In the MLM, the bandpass filters designed adapt to the data given, and they differ depending on the center frequency of the bandpass filter.

Equations (6.93) and (6.94) can be expressed more compactly by using matrix notation. Specifically, let M denote the total number of points in the region B in (6.92), and let these M points be ordered arbitrarily. We will denote the ith value of (n_1, n_2) in B as $(n_1, n_2)^i$ or (n_{1i}, n_{2i}). Equations (6.93) and (6.94) can then be stated as

$$\text{Minimize} \quad \mathbf{b}^H R \mathbf{b} \atop \mathbf{b} \tag{6.95}$$

$$\text{Subject to the constraint} \quad \mathbf{e}^H \mathbf{b} = 1 \tag{6.96}$$

where $\mathbf{b}$ is an $M \times 1$ column vector whose element b_i is given by

$$b_i = b(n_1, n_2)^i \tag{6.97}$$

R is an $M \times M$ Hermitian symmetric matrix whose (i, j)th element R_{ij} is given by

$$R_{ij} = E[x(n_1, n_2)^i x^*(n_1, n_2)^j] \tag{6.98}$$

$\mathbf{e}$ is an $M \times 1$ column vector whose ith element e_i is given by

$$e_i = e^{-j\omega_1' n_{1i} - j\omega_2' n_{2i}} \tag{6.99}$$

and $\mathbf{b}^H$ and $\mathbf{e}^H$ represent the Hermitian transpose of $\mathbf{b}$ and $\mathbf{e}$, respectively.

Since (6.96) is a linear constraint for $b(n_1, n_2)$, and $E[|y(n_1, n_2)|^2]$ is a quadratic form of $b(n_1, n_2)$, the constrained optimization problem of (6.95) and (6.96) is a linear problem. The solution is given by

$$\mathbf{b} = \frac{R^{-1}\mathbf{e}}{\mathbf{e}^H R^{-1}\mathbf{e}} \tag{6.100}$$

and the resulting spectral estimate $\hat{P}_x(\omega_1', \omega_2')$ is given by

$$\hat{P}_x(\omega_1', \omega_2') = \frac{1}{\mathbf{e}^H R^{-1}\mathbf{e}}. \tag{6.101}$$

The elements in R in (6.101) are the correlation points of the random process x. Since R does not depend on (ω_1, ω_2), R^{-1} in (6.101) needs to be computed only once. Furthermore, $\hat{P}_x(\omega_1, \omega_2)$ in (6.101) can be computed [Musicus] by using FFT algorithms in some cases by exploiting special structures in the matrix R.

In (6.92), we wish to choose as large a region B as possible, since we can minimize $E[|y(n_1, n_2)|^2]$ better with a larger B. A larger B, however, corresponds to a larger matrix R. Since all the elements in R are assumed known in (6.101), the size of B can be increased only to the extent that all the elements in R can be reliably estimated from the available data. The elements in R are $R_x(n_1 - l_1, n_2 - l_2)$ for $(n_1, n_1) \in B$ and $(l_1, l_2) \in B$. If the data $x(n_1, n_2)$ or its shifted version is available for a region that includes B, then all the elements in R may be estimated from the available data.

In array processing applications, the data along the time dimension are typically processed first by using 1-D conventional methods. The available data $x(n_1, n_2)$ typically represent the data at the sensor locations (n_1, n_2) at a certain temporal frequency. If the region D represents all the sensor locations, some elements in the matrix R in (6.101) must be estimated from only one sample; that is, $R_x(n_1 - l_1, n_2 - l_2) \approx x(n_1, n_2) x^*(l_1, l_2)$. This is a very poor estimate of $R_x(n_1 - l_1, n_2 - l_2)$ and can degrade the performance of the MLM. This problem is more serious when the sensors are nonuniformly located. Spacing the sensors nonuniformly leads to a larger number of distinct elements in $R_x(n_1 - l_1, n_2 - l_2)$ for a given number of sensors. This trade-off should be considered in designing arrays.

The MLM generally performs better than the Fourier-transform-based conventional methods. This is illustrated in the following example. Figure 6.5(a) shows the spectral estimate obtained from the correlation windowing method using the 5×5-point segment of the exact correlation given by (6.85) with $M = 2$, $a_1 = 1$, $a_2 = 1$, $(\omega_{11}/2\pi, \omega_{12}/2\pi) = (0.1, 0.1)$, $(\omega_{21}/2\pi, \omega_{22}/2\pi) = (0.25, 0.35)$, $\sigma^2 = 0.5$, and SNR = 6 dB. Figure 6.5(b) shows the spectral estimate obtained by the MLM using exactly the same correlation segment used in Figure 6.5(a). The two figures illustrate the higher resolution characteristics of the MLM relative to a conventional spectral estimation technique.

As will be discussed in Section 6.2.6, the MLM does not perform as well as the maximum entropy method (MEM) in resolving peaked spectra. Nevertheless, in practice the MLM is used more often than the MEM because of its computational simplicity. One method for retaining the computational simplicity of the MLM while improving resolution properties was developed by heuristically extending the relationship that exists between the MLM and the MEM in 1-D to 2-D signals.

Specifically, for 1-D signals whose correlation $R_x(n)$ is either known or estimated for $-N \le n \le N$, Burg (1972) has shown the following relationship:

$$\frac{1}{\text{MLM}(\omega : N)} = \sum_{p=0}^{N} \frac{1}{\text{MEM}(\omega : p)} \tag{6.102}$$

where $\text{MLM}(\omega : N)$ represents the MLM spectral estimate, and $\text{MEM}(\omega : p)$ represents the MEM spectral estimate based on $R_x(n)$ for $-p \le n \le p$. By rewriting (6.102), we can express $\text{MEM}(\omega : N)$ in terms of $\text{MLM}(\omega : N)$ and $\text{MLM}(\omega : N - 1)$ as

$$\text{MEM}(\omega : N) = \frac{1}{\dfrac{1}{\text{MLM}(\omega : N)} - \dfrac{1}{\text{MLM}(\omega : N - 1)}}. \tag{6.103}$$

Equation (6.103) is the basis for the algorithm.

Now suppose that $R_x(n_1, n_2)$, the correlation of a 2-D signal, is either known or estimated for $(n_1, n_2) \in C$. A straightforward extension of (6.103) to 2-D signals

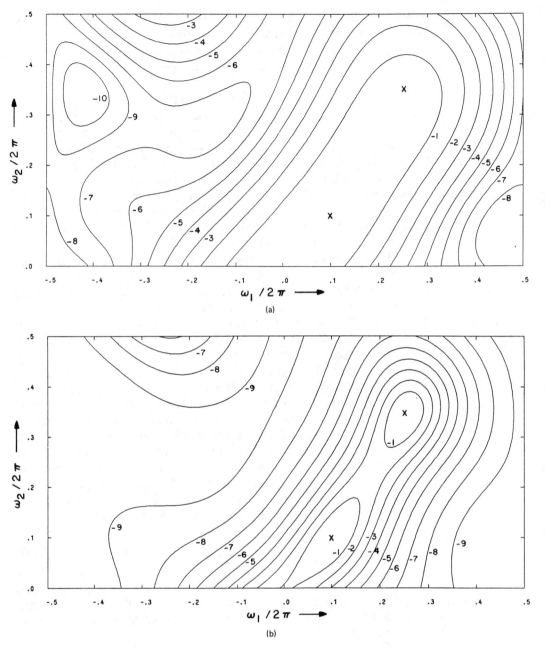

Figure 6.5 Correlation windowing and MLM spectral estimates based on exact correlation points. Parameters used for correlation generation: $M = 2$; $a_1 = 1$; $a_2 = 1$; $(\omega_{11}/2\pi, \omega_{12}/2\pi)$ = $(0.1, 0.1)$; $(\omega_{21}/2\pi, \omega_{22}/2\pi)$ = $(0.25, 0.35)$; $\sigma^2 = 0.5$; SNR = 6 dB; correlation size, 5×5; CINC = 1 dB. (a) Correlation windowing; (b) MLM.

can be expressed as

$$IMLM(\omega_1, \omega_2 : C, C') = \cfrac{1}{\cfrac{1}{MLM(\omega_1, \omega_2 : C)} - \cfrac{1}{MLM(\omega_1, \omega_2 : C')}} \qquad (6.104)$$

where $MLM(\omega_1, \omega_2 : C)$ represents the MLM spectral estimate based on all the known correlation points, $MLM(\omega_1, \omega_2 : C')$ represents the MLM spectral estimate with C' chosen as a subset of C, and $IMLM(\omega_1, \omega_2 : C, C')$ is the new 2-D spectral estimate. Since the new spectral estimate is not the same as the MEM spectral estimate and, with a proper choice of C', has better resolution properties than the MLM spectral estimate, the new spectral estimate is referred to as the "improved MLM" (IMLM). From (6.104) it appears that the IMLM spectral estimate requires about twice the computation required for the MLM. However, when the two terms in the right-hand side of (6.104) are combined appropriately, the amount of computation required for the IMLM can be shown [Dowla] to be essentially the same as that required for the MLM.

In using (6.104) as a method for spectral estimation, we must ensure that the resulting $IMLM(\omega_1, \omega_2 : C, C')$ will be nonnegative. By rewriting (6.104) in matrix form and with some algebraic manipulation, the $IMLM(\omega_1, \omega_2 : C, C')$ can be shown to always be nonnegative as long as C' is a subset of C. Another issue to consider is the choice of C'. From (6.103), it is reasonable to require the size of C' to be less than but close to the size of C. In addition, it has been empirically observed that the correlation points that are in C but not in C' should be chosen to be as symmetric as possible with respect to the origin. For example, when $R_x(n_1, n_2)$ is given for $-N \le n_1 \le N$, $-N \le n_2 \le N$ (Region C), the choice of C' that excludes $R_x(N, N)$, $R_x(-N, N)$, $R_x(-N, -N)$, and $R_x(N, -N)$ from Region C is a good one and meets both of the above requirements.

The performance of the IMLM in resolving peaked spectra appears to be somewhere between the MLM and MEM, closer to the MLM for a small C (below 7×7) and closer to the MEM for a large C (above 7×7).

The MLM and the IMLM are compared in the example in Figure 6.6. Figure 6.6(a) shows the spectral estimate obtained from the MLM by using the 5×5-point segment (Region C) of the exact correlation given by (6.85) with $M = 2$, $a_1 = 1$, $a_2 = 1$, $(\omega_{11}/2\pi, \omega_{12}/2\pi) = (0.2, 0.2)$, $(\omega_{21}/2\pi, \omega_{22}/2\pi) = (0.3, 0.3)$, $\sigma^2 = 0.25$, and SNR $= 9$ dB. Figure 6.6(b) shows the spectral estimate obtained by the IMLM using exactly the same correlation segment used in Figure 6.6(a). The choice of C' was made by excluding $R_x(2, 2)$, $R_x(-2, 2)$, $R_x(-2, -2)$, and $R_x(2, -2)$ from Region C. The two figures illustrate the higher resolution characteristics of the IMLM relative to the MLM.

6.2.4 Spectral Estimation Based on Autoregressive Signal Modeling

If the random process x can be modeled as the response of a parametric system excited by white noise, one approach to estimating $P_x(\omega_1, \omega_2)$ is to estimate the system model parameters from which $P_x(\omega_1, \omega_2)$ can be determined. The model

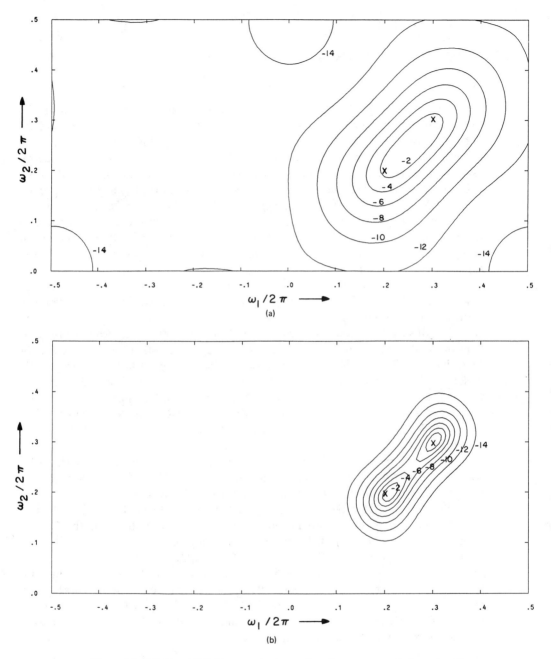

Figure 6.6 MLM and IMLM spectral estimates based on exact correlation points. Parameters used for correlation generation: $M = 2$; $a_1 = 1$; $a_2 = 1$; $(\omega_{11}/2\pi, \omega_{12}/2\pi) = (0.2, 0.2)$; $(\omega_{21}/2\pi, \omega_{22}/2\pi) = (0.3, 0.3)$; $\sigma^2 = 0.25$; SNR = 9 dB; correlation size, 5 × 5; CINC = 2 dB. (a) MLM; (b) IMLM.

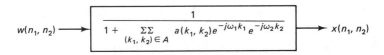

Figure 6.7 Two-dimensional autoregressive random process.

most often used is autoregressive (AR), since estimating the AR model parameters is a simple linear problem for many cases of interest.

In AR signal modeling, the random process x is considered to be the response of an AR model excited by white noise $w(n_1, n_2)$ with variance σ^2, as shown in Figure 6.7. From $x(n_1, n_2)$, $a(n_1, n_2)$ for $(n_1, n_2) \in A$ and σ^2 are estimated. Since the input random process has a constant power spectrum of amplitude σ^2, from (6.55) the spectral estimate $\hat{P}_x(\omega_1, \omega_2)$ is given by

$$\hat{P}_x(\omega_1, \omega_2) = \frac{\sigma^2}{|1 + \sum\limits_{(n_1,n_2) \in A} \sum a(n_1, n_2)e^{-j\omega_1 n_1}e^{-j\omega_2 n_2}|^2}. \tag{6.105}$$

To estimate $a(n_1, n_2)$, we express the system by a difference equation as

$$x(n_1, n_2) = -\sum_{(k_1,k_2) \in A} \sum a(k_1, k_2)x(n_1 - k_1, n_2 - k_2) + w(n_1, n_2). \tag{6.106}$$

Suppose the difference equation in (6.106) is recursively computable, so that $x(n_1, n_2)$ can always be computed from previously computed values and an appropriate set of boundary conditions. Multiplying both sides of (6.106) with $x^*(n_1 - l_1, n_2 - l_2)$ and taking the expectation, we have

$$R_x(l_1, l_2) = -\sum_{(k_1,k_2) \in A} \sum a(k_1, k_2)R_x(l_1 - k_1, l_2 - k_2)$$

$$+ E[w(n_1, n_2)x^*(n_1 - l_1, n_2 - l_2)]. \tag{6.107}$$

If we choose (l_1, l_2) such that $x^*(n_1 - l_1, n_2 - l_2)$ represents a previously computed point relative to $x(n_1, n_2)$, $x^*(n_1 - l_1, n_2 - l_2)$ is uncorrelated with $w(n_1, n_2)$. Since $w(n_1, n_2)$ is assumed to be white noise with zero mean, for such values of (l_1, l_2), (6.107) reduces to

$$R_x(l_1, l_2) = -\sum_{(k_1,k_2) \in A} \sum a(k_1, k_2)R_x(l_1 - k_1, l_2 - k_2). \tag{6.108}$$

Equation (6.108) is a linear set of equations for $a(n_1, n_2)$. Once $a(n_1, n_2)$ is determined, σ^2 can be obtained from (6.107) with $l_1 = l_2 = 0$. When $l_1 = l_2 = 0$, $E[w(n_1, n_2)x^*(n_1 - l_1, n_2 - l_2)]$ in (6.107) becomes σ^2. Equation (6.108) is based on the recursive computability assumption, which limits the shape of Region A. Computing $a(n_1, n_2)$ for a nonrecursively computable system is a nonlinear problem. Partly for this reason, (6.107) is sometimes used for any shape of Region A.

Equation (6.108) is called the normal equation and can be very easily related to (5.21), which is used in designing an IIR filter, as discussed in Section 5.2.1. However, there are two major differences between the filter design problem and AR modeling for spectral estimation. The filter design problem is a deterministic problem in the sense that the input to the system is a fixed known sequence such as $\delta(n_1, n_2)$, so the parameters (filter coefficients) completely determine the output of the system, which is compared to a fixed known desired output sequence. In AR modeling for spectral estimation, the input is assumed to be a random signal, and the output of the system is characterized by the model parameters only in a stochastic sense. Another difference is that the designed filter is required to be recursively computable and stable, since it is used to filter the data. In spectral estimation, using the system itself to filter the data is not an objective, so recursive computability is not required except in those cases when the system is used to extend the available data or correlation.

To consider (6.108) as a linear set of equations for $a(n_1, n_2)$, the $R_x(l_1, l_2)$ and $R_x(l_1 - k_1, l_2 - k_2)$ used in (6.108) must be known. There are many different sets of $R_x(n_1, n_2)$, from which $a(n_1, n_2)$ can be determined, depending on the (l_1, l_2) used in (6.108). Typically, (l_1, l_2) is chosen such that $(l_1, l_2) \in A$. An example of the region A and the region of (n_1, n_2) for which $R_x(n_1, n_2)$ must be known is shown in Figure 6.8.

From Figure 6.8, the number of independent correlation points required to determine the unknown coefficients [eight points in this case, since $R_x(n_1, n_2) = R_x^*(-n_1, -n_2)$] is greater than the number of unknown coefficients (six) in $a(n_1, n_2)$ and σ^2. This is the case, in general, in 2-D AR modeling. If the correlation points were exact and had truly come from the assumed AR model, then the correlation sequence obtained by inverse Fourier transforming $\hat{P}_x(\omega_1, \omega_2)$ in (6.105) would agree with the correlation points used in (6.108). This is called the *correlation matching property*. In practice, however, the correlation points used in (6.108) must be estimated from the data and the data may not have come from the assumed model. As a consequence, since the number of independent correlation points is greater than the number of system parameters, and since the system parameters completely determine $\hat{P}_x(\omega_1, \omega_2)$, the correlation sequence obtained by inverse Fourier transforming $\hat{P}_x(\omega_1, \omega_2)$ will not agree with the correlation points used in (6.108). This differs fundamentally from 1-D AR modeling, where the number of independent correlation points used in the normal equations is the same as the number of system parameters, and the correlation matching property holds.

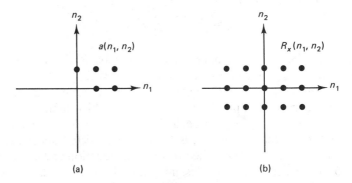

Figure 6.8 (a) AR model coefficients to be estimated; (b) correlation points required to estimate the AR model coefficients in (a), using (6.108).

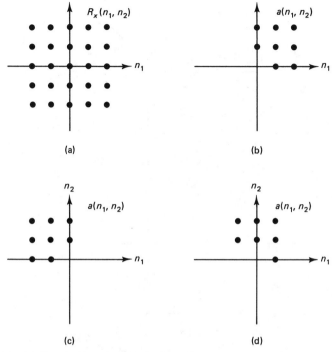

Figure 6.9 (a) Correlation points required to estimate the AR model coefficients in (b), (c), or (d) by using (6.108); (b)–(d) examples of AR model coefficients that can be estimated using (6.108).

As discussed in Section 6.2.6, the absence of the correlation matching property explains why AR modeling differs from the maximum entropy method in 2-D.

We assumed above that the region A for which $a(n_1, n_2)$ is nonzero is assumed known. In practice, the data may not have come from an AR system, or the available data may not be sufficient to estimate the correlation points needed for (6.108). As a result, the shape and size of A are generally unknown and are constrained by the correlation points that can be estimated from the data. If $R_x(n_1, n_2)$ can be estimated for $(n_1, n_2) \in C$, for any given C, several choices of A can be made in using (6.108). Figure 6.9 shows examples of the shapes and sizes of A that can be used in (6.108), if the region C corresponds to $-2 \leq n_1 \leq 2$, $-2 \leq n_2 \leq 2$, as shown in Figure 6.9(a).

For a given set of correlation points, the spectral estimate $\hat{P}_x(\omega_1, \omega_2)$ based on AR modeling can vary considerably, depending on the shape and size of A used. This is illustrated in Figure 6.10. Figure 6.10(a) shows the spectral estimate obtained by AR modeling using the exact correlation given by (6.85) with $M = 2$, $a_1 = 1$, $a_2 = 1$, $(\omega_{11}/2\pi, \omega_{12}/2\pi) = (-0.1, 0.22)$, $(\omega_{21}/2\pi, \omega_{22}/2\pi) = (0.1, 0.28)$, $\sigma^2 = 1$, and SNR = 3 dB. For the estimate in Figure 6.10(a), the region C [the region for which $R_x(n_1, n_2)$ is assumed known] used is the one shown in Figure 6.9(a). The region A [the region for which $a(n_1, n_2)$ is estimated] shown in Figure

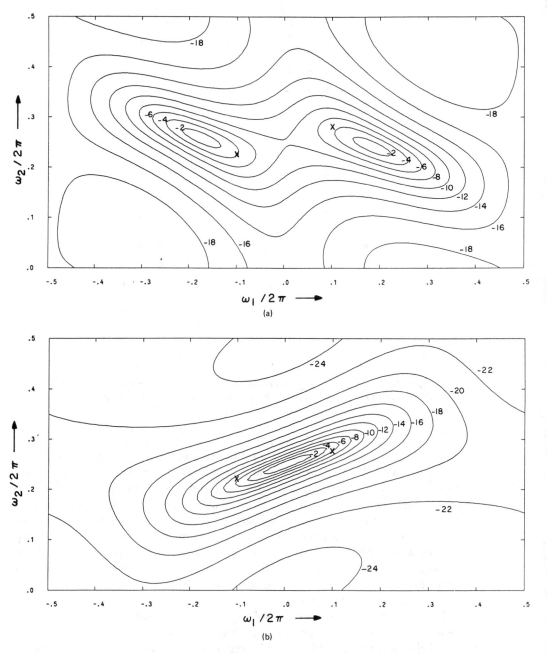

Figure 6.10 Spectral estimates by AR signal modeling based on exact correlation points. Parameters used for correlation generation: $M = 2$; $a_1 = 1$; $a_2 = 1$; $(\omega_{11}/2\pi, \omega_{12}/2\pi) = (-0.1, 0.22)$; $(\omega_{21}/2\pi, \omega_{22}/2\pi) = (0.1, 0.28)$; $\sigma^2 = 1$; SNR = 3 dB; correlation size, 5×5; CINC = 2 dB. (a) AR model coefficients have the region of support shape shown in Figure 6.9(b); (b) same as (a) with the shape shown in Figure 6.9(c).

6.9(b) was used in this estimate. Figure 6.10(b) is identical to Figure 6.10(a), except that the region A is the one shown in Figure 6.9(c). Figure 6.10 illustrates the considerable variance that exists in the spectral estimates obtained from AR modeling, depending on the shape and size of the A used, even when exactly the same correlation points have been used in solving for the AR model coefficients. The shape and size of A are important choices in AR modeling for spectral estimation, since they are typically not known in advance.

The reason for the spectral distortion illustrated in this example is not well understood. Some heuristic methods have been developed, however, which take spectral distortion into account. One method would be to form some combination of the spectral estimates. The combination that appears to reduce the spectral distortion problem is the "parallel resistor" type of average given by

$$\hat{P}_x(\omega_1, \omega_2) = \cfrac{1}{\cfrac{1}{\hat{P}_{x1}(\omega_1, \omega_2)} + \cfrac{1}{\hat{P}_{x2}(\omega_1, \omega_2)}} \tag{6.109}$$

where $\hat{P}_{x1}(\omega_1, \omega_2)$ and $\hat{P}_{x2}(\omega_1, \omega_2)$ are spectral estimates obtained from two different shapes of the region A. This is illustrated in the following example. Figure 6.11 shows the spectral estimate obtained by using (6.109), with $\hat{P}_{x1}(\omega_1, \omega_2)$ and $\hat{P}_{x2}(\omega_1, \omega_2)$ being the spectral estimates in Figures 6.10(a) and 6.10(b), respectively. The spectral distortions evident in Figures 6.10(a) and 6.10(b) have been significantly reduced in Figure 6.11.

When spectral distortion is reduced by such methods, spectral estimation based on AR modeling appears to have resolution properties better than the MLM

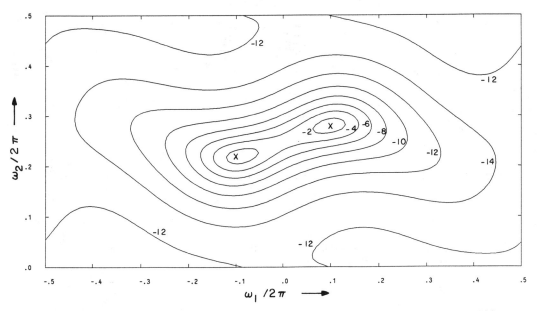

Figure 6.11 Spectral estimate obtained by combining the spectral estimates in Figures 6.10(a) and (b), using (6.109).

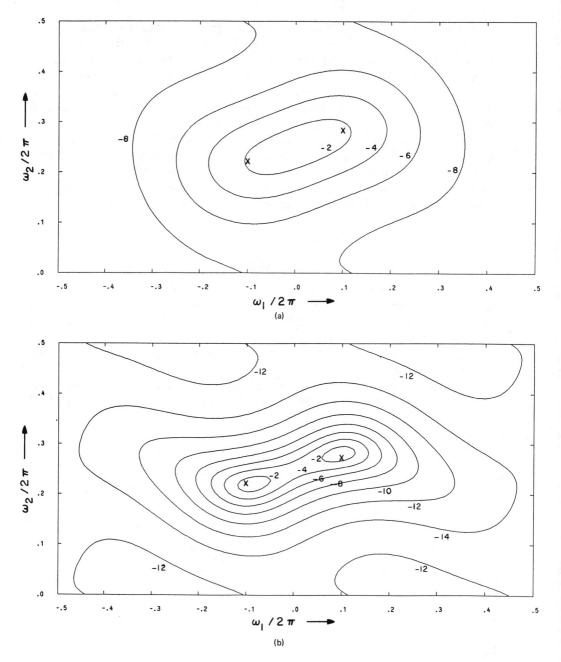

Figure 6.12 MLM and IMLM spectral estimates. Parameters used are the same as those in Figure 6.10: (a) MLM; (b) IMLM.

and similar to the IMLM. Figures 6.12(a) and 6.12(b) show the spectral estimates obtained from the MLM and IMLM, respectively, for the correlation set used in Figures 6.10 and 6.11.

AR signal modeling is not the only signal modeling approach to spectral estimation. Other models, such as the ARMA (autoregressive moving average) model, can also be used for spectral estimation.

6.2.5 Data or Correlation Extension

In the previous section, a random process was modeled by a parametric model, and the parameters of the model were then estimated from the data. The power spectrum was directly determined from the estimated model parameters. An alternative approach is to extend the data or correlation sequence using the model and then use Fourier-transform-based conventional techniques to estimate the power spectrum. In this case, it is desirable for the system to be recursively computable and stable. For AR modeling, the data or correlation can be extended by using the following relations:

$$x(n_1, n_2) = - \sum_{(k_1, k_2) \in A} \sum a(k_1, k_2) x(n_1 - k_1, n_2 - k_2) \qquad (6.110)$$

$$R_x(n_1, n_2) = - \sum_{(k_1, k_2) \in A} \sum a(k_1, k_2) R_x(n_1 - k_1, n_2 - k_2). \qquad (6.111)$$

Data or correlation extension has an important advantage over the signal modeling approach discussed in Section 6.2.4. When the data do not follow the model, the data or correlation extension method tends to be less sensitive to model inaccuracy than the signal modeling method. This is because the original data or correlation is part of the overall data or correlation used in the Fourier-transform-based conventional techniques. Disadvantages of the data or correlation extension method include the greater computational complexity and the requirement of recursive computability and stability in the system. If the system is unstable, the extended data can grow unbounded, and the resulting spectral estimate can be dominated by a small portion of the extended data. Testing the stability of a 2-D system based on models such as an AR model is much more difficult than in the 1-D case.

This method has been used successfully in 1-D when the data or correlation sequence has scattered missing points. In such a case, the missing points are first filled in by the data or correlation extension method, and then the spectral estimate is obtained from the extended data set. The data and correlation extension methods would probably be useful in 2-D as well when the data or correlation sequence has scattered missing points.

6.2.6 The Maximum Entropy Method

Because of its high resolution characteristics, the maximum entropy method (MEM) for spectral estimation has been studied extensively. For 1-D signals with no missing correlation points, the MEM is equivalent to spectral estimation based on

AR signal modeling and the MEM spectral estimate can be obtained by solving a set of linear equations. The 2-D MEM, in contrast, is a highly nonlinear problem, and a closed-form solution has not yet been found. The properties of the MEM, however, extend straightforwardly from 1-D to 2-D.

Suppose the correlation $R_x(n_1, n_2)$ is known (or estimated from data) for $(n_1, n_2) \in C$. In the MEM, $R_x(n_1, n_2)$ is estimated for $(n_1, n_2) \notin C$ by maximizing the entropy, and the power spectrum is determined from the known and extended correlation. The MEM spectral estimation problem can be stated as

Given $R_x(n_1, n_2)$ for $(n_1, n_2) \in C$, determine $\hat{P}_x(\omega_1, \omega_2)$ such that the entropy* H, which is given by

$$H = \frac{1}{(2\pi)^2} \int_{\omega_1 = -\pi}^{\pi} \int_{\omega_2 = -\pi}^{\pi} \log \hat{P}_x(\omega_1, \omega_2) \, d\omega_1 \, d\omega_2 \qquad (6.112)$$

is maximized and

$$R_x(n_1, n_2) = F^{-1}[\hat{P}_x(\omega_1, \omega_2)] \quad \text{for } (n_1, n_2) \in C. \qquad (6.113)$$

Rewriting $\hat{P}_x(\omega_1, \omega_2)$ in terms of the given $R_x(n_1, n_2)$ for $(n_1, n_2) \in C$ and $\hat{R}_x(n_1, n_2)$ for $(n_1, n_2) \notin C$, we have

$$\hat{P}_x(\omega_1, \omega_2) = \sum_{(n_1, n_2) \in C} \sum R_x(n_1, n_2) e^{-j\omega_1 n_1} e^{-j\omega_2 n_2}$$

$$+ \sum_{(n_1, n_2) \notin C} \sum \hat{R}_x(n_1, n_2) e^{-j\omega_1 n_1} e^{-j\omega_2 n_2}. \qquad (6.114)$$

To maximize H in (6.112) with respect to $\hat{R}_x(n_1, n_2)$ in (6.114),

$$\frac{\partial H}{\partial \hat{R}_x(n_1, n_2)} = \frac{1}{(2\pi)^2} \int_{\omega_1 = -\pi}^{\pi} \int_{\omega_2 = -\pi}^{\pi} \frac{1}{\hat{P}_x(\omega_1, \omega_2)} \frac{\partial \hat{P}_x(\omega_1, \omega_2)}{\partial \hat{R}_x(\omega_1, \omega_2)} \, d\omega_1 \, d\omega_2$$

$$= \frac{1}{(2\pi)^2} \int_{\omega_1 = -\pi}^{\pi} \int_{\omega_2 = -\pi}^{\pi} \frac{1}{\hat{P}_x(\omega_1, \omega_2)} e^{-j\omega_1 n_1} e^{-j\omega_2 n_2} \, d\omega_1 \, d\omega_2 \qquad (6.115)$$

$$= 0 \text{ for } (n_1, n_2) \notin C$$

From (6.115), and noting that $\hat{P}_x(\omega_1, \omega_2) = \hat{P}_x^*(\omega_1, \omega_2)$ from (6.31), we obtain

$$\frac{1}{(2\pi)^2} \int_{\omega_1 = -\pi}^{\pi} \int_{\omega_2 = -\pi}^{\pi} \frac{1}{\hat{P}_x(\omega_1, \omega_2)} e^{j\omega_1 n_1} e^{j\omega_2 n_2} \, d\omega_1 \, d\omega_2 = 0 \text{ for } (n_1, n_2) \notin C.$$

$$(6.116)$$

The expression in (6.116) is the inverse Fourier transform of $1/\hat{P}_x(\omega_1, \omega_2)$, and therefore from (6.116),

$$\lambda(n_1, n_2) = F^{-1}\left[\frac{1}{\hat{P}_x(\omega_1, \omega_2)}\right] = 0 \quad \text{for } (n_1, n_2) \notin C. \qquad (6.117)$$

*The entropy expression in (6.112) is valid for a stationary Gaussian random process [Burg (1975)].

From (6.117), $\lambda(n_1, n_2)$ is a finite-extent sequence which is zero for $(n_1, n_2) \notin C$ so $\hat{P}_x(\omega_1, \omega_2)$ can be expressed as

$$\hat{P}_x(\omega_1, \omega_2) = \frac{1}{\displaystyle\sum_{(n_1,n_2) \in C} \lambda(n_1, n_2) e^{-j\omega_1 n_1} e^{-j\omega_2 n_2}}. \tag{6.118}$$

Equation (6.118) states that the finite-extent sequence $\lambda(n_1, n_2)$ completely specifies $\hat{P}_x(\omega_1, \omega_2)$. From (6.118), the MEM spectral estimation problem in (6.112) and (6.113) can be stated as

Given $R_x(n_1, n_2)$ for $(n_1, n_2) \in C$, determine $\hat{P}_x(\omega_1, \omega_2)$ such that $\hat{P}_x(\omega_1, \omega_2)$ is in the form of

$$\hat{P}_x(\omega_1, \omega_2) = \frac{1}{\displaystyle\sum_{(n_1,n_2) \in C} \lambda(n_1, n_2) e^{-j\omega_1 n_1} e^{-j\omega_2 n_2}} \tag{6.119}$$

and $\quad R_x(n_1, n_2) = F^{-1}[\hat{P}_x(\omega_1, \omega_2)] \quad$ for $(n_1, n_2) \in C$. $\tag{6.120}$

The MEM problem above generally has a unique solution if the given $R_x(n_1, n_2)$ for $(n_1, n_2) \in C$ is *extendable*, that is, is a part of some positive definite correlation function (meaning that its Fourier transform is positive for all (ω_1, ω_2)). In general, it is difficult to determine if the given segment of the correlation is part of some positive definite correlation sequence, although this is often the case in practice. In the following discussion, we will assume that the given $R_x(n_1, n_2)$ for $(n_1, n_2) \in C$ is extendable.

Even though the MEM problem statement of (6.119) and (6.120) applies, with appropriate dimensionality changes, to signals of any dimensionality, the solutions depend strongly on dimensionality. For 1-D signals with no missing correlation points, the spectral estimate obtained from AR signal modeling is in the form of (6.119) and satisfies the correlation matching property of (6.120), so it is the same as the spectral estimate given by the MEM. This is not the case for 2-D signals. In the 2-D case, the spectral estimate obtained from AR signal modeling is in the form of (6.119), but does not satisfy the correlation matching property of (6.120). As discussed in Section 6.2.4, this is because the number of independent correlation points needed in solving the 2-D normal equation of (6.108) is greater than the number of AR model parameters, and because the spectral estimate is completely determined by the AR model parameters. As a result, the spectral estimate obtained from AR signal modeling does not have enough degrees of freedom to satisfy (6.120). Because of this difficulty, no closed-form solution for the 2-D MEM problem has yet been found.

Many attempts have been made to solve the 2-D MEM problem. In all cases, the resulting algorithms are iterative ones which attempt to improve the spectral estimate in each iteration. Burg (1975) has proposed an iterative solution which requires the inversion of a matrix in each iteration, with the matrix dimension being on the order of the number of the given correlation points. No experimental results using this technique have yet been reported. Wernecke and D'Addario have proposed a scheme in which the entropy is numerically maximized. The maximization is done by continuously adjusting the spectral estimate and evaluating the expression for the entropy and its gradient. The procedure is computationally expensive and is not guaranteed to have a solution. Woods expresses the MEM spectral estimate as a power series in the frequency domain and attempts to approximate the MEM spectral estimate by truncating the power series expansion. This method is restricted to that class of signals for which power series expansion is possible. Lang and McClellan have shown that the MEM spectral estimation problem involves the optimization of a convex functional over a convex set. A number of standard algorithms that are guaranteed to converge can thus be used to solve the MEM spectral estimation problem.

The iterative procedure proposed by Lim and Malik was used* to generate the 2-D MEM spectral estimates in this and following sections, although it has not been shown to converge. To develop this algorithm, we express a power spectrum $P_y(\omega_1, \omega_2)$ as

$$P_y(\omega_1, \omega_2) = F[R_y(n_1, n_2)] = \sum_{n_1=-\infty}^{\infty} \sum_{n_2=-\infty}^{\infty} R_y(n_1, n_2)e^{-j\omega_1 n_1}e^{-j\omega_2 n_2} \qquad (6.121)$$

and

$$\frac{1}{P_y(\omega_1, \omega_2)} = F[\lambda(n_1, n_2)] = \sum_{n_1=-\infty}^{\infty} \sum_{n_2=-\infty}^{\infty} \lambda(n_1, n_2)e^{-j\omega_1 n_1}e^{-j\omega_2 n_2}. \qquad (6.122)$$

From (6.121) and (6.122), it is clear that $R_y(n_1, n_2)$ can be obtained from $\lambda(n_1, n_2)$, and vice versa, through Fourier transform operations. Now from (6.119) and (6.120), $P_y(\omega_1, \omega_2)$ is the MEM spectral estimate if and only if $\lambda(n_1, n_2) = 0$ for $(n_1, n_2) \notin C$ and $R_y(n_1, n_2) = R_x(n_1, n_2)$ for $(n_1, n_2) \in C$. Thus, we see that for $P_y(\omega_1, \omega_2)$ to be the desired MEM spectral estimate, we have constraints on $R_y(n_1, n_2)$ and $\lambda(n_1, n_2)$. Recognizing this, we can develop a simple iterative algorithm by going back and forth between $R_y(n_1, n_2)$ (the correlation domain) and $\lambda(n_1, n_2)$ (the coefficient domain), each time imposing the constraints on $R_y(n_1, n_2)$ and $\lambda(n_1, n_2)$. Thus, starting with some initial estimates of $\lambda(n_1, n_2)$, we obtain an estimate of $R_y(n_1, n_2)$. This estimate is then corrected by the given $R_x(n_1, n_2)$ over Region C and is used to generate a new $\lambda(n_1, n_2)$. The new $\lambda(n_1, n_2)$ is then truncated to the desired limits, and the procedure is repeated. This iterative procedure is illustrated in Figure 6.13.

The procedure discussed above is very similar in form to other iterative techniques discussed in Section 1.4.1 for reconstructing a signal from its Fourier transform phase. The conditions under which the algorithm converges are not yet known, but if the algorithm does converge, the converging solution will satisfy both (6.119) and (6.120), and consequently will be the desired MEM spectral estimate. The major computations in each iteration of the algorithm are Fourier transform

---

*The use of this method for generating examples in this chapter reflects the algorithm's accessibility to the author, and does not imply that this is the most computationally efficient method. A comparison of different methods of computing the MEM spectral estimates is not available in the open literature.

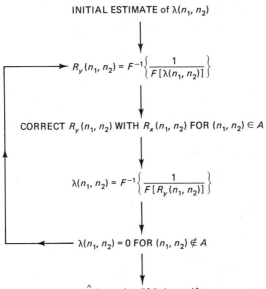

INITIAL ESTIMATE of $\lambda(n_1, n_2)$

$$R_y(n_1, n_2) = F^{-1}\left\{\frac{1}{F[\lambda(n_1, n_2)]}\right\}$$

CORRECT $R_y(n_1, n_2)$ WITH $R_x(n_1, n_2)$ FOR $(n_1, n_2) \in A$

$$\lambda(n_1, n_2) = F^{-1}\left\{\frac{1}{F[R_y(n_1, n_2)]}\right\}$$

$\lambda(n_1, n_2) = 0$ FOR $(n_1, n_2) \notin A$

$$\hat{P}_x(\omega_1 \omega_2) = F[R_y(n_1, n_2)]$$

Figure 6.13 Iterative procedure for solving 2-D MEM problem.

operations, which can be computed using FFT algorithms. This algorithm has been successfully used to determine a number of 2-D spectral estimates when the size of C is small (on the order of 7×7) and the SNR is low (typically less than 5 dB). For a large C or a high SNR, the DFT used to approximate the Fourier transform operation is large, and the algorithm has been observed to have difficulty in converging to the MEM spectral estimate.

The properties of the MEM, unlike its computational aspects, have been observed to extend straightforwardly from 1-D to 2-D. For example, the MEM's resolution properties have been observed to be significantly higher than those of conventional methods or the MLM. This is illustrated in the following example. Figure 6.14(a) shows the spectral estimates obtained with the correlation windowing method by using a 5×5-point segment (Region C) of the exact correlation given by (6.85), with $M = 3$, $a_1 = 1$, $a_2 = 1$, $a_3 = 1$, $(\omega_{11}/2\pi, \omega_{12}/2\pi) = (0.3, 0.1)$, $(\omega_{21}/2\pi, \omega_{22}/2\pi) = (0.1, 0.1)$, $(\omega_{31}/2\pi, \omega_{32}/2\pi) = (0.2, 0.2)$, $\sigma^2 = 6$, and SNR $= -3$ dB. Figures 6.14(b), (c), and (d) show the spectral estimates obtained from the MLM, IMLM, and MEM, respectively. The spectral estimates in Figure 6.14 show that the three spectral peaks are clearly resolved in the MEM spectral estimate, while they are not in the other three spectral estimates. A more quantitative comparison of the resolution characteristics of different spectral estimation techniques is given in the next section.

For 1-D signals, it is well known that the power of a sinusoid is linearly proportional to the area under the peak corresponding to the sinusoid in the MEM spectral estimate. For 2-D signals, the power of a 2-D sinusoid has been observed to be linearly proportional to the volume under the peak in the MEM spectral estimate. This is another property of the MEM that extends straightforwardly from 1-D to 2-D.

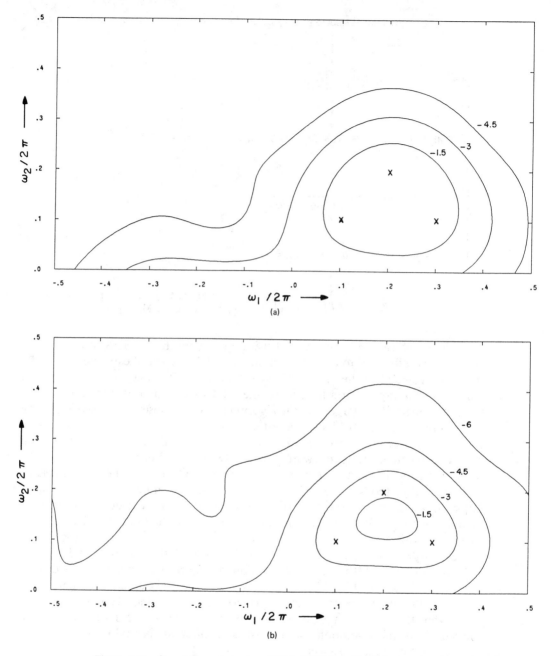

Figure 6.14 Correlation windowing, MLM, IMLM, and MEM spectral estimates based on exact correlation points. Parameters used for correlation generation: $M = 3$; $a_1 = 1$; $a_2 = 1$; $a_3 = 1$; $(\omega_{11}/2\pi, \omega_{12}/2\pi) = (0.3, 0.1)$; $(\omega_{21}/2\pi, \omega_{22}/2\pi) = (0.1, 0.1)$; $(\omega_{31}/2\pi, \omega_{32}/2\pi) = (0.2, 0.2)$; $\sigma^2 = 6$; SNR $= -3$ dB; correlation size, 5×5; CINC $= 1.5$ dB. (a) Correlation windowing; (b) MLM; (c) IMLM; (d) MEM.

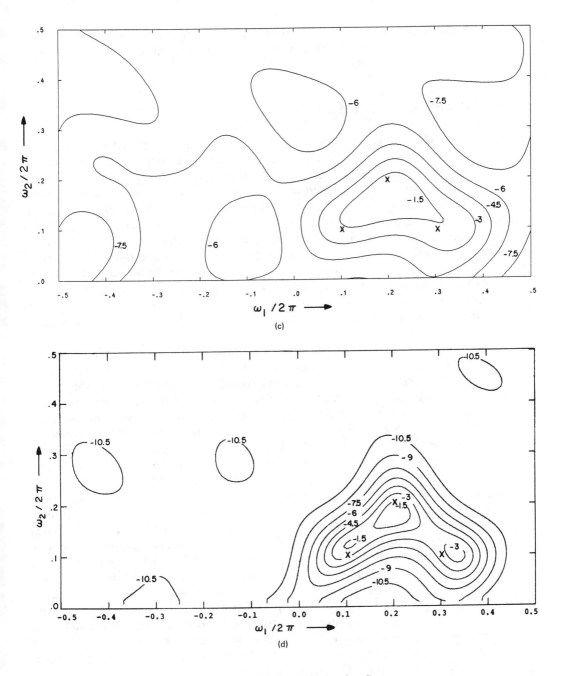

Figure 6.14 (continued)

In this section, we have discussed different approaches to 2-D estimation. For additional methods of 2-D spectral estimation not discussed in this section, see [Davis and Regier; Pisarenko; Schmidt].

6.3 PERFORMANCE COMPARISON

Only certain aspects of the spectral estimation techniques discussed in the previous section have been quantitatively compared [Malik and Lim]. In this section, we discuss the limited results available in the literature.

The major impetus behind the development of high resolution spectral estimation techniques is the improvement in resolution they offer over conventional techniques. Experimental studies of two aspects of resolution have been reported. The first is the resolvability of two sinusoids in the presence of noise. The other is the accuracy of the frequency estimation of a sinusoid when the sinusoid is well resolved. In both cases, complex data with exact correlation values [Equation (6.84)] were used to separate the issue of resolution from the issue of correlation estimation from data. The correlation windowing method, the MLM, and the MEM were compared. For correlation windowing, the 2-D separable triangular window was used.

To compare different spectral estimation algorithms in their ability to resolve two sinusoids, a quantitative measure of resolvability was developed based on empirical observations. It has been observed that for a given size and shape of C (the region for which the exact correlation sequence is assumed to be known) and a given SNR, the spectral estimates for the correlation windowing method, the MLM, and the MEM do not depend on the absolute location of the peaks in the 2-D frequency plane. That is, the shape and size of the estimated spectral peaks remain the same regardless of the complex sinusoids' locations, if the same relative distance and orientation of the peaks are maintained. Figure 6.15 illustrates this phenomenon for MEM spectral estimates. In these cases, the frequency separation between the peaks is held constant, and the orientation of the peaks is kept horizontal. The results clearly show the invariance of the MEM spectral estimates under these conditions. Other examples support this conclusion, and similar results have been observed for the MLM and the correlation windowing method. In addition, in all three methods, a larger separation in the frequencies of two sinusoids has been observed to always produce "more resolved" spectral estimates at a given SNR, size and shape of Region C, and orientation of the peaks. Based on the above, a reasonable quantitative measure of the resolvability would be the minimum frequency separation, denoted by d, above which the two sinusoids are resolved and below which they are not. For this measure d, smaller values imply higher resolution, while larger values imply lower resolution.

Many cases have been studied in an effort to determine the minimum separation distance d required to resolve two peaks in the sense that the spectral estimates will display two distinct peaks. One peak's location is held constant while the second peak's location is varied over a range such that the peaks are not

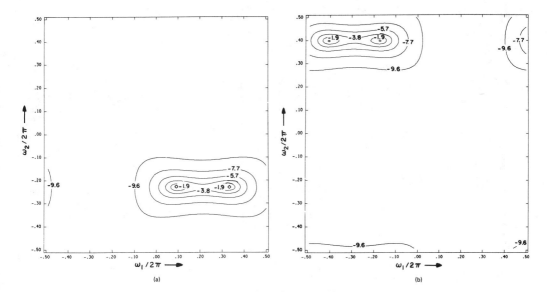

Figure 6.15 Spectral estimates do not depend on absolute peak locations for complex signals. Parameters used for these MEM spectral estimates: $M = 2$; $a_1 = 1$; $a_2 = 1$; $\sigma^2 = 2$; SNR = 0 dB; correlation size, 5 × 5; CINC = 1.915 dB; ($\diamond$) 0-dB (maximum) points. (a) $(\omega_{11}/2\pi, \omega_{12}/2\pi) = (0.1, -0.23)$; $(\omega_{21}/2\pi, \omega_{22}/2\pi) = (0.325, -0.23)$; (b) $(\omega_{11}/2\pi, \omega_{12}/2\pi) = (-0.4, 0.4)$; $(\omega_{21}/2\pi, \omega_{22}/2\pi) = (-0.175, 0.4)$.

initially resolved, but become resolved as the distance between them is increased. Figure 6.16 shows the results obtained by the MEM as the separation between peaks is increased. Initially, the two peaks are not resolved, and the spectral estimate consists of a single spectral peak located approximately at the midpoint of the line joining the true peak locations. As the distance between the peaks increases, the spectral estimate shows a distortion or stretching in the direction of the peaks, and eventually the two peaks are resolved. Figure 6.17 summarizes the resolution performance of the three techniques. It is clear that in the 2-D case, as in the 1-D case, the MEM provides higher resolution than the other two methods. Note that the resolution performance of the correlation windowing method is determined only by the size of the correlation available for analysis, and is independent of the SNR as far as the resolution d is concerned. This is because changing noise level adds a constant independent of frequency to the spectral estimate when exact correlation values are used. The minimum distance for the peaks to be resolved in the MEM and MLM estimates decreases with increasing SNR, with the MEM consistently outperforming the MLM.

The measure adopted for the resolution performance evaluation is fairly arbitrary, and is used only to study the relative performance of the different techniques under the same set of conditions. The minimum resolution distance between two peaks also depends on their orientation in the 2-D frequency plane and

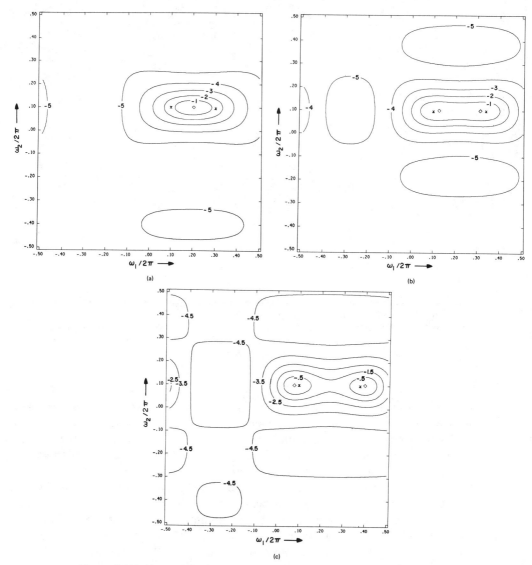

Figure 6.16 Change in MEM spectral estimates as the separation between two peaks is increased. Parameters used for the MEM spectral estimates: $M = 2$; $a_1 = 1$; $a_2 = 1$; $\sigma^2 = 6.32$; SNR = -5 dB; correlation size, 3×3; ($\diamond$) 0-dB (maximum) point in the spectral estimate. CINC = 1 dB. (a) $(\omega_{11}/2\pi, \omega_{12}/2\pi) = (0.1, 0.1)$; $(\omega_{21}/2\pi, \omega_{22}/2\pi) = (0.3, 0.1)$. (b) $(\omega_{11}/2\pi, \omega_{12}/2\pi) = (0.1, 0.1)$; $(\omega_{21}/2\pi, \omega_{22}/2\pi) = (0.34, 0.1)$. (c) $(\omega_{11}/2\pi, \omega_{12}/2\pi) = (0.1, 0.1)$; $(\omega_{21}/2\pi, \omega_{22}/2\pi) = (0.38, 0.1)$.

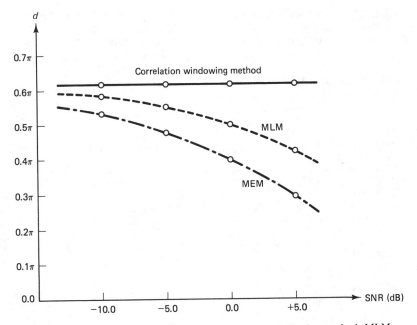

Figure 6.17 Resolution measure d of the correlation windowing method, MLM, and MEM.

on the shape of C. Thus, the resolution measure is only an indicator of the relative performance, and should not be considered meaningful as an absolute measure.

Another set of experiments has been directed toward studying the accuracy of the peak location resulting when the number of sinusoids present is accurately estimated. The quantitative measure of the error in the location of the spectral peak (LOSP) is defined as

$$\text{Error (LOSP)} = \sum_{i=1}^{M} \sqrt{\left(\frac{\omega_{i1e}}{2\pi} - \frac{\omega_{i1t}}{2\pi}\right)^2 + \left(\frac{\omega_{i2e}}{2\pi} - \frac{\omega_{i2t}}{2\pi}\right)^2} \qquad (6.123)$$

where the number of sinusoids is M, ω_{i1e} and ω_{i2e} represent the estimated frequency location of the ith peak, and ω_{i1t} and ω_{i2t} represent the true peak location. Table 6.1 shows the LOSP error of some representative one sinusoid and two sinusoid cases. For the one sinusoid case with 3×3 correlation values, the correlation windowing method, MLM, and MEM estimates all show LOSP errors very close to zero. For two sinusoids with 5×5 correlation values, all methods show some finite LOSP error. Although the MEM estimates exhibit much sharper peaks than do the other two, the table shows that the MEM gives the worst LOSP estimate for the two sinusoid cases. The correlation windowing method and the MLM give LOSP estimates of approximately the same magnitude.

TABLE 6.1 COMPARISON OF THE MEM, MLM, AND CORRELATION-WINDOWING METHOD FOR PEAK-LOCATION ACCURACY USING EXACT CORRELATION VALUES. THE PEAK LOCATIONS ARE LISTED AS PAIRS ($\omega_1/2\pi$, $\omega_2/2\pi$), AND THE PEAK LOCATION ERROR (LOSP) IS ALSO IN UNITS OF 2π. $S/N = 5$ dB. ONE-SINUSOID CASES: CORRELATION OF SIZE 3×3. TWO-SINUSOID CASES: CORRELATION OF SIZE 5×5.

	True Location	Maximum Entropy		Maximum Likelihood		Correlation Windowing	
		Estimated Location	LOSP Error	Estimated Location	LOSP Error	Estimated Location	LOSP Error
ONE SINUSOID	−.4000,0.4000	−.4000,0.4000	0.0000	−.4000,0.4000	0.0000	−.4000,0.4000	0.0000
	0.0745,−.4456	0.0745,−.4456	0.0000	0.0745,−.4456	0.0000	0.0745,−.4456	0.0000
	−.3000,−.3000	−.3000,−.3000	0.0000	−.3000,−.3000	0.0000	−.3000,−.3000	0.0000
	−.0500,−.0500	−.0500,−.0500	0.0000	−.0500,−.0500	0.0000	−.0500,−.0500	0.0000
	−.3125,0.3000	−.3125,0.3000	0.0000	−.3125,0.3000	0.0000	−.3125,0.3000	0.0000
TWO SINUSOIDS	−.4000,0.0000	−.4010,0.0040	0.0082	−.4010,0.0000	0.0020	−.4010,0.0020	0.0045
	0.0745,−.4456	0.0755,−.4496		0.0755,−.4456		0.0755,−.4476	
	0.3000,−.3000	0.2670,−.3000	0.0480	0.2970,−.3000	0.0059	0.2810,−.3000	0.0379
	−.3000,−.3000	−.2670,−.3000		−.2970,−.3000		−.2810,−.3000	
	0.3000,0.4120	0.3050,0.4060	0.0156	0.2990,0.4110	0.0028	0.3010,0.4120	0.0019
	−.0500,−.0500	−.0550,−.0440		−.0490,−.0490		−.0510,−.0500	
	0.1234,0.3456	0.1374,0.3396	0.0304	0.1304,0.3476	0.0146	0.1374,0.3416	0.0291
	−.3125,0.3000	−.2265,0.3060		−.3195,0.2980		−.3265,0.3040	
	0.2000,0.3125	0.1950,0.3135	0.0102	0.1990,0.3115	0.0028	0.1990,0.3125	0.0019
	−.1125,0.0330	−.1075,0.0320		−.1115,0.0340		−.1115,0.0330	
	0.3300,0.0000	0.3230,0.0070	0.0197	0.3300,0.0000	0.0000	0.3300,0.0000	0.0000
	0.0000,0.3333	0.0070,0.3263		0.0000,0.3333		0.0000,0.3333	
	−.3000,−.2000	−.2900,−.2040	0.0215	−.3000,−.2000	0.0000	−.3000,−.2000	0.0000
	0.1000,0.4430	0.0900,0.4470		0.1000,0.4430		0.1000,0.4430	
	−.1000,−.1000	−.1010,−.1000	0.0019	−.1000,−.1000	0.0000	−.1000,−.1000	0.0000
	0.3900,0.4000	0.3910,0.4000		0.3900,0.4000		0.3900,0.4000	

6.4 FURTHER COMMENTS

This section deals with estimating the correlation from data. Some guidelines for applying existing methods to practical problems are also given.

6.4.1 Estimating Correlation

Spectral estimation techniques such as the MLM and MEM are based on the assumption that the correlation sequence is available. To apply these methods, therefore, the correlation must be estimated from the available data. The spectral estimate can vary significantly depending on the method by which the correlation sequence is estimated.

We will describe the two most common of the many available methods for estimating the correlation sequence from the available data. Suppose the data $x(n_1, n_2)$ is available for $(n_1, n_2) \in D$. In one method, the correlation $R_x(n_1, n_2)$ for $(n_1, n_2) \in C$ is estimated by

$$\hat{R}_x(n_1, n_2) = \frac{1}{N} \sum_{k_1=-\infty}^{\infty} \sum_{k_2=-\infty}^{\infty} x(k_1, k_2)x^*(k_1 - n_1, k_2 - n_2) \text{ for } (n_1, n_2) \in C$$

(6.124)

where $x(n_1, n_2)$ is assumed to be zero for $(n_1, n_2) \notin D$ and N is the number of points in D. Estimating the correlation sequence in this way is identical to performing the inverse Fourier transform operation of the periodogram. The Fourier transform of $\hat{R}_x(n_1, n_2)$ in (6.124) is, therefore, always nonnegative. Equation (6.124) leads to a biased estimate of $R_x(n_1, n_2)$, since the normalization factor N is not the same as the number of terms that contribute to $\hat{R}_x(n_1, n_2)$. In the second method, $R_x(n_1, n_2)$ for $(n_1, n_2) \in C$ is estimated from $x(n_1, n_2)$ by

$$\hat{R}_x(n_1, n_2) = \frac{1}{N(n_1, n_2)} \sum_{k_1=-\infty}^{\infty} \sum_{k_2=-\infty}^{\infty} x(k_1, k_2)x^*(k_1 - n_1, k_2 - n_2)$$

$$\text{for } (n_1, n_2) \in C \tag{6.125}$$

where $x(n_1, n_2)$ is assumed to be zero for $(n_1, n_2) \notin D$ and $N(n_1, n_2)$ is the number of terms in the summation that contribute to the estimate. Equation (6.125) leads to an unbiased estimate of $R_x(n_1, n_2)$. The Fourier transform of $R_x(n_1, n_2)$ estimated by (6.125) is not guaranteed to be nonnegative, and partly for this reason, (6.124) is more often used.

Figure 6.18 shows that a given spectral estimation method will produce markedly different results, depending on the method by which the correlation is estimated from the available data. Figure 6.18(a) and (b) show the results of the MLM when the correlation sequence is obtained from (6.124) and (6.125), respectively. In both cases, synthetic data were generated by using (6.83) with $M = 2$, $a_1 = 1$, $a_2 = 1$, $(\omega_{11}/2\pi, \omega_{12}/2\pi) = (0.1, 0.1)$, $(\omega_{21}/2\pi, \omega_{22}/2\pi) = (0.4, 0.4)$, $\phi_i = 0$, $\sigma^2 = 1$, and SNR = 3 dB. The size of the data generated is 6×6, and the size of the correlation sequence used is 5×5.

The more data we have available, the more accurately we can estimate the correlation points. This generally leads to a better spectral estimate. To illustrate this point, Figure 6.19(a) shows an MLM spectral estimate in which 3×3-point correlation was estimated from 4×4-point synthetic data generated by using (6.82) with $M = 2$, $a_1 = 1$, $a_2 = 1$, $(\omega_{11}/2\pi, \omega_{12}/2\pi) = (0.1, 0.1)$, $(\omega_{21}/2\pi, \omega_{22}/2\pi) = (0.35, 0.4)$, $\phi_i = 0$, $\sigma^2 = 0.632$, and SNR = 5 dB. Figure 6.19(b) is identical to Figure 6.19(a) except that the 3×3-point correlation was estimated from 60×60-point synthetic data. The spectral estimate in Figure 6.19(b) clearly corresponds better to the two sinusoids buried in noise than does the spectral estimate in Figure 6.19(a).

Estimating more correlation points from a given amount of available data does not necessarily lead to a better spectral estimate. If we were to estimate too

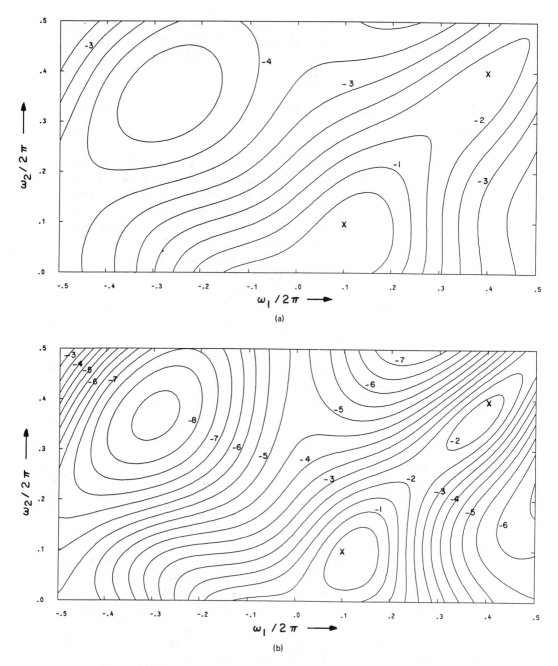

Figure 6.18 Spectral estimates are affected by how correlation points are estimated from data. MLM spectral estimates. Parameters used for data generation: $M = 2$; $a_1 = 1$; $a_2 = 1$; $(\omega_{11}/2\pi, \omega_{12}/2\pi) = (0.1, 0.1)$; $(\omega_{21}/2\pi, \omega_{22}/2\pi) = (0.4, 0.4)$; $\phi_i = 0$, $\sigma^2 = 1$; SNR $= 3$ dB; data size, 6×6; correlation size, 5×5; CINC $= 0.5$ dB. (a) Correlation points estimated by using (6.124); (b) correlation points estimated by using (6.125).

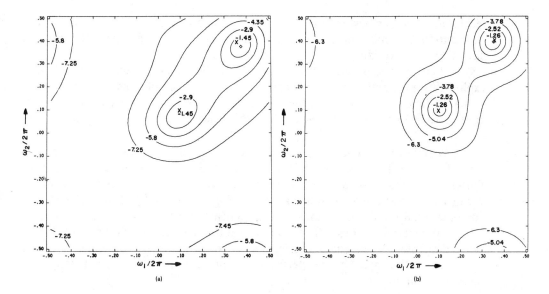

Figure 6.19 Spectral estimates are affected by the data size used in estimating correlation points of a given size. MLM spectral estimates. Parameters used for data generation: $M = 2$; $a_1 = 1$; $a_2 = 1$; $(\omega_{11}/2\pi, \omega_{12}/2\pi) = (0.1, 0.1)$; $(\omega_{21}/2\pi, \omega_{22}/2\pi) = (0.35, 0.4)$; $\phi_i = 0$, $\sigma^2 = 0.632$; SNR = 5 dB; correlation size, 3 × 3. (a) Data length, 4 × 4; CINC = 1.45 dB; (b) data length 60 × 60; CINC = 1.26 dB.

many correlation points from the data, those correlation points with large lags would be estimated from a small amount of data and consequently might not be reliable. Using unreliable correlation points can sometimes do more harm than good to the resulting spectral estimate. Determining the optimum size and shape of the correlation points to be estimated depends on many factors, including the type of data, the specific spectral estimation technique used, and the method by which the correlation points are estimated from available data. Despite some studies [Burg et al.] on this problem, few concrete results are available.

In summary, a spectral estimate can be significantly affected by the way in which the correlation points are estimated from available data. Estimating the correlation points to optimize the performance of a given spectral estimation method remains an area in need of further research.

6.4.2 Application of Spectral Estimation Methods

Determining which method is best for any given application is an important issue in using spectral estimation techniques. Since the various methods were developed on the basis of different assumptions, and since only limited comparisons of the methods' performance are available, choosing the best method for a given application problem is a difficult task. We offer here some general guidelines for choosing a spectral estimation method.

If enough data are available so that resolution is not a problem, conventional techniques, such as periodogram averaging or smoothing, may be a good choice. Conventional techniques are computationally simple and the spectral estimates have very good variance characteristics.

When some structural information is available about the data, methods that exploit this information should be considered. For example, if we know that the available data are a segment of the response of an AR process excited by random noise, spectral estimation based on AR signal modeling will generally be a good choice.

When not much data are available and high resolution is required, the MLM, IMLM, or MEM should be considered. The MEM has significantly better resolution characteristics than the MLM and IMLM, but its computational requirements are orders of magnitude greater than those of the MLM and IMLM. The MEM's computational requirements make it impractical for most real time applications.

If the characteristics of available data differ significantly in different dimensions, dimension-dependent processing should be considered. The guidelines above apply also to the choice of the spectral estimation technique to be used along each individual dimension.

6.5 APPLICATION EXAMPLE

In this section, we present a typical application of a multidimensional spectral estimation technique. We will consider the problem of determining the direction of low-flying aircraft by processing acoustic signals on a sensor array having a planar spatial distribution. The two directional parameters of interest are the azimuthal (or bearing) angle θ, defined in the interval

$$0° \leq \theta \leq 360° \tag{6.126}$$

and the elevation angle ϕ, defined in the interval

$$0° \leq \phi \leq 90°. \tag{6.127}$$

These two parameters are illustrated in Figure 6.20. To illustrate how a spectral estimation technique can be used in determining the directions of aircraft, we first consider the case of a single aircraft as an acoustic source. The acoustic source generates a space-time wavefield $s(t, x, y, z)$, where t represents time and (x, y, z) represent the Cartesian spatial coordinates. If the array is in the far field of the acoustic source, the signal in the region of the array can be approximated as a plane wave. For any plane wave, the wavefield $s(t, x, y, z)$ is constant along the plane perpendicular to the directon of wave propagation, and therefore $s(t, x, y, z)$ can be expressed as

$$s(t, x, y, z) = s\left(t - \frac{x \cos \theta \cos \phi}{c} - \frac{y \sin \theta \cos \phi}{c} - \frac{z \sin \phi}{c}\right) \tag{6.128}$$

where $s(t)$ is the acoustic signal received at a particular spatial location (for example, $x = y = z = 0$) and c is the speed of wave propagation. Sound travels in air at

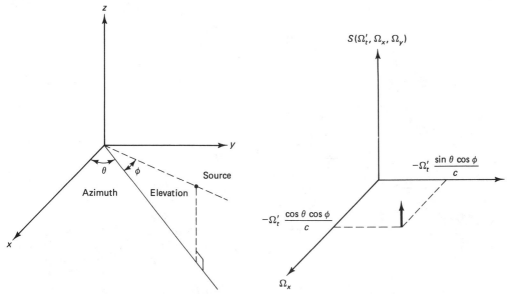

Figure 6.20 Direction parameters θ (azimuth) and ϕ (elevation) of a source.

Figure 6.21 $S(\Omega'_t, \Omega_x, \Omega_y)$ as a function of Ω_x and Ω_y at a particular temporal frequency Ω'_t for one source.

a speed of approximately 340 m/sec. For a planar array parallel to the ground surface, we can assume $z = 0$ without any loss of generality. Equation (6.128) then becomes

$$s(t, x, y) = s\left(t - \frac{x \cos \theta \cos \phi}{c} - \frac{y \sin \theta \cos \phi}{c} \right). \tag{6.129}$$

Denoting the 3-D analog Fourier transform of $s(t, x, y)$ by $S(\Omega_t, \Omega_x, \Omega_y)$, we have

$$S(\Omega_t, \Omega_x, \Omega_y) = (2\pi)^2 S(\Omega_t) \delta\left(\Omega_x + \Omega_t \frac{\cos \theta \cos \phi}{c} \right) \delta\left(\Omega_y + \Omega_t \frac{\sin \theta \cos \phi}{c} \right) \tag{6.130}$$

where $S(\Omega_t)$ is the 1-D analog Fourier transform of $s(t)$, Ω_t is the temporal frequency, and Ω_x and Ω_y are the two spatial frequencies. Now consider a particular temporal frequency Ω'_t. From (6.130), as shown in Figure 6.21, the spectrum $S(\Omega'_t, \Omega_x, \Omega_y)$ should have a strong spectral peak at the following values of (Ω_x, Ω_y):

$$\Omega_x = -\Omega'_t \frac{\cos \theta \cos \phi}{c} \tag{6.131}$$

$$\Omega_y = -\Omega'_t \frac{\sin \theta \cos \phi}{c}. \tag{6.132}$$

By looking for a spectral peak in the function $S(\Omega'_t, \Omega_x, \Omega_y)$, then, we can determine the two direction parameters θ and ϕ using (6.131) and (6.132). For SNR considerations, Ω'_t is chosen such that $S(\Omega'_t, \Omega_x, \Omega_y)$ is large.

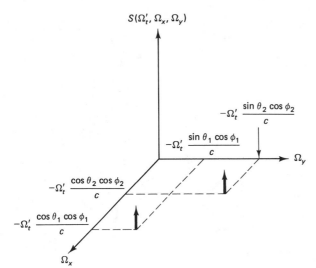

$S(\Omega_t', \Omega_x, \Omega_y)$

$-\Omega_t' \dfrac{\sin\theta_2 \cos\phi_2}{c}$

$-\Omega_t' \dfrac{\sin\theta_1 \cos\phi_1}{c}$

Ω_y

$-\Omega_t' \dfrac{\cos\theta_2 \cos\phi_2}{c}$

$-\Omega_t' \dfrac{\cos\theta_1 \cos\phi_1}{c}$

Ω_x

Figure 6.22 $S(\Omega_t', \Omega_x, \Omega_y)$ as a function of Ω_x and Ω_y at a particular temporal frequency Ω_t' for two sources.

If there are N sources at different azimuths and elevations, $s(t, x, y)$ can be expressed as

$$s(t, x, y) = \sum_{i=1}^{N} s_i\left(t - \frac{x \cos\theta_i \cos\phi_i}{c} - \frac{y \sin\theta_i \cos\phi_i}{c}\right). \qquad (6.133)$$

At a particular temporal frequency Ω_t', the spectrum $S(\Omega_t', \Omega_x, \Omega_y)$ should have N strong spectral peaks. An example of $S(\Omega_t', \Omega_x, \Omega_y)$ for $N = 2$ is shown in Figure 6.22. For each of these spectral peaks, the directional parameters can be determined by using (6.131) and (6.132). It is clear from the above discussion that the number of sources and the direction parameters for each source can be determined through spectral analysis of $s(t, x, y)$.

In the above discussion, determining the number of sources and their direction parameters was a straightforward task that involved evaluating the Fourier transform of $s(t, x, y)$. This is because $s(t, x, y)$ was assumed to be available for all (t, x, y). In practice, $s(t, x, y)$ is sampled in both the temporal and spatial dimensions and only a finite number of samples are available. Along the spatial dimensions, the number of array sensors limits the number of spatial samples. Along the temporal dimension, an aircraft can be assumed to be coming from a constant direction only for a limited amount of time, and this limits the number of samples. Sampling $s(t, x, y)$ repeats $S(\Omega_t, \Omega_x, \Omega_y)$ periodically in the frequency domain, and sampling periods should be chosen to avoid aliasing effects. The limitation on the number of samples can be viewed as signal windowing, which smoothes the spectrum in the frequency domain. As a result, the spectral peaks in $S(\Omega_t', \Omega_x, \Omega_y)$ are no longer impulses but have finite widths. This will clearly make it more difficult to resolve two aircraft coming from similar directions, particularly when the number of samples within the window is small. Another potential problem area is noise. The wavefront $s(t, x, y)$ may be degraded by noise, further reducing resolution. Simply computing the discrete Fourier transform of samples of $s(t, x, y)$

may not produce adequate resolution. This is one reason for the application of high resolution spectral estimation techniques to array processing problems.

Real data for this aircraft resolution problem was gathered from a 2-D array of sensors. The sensor array consisted of nine microphones (channels), equally spaced one meter apart in a 3×3 format on a square grid. The data gathered were the sound of a helicopter flying past the array for two seconds, during which period the angles θ and ϕ were assumed to be approximately constant. Each of the nine sensors was used to record a time series. The wavefront $s(t, x, y)$ was sampled along the temporal dimension with a sampling rate of 2048 Hz, and the total number of data points in each channel was 4096. Since the temporal spectral contents of acoustic sounds from a helicopter typically consist of frequency contents below 200 Hz, this sampling rate was more than sufficient to avoid aliasing. The time series obtained were three-dimensional, with one temporal and two spatial dimensions. Since much more data were available in the temporal than in the spatial dimensions, dimension-dependent processing was used. A Fourier-transform-based conventional technique was first used along the time dimension. Along the two spatial dimensions, the MLM was used, because it is relatively simple computationally, while offering higher resolution than Fourier-transform-based conventional techniques.

Taking the time series at each of the microphones, we first determine the temporal frequency at which there was a strong spectral peak. This is accomplished by looking at a single channel and performing periodogram averaging. Once the temporal frequency is chosen, the data from each of the nine channels is divided into 512-point sections. The nine channels are then correlated spatially at the chosen temporal frequency. This is illustrated in Figure 6.23. Note that the result

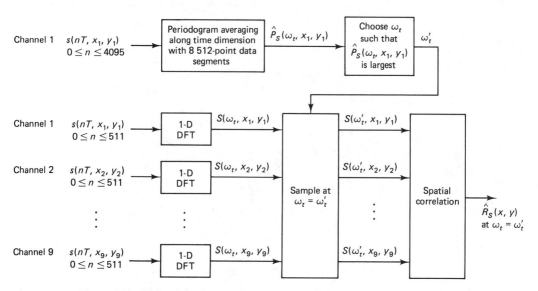

Figure 6.23 Estimation of spatial correlation at a particular temporal frequency from nine array sensors, with 512 data points in each channel.

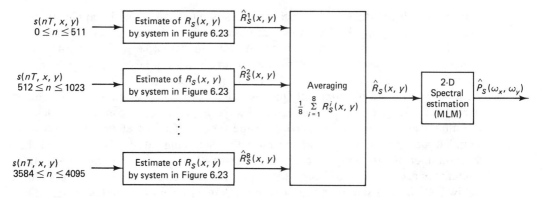

Figure 6.24 Estimation of spatial correlation at a particular temporal frequency from nine array sensors with 4096 data points in each channel. The estimate is obtained by segmenting the data into eight blocks of 512 data points each and by averaging the result of applying the system in Figure 6.23 to each of the eight data blocks.

of the 1-D DFT along the time dimension in each of the channels is complex-valued, and we do not perform the magnitude squared operation. This is because the phase information must be preserved in the intermediate stages of dimension-dependent processing, as discussed in Section 6.2.2.

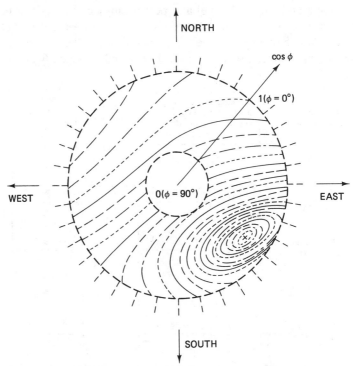

Figure 6.25 MLM spectral estimate for real data gathered from nine microphones (3×3-size grid with 1-m spacing). The spectral peak shows the estimated direction parameters of a flying helicopter.

This process is repeated for eight consecutive 512-point sections, and the results are averaged to obtain the spatial correlation values:

$$\hat{R}_s(x, y) = \tfrac{1}{8} \sum_{i=1}^{8} R_s^i(x, y). \tag{6.134}$$

The spatial correlation estimate $\hat{R}_s(x, y)$ is used along the two spatial dimensions by the MLM. This is shown in Figure 6.24. The spectral estimate obtained as a function of θ and ϕ is shown in Figure 6.25. Conversion from the spatial frequencies to θ and ϕ is made by using (6.131) and (6.132). The spectral estimate clearly shows one strong spectral peak representing the presence of an acoustic source. The location of the spectral peak is well within the tolerance of the experiment in this example.

We have discussed one particular way to process the data from a sensor array. There are many variations of this method, in addition to methods that are significantly different from this one. A method that is useful for determining the azimuth angle θ alone is discussed in Problem 6.24.

REFERENCES

For books on probability, random processes, and estimation, see [Papoulis; Van Trees].

For reprint books of journal articles on spectral estimation, see [Childers; Kesler]. For a special journal issue on the topic, see [Haykin and Cadzow]. For books on conventional spectral estimation techniques, see [Blackman and Tukey; Jenkins and Watts]. For articles on conventional methods, see [Welch; Nuttal and Carter]. For a tutorial article on 1-D high resolution spectral estimation techniques, see [Kay and Marple]. For a review article on multidimensional spectral estimation, see [McClellan].

For readings on the MLM for spectral estimation, see [Capon et al.; Capon]. For the IMLM, see [Burg (1972); Lim and Dowla]. For a computationally efficient method to compute the MLM spectral estimate, see [Musicus].

For the relationship between maximum entropy method and autoregressive signal modeling in 1-D, see [Ulrych and Bishop]. For a method to reduce spectral distortion caused by 2-D autoregressive (AR) modeling, see [Jackson and Chien]. For a method to improve the performance of the AR signal modeling method in the frequency estimation of multiple sinusoids, see [Tufts and Kumaresan]. For spectral estimation based on autoregressive moving average (ARMA) signal modeling, see [Cadzow and Ogino]. For spectral estimation based on a Markov model, see [Chellappa et al.].

For a reading on the original development of the MEM for spectral estimation, see [Burg (1975)]. For a reading on extendability of a given correlation sequence, see [Dickinson]. For the development of the MEM algorithms, see [Woods; Wernecke and D'Addario; Lim and Malik; Lang and McClellan (1982)].

Many methods for spectral estimation have been proposed in addition to the standard spectral estimation methods discussed in this chapter. The data adaptive spectral estimation (DASE) method, which is a generalization of the MLM discussed in Section 6.2.3, is discussed in [Davis and Regier]. Pisarenko's method, which is based on a parametric model of the spectrum consisting of impulses and a noise spectrum with a known shape is discussed in [Pisarenko; Lang and McClellan (1983)]. This method requires eigenanalysis of the correlation. A description of multiple signal classification (MUSIC), which is related to Pisarenko's method, can be found in [Schmidt].

For readings on performance comparison of different spectral estimation methods, see [Lacoss (1971); Cox; Malik and Lim]. For the effect of correlation estimation on spectral estimation, see [Burg, et al.].

For applications of spectral estimation to sonar and radar signal processing, see [Oppenheim]. For a tutorial article on array processing, which is one of the main applications of 2-D spectral estimation, see [Dudgeon]. For estimation of the bearing angle and elevation angle for array sensors through dimension-dependent spectral estimation, see [Lacoss, et al.]. For methods to estimate the bearing angle through spectral estimation, see [Johnson; Nawab, et al.].

R. B. Blackman and J. W. Tukey, *The Measurement of Power Spectra*. New York: Dover, 1958.

J. P. Burg, Maximum entropy spectral analysis, Ph.D. Thesis. Stanford, CA: Stanford University, 1975.

J. P. Burg, The relationship between maximum entropy spectra and maximum likelihood spectra, *Geophysics*, Vol. 37, 1972, p. 375.

J. P. Burg, D. G. Luenberger, and D. L. Wenger, Estimation of structured covariance matrices, *Proc. IEEE*, Vol. 70, September 1982, pp. 963–974.

J. A. Cadzow and K. Ogino, Two-dimensional signal estimation, *IEEE Trans. Acoust. Speech Sig. Proc.*, Vol. ASSP-29, June 1981, pp. 396–401.

J. Capon, High-resolution frequency-wavenumber spectrum analysis, *Proc. IEEE*, Vol. 57, August 1969, pp. 1408–1419.

J. Capon, R. J. Greenfield, and R. J. Kolken, Multi-dimensional maximum likelihood processing of a large aperture seismic array, *Proc. IEEE*, Vol. 55, February 1967, pp. 192–211.

R. Chellappa, Y. H. Hu, and S. Y. Kung, On two-dimensional Markov spectral estimation, *IEEE Trans. Acoust. Speech Sig. Proc.*, Vol. ASSP-31, August 1983, pp. 836–841.

D. G. Childers, ed., *Modern Spectral Analysis*. New York: IEEE Press, 1978.

H. Cox, Resolving power and sensitivity to mismatch of optimum array processors, *J. Acoust. Soc. Am.*, Vol. 54, 1973, pp. 771–785.

R. E. Davis and L. A. Regier, Methods for estimating directional wave spectra from multi-element arrays, *J. Marine Res.*, Vol. 35, 1977, pp. 453–477.

B. W. Dickinson, Two-dimensional Markov spectrum estimates need not exist, *IEEE Trans. Inf. Theory*, Vol. IT-26, January 1980, pp. 120–121.

F. U. Dowla, Bearing estimation of wideband signals by multidimensional spectral estimation, Ph.D. Thesis, Cambridge, MA: M.I.T., 1984.

D. E. Dudgeon, Fundamentals of digital array processing, *Proc. IEEE*, Vol. 65, June 1977, pp. 898–904.

S. Haykin and J. A. Cadzow, eds., Special issue on *Spectral Estimation*, *Proc. IEEE*, September 1982.

L. B. Jackson and H. C. Chien, Frequency and bearing estimation by two-dimensional linear prediction, *Proc. 1979 Int. Conf. Acoust. Speech Sig. Proc.*, April 1979, pp. 665–668.

G. M. Jenkins and D. G. Watts, *Spectral Analysis and Its Applications.* San Francisco: Holden-Day, 1968.

D. H. Johnson, The application of spectral estimation methods to bearing estimation problems, *Proc. IEEE*, Vol. 70, September 1982, pp. 1018–1028.

S. M. Kay and S. L. Marple, Jr., Spectrum analysis—a modern perspective, *Proc. IEEE*, Vol. 69, November 1981, pp. 1380–1419.

S. B. Kesler, ed., *Modern Spectrum Analysis, II.* New York: IEEE Press, 1986.

R. T. Lacoss, Data-adaptive spectral analysis methods, *Geophysics*, Vol. 36, August 1971, pp. 661–675.

R. T. Lacoss, et al., Distributed sensor networks, *Semiannual Tech. Rep., M.I.T. Lincoln Lab*, May 1980.

S. W. Lang and J. H. McClellan, Multi-dimensional MEM spectral estimation, *IEEE Trans. Acoust. Speech, Sig. Proc.*, Vol. ASSP-30, December 1982, pp. 880–887.

S. W. Lang and J. H. McClellan, Spectral estimation for sensor arrays, *IEEE Trans. Acoust. Speech, Sig. Proc.*, Vol. ASSP-31, April 1983, pp. 349–358.

J. S. Lim and F. U. Dowla, A new algorithm for high-resolution two-dimensional spectral estimation, *Proc. IEEE*, Vol. 71, February 1983, pp. 284–285.

J. S. Lim and N. A. Malik, A new algorithm for two-dimensional maximum entropy power spectrum estimation, *IEEE Trans. Acoust. Speech, Sig. Proc.*, Vol. ASSP-29, June 1981, pp. 401–413.

N. A. Malik and J. S. Lim, Properties of two-dimensional maximum entropy power spectrum estimates, *IEEE Trans. Acoust. Speech, Sig. Proc.*, Vol. ASSP-30, October 1982, pp. 788–798.

J. H. McClellan, Multi-dimensional spectral estimation, *Proc. IEEE*, Vol. 70, September 1982, pp. 1029–1039.

B. R. Musicus, Fast MLM power spectrum estimation from uniformly spaced correlations, *IEEE Trans. Acoust. Speech, Sig. Proc.*, Vol. ASSP-33, October 1985, pp. 1333–1335.

S. H. Nawab, F. U. Dowla, and R. T. Lacoss, Direction determination of wideband signals, *IEEE Trans. Acoust. Speech, Sig. Proc.*, Vol. ASSP-33, October 1985, pp. 1114–1122.

S. H. Nuttal and G. C. Carter, Spectral estimation using combined time and lag weighting, *Proc. IEEE*, Vol. 70, September 1982, pp. 1115–1125.

A. V. Oppenheim, ed., *Applications of Digital Signal Processing.* Englewood Cliffs, NJ: Prentice-Hall, 1978, pp. 331–428.

A. Papoulis, *Probability, Random Variables, and Stochastic Processes.* New York: McGraw-Hill, 1965.

V. F. Pisarenko, On the estimation of spectra by means of nonlinear functions of the covariance matrix, *Geophys. J. Roy. Astron. Soc.*, Vol. 28, 1972, pp. 511–531.

R. O. Schmidt, A signal subspace approach to emitter location and spectral estimation, Ph.D thesis. Stanford, CA: Stanford University, August 1981.

D. W. Tufts and R. Kumaresan, Estimation of frequencies of multiple sinusoids: making linear prediction perform like maximum likelihood, *Proc. IEEE*, Vol. 70, September 1982, pp. 975–989.

T. J. Ulrych and T. N. Bishop, Maximum entropy spectral analysis and autoregressive decomposition, *Rev. Geophys. Space, Phys.*, Vol. 13, February 1975, pp. 183–200.

H. L. Van Trees, *Detection, Estimation and Modulation Theory*. New York: Wiley, 1968.

P. D. Welch, The use of the fast Fourier transform for the estimation of power spectra: a method based on time averaging over short, modified periodograms, *IEEE Trans. Audio Electron.*, Vol. AU-15, June 1967, pp. 70–73.

S. J. Wernecke and L. R. D'Addario, Maximum entropy image reconstruction, *IEEE Trans. Comput.*, Vol. C-26, April 1979, pp. 351–364.

J. W. Woods, Two-dimensional Markov spectral estimation, *IEEE Trans. Inf. Theory*, Vol. IT-22, September 1976, pp. 552–559.

PROBLEMS

6.1. Let x and y be uncorrelated real random variables. Define a new random variable w by $w = x + y$.
   (a) Show that $E[w] = E[x] + E[y]$.
   (b) Show that Var $[w]$ = Var $[x]$ + Var $[y]$, where Var $[w]$ represents variance of w and Var $[x]$ and Var $[y]$ are similarly defined. You may use the result of (a).

6.2. Let x denote a real random variable whose probability density function $p_x(x_0)$ is given by

$$p_x(x_0) = \begin{cases} k, & 0 \le x_0 \le 4 \\ 0, & \text{otherwise.} \end{cases}$$

   (a) Determine k.
   (b) What is the probability that x will lie between 2 and 3?
   (c) Determine $E[x]$, the mean of x.
   (d) Determine Var $[x]$, the variance of x.

6.3. Let x and y denote two real random variables with a joint probability density function $p_{x,y}(x_0, y_0)$ given by

$$p_{x,y}(x_0, y_0) = \begin{cases} \frac{1}{2}, & \text{in the shaded region} \\ 0, & \text{otherwise.} \end{cases}$$

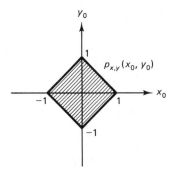

Figure P6.3

   (a) Are x and y statistically independent?
   (b) Are x and y linearly independent?
   (c) Are the results in (a) and (b) consistent with the fact that statistical independence implies linear independence, but linear independence does not imply statistical independence?

6.4. Let x and y denote two real random variables with a joint probability density function given by

$$p_{x,y}(x_0, y_0) = \begin{cases} 1, & \text{in the shaded region} \\ 0, & \text{otherwise.} \end{cases}$$

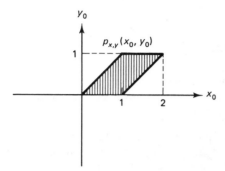

Figure P6.4

(a) Determine $p_{x|y}(x_0|y_0)|_{y_0=1/2} = p_{x|y}(x_0|\tfrac{1}{2})$.

(b) Determine $p_y(y_0)|_{y_0=1/2} = p_y(\tfrac{1}{2})$.

(c) From the results of (a) and (b), show that

$$p_{x|y}(x_0|y_0)|_{y_0=1/2} = \left.\frac{p_{x,y}(x_0,y_0)}{p_y(y_0)}\right|_{y_0=1/2}$$

(d) Show that $p_{x|y}(x_0|y_0) = p_{x,y}(x_0, y_0)/p_y(y_0)$ for all y_0.

6.5. Let w denote a complex random variable given by

$$w = x + jy.$$

The joint probability density function of x and y is given by

$$p_{x,y}(x_0, y_0) = \begin{cases} 1, & \text{in the shaded region} \\ 0, & \text{otherwise.} \end{cases}$$

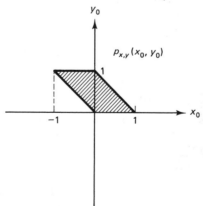

Figure P6.5

(a) Determine $E[w]$.

(b) Determine Var $[w]$.

6.6. Let $\mathbf{x}$ denote an $N \times 1$-column vector that consists of N real random variables. The N random variables are said to be jointly Gaussian if their joint probability density function $p_x(\mathbf{x}_0)$ is given by

$$p_x(\mathbf{x}_0) = \frac{1}{(2\pi)^{N/2}|\Lambda|^{1/2}} e^{-(\mathbf{x}_0 - E[\mathbf{x}])^T\Lambda^{-1}(\mathbf{x}_0 - E[\mathbf{x}])/2}$$

where Λ is an $N \times N$ covariance matrix and $|\Lambda|$ is the determinant of Λ. The ijth element of Λ, λ_{ij}, is given by

$$\lambda_{ij} = E[(x_i - E[x_i])(x_j - E[x_j])]$$

where x_i is the ith element in $\mathbf{x}$.

(a) Let a and b denote two independent real random variables whose probability density functions are Gaussian with means of 5 and 0, respectively, and standard deviations of 2 and 3, respectively. We define a new random variable r by

$$r = 2a + 3b.$$

Determine $p_r(r_0)$, the probability density function of r.

(b) Show that a linear combination of jointly Gaussian random variables is Gaussian. Is your result in (a) consistent with this result?

6.7. Let $R_x(n_1, n_2)$ denote the correlation sequence of a stationary random process. Show that

$$R_x(0, 0) \geq |R_x(n_1, n_2)| \quad \text{for all } n_1 \text{ and } n_2.$$

6.8. Let $x(n_1, n_2)$ denote a stationary white noise process whose power spectrum $P_x(\omega_1, \omega_2)$ satisfies

$$\frac{1}{(2\pi)^2} \int_{\omega_1 = -\pi}^{\pi} \int_{\omega_2 = -\pi}^{\pi} P_x(\omega_1, \omega_2) \, d\omega_1 \, d\omega_2 = 1.$$

Determine $R_x(n_1, n_2)$, the correlation sequence of $x(n_1, n_2)$.

6.9. Let $x(n_1, n_2)$ denote a stationary complex random process.

(a) Suppose $R_x(n_1, n_2)$ is known for $n_2 \geq 0$. Is it possible to determine $R_x(n_1, n_2)$ for all (n_1, n_2)?

(b) Suppose $P_x(\omega_1, \omega_2)$ is known for $-\pi \leq \omega_1 \leq \pi, 0 \leq \omega_2 \leq \pi$. Is it possible to determine $P_x(\omega_1, \omega_2)$ for all (ω_1, ω_2)?

6.10. Suppose the correlation function $R_x(n_1, n_2)$ of a stationary random process is given by

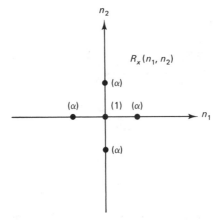

Figure P6.10

For what values of α is $R_x(n_1, n_2)$ a valid correlation function? Assume α is real.

6.11. Let $f(n_1, n_2)$ denote a deterministic signal. We have two observations of $f(n_1, n_2)$, $r_1(n_1, n_2)$ and $r_2(n_1, n_2)$, given by

$$r_1(n_1, n_2) = f(n_1, n_2) + w_1(n_1, n_2)$$

$$r_2(n_1, n_2) = f(n_1, n_2) + w_2(n_1, n_2).$$

The sequences $w_1(n_1, n_2)$ and $w_2(n_1, n_2)$ are samples of zero-mean white Gaussian noise processes each with variance of σ^2. One simple way to estimate $f(n_1, n_2)$ from $r_1(n_1, n_2)$ and $r_2(n_1, n_2)$ is

$$\hat{f}(n_1, n_2) = \frac{r_1(n_1, n_2) + r_2(n_1, n_2)}{2}.$$

Assume that $f(n_1, n_2)$, $w_1(n_1, n_2)$ and $w_2(n_1, n_2)$ are all real. Assume also that $w_1(n_1, n_2)$ and $w_2(n_1, n_2)$ are independent.
(a) Determine $E[\hat{f}(n_1, n_2)]$.
(b) Suppose we express $\hat{f}(n_1, n_2)$ as

$$\hat{f}(n_1, n_2) = f(n_1, n_2) + w(n_1, n_2).$$

Determine $E[w^2(n_1, n_2)]$ and compare it with $E[w_1^2(n_1, n_2)]$.
(c) Is $w(n_1, n_2)$ defined in (b) a zero-mean white Gaussian noise?
(d) Determine $P_w(\omega_1, \omega_2)$, the power spectrum of $w(n_1, n_2)$. The sequence $w(n_1, n_2)$ is defined in (b).

6.12. Consider a linear shift-invariant system whose impulse response $h(n_1, n_2)$ is as shown below.

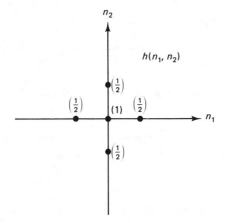

Figure P6.12

Suppose the response of the system to a real, stationary, zero-mean, and white noise process with variance of 1 is denoted by $x(n_1, n_2)$, as shown below.

zero-mean
white noise $\longrightarrow$ $\boxed{h(n_1, n_2)}$ $\longrightarrow x(n_1, n_2)$.
$w(n_1, n_2)$

(a) Determine $P_x(\omega_1, \omega_2)$, the power spectrum of $x(n_1, n_2)$.
(b) Determine $R_x(n_1, n_2)$, the correlation sequence of $x(n_1, n_2)$.
(c) Determine $E[x(n_1, n_2)]$.
(d) Determine $E[x^2(n_1, n_2)]$.

6.13. Consider a linear shift-invariant system whose impulse response is real and is given by $h(n_1, n_2)$. Suppose the responses of the system to the two inputs $x(n_1, n_2)$ and $v(n_1, n_2)$ are $y(n_1, n_2)$ and $z(n_1, n_2)$, as shown in the figure below.

$$x(n_1, n_2) \rightarrow \boxed{h(n_1, n_2)} \rightarrow y(n_1, n_2)$$

$$v(n_1, n_2) \rightarrow \boxed{h(n_1, n_2)} \rightarrow z(n_1, n_2)$$

The inputs $x(n_1, n_2)$ and $v(n_1, n_2)$ in the figure represent real stationary zero-mean random processes with autocorrelation functions $R_x(n_1, n_2)$ and $R_v(n_1, n_2)$, the power spectra $P_x(\omega_1, \omega_2)$ and $P_v(\omega_1, \omega_2)$, cross-correlation function $R_{xv}(n_1, n_2)$, and cross-power spectrum $P_{xv}(\omega_1, \omega_2)$.

(a) Given $R_x(n_1, n_2)$, $R_v(n_1, n_2)$, $R_{xv}(n_1, n_2)$, $P_x(\omega_1, \omega_2)$, $P_v(\omega_1, \omega_2)$, and $P_{xv}(\omega_1, \omega_2)$, determine $P_{yz}(\omega_1, \omega_2)$, the cross-power spectrum of $y(n_1, n_2)$ and $z(n_1, n_2)$.

(b) Is the cross-power spectrum $P_{xv}(\omega_1, \omega_2)$ always nonnegative? That is, is $P_{xv}(\omega_1, \omega_2) \geq 0$ for all (ω_1, ω_2)? Justify your answer.

6.14. Let $s(n_1, n_2)$ and $w(n_1, n_2)$ denote two stationary zero-mean random processes. The power spectra of the two processes $P_s(\omega_1, \omega_2)$ and $P_w(\omega_1, \omega_2)$ are given by

$$P_s(\omega_1, \omega_2) = \begin{cases} 1 - \sqrt{\dfrac{\omega_1^2}{(2\pi)^2} + \dfrac{\omega_2^2}{(2\pi)^2}}, & \sqrt{\omega_1^2 + \omega_2^2} \leq \pi \\ 0, & \text{otherwise} \end{cases}$$

and $P_w(\omega_1, \omega_2) = 1$ for all (ω_1, ω_2). The observation process is given by

$$r(n_1, n_2) = s(n_1, n_2) + w(n_1, n_2).$$

Determine the frequency response of the noncausal Wiener filter that can be used to estimate $s(n_1, n_2)$ from $r(n_1, n_2)$. Assume that $s(n_1, n_2)$ and $w(n_1, n_2)$ are uncorrelated with each other.

6.15. Let r denote an observation variable that can be modeled by

$$r = s + w$$

where s is the variable we wish to estimate and w is a Gaussian random variable with mean of 0 and variance of σ_w^2; that is,

$$p_w(w_0) = \frac{1}{\sqrt{2\pi}\,\sigma_w} e^{-w_0^2/2\sigma_w^2}.$$

(a) Assuming that s is nonrandom, determine $\hat{s}_{\text{ML}}$, the maximum likelihood estimate of s.

(b) Suppose s is a random variable which has a Gaussian probability density function with mean of m_s and variance of σ_s^2. Determine $\hat{s}_{\text{MAP}}$, the maximum a posteriori estimate of s. Discuss how $\hat{s}_{\text{MAP}}$ is affected by r, m_s, σ_s^2, and σ_w^2. Does your answer make sense?

(c) Suppose s is again assumed to be a random variable which has a Gaussian probability density function with mean of m_s and variance of σ_s^2. Determine $\hat{s}_{\text{MMSE}}$, the minimum mean square error estimate of s. You do not have to evaluate your expression explicitly. What is $\hat{s}_{\text{MMSE}}$ when $r = m_s$?

6.16. Signals such as sinusoids buried in noise are not samples of stationary random processes. Consider $x(n_1, n_2)$ given by

$$x(n_1, n_2) = \sum_{i=1}^{M} a_i e^{j(\omega_{i1}n_1 + \omega_{i2}n_2 + \phi_i)} + w(n_1, n_2)$$

where the sinusoidal components are deterministic and $w(n_1, n_2)$ is zero-mean white noise with variance of σ_w^2. Clearly, $E[x(k_1, k_2)x^*(k_1 - n_1, k_2 - n_2)]$ depends not only on n_1 and n_2 but also on k_1 and k_2, so $x(n_1, n_2)$ is not a stationary random process. One way to eliminate the dependence on k_1 and k_2 is to define $\theta_x(n_1, n_2)$ by

$$\theta_x(n_1, n_2) = \lim_{N \to \infty} \frac{1}{(2N+1)^2} \sum_{k_1=-N}^{N} \sum_{k_2=-N}^{N} x(k_1, k_2)x^*(k_1 - n_1, k_2 - n_2)$$

and assume that $\theta_x(n_1, n_2)$ is the correlation function. Show that $\theta_x(n_1, n_2)$ for the above $x(n_1, n_2)$ is given by

$$\theta_x(n_1, n_2) = \sum_{i=1}^{M} a_i^2 e^{j(\omega_{i1}n_1 + \omega_{i2}n_2)} + \sigma_w^2 \, \delta(n_1, n_2).$$

6.17. Let $x(n_1, n_2)$ be a sample of a 2-D stationary random process. The sequence $x(n_1, n_2)$ is given for $(n_1, n_2) \in D$. In spectral estimation based on a periodogram, $x(n_1, n_2)$ is assumed to be zero for $(n_1, n_2) \notin D$, and the spectral estimate $\hat{P}_x(\omega_1, \omega_2)$ is obtained by

$$\hat{P}_x(\omega_1, \omega_2) = \frac{1}{N} |X(\omega_1, \omega_2)|^2$$

where N represents the number of points in the region D. Show that the periodogram $\hat{P}_x(\omega_1, \omega_2)$ is the Fourier transform of $\hat{R}_x(n_1, n_2)$ given by

$$\hat{R}_x(n_1, n_2) = \frac{1}{N} \left[\sum_{k_1=-\infty}^{\infty} \sum_{k_2=-\infty}^{\infty} x(k_1, k_2)x^*(k_1 - n_1, k_2 - n_2) \right].$$

6.18. Let $x(n_1, n_2)$ be a sample of a 2-D stationary random process. The sequence $x(n_1, n_2)$ is given for $0 \le n_1 \le N_1 - 1, 0 \le n_2 \le N_2 - 1$. In spectral estimation based on a periodogram, $x(n_1, n_2)$ is assumed to be zero outside $0 \le n_1 \le N_1 - 1, 0 \le n_2 \le N_2 - 1$ and the spectral estimate $\hat{P}_x(\omega_1, \omega_2)$ is obtained by

$$\hat{P}_x(\omega_1, \omega_2) = \frac{1}{N_1 N_2} |X(\omega_1, \omega_2)|^2.$$

The periodogram $\hat{P}_x(\omega_1, \omega_2)$ is often computed by using FFT algorithms. Suppose we compute the periodogram by computing an $M_1 \times M_2$-point DFT. If $M_1 > N_1$ or $M_2 > N_2$, we can pad $x(n_1, n_2)$ with enough zeros before we compute its DFT. Does the choice of M_1 and M_2 affect the resolution characteristic of the periodogram?

6.19. In this problem, we show that a periodogram can be viewed as filtering the data with a bank of bandpass filters that are identical except for their passband center frequencies. Let $x(n_1, n_2)$ be a sample of a 2-D stationary random process. The sequence $x(n_1, n_2)$ is assumed to be given for $0 \le n_1 \le N_1 - 1, 0 \le n_2 \le N_2 - 1$. In computing the periodogram, $x(n_1, n_2)$ is assumed to be zero outside $0 \le n_1 \le N_1 - 1, 0 \le n_2 \le N_2 - 1$. This is equivalent to windowing $x(n_1, n_2)$ with a rectangular window $w(n_1, n_2)$ given by

$$w(n_1, n_2) = \begin{cases} 1, & 0 \le n_1 \le N_1 - 1, 0 \le n_2 \le N_2 - 1 \\ 0, & \text{otherwise.} \end{cases}$$

Let $x_w(n_1, n_2)$ denote the windowed signal so that

$$x_w(n_1, n_2) = x(n_1, n_2)w(n_1, n_2).$$

The periodogram $\hat{P}_x(\omega_1, \omega_2)$ is given by

$$\hat{P}_x(\omega_1, \omega_2) = \frac{1}{N_1 N_2} |X_w(\omega_1, \omega_2)|^2.$$

Consider $X_w(\omega_1, \omega_2)$. From the Fourier transform definition,

$$X_w(\omega_1, \omega_2) = \sum_{n1=-\infty}^{\infty} \sum_{n2=-\infty}^{\infty} x_w(n_1, n_2)e^{-j\omega_1 n_1}e^{-j\omega_2 n_2}.$$

(a) Show that $X_w(\omega_1, \omega_2)$ can be interpreted as

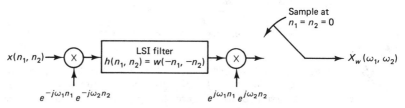

Figure P6.19

where $h(n_1, n_2) = w(-n_1, -n_2)$ is the impulse response of the filter.

(b) From the result in (a), show that the periodogram $\hat{P}_x(\omega_1, \omega_2)$ can be viewed as the result of filtering the data with a bank of bandpass filters that are identical except the passband center frequencies. Sketch the shape of one bandpass filter. The result in (b) shows that the characteristics of the bandpass filter do not change for different frequencies. This contrasts sharply with the maximum likelihood method, in which a different bandpass filter is designed at each frequency.

6.20. Let $x(n_1, n_2)$ denote a stationary zero-mean random process. The correlation function $R_x(n_1, n_2)$ is given only for $-1 \le n_1 \le 1$, $-1 \le n_2 \le 1$, as shown in the following figure.

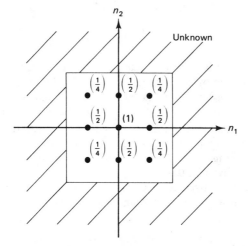

Figure P6.20

Determine $\hat{P}_x(\omega_1, \omega_2)$, the MLM spectral estimate of $x(n_1, n_2)$.

6.21. Let $x(n_1, n_2)$ denote a stationary random process whose correlation function $R_x(n_1, n_2)$ is assumed to be known only for $-N \leq n_1 \leq N$, $-N \leq n_2 \leq N$. To estimate the power spectrum, we model $x(n_1, n_2)$ as the response of an auto-regressive model excited by white noise with variance of σ^2, as shown in the following figure.

$$w(n_1, n_2) \longrightarrow \boxed{\frac{1}{1 + \sum\limits_{(k_1, k_2) \in A} \sum a(n_1, n_2) e^{-j\omega_1 n_1} e^{-j\omega_2 n_2}}} \longrightarrow x(n_1, n_2)$$

(a) Suppose the region A is chosen as shown in Figure P6.21a. Determine and sketch the region of support of the correlation points that are used in estimating the model parameters $a(n_1, n_2)$ with the smallest possible choice of N.

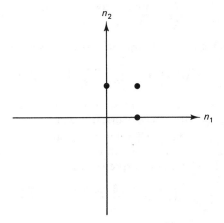

Figure P6.21a

(b) Suppose the region A is chosen as shown in Figure P6.21b. Determine and sketch the region of support of the correlation points that are used in estimating the model parameters $a(n_1, n_2)$ with the smallest possible value of N.

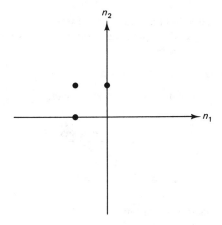

Figure P6.21b

(c) Suppose we conjecture that the resolution of the spectral estimate is better along the direction where more correlation points are used. Using this conjecture and the results of (a) and (b), explain the spectral distortions in Figures 6.10(a) and 6.10(b).

(d) Assuming that the conjecture in (c) is correct, approximately sketch the spectral estimate obtained from one sinusoid buried in noise using a first-quadrant support $a(n_1, n_2)$.

(e) Answer (d), using a second-quadrant support $a(n_1, n_2)$.

6.22. The correlation function $R_x(n_1, n_2)$ of a stationary random process $x(n_1, n_2)$ is assumed known only for $-1 \le n_1 \le 1$ and $-1 \le n_2 \le 1$. Determine one possible maximum entropy spectral estimate. You may choose any nonzero $R_x(n_1, n_2)$ within the above constraints on $R_x(n_1, n_2)$.

6.23. Consider a 3-D analog signal $s(t, x, y)$ given by

$$s(t, x, y) = s\left(t - \frac{x \cos \theta \cos \phi}{c} - \frac{y \sin \theta \cos \phi}{c}\right).$$

Show that the 3-D analog Fourier transform $S(\Omega_t, \Omega_x, \Omega_y)$ of $s(t, x, y)$ is given by

$$S(\Omega_t, \Omega_x, \Omega_y) = (2\pi)^2 S(\Omega_t) \, \delta\left(\Omega_x + \Omega_t \frac{\cos \theta \cos \phi}{c}\right) \delta\left(\Omega_y + \Omega_t \frac{\sin \theta \cos \phi}{c}\right)$$

where $S(\Omega_t)$ is the 1-D analog Fourier transform of $s(t)$, Ω_t is the temporal frequency, and (Ω_x, Ω_y) are two spatial frequencies corresponding to the spatial variables (x, y).

6.24. In this problem, we discuss a method developed for estimating the bearing angle from the data for a planar array. For a planar array, a space-time wavefield $s(t, x, y)$ based on the plane wave assumption can be expressed as

$$s(t, x, y) = s\left(t - \frac{x \cos \theta \cos \phi}{c} - \frac{y \sin \theta \cos \phi}{c}\right)$$

where $s(t)$ is the wavefield received at $x = y = 0$, θ is the bearing angle, and ϕ is the elevation angle, as shown in Figure 6.20. The Fourier transform of $s(t, x, y)$ is given by

$$S(\Omega_t, \Omega_x, \Omega_y) = (2\pi)^2 S(\Omega_t) \, \delta\left(\Omega_x + \Omega_t \frac{\cos \theta \cos \phi}{c}\right) \delta\left(\Omega_y + \Omega_t \frac{\sin \theta \cos \phi}{c}\right)$$

where $S(\Omega_t)$ is the 1-D analog Fourier transform of $s(t)$, Ω_t is the temporal frequency, and (Ω_x, Ω_y) are the two spatial frequencies. Suppose we estimate $P_s(\Omega_t, \Omega_x, \Omega_y)$, the power spectrum of $s(t, x, y)$, by

$$P_s(\Omega_t, \Omega_x, \Omega_y) = k|S(\Omega_t, \Omega_x, \Omega_y)|^2$$

where k is a normalization constant. As sketched in Figure P6.24a, $P_s(\Omega_t, \Omega_x, \Omega_y)|_{\Omega_t = \Omega_t'}$ has a strong spectral peak at

$$\Omega_x = -\Omega_t' \frac{\cos \theta \cos \phi}{c}$$

$$\Omega_y = -\Omega_t' \frac{\sin \theta \cos \phi}{c}$$

where c is the speed of wave propagation.

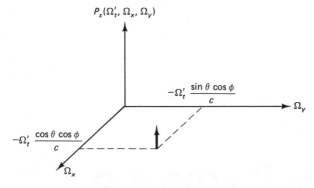

$P_s(\Omega'_t, \Omega_x, \Omega_y)$

$-\Omega'_t \dfrac{\sin\theta\,\cos\phi}{c}$

Ω_y

$-\Omega'_t \dfrac{\cos\theta\,\cos\phi}{c}$

Ω_x

Figure P6.24a

(a) Suppose we consider a different temporal frequency Ω''_t. Sketch $P_s(\Omega''_t, \Omega_x, \Omega_y)$ and determine the (Ω_x, Ω_y) at which the spectral peak occurs.

(b) Suppose $|S(\Omega_t)|^2$ is given by Figure 6.24b. Sketch $P_{ZD}(\Omega_x, \Omega_y) = \displaystyle\int_{\Omega_t=-\infty}^{\infty}$
$P_s(\Omega_t, \Omega_x, \Omega_y)\,d\Omega_t$. The subscript ZD in $P_{ZD}(\Omega_x, \Omega_y)$ refers to "zero delay," for reasons to be discussed shortly.

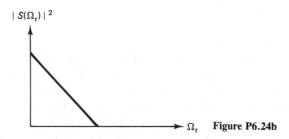

$|S(\Omega_t)|^2$

Ω_t Figure P6.24b

(c) Develop a method of estimating the bearing angle θ from $P_{ZD}(\Omega_x, \Omega_y)$.
(d) Explain why the elevation angle ϕ cannot be estimated from $P_{ZD}(\Omega_x, \Omega_y)$.
(e) Let $R_s(t, x, y)$ denote the correlation function of $s(t, x, y)$ defined by

$$R_s(t, x, y) = F^{-1}[P_s(\Omega_t, \Omega_x, \Omega_y)].$$

We define $R_{ZD}(x, y)$ by

$$R_{ZD}(x, y) = R_s(t, x, y)|_{t=0}.$$

What is the relationship between $R_{ZD}(x, y)$ and $P_{ZD}(\Omega_x, \Omega_y)$? Since $R_{ZD}(x, y)$ does not require correlation along the temporal dimension, it is said to be a zero delay correlation function.

(f) Since $R_{ZD}(x, y)$ does not require correlation along the temporal dimension, estimating $R_{ZD}(x, y)$ or $P_{ZD}(x, y)$ is considerably simpler computationally than estimating $R_s(t, x, y)$ or $P_s(t, x, y)$. Discuss the circumstances under which it is better to estimate the bearing angle from $P_{ZD}(\Omega_x, \Omega_y)$ than from $P_s(\Omega'_t, \Omega_x, \Omega_y)$ at some particular temporal frequency Ω'_t.

7

Image Processing Basics

7.0 INTRODUCTION

In the previous six chapters, we studied the fundamentals of 2-D digital signal processing. In the next four chapters, we will study digital image processing, an important application of 2-D digital signal processing theories.

Digital image processing has many practical applications. One of the earliest applications was processing images from the Ranger 7 mission at the Jet Propulsion Laboratory in the early 1960s. The imaging system mounted on the spacecraft had a number of constraints imposed on it, such as size and weight, and images received had such degradations as blurring, geometric distortions, and background noise. These images were successfully processed by digital computers, and since then images from space missions have been routinely processed by digital computers. The striking pictures of the moon and the planet Mars we see in magazines have all been processed by digital computers.

The image processing application which probably has had the greatest impact on our lives is in the field of medicine. Computed tomography, which has its basis in the projection-slice theorem discussed in Section 1.4.3, is used routinely in many clinical situations, for example, detecting and identifying brain tumors. Other medical applications of digital image processing include enhancement of x-ray images and identification of blood vessel boundaries from angiograms.

Another application, much closer to home for the average person, is the improvement of television images.* The image that we view on a television monitor has flickering, limited resolution, a ghost image, background noise, and motion

*It may be argued that improving the quality of television programs is a much more urgent problem. We point out, though, that improving image quality probably will not make mediocre programs any worse.

crawling due to line interlace. Digital televisions are not far from realization, and digital image processing will have a major impact on improving the image quality of existing television systems and on developing such new television systems as high definition television.

One major problem of such video communications as video conferencing and video telephone has been the enormous bandwidth required. A straightforward coding of broadcast quality video requires on the order of 100 million bits per second. By sacrificing quality and using digital image coding schemes, systems that transmit intelligible images at bit rates lower than 100 thousand bits per second have become commercially available.

Robots are expected to play an increasingly important role in industries and homes. They will perform jobs that are very tedious or dangerous and jobs that require speed and accuracy beyond human ability. As robots become more sophisticated, computer vision will play an increasingly important role. Robots will be asked not only to detect and identify industrial parts, but also to "understand" what they "see" and take appropriate actions. Digital image processing will have a major impact on computer vision.

In addition to these well-established application areas of digital image processing, there are a number of less obvious ones. Law enforcement agents often take pictures in uncooperating environments, and the resulting images are often degraded. For example, snapshots of moving cars' license plates are often blurred; reducing blurring is essential in identifying the car. Another potential application is the study of whale migration. When people study the migratory behavior of lions, tigers, and other land animals, they capture animals and tag them on a convenient tail or ear. When the animals are recaptured at another location, the tags serve as evidence of migratory behavior. Whales, however, are quite difficult to capture and tag. Fortunately, whales like to show their tails, which have features that can be used to distinguish them. To identify a whale, a snapshot of its tail, taken from shipboard, is compared with a reference collection of photographs of thousands of different whales' tails. Successive sightings and identifications of an individual whale allow its migration to be tracked. Comparing photographs, though, is extremely tedious, and digital image processing may prove useful in automating the task.

The potential applications of digital image processing are limitless. In addition to the applications discussed above, they include home electronics, astronomy, biology, physics, agriculture, geography, defense, anthropology, and many other fields. Vision and hearing are the two most important means by which humans perceive the outside world, so it is not surprising that digital image processing has potential applications not only in science and engineering, but also in any human endeavor.

Digital image processing can be classified broadly into four areas, depending on the nature of the task. These are image enhancement, restoration, coding, and understanding. In image enhancement, images either are processed for human viewers, as in television, or are preprocessed to aid machine performance, as in object identification by machine. Image enhancement is discussed in Chapter 8. In image restoration, an image has been degraded in some manner such as blurring, and the objective is to reduce or eliminate the effect of degradation. Image restoration is closely related to image enhancement. When an image is degraded, reducing the image degradation often results in enhancement. There are, however, some important differences between restoration and enhancement. In image restoration, an ideal image has been degraded and the objective is to make the processed image resemble the original as much as possible. In image enhancement, the objective is to make the processed image look better in some sense than the unprocessed image. To illustrate this difference, note that an original, undegraded image cannot be further restored, but can be enhanced by increasing sharpness. Image restoration is discussed in Chapter 9. In image coding, one objective is to represent an image with as few bits as possible, preserving a certain level of image quality and intelligibility acceptable for a given application such as video conferencing. Image coding is related to image enhancement and restoration. If we can enhance the visual appearance of the reconstructed image, or if we can reduce degradation from such sources as quantization noise from an image coding algorithm, then we can reduce the number of bits required to represent an image at a given level of image quality and intelligibility. Image coding is discussed in Chapter 10.

In image understanding, the objective is to symbolically represent the contents of an image. Applications of image understanding include computer vision, robotics, and target identification. Image understanding differs from the other three areas in one major respect. In image enhancement, restoration, and coding, both the input and the output are images, and signal processing has been the backbone of many successful systems in these areas. In image understanding, the input is an image, but the output is typically some symbolic representation of the contents of the input image. Successful development of systems in this area involves both signal processing and artificial intelligence concepts. In a typical image understanding system, signal processing is used for such lower-level processing tasks as reduction of degradation and extraction of edges or other image features, and artificial intelligence is used for such higher-level processing tasks as symbol manipulation and knowledge base management. We treat some of the lower-level processing techniques useful in image understanding as part of our general discussion of image enhancement, restoration, and coding. A more complete treatment of image understanding is beyond the scope of this book.

The theoretical results we studied in the first six chapters are generally based on a set of assumptions. In practice, these assumptions rarely are satisfied exactly. Some results may not be useful in image processing; others may have to be modified. We need to know the basics of image processing if we are to understand the theories' applicability and limitations and to modify them when necessary to adapt to real-world problems. Moreover, the first six chapters have focused on general theories that apply not only to image processing, but also to other 2-D signal processing problems such as geophysical data processing. Some important theoretical results specific to image processing have not yet been discussed. Some basic knowledge of image processing is needed to understand these theories. In this chapter, we present the basics of image processing. These basics will lay a foundation for later chapters' discussion of image enhancement, restoration, and coding. In Section

7.1, we discuss basics of the images we process. In Sections 7.2 and 7.3, we discuss the basics of the human visual system. In Section 7.4, we discuss the basics of a typical image processing environment.

7.1 LIGHT

7.1.1 Light as an Electromagnetic Wave

Everything that we view is seen with light. There are two types of light sources. One type, called a *primary light source*, emits its own light. Examples of primary light sources include the sun, lamps, and candles. The other type, called a *secondary light source*, only reflects or diffuses the light emitted by another source. Examples of secondary light sources include the moon, clouds, and apples.

Light is part of a vast, continuous spectrum of electromagnetic radiation. An electromagnetic wave carries energy, and the energy distribution of the wave passing through a spatial plane can be represented by $c(x, y, t, \lambda)$, where x and y are two spatial variables, t is the time variable, and λ is the wavelength. The function $c(x, y, t, \lambda)$ is called *radiant flux* per (area × wavelength) or *irradiance* per wavelength. The wavelength λ is related to the frequency f by

$$\lambda = c/f \tag{7.1}$$

where c is the speed* of an electromagnetic wave, approximately 3×10^8 m/sec in vacuum and air. Although the function $c(x, y, t, \lambda)$ can be expressed in terms of the frequency f, it is more convenient to use the wavelength λ. The unit associated with $c(x, y, t, \lambda)$ is energy per (area × time × wavelength) and is joules/(m^3 sec) in the MKS (meter, kg, second) system. If we integrate $c(x, y, t, \lambda)$ with respect to λ, we obtain irradiance that has the unit of joules/(m^2 sec) or watts/m^2. Radiation from the sun that passes through a spatial plane perpendicular to the rays has 1350 watts/m^2 of irradiance in the absence of atmospheric absorption. If we integrate $c(x, y, t, \lambda)$ with respect to all four variables x, y, t and λ, we obtain the total energy (in joules) of the electromagnetic wave that passes through the spatial plane.

Light is distinguished from other electromagnetic waves—for instance, radio transmission waves—by the fact that the eye is sensitive to it. Suppose we consider a fixed spatial point (x', y') and a fixed time t'. The function $c(x, y, t, \lambda)$ can be viewed as a function of λ only. We can express it as $c(x', y', t', \lambda)$, or $c(\lambda)$ for convenience. An example of $c(\lambda)$ for the radiation from the sun is shown in Figure 7.1. The eye is sensitive to electromagnetic waves over an extremely narrow range of λ, that is, approximately from 350 nm to 750 nm. (1 nm = 10^{-9} meter). Figure 7.2 shows different types of electromagnetic waves as a function of the wavelength λ. Electromagnetic radiation with large λ, from a few centimeters to several thousand meters, can be generated by electrical circuits. Such radiation is used

---

*The variable c is used both as the speed and as the energy distribution function of an electromagnetic wave. Which is meant will be apparent from the context.

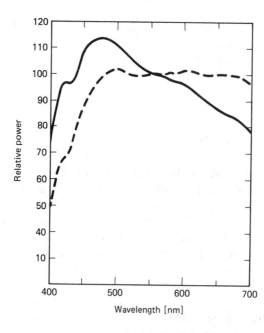

Figure 7.1 Spectral contents of the sun's radiation, above the earth's atmosphere (solid line) and on the ground at noon in Washington (dotted line). After [Hardy].

for radio transmission and radar. Radiation with λ just above the visible range is called *infrared*; with λ just below the visible range, it is called *ultraviolet*. Both infrared and ultraviolet radiations are emitted by typical light sources such as the sun. Radiation with λ far below the visible range includes X rays, γ rays, and cosmic rays; for cosmic rays, λ is less than 10^{-5} nm or 10^{-14} m.

7.1.2 Brightness, Hue, and Saturation

Human perception of light with $c(\lambda)$ is generally described in terms of brightness, hue, and saturation. *Brightness* refers to how bright the light is. *Hue* refers to the color, such as red, orange, or purple. *Saturation*, sometimes called *chroma*, refers to how vivid or dull the color is. Brightness, hue, and saturation are perceptual terms, and they depend on a number of factors, including the detailed shape of $c(\lambda)$, the past history of the observer's exposure to visual stimuli, and the specific environment in which the light is viewed. Nevertheless, it is possible to relate them very approximately to specific features of $c(\lambda)$.

To relate the human perception of brightness to $c(\lambda)$, it is useful to define photometric quantities. The quantities associated with $c(\lambda)$, such as radiant flux, irradiance, and watts/m^2, are called *radiometric units*. These physical quantities can be defined independent of a specific observer. The contributions that $c(\lambda_1)$ and $c(\lambda_2)$ make to human perception of brightness are in general quite different for $\lambda_1 \neq \lambda_2$ even though $c(\lambda_1)$ may be the same as $c(\lambda_2)$. For example, an electromagnetic wave with $c(\lambda)$ is invisible to a human observer as long as $c(\lambda)$ is zero in the visible range of λ, no matter how large $c(\lambda)$ may be outside the visible range. Even within the visible range, the brightness depends on λ. For this reason, a simple integral of $c(\lambda)$ over the variable λ does not relate well to the

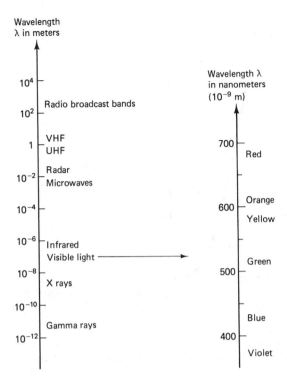

Wavelength
λ in meters

10^4

Radio broadcast bands

10^2

VHF
1
UHF

Radar
10^{-2}
Microwaves

10^{-4}

10^{-6} Infrared
Visible light

10^{-8}

X rays

10^{-10}

Gamma rays

10^{-12}

Wavelength λ
in nanometers
$(10^{-9}$ m)

700
Red

Orange
600
Yellow

Green
500

Blue

400

Violet

Figure 7.2 Different types of electro-magnetic waves as a function of the wavelength λ.

perception of brightness. The quantities which take the human observer's characteristics into account, thus relating to brightness better than the integral of $c(\lambda)$, are called *photometric quantities*.

The basic photometric quantity is luminance, adopted in 1948 by the C.I.E. (Commission Internationale de l'Eclairage), an international body concerned with standards for light and color. Consider a light with $c(\lambda)$ that is zero everywhere except at $\lambda = \lambda_r$, where λ_r denotes a fixed reference wavelength. A light that consists of only one spectral component (one wavelength) is called a *monochromatic light*. Suppose we ask a human observer to compare the brightness from a monochromatic light with $c(\lambda_r)$ with that from another monochromatic light with $c'(\lambda_t)$ where λ_t is a test wavelength. Suppose further that the observer says that $c(\lambda_r)$ matches $c'(\lambda_t)$ in brightness. The equal brightness points $c(\lambda_r)$ and $c'(\lambda_t)$ can be obtained by such experiments as showing two patches of light with a fixed $c(\lambda_r)$ and a variable $c'(\lambda_t)$ and asking the observer to decrease or increase the amplitude of $c'(\lambda_t)$ until they match in brightness. The ratio $c(\lambda_r)/c'(\lambda_t)$ where $c(\lambda_r)$ and $c'(\lambda_t)$ match in brightness, is called the *relative luminous efficiency* of a monochromatic light with λ_t relative to λ_r, and is approximately independent of the amplitude of $c(\lambda_r)$ under normal viewing conditions. The wavelength λ_r used is 555 nm (yellow-green light), at which a typical observer has maximum brightness sensitivity. For this choice of λ_r, the relative luminous efficiency $c(\lambda_r)/c'(\lambda_t)$ is always less than or equal to 1, since $c(\lambda_r)$ is not greater than $c'(\lambda_t)$; that is, it takes less energy at λ_r to produce the same brightness. The relative luminous efficiency as a function of

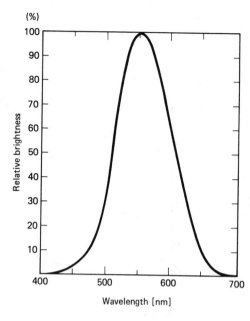

Figure 7.3 C.I.E. relative luminous efficiency function. After [Hardy].

λ is called the *relative luminous efficiency function* and is denoted by $v(\lambda)$. Two monochromatic lights with $c_1(\lambda_1)$ and $c_2(\lambda_2)$ appear equally bright to an observer when*

$$c_1(\lambda_1)v(\lambda_1) = c_2(\lambda_2)v(\lambda_2). \tag{7.2}$$

The relative luminous efficiency function $v(\lambda)$ depends on the observer. Even with a single observer, a slightly different $v(\lambda)$ is obtained when measured at different times. To eliminate this variation, the *C.I.E. standard observer* was defined in 1929, based on experimental results obtained from a number of different observers. The resulting function $v(\lambda)$ is called the *C.I.E. relative luminous efficiency function* and is shown in Figure 7.3. The C.I.E. function is roughly bell-shaped with a maximum of 1 at $\lambda = 555$ nm.

The basic unit of luminance is the lumen (lm). The luminance per area l of a light with $c(\lambda)$ can be defined by

$$l = k \int_{\lambda=0}^{\infty} c(\lambda)v(\lambda) \, d\lambda. \tag{7.3}$$

In (7.3), the quantity l is in units of lumens/m², k is 685 lumens/watt, $c(\lambda)$ is in units of watts/m³, $v(\lambda)$ is unitless, and λ has a unit of m. A monochromatic light with an irradiance of 1 watt/m² produces 685 lumens/m² when $v(\lambda) = 1$. This

---

*Our discussions in this section are brief and some reasonable assumptions are made. For example, (7.2) is based on the transmittivity law, which states that if A and B are equally bright, and B and C are equally bright, then A and C are equally bright. This transmittivity law has been approximately verified experimentally.

occurs when $\lambda = 555$ nm. At other wavelengths, $v(\lambda) < 1$, so the irradiance of the monochromatic light must be greater than 1 watt/m^2 to generate a luminance per area of 685 lumens/m^2. Many other photometric units such as footcandle (lumen/ft^2) and phot (lumen/cm^2) can be defined in terms of lumen.

It is important to note that the luminance or luminance per area does not measure human perception of brightness. For example, a light with 2 lumens/m^2 does not appear twice as bright to a human observer as a light with 1 lumen/m^2. It is also possible to create an environment where a light with a smaller luminance per area looks brighter than a light with a larger luminance per area. However, luminance per area is related more directly than an integral of $c(\lambda)$ to human perception of brightness. Furthermore, in typical viewing conditions (light neither too weak nor excessive), a light with larger luminance per area is perceived to be brighter than a light with smaller luminance per area.

Hue is defined as that attribute of color which allows us to distinguish red from blue. In some cases, the hue of a color can be related to simple features of $c(\lambda)$. Light with approximately constant $c(\lambda)$ in the visible range appears white or colorless. Under normal viewing conditions, a monochromatic light appears colored and its color depends on λ. When an observer is shown a succession of monochromatic lights side by side, the color changes smoothly from one hue to another. Light can be split into a succession of monochromatic lights by a prism, as shown in Figure 7.4. This was first done in 1666 by Newton. Newton divided the color spectrum in the visible range into seven broad categories: red, orange, yellow, green, blue, indigo, and violet, in the order of longer to shorter λ. These are known as the seven colors of the rainbow. Newton originally began with only red, yellow, green, blue and violet. He later added orange and indigo to bring the number to seven (in keeping with the tradition of dividing the week into seven days, the musical notes into seven, and so on).

When a light is not monochromatic but its $c(\lambda)$ is narrow band in the sense that most of its energy is concentrated in $\lambda' - \Delta\lambda < \lambda < \lambda' + \Delta\lambda$ for small $\Delta\lambda$, the perceived hue roughly corresponds to monochromatic light with $\lambda = \lambda'$. The color will appear less pure, however, than a monochromatic light of a similar hue. When $c(\lambda)$ is some arbitrary function, it is difficult to relate the hue to some simple features of $c(\lambda)$. By proper choice of $c(\lambda)$, it is possible to produce hues that do not correspond to any monochromatic light. By mixing red and blue lights, for example, it is possible to produce purple light.

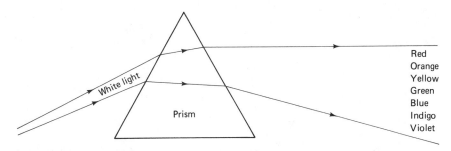

Figure 7.4 White light split into a succession of monochromatic lights by a prism.

Saturation refers to the purity or vividness of the color. A monochromatic light has very pure spectral contents, and looks very vivid and pure. It is said to be highly saturated. As the spectral content of $c(\lambda)$ widens, the color is perceived as less vivid and pure, and the color is said to be less saturated. Color saturation is related very approximately to the effective width of $c(\lambda)$.

7.1.3 Additive and Subtractive Color Systems

When two lights with $c_1(\lambda)$ and $c_2(\lambda)$ are combined, the resulting light has $c(\lambda)$ given by

$$c(\lambda) = c_1(\lambda) + c_2(\lambda). \tag{7.4}$$

Since the lights add in (7.4), this is called an *additive color system*. By adding light sources with different wavelengths, many different colors can be generated. For example, the lighted screen of a color television tube is covered with small, glowing phosphor dots arranged in groups of three. Each of these groups contains one red, one green, and one blue dot. These three colors are used because by proper combination they can produce a wider range of colors than any other combination of three colors; they are the primary colors of the additive color system. Colors of monochromatic lights change gradually, and it is difficult to pinpoint the specific wavelength corresponding to red (R), green (G), and blue (B). The C.I.E. has chosen $\lambda = 700$ nm for red, $\lambda = 546.1$ nm for green, and $\lambda = 435.8$ nm for blue.

The three primary colors of the additive color system are shown in Figure 7.5. In the additive color system, a mixture of equal amounts of blue and green produces cyan. A mixture of equal amounts of red and blue produces magenta, and a mixture of equal amounts of red and green produces yellow. The three colors yellow (Y), cyan (C), and magenta (M) are called the secondary colors of the additive color system. When roughly equal amounts of all three colors R, G,

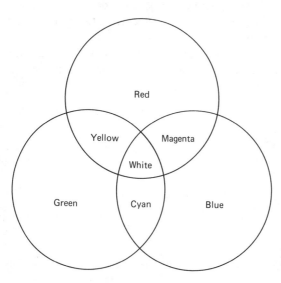

Figure 7.5 Primary colors of the additive color system.

and B are combined, the result is white. When roughly equal amounts of R, G, and B components are used in a color TV monitor, therefore, the result is a black-and-white image. By combining different amounts of the R, G, and B components, other colors can be obtained. A mixture of a red light and a weak green light with no blue light, for example, produces a brown light.

Nature often generates color by filtering out, or subtracting, some wavelengths and reflecting others. This process of wavelength subtraction is accomplished by molecules called *pigments*, which absorb particular parts of the spectrum. For example, when sunlight, which consists of many different wavelengths, hits a red apple, the billions of pigment molecules on the surface of the apple absorb all the wavelengths except those corresponding to red. As a result, the reflected light has a $c(\lambda)$ which is perceived as red. The pigments subtract out certain wavelengths, and a mixture of two different types of pigments will result in a reflected light whose wavelengths are further reduced. This is called a *subtractive color system*. When two inks of different colors are combined to produce another color on paper, the subtractive color system applies.

The three primary colors of a subtractive color system are yellow (Y), cyan (C), and magenta (M), which are the secondary colors of the additive color system. The three colors are shown in Figure 7.6. By mixing the proper amounts of these colors (pigments), a wide range of colors can be generated. A mixture of yellow and cyan produces green. A mixture of yellow and magenta produces red. A mixture of cyan and magenta produces blue. Thus, the three colors, red, green, and blue, the primary colors of the additive color system, are the secondary colors of the subtractive color system. When all three primary colors Y, C, and M are combined, the result is black; the pigments absorb all the visible wavelengths.

It is important to note that the subtractive color system is fundamentally different from the additive color system. In the additive color system, as we add colors (lights) with different wavelengths, the resulting light consists of more wavelengths. We begin with black, corresponding to no light. As we then go from

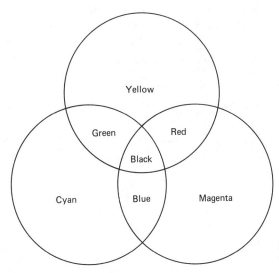

Figure 7.6 Primary colors of the subtractive color system.

the primary colors (RGB) to the secondary colors (YCM) and then to white, we increase the spread of the wavelengths in the resulting light. In a subtractive color system, we begin with white, corresponding to no pigments. As we go from the primary colors (YCM) to the secondary colors (RGB) and then to black, we decrease the spread of the wavelengths in the resulting reflected light.

In an additive color system, we can think of red, green, and blue light as the result of passing white light through three different bandpass filters. Mixing two colors can be viewed as passing white light through a filter which is a parallel combination of the two corresponding bandpass filters. In a subtractive color system, we can think of yellow, cyan, and magenta as the result of passing white light through three different bandstop filters. Mixing two colors can be viewed as passing white light through a cascade of the two corresponding bandstop filters.

7.1.4 Representation of Monochrome and Color Images

For a black-and-white image, a light with $c(\lambda)$ can be represented by one number I given by

$$I = k \int_{\lambda=0}^{\infty} c(\lambda)s_{\mathrm{BW}}(\lambda)\,d\lambda \tag{7.5}$$

where $s_{\mathrm{BW}}(\lambda)$ is the spectral characteristic of the sensor used and k is some scaling constant. Since brightness perception is the primary concern with a black-and-white image, $s_{\mathrm{BW}}(\lambda)$ is typically chosen to resemble the relative luminous efficiency function discussed in Section 7.1.2. The value I is often referred to as the *luminance*, *intensity*, or *gray level* of a black-and-white image. Since I in (7.5) represents power per unit area, it is always nonnegative and finite; that is,

$$0 \leq I \leq I_{\mathrm{MAX}} \tag{7.6}$$

where I_{MAX} is the maximum I possible. In image processing, I is typically scaled such that it lies in some convenient arbitrary range, for example, $0 \leq I \leq 1$ or $0 \leq I \leq 255$. In these cases 0 corresponds to the darkest possible level and 1 or 255 corresponds to the brightest possible level. Because of this scaling, the specific radiometric or photometric units associated with I become unimportant. A black-and-white image has, in a sense, only one color. Thus, it is sometimes called a monochrome image.

A color image can be viewed as three monochrome images. For a color image, a light with $c(\lambda)$ is represented by three numbers which are called *tristimulus values*. One three-number set that is frequently used in practice is R, G, and B, representing the intensity of the red, green, and blue components. The tristimulus values R, G, and B are obtained by

$$\mathrm{R} = k \int_{\lambda=0}^{\infty} c(\lambda)s_{\mathrm{R}}(\lambda)\,d\lambda \tag{7.7a}$$

$$\mathrm{G} = k \int_{\lambda=0}^{\infty} c(\lambda)s_{\mathrm{G}}(\lambda)\,d\lambda \tag{7.7b}$$

and

$$B = k \int_{\lambda=0}^{\infty} c(\lambda) s_B(\lambda) \, d\lambda \qquad (7.7c)$$

where $s_R(\lambda)$, $s_G(\lambda)$, and $s_B(\lambda)$ are spectral characteristics of the red, green, and blue sensors (filters) respectively. Like the gray level I in a monochrome image, R, G, and B are nonnegative and finite. One possible set of $s_R(\lambda)$, $s_G(\lambda)$, and $s_B(\lambda)$ is shown in Figure 7.7. Examples of $f_R(x, y)$, $f_G(x, y)$ and $f_B(x, y)$, which represent the red, green and blue components of a color image, are shown in Figures 7.8(a), (b), and (c), respectively (see color insert). The color image that results when the three components are combined by a color television monitor is shown in Figure 7.8(d).

One approach to processing a color image is to process three monochrome images, R, G, and B, separately and then combine the results. This approach is simple and is often used in practice. Since brightness, hue, and saturation each depends on all three monochrome images, independent processing of R, G, and B may affect hue and saturation, even though the processing objective may be only modifying the brightness.

The three tristimulus values R, G, and B can be transformed into a number of other sets of tristimulus values. One particular set, known as *luminance-chrominance*, is quite useful in practice. When R, G, and B are the values used on a

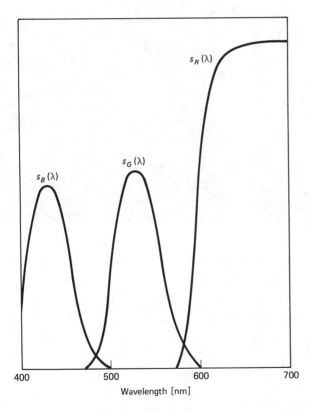

Figure 7.7 Example of spectral characteristics of red, green, and blue color sensors.

(a)

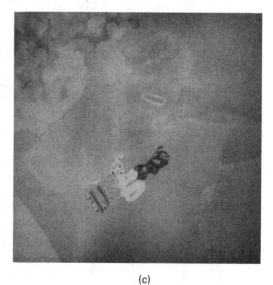

(b)   (c)

Figure 7.9 *Y*, *I*, and *Q* components of the color image in Figure 7.8(d). (a) *Y* component; (b) *I* component; (c) *Q* component.

TV monitor (in the NTSC* color system), the corresponding luminance-chrominance values Y, I, and Q are related to R, G, and B by

$$\begin{bmatrix} Y \\ I \\ Q \end{bmatrix} = \begin{bmatrix} 0.299 & 0.587 & 0.114 \\ 0.596 & -0.274 & -0.322 \\ 0.211 & -0.523 & 0.312 \end{bmatrix} \begin{bmatrix} R \\ G \\ B \end{bmatrix} \qquad (7.8a)$$

*NTSC stands for National Television Systems Committee.

$$\text{and} \qquad \begin{bmatrix} R \\ G \\ B \end{bmatrix} = \begin{bmatrix} 1.000 & 0.956 & 0.621 \\ 1.000 & -0.273 & -0.647 \\ 1.000 & -1.104 & 1.701 \end{bmatrix} \begin{bmatrix} Y \\ I \\ Q \end{bmatrix}. \qquad (7.8b)$$

The Y component is called the luminance component, since it roughly reflects the luminance l in (7.3). It is primarily responsible for the perception of the brightness of a color image, and can be used as a black-and-white image. The I and Q components are called chrominance components, and they are primarily responsible for the perception of the hue and saturation of a color image. The $f_Y(x, y)$, $f_I(x, y)$, and $f_Q(x, y)$ components, corresponding to the color image in Figure 7.8, are shown as three monochrome images in Figures 7.9(a), (b), and (c), respectively. Since $f_I(x, y)$ and $f_Q(x, y)$ can be negative, a bias has been added to them for display. The mid-gray intensity in Figures 7.9(b) and (c) represents the zero amplitude of $f_I(x, y)$ and $f_Q(x, y)$. One advantage of the YIQ tristimulus set relative to the RGB set is that we can process the Y component only. The processed image will tend to differ from the unprocessed image in its appearance of brightness. Another advantage is that most high-frequency components of a color image are primarily in the Y component. Therefore, significant spatial lowpass filtering of I and Q components does not significantly affect the color image. This feature can be exploited in the coding of a digital color image or in the analog transmission of a color television signal.

When the objective of image processing goes beyond accurately reproducing the "original" scene as seen by human viewers, we are not limited to the range of wavelengths visible to humans. Detecting objects that generate heat, for example, is much easier with an image obtained using a sensor that responds to infrared light than with a regular color image. Infrared images can be obtained in a manner similar to (7.7) by simply changing the spectral characteristics of the sensor used.

7.2 THE HUMAN VISUAL SYSTEM

7.2.1 The Eye

The human visual system is one of the most complex in existence. Our visual system allows us to organize and understand the many complex elements of our environment. For nearly all animals, vision is just an instrument of survival. For humans, vision is not only an aid to survival, but an instrument of thought and a means to a richer life.

The visual system consists of an eye that transforms light to neural signals, and the related parts of the brain that process the neural signals and extract necessary information. The eye, the beginning of the visual system, is approximately spherical with a diameter of around 2 cm. From a functional point of view, the eye is a device that gathers light and focuses it on its rear surface.

A horizontal cross section of an eye is shown in Figure 7.10. At the very front of the eye, facing the outside world, is the cornea, a tough, transparent membrane. The main function of the cornea is to refract (bend) light. Because

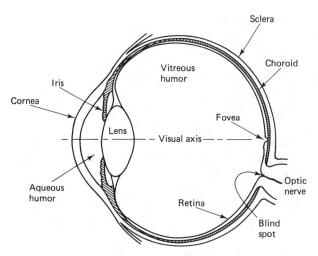

Figure 7.10 Horizontal cross section of a right human eye, seen from above.

of its rounded shape, it acts like the convex lens of a camera. It accounts for nearly two-thirds of the total amount of light bending needed for proper focusing.

Behind the cornea is the aqueous humor, which is a clear, freely flowing liquid. Through the cornea and the aqueous humor, we can see the iris. By changing the size of the pupil, a small round hole in the center of the iris, the iris controls the amount of light entering the eye. Pupil diameter ranges between 1.5 mm ~ 8 mm, with smaller diameter corresponding to exposure to brighter light. The color of the iris determines the color of the eye. When we say that a person has blue eyes, we mean blue irises. Iris color, which has caught the attention of so many lovers and poets, is not functionally significant to the eye.

Behind the iris is the lens. The lens consists of many transparent fibers encased in a transparent elastic membrane about the size and shape of a small bean. The lens grows throughout a person's lifetime. Thus, the lens of an eighty-year-old man is more than fifty percent larger than that of a twenty-year old. As with an onion, cells in the oldest layer remain in the center, and cells in newer layers grow further from the center. The lens has a bi-convex shape and a refractive index of 1.4, which is higher than any other part of the eye through which light passes. However, the lens is surrounded by media that have refractive indices close to its own. For this reason, much less light-bending takes place at the lens than at the cornea. The cornea has a refractive index of 1.38, but faces the air, which has a refractive index of 1. The main function of the lens is to accurately focus the incoming light on a screen at the back of the eye called the retina. For a system with a fixed lens and a fixed distance between the lens and the screen, it is possible to focus objects at only one particular distance. If faraway objects are in sharp focus, for example, close objects will be focused behind the screen. To be able to focus close objects at one time and distant objects at some other time, a camera changes the distance between the fixed lens and the screen. This is what the eyes of many fish do. In the case of the human eye, the shape of the lens, rather than the distance between the lens and screen, is changed. This process of changing shape to meet the needs of both near and far vision is called *accommo-*

dation. This adjustability of the shape is the most important feature of the lens. Accommodation takes place almost instantly and is controlled by the ciliary body, a group of muscles surrounding the lens.

Behind the lens is the vitreous humor, which is a transparent jelly-like substance. It is optically matched so that light which has been sharply focused by the lens keeps the same course. The vitreous humor fills the entire space between the lens and the retina and occupies about two-thirds of the eye's volume. One of its functions is to support the shape of the eye.

Behind the vitreous humor is the retina, which covers about 65% of the inside of the eyeball. This is the screen on which the entering light is focused and light-receptive cells convert light to neural signals. All of the other eye parts we have discussed so far serve the function of placing a sharp image on this receptor surface. The fact that an image is formed on the retina, so the eye is simply an image catching device, was not known until the early seventeenth century. Even though the ancient Greeks knew the structure of an eye accurately and performed delicate surgery on it, they theorized that light-like rays emanate from the eye, touch an object, and make it visible. After all, things appear "out there." In 1625, Scheiner demonstrated that light enters the eye and vision stems from the light that enters the eye. By exposing the retina of an animal and looking through it from behind, he was able to see miniature reproductions of the objects in front of the eyeball.

There are two types of light-receptive cells in the retina. They are called cones and rods because of their shape. The cones, which number about 7 million, are less sensitive to light than rods, and are primarily for day (photopic) vision. They are also responsible for seeing color. The three types of cones are most sensitive to red, green, and blue light, respectively. This is the qualitative physiological basis for representing a color image with red, green, and blue monochrome images. The rods, which number about 120 million, are more sensitive to light than cones, and are primarily for night (scotopic) vision. Since the cones responsible for color vision do not respond to dim light, we do not see color in very dim light.

Rods and cones are distributed throughout the retina. However, their distribution is highly uneven. The distribution of the rods and cones in the retina is shown in Figure 7.11. Directly behind the middle point of the pupil is a small

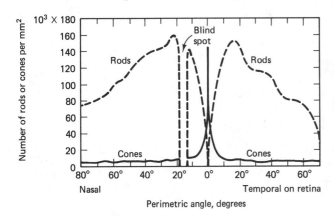

Figure 7.11 Distribution of rods (dotted line) and cones (solid line) on the retina. After [Pirenne].

depressed dimple on the retina called the fovea. There are no rods in this small region, and most of the cones are concentrated here. Therefore, this is the region for the most accurate vision in bright light. When we look straight ahead at an object, the object is focused on the fovea. Since the fovea is very small, we constantly move our attention from one region to another when studying a larger region in detail. The rods, which function best in night vision, are concentrated away from the fovea. Since there are no rods in the fovea, an object focused in the fovea is not visible in dim light. To see objects at night, therefore, we look at them slightly sideways.

There are many thin layers in the retina. Even though cones and rods are light-receptive cells, so that it would be reasonable for them to be located closer to the vitreous humor, they are located farther away from it. Therefore, light has to pass through other layers of the retina, such as nerve fibers, to reach the cones and rods. This is shown in Figure 7.12. It is not clear why nature chose to do it this way, but the arrangement works. In the fovea, at least, the nerves are pushed aside so that the cones are directly exposed to light. Due to this particular arrangement, the optic nerve fibers have to pass through the light-receptive cell layers

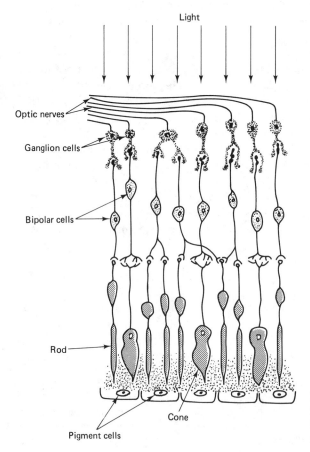

Light

Optic nerves

Ganglion cells

Bipolar cells

Rod

Cone

Pigment cells

Figure 7.12 Layers in the retina. Note that light has to travel through several layers before it reaches light-sensitive cells.

426　　　　　　　　　　　　　　　　　　　　Image Processing Basics　　Chap. 7

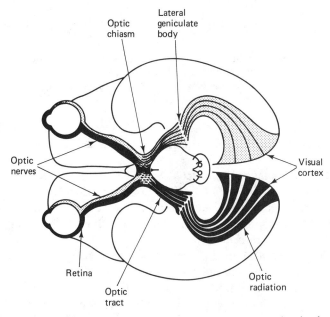

Figure 7.13 Path that neural signals travel from the retina to the visual cortex.

on the way to the brain. Instead of crossing the light-receptive cell layers throughout the retina, they bundle up at one small region the size of a pinhead in the retina, known as the blind spot. Since there are no light receptive cells in this region, we cannot see light focused on the blind spot.

When light hits cones and rods, a complex electrochemical reaction takes place, and light is converted to neural impulses, which are transmitted to the brain through the optic nerve fibers. There are about 130 million light-receptive cells (cones and rods), but only about 1 million nerve fibers. This means that one nerve fiber, on the average, serves more than 100 light-receptive cells. The nerve fibers are not shared equally. Some cones in the fovea are served by one nerve fiber each, increasing the visual acuity in this region. The rods, however, always share nerve fibers. This is one reason why visual acuity at night is not as good as it is during the day even though there are many more rods than cones.

After the optic nerve bundles leave the two eyes, the two bundles meet at an intersection called the optic chiasm. This is shown in Figure 7.13. Each of the two bundles is divided into two branches. Two branches, one from each of the two bundles, join together to form a new bundle. The remaining two branches form another bundle. This crossing of the nerve fibers from two eyes is partly responsible for our stereoscopic vision, which mixes the images from each eye to allow the visual field to be perceived as a 3-D space. These two new bundles go to the left and right lateral geniculate bodies, respectively. The original fibers end here and new fibers continue to the visual cortex, where the neural signals are processed and vision takes place. The visual cortex is a small part of the cortex, a mass of gray matter forming two hemispheres in the back of the brain. Little is known about how the visual neural signals are processed in the visual cortex.

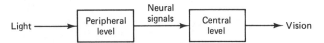

Figure 7.14 Human visual system as a cascade of two systems. The first system represents the peripheral level of the visual system and converts light to neural signals. The second system represents the central level and processes neural signals to extract necessary information.

7.2.2 Model for Peripheral Level of Visual System

The human visual system discussed in Section 7.2.1 can be viewed as a cascade of two systems, as shown in Figure 7.14. The first system, which represents the peripheral level of the visual system, converts light to a neural signal. The second system, which represents the central level of the visual system, processes the neural signal to extract information.

Unlike central level processing, about which little is known, peripheral level processing is fairly well understood, and many attempts have been made to model it. One very simple model [Stockham] for a monochrome image that is consistent with some well-known visual phenomena is shown in Figure 7.15. In this model, the monochrome image intensity $I(x, y)$ is modified by a nonlinearity, such as a logarithmic operation, that compresses the high level intensities but expands the low level intensities. The result is then filtered by a linear shift-invariant (LSI) system with spatial frequency response $H(\Omega_x, \Omega_y)$. The nonlinearity is motivated by the results of some psychophysical experiments that will be discussed in the next section. The LSI system $H(\Omega_x, \Omega_y)$, which is bandpass in character, is motivated by the finite size of the pupil, the resolution limit imposed by a finite number of light-receptive cells, and the lateral inhibition process. The finite size of the pupil and the resolution limit due to a finite number of light receptive cells are responsible for the lowpass part of the bandpass nature of $H(\Omega_x, \Omega_y)$. The lateral inhibition process stems from the observation that one neural fiber responds to many cones and rods. The response of the neural fiber is some combination of the signals from the cones and rods. While some cones and rods contribute positively, others contribute negatively (inhibition). This lateral inhibition process is the rationale for the highpass part of the bandpass character of $H(\Omega_x, \Omega_y)$. Even though the model in Figure 7.15 is very simple and applies only to the peripheral level of the human visual system, it is consistent with some of the visual phenomena that are discussed in the next section.

One way to exploit a model such as the one in Figure 7.15 is to process an image in a domain closer to where vision takes place. This can be useful in some applications. In image coding, for example, the information that is in the image but is discarded by the visual system does not need to be coded. By processing an image in a domain closer to where vision takes place, more emphasis can be

Figure 7.15 Simple model of peripheral level of human visual system.

placed on what is important to the visual system. This is one reason why some image processing operations are performed in the log intensity domain rather than the intensity domain.

In addition to the model in Figure 7.15, more sophisticated models for monochrome images and models for color images have also been proposed in the literature [Budrikis; Mannos and Sakrison; Hall and Hall; Faugeras; Granrath].

7.3 VISUAL PHENOMENA

7.3.1 Intensity Sensitivity

One way to quantify our ability to resolve two visual stimuli which are the same except for their intensities or luminances is by measuring the just-noticeable difference (j.n.d.). The j.n.d. can be defined and measured in a variety of ways. One way is through a psychophysical experiment called *intensity discrimination*. Suppose we present the visual stimulus in Figure 7.16 to an observer. The inside region is a monochrome image of uniform intensity I_{in}, which is randomly chosen to be either I or $I + \Delta I$. The outside region is a monochrome image of intensity I_{out}, which is chosen to be $I + \Delta I$ when $I_{in} = I$ and I when $I_{in} = I + \Delta I$. The observer is asked to make a forced choice as to which of the two intensities I_{in} and I_{out} is brighter. When ΔI is very large, the observer will give a correct answer most of the time, correct in the sense that the region with $I + \Delta I$ is chosen. When ΔI is very small, the observer will give a correct answer about half of the time. As we move away from a very large ΔI to a very small ΔI, the percentage of the observer's correct responses decreases continuously, and we can define ΔI at which the observer gives correct responses 75% of the time as the j.n.d. at the intensity I.

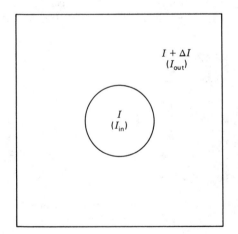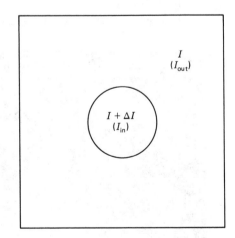

Figure 7.16   Two stimuli used in an intensity discrimination experiment. Each trial consists of showing one of the two stimuli to an observer and asking the observer to make a forced choice of which between I_{in} and I_{out} appears brighter. The stimulus used in a trial is chosen randomly from the two stimuli. Results of this experiment can be used to measure the just noticeable difference ΔI as a function of I.

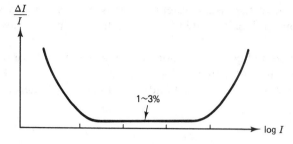

The result of plotting $\Delta I/I$ as a function of I where ΔI is the j.n.d. is shown in Figure 7.17. From the figure, for a wide range of I,

$$\frac{\Delta I}{I} \approx \text{constant.} \tag{7.9}$$

This relationship is called *Weber's law*. Weber's law states that the j.n.d. ΔI is proportional to I. As we increase I, we need a larger ΔI to make $I + \Delta I$ noticeably different from I. This is one way in which the visual system remains sensitive to a wide dynamic range of the visual stimulus intensity. Weber's law holds approximately not only for vision, but also for all other human sense modalities: hearing, smell, taste, and touch.

As we let ΔI approach 0, (7.9) can be written as

$$\frac{dI}{I} = d(\log I) \approx \text{constant.} \tag{7.10}$$

From (7.10), the j.n.d. is constant in the log I domain for a wide range of I. This is consistent with the notion that a nonlinear operation such as the log is applied to the image intensity in the simple model in Figure 7.15. The intensity discrimination experiment involves a very simple task on the part of the observer, and

Figure 7.18 Image of 512 × 512 pixels degraded by zero-mean white noise with a uniform probability density function. Same level of noise is more visible in a dark region relative to a bright region. Same level of noise is more visible in a uniform background region than in a region with edges.

(a)

(b)

Figure 7.8 Red, green, and blue components of a color image. (a) Red component; (b) green component; (c) blue component; (d) color image of 512 × 512 pixels.

(c)

(d)

Figure 8.7 Example of gray scale modification. (a) Original color image of 512 × 512 pixels; (b) processed color image.

(a)

(b)

Figure 8.49 Creation of a color image from a black-and-white image by mapping the result of lowpass filtering the black-and-white image to the blue component, the result of bandpass filtering to the green component, and the result of highpass filtering to the red component.

(a)

(b)

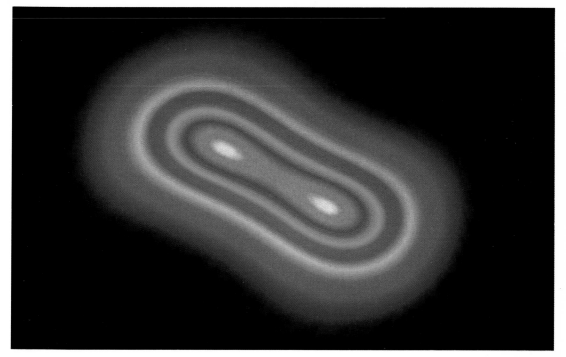

Figure 8.50 (b)

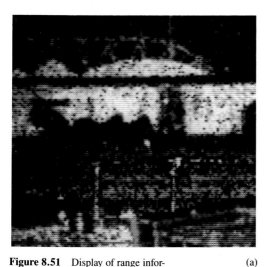

Figure 8.51 Display of range information with color. After [Sullivan et al.]. (a) Intensity image from an infrared radar imaging system with range information discarded; (b) image in (a) with range information displayed with color.

(a)

(b)

 (a)  (b)

Figure 10.58 Example of color image coding using the scene adaptive coder. Courtesy of Wen-Hsiung Chen. (a) Original image of 512 × 512 pixels; (b) reconstructed image at 0.4 bit/pixel.

 (a)  (b)

Figure 10.59 Another example of color image coding using the scene adaptive coder. Courtesy of Wen-Hsiung Chen. (a) Original image of 512 × 512 pixels; (b) reconstructed color image of 0.4 bit/pixel.

complicated central level processing probably is not needed. Consequently, the result of the intensity discrimination experiment can be related to peripheral level processing in the visual system.

The result of the intensity discrimination experiment states that the j.n.d. ΔI increases as I increases. This partly explains why a uniform level of random noise is more visible in a darker region than in a brighter region. This is illustrated in Figure 7.18. The image in Figure 7.18 is the result of adding zero-mean white noise with a uniform probability density to an original undegraded image. The grainy appearance, which is due to noise, is more pronounced in the darker uniform background region than in the brighter uniform background region. Since the j.n.d. ΔI is larger for a larger I, a higher level of noise in the brighter region is required for it to be equally visible as a lower level of noise in the darker region. The implication of this observation for image processing is that reducing noise in the darker region is more important than reducing noise in the brighter region.

7.3.2 Adaptation

In the intensity discrimination experiment discussed above, the intensities shown at any given time are I and $I + \Delta I$. If the observer is assumed to spend some time before making a decision, the result is obtained when the observer is adapted to the intensity level I. When the intensity level the observer is adapted to is different from I, the observer's intensity resolution ability decreases. Suppose we run the same intensity discrimination experiment discussed in Section 7.3.1, but with I and $I + \Delta I$ surrounded by a much larger region with intensity I_0, as shown in Figure 7.19. The result of plotting $\Delta I/I$ as a function of I and I_0 is shown in

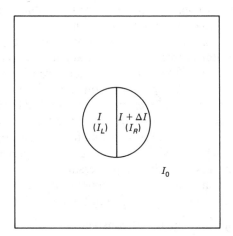 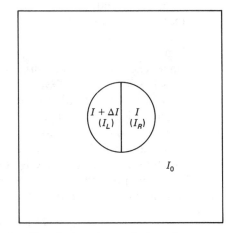

Figure 7.19 Two stimuli used in studying the effect of adaptation on the intensity sensitivity. Each trial consists of showing one of the two stimuli to an observer and asking the observer to make a forced choice of which between I_R or I_L appears brighter. The stimulus used in a trial is chosen randomly from the two stimuli. Results of this experiment can be used to measure the just noticeable difference ΔI as a function of I and I_0.

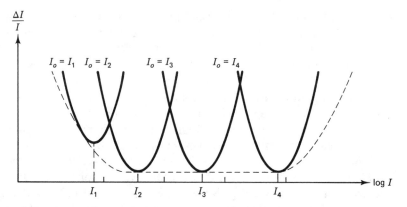

Figure 7.20 Plot of $\Delta I / I$ as a function of I and I_0. When I_0 equals I, $\Delta I / I$ is the same as that in Figure 7.17 (dotted line in this figure). When I_0 is different from I, $\Delta I / I$ increases relative to the case $I_0 = I$. This means that the observer's sensitivity to intensity decreases.

Figure 7.20. When I_0 is equal to I, the result is the same as that in Figure 7.17. When I_0 is different from I, however, the j.n.d. ΔI increases relative to the case $I_0 = I$, indicating that the observer's sensitivity to intensity decreases. The result shows that sensitivity to intensity is highest near the level that the observer is adapted to. This is another way in which the visual system responds to a wide range of intensities at different times.

7.3.3 Mach Band Effect and Spatial Frequency Response

Consider an image whose intensity is constant along the vertical dimension but increases in a staircase manner along the horizontal dimension, as shown in Figure 7.21(a). The intensities along the horizontal direction are sketched in Figure 7.21(b). Even though the intensity within each rectangular region is constant, each region looks brighter towards the left and darker towards the right. This is known as the *Mach band effect*. This phenomenon is consistent with the presence of spatial filtering in the peripheral-level model of the visual system in Figure 7.15. When a filter is applied to a signal with sharp discontinuities, an overshoot and undershoot occur. This is partly responsible for uneven brightness perception within the region of uniform intensity. This suggests that precise preservation of the edge shape is not necessary in image processing.

The presence in the visual system of a spatial bandpass filter can be seen by looking at the image in Figure 7.22. The image $I(x, y)$ in Figure 7.22 is given by

$$I(x, y) = I_0(y) \cos (\omega(x)x) + \text{constant} \tag{7.11}$$

where the constant is chosen such that $I(x, y)$ is positive for all (x, y). As we move in the horizontal direction from left to right, the spatial frequency $\omega(x)$ increases. As we move in the vertical direction from top to bottom, the amplitude $I_0(y)$ increases. If the spatial frequency response were constant across the frequency range, sensitivity to intensity would be constant along the horizontal di-

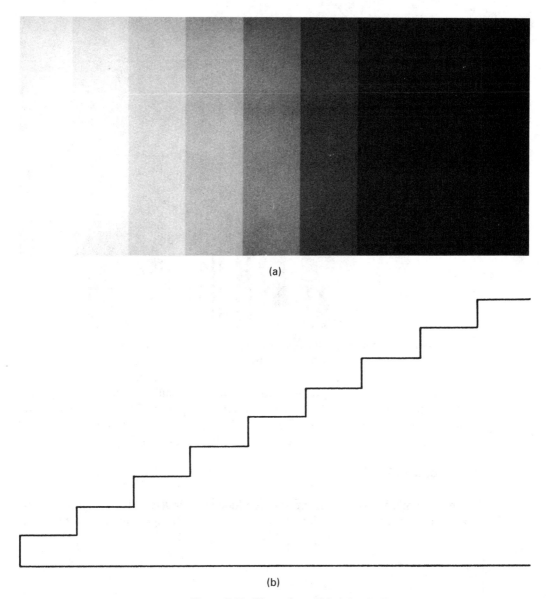

(a)

(b)

Figure 7.21 Illustration of Mach band effect.

rection. In Figure 7.22, we are more sensitive to the contrast in mid-frequency regions than in low- and high-frequency regions, indicating the bandpass character of the visual system. A spatial filter frequency response $H(\Omega_x, \Omega_y)$, which is more accurately measured by assuming the model in Figure 7.15 is correct, is shown in Figure 7.23. The horizontal axis is the spatial frequency/angle of vision. The perceived spatial frequency of an image changes as a function of the distance between the eye and the image. As the distance increases, the perceived spatial frequency increases. To take this effect into account, the spatial frequency/angle

Figure 7.22 Modulated sinewave grating that illustrates the bandpass character of the peripheral level of the human visual system.

of vision (spatial frequency relative to the spatial domain in the retina) is often used in determining $H(\Omega_x, \Omega_y)$. The frequency response $H(\Omega_x, \Omega_y)$ is maximum at the spatial frequency in the range of approximately $5 \sim 10$ cycles/degree and decreases as the spatial frequency increases or decreases from $5 \sim 10$ cycles/degree.

7.3.4 Spatial Masking

When random noise of a uniform level is added to an image, it is much more visible in a uniform background region than in a region with high contrast. This effect

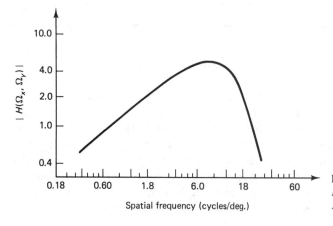

Figure 7.23 Frequency response $H(\Omega_x, \Omega_y)$ in the model in Figure 7.15. After [Davidson].

is much more pronounced than the effect of the brightness level on noise visibility discussed in Section 7.3.1. Consider the image in Figure 7.18, which illustrates the effect of overall brightness level on noise visibility. In the figure, the noise is much less visible in the edge regions than in the uniform background regions. In addition, the noise in the dark edge regions is less visible than the noise in the bright uniform background regions. One way to explain this is to consider the local signal-to-noise ratio (SNR). If we consider the local SNR to be the ratio of the signal variance to the noise variance in a local region, at the same level of noise, the SNR is higher in the high-contrast region than in the uniform background. Another related view is spatial masking. In a high-contrast region, signal level is high and tends to mask the noise more.

The spatial masking effect can be exploited in image processing. For example, attempting to reduce background noise by spatial filtering typically involves some level of image blurring. In high-contrast regions, where the effect of blurring due to spatial filtering is more likely to be pronounced, the noise is not as visible, so little spatial filtering may be needed.

7.3.5 Other Visual Phenomena

It is well known that a sharper image generally looks even more pleasant to a human viewer than an original image. This is often exploited in improving the appearance of an image for a human viewer. It is also a common experience that an unnatural aspect catches a viewer's attention. A positive aspect of this phenomenon is that it can be exploited in such applications as production of television commercials. A negative aspect is that it sometimes makes it more difficult to develop a successful algorithm using computer processing techniques. For example, some image processing algorithms are capable of reducing a large amount of background noise. In the process, however, they introduce noise that has an artificial tone to it. Even when the amount of the artificial noise introduced is much less than the amount by which the background noise is reduced, the artificial noise may catch a viewer's attention more, and a viewer may prefer the unprocessed image over the processed image.

The visual phenomena discussed in the previous sections can be explained simply, at least at a qualitative level; however, many other visual phenomena cannot be explained simply, in part due to our lack of knowledge. For example, a visual phenomenon that involves a fair amount of central level processing cannot be explained in a satisfactory manner. Figure 7.24 shows a sketch consisting of just a small number of lines. How we can associate this image with Einstein is not clear. The example does demonstrate, however, that simple outlines representing the gross features of an object are important for its identification. This can be exploited in such applications as object identification in computer vision and the development of a very low bit-rate video telephone system for the deaf.

The visual phenomena discussed above relate to the perception of light that shines continuously. When light shines intermittently, our perception depends a great deal on its frequency. Consider a light that flashes on for a brief duration N times per second. When N is small, the light flashes are perceived to be separate.

Figure 7.24 Sketch of Einstein's face with a small number of lines.

As we increase N, an unsteady flicker that is quite unpleasant to the human viewer occurs. As we increase N further, the flicker becomes less noticeable, and eventually the observer can no longer detect that the light intensity is changing as a function of time. The frequency at which the observer begins perceiving light flashes as continuous light is called the *critical flicker frequency* or *fusion frequency*. The fusion frequency increases as the size and overall intensity of the flickering object increase. The fusion frequency can be as low as a few cycles/sec for a very dim, small light and may exceed 100 cycles/sec for a very bright, larger light. When a flicker is perceived, visual acuity is at its worst.

Intermittent light is common in everyday vision. Fluorescent lights do not shine continuously, as they appear to, but flicker at a sufficiently high rate (over 100 times/sec) that fusion is reached in typical viewing conditions. Avoiding the perception of flicker is an important consideration in deciding the rate at which a CRT (cathode ray tube) display monitor is refreshed. As is discussed further in Section 7.4, CRT display monitors are illuminated only for a short period of time. For an image to be displayed continuously without the perception of flicker, the monitor has to be refreshed at a sufficiently high rate. Typically, a CRT display monitor is refreshed 60 times per second. With 2:1 interlace, which is discussed further in Section 7.4, this corresponds to 30 frames/sec. The current NTSC (National Television Systems Committee)* television system employs 30 frames/

*The NTSC is a group formed in the United States in 1940 to formulate standards for monochrome television broadcast service. It was reconvened in 1950 to study color television systems and recommended a standard system for the United States. The NTSC system is used in North America and Japan. Both the SECAM (Sequential Couleur a Memoire) System used in France, the USSR and Eastern Europe, and the PAL (Phase Alternating Line) system used in South America, Africa, and Western Europe except France employ 25 frames/sec with 2:1 interlace.

sec with 2:1 interlace. For motion pictures, 24 frames per second are shown, with one frame shown twice. The effective flicker rate is therefore 48 frames/sec. In addition, the typical viewing condition in a cinema is very dark, decreasing the fusion frequency to below 40 cycles/sec. For this reason, flickering is not visible in a motion picture, even though the screen is dark for approximately half of the time.

Even though each frame of a motion picture or television broadcast is actually still, and only a finite number of frames are shown in a second, the objects in the scene appear to be moving in a continuous manner. This effect, known as *motion rendition*, is closely related to the phi phenomenon [Kinchla et al]. Consider two pulsating light sources separated by approximately 1 degree of an observer's viewing angle. When the lights shine for one msec each with a separation of 10 msec, the light is perceived to move continuously from one source to the other. When the time difference between the two lights is on the order of 1 msec, they appear to flash simultaneously. When the time difference is more than 1 second, they are perceived as two separate flashes. This is known as the *phi phenomenon*.

In general, frame rates that are sufficiently high to avoid flicker are adequate for motion rendition. The fact that an object appears to move in a continuous manner in a motion picture or television broadcast does not necessarily imply that the sampling rate along the temporal dimension is above the Nyquist rate. For objects with sufficiently rapid motion, sampling the temporal dimension 24 times/sec or 30 times/sec is far lower than the Nyquist rate, and temporal aliasing occurs. Temporal aliasing does not always cause motion discontinuity. In a movie, we sometimes see a wheel that moves continuously, but backwards. In this case, motion rendition is present, but significant temporal aliasing has occurred. Our current knowledge of the flicker effect, motion rendition, the temporal aliasing effect, and their interrelations is far from complete. A comprehensive understanding of this topic would be useful in a number of applications, such as bit rate reduction by frame elimination in a sequence of image frames.

7.4 IMAGE PROCESSING SYSTEMS

7.4.1 Overview of an Image Processing System

A typical image processing system that involves digital signal processing is shown in Figure 7.25. The input image source $I(x, y)$ is generally an object or a natural scene, but it may be an image produced by another system, such as a filter, a cathode ray tube (CRT) display monitor, or a video cassette recorder (VCR). The digitizer converts the input source to an electrical signal whose amplitude represents the image intensity and digitizes the electrical signal using an analog-to-digital (A/D) converter.

The sequence $f(n_1, n_2)$ that results from the digitizer is then processed by a digital image processing algorithm. The algorithm may be implemented on a general purpose computer, a microprocessor, or a special purpose hardware. The specific algorithm used depends on the objective, which may involve image en-

<figure>Input image source $\xrightarrow{I(x, y)}$ Digitizer $\xrightarrow{f(n_1, n_2)}$ Digital processing $\xrightarrow{g(n_1, n_2)}$ Display $\xrightarrow{I'(x, y)}$ Output image</figure>

Figure 7.25 Typical overall image processing system.

hancement, restoration, coding, understanding, or any combination of them. The result of processing is then displayed, generally for human viewing, but sometimes as an input to another system. The display typically used is a CRT monitor, but may be a photograph or VCR tape. If the result is some symbolic representation, as in image understanding, the display used can also be just a printer.

7.4.2 The Digitizer

The digitizer converts the input image source to an electrical signal and then samples the electrical signal, using an A/D converter. The specific functions performed by a digitizer depend on the input image source. When the input is already in the form of an electrical signal, as in the case of a VCR tape, the digitizer is interfaced to the input source and is used in sampling the electrical signal following the format used in converting the input source to an electrical signal.

When the input source is in the form of an image, an electronic camera converts the image to an electrical signal, and the result is digitized by using an A/D converter. In some camera systems, parallel paths allow light intensities at many spatial points to be measured simultaneously. In typical systems, however, there is only one path, and the light intensity at only one spatial point can be measured at a given time. In this case, the entire image is covered by scanning. In most scanning systems, a small aperture searches the image following a certain pattern called a *raster*. The light intensity integrated over a small aperture is measured, is converted to an electrical signal, and is considered to be the image intensity at that spatial point. This process can be viewed as convolving the input image intensity $I(x, y)$ with the aperture and then sampling the result of the convolution. The effect of the aperture is, therefore, lowpass filtering $I(x, y)$. This limits the spatial resolution of $I(x, y)$ and can be used for antialiasing necessary in an A/D converter. For a still image, the image is generally scanned once, but may be scanned more times for noise reduction through frame averaging. For a moving scene, the image is scanned at periodic time intervals.

When the input is a film or a photograph, a common device used to convert the image intensity to an electrical signal is a flying spot scanner. In this arrangement, a small spot of light scans the input source, and the light that is reflected by the photograph or transmitted through the film is collected by wide-area photodetectors. The source of the small light spot is a CRT screen. The CRT is discussed in greater detail in Section 7.4.3.

When the input image source is an object or a natural scene, the most common device for converting light intensity to an electrical signal has been the *vidicon* and its relatives such as a Saticon and Newvicon. The vidicon and its relatives were employed until the early 1980s in practically all TV applications, including broad-

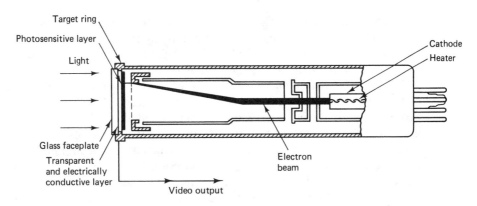

Figure 7.26 Construction of a Vidicon camera.

casting, small portable video cameras, and surveillance cameras. The construction of a vidicon camera tube is shown in Figure 7.26. At one end (the left end in the figure) of the tube inside the glass envelope is an image plate. The plate has two layers. Facing the light from the input source is a thin layer of tin oxide coating that is transparent to light but is electrically conductive. Facing the electron gun is the second layer. This second layer has a coating of photosensitive material, antimony trisulfide for a basic vidicon. The light from the input source passes through an optical lens that is the focusing mechanism, through an optically flat glass plate, and through the first layer of the image plate. The light is then focused on the second layer. The photosensitive image plate (the second layer) is scanned by an electron gun, and the resulting electrical current is the camera signal that is digitized by an A/D converter. The scanning pattern used is from right to left, bottom to top. Since the input source is inverted by the lens, this scanning pattern is equivalent to scanning left to right, top to bottom, in the input image plane.

The photosensitive layer is a semiconductor, which acts as an insulator when no light is present. As light hits this layer, electrons move to the electrically conductive tin oxide layer, creating positive charges on the image-plate surface facing the electron gun. The number of electrons that move, or alternatively the number of positive charges facing the electron gun, represents the image intensity at that spatial point. As the low-energy electron beam from the electron gun scans the image plate, it drops enough electrons to neutralize the positive charges. This discharge current is collected at a metal target ring that is electrically attached to the tin oxide layer. The current at the metal target ring is the camera signal. The electrons originate at the cathode, which is at the other end of the vidicon camera tube. Electrons converge to a narrow beam by means of the electrostatic lens and magnetic focusing.

The spectral response of a basic vidicon for a black and white image is very similar to the C.I.E. relative luminous efficiency function discussed in Section 7.1.2. For color images, a color camera optically separates the incoming light into red, green, and blue components. Each component is the input to a vidicon camera tube. A color camera, therefore, houses three separate tubes.

The camera signal, which represents the intensity of the input image source, is sampled by an A/D converter to obtain digital images. Common digital image sizes are 128×128, 256×256, 512×512, and 1024×1024 pixels. As we reduce the number of pixels, the spatial resolution, which is also referred to as *definition*, is decreased and the details in the image begin to disappear. Examples of images of different sizes can be found in Figure 1.11. The amplitude of each pixel is typically quantized to 256 levels (represented by 8 bits). Often, each level is denoted by an integer, with 0 corresponding to the darkest level and 255 corresponding to the brightest. As we decrease the number of amplitude quantization levels, the signal-dependent quantization noise begins to appear first as random noise and then as false contours. Examples of images quantized at different numbers of quantization levels can be found in Figure 1.10. For a color image, each of the red, green, and blue components is typically quantized to 8 bits/pixel, a total of 24 bits/pixel.

The vidicon and its relatives are called *photo-conductive sensors* or *tube sensors*, and were employed until the early 1980s in practically all TV applications. Since the mid-1980s, however, there has been a rapid growth in *solid-state sensors*. In a typical solid-state sensor, a 2-D array of sensor elements is integrated on a chip. One sensor element is located spatially at each pixel location and senses the light intensity at the pixel, the value of which is then read by a scanning mechanism.

The charge coupled device (CCD) is an example of a solid-state sensor element. When a CCD array is exposed to light, charge packets proportional to the light intensity develop. The stored charge packets are shifted to the storage CCD array which is not exposed to light. The light intensity values are then read from the storage array. Depending on how the imaging and storage CCD arrays are configured, different methods [Flory] have been developed to read the light intensity values from the storage array.

Solid-state sensors have many advantages over photo-conductive sensors. They are inherently more stable and compact. A solid-state sensor also has a well-defined structure and the location of every pixel is known accurately in both space and time. As a result, color extraction requires simpler signal processing and better color uniformity can be obtained. Solid-state sensors also have the potential for much higher sensitivity of light detection. In a typical photo-conductive sensor, each pixel is examined one at a time by a single light sensor. The time for light detection is measured in microseconds in a typical TV application and the sensitivity of light detection is low. In a typical solid-state sensor, an array of sensor elements, one for each pixel, is used. Each sensor element, therefore, needs to be checked once per picture. The light energy can be integrated over the time of one frame rather than one pixel, increasing the potential for light sensitivity by orders of magnitude. A solid-state sensor also has a lower lag factor than a photo-conductive sensor. The *lag* is the residual output of a sensor after the light intensity is changed or removed.

Solid-state sensors have some disadvantages in comparison with photo-conductive sensors. One is the spatial resolution. A higher spatial resolution requires

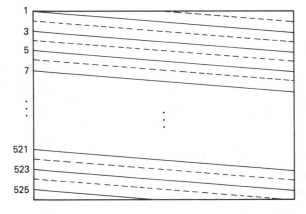

Figure 7.27 Odd (solid line) and even (dotted line) fields in a 2:1 interlace.

a larger number of pixels, meaning a larger number of sensor elements that need to be integrated on a chip. Another is the relatively low signal-to-noise ratio. Despite some of these disadvantages, the solid-state technology is advancing rapidly, and it is expected that solid-state sensors will replace photo-conductive sensors in almost all TV applications in the near future. More details on solid-state sensors can be found in [Baynon and Lamb; Kosonocky; Flory; Declerck].

In the NTSC television broadcasting, 30 frames are transmitted every second. Each frame consists of 525 horizontal scan lines, which are divided into two fields, the odd field and the even field. Each of the two fields is scanned from left to right and from top to bottom, and covers $262\frac{1}{2}$ horizontal lines. The odd field consists of odd-numbered lines, and the even field, of even-numbered lines. The horizontal lines in the odd and even fields are interlaced to form a frame. This is shown in Figure 7.27 and is called a 2:1 interlace. The 2:1 interlace is used so that the vertical resolution will be 525 lines per frame at the rate of 30 frames/sec, but the flickering frequency will be 60 cycles/sec, reducing the perception of flicker at the receiver monitor. Without the interlace, a frame would have to be displayed twice to achieve a flicker frequency of 60 cycles/sec, and this requires frame storage or more transmission bandwidth. The spatial resolution of a television frame displayed as a still image is similar to that of a 512×512-pixel digital image. In the television broadcast, the signal remains in the same analog form both at the transmitter and receiver. Digital processing of these signals involves sampling using an A/D converter.

The output of the digitizer is a sequence of numbers. Although the output is represented by a 2-D sequence $f(n_1, n_2)$ in Figure 7.25, the output may be three sequences, $f_R(n_1, n_2)$, $f_G(n_1, n_2)$, and $f_B(n_1, n_2)$, corresponding to the red, green, and blue components for a color image. The output may also be a 3-D sequence $f(n_1, n_2, n_t)$, which is a function of two spatial variables and a time variable for a sequence of frames. These signals are then processed by digital image processing algorithms, which may be implemented on a general purpose computer, a microprocessor, or a special purpose hardware. Digital processing of these signals is the topic of Chapters 8, 9, and 10.

7.4.3 Display

The most common display device in an image processing environment is a CRT. A CRT consists of an electron gun and a phosphor screen, as shown in Figure 7.28. The electron gun produces a beam of electrons that are focused in a narrow region on the phosphor screen through the electrostatic lens, which involves the use of an electrostatic and a magnetic field. The electron beam excites the phosphor to generate light. For monochrome tubes, one beam is used. For color tubes, three separate beams are used to excite the phosphor for each of the three colors. The screen is typically scanned from left to right and from top to bottom to cover the whole screen.

The phosphor chemicals are generally light metals, such as zinc, in the form of sulfide and sulfate. The phosphor material is processed to obtain very fine particles, which are applied to the inside of the glass plate. For monochrome tubes, the phosphor coating is a uniform layer. For color tubes, the phosphor is deposited in dots or vertical layers for each color.

When the high-velocity beams excite the phosphor, electrons in the atoms of the phosphor move to a high energy level. As the electron beams move to a different spot on the screen in raster scanning, the electrons move back to a lower energy level and emit light. The radiation of light from the screen is called *luminescence*. When the light is extinguished, the screen fluoresces. The time it takes for light emitted from the screen to decay to 1% of the maximum value is called the *screen persistence*. For medium to short persistence, generally used for television monitors, the decay time is about 5 msec. Long persistence phosphor exists, but can cause merging of two frames, resulting in substantial blurring for a sequence of images with motion.

Since their screen persistence is short, most CRTs used for display are brightened for short periods of time as the electron beams scan the screen. To display a still image or a sequence of images, the screen must be refreshed continuously. Display monitors often have semiconductor random access memory (RAM) for

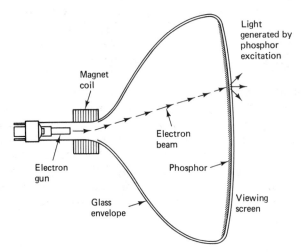

Figure 7.28 Construction of a cathode ray tube.

refresh. The refresh memory, which can typically hold more than one image frame of a reasonable size, is used to refresh the screen.

The display monitor used in an image processing environment is typically a high quality industrial type. A typical monitor has an aspect ratio (width to height) of 1* and is typically calibrated to display a square frame. Standard display sizes are 512×512 and 1024×1024, but other sizes, such as 640×480 and 1024×1280, are available. The refresh rate has traditionally been 30 frames/sec with a 2:1 interlace, the same as for television broadcasting. Individual scan lines are sharply defined on high quality monitors. A sharp horizontal line, therefore, may appear only in the odd field or only in the even field, so the flicker rate for this scan line would become 30 cycles/sec. At this rate, the flicker may be visible. To avoid this problem, displays which refresh the screen at 50 frames/sec or 60 frames/sec without interlace are available. This provides a display that is essentially flicker-free and is more pleasant to the human viewer.

In many displays, the refresh memory is used not only for refreshing the screen but also for processing. The displayed image can be zoomed, scrolled, or processed with a simple filter in the refresh memory in real time. As hardware advances continue, the display unit will gain even more sophisticated capabilities.

An image displayed on a CRT screen is considered soft since it is in a sense a temporary image. One approach to making a hard copy is to photograph the CRT screen on film. Special camera systems that have their own small CRT screens are available for such purposes. There are many additional hard copy devices [Schreiber]. To obtain multiple copies, for example, methods such as letterpress, lithography (offset), and gravure exist.

REFERENCES

For books on digital image processing, see [Rosenfeld (1969); Lipkin and Rosenfeld; Andrews (1970); Huang (1975); Andrews and Hunt; Pratt (1978); Hall (1979); Castleman; Pratt (1979); Moik; Huang (1981); Rosenfeld and Kak; Green; Braddick and Sleigh; Ekstrom; Rosenfeld (1984); Yaroslavsky; Niblack; Jensen; Huang (1986); Young and Fu; Gonzalez and Wintz]. For collections of selected papers on digital image processing, see [Andrews (1978); Chellapa and Sawchuck]. For special journal issues on digital image processing, see [Andrews and Enloe; Andrews (1974); Netravali and Habibi; Hunt (1981); Hwang (1983); Netravali and Prasada (1985)].

For readings on light and human vision, see [Graham, et al.; Mueller, et al.; Le Grande; Gregory; Cornsweet; Kupchella]. For a computational view of visual perception, see [Barrow and Tenenbaum]. For readings on color and color vision, see [Wyszecki and Stiles; Yule; Boynton]. On models of human visual system, see [Stockham; Budrikis; Mannos and Sakrison; Hall and Hall; Faugeras; Gran-

---

*Currently, television receivers have an aspect ratio of 4:3, which is the same ratio used in conventional motion pictures. In high resolution or high definition television systems, the aspect ratio will increase to approximately 5:3 or 16:9.

rath]. On mathematical models of images, see [Jain]. For readings on image processing systems, including cameras and displays, see [Hunt and Breedlove; Weimer; Reader and Hubble; Grob; Flory; Schreiber; Declerck].

H. C. Andrews, *Computer Techniques in Image Processing*. New York: Academic Press, 1970.

H. C. Andrews, ed., *Special Issue on Digital Image Processing, Computer*, Vol. 7, May 1974.

H. C. Andrews, ed., *Tutorial and Selected Papers in Digital Image Processing*. New York: IEEE Press, 1978.

H. C. Andrews and L. H. Enloe, eds., *Special Issue on Digital Picture Processing, Proc. IEEE*, Vol. 60, July 1972.

H. C. Andrews and B. R. Hunt, *Digital Image Restoration*. Englewood Cliffs, NJ: Prentice-Hall, 1977.

H. C. Barrow and J. M. Tenenbaum, Computational vision, *Proc. IEEE*, Vol. 69, May 1981, pp. 572–595.

J. D. Baynon and D. R. Lamb, *Charge Coupled Devices and Their Applications*. London: McGraw-Hill, 1980.

R. M. Boynton, *Human Color Vision*. New York: Holt, Rinehart and Winston, 1979.

O. J. Braddick and A. C. Sleigh, eds., *Physical and Biological Processing of Images*. Berlin: Springer-Verlag, 1983.

Z. L. Budrikis, Visual fidelity criterion and modeling, *Proc. IEEE*, Vol. 60, 1972, pp. 771–779.

K. R. Castleman, *Digital Image Processing*. Englewood Cliffs, NJ: Prentice-Hall, 1979.

R. Chellapa and A. A. Sawchuck, eds., *Digital Image Processing and Analysis: Vol. 1, Digital Image Processing*. New York: IEEE Press, 1985.

T. N. Cornsweet, *Visual Perception*. New York: Academic Press, 1970.

M. L. Davidson, Perturbation approach to spatial brightness interaction in human vision, *J. Opt. Soc. of Am.*, Vol. 58, Sept. 1968, pp. 1300–1308.

G. J. Declerck, ed., *Solid State Imagers and Their Applications, SPIE*, Vol. 591, 1986.

M. P. Ekstrom, ed., *Digital Image Processing Techniques*. Orlando, FL: Academic Press, 1984.

O. D. Faugeras, Digital color image processing within the framework of a human visual model, *IEEE Trans. Acoust. Speech, and Sig. Proc.*, Vol. ASSP-27, August 1979, pp. 380–393.

R. E. Flory, Image Acquisition Technology, *Proc. IEEE*, Vol. 73, April 1985, pp. 613–637.

R. C. Gonzalez and P. Wintz, *Digital Image Processing*, 2nd ed. Reading, MA: Addison-Wesley, 1987.

C. H. Graham et al., eds., *Vision and Visual Perception*. New York: Wiley, 1965.

D. J. Granrath, The role of human visual models in image processing, *Proc. IEEE*, Vol. 69, May 1981, pp. 552–561.

W. B. Green, *Digital Image Processing, A Systems Approach*. New York: Van Nostrand and Reinhold Comp., 1983.

R. L. Gregory, Visual Illusions, *Scientific American*, November 1968, pp. 66–76.

B. Grob, *Basic Television and Video Systems*. 5th ed. New York: McGraw-Hill, 1984.

E. L. Hall, *Computer Image Processing and Recognition*. New York: Academic Press, 1979.

C. F. Hall and E. L. Hall, A nonlinear model for the spatial characteristics of the human visual system, *IEEE Trans. Syst., Man. and Cybern.*, Vol. SMC-7, March 1977, pp. 161–170.

A. C. Hardy, *Handbook of Colorimetry*. Cambridge, MA: MIT Press, 1936.

T. S. Huang, ed., *Picture Processing and Digital Filtering*. Berlin: Springer-Verlag, 1975.

T. S. Huang, ed., *Image Sequence Analysis*. Berlin: Springer-Verlag, 1981.

T. S. Huang, ed., *Advances in Computer Vision and Image Processing, Vol. 2*. New York: JAI Press, 1986.

B. R. Hunt, ed., *Special Issue on Image Processing*, Proc. IEEE, May 1981.

B. R. Hunt and J. R. Breedlove, Scan and display considerations in processing images by digital computer, *IEEE Trans. Computers*, Vol. C-24, Aug. 1975, pp. 848–853.

K. Hwang, ed., *Special Issue on Computer Architectures for Image Processing, Computer*, Vol. 16, January 1983.

A. K. Jain, Advances in mathematical models for image processing, *Proc. IEEE*, Vol. 69, March 1981, pp. 502–528.

J. R. Jensen, *Introductory Digital Image Processing. A Remote Sensing Perspective*. Englewood Cliffs, NJ: Prentice-Hall, 1986.

D. H. Kelly, Theory of flicker and transient responses, I: Uniform Fields, *J. Opt. Soc. Am.*, Vol. 61, April 1971, pp. 537–546.

R. A. Kinchla et al., A theory of visual movement perception, *Psych. Rev.*, Vol. 76, 1968, pp. 537–558.

W. F. Kosonocky, Visible and infra-red solid-state sensors, *1983 IEDM Tech. Dig.*, Int. Elec. Dev. Meeting, Dec. 1983, pp. 1–7.

C. E. Kupchella, *Sights and Sounds*. Indianapolis, IN: The Bobbs-Merrill Company, 1976.

Y. LeGrand, *Light, Color, and Vision*, translated from *Lumiere et Couleurs*. Chapman and Hall, 1968.

B. S. Lipkin and A. Rosenfeld, eds., *Picture Processing and Psychopictorics*. New York: Academic Press, 1970.

J. L. Mannos and D. J. Sakrison, The effects of a visual fidelity criterion on the encoding of images, *IEEE Trans. Inf. Theory*, Vol. IT-20, July 1974, pp. 525–536.

J. K. Moik, *Digital Processing of Remotely Sensed Images*. NASA, 1980.

C. G. Mueller, M. Rudolph, and the editors of LIFE, *Light and Vision*. New York: Time, 1966.

W. Niblack, *An Introduction to Digital Image Processing*. Englewood Cliffs, NJ: Prentice-Hall, 1986.

A. N. Netravali and A. Habibi, *Special Issue on Picture Communication Systems, IEEE Trans. Commun.*, Vol. COM-29, Dec. 1981.

A. N. Netravali and B. Prasada, *Special Issue on Visual Communications Systems, Proc. IEEE*, Vol. 73, April 1985.

M. H. Pirenne, *Vision and the Eye*, 2nd ed. London: Associated Book Publishers, 1967.

W. K. Pratt, *Digital Image Processing*. New York: Wiley, 1978.

W. K. Pratt, ed., *Image Transmission Techniques*. New York: Academic Press, 1979.

C. Reader and L. Hubble, Trends in image display systems, *Proc. IEEE*, Vol. 69, May 1981, pp. 606–614.

A. Rosenfeld, *Picture Processing by Computer*. New York: Academic Press, 1969.

A. Rosenfeld, ed., *Multiresolution Image Processing and Analysis*. Berlin: Springer-Verlag, 1984.

A. Rosenfeld and A. C. Kak, *Digital Picture Processing*, 2nd ed. New York: Academic Press, 1982.

W. F. Schreiber, *Fundamentals of Electronic Imaging Systems*. Berlin: Springer-Verlag, 1986.

T. G. Stockham, Jr., Image processing in the context of a visual model, *Proc. IEEE*, Vol. 60, July 1972, pp. 828–842.

P. K. Weimer, A historical review of the development of television pickup devices (1930–1976), *IEEE Trans. Electron. Devices*, Vol. ED-23, July 1976, pp. 739–752.

G. W. Wyszecki and W. S. Stiles, *Color Science*. New York: Wiley, 1967.

L. P. Yaroslavsky, *Digital Picture Processing: An Introduction*. Berlin: Springer-Verlag, 1985.

T. Y. Young and K. S. Fu, eds., *Handbook of Pattern Recognition and Image Processing*. Orlando, FL: Academic Press, 1986.

J. A. C. Yule, *Principles of Color Reproduction*. New York: Wiley, 1967.

PROBLEMS

7.1. Let λ denote the wavelength of an electromagnetic wave propagating in a vacuum and let f denote its frequency. For each of the following wavelengths, determine the corresponding frequency f.
(a) $\lambda = 100$ m (radio broadcast)
(b) $\lambda = 1$ m (VHF, UHF)
(c) $\lambda = 10^{-1}$ m (radar)
(d) $\lambda = 5 \cdot 10^{-7}$ m (visible light)
(e) $\lambda = 10^{-9}$ m (X rays)
(f) $\lambda = 10^{-11}$ m (gamma rays)

7.2. Let $c(\lambda)$ denote the energy distribution of a light as a function of its wavelength λ. Let A and B denote two lights with energy distributions of $c_A(\lambda)$ and $c_B(\lambda)$, respectively, as shown below.

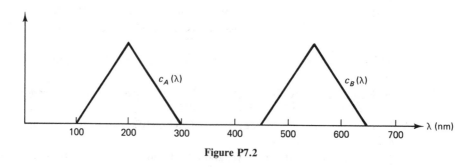

Figure P7.2

Under identical normal viewing conditions, which light will appear brighter to a human viewer?

7.3. Consider a monochromatic light with wavelength of 555 nm which has irradiance of

50 watts/m². We wish to have another monochromatic light with wavelength of 450 nm to appear to the C.I.E. standard observer to be as bright, under identical normal viewing conditions, as the monochromatic light with wavelength of 555 nm. Determine the irradiance of the monochromatic light with wavelength of 450 nm.

7.4. Let $c(\lambda)$ denote the energy density of a light as a function of the wavelength λ. Let A and B denote two lights with energy density of $c_A(\lambda)$ and $c_B(\lambda)$, respectively, as shown below.

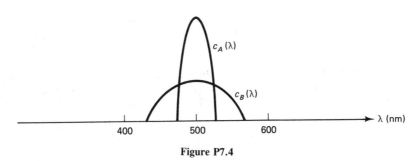

Figure P7.4

Under identical normal viewing conditions, which light will appear to be more saturated (more pure or vivid) to a human viewer?

7.5. Let the energy densities of two lights be denoted by $c_1(\lambda)$ and $c_2(\lambda)$, as shown in the following figure.

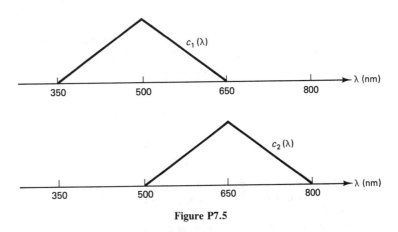

Figure P7.5

(a) Suppose we mix the two lights. Let $c(\lambda)$ denote the energy density of the combined light. Sketch $c(\lambda)$.

(b) Suppose that $c_1(\lambda)$ denotes the energy density of light reflected by a particular ink illuminated by white light. Suppose further that $c_2(\lambda)$ denotes the energy density of light reflected by a second ink illuminated by the same white light. We now mix the two inks in equal amounts and illuminate the mixture with the same white light. Sketch $c(\lambda)$, the light reflected by the mixture of the two inks.

7.6. A color image can be viewed as three monochrome images, which we will denote by $f_R(n_1, n_2), f_G(n_1, n_2),$ and $f_B(n_1, n_2)$, representing the red, green, and blue components, respectively. Alternatively, we can transform the $f_R(n_1, n_2), f_G(n_1, n_2),$ and $f_B(n_1, n_2)$

components to the luminance component $f_Y(n_1, n_2)$ and the two chrominance components $f_I(n_1, n_2)$ and $f_Q(n_1, n_2)$. Consider the system shown in the following figure.

$$f_R(n_1, n_2) \longrightarrow \boxed{h(n_1, n_2)} \longrightarrow \hat{f}_R(n_1, n_2)$$

$$f_G(n_1, n_2) \longrightarrow \boxed{h(n_1, n_2)} \longrightarrow \hat{f}_G(n_1, n_2)$$

$$f_B(n_1, n_2) \longrightarrow \boxed{h(n_1, n_2)} \longrightarrow \hat{f}_B(n_1, n_2)$$

In the above system, each of the three components $f_R(n_1, n_2), f_G(n_1, n_2)$, and $f_B(n_1, n_2)$ is filtered by a linear shift-invariant system with impulse response $h(n_1, n_2)$. Now consider the system shown in the following figure.

In the figure, $h(n_1, n_2)$ is the same as in the previous system. How are $\hat{f}'_R(n_1, n_2)$, $\hat{f}'_G(n_1, n_2)$, and $\hat{f}'_B(n_1, n_2)$, related to $\hat{f}_R(n_1, n_2)$, $\hat{f}_G(n_1, n_2)$, and $\hat{f}_B(n_1, n_2)$?

7.7. Consider a light represented in terms of its three components, red (R), green (G), and blue (B). Suppose the intensities R, G, and B are in the following ranges:

$$0 \le R \le R_{MAX}$$

$$0 \le G \le G_{MAX}$$

$$0 \le B \le B_{MAX}$$

Note that R, G, and B are nonnegative. Another way of representing the same light is in terms of the luminance component Y and the two chrominance components I and Q.

(a) The possible ranges of Y, I, and Q are

$$Y_{MIN} \le Y \le Y_{MAX}$$

$$I_{MIN} \le I \le I_{MAX}$$

$$Q_{MIN} \le Q \le Q_{MAX}$$

Determine $Y_{MIN}, Y_{MAX}, I_{MIN}, I_{MAX}, Q_{MIN}$, and Q_{MAX}.

(b) Does any set of Y, I, and Q in the ranges obtained in (a) correspond to a valid light, valid in the sense that it corresponds to a set of nonnegative R, G, and B?

7.8 Figure 7.22 illustrates that the peripheral level of the human visual system is approximately bandpass in character. Design an experiment that can be used in more accurately determining the frequency response of the bandpass filter. State the assumptions that you make in designing your experiment.

7.9. Weber's law, which states that the just-noticeable intensity difference ΔI at a given intensity I is proportional to I, appears to hold over a wide range of I not only for human vision, but for all other sense modalities, including hearing, smell, taste, and touch. Discuss why evolution may have favored such a mechanism rather than a system whereby the just-noticeable intensity difference ΔI is constant independent of I.

7.10. Supppose Weber's law holds strictly. The just-noticeable intensity difference ΔI when $I = 100$ has been measured to be 1. How many just-noticeable differences are there between $I = 10$ and $I = 1000$?

7.11. A physical unit of spatial frequency is cycles/cm. A unit of spatial frequency more directly related to our perception of spatial frequency is cycles/viewing angle. Suppose we have a sinusoidal intensity grating at a horizontal frequency of 10 cycles/cm. In answering the following questions, assume that the image is focused on the retina, the sinusoidal grating is directly displayed before an eye, and the eye looks at the image straight on.

(a) When the image is viewed at a distance of 50 cm away from the eye, what is the horizontal spatial frequency in units of cycles/degree in viewing angle?

(b) Answer (a) when the image is viewed at a distance of 100 cm from the eye.

(c) From the results of (a) and (b), and the spatial frequency response of a human eye, discuss why the details of a distant scene are less visible than those of nearby objects.

7.12. The lens of a human eye is convex. As a result, the image formed on the retina is upside down, as shown in the following figure. Nevertheless, the object is perceived to be right side up. Discuss possible explanations for this phenomenon.

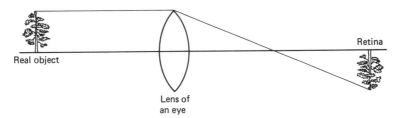

Figure P7.12

7.13. In the region on the retina known as the blind spot, there are no light-sensitive cells, so light focused on this region is not visible. The blind spot often goes unnoticed in our everyday experience. Design an experiment that can be used to perceive the blind spot.

7.14. In a typical image digitization system, a small aperture $A(x, y)$ searches an image following a certain pattern called a raster. Let $A(x, y)$ be given by

$$A(x, y) = \begin{cases} 1, & \sqrt{x^2 + y^2} < r \\ 0, & \text{otherwise} \end{cases}$$

The light intensity integrated over the small aperture is measured and is considered the image intensity at the spatial point.

(a) Discuss how this process can be viewed as convolving the input intensity $I(x, y)$ with the aperture $A(x, y)$ and then sampling the result of convolution.

(b) Suppose we have an image of 5 cm × 5 cm, and we wish to sample it with uniform sampling on a Cartesian grid at 512 × 512 points. What is a reasonable size of r for the aperture? What is the effect of choosing a very small r? a very large r?

7.15. In television broadcasting, a horizontal line time is the time interval during which the scanner traces one horizontal line and retraces to the beginning of the next horizontal line.

(a) What is the approximate horizontal line time in an NTSC television system? Assume that the time it takes the scanner to retrace from the end of one field to the beginning of the next is negligible.

(b) Approximately 16% of the horizontal line time is used to allow the retrace from the end of one horizontal line to the beginning of the next horizontal line. The picture intensity information is blanked out during this period, which is called a *horizontal blanking period*. What is the approximate time used as horizontal blanking periods during the period the scanner scans one complete field in an NTSC television system?

7.16. In NTSC television broadcasting, the frame rate used is 30 frames/sec. Each frame consists of 525 horizontal lines which are divided into two fields, odd and even. The field rate used is, therefore, 60 fields/sec. The spacing in time between any two consecutive fields is the same. The 2:1 interlace is used to give a vertical resolution of 525 lines/frame at the rate of 30 frames/sec, but with a flicker frequency of 60 cycles/sec to reduce the perception of flicker. The 2:1 interlace, however, causes some artifacts. By considering a solid circle moving in the horizontal and vertical directions, discuss what distortion in the circle may be visible. Assume that all the lines in a given field are displayed at the same time.

8

Image Enhancement

8.0 INTRODUCTION

Image enhancement is the processing of images to improve their appearance to human viewers or to enhance other image processing systems' performance. Methods and objectives vary with the application. When images are enhanced for human viewers, as in television, the objective may be to improve perceptual aspects: image quality, intelligibility, or visual appearance. In other applications, such as object identification by machine, an image may be preprocessed to aid machine performance. Because the objective of image enhancement is dependent on the application context, and the criteria for enhancement are often subjective or too complex to be easily converted to useful objective measures, image enhancement algorithms tend to be simple, qualitative, and ad hoc. In addition, in any given application, an image enhancement algorithm that performs well for one class of images may not perform as well for other classes.

Image enhancement is closely related to image restoration, which will be discussed in Chapter 9. When an image is degraded, restoration of the original image often results in enhancement. There are, however, some important differences between restoration and enhancement. In image restoration, an ideal image has been degraded, and the objective is to make the processed image resemble the original image as much as possible. In image enhancement, the objective is to make the processed image better in some sense than the unprocessed image. In this case, the ideal image depends on the problem context and often is not well defined. To illustrate this difference, note that an original, undegraded image cannot be further restored but can be enhanced by increasing sharpness through highpass filtering.

Image enhancement is desirable in a number of contexts. In one important class of problems, an image is enhanced by modifying its contrast and/or dynamic

Some part of this chapter has been adapted from "Image Enhancement" by Jae S. Lim in *Digital Image Processing Techniques*, edited by Michael P. Ekstrom. Copyright © 1984 by Academic Press, Inc. Reprinted by permission of the publisher.

range. For example, a typical image, even if undegraded, will often appear better when its edges are sharpened. Also, if an image with a large dynamic range is recorded on a medium with a small dynamic range, such as film or paper, the contrast and therefore the details of the image are reduced, particularly in the very bright and dark regions. Contrast in an image taken from an airplane is reduced when the scenery is covered by cloud or mist. Increasing the local contrast and reducing the overall dynamic range can significantly enhance the quality of such an image.

In another class of enhancement problems, a degraded image may be enhanced by reducing the degradation. Examples of image degradation are blurring, random background noise, speckle noise, and quantization noise. This area of image enhancement overlaps with image restoration. An algorithm that is simple and ad hoc, and does not attempt to exploit the characteristics of the signal and degradation, is generally considered an enhancement algorithm. An algorithm that is more mathematical and complex, and exploits the characteristics of the signal and degradation with an explicit error criterion that attempts to compare the processed image with the original undegraded image, is generally regarded as a restoration algorithm. This distinction is admittedly somewhat vague and arbitrary. Some arbitrary decisions have been necessary in dividing certain topics between this chapter and the next chapter, which deals with the image restoration problem.

It is well known that the contours or edges in an object contain very important information that may be used in image understanding applications. The first step in such an application may be to preprocess an image into an edge map that consists of only edges. Since more accurate detection of edges in an image can enhance the performance of an image understanding system that exploits such information, converting an image to its corresponding edge map may be viewed as an enhancement process.*

Another important class of image enhancement problems is the display of 2-D data that may or may not represent the intensities of an actual image. A low-resolution image of 128×128 pixels may be made more visually pleasant to a human observer by interpolating it to generate a larger image, say 256×256 pixels. In 2-D spectral estimation, the spectral estimates have traditionally been displayed as contour plots. Although such 2-D data are not images in the conventional sense, they can be presented as images. We can display them as black-and-white images, or we can enhance them with color so that their appearance may be improved and information conveyed more clearly. In other applications, such as infrared radar imaging, range information as well as image intensities may be available. By displaying the range information with color, relative distances of objects in an image can be highlighted. Even good-quality images may be enhanced by certain types of distortion. For example, when an object in an image is displayed with false color, the object may stand out more clearly to a human viewer.

---

*Edge detection is useful in a variety of image processing applications including image enhancement, restoration, coding, and understanding. We have chosen to discuss the topic in this chapter.

In this chapter, we study methods of solving the image enhancement problems discussed above. In Section 8.1, we discuss modification of the contrast and dynamic range. In Section 8.2, we discuss noise smoothing. In Section 8.3, the problem of detecting edges of an image is discussed. In Section 8.4, we discuss the problem of image interpolation and motion estimation which can be used for image interpolation. In Section 8.5, we discuss enhancement of images by means of pseudocolor and false color.

Throughout this chapter and Chapters 9 and 10, the performance of various algorithms is illustrated using examples. These examples are included only for illustrative purposes and should not be used for comparing the performance of different algorithms. The performance of an image processing algorithm depends on many factors, such as the objective of the processing and the type of image used. One or two examples do not adequately demonstrate the performance of an algorithm. Unless specified otherwise, all images used are quantized at 8 bits/pixel for monochrome images and at 24 bits/pixel (8 bits/pixel for each of the red, green, and blue components) for color images.

8.1 CONTRAST AND DYNAMIC RANGE MODIFICATION

8.1.1 Gray Scale Modification

Gray scale modification is a simple and effective way of modifying an image's dynamic range (the range of image intensities) or contrast. In this method, the gray scale or intensity level of an input image $f(n_1, n_2)$ is modified according to a specific transformation. The transformation $g = T[f]$ that relates an input intensity f to an output intensity g is often represented by a plot or a table. Consider a simple illustration of this method. Figure 8.1(a) shows an image of 4×4 pixels, with each pixel represented by three bits, so that there are eight levels; that is, $f = 0$ (darkest level), 1, 2, . . . , 7 (brightest level). The transformation that relates the input intensity to the output intensity is shown as a plot and as a table in Figure 8.1(b). For each pixel in the input image, the corresponding output intensity is obtained from the plot or the table in Figure 8.1(b). The result is shown in Figure 8.1(c). By properly choosing the specific transformation, contrast or dynamic range can be modified.

The specific transformation desired depends on the application. In some applications, physical considerations determine the transformation selected. For example, when a display system has nonlinear characteristics, the objective of the modification may be to compensate for the nonlinearities. In such a case, the most suitable transformation can be determined from the nonlinearity of the display system.

A good transformation in typical applications can be identified by computing the histogram of the input image and studying its characteristics. The *histogram* of an image, denoted by $p(f)$, represents the number of pixels that have a specific intensity f as a function of f. For example, the 4×4-pixel image shown in Figure 8.1(a) has the histogram shown in Figure 8.2(a). The histogram displays some

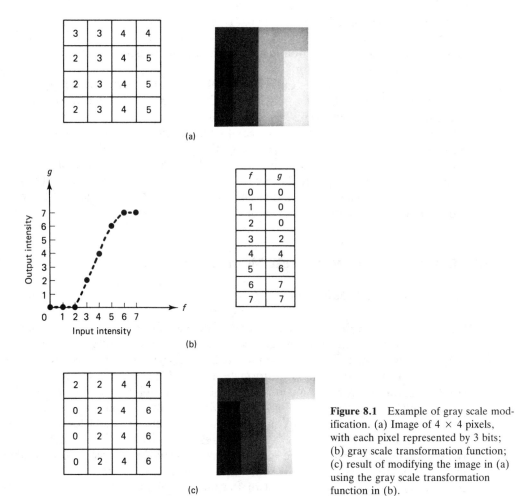

Figure 8.1 Example of gray scale modification. (a) Image of 4 × 4 pixels, with each pixel represented by 3 bits; (b) gray scale transformation function; (c) result of modifying the image in (a) using the gray scale transformation function in (b).

important image features that help determine which particular gray scale transformation is desirable. In Figure 8.2(a), the image's intensities are clustered in a small region, and the available dynamic range is not very well utilized. In such a case, a transformation of the type shown in Figure 8.1(b) would increase the overall dynamic range, and the resulting image would appear to have greater contrast. This is evidenced by Figure 8.2(b), which is the histogram of the processed image shown in Figure 8.1(c).

Because computing the histogram of an image and modifying its gray scale for a given gray scale transformation requires little computation, the desirable gray scale transformation can be determined by an experienced human operator in real

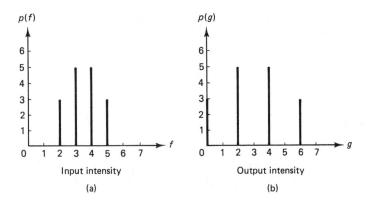

Figure 8.2 Histogram of 4 × 4-pixel image (a) in Figure 8.1(a); (b) in Figure 8.1(c).

time. On the basis of the initial histogram computation, the operator chooses a gray scale transformation to produce a processed image. By looking at the processed image and its histogram, the operator can choose another gray scale transformation, obtaining a new processed image. These steps can be repeated until the output image satisfies the operator.

In such circumstances as when there are too many images for individual attention by a human operator, the gray scale transformation must be chosen automatically. A method known as *histogram modification* is useful in this case. In this method, the gray scale transformation that produces a desired histogram is chosen for each individual image. The desired histogram of the output image, denoted by $p_d(g)$, that is useful for typical images has a maximum around the middle of the dynamic range and decreases slowly as the intensity increases or decreases. For a given image, we wish to determine the transformation function so that the resulting output image has a histogram similar to $p_d(g)$. This problem can be phrased in terms of a problem in elementary probability theory. Specifically, the histograms $p(f)$ and $p_d(g)$ can be viewed as scaled probability density functions of random variables f and g, respectively. For example, $p(3)/16$ in Figure 8.2(a) is the probability that a randomly chosen pixel in the 4 × 4-pixel image in Figure 8.1(a) will have an intensity level of 3. We wish to find a transformation $g = T[f]$ with the constraint that $T[f]$ must be a monotonically nondecreasing function of f such that $p(g)$ is equal to or close to $p_d(g)$. One approach to solving this probability problem is to obtain the probability distribution functions $P(f)$ and $P_d(g)$ by integrating the probability density functions $p(f)$ and $p_d(g)$ and then choosing the transformation function such that $P(f)$ will be equal to or close to $P_d(g)$ at $g = T[f]$. Imposing the constraint that $T[f]$ must be a monotonically nondecreasing function ensures that a pixel with a higher intensity than another pixel will not become a pixel with a lower intensity in the output image.

Applying this approach to the histogram modification problem which involves discrete variables f and g, we first compute the cumulative histograms $P(f)$ and $P_d(g)$ from $p(f)$ and $p_d(g)$ by

$$P(f) = \sum_{k=0}^{f} p(k) = P(f-1) + p(f) \qquad (8.1a)$$

$$P_d(g) = \sum_{k=0}^{g} p_d(k) = P_d(g-1) + p_d(g). \qquad (8.1b)$$

An example of the cumulative histograms is shown in Figure 8.3. Figures 8.3(a) and (b) show an example of $p(f)$ and $p_d(g)$, and Figures 8.3(c) and (d) show $P(f)$ and $P_d(g)$ obtained by using (8.1). From $P(f)$ and $P_d(g)$, the gray scale transformation $g = T[f]$ can be obtained by choosing g for each f such that $P_d(g)$ will be closest to $P(f)$. The gray scale transformation function obtained from Figure 8.3 is shown in Figure 8.4(a), and the histogram of the image obtained by using this transformation function is shown in Figure 8.4(b). If the desired histogram $p_d(g)$ remains the same for different input images, $P_d(g)$ needs to be computed only once from $p_d(g)$.

In the example we considered above, note that the histogram of the processed image is not the same as the given desired histogram. This is in general the case when f and g are discrete variables and we require that all the pixels with the same input intensity be mapped to the same output intensity. Note also that the desired cumulative histogram $P_d(g)$ is close to a straight line. In the special case of the

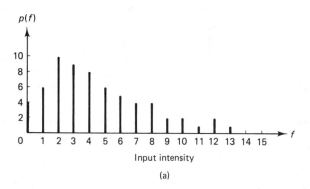

(a)

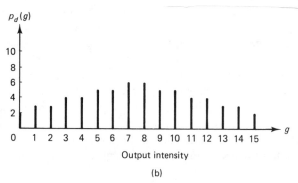

(b)

Figure 8.3 Histograms and cumulative histograms. (a) Histogram of an 8×8-pixel image; (b) desired histogram; (c) cumulative histogram derived from (a); (d) cumulative histogram derived from (b).

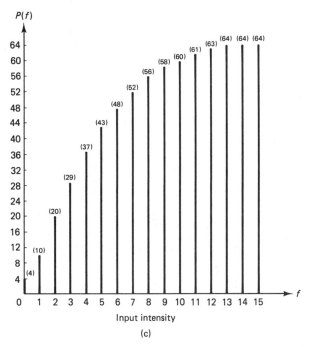

Input intensity

(c)

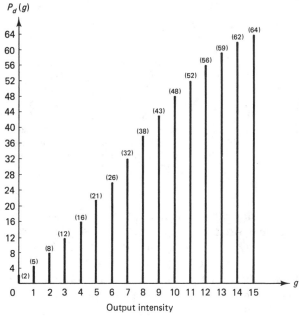

Output intensity

(d)

Figure 8.3 (continued)

Sec. 8.1 Contrast and Dynamic Range Modification

457

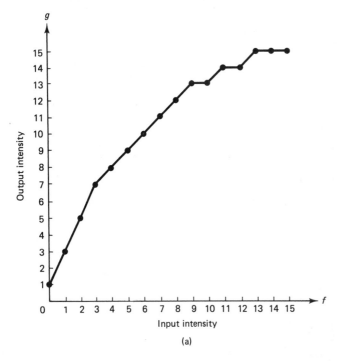

(a)

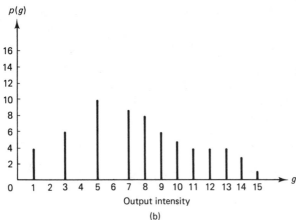

(b)

Figure 8.4 (a) Gray scale transformation function that approximately transforms the histogram in Figure 8.3(a) to the desired histogram in Figure 8.3(b); (b) histogram of gray-scale-transformed image obtained by applying the transformation function in (a) to an image with the histogram shown in Figure 8.3(a).

histogram modification known as *histogram equalization*, the desired histogram is assumed constant. In this case, the desired cumulative histogram would be exactly a straight line. Images processed by histogram equalization typically have more contrast than unprocessed images, but they tend to appear somewhat unnatural.

Even though gray scale modification is conceptually and computationally simple, it can often provide significant improvement in image quality or intelligibility to the human observer, and is therefore used routinely in many image processing applications. This is illustrated in the following two examples. Figure 8.5(a) shows an original image of 512×512 pixels, with each pixel represented by eight bits. Figure 8.5(b) shows the histogram of the image in Figure 8.5(a). The histogram clearly shows that a large number of the image's pixels are concentrated in the lower intensity levels of the dynamic range, suggesting that the image will appear very dark with a loss of contrast in the dark regions. By increasing the contrast in the dark regions, the details can be made more visible. This can be accomplished by using the transformation function shown in Figure 8.5(c). The processed image using the function in Figure 8.5(c) is shown in Figure 8.5(d), and the histogram of the processed image is shown in Figure 8.5(e). Another example is shown in Figure 8.6. The unprocessed image is shown in Figure 8.6(a) and the image processed by gray scale modification is shown in Figure 8.6(b).

The histogram modification method discussed above can also be applied to color images. To improve the image contrast with only a relatively small effect on the hue or saturation, we can transform RGB images $f_R(n_1, n_2)$, $f_G(n_1, n_2)$, and $f_B(n_1, n_2)$ to YIQ images $f_Y(n_1, n_2)$, $f_I(n_1, n_2)$, and $f_Q(n_1, n_2)$ by using the transformation in (7.8). Gray scale modification can be applied to only the Y image $f_Y(n_1, n_2)$, and the result can be combined with the unprocessed $f_I(n_1, n_2)$ and $f_Q(n_1, n_2)$. Again using the transformation in (7.8), the processed RGB images $g_R(n_1, n_2)$, $g_G(n_1, n_2)$, and $g_B(n_1, n_2)$ can be obtained. Figure 8.7(a) (see color insert) shows an original color image of 512×512 pixels, and Figure 8.7(b) shows the image processed by the gray scale transformation discussed above.

8.1.2 Highpass Filtering and Unsharp Masking

Highpass filtering emphasizes the high-frequency components of a signal while reducing the low-frequency components. Because edges or fine details of an image are the primary contributors to the high-frequency components of an image, highpass filtering often increases the local contrast and sharpens the image.

Unsharp masking, which has been known to photographic artists for a long time, is closely related to highpass filtering. In unsharp masking, the original image is blurred (unsharpened) and a fraction of the unsharp image is subtracted from, or masks, the original. The subtraction is performed by adding the negative of the unsharp image to the original. The image processed by unsharp masking can be expressed as

$$g(n_1, n_2) = af(n_1, n_2) - bf_L(n_1, n_2) \qquad (8.2)$$

where $f(n_1, n_2)$ is the original image, $f_L(n_1, n_2)$ is the lowpass filtered or unsharp image, a and b are positive scalars with $a > b$, and $g(n_1, n_2)$ is the processed image. Rewriting $f(n_1, n_2)$ as a sum of the lowpass filtered image $f_L(n_1, n_2)$ and the highpass filtered image $f_H(n_1, n_2)$, we can write (8.2) as

$$g(n_1, n_2) = (a - b) f_L(n_1, n_2) + af_H(n_1, n_2). \qquad (8.3)$$

From (8.3), it is clear that high-frequency components are emphasized over low-

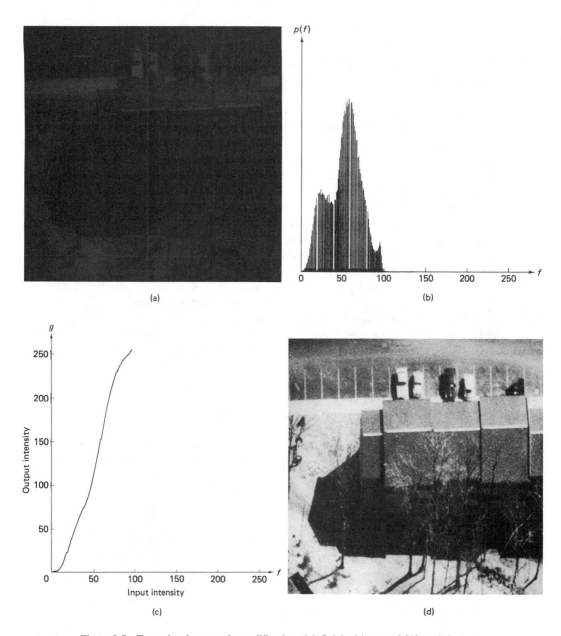

(a)

(b)

(c)

(d)

Figure 8.5 Example of gray scale modification. (a) Original image of 512 × 512 pixels; (b) histogram of the image in (a); (c) transformation function used in the gray scale modification; (d) processed image; (e) histogram of the processed image in (d).

Image Enhancement   Chap. 8

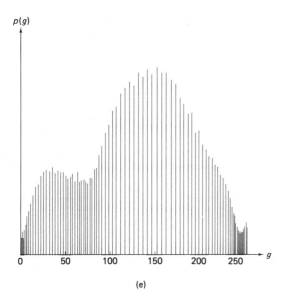

(e)

Figure 8.5 (continued)

(a)   (b)

Figure 8.6 Example of gray scale modification. (a) Original image of 512 × 512 pixels; (b) processed image.

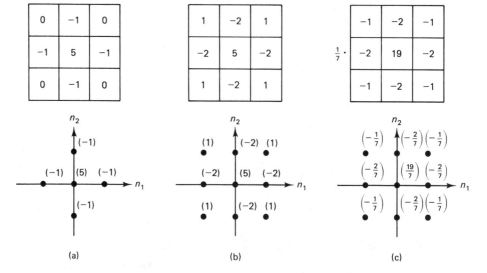

Figure 8.8 Impulse responses of highpass filters useful for image enhancement. Two representations are shown for each filter.

frequency components and that unsharp masking is some form of highpass filtering.*

Some typical examples of the impulse response of a highpass filter used for contrast enhancement are shown in Figure 8.8. One characteristic of all the filters in Figure 8.8 is that the sum of all amplitudes of each impulse response is one, so that the filter frequency response $H(\omega_1, \omega_2)$ is one at $\omega_1 = \omega_2 = 0$ and passes the DC component unaltered. This characteristic has the effect of preserving the average intensity of the original image in the processed image. It should be noted that this characteristic does not itself guarantee that the intensity of the processed image will remain in the range between 0 and 255. If intensities of some pixels in the processed image lie outside this range, they can be clipped to 0 and 255, or the image can be rescaled so that the intensities of all pixels in the processed image lie in the range between 0 and 255.

Figure 8.9 illustrates the performance of highpass filtering. Figure 8.9(a) shows an original image of 256×256 pixels, and Figure 8.9(b) shows the result of highpass filtering using the filter in Figure 8.8(a). Although the original image is not degraded, some highpass filtering increases the local contrast and thus gives it a sharper visual appearance. However, because a highpass filter emphasizes high-frequency components, and background noise typically has significant high-frequency components, highpass filtering tends to increase the background noise

---

*Unsharp masking is performed by a photographic artist in the domain of film density, which can be approximated by log intensity. Unsharp masking, therefore, is highpass filtering in the log intensity domain, which is related to homomorphic processing discussed in Section 8.1.3.

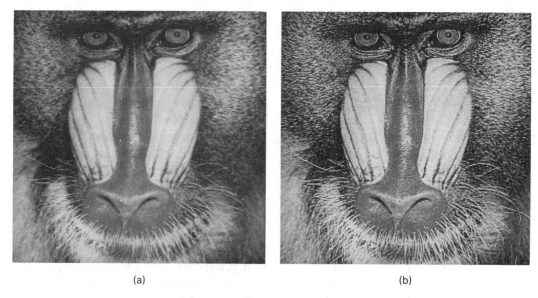

(a)   (b)

Figure 8.9 Example of highpass filtering. (a) Original image of 256×256 pixels; (b) highpass filtered image.

power. A comparison of the background regions of Figures 8.9(a) and 8.9(b) shows that the highpass filtered image appears more noisy than the unprocessed image. The accentuation of background noise is a limitation of any algorithm that attempts to increase the local contrast and sharpen the visual appearance of an image.

8.1.3 Homomorphic Processing

When an image with a large dynamic range, for instance, a natural scene on a sunny day, is recorded on a medium with a small dynamic range, such as film or paper, the image's contrast is significantly reduced, particularly in the dark and bright regions. One approach to enhancing the image is to reduce its dynamic range and increase its local contrast prior to recording it on a medium with a small dynamic range.

One method developed to reduce the dynamic range and increase the local contrast is based on applying a homomorphic system for multiplication to an image formation model. An image is typically formed by recording the light reflected from an object that has been illuminated by some light source. Based on this observation, one simple model of an image is

$$f(n_1, n_2) = i(n_1, n_2)r(n_1, n_2) \tag{8.4}$$

where $i(n_1, n_2)$ represents the illumination and $r(n_1, n_2)$ represents the reflectance. In developing a homomorphic system for image enhancement, the illumination component $i(n_1, n_2)$ is assumed to be the primary contributor to the dynamic range of an image and is assumed to vary slowly, while the reflectance component $r(n_1, n_2)$ that represents the details of an object is assumed to be the primary contributor to local contrast and is assumed to vary rapidly. To reduce the dynamic range and increase the local contrast, then, we need to reduce $i(n_1, n_2)$ and increase $r(n_1, n_2)$. To separate $i(n_1, n_2)$ from $r(n_1, n_2)$ in (8.4), a logarithmic operation is applied to (8.4) and the result is

$$\log f(n_1, n_2) = \log i(n_1, n_2) + \log r(n_1, n_2). \tag{8.5}$$

If we assume that $\log i(n_1, n_2)$ remains slowly varying and $\log r(n_1, n_2)$ remains rapidly varying, lowpass filtering $\log f(n_1, n_2)$ will result in $\log i(n_1, n_2)$ and highpass filtering $\log f(n_1, n_2)$ will result in $\log r(n_1, n_2)$. Once $\log i(n_1, n_2)$ and $\log r(n_1, n_2)$ have been separated at least approximately, $\log i(n_1, n_2)$ is attenuated to reduce the dynamic range while $\log r(n_1, n_2)$ is emphasized to increase the local contrast. The processed $\log i(n_1, n_2)$ and $\log r(n_1, n_2)$ are then combined and the result is exponentiated to get back to the image intensity domain. This is shown in Figure 8.10(a). The system in Figure 8.10(a) can be simplified by replacing the system inside the dotted line in the figure with the corresponding highpass filter. The simplified system is shown in Figure 8.10(b). An example illustrating the performance of the system in Figure 8.10(b) is shown in Figure 8.11. Figure 8.11(a)

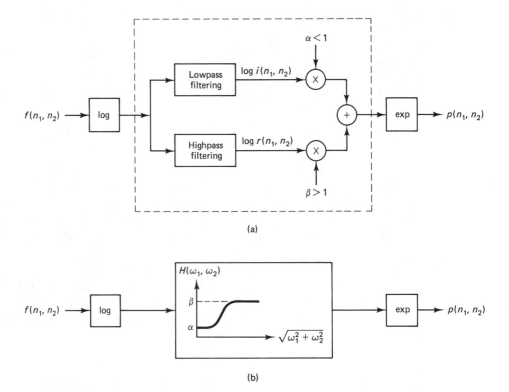

(a)

(b)

Figure 8.10 Homomorphic system for image enhancement. (a) Homomorphic system for contrast enhancement and dynamic range modification; (b) simplification of system in (a).

Image Enhancement   Chap. 8

shows an original image of 256×256 pixels and Figure 8.11(b) shows the processed image using the system in Figure 8.10(b).

A system like that in Figure 8.10, which performs a logarithmic operation followed by a linear operation followed by an exponentiation operation, is called a *homomorphic system for multiplication*. This is the origin of the terms *homomorphic processing* and *homomorphic filtering*. The logarithmic operation transforms multiplicative components to additive components. The linear operation performs the separation by exploiting characteristics of the resulting additive components and processes the separated components. The exponentiation operation brings the processed signal back to the original signal domain.

Although the system in Figure 8.10 was developed from a model of image formation and a homomorphic system, the system can be viewed simply as highpass filtering in the log intensity domain. Performing highpass filtering in the log intensity domain is also reasonable in light of human visual system. As discussed in Section 7.2.2, the image intensity appears to be modified at the peripheral level of a human visual system by some form of nonlinearity such as a logarithmic operation. Thus, the log intensity domain is, in a sense, more central to the human visual system than is the intensity domain.

8.1.4 Adaptive Modification of Local Contrast and Local Luminance Mean

In some applications, it is desirable to modify the local contrast and local luminance mean as the local characteristics of an image vary. In such applications, it is reasonable to process an image adaptively.

(a)   (b)

Figure 8.11 Example of homomorphic processing for image enhancement. (a) Original image of 256×256 pixels; (b) processed image by homomorphic system for multiplication. After [Oppenheim et al.].

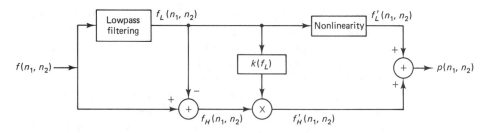

Figure 8.12 System for the modification of local contrast and local luminance mean as a function of luminance mean.

One application in which adaptive modification of the local contrast and local luminance mean is used is enhancement of an image taken from an airplane through varying amounts of cloud cover. According to one simple model of image degradation due to cloud cover, regions of an image covered by cloud have increased local luminance mean due to direct reflection of the sunlight by cloud cover and decreased local contrast due to attenuation of the signal from the ground when it passes through the cloud cover. One approach to enhancing the image, then, is to increase the local contrast and decrease the local luminance mean whenever cloud cover is detected. One way to detect the presence of cloud cover is by measuring the local luminance mean. When the local luminance mean is high, it is likely that cloud cover is present.

One system developed to reduce the effect of the cloud cover is shown in Figure 8.12. This system modifies the local contrast and the local luminance mean. In the figure, $f(n_1, n_2)$ denotes the unprocessed image. The sequence $f_L(n_1, n_2)$ which denotes the local luminance mean of $f(n_1, n_2)$ is obtained by lowpass filtering $f(n_1, n_2)$. The sequence $f_H(n_1, n_2)$, which denotes the local contrast, is obtained by subtracting $f_L(n_1, n_2)$ from $f(n_1, n_2)$. The local contrast is modified by multiplying $f_H(n_1, n_2)$ with $k(f_L)$, a scalar that is a function of $f_L(n_1, n_2)$. The modified contrast is denoted by $f_H'(n_1, n_2)$. If $k(f_L)$ is greater than one, the local contrast is increased, while $k(f_L)$ less than one represents local contrast decrease. The local luminance mean is modified by a point nonlinearity, and the modified local luminance mean is denoted by $f_L'(n_1, n_2)$. The modified local contrast and local luminance mean are then combined to obtain the processed image, $p(n_1, n_2)$. To increase the local contrast and decrease the local luminance mean when the local luminance mean is high, we choose a larger $k(f_L)$ for a larger f_L, and we choose the nonlinearity, taking into account the local luminance mean change and the contrast increase. Figure 8.13 shows the result of applying the system in Figure 8.12 to enhance an image taken from an airplane through varying amounts of cloud cover. Figure 8.13(a) shows the original image of 256×256 pixels. Figure 8.13(b) shows the processed image. The function $k(f_L)$ and the nonlinearity used are shown in Figures 8.13(c) and 8.13(d). The lowpass filtering operation was performed by using an FIR filter whose impulse response is an 8×8-point rectangular window.

The system in Figure 8.12 can be viewed as a special case of a *two-channel process*. In the two-channel process, the image to be processed is divided into

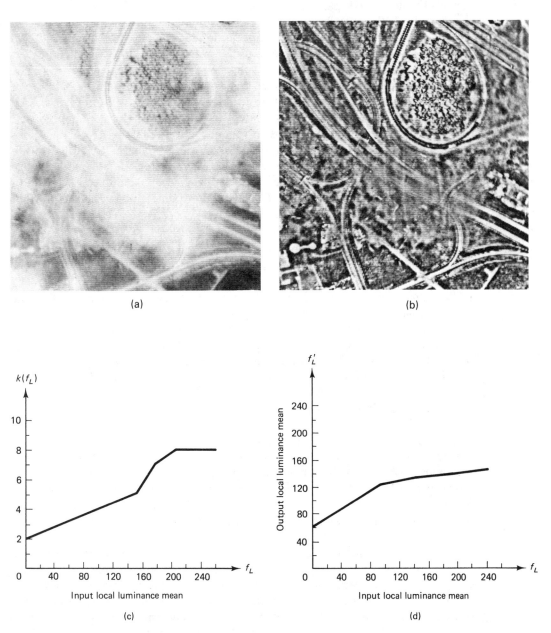

(a)

(b)

$k(f_L)$

10

8

6

4

2

0   40   80   120   160   200   240 f_L

Input local luminance mean

(c)

f_L'

Output local luminance mean

240

200

160

120

80

40

0   40   80   120   160   200   240 f_L

Input local luminance mean

(d)

Figure 8.13 Example of image enhancement by adaptive filtering. (a) Image of 256×256 pixels taken from an airplane through varying amounts of cloud cover; (b) result of processing the image in (a) with the system in Figure 8.12; (c) function $k(f_L)$ used in the processing; (d) nonlinearity used in the processing. After [Peli and Lim].

Sec. 8.1   Contrast and Dynamic Range Modification   **467**

two components, the local luminance mean and the local contrast. The two components are modified separately and then the results are combined. In the system in Figure 8.12, the local luminance mean is modified by a nonlinearity, and the local contrast is modified by the multiplication factor $k(f_L)$. As Chapters 9 and 10 show, a two-channel process is also useful in image restoration and coding.

The notion of adapting an image enhancement system to changing local characteristics is generally a very useful idea that can be applied in a number of different contexts. For example, gray scale transformation and highpass filtering, discussed earlier, can be modified so that they adapt to some varying local characteristics. Even though an adaptive system often requires considerably more computations than a nonadaptive system, the adaptive system's performance is generally considerably better. It is worthwhile to explore adaptive systems in solving an image enhancement problem that requires high performance. Adaptive processing of images is also very useful in image restoration and coding, and is discussed further in Chapters 9 and 10.

8.2 NOISE SMOOTHING

In addition to enhancement of images by contrast and dynamic range modification, images can also be enhanced by reducing degradations that may be present. This area of image enhancement overlaps with image restoration. In this section, we discuss very simple algorithms that attempt to reduce random noise and salt-and-pepper type of noise. Algorithms that are more mathematical and complex and algorithms that address other types of degradation will be discussed in Chapter 9.

8.2.1 Lowpass Filtering

The energy of a typical image is primarily concentrated in its low-frequency components. This is due to the high spatial correlation among neighboring pixels. The energy of such forms of image degradation as wideband random noise is typically more spread out over the frequency domain. By reducing the high-frequency components while preserving the low-frequency components, lowpass filtering reduces a large amount of noise at the expense of reducing a small amount of signal.

Lowpass filtering can also be used together with highpass filtering in processing an image prior to its degradation by noise. In applications such as image coding, an original undegraded image is available for processing prior to its degradation by noise, for instance, quantization noise. In such applications, the undegraded image can be highpass filtered prior to its degradation and then lowpass filtered after degradation. This may result in some improvement in the quality or intelligibility of the resulting image. For example, when the degradation is due to wideband random noise, the effective SNR (signal-to-noise ratio) of the degraded image is much lower in the high-frequency components than in the low-frequency components, due to the lowpass character of a typical image. Highpass filtering prior to the degradation significantly improves the SNR in the high-frequency components at the expense of small SNR decrease in the low-frequency components.

Examples of impulse responses of lowpass filters typically used for image enhancement are shown in Figure 8.14. To illustrate the performance of lowpass filtering for image enhancement, two examples are considered. Figure 8.15(a) shows an original noise-free image of 256 × 256 pixels, and Figure 8.15(b) shows an image degraded by wideband Gaussian random noise at an SNR of 15 dB. The SNR is defined as $10 \log_{10}$ (image variance/noise variance). Figure 8.15(c) shows the result of lowpass filtering the degraded image. The lowpass filter used is shown in Figure 8.14(c). In Figure 8.15, lowpass filtering clearly reduces the additive noise, but at the same time it blurs the image. Blurring is a primary limitation of lowpass filtering. Figure 8.16(a) shows an original image of 256 × 256 pixels with 8 bits/pixel. Figure 8.16(b) shows the image coded by a PCM system with Roberts's pseudonoise technique at 2 bits/pixel. Roberts's pseudonoise technique is discussed in Chapter 10. Figure 8.16(c) shows the result of highpass filtering before coding and lowpass filtering after coding. The highpass and lowpass filters used in this example are those in Figure 8.8(c) and Figure 8.14(c), respectively.

8.2.2 Median Filtering

Median filtering is a nonlinear process useful in reducing impulsive, or salt-and-pepper noise. It is also useful in preserving edges in an image while reducing random noise. Impulsive or salt-and-pepper noise can occur due to a random bit error in a communication channel. In a median filter, a window slides along the image, and the median intensity value of the pixels within the window becomes the output intensity of the pixel being processed. For example, suppose the pixel values within a window are 5, 6, 55, 10, and 15, and the pixel being processed has

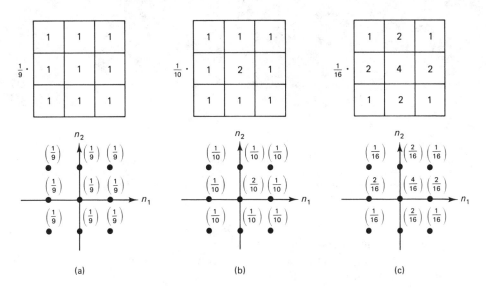

Figure 8.14   Impulse responses of lowpass filters useful for image enhancement.

(a)

(b)

(c)

Figure 8.15 Example of noise reduction by lowpass filtering. (a) Original image of 256 × 256 pixels; (b) original image degraded by wideband Gaussian random noise at SNR of 15 dB; (c) result of processing the image in (b) with a lowpass filter.

a value of 55. The output of the median filter at the current pixel location is 10, which is the median of the five values.

Like lowpass filtering, median filtering smoothes the image and is thus useful in reducing noise. Unlike lowpass filtering, median filtering can preserve discontinuities in a step function and can smooth a few pixels whose values differ signif-

(a)

(b)   (c)

Figure 8.16 Application of lowpass filtering in image coding. (a) Original image of 256 × 256 pixels; (b) image in (a) coded by a PCM system with Roberts's pseudonoise technique at 2 bits/pixel; (c) result of highpass filtering the image in (a) before coding and lowpass filtering after coding with a PCM system with Roberts's pseudonoise technique at 2 bits/pixel.

icantly from their surroundings without affecting the other pixels. Figure 8.17(a) shows a 1-D step sequence degraded by a small amount of random noise. Figure 8.17(b) shows the result after filtering with a lowpass filter whose impulse response is a 5-point rectangular window. Figure 8.17(c) shows the result after filtering with a 5-point median filter. It is clear from the figure that the step discontinuity is better preserved by the median filter. Figure 8.18(a) shows a 1-D sequence with two values that are significantly different from the surrounding points. Figures 8.18(b) and (c) show the result of a lowpass filter and a median filter, respectively. The filters used in Figure 8.18 are the same as those used in Figure 8.17. If the two impulsive values are due to noise, the result of using a median filter will be to reduce the noise. If the two values are part of the signal, however, using the median filter will distort the signal.

An important parameter in using a median filter is the size of the window. Figure 8.19 illustrates the result of median filtering the signal in Figure 8.18(a) as a function of the window size. If the window size is less than 5, the two pixels with impulsive values will not be significantly affected. For a larger window, they will be. Thus, the choice of the window size depends on the context. Because it is difficult to choose the optimum window size in advance, it may be useful to try several median filters of different window sizes and choose the best of the resulting images.

In the above, we discussed 1-D median filtering. The task involved in performing a median filtering operation extends straightforwardly from the 1-D case to the 2-D case. However, not all properties of a 1-D median filter apply to a

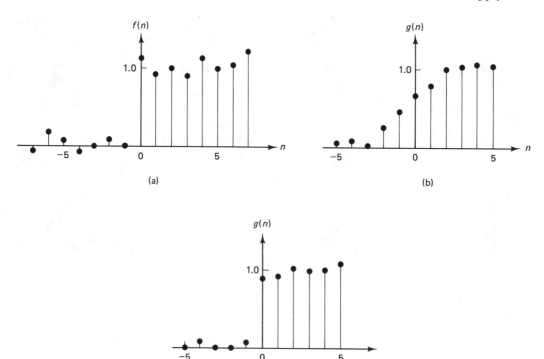

Figure 8.17 Illustration of median filter's tendency to preserve step discontinuities. (a) One-dimensional step sequence degraded by random noise; (b) result of lowpass filtering the sequence in (a) with a 5-point rectangular impulse response; (c) result of applying a 5-point median filter to the sequence in (a).

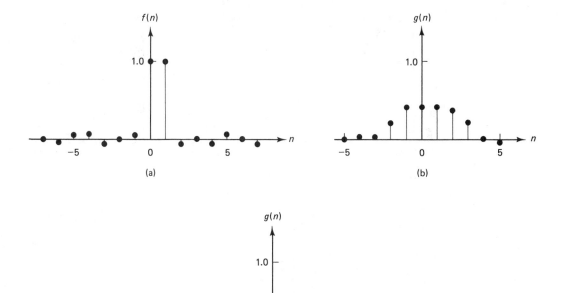

Figure 8.18 Illustration of a median filter's capability to remove impulsive values. (a) One-dimensional sequence with two consecutive samples significantly different from surrounding samples; (b) result of lowpass filtering the sequence in (a) with a 5-point rectangular impulse response; (c) result of applying a 5-point median filter to the sequence in (a).

2-D median filter. For example, median filtering a 1-D unit step sequence $u(n)$ preserves the step discontinuity and does not affect the signal $u(n)$ at all. Suppose we filter a 2-D step sequence $u(n_1, n_2)$ with a 2-D $N \times N$-point median filter. Figure 8.20(a) shows $u(n_1, n_2)$ and Figure 8.20(b) shows the result of filtering $u(n_1, n_2)$ with a 2-D 5×5-point median filter. From Figure 8.20(b), the intensity discontinuities which can be viewed as 1-D steps (for large n_1 at $n_2 = 0$ and large n_2 at $n_1 = 0$) are not affected. However, the discontinuities which are truly 2-D steps ($n_1 = n_2 = 0$) are seriously distorted. One method that tends to preserve 2-D step discontinuities well is to filter a 2-D signal along the horizontal direction with a 1-D median filter and then filter the result along the vertical direction with another 1-D median filter. This method is called *separable median filtering*, and is often used in 2-D median filtering applications. When a separable median filter is applied to $u(n_1, n_2)$, the signal $u(n_1, n_2)$ is not affected.

A median filter is a nonlinear system, and therefore many theoretical results on linear systems are not applicable. For example, the result of separable median filtering depends on the order in which the 1-D horizontal and vertical median filters are applied. Despite this difficulty, some theoretical results have been developed [Gallagher and Wise; Nodes and Gallagher; Arce and McLoughlin] on

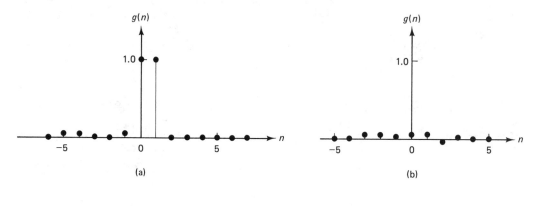

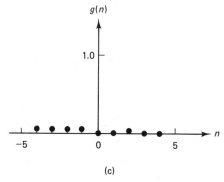

Figure 8.19 Results of applying a median filter to the sequence in Figure 8.18(a) as a function of window size. This illustrates that removal of impulsive values by a median filter depends on the window size. (a) Window size = 3; (b) window size = 5; (c) window size = 7.

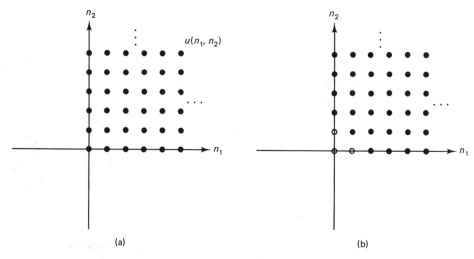

Figure 8.20 Illustration that a 2-D $N \times N$-point median filter distorts 2-D step discontinuities. (a) Unit step sequence $u(n_1, n_2)$; (b) result of filtering $u(n_1, n_2)$ with a 5×5-point median filter.

median filtering. One result states that repeated application of a 1-D median filter to a 1-D sequence eventually leads to a signal called a *root signal*, which is invariant under further applications of the 1-D median filter.

Two examples are given to illustrate the performance of a median filter. In the first, the original image of 512×512 pixels shown in Figure 8.21(a), is degraded

(a)

(b)   (c)

Figure 8.21 Example of wideband random noise reduction by median filtering. (a) Original image of 512×512 pixels; (b) image degraded by wideband Gaussian random noise at SNR of 7 dB; (c) processed image by a separable median filter with window size of 3 for both the horizontal and vertical 1-D median filter.

by wideband Gaussian random noise at an SNR of 7 dB. The degraded image is shown in Figure 8.21(b). Figure 8.21(c) shows the image processed by a separable median filter with a window size of 3 for both the horizontal and vertical 1-D median filters. Although the very sharp edges are not blurred, median filtering blurs the image significantly. In the second example, the original image from Figure 8.21(a) is degraded by salt-and-pepper noise. The degraded image is shown in Figure 8.22(a) and the image processed by the same separable median filter used in Figure 8.21 is shown in Figure 8.22(b). This example shows that median filtering is quite effective in removing salt-and-pepper noise.

8.2.3 Out-Range Pixel Smoothing

Like median filtering, *out-range pixel smoothing* is a nonlinear operation and is useful in reducing salt-and-pepper noise. In this method, a window slides along the image, and the average of the pixel values, excluding the pixel being processed, is obtained. If the difference between the average and the value of the pixel processed is above some threshold, then the current pixel value is replaced by the average. Otherwise, the value is not affected. Because it is difficult to determine the best parameter values in advance, it may be useful to process an image using several different threshold values and window sizes and select the best result.

Figure 8.23 illustrates the performance of out-range pixel smoothing. The image in Figure 8.23 is the result after processing the image in Figure 8.22(a) using out-range pixel smoothing with a threshold value of 50 and a 3×3-point window.

8.3 EDGE DETECTION

An *edge* in an image is a boundary or contour at which a significant change occurs in some physical aspect of an image, such as the surface reflectance, illumination, or the distances of the visible surfaces from the viewer. Changes in physical aspects manifest themselves in a variety of ways, including changes in intensity, color, and texture. In our discussion, we are concerned only with the changes in image intensity.

Detecting edges is very useful in a number of contexts. For example, in a typical image understanding task such as object identification, an essential step is to segment an image into different regions corresponding to different objects in the scene. Edge detection is often the first step in image segmentation. As another example, one approach to the development of a low bit-rate image coding system is to code only the detected edges. It is well known that an image that consists of only edges is highly intelligible.

The significance of a physical change in an image depends on the application; an intensity change that would be classified as an edge in some applications might not be considered an edge in other applications. In an object identification system, an object's boundaries may be sufficient for identification, and contours that rep-

(a)

(b)

Figure 8.22 Example of salt-and-pepper noise reduction by median filtering. (a) Image in Figure 8.21(a) degraded by salt-and-pepper noise; (b) processed image by the same separable median filter used in Figure 8.21.

resent additional details within the object may not be considered edges. An edge cannot be defined, then, outside of the context of an application. Nevertheless, edge detection algorithms that detect edges that are useful in a broad set of applications have been developed. In this section, we discuss some of the more representative edge detection algorithms.

Figure 8.23 Example of salt-and-pepper noise reduction by outrange pixel smoothing. Image in Figure 8.22(a) processed by outrange pixel smoothing with threshold value of 50 and window size of 3×3.

8.3.1 Gradient-Based Methods

Consider an analog* function $f(x)$ which represents a typical 1-D edge, as shown in Figure 8.24(a). In typical problems, it is reasonable to consider the value x_0 in the figure an edge point. One way to determine x_0 is to compute the first derivative $f'(x)$ or the second derivative $f''(x)$. Figures 8.24(b) and (c) show $f'(x)$ and $f''(x)$. From the figure, the value x_0 can be determined by looking for the local extremum (maximum or minimum) of $f'(x)$ or by looking for a zero crossing of $f''(x)$ where $f''(x)$ changes its sign. In this section, we discuss methods that exploit the characteristics of $f'(x)$. In the next section, we discuss methods that exploit the characteristics of $f''(x)$.

In addition to determining the possible edge point x_0, $f'(x)$ can also be used in estimating the strength and direction of the edge. If $|f'(x)|$ is very large, $f(x)$ is changing very rapidly and a rapid change in intensity is indicated. If $f'(x)$ is positive, $f(x)$ is increasing. Based on the above observations, one approach to detecting edges is to use the system shown in Figure 8.25. In the system, first $|f'(x)|$ is computed from $f(x)$. If $|f'(x)|$ is greater than some threshold, it is a candidate to be an edge. If all values of x such that $|f'(x)|$ is greater than a certain threshold are detected to be edges, an edge will appear as a line rather than a point. To avoid this problem, we further require $|f'(x)|$ to have a local maximum at the edge points. It may be desirable to determine whether $f(x)$ is increasing or decreasing at $x = x_0$. The necessary information is contained in $f'(x)$ at $x = x_0$. The choice of the threshold depends on the application. As the threshold increases, only the values of x where $f(x)$ changes rapidly will be registered as candidate edges. Since it is difficult to choose the threshold optimally, some trial and error is usually involved. It is also possible to choose the threshold adaptively. The system in Figure 8.25 is based on the particular type of edge shown in Figure 8.24(a), but it is generally applicable to detecting various other types of edges.

The generalization of $f'(x)$ to a 2-D function $f(x, y)$ is the gradient $\nabla f(x, y)$ given by

$$\nabla f(x, y) = \frac{\partial f(x, y)}{\partial x}\, \hat{i}_x + \frac{\partial f(x, y)}{\partial y}\, \hat{i}_y \tag{8.6}$$

where $\hat{i}_x$ is the unit vector in the x-direction and $\hat{i}_y$ is the unit vector in the y-direction. A generalization of the edge detection system in Figure 8.25 based on $\nabla f(x, y)$ is shown in Figure 8.26. The magnitude of $\nabla f(x, y)$ is first computed and is then compared with a threshold to determine candidate edge points. If all values of (x, y) such that $|\nabla f(x, y)|$ is greater than a certain threshold are detected to be edges, the edges will appear as strips rather than lines. The process of determining an edge line from a strip of candidate edge points is called *edge thinning*. In one simple edge thinning algorithm, the edge points are selected by

---

*Sometimes, it is more convenient to develop results in the analog domain. In such instances, we will begin the development of results in the analog domain and then discretize the results at some later point in the development.

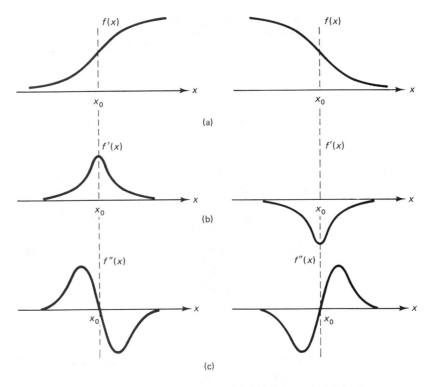

Figure 8.24 (a) $f(x)$, (b) $f'(x)$, and (c) $f''(x)$ for a typical 1-D edge.

checking if $|\nabla f(x, y)|$ is a local maximum in at least one direction. The property that $|\nabla f(x, y)|$ achieves its local maximum in at least one direction is usually checked along a few specified directions. In most cases, it is sufficient to check for local maxima in only the horizontal and vertical directions. If $|\nabla f(x, y)|$ is a local maximum along any one of the specified directions at a potential edge point, the potential edge point is considered to be an edge point. One difficulty with this simple edge thinning algorithm is that it creates a number of minor false edge lines in the vicinity of strong edge lines. One simple method to remove most of these minor false edge lines is to impose the following additional constraints:

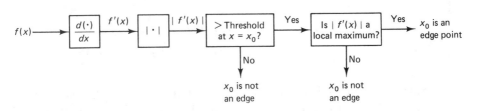

Figure 8.25 System for 1-D edge detection.

Sec. 8.3 Edge Detection

479

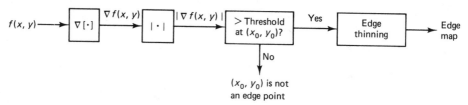

Figure 8.26 System for 2-D edge detection.

(a) If $|\nabla f(x, y)|$ has a local maximum at (x_0, y_0) in the horizontal direction but not in the vertical direction, (x_0, y_0) is an edge point when

$$\left|\frac{\partial f(x, y)}{\partial x}\right|_{x=x_0, y=y_0} > k \left|\frac{\partial f(x, y)}{\partial y}\right|_{x=x_0, y=y_0} \quad \text{with } k \text{ typically chosen around 2.}$$

(b) If $|\nabla f(x, y)|$ has a local maximum at (x_0, y_0) in the vertical direction but not in the horizontal direction, (x_0, y_0) is an edge point when

$$\left|\frac{\partial f(x, y)}{\partial y}\right|_{x=x_0, y=y_0} > k \left|\frac{\partial f(x, y)}{\partial x}\right|_{x=x_0, y=y_0} \quad \text{with } k \text{ typically chosen around 2.}$$

When $|\nabla f(x, y)|$ has a local maximum at (x_0, y_0) in the horizontal direction but not in the vertical direction, Condition (a) requires that the rate of intensity change along the horizontal direction is significantly larger than that along the vertical direction. Condition (b) is the same as Condition (a) with the roles of x and y reversed. Why these additional constraints remove most of the minor false edges in the vicinity of major edges is discussed in Problem 8.17.

An edge detection system that is based on a function such as $|\nabla f(x, y)|$ is called a *nondirectional edge detector*, since such functions do not have a bias toward any particular direction. If an edge detection system is based on a function that has a bias toward one particular direction, it is called a *directional edge detector*. If we use $|\partial f(x, y)/\partial x|$ instead of $|\nabla f(x, y)|$ in the system in Figure 8.26, for example, the system will detect edges in the vertical direction, but will not respond to edges in the horizontal direction.

For a 2-D sequence $f(n_1, n_2)$, the partial derivatives $\partial f(x, y)/\partial x$ and $\partial f(x, y)/\partial y$ can be replaced by some form of difference. For example, $\partial f(x, y)/\partial x$ may be replaced by

$$\frac{\partial f(x, y)}{\partial x} \leftrightarrow [f(n_1, n_2) - f(n_1 - 1, n_2)]/T, \tag{8.7a}$$

$$[f(n_1 + 1, n_2) - f(n_1, n_2)]/T, \tag{8.7b}$$

$$\text{or} \quad [f(n_1 + 1, n_2) - f(n_1 - 1, n_2)]/(2T). \tag{8.7c}$$

Since the computed derivatives are compared later with a threshold and the threshold can be adjusted, the scaling factors $1/T$ and $1/2T$ can be omitted. Typically the expressions in (8.7) are averaged over several samples to improve the reliability and continuity of the estimate of $\partial f(x, y)/\partial x$. Examples of "improved" estimates of $\partial f(x, y)/\partial x$ are

$$\frac{\partial f(x,y)}{\partial x} \leftrightarrow [f(n_1 + 1, n_2 + 1) - f(n_1 - 1, n_2 + 1)] + [f(n_1 + 1, n_2) - f(n_1 - 1, n_2)]$$

$$+ [f(n_1 + 1, n_2 - 1) - f(n_1 - 1, n_2 - 1)] \qquad (8.8a)$$

or

$$[f(n_1 + 1, n_2 + 1) - f(n_1 - 1, n_2 + 1)] + 2[f(n_1 + 1, n_2) - f(n_1 - 1, n_2)]$$

$$+ [f(n_1 + 1, n_2 - 1) - f(n_1 - 1, n_2 - 1)]. \qquad (8.8b)$$

The unnecessary scaling factors have been omitted in (8.8).

The differencing operation in (8.7) and (8.8) can be viewed as the convolution of $f(n_1, n_2)$ with the impulse response of a filter $h(n_1, n_2)$. Examples of impulse responses that can be used in developing directional edge detectors are shown in Figure 8.27. The filters $h(n_1, n_2)$ in Figures 8.27(a) and (b) detect edges in the vertical and horizontal directions and can be viewed as approximating $\partial f(x, y)/\partial x$ and $\partial f(x, y)/\partial y$ respectively. The filters $h(n_1, n_2)$ in Figures 8.27(c) and (d) detect edges along the two diagonal directions. The gradient $\nabla f(x, y)$ in (8.6) can also be expressed in terms of the first-order partial derivatives in a rotated coordinate system. When the rotation is 45 degrees, the directions of partial derivatives are along the two diagonal directions.

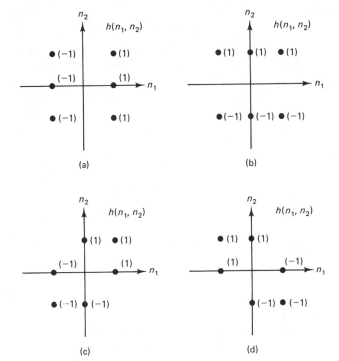

Figure 8.27 Impulse responses of filters that can be used for directional edge detection. (a) Vertical edge detection; (b) horizontal edge detection; (c) and (d) diagonal edge detection.

Nondirectional edge detectors can be developed by discrete approximation of $|\nabla f(x, y)|$ in the system in Figure 8.26. From (8.6),

$$|\nabla f(x, y)| = \sqrt{\left(\frac{\partial f(x, y)}{\partial x}\right)^2 + \left(\frac{\partial f(x, y)}{\partial y}\right)^2} \qquad (8.9)$$

From (8.9), nondirectional edge detectors can be developed by nonlinear combination of the terms used in the development of directional edge detectors. An example of discrete approximation of (8.9) that can be used for nondirectional edge detectors is given by

$$|\nabla f(x, y)| \rightarrow \sqrt{(f_x(n_1, n_2))^2 + (f_y(n_1, n_2))^2} \qquad (8.10)$$

where

$$f_x(n_1, n_2) = f(n_1, n_2) * h_x(n_1, n_2)$$

$$f_y(n_1, n_2) = f(n_1, n_2) * h_y(n_1, n_2)$$

and $h_x(n_1, n_2)$ and $h_y(n_1, n_2)$ are shown in Figure 8.28. The method developed by Sobel [Duda and Hart] is based on (8.10) with $h_x(n_1, n_2)$ and $h_y(n_1, n_2)$ in Figure 8.28. Another example is the method developed by [Roberts], which is based on (8.10) with $h_x(n_1, n_2)$ and $h_y(n_1, n_2)$ shown in Figure 8.29. Depending on exactly how $|\nabla f(x, y)|$ is approximated in the discrete domain, many other variations can be developed.

Figure 8.30 shows the result of edge detection using a directional edge detector. Figure 8.30(a) shows an image of 512×512 pixels. Figure 8.30(b) and (c) show the results of a vertical edge detector and a horizontal edge detector, respectively, applied to the image in Figure 8.30(a). The vertical and horizontal edge detectors are based on $h(n_1, n_2)$ in Figures 8.27(a) and (b). Figures 8.31(a) and (b) show the results of applying the Sobel edge detector and Roberts's edge detector to the image in Figure 8.30(a). Both belong to the class of nondirectional edge detectors, and the specific method of determining the threshold value and checking the local maximum property of an edge used is the same as that used in Figure 8.30.

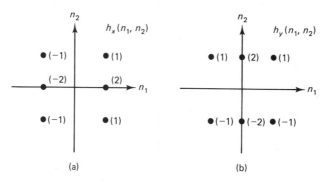

(a)   (b)

Figure 8.28 Approximation of (a) $\partial f(x, y)/\partial x$ with $f(n_1, n_2) * h_x(n_1, n_2)$; (b) $\partial f(x, y)/\partial y$ with $f(n_1, n_2) * h_y(n_1, n_2)$. Sobel's edge detection method is based on comparison of $\sqrt{(f(n_1, n_2) * h_x(n_1, n_2))^2 + (f(n_1, n_2) * h_y(n_1, n_2))^2}$ with a threshold.

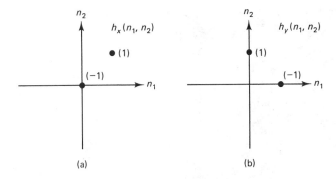

Figure 8.29 Impulse responses of filters used in Roberts's edge detection method. The method is based on comparison of

$$\sqrt{(f(n_1, n_2) * h_x(n_1, n_2))^2 + (f(n_1, n_2) * h_y(n_1, n_2))^2}$$

with a threshold.

There are many variations of the edge detection methods discussed in this section. For example, we could use a different nonlinear combination of $\partial f(x, y)/\partial x$ and $\partial f(x, y)/\partial y$ instead of

$$\sqrt{(\partial f(x, y)/\partial x)^2 + (\partial f(x, y)/\partial y)^2}$$

in the system in Figure 8.26. Many different methods can also be developed for edge thinning.

The edge detection methods discussed in this section can be improved in various ways. Methods based on computing some form of gradient or differencing are typically sensitive to noise. A certain number of isolated edge points which appear randomly distributed throughout the edge maps in Figure 8.31 are most likely the result of some background noise or very fine image details. Some noise smoothing using the methods discussed in Section 8.2 or more sophisticated noise reduction methods that we will discuss in Chapter 9 may be desirable prior to applying an edge detection algorithm. Isolated random edge points may also be removed by some simple processing of the edge maps. Gradient-based edge detection methods also cause some discontinuities in the detected edge contours, as can be seen from the edge maps in Figure 8.31. Methods that impose continuity constraints in the detected edge contours can also be developed [Roberts].

8.3.2 Laplacian-Based Methods

The objective of an edge detection algorithm is to locate the regions where the intensity is changing rapidly. In the case of a 1-D function $f(x)$, searching for regions of rapidly changing intensity corresponds to searching for regions where $f'(x)$ is large. For gradient-based methods, $f'(x)$ is considered large when its magnitude $|f'(x)|$ is greater than a threshold. Another possible way is to assume that $f'(x)$ is large whenever it reaches a local extremum, that is, whenever the second derivative $f''(x)$ has a zero crossing. This is illustrated in Figure 8.24. Declaring zero-crossing points as edges results in a large number of points being declared to be edge points. Since there is no check on the magnitude of $f'(x)$,

(a)

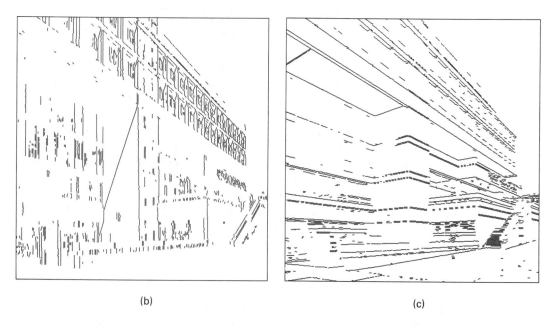

(b)   (c)

Figure 8.30 Edge maps obtained by directional edge detectors. (a) Image of 512 × 512 pixels; (b) result of applying a vertical edge detector; (c) result of applying a horizontal edge detector.

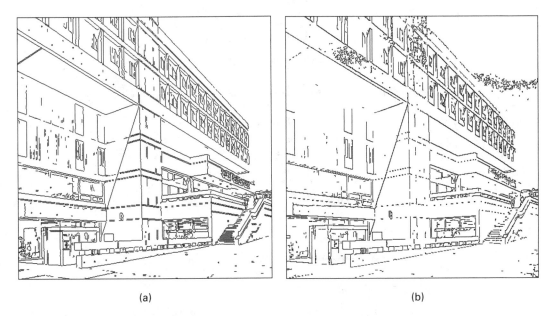

Figure 8.31 Result of applying (a) Sobel edge detector and (b) Roberts's edge detector to the image in Figure 8.30(a).

any small ripple in $f(x)$ is enough to generate an edge point. Due to this sensitivity to noise, the application of a noise reduction system prior to edge detection is very desirable in processing images with background noise.

A generalization of $\partial^2 f(x)/\partial x^2$ to a 2-D function $f(x, y)$ for the purpose of edge detection (see Problem 8.19) is the Laplacian $\nabla^2 f(x, y)$ given by

$$\nabla^2 f(x, y) = \nabla(\nabla f(x, y)) = \frac{\partial^2 f(x, y)}{\partial x^2} + \frac{\partial^2 f(x, y)}{\partial y^2}. \tag{8.11}$$

For a 2-D sequence $f(n_1, n_2)$, the partial second derivatives $\partial^2 f(x, y)/\partial x^2$ and $\partial^2 f(x, y)/\partial y^2$ can be replaced by some form of second-order differences. Second-order differences can be represented by convolution of $f(n_1, n_2)$ with the impulse response of a filter $h(n_1, n_2)$. Examples of $h(n_1, n_2)$ that may be used are shown in Figure 8.32. To illustrate that $f(n_1, n_2) * h(n_1, n_2)$ may be viewed as a discrete approximation of $\nabla^2 f(x, y)$, let us consider $h(n_1, n_2)$ in Figure 8.32(a). Suppose we approximate $\partial f(x, y)/\partial x$ by

$$\frac{\partial f(x, y)}{\partial x} \to f_x(n_1, n_2) = f(n_1 + 1, n_2) - f(n_1, n_2). \tag{8.12}$$

We again omitted the scaling factor, since it does not affect zero-crossing points. Since the forward difference is used in (8.12), we can use the backward difference in approximating $\partial^2 f(x, y)/\partial x^2$:

$$\frac{\partial^2 f(x)}{\partial x^2} \to f_{xx}(n_1, n_2) = f_x(n_1, n_2) - f_x(n_1 - 1, n_2). \tag{8.13}$$

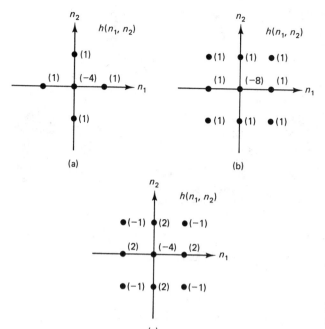

Figure 8.32 Examples of $h(n_1, n_2)$ that may be used in approximating $\nabla^2 f(x, y)$ with $f(n_1, n_2) * h(n_1, n_2)$.

From (8.12) and (8.13),

$$\frac{\partial^2 f(x, y)}{\partial x^2} \rightarrow f_{xx}(n_1, n_2) = f(n_1 + 1, n_2) - 2f(n_1, n_2) + f(n_1 - 1, n_2). \quad (8.14)$$

From (8.11) and (8.14), and approximating $\partial^2 f(x, y)/\partial y^2$ in a similar manner, we obtain

$$\nabla^2 f(x, y) \rightarrow \nabla^2 f(n_1, n_2) = f_{xx}(n_1, n_2) + f_{yy}(n_1, n_2)$$

$$= f(n_1 + 1, n_2) + f(n_1 - 1, n_2) + f(n_1, n_2 + 1)$$

$$+ f(n_1, n_2 - 1) - 4f(n_1, n_2). \quad (8.15)$$

The resulting $\nabla^2 f(n_1, n_2)$ is $f(n_1, n_2) * h(n_1, n_2)$ with $h(n_1, n_2)$ in Figure 8.32(a). Depending on how the second-order derivatives are approximated, it is possible to derive many other impulse responses $h(n_1, n_2)$, including those shown in Figures 8.32(b) and (c).

Figure 8.33 shows an example where edges were detected by looking for zero-crossing points of $\nabla^2 f(n_1, n_2)$. Figure 8.33(a) shows an image of 512×512 pixels. Figure 8.33(b) shows the zero-crossing points of $\nabla^2 f(n_1, n_2)$, obtained from (8.15) and using the image in Figure 8.33(a) as $f(n_1, n_2)$. Since zero-crossing contours are boundaries between regions, they tend to be continuous lines. As a result, edge thinning necessary in gradient-based methods is not needed in Laplacian-based methods. In addition, algorithms that force continuity in edge contours are not as useful in Laplacian-based methods as in gradient-based methods. As is clear from Figure 8.33(b), however, choosing all zero-crossing points as edges tends to generate a large number of edge points.

The Laplacian-based methods discussed above generate many "false" edge contours, which typically appear in regions where the local variance of the image is small. As a special case, consider a uniform background region so that $f(n_1, n_2)$ is constant. Since $\nabla^2 f(n_1, n_2)$ is zero and we detect edges from zero-crossing points of $\nabla^2 f(n_1, n_2)$, any small perturbation of $f(n_1, n_2)$ is likely to cause false edge contours. One method to remove many of these false contours is to require that the local variance is sufficiently large at an edge point, as shown in Figure 8.34. The local variance $\sigma_f^2(n_1, n_2)$ can be estimated by

$$\sigma_f^2(n_1, n_2) = \frac{1}{(2M + 1)^2} \sum_{k_1 = n_1 - M}^{n_1 + M} \sum_{k_2 = n_2 - M}^{n_2 + M} \left[f(k_1, k_2) - m_f(k_1, k_2) \right]^2 \qquad (8.16a)$$

where

$$m_f(n_1, n_2) = \frac{1}{(2M + 1)^2} \sum_{k_1 = n_1 - M}^{n_1 + M} \sum_{k_2 = n_2 - M}^{n_2 + M} f(k_1, k_2) \qquad (8.16b)$$

with M typically chosen around 2. Since $\sigma_f^2(n_1, n_2)$ is compared with a threshold, the scaling factor $1/(2M + 1)^2$ in (8.16a) can be eliminated. In addition, the local variance σ_f^2 needs to be computed only for (n_1, n_2) which are zero-crossing points of $\nabla^2 f(n_1, n_2)$. Figure 8.35 shows the result of applying the system in Figure 8.34 to the image in Figure 8.33(a). Comparison of Figures 8.33(b) and 8.35 shows considerable reduction in the "false" edge contours.

The system in Figure 8.34 can be interpreted as a gradient-based method.

(a)

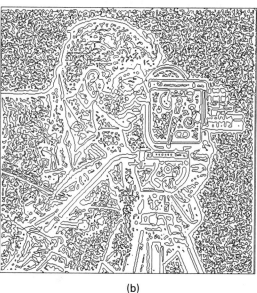

(b)

Figure 8.33 Edge map obtained by a Laplacian-based edge detector. (a) Image of 512 × 512 pixels; (b) result of convolving the image in (a) with $h(n_1, n_2)$ in Figure 8.32(a) and then finding zero-crossing points.

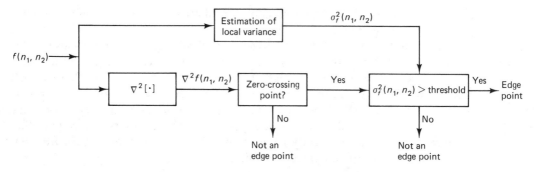

Figure 8.34 Laplacian-based edge detection system that does not produce many false edge contours.

The local variance $\sigma_f^2(n_1, n_2)$ is closely related to the gradient magnitude. Comparing $\sigma_f^2(n_1, n_2)$ with a threshold is similar to comparing the gradient magnitude with a threshold. Requiring that $\nabla^2 f(n_1, n_2)$ crosses zero at an edge can be interpreted as edge thinning. With this interpretation, we can implement the system in Figure 8.34 by computing $\sigma_f^2(n_1, n_2)$ first and then by detecting the zero-crossing points of $\nabla^2 f(n_1, n_2)$ only at those points where $\sigma_f^2(n_1, n_2)$ is above the chosen threshold.

8.3.3 Edge Detection by Marr and Hildreth's Method

In the previous two sections, we discussed edge detection algorithms that produce one edge map from an input image. Marr and Hildreth [Marr and Hildreth; Marr] observed that significant intensity changes occur at different scales (resolution) in an image. For example, blurry shadow regions and sharply focused fine-detail

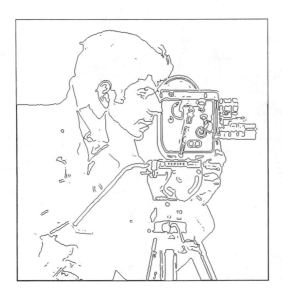

Figure 8.35 Edge map obtained by applying the system in Figure 8.34 to the image in Figure 8.33(a).

Image Enhancement   Chap. 8

regions may be present in the same image. "Optimal" detection of significant intensity changes, therefore, generally requires the use of operators that respond to several different scales. Marr and Hildreth suggested that the original image be band-limited at several different cutoff frequencies and that an edge detection algorithm be applied to each of the images. The resulting edge maps have edges corresponding to different scales.

Marr and Hildreth argue that edge maps of different scales contain important information about physically significant parameters. The visual world is made of elements such as contours, scratches, and shadows, which are highly localized at their own scale. This localization is also reflected in such physically important changes as reflectance change and illumination change. If the same edge is present in a set of edge maps of different scale, it represents the presence of an image intensity change due to a single physical phenomenon. If an edge is present in only one edge map, one reason may be that two independent physical phenomena are operating to produce intensity changes in the same region of the image.

To bandlimit an image at different cutoff frequencies, the impulse response $h(x, y)$ and frequency response $H(\Omega_x, \Omega_y)$ of the lowpass filter proposed [Marr and Hildreth; Canny] is Gaussian-shaped and is given by

$$h(x, y) = e^{-(x^2 + y^2)/(2\pi\sigma^2)} \tag{8.17a}$$

$$H(\Omega_x, \Omega_y) = 2\pi^2\sigma^2 e^{-\pi\sigma^2(\Omega_x^2 + \Omega_y^2)/2} \tag{8.17b}$$

where σ determines the cutoff frequency with larger σ corresponding to lower cutoff frequency. The choice of Gaussian shape is motivated by the fact that it is smooth and localized in both the spatial and frequency domains. A smooth $h(x, y)$ is less likely to introduce any changes that are not present in the original shape. A more localized $h(x, y)$ is less likely to shift the location of edges.

From the smoothed images, edges can be detected by using the edge detection algorithms discussed in the previous two sections. Depending on which method is used, the lowpass filtering operation in (8.17) and the partial derivative operation used for edge detection may be combined. For example, noting that $\nabla^2[\cdot]$ and convolution $*$ are linear, we obtain

$$\nabla^2(f(x, y) * h(x, y)) = f(x, y) * [\nabla^2 h(x, y)]$$
$$= f(x, y) * \left[\frac{\partial^2 h(x, y)}{\partial x^2} + \frac{\partial^2 h(x, y)}{\partial y^2} \right]. \tag{8.18}$$

For the Gaussian function $h(x, y)$ in (8.17), $\nabla^2 h(x, y)$ and its Fourier transform are given by

$$\nabla^2 h(x, y) = \frac{e^{-(x^2 + y^2)/(2\pi\sigma^2)}}{(\pi\sigma^2)^2} (x^2 + y^2 - 2\pi\sigma^2) \tag{8.19a}$$

$$F[\nabla^2 h(x, y)] = -2\pi^2\sigma^2 e^{-\pi\sigma^2(\Omega_x^2 + \Omega_y^2)/2}(\Omega_x^2 + \Omega_y^2). \tag{8.19b}$$

Marr and Hildreth chose, for simplicity, to detect edges by looking for zero-crossing points of $\nabla^2 f(x, y)$. Bandlimiting $f(x, y)$ tends to reduce noise, thus reducing the noise sensitivity problem associated with detecting zero-crossing points. The functions $\nabla^2 h(x, y)$ and $-F[\nabla^2 h(x, y)]$ in (8.19) are sketched in Figure 8.36. Clearly, computing $f(x, y) * \nabla^2 h(x, y)$ is equivalent to bandpass filtering $f(x, y)$ where σ^2 in (8.19) is a parameter that controls the bandwidth of the bandpass filter. For a sequence $f(n_1, n_2)$, one approach is to simply replace x and y in (8.19) with n_1 and n_2.

Figure 8.37 shows an example of the approach under discussion. Figures 8.37(a), (b), and (c) show three images obtained by blurring the original image in Figure 8.33(a) with $h(n_1, n_2)$ obtained by replacing x and y of $h(x, y)$ in (8.17) with n_1 and n_2 with $\sigma^2 = 4$, 16, and 36, respectively. Figures 8.37(d), (e), and (f) show the images obtained by detecting zero crossings of $f(n_1, n_2) * \nabla^2 h(x, y)|_{x=n_1, y=n_2}$, with $\nabla^2 h(x, y)$ given by (8.19a) for $\sigma^2 = 4$, 16, and 36, respectively. Marr and Hildreth used the edge maps of different scales, such as those in Figures 8.37(d), (e), and (f) for object representation in their image understanding work.

8.3.4 Edge Detection Based on Signal Modeling

The edge detection algorithms discussed above are general methods, in that they are developed independent of an application context. An alternative approach is to develop an edge detection algorithm specific to a particular application problem. If we know the shape of an edge, for example, this information can be incorporated in the development of an edge detection algorithm. To illustrate how an edge detection algorithm specific to an application problem may be developed, we consider the problem of detecting boundaries of coronary arteries from an angiogram [Abrams].

The coronary arteries are the blood vessels which encircle the heart and supply blood to the heart muscle. Narrowing of the coronary arteries prevents adequate blood supply from reaching the heart, causing pain and damage to the heart muscle. Such damage is called coronary disease. To determine the severity of coronary disease, a coronary angiogram is used. An angiogram is an X ray picture of arteries taken after a contrast agent, typically iodine, has been injected into the vessels. The contrast agent is injected directly into the arteries through a catheter in order to achieve high concentrations. An example of a coronary angiogram is shown in Figure 8.38. Different observers making conventional visual evaluations of an angiogram will give widely varying evaluations of the severity of the disease.

The most commonly used measure of an obstruction is percentage of stenosis, which is defined as the maximum percentage of arterial narrowing within a specified length of the vessel. One approach to estimating the percentage of stenosis begins with determining the vessel boundaries from an angiogram. We will be concerned with the problem of detecting the vessel boundaries.

One reasonable model of an angiogram $f(n_1, n_2)$ is given by

$$f(n_1, n_2) = (v(n_1, n_2) + p(n_1, n_2)) * g(n_1, n_2) + w(n_1, n_2) \qquad (8.20)$$

where $v(n_1, n_2)$ denotes the vessel, $p(n_1, n_2)$ denotes the background, $g(n_1, n_2)$ denotes blurring, and $w(n_1, n_2)$ denotes the background noise. The vessel function $v(n_1, n_2)$ is derived from a generalized cone model of a 3-D vessel which is con-

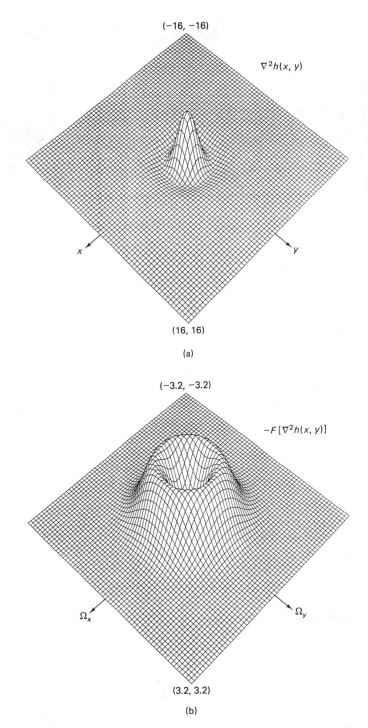

$(-16, -16)$

$\nabla^2 h(x, y)$

x

y

$(16, 16)$

(a)

$(-3.2, -3.2)$

$-F[\nabla^2 h(x, y)]$

Ω_x

Ω_y

$(3.2, 3.2)$

(b)

Figure 8.36 Sketch of (a) $\nabla^2 h(x, y)$ and (b) $-F[\nabla^2 h(x, y)]$ in Equation (8.19) for $\sigma^2 = 1$.

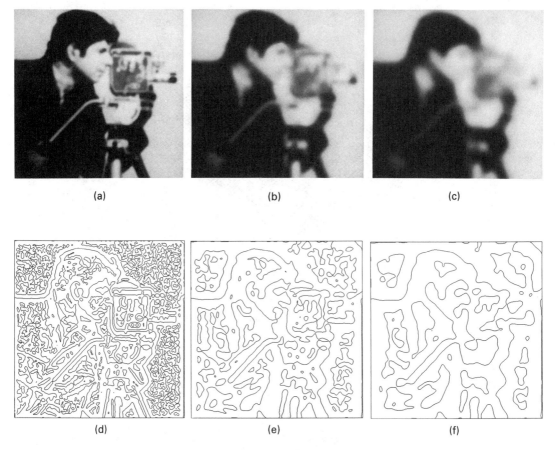

Figure 8.37 Edge maps obtained from lowpass filtered image. Blurred image with (a) $\sigma^2 = 4$; (b) $\sigma^2 = 16$; (c) $\sigma^2 = 36$. Result of applying Laplacian-based algorithm to the blurred image; (d) $\sigma^2 = 4$; (e) $\sigma^2 = 16$; (f) $\sigma^2 = 36$.

tinuous and has elliptical cross sections. The elliptical shape is chosen because of the small number of parameters involved in its characterization and because of some empirical evidence that it leads to a good estimate of percentage of stenosis. The 1-D cross section of $v(n_1, n_2)$, which consists of one blood vessel, is totally specified by three parameters, two representing the blood vessel boundaries and one related to the x-ray attenuation coefficient of iodine. The continuity of the vessel is guaranteed by fitting a cubic spline function to the vessel boundaries. The background $p(n_1, n_2)$ is modeled by a 2-D low-order polynomial. Low-order polynomials are very smooth functions, and their choice is motivated by the observation that objects in the background, such as tissue and bone, are much bigger than the blood vessels. The blurring function $g(n_1, n_2)$ is modeled by a known 2-D Gaussian function that takes into account the blurring introduced at various stages of the imaging process. The noise $w(n_1, n_2)$ is random background noise and is assumed to be white. The parameters in the model of $f(n_1, n_2)$ are the

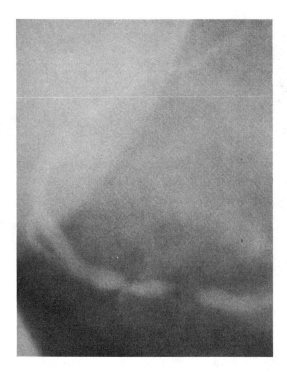

Figure 8.38 Coronary angiogram.

vessel parameters, the polynomial coefficients of $p(n_1, n_2)$, and the noise variance.

The vessels, tissues, bones, and the radiographic imaging process are much more complicated than suggested by the simple model presented above. Nevertheless, the model has been empirically observed to lead to good estimates of the vessel boundaries and corresponding percentage of stenosis. The model parameters may be estimated by a variety of different procedures. One possibility is the maximum likelihood (ML) parameter estimation method discussed in Section 6.1.5. In the ML method, the unknown parameters denoted by θ are estimated by maximizing the probability density function $p_{f(n_1,n_2)|\theta}(f_0(n_1, n_2)|\theta_0)$ where $f(n_1, n_2)$ is the angiogram observation and θ represents all the unknown parameters to be estimated. The ML method applied to vessel boundary detection is a nonlinear problem, but has been solved approximately [Pappas and Lim]. Figures 8.39 and 8.40 illustrate the results of applying the ML parameter estimation method to the detection of the blood vessels using the 1-D version of the 2-D model in (8.20). In the 1-D version, $f(n_1, n_2)$ in (8.20) is considered a 1-D sequence with variable n_1 for each n_2. Computations simplify considerably when the 1-D model is used. Figure 8.39(a) shows the original angiogram of 80×80 pixels, and Figure 8.39(b) shows the detected vessel boundaries superimposed on the original image. Figure 8.40 is another example. Developing an edge detection algorithm specific to an application problem is considerably more complicated than applying the general edge detection algorithms discussed in previous sections. However, it has the potential of leading to much more accurate edge detection.

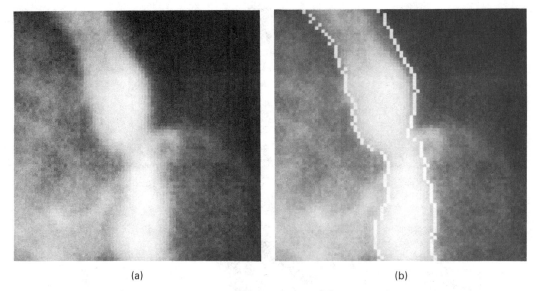

(a)   (b)

Figure 8.39 Example of vessel boundary detection from an angiogram by signal modeling. (a) Angiogram of 80 × 80 pixels; (b) vessel boundary detection.

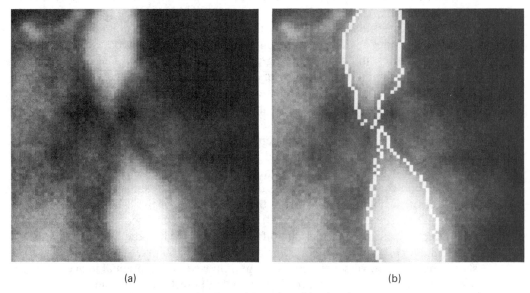

(a)   (b)

Figure 8.40 Another example of vessel boundary detection from an angiogram by signal modeling. (a) Angiogram of 80 × 80 pixels; (b) vessel boundary detection.

8.4 IMAGE INTERPOLATION AND MOTION ESTIMATION

In signal interpolation, we reconstruct a continuous signal from samples. Image interpolation has many applications. It can be used in changing the size of a digital image to improve its appearance when viewed on a display device. Consider a digital image of 64 × 64 pixels. If the display device performs zero-order hold, each individual pixel will be visible and the image will appear blocky. If the image size is increased through interpolation and resampling prior to its display, the resulting image will appear smoother and more visually pleasant. A sequence of image frames can also be interpolated along the temporal dimension. A 24 frames/sec motion picture can be converted to a 60 fields/sec NTSC signal for TV through interpolation. Temporal interpolation can also be used to improve the appearance of a slow-motion video.

Interpolation can also be used in other applications such as image coding. For example, a simple approach to bit rate reduction would be to discard some pixels or some frames and recreate them from the coded pixels and frames.

8.4.1 Spatial Interpolation

Consider a 2-D sequence $f(n_1, n_2)$ obtained by sampling an analog signal $f_c(x, y)$ through an ideal A/D converter:

$$f(n_1, n_2) = f_c(x, y)|_{x = n_1 T_1, y = n_2 T_2}. \qquad (8.21)$$

If $f_c(x, y)$ is bandlimited and the sampling frequencies $1/T_1$ and $1/T_2$ are higher than the Nyquist rate, from Section 1.5 $f_c(x, y)$ can be reconstructed from $f(n_1, n_2)$ with an ideal D/A converter by

$$f_c(x, y) = \sum_{n_1 = -\infty}^{\infty} \sum_{n_2 = -\infty}^{\infty} f(n_1, n_2) h(x - n_1 T_1, y - n_2 T_2) \qquad (8.22)$$

where $h(x, y)$ is the impulse response of an ideal separable analog lowpass filter given by

$$h(x, y) = \frac{\sin \dfrac{\pi}{T_1} x}{\dfrac{\pi}{T_1} x} \frac{\sin \dfrac{\pi}{T_2} y}{\dfrac{\pi}{T_2} y}. \qquad (8.23)$$

There are several difficulties in using (8.22) and (8.23) for image interpolation. The analog image $f_c(x, y)$, even with an antialiasing filter, is not truly bandlimited, so aliasing occurs when $f_c(x, y)$ is sampled. In addition, $h(x, y)$ in (8.23) is an infinite-extent function, so evaluation of $f_c(x, y)$ using (8.22) cannot be carried out in practice. To approximate the interpolation by (8.22) and (8.23), one approach is to use a lowpass filter $h(x, y)$ that is spatially limited. For a spatially limited

$h(x, y)$, the summation in (8.22) has a finite number of nonzero terms. If $h(x, y)$ is a rectangular window function given by

$$h(x, y) = 1, \qquad -\frac{T_1}{2} \le x \le \frac{T_1}{2}, \qquad -\frac{T_2}{2} \le y \le \frac{T_2}{2} \qquad (8.24)$$

then it is called *zero-order interpolation*. In zero-order interpolation, $\hat{f}_c(x, y)$ is chosen as $f(n_1, n_2)$ at the pixel closest to (x, y). Other examples of $h(x, y)$ which are more commonly used are functions of smoother shape, such as the spatially limited Gaussian function or the windowed ideal lowpass filter.

Another simple method widely used in practice is *bilinear interpolation*. In this method, $f_c(x, y)$ is evaluated by a linear combination of $f(n_1, n_2)$ at the four closest pixels. Suppose we wish to evaluate $f(x, y)$ for $n_1 T_1 \le x \le (n_1 + 1)T_1$ and $n_2 T_2 \le y \le (n_2 + 1)T_2$, as shown in Figure 8.41. The interpolated $\hat{f}_c(x, y)$ in the bilinear interpolation method is

$$\hat{f}_c(x, y) = (1 - \Delta_x)(1 - \Delta_y)f(n_1, n_2) + (1 - \Delta_x)\Delta_y f(n_1, n_2 + 1) \qquad (8.25a)$$
$$+ \Delta_x(1 - \Delta_y)f(n_1 + 1, n_2) + \Delta_x\Delta_y f(n_1 + 1, n_2 + 1)$$

where
$$\Delta_x = (x - n_1 T_1)/T_1 \qquad (8.25b)$$

and
$$\Delta_y = (y - n_2 T_2)/T_2. \qquad (8.25c)$$

Another method is *polynomial interpolation*. Consider a local spatial region, say 3×3 or 5×5 pixels, over which $f(x, y)$ is approximated by a polynomial. The interpolated image $\hat{f}_c(x, y)$ is

$$\hat{f}_c(x, y) = \sum_{i=1}^{N} S_i \phi_i(x, y) \qquad (8.26)$$

where $\phi_i(x, y)$ is a term in a polynomial. An example of $\phi_i(x, y)$ when $N = 6$ is

$$\phi_i(x, y) = 1, x, y, x^2, y^2, xy. \qquad (8.27)$$

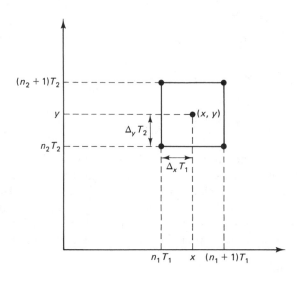

Figure 8.41 Region where $f_c(x, y)$ is interpolated from the four neighboring pixels $f_c(n_1 T_1, n_2 T_2)$, $f_c((n_1 + 1)T_1, n_2 T_2)$, $f_c(n_1 T_1, (n_2 + 1)T_2)$, and $f_c((n_1 + 1)T_1, (n_2 + 1)T_2)$.

The coefficients S_i can be determined by minimizing

$$\text{Error} = \sum_{(n_1, n_2) \in \psi} \sum \left[f(x, y) - \sum_{i=1}^{N} S_i \phi_i(x, y) \right]^2 \Bigg|_{x = n_1 T_1, y = n_2 T_2} \tag{8.28}$$

where ψ denotes the pixels over which $f(x, y)$ is approximated. Solving (8.28) is a simple linear problem, since the $\phi_i(x, y)$ are fixed. Advantages of polynomial interpolation include the smoothness of $\hat{f}_c(x, y)$ and the simplicity of evaluating $\partial \hat{f}_c(x, y)/\partial x$ and $\partial \hat{f}_c(x, y)/\partial y$, partial derivatives which are used in such applications as edge detection and motion estimation. In addition, by fitting a polynomial with fewer coefficients than the number of pixels in the region ψ in (8.28), some noise smoothing can be accomplished. Noise smoothing is particularly useful in applications where the partial derivatives $\partial \hat{f}_c(x, y)/\partial x$ and $\partial \hat{f}_c(x, y)/\partial y$ are used.

Spatial interpolation schemes can also be developed using motion estimation algorithms discussed in the next section. One example, where an image frame that consists of two image fields is constructed from a single image field, is discussed in Section 8.4.4.

Figure 8.42 shows an example of image interpolation. Figure 8.42(a) shows an image of 256 × 256 pixels interpolated by zero-order hold from an original image of 64 × 64 pixels. Figure 8.42(b) shows an image of 256 × 256 pixels obtained by bilinear interpolation of the same original image of 64 × 64 pixels.

8.4.2 Motion Estimation

New image frames can be created from existing ones through temporal interpolation. Unlike spatial interpolation, temporal interpolation requires a large amount

(a)   (b)

Figure 8.42 Example of spatial interpolation. (a) Image of 256 × 256 pixels interpolated by zero-order hold from an original image of 64 × 64 pixels; (b) image of 256 × 256 pixels obtained by bilinear interpolation of the same original image used in (a).

of storage. Therefore a new frame is usually created from two adjacent frames, one in the past and one in the future relative to the frame being created.

The simplest method, often used in practice, is the zero-order hold method, in which a new frame is created by repeating the existing frame which is closest in time. In transforming a 24 frames/sec motion picture to a 60 fields/sec NTSC signal, three consecutive fields are created from a single motion picture frame, and the next two consecutive fields are created from the next motion picture frame. This process is repeated for the entire motion picture. This is known as the 3:2 pull-down method. For most scenes without large global motion, the results are quite good. If there is large global motion, however, zero-order hold temporal interpolation can cause noticeable jerkiness of motion. One method to improve the performance of temporal interpolation is through motion compensation.

A motion picture or television broadcast is a sequence of still frames that are displayed in rapid succession. The frame rate necessary to achieve proper motion rendition is usually high enough to ensure a great deal of temporal redundancy among adjacent frames. Much of the variation in intensity from one frame to the next is due to object motion. The process of determining the movement of objects within a sequence of image frames is known as *motion estimation*. Processing images accounting for the presence of motion is called *motion-compensated image processing*.

Motion-compensated image processing has a variety of applications. One application is image interpolation. By estimating motion parameters, we can create a new frame between two adjacent existing frames. The application of motion-compensated processing to image interpolation is discussed in the next section. Another application is image restoration. If we can estimate the motion parameters and identify regions in different frames where image intensities are expected to be the same or similar, temporal filtering can be performed in those regions. Application to image restoration is discussed in Chapter 9. Motion-compensated image processing can also be applied to image coding. By predicting the intensity of the current frame from the previous frames, we can limit our coding to the difference in intensities between the current frame and the predicted current frame. In addition, we may be able to discard some frames and reconstruct the discarded frames through interpolation from the coded frames. Application to image coding is discussed in Chapter 10.

The motion estimation problem we consider here is the translational motion of objects. Let $f(x, y, t_{-1})$ and $f(x, y, t_0)$ denote the image intensity at times t_{-1} and t_0, respectively. We will refer to $f(x, y, t_{-1})$ and $f(x, y, t_0)$ as the past and current frame. We assume that

$$f(x, y, t_0) = f(x - d_x, y - d_y, t_{-1}) \qquad (8.29)$$

where d_x and d_y are the horizontal and vertical displacement between t_{-1} and t_0. An example of $f(x, y, t_{-1})$ and $f(x, y, t_0)$ which satisfy (8.29) is shown in Figure 8.43. If we assume uniform motion between t_{-1} and t_0,

$$f(x, y, t) = f(x - v_x(t - t_{-1}), y - v_y(t - t_{-1}), t_{-1}), \, t_{-1} \le t \le t_0 \qquad (8.30)$$

where v_x and v_y are the uniform horizontal and vertical velocities.

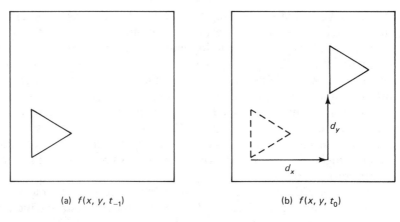

(a) $f(x, y, t_{-1})$ (b) $f(x, y, t_0)$

Figure 8.43 Image translated with displacement of (d_x, d_y). (a) $f(x, y, t_{-1})$; (b) $f(x, y, t_0)$.

A direct consequence of (8.30) is a differential equation which relates v_x and v_y to $\partial f(x, y, t)/\partial x$, $\partial f(x, y, t)/\partial y$, and $\partial f(x, y, t)/\partial t$, which is valid in the spatio-temporal region over which uniform translational motion is assumed. To derive the relationship, let $f(x, y, t_{-1})$ be denoted by $s(x, y)$:

$$s(x, y) = f(x, y, t_{-1}). \tag{8.31}$$

From (8.30) and (8.31),

$$f(x, y, t) = s(\alpha(x, y, t), \beta(x, y, t)), \qquad t_{-1} \le t \le t_0 \tag{8.32a}$$

where

$$\alpha(x, y, t) = x - v_x(t - t_{-1}) \tag{8.32b}$$

and

$$\beta(x, y, t) = y - v_y(t - t_{-1}). \tag{8.32c}$$

From (8.32), assuming $\partial f(x, y, t)/\partial x$, $\partial f(x, y, t)/\partial y$, and $\partial f(x, y, t)/\partial t$ exist, we obtain

$$\frac{\partial f(x, y, t)}{\partial x} = \frac{\partial s}{\partial \alpha} \frac{\partial \alpha}{\partial x} + \frac{\partial s}{\partial \beta} \frac{\partial \beta}{\partial x} = \frac{\partial s}{\partial \alpha} \tag{8.33a}$$

$$\frac{\partial f(x, y, t)}{\partial y} = \frac{\partial s}{\partial \alpha} \frac{\partial \alpha}{\partial y} + \frac{\partial s}{\partial \beta} \frac{\partial \beta}{\partial y} = \frac{\partial s}{\partial \beta} \tag{8.33b}$$

$$\frac{\partial f(x, y, t)}{\partial t} = \frac{\partial s}{\partial \alpha} \frac{\partial \alpha}{\partial t} + \frac{\partial s}{\partial \beta} \frac{\partial \beta}{\partial t} = -v_x \frac{\partial s}{\partial \alpha} - v_y \frac{\partial s}{\partial \beta}. \tag{8.33c}$$

From (8.33),

$$v_x \frac{\partial f(x, y, t)}{\partial x} + v_y \frac{\partial f(x, y, t)}{\partial y} + \frac{\partial f(x, y, t)}{\partial t} = 0. \tag{8.34}$$

Equation (8.34) is called a *spatio-temporal constraint equation* and can be generalized to incorporate other types of motion, such as zooming [Martinez].

The assumption of simple translation that led to (8.29) and the additional assumption of translation with uniform velocity that led to (8.34) are highly restrictive. For example, they do not allow for object rotation, camera zoom, regions uncovered by translational object motion, or multiple objects moving with different v_x and v_y. However, by assuming uniform translational motion only locally and estimating the two motion parameters (d_x, d_y) or (v_x, v_y) at each pixel or at each small subimage, (8.29) and (8.34) are valid for background regions that are not affected by object motion and for regions occupied by objects which do indeed translate with a uniform velocity. Such regions occupy a significant portion of a typical sequence of image frames. In addition, if we identify the regions where motion estimates are not accurate, we can suppress motion-compensated processing in those regions. In image interpolation, for example, we can assume that v_x and v_y are zero.

Motion estimation methods can be classified broadly into two groups, that is, region matching methods and spatio-temporal constraint methods. Region matching methods are based on (8.29), and spatio-temporal constraint methods are based on (8.34). We first discuss region matching methods.

Region matching methods. Region matching methods involve considering a small region in an image frame and searching for the displacement which produces the "best match" among possible regions in an adjacent frame. In region matching methods, the displacement vector (d_x, d_y) is estimated by minimizing

$$\text{Error} = \iint\limits_{(x,y)\in R} C[f(x, y, t_0), f(x - d_x, y - d_y, t_{-1})] \, dx \, dy \qquad (8.35)$$

where R is the local spatial region used to estimate (d_x, d_y) and $C[\cdot,\cdot]$ is a metric that indicates the amount of dissimilarity between two arguments. The integrals in (8.35) can be replaced by summation if $f(x, y, t)$ is sampled at the spatial variables (x, y). If we estimate (d_x, d_y) at time t_0, the region R is the local spatial region that surrounds the particular spatial position at which (d_x, d_y) is estimated. The size of R is dictated by several considerations. If it is chosen too large, the assumption that (d_x, d_y) is approximately constant over the region R may not be valid and evaluation of the error expression requires more computations. If it is chosen too small, the estimates may become very sensitive to noise. One reasonable choice based on these considerations is a 5×5-pixel region. There are many possible choices for the dissimilarity function $C[\cdot, \cdot]$. Two commonly used choices are the squared difference and absolute difference between the two arguments. With these choices of $C[\cdot, \cdot]$, (8.35) reduces to

$$\text{Error} = \iint\limits_{(x,y)\in R} [f(x, y, t_0) - f(x - d_x, y - d_y, t_{-1})]^2 \, dx \, dy \qquad (8.36)$$

$$\text{Error} = \iint\limits_{(x,y)\in R} |f(x, y, t_0) - f(x - d_x, y - d_y, t_{-1})| \, dx \, dy. \qquad (8.37)$$

The function $f(x, y, t_0) - f(x - d_x, y - d_y, t_{-1})$ is called the *displaced frame difference*. In typical applications of motion-compensated processing, the system performance is not very sensitive to the specific choice of the dissimilarity function.

To the extent that (8.29) is valid, the error expression in (8.36) or (8.37) is zero at the correct (d_x, d_y).

Minimizing the Error in (8.36) or (8.37) is a nonlinear problem. Attempts to solve this nonlinear problem have produced many variations, which can be grouped into block matching and recursive methods. We discuss block matching methods first.

One straightforward approach to solve the above minimization problem is to evaluate the Error for every possible (d_x, d_y) within some reasonable range and choose (d_x, d_y) at which the Error is minimum. In this approach, a block of pixel intensities at time t_0 is matched directly to a block at time t_{-1}. This is the basis of *block matching methods*. Since the error expression has to be evaluated at many values of (d_x, d_y), this method of estimating (d_x, d_y) is computationally very expensive and many methods have been developed to reduce computations. In one simplification, we assume that (d_x, d_y) is constant over a block, say of 7×7 pixels. Under this assumption, we divide the image into many blocks and we estimate (d_x, d_y) for each block. Even though we generally choose the block size to be the same as the size of R in (8.35), it is not necessary to do so. In another simplification, we can limit the search space to integer values of (d_x, d_y). In addition to reducing the search space from continuous variables (d_x, d_y) to discrete variables, limiting the search space to integer values allows us to determine $f(n_1 - d_x, n_2 - d_y, t_{-1})$, necessary in the evaluation of the error expression, without interpolation. However, the estimates of (d_x, d_y) are restricted to discrete values.

We can reduce the computational requirements in block matching methods further by using a more efficient search procedure than a brute force search. One such method is called a three-step search method, illustrated in Figure 8.44. In the first step of this method, the error expression is evaluated at nine values of (d_x, d_y) which are marked by "1" and filled circles. Among these nine values, we choose (d_x, d_y) with the smallest Error. Suppose the smallest Error is at $(d_x = 3, d_y = -3)$. In the second step, we evaluate the error expression at eight additional values of (d_x, d_y) which are marked by "2" and filled squares. We now choose (d_x, d_y) from nine values [eight new values and $(d_x = 3, d_y = -3)$]. This procedure is repeated one more time. At the end of the third step, we have an estimate of (d_x, d_y). This procedure can be easily generalized to more than three steps to increase the range of possible (d_x, d_y). Another search method is to estimate d_x first by searching $(d_x, 0)$. Once d_x is estimated, say $\hat{d}_x$, d_y is estimated by searching $(\hat{d}_x, d_y)$. If we wish to improve the estimate further, we can reestimate d_x by searching $(d_x, \hat{d}_y)$ where $\hat{d}_y$ is the estimate of d_y obtained in the previous step. At each step in this procedure, we estimate only one parameter, which is considerably simpler than estimating two parameters jointly. These heuristic methods reduce the number of computations by reducing the number of values of (d_x, d_y) at which the error expression is evaluated. However, the Error at the estimated (d_x, d_y) may not be the global minimum.

In block matching methods, we estimate (d_x, d_y) by explicitly evaluating the Error at some specified set of (d_x, d_y). As an alternative, we can use descent algorithms such as the steepest descent, Newton-Raphson, and Davidon-Fletcher-Powell methods discussed in Section 5.2.3, to solve the nonlinear problem of

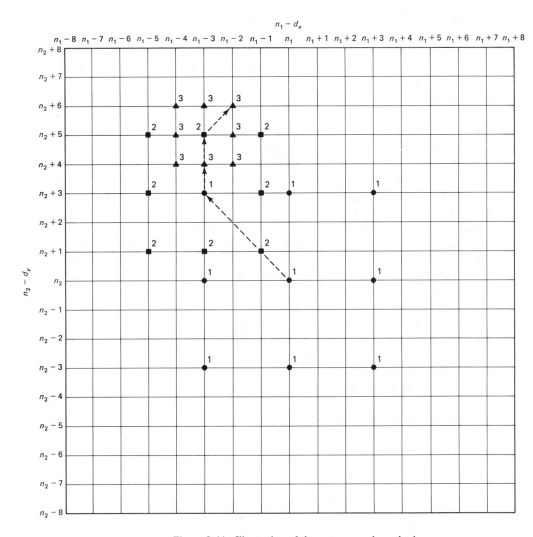

Figure 8.44 Illustration of three-step search method.

minimizing the Error with respect to (d_x, d_y). In this class of algorithms, a recursive (iterative) procedure is used to improve the estimate in each iteration. For this reason, they are called *recursive methods*.

Let $(\hat{d}_x(k), \hat{d}_y(k))$ denote the estimate of (d_x, d_y) after the kth iteration. In recursive methods, the estimate of (d_x, d_y) after the $k + 1$th iteration, $(\hat{d}_x(k + 1), \hat{d}_y(k + 1))$, is obtained by

$$\hat{d}_x(k + 1) = \hat{d}_x(k) + u_x(k) \tag{8.38a}$$

$$\hat{d}_y(k + 1) = \hat{d}_y(k) + u_y(k) \tag{8.38b}$$

where $u_x(k)$ and $u_y(k)$ are the update or correction terms. The update terms vary depending on the descent method used. If we use the steepest descent method, for example, (8.38) becomes

$$\hat{d}_x(k + 1) = \hat{d}_x(k) - \epsilon \left. \frac{\partial \text{Error } (d_x, d_y)}{\partial d_x} \right|_{d_x = \hat{d}_x(k), d_y = \hat{d}_y(k)} \tag{8.39a}$$

$$\hat{d}_y(k + 1) = \hat{d}_y(k) - \epsilon \left. \frac{\partial \text{Error } (d_x, d_y)}{\partial d_y} \right|_{d_x = \hat{d}_x(k), d_y = \hat{d}_y(k)} \tag{8.39b}$$

where ϵ is a step size that can be adjusted and $\text{Error}(d_x, d_y)$ is the Error in (8.35) as a function of d_x and d_y for a given R. Recursive methods typically involve partial derivatives and tend to be sensitive to the presence of noise or fine details in an image. Smoothing the image before motion estimation often improves the performance of recursive methods.

In recursive methods, (d_x, d_y) is not limited to integer values and can be estimated within subpixel accuracy. Update terms typically include evaluation of partial derivatives of $\text{Error}(d_x, d_y)$, which involves evaluation of $f(x, y, t_{-1})$ and its partial derivatives at an arbitrary spatial point. In practice, $f(x, y, t_{-1})$ is known only at $x = n_1 T_1$ and $y = n_2 T_2$. To evaluate the necessary quantities at an arbitrary spatial point (x, y), we can use the spatial interpolation techniques discussed in Section 8.4.1.

In recursive methods, (d_x, d_y) is typically estimated at each pixel. In using the recursion relationship in (8.38), $(\hat{d}_x(0), \hat{d}_y(0))$ is typically obtained from the estimate at the adjacent pixel in the same horizontal scan line, in the adjacent scan line, or in the adjacent frame. These methods are called pel (picture element) recursive estimation with horizontal, vertical, and temporal recursion, respectively. Given $(\hat{d}_x(k), \hat{d}_y(k))$, we can use the recursion relation in (8.38) only once for a pixel and then move on to the next pixel. Alternatively, we can use the recursion relation more than once for a more accurate estimate of (d_x, d_y) before we move on to the next pixel.

Although we classified region matching methods into block matching methods and recursive methods to be consistent with the literature, the boundary between the two classes of methods is fuzzy. By choosing a finer grid at which the error expression is evaluated, we can also estimate (d_x, d_y) within subpixel accuracy with block matching methods. In addition, the three-step search procedure discussed as a block matching method can be viewed as a recursive method in which the estimate is improved iteratively. A major disadvantage of region matching methods is in the amount of computation required. Even though only two parameters, d_x and d_y, must be estimated, solving the nonlinear problem at each pixel or at each small subimage can be computationally very expensive.

Spatio-temporal constraint methods. Algorithms of this class are based on the spatio-temporal constraint equation (8.34), which can be viewed as a linear equation for two unknown parameters v_x and v_y under the assumption that $\partial f(x, y, t)/\partial x$, $\partial f(x, y, t)/\partial y$, and $\partial f(x, y, t)/\partial t$ are given. By evaluating $\partial f(x, y, t)/\partial x$,

$\partial f(x, y, t)/\partial y$ and $\partial f(x, y, t)/\partial t$ at many points (x_i, y_i, t_i) for $1 \leq i \leq N$ at which v_x and v_y are assumed constant, we can obtain an overdetermined set of linear equations:

$$v_x \left.\frac{\partial f(x, y, t)}{\partial x}\right|_{(x_i,y_i,t_i)} + v_y \left.\frac{\partial f(x, y, t)}{\partial y}\right|_{(x_i,y_i,t_i)} + \left.\frac{\partial f(x, y, t)}{\partial t}\right|_{(x_i,y_i,t_i)} \simeq 0, \ 1 \leq i \leq N.$$

(8.40)

The velocity estimates can be obtained by minimizing

$$\text{Error} = \sum_{i=1}^{N} \left[v_x \left.\frac{\partial f(x, y, t)}{\partial x}\right|_{(x_i,y_i,t_i)} + v_y \left.\frac{\partial f(x, y, t)}{\partial y}\right|_{(x_i,y_i,t_i)} + \left.\frac{\partial f(x, y, t)}{\partial t}\right|_{(x_i,y_i,t_i)} \right]^2.$$

(8.41)

Since the error expression in (8.41) is a quadratic form of the unknown parameters v_x and v_y, the solution requires solving two linear equations. More generally, suppose (8.34) is valid in a local spatio-temporal region denoted by ψ. To estimate v_x and v_y, we minimize

$$\text{Error} = \iiint_{(x,y,t) \in \psi} \left(v_x \frac{\partial f(x, y, t)}{\partial x} + v_y \frac{\partial f(x, y, t)}{\partial y} + \frac{\partial f(x, y, t)}{\partial t} \right)^2 dx \, dy \, dt. \qquad (8.42)$$

The integrals in (8.42) may be replaced by summations. One such example is (8.41). Differentiating the Error in (8.42) with respect to v_x and v_y and setting the results to zero leads to

$$W\mathbf{v} = \boldsymbol{\gamma} \qquad (8.43a)$$

where

$$W = \begin{bmatrix} \displaystyle\iiint_{(x,y,t) \in \psi} \left(\frac{\partial f(x, y, t)}{\partial x}\right)^2 dx \, dy \, dt & \displaystyle\iiint_{(x,y,t) \in \psi} \frac{\partial f(x, y, t)}{\partial x} \frac{\partial f(x, y, t)}{\partial y} dx \, dy \, dt \\[2em] \displaystyle\iiint_{(x,y,t) \in \psi} \frac{\partial f(x, y, t)}{\partial x} \frac{\partial f(x, y, t)}{\partial y} dx \, dy \, dt & \displaystyle\iiint_{(x,y,t) \in \psi} \left(\frac{\partial f(x, y, t)}{\partial y}\right)^2 dx \, dy \, dt \end{bmatrix}$$

(8.43b)

$$\mathbf{v} = [v_x, v_y]^T \qquad (8.43c)$$

$$\boldsymbol{\gamma} = - \begin{bmatrix} \displaystyle\iiint_{(x,y,t) \in \psi} \frac{\partial f(x, y, t)}{\partial x} \frac{\partial f(x, y, t)}{\partial t} dx \, dy \, dt \\[2em] \displaystyle\iiint_{(x,y,t) \in \psi} \frac{\partial f(x, y, t)}{\partial y} \frac{\partial f(x, y, t)}{\partial t} dx \, dy \, dt \end{bmatrix}. \qquad (8.43d)$$

The two linear equations in (8.43) may have multiple solutions. Suppose $f(x, y, t)$ is constant in the spatio-temporal region ψ. Then $\partial f(x, y, t)/\partial x$, $\partial f(x, y, t)/\partial y$, and $\partial f(x, y, t)/\partial t$ are all zero, and all the elements in W and γ in (8.43) are zero. Therefore, any (v_x, v_y) will satisfy (8.43a). Any velocity in a uniform intensity region will not affect $f(x, y, t)$, so the true velocity cannot be estimated from $f(x, y, t)$. Suppose $f(x, y, t)$ is a perfect step edge. The velocity along the direction parallel to the step edge will not affect $f(x, y, t)$ and therefore cannot be estimated. These problems have been studied, and a solution has been developed [Martinez]. Let λ_1 and λ_2 denote the eigenvalues of W, and let α_1 and α_2 denote the corresponding orthonormal eigenvectors. A reasonable solution to (8.43) is given by (see Problem 8.24)

Case 1. $\qquad\qquad\qquad$ $\mathbf{v} = 0, \qquad \lambda_1, \lambda_2 < \text{threshold}$ $\qquad\qquad$ (8.44a)

Case 2.

$$\mathbf{v} = \frac{[\alpha_1^T \gamma]\alpha_1}{\lambda_1}, \qquad \lambda_1 \gg \lambda_2 \qquad\qquad (8.44b)$$

Case 3. $\qquad\qquad\qquad$ $\mathbf{v} = W^{-1}\gamma, \qquad \text{otherwise}$ $\qquad\qquad\qquad$ (8.44c)

Case 1 includes the uniform intensity region, where the velocity is set to zero. Case 2 includes the perfect step edge region, and the velocity estimate in (8.44b) is along the direction perpendicular to the step edge.

Solving the linear equations in (8.43) requires evaluation of $\partial f(x, y, t)/\partial x$, $\partial f(x, y, t)/\partial y$, and $\partial f(x, y, t)/\partial t$ at arbitrary spatio-temporal positions. This can be accomplished by extending the spatial polynomial interpolation method to 3-D, which has the advantage over other approaches in computational simplicity and robustness to noise. In the 3-D polynomial interpolation, the interpolated $\hat{f}(x, y, t)$ is

$$\hat{f}(x, y, t) = \sum_{i=1}^{N} S_i \, \phi_i(x, y, t). \qquad\qquad (8.45)$$

An example of $\phi_i(x, y, t)$ for $N = 9$ is

$$\phi_i(x, y, t) = 1, x, y, t, x^2, y^2, xy, xt, yt. \qquad\qquad (8.46)$$

The coefficients S_i can be determined by minimizing

$$\text{Error} = \sum\sum\sum_{(n_1,n_2,n_3)\in\psi} \left(f(x, y, t) - \sum_{i=1}^{N} S_i\phi_i(x, y, t) \right)^2 \Bigg|_{x=n_1T_1, y=n_2T_2, t=n_3T_3} \qquad (8.47)$$

One reasonable choice of the region ψ typically contains 50 pixels: 5 for n_1, 5 for n_2, and 2 for t. Minimizing the error expression in (8.47) with respect to S_i requires solving a set of linear equations. Note that the partial derivatives $\partial f(x, y, t)/\partial x$, $\partial f(x, y, t)/\partial y$, and $\partial f(x, y, t)/\partial t$ used in (8.43) can be precomputed in terms of S_i.

The motion estimation algorithms discussed above require determination of the spatio-temporal regions denoted by ψ over which the uniform translational motion can be assumed. Since a local spatial region in a frame is on the order of 5×5 pixels in size, determining a reasonable ψ requires an initial displacement estimate within a few pixels of the true displacement. In practice, it is not uncommon for the displacement between two adjacent frames to be on the order of 10 pixels. One approach to obtaining an initial displacement (or velocity) is to use a previously computed velocity in the neighborhood and then determine the appropriate ψ as shown in Figure 8.45. Another approach is the hierarchical method [Bierling and Thoma] or the multigrid method [Martinez]. The multigrid method begins with the velocity estimated on a coarse grid. The coarse grid has been obtained from the original frame by lowpass filtering and down-sampling. Down-sampling contracts the displacement (or velocity). A large velocity in the original frame is scaled down by the down-sampling factor. The velocities in the down-sampled frames can be estimated by using the spatio-temporal constraint method with an assumed initial velocity of zero. The velocities estimated on the coarse grid are interpolated to generate initial estimates of the velocities at a finer grid. The bilinear interpolation may be used in the velocity interpolation. The multigrid method can be viewed as an example of *pyramid processing*, which exploits a data structure called a pyramid. A pyramid provides successively condensed information of an image. An example of a pyramid is a higher resolution image and successively lower resolution images. Pyramid processing is useful in various applications including image coding and is discussed further in Chapter 10.

A major advantage of the spatio-temporal constraint methods over region matching methods is their computational simplicity. In addition, preliminary studies based on both synthetic and real data indicate that a spatio-temporal constraint method with polynomial interpolation for $f(x, y, t)$ performs as well as a region matching method for both noisy and noise-free image frames [Martinez].

Both the region matching and spatio-temporal constraint methods can produce substantial errors in motion estimation for some regions, either because the signal $f(x, y, t)$ cannot be modeled by uniform translational motion, or because the motion estimation algorithm does not perform well, perhaps due to the presence of noise. One means of detecting large errors is to compare the error expression in (8.35) for region matching methods or the error expression in (8.42) for spatio-

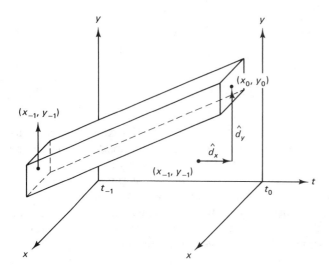

Figure 8.45 Region ψ used in (8.42) to estimate (d_x, d_y) at the spatial location (x_0, y_0) and time t_0. The estimate $(\hat{d}_x, \hat{d}_y)$ represents a previously computed displacement in the neighborhood.

temporal constraint methods with some threshold. The regions where the error is above the threshold can be declared to have unreliable motion estimates, and motion-compensated processing can be suppressed in such regions.

8.4.3 Motion-Compensated Temporal Interpolation

Suppose we have two consecutive image frames $f(n_1, n_2, t_{-1})$ and $f(n_1, n_2, t_0)$, as shown in Figure 8.46. We wish to create a new frame $f(n_1, n_2, t)$ where $t_{-1} < t < t_0$. A simple approach is to choose the original frame that is closer in time to the desired frame. One problem with this approach is the jerkiness that results if the sequence has a large global motion.

An alternative is motion-compensated interpolation using the motion estimation algorithms discussed in the previous section. In motion-compensated interpolation, we assume the uniform translational motion model within a local spatio-temporal region. From $f(n_1, n_2, t_{-1})$ and $f(n_1, n_2, t_0)$, we compute the velocities at $f(n_1, n_2, t)$. We then project the velocities to the frame at t_{-1} or t_0 closer in time to the desired time t, as shown in Figure 8.46. Since the projected spatial point generally does not lie on the original sampling grid, spatial interpolation is needed to obtain the interpolated frame. If the velocity estimated at a particular pixel in $f(n_1, n_2, t_{-1})$ is not considered accurate enough, the velocity is assumed to be zero. In this case, the interpolated pixel value is identical to that at the same pixel location in $f(n_1, n_2, t_{-1})$ or $f(n_1, n_2, t_0)$, whichever is closer in time to the desired time t.

It is not possible to illustrate the motion rendition characteristics of motion-compensated frame interpolation by using only still pictures. However, we can get a rough idea by looking at a still frame created from two image frames by this method. Figure 8.47 shows a set of four frames: two original frames, shown in Figures 8.47(a) and (d), and the two interpolated frames, shown in Figures 8.47(b) and (c). The interpolated frame in (b) is by motion compensation. The frame in (c) is obtained by simply averaging the two key frames. This frame shows the amount of motion that occured between the two key frames. The four frames have a spatial resolution of 512×512 pixels. The interpolated frame corresponds to the time instant midway between the two key original frames. Note that the

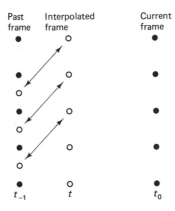

Past frame Interpolated frame Current frame

t_{-1} t t_0

Figure 8.46 Creation of $f(n_1, n_2, t)$ by interpolation of $f(n_1, n_2, t_{-1})$ and $f(n_1, n_2, t_0)$. In this example, the displacement (d_x, d_y) is obtained at each pixel (n_1, n_2) at time t from $f(n_1, n_2, t_{-1})$ and $f(n_1, n_2, t_0)$. Each pixel at time t is projected to the corresponding spatial location at time t_{-1} (t closer to t_{-1} than to t_0 in this example) and the pixel intensity is determined from $f(n_1, n_2, t_{-1})$ at the projected pixel location. Note that spatial interpolation of $f(n_1, n_2, t_{-1})$ is usually necessary to perform this operation.

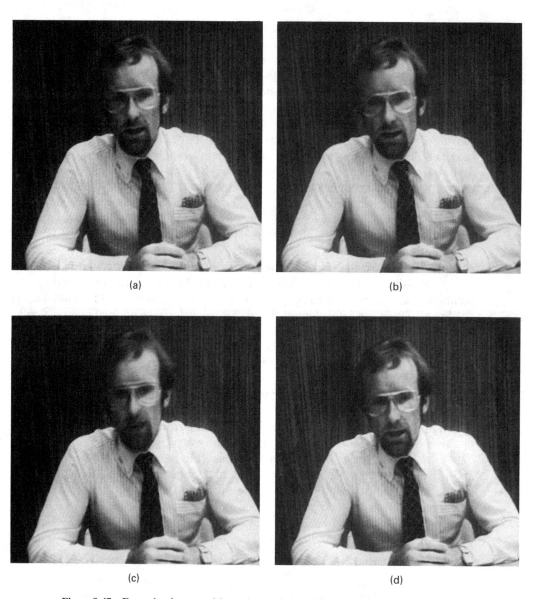

(a)

(b)

(c)

(d)

Figure 8.47 Example of temporal frame interpolation. (a) Original frame 1; (b) interpolated frame by motion compensation; (c) interpolated frame by averaging two key frames. This shows the amount of movement between the two key frames; (d) original frame 2.

interpolated image's quality in this example is essentially the same as that of the two key frames when motion compensation is used. The motion estimation method used here is the spatio-temporal constraint method with polynomial interpolation, discussed in Section 8.4.2.

Motion-compensated interpolation has been used in modifying the frame rate. Frame rate modification can be combined with time-scale modification of audio

[Lim] to modify the length of a motion picture or a TV program. Experience with typical scenes indicates that frame rate modification of video through motion-compensated interpolation can produce video of quality comparable to the original, except for somewhat unnatural motion rates for such actions as walking and talking [Martinez] which occur when the rate modification factor is sufficiently high.

8.4.4 Application of Motion Estimation Methods to Spatial Interpolation

The general idea behind motion-compensated temporal interpolation can be used to develop new algorithms for spatial interpolation. To examine these new algorithms, let us consider a specific spatial interpolation problem. As discussed in Section 7.4, an NTSC television system uses a 2:1 interlaced format, scanned at a rate of 30 frames/sec. A frame consists of 525 horizontal scan lines which are divided into two fields, the odd field consisting of odd-numbered lines and the even field consisting of even-numbered ones. Creating a frame at time t from a field at time t through spatial interpolation can be useful in many applications, including a 60 frames/sec television without bandwidth increase and improved vertical resolution of frozen frames.

The spatial interpolation techniques discussed in Section 8.4.1 may be used in creating a frame from a field, but incorporating some additional knowledge about images may improve the performance of spatial interpolation algorithms. Many elements in the visual world, such as contours and scratches, are spatially continuous. This information can be exploited in creating a frame from a field. Let $f(x, y_{-1})$ and $f(x, y_0)$ denote image intensities of two adjacent horizontal scan lines of a field. We wish to create a new horizontal scan line between $f(x, y_{-1})$ and $f(x, y_0)$. One model that takes into account the spatial continuity of such elements as contours and scratches is

$$f(x, y_0) = f(x - d_x, y_{-1}) \tag{8.48}$$

where d_x is a horizontal displacement between y_{-1} and y_0. Equation (8.48) can be viewed as a special case of the uniform translational velocity model of (8.29). The spatial variable y in (8.48) has a function very similar to the time variable t in (8.29), and there is only one spatial variable x in (8.48), while there are two spatial variables x and y in (8.29). As a result, all our discussions in Section 8.4.2 apply to the problem of estimating d_x in (8.48). For example, under the assumption of uniform velocity, (8.48) can be expressed as

$$f(x, y) = f(x - v_x(y - y_{-1}), y_{-1}), \, y_{-1} \le y \le y_0 \tag{8.49}$$

which leads directly to

$$v_x \frac{\partial f(x, y)}{\partial x} + \frac{\partial f(x, y)}{\partial y} = 0. \tag{8.50}$$

Equation (8.48) can be used to develop region matching methods and (8.50) can be used to develop spatio-temporal constraint methods for estimating d_x or v_x.

Once the horizontal displacement (or velocity) is estimated, it can be used in spatial interpolation in a manner analogous to the temporal interpolation discussed in Section 8.4.3. Figure 8.48 illustrates the performance of a spatial interpolation algorithm based on (8.50). Figure 8.48(a) shows a frame of 256 × 256 pixels obtained by repeating each horizontal line of a 256 × 128-pixel image. Figure

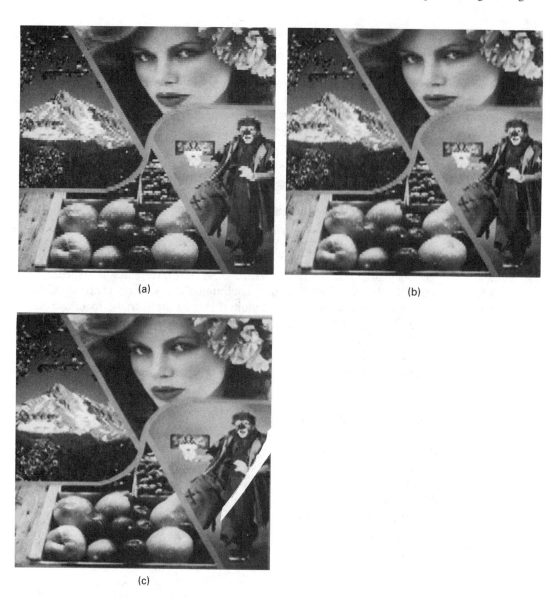

(a)

(b)

(c)

Figure 8.48 Creation of a frame from a field by spatial interpolation. (a) Image of 256 × 256 pixels obtained from an image of 256 × 128 pixels by zero-order hold interpolation; (b) same as (a) obtained by bilinear interpolation; (c) same as (a) obtained by application of a motion estimation algorithm.

8.48(b) shows the frame obtained by bilinear spatial interpolation. Figure 8.48(c) shows the frame obtained by estimating the horizontal displacement based on (8.50) and then using the estimate for spatial interpolation. Spatial continuity of lines and contours is preserved better in the image in Figure 8.48(c) than in the other two images in Figures 8.48(a) and (b).

8.5 FALSE COLOR AND PSEUDOCOLOR

It is well known that the human visual system is quite sensitive to color. The number of distinguishable intensities, for example, is much smaller than the number of distinguishable colors and intensities. In addition, color images are generally much more pleasant to view than black-and-white images. The aesthetic aspect of color can be used for image enhancement. In some applications, such as television commercials, *false color* can be used to emphasize a particular object in an image. For example, a red banana in a surrounding of other fruits of natural color will receive more of a viewer's attention. In other applications, data that do not represent an image in the conventional sense can be represented by a color image. In this case, the color used is called *pseudocolor*. As an example, a speech spectrogram showing speech energy as a function of time and frequency can be represented by a color image, with silence, voiced segments, and unvoiced segments distinguished by different colors and energy represented by color brightness.

The use of color in image enhancement is limited only by artistic imaginations, and there are no simple guidelines or rules to follow. In this section, therefore, we will concentrate on three examples that illustrate the type of image enhancement that can be achieved by using color. In the first example, we transform a monochrome (black and white) image to a color image by using a very simple rule. To obtain a color image from a monochrome image, the monochrome image is first filtered by a lowpass filter, a bandpass filter, and a highpass filter. The lowpass filtered image is considered to be the blue component of the resulting color image. The bandpass filtered image is considered the green component, and the highpass filtered image is considered the red component. The three components—red, green, and blue—are combined to form a color image. Figure 8.49(a) (see color insert) shows an original monochrome image of 512×512 pixels. Figure 8.49(b) shows the color image obtained by using this procedure. The color is pleasant, but this arbitrary procedure does not generate a natural-looking color image. Changing classic black-and-white movies such as *Casablanca* or *It's A Wonderful Life* to color movies requires much more sophisticated processing and a great deal of human intervention.

In the second example, we consider the display of a 2-D spectral estimate on a CRT. The 2-D spectral estimate, represented by $P_x(\omega_1, \omega_2)$ in dB, is typically displayed by using a contour plot. An example of a 2-D maximum likelihood spectral estimate for the data of two sinusoids in white noise is shown in Figure 8.50(a). The maximum corresponds to 0 dB and the contours are in increments of 0.5 dB downward from the maximum point. As we discussed in Chapter 6, in such applications as detection of low-flying aircraft by an array of microphone sensors, we wish to determine the number of sinusoids present and their frequen-

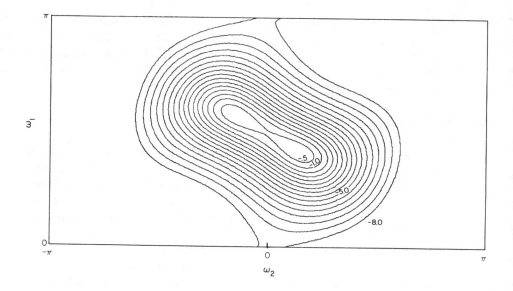

Figure 8.50 Display of spectral estimate using pseudocolor. (a) 2-D maximum likelihood spectral estimate represented by a contour plot; (b) spectral estimate in (a) represented by a color image (see color insert).

cies. An alternative way of representing the spectral estimate is to use pseudocolor. Figure 8.50(b) (see color insert) gives an example, where different amplitudes of $P_x(\omega_1, \omega_2)$ have been mapped to different colors. Comparing the two figures shows that the two peaks and their locations in the spectral estimate stand out more clearly in Figure 8.50(b)

The third example is the display of range information using color [Sullivan et al.]. In such applications as infrared radar imaging systems, range information and image intensity are available. Figure 8.51(a) (see color insert) shows an intensity image of several buildings located two to four kilometers away from the radar; the range information has been discarded. Figure 8.51(b) shows an image that uses color to display range information. The range value determines the hue, and the intensity determines the brightness level of the chosen hue. The most striking aspect of this technique is demonstrated by the observation that a horizontal line seen at close range (actually a telephone wire) is visible in Figure 8.51(b), but is completely obscured in Figure 8.51(a).

REFERENCES

For readings on gray scale modification, see [Hall et al. (1971); Troy et al.; Gonzales and Fittes; Woods and Gonzales]. For image enhancement by lowpass filtering and highpass filtering, see [O'Handley and Green; Hall and Awtrey]. For unsharp masking see [Schreiber (1970)]. For readings on homomorphic processing for image enhancement, see [Oppenheim et al.; Schreiber (1978)]. See [Peli and

Lim] for adaptive image enhancement. For an overview of median filtering, see [Arce et al.]. For readings on theoretical results of median filtering, see [Gallagher and Wise; Nodes and Gallagher; Arce and McLoughlin]. For fast algorithms for median filtering, see [Huang et al.; Ataman et al.].

For a survey of edge detection methods, see [Davis; Shaw; Peli and Malah]. For readings on signal and image interpolation, see [Crochiere and Rabiner; Dubois]. For a survey of motion estimation methods, see [Musmann et al.; Aggarwal and Nandhakumar]. For region matching methods, see [Netravali and Robbins; Huang and Tsai; Srinivasan and Rao]. For spatio-temporal constraint methods, see [Paquin and Dubois; Martinez]. For readings on false color and pseudocolor, see [Gazley et al.; Sheppard; Kreins and Allison; Sullivan et al.].

H. L. Abrams, ed., *Coronary Artiography*. Boston: Little, Brown and Company, 1983.

J. K. Aggarwal and N. Nandhakumar, On the computation of motion from sequences of images—a review, *Proc. IEEE*, Vol. 76, August 1988, pp. 917–935.

G. R. Arce, N. C. Gallagher, and T. A. Nodes, Median filters: Theory for one or two dimensional filters, *Advances in Computer Vision and Image Processing*, T. S. Huang, ed., Greenwich, CT: JAI Press, 1986.

G. R. Arce and M. P. McLoughlin, Theoretical analysis of the max/median filter, *IEEE Trans. Acoust., Speech and Sig. Proc.*, Vol. ASSP-35, January 1987, pp. 60–69.

E. Ataman, V. K. Aatre, and K. M. Wong, A fast method for real-time median filtering, *IEEE Trans. Acoust., Speech and Sig. Proc.*, Vol. ASSP-28, August 1980, pp. 415–421.

M. Bierling and R. Thoma, Motion compensating field interpolation using a hierarchically structured displacement estimator, *Signal Processing*, Vol. 11, December 1986, pp. 387–404.

J. Canny, A computational approach to edge detection, *IEEE Trans. on Patt. Ana. Mach. Intell.*, Vol. PAMI-8, November 1986, pp. 679–698.

R. E. Crochiere and L. R. Rabiner, Interpolation and decimation of digital signals—a tutorial review, *Proc. IEEE*, Vol. 69, March 1981, pp. 300–331.

L. S. Davis, A survey of edge detection techniques, *Computer Graphics and Image Processing*, Vol. 4, 1975, pp. 248–270.

L. S. Davis and A. Mitiche, Edge detection in textures, *Computer Graphics and Image Processing*, Vol. 12, 1980, pp. 25–39.

E. Dubois, The sampling and reconstruction of time-varying imagery with application in video systems, *Proc. IEEE*, Vol. 73, April 1985, pp. 502–522.

R. O. Duda and P. E. Hart, *Pattern Classification and Scene Analysis*. New York: Wiley, 1973.

N. C. Gallagher, Jr. and G. L. Wise, A theoretical analysis of the properties of median filters, *IEEE Trans. Acoust., Speech and Sig. Proc.*, Vol. ASSP-29, December 1981, pp. 1136–1141.

C. Gazley, J. E. Reiber, and R. H. Stratton, Computer works a new trick in seeing pseudocolor processing, *Aeronaut. Astronaut.*, Vol. 4, 1967, p. 76.

R. C. Gonzales and B. A. Fittes, Gray-level transformation for interactive image enhancement, *Mech. Mach. Theory*, Vol. 12, 1977, pp. 111–122.

J. E. Hall and J. D. Awtrey, Real-time image enhancement using 3×3 pixel neighborhood operator functions, *Opt. Eng.*, Vol. 19, May/June 1980, pp. 421–424.

E. L. Hall, R. P. Kruger, S. J. Dwyer, III, D. L. Hall, R. W. McLaren, and G. S. Lodwick, A survey of preprocessing and feature extraction techniques for radiographic images, *IEEE Trans. Computer*, Vol. C-20, 1971, pp. 1032–1044.

T. S. Huang and R. Y. Tsai, *Image Sequence Analysis.* Berlin: Springer-Verlag, 1981, Chapter 1.

T. S. Huang, G. J. Yang, and G. Y. Tang, A fast two-dimensional median filtering algorithm, *IEEE Trans. Acoust., Speech and Sig. Proc.*, Vol. ASSP-27, February 1979, pp. 13–18.

E. R. Kreins and L. J. Allison, Color enhancement of nimbus high resolution infrared radiometer data, *Appl. Opt.*, Vol. 9, 1970, p. 681.

J. S. Lim, ed., *Speech Enhancement.* Englewood Cliffs, NJ: Prentice-Hall, 1983.

D. Marr, *Vision, A Computational Investigation into the Human Representation of Visual Information.* New York: W. H. Freeman and Company, 1982.

D. Marr and E. Hildreth, Theory of edge detection, *Proc. R. Soc.*, London, Vol. B207, 1980, pp. 187–217.

D. M. Martinez, Model-based motion estimation and its application to restoration and interpolation of motion pictures, Ph.D. Thesis, M.I.T., Dept. of Elec. Eng. Comp. Sci., 1986.

H. G. Musmann, P. Pirsch, and H. Grallert, Advances in picture coding, *Proc. IEEE*, Vol. 73, April 1985, pp. 523–548.

A. N. Netravali and J. D. Robbins, Motion-compensated coding: some new results, *The Bell System Tech. J.*, Vol. 59, November 1980, pp. 1735–1745.

T. A. Nodes and N. C. Gallagher, Jr., Two-dimensional root structures and convergence properties of the separable median filter, *IEEE Trans. Acoust., Speech and Sig. Proc.*, December 1983, pp. 1350–1365.

D. A. O'Handley and W. B. Green, Recent developments in digital image processing at the image processing laboratory at the Jet Propulsion Laboratory, *Proc. IEEE*, Vol. 60, July 1972, pp. 821–828.

A. V. Oppenheim, R. W. Schafer, and T. G. Stockham, Jr., Nonlinear filtering of multiplied and convolved signals, *Proc. IEEE*, Vol. 56, 1968, pp. 1264–1291.

T. N. Pappas and J. S. Lim, A new method for estimation of coronary artery dimensions in angiograms, *IEEE Trans. Acoust., Speech and Sig. Proc.*, Vol. ASSP-36, September 1988, pp. 1501–1513.

R. Paquin and E. Dubois, A spatio-temporal gradient method for estimating the displacement field in time-varying imagery, *Computer Vision, Graphics, and Image Processing*, Vol. 21, 1983, pp. 205–221.

T. Peli and J. S. Lim, Adaptive filtering for image enhancement, *J. Opt. Eng.*, Vol. 21, January/February 1982, pp. 108–112.

T. Peli and D. Malah, A study of edge detection algorithms, *Computer Graphics and Image Processing*, Vol. 20, 1982, pp. 1–20.

L. G. Roberts, Machine perception of three-dimensional solids, in *Optical and Electro-Optical Information Processing*, J. T. Tippett et al., eds., Cambridge, MA: MIT Press, 1965, pp. 159–197.

W. F. Schreiber, Image processing for quality improvement, *Proc. IEEE*, Vol. 66, December 1978, pp. 1640–1651.

W. F. Schreiber, Wirephoto quality improvement by unsharp masking, *J. Pattern Recognition*, Vol. 2, 1970, pp. 117–121.

G. B. Shaw, Local and regional edge detectors: some comparisons, *Computer Graphics and Image Processing*, Vol. 9, 1979, pp. 135–149.

J. J. Sheppard, Jr., Pseudocolor as a means of image enhancement, *Am. J. Opthalmol. Arch. Am. Acad. Optom.*, Vol. 46, 1969, pp. 735–754.

R. Srinivasan and K. R. Rao, Predictive coding based on efficient motion estimation, *IEEE Trans. on Comm.*, Vol. COM-33, August 1985, pp. 888–896.

D. R. Sullivan, R. C. Harney, and J. S. Martin, Real-time quasi-three-dimensional display of infrared radar images, *SPIE*, Vol. 180, 1979, pp. 56–64.

E. G. Troy, E. S. Deutsch, and A. Rosenfeld, Gray-level manipulation experiments for texture analysis, *IEEE Trans. Sys. Man. Cybernet.*, Vol. SMC-3, 1973, pp. 91–98.

R. E. Woods and R. C. Gonzales, Real-time digital image enhancement, *Proc. IEEE*, Vol. 69, May 1981, pp. 643–654.

PROBLEMS

8.1. Let $f(n_1, n_2)$ denote an image of 256×256 pixels. The histogram of $f(n_1, n_2)$ is sketched below.

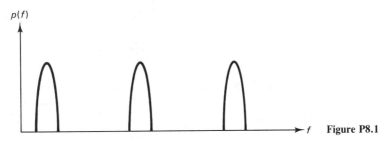

$p(f)$

f **Figure P8.1**

What can we say about $f(n_1, n_2)$? Approximately sketch a transformation function which is likely to improve the contrast of the image when it is used to modify the gray scale of the image.

8.2. Suppose we have an image of 8×8 pixels as shown below.

0	0	0	0	0	0	0	0
1	1	1	1	1	1	1	1
1	1	1	2	2	2	3	4
1	1	1	2	2	2	3	4
1	1	1	2	2	3	3	5
1	1	1	2	2	3	3	5
1	1	2	2	2	3	4	6
1	1	2	2	2	3	4	7

Figure P8.2

We wish to modify the gray scale of this image such that the histogram of the processed image is as close as possible to being constant in the range between 0 and 7. This is known as histogram equalization.

(a) Determine a transformation function that will achieve the above objective.

(b) Determine the processed image based on the transformation function you obtained in (a).

8.3. Modifying a histogram so that the output image has a histogram which has a maximum around the middle of the dynamic range and decreases slowly as the intensity increases or decreases does not always result in an output image more useful than the original unprocessed image. Discuss one such example.

8.4. In this problem, we consider an elementary probability problem closely related to the histogram modification problem. Let f denote a random variable with probability density function $p_f(f_0)$. We define g as $g = T[f]$. The function $T[\cdot]$ is a deterministic monotonically increasing function. The variable g is a random variable with probability density function $p_g(g_0)$.

(a) Let $p_f(f_0)$ be a uniform probability density function given by

$$p_f(f_0) = \begin{cases} 1, & 0 \le f_0 \le 1 \\ 0, & \text{otherwise.} \end{cases}$$

Suppose $T[f]$ is given by

$$g = T[f] = e^f.$$

Determine $p_g(g_0)$.

(b) More generally, develop a method to determine $p_g(g_0)$ given $p_f(f_0)$ and $T[\cdot]$.

(c) Let $p_f(f_0)$ be the same uniform probability density function as in (a). Suppose $p_g(g_0)$ is given by

$$p_g(g_0) = e^{-g_0}u(g_0)$$

where $u(g_0)$ is the unit step function. Determine $T[\cdot]$.

(d) More generally, develop a method to determine $T[\cdot]$, given $p_f(f_0)$ and $p_g(g_0)$.

(e) Discuss how the solution to (d) can be used as a basis for developing a histogram modification method.

8.5. Consider a color image represented by $f_R(n_1, n_2)$, $f_G(n_1, n_2)$, and $f_B(n_1, n_2)$, the red, green, and blue components. We modify the gray scale of each of the three components using histogram modification.

(a) Suppose the desired histogram used in modifying the gray scale is the same for each of the three components. Does the modification affect the hue and saturation of the color image?

(b) Suppose we filter $f_R(n_1, n_2)$ with a system whose impulse response is $h(n_1, n_2)$ and then modify the gray scale of the filtered signal, as shown in the following figure.

Suppose we change the order of the filtering and gray scale modification, as shown below.

If the gray scale is modified by histogram modification by using the same desired histograms in both systems, are the results the same; that is, $f'_R(n_1, n_2) = f''_R(n_1, n_2)$?

8.6. Let V_{IN} denote the input voltage to a display monitor and I_{OUT} denote the output intensity actually displayed on the display monitor. Ideally, we wish to have I_{OUT} proportional to V_{IN}. In practice, however, the relationship between V_{IN} and I_{OUT} is nonlinear and is called *gamma*. Suppose the nonlinear relationship is given approximately by

$$I_{OUT} \propto V_{IN}^2$$

(a) What effect does the nonlinear relation have on a black-and-white image?

(b) What effect does the nonlinear relation have on a color image?

(c) One way to compensate for the gamma effect is to process V_{IN} prior to its input to the display monitor. Discuss a method of processing V_{IN} so that the gamma effect can be taken into account. In practice, such processing is not necessary, since the correction can be incorporated in the camera design.

8.7. Determine and sketch the frequency response of each of the filters in Figure 8.8. In sketching the frequency response $H(\omega_1, \omega_2)$, you need to sketch only $H(\omega_1, \omega_2)|_{\omega_1 = 0}$.

8.8. In homomorphic processing for contrast enhancement, an image $f(n_1, n_2)$ is logarithmically transformed, highpass filtered, and then exponentiated, as shown in the following figure.

$$f(n_1, n_2) \longrightarrow \boxed{\log} \longrightarrow \boxed{\begin{array}{c} \text{Highpass} \\ \text{filtering} \end{array}} \longrightarrow \boxed{\exp} \longrightarrow p(n_1, n_2)$$

We assume in this problem that the highpass filter used is a linear shift-invariant system.

(a) In practice, the scale of $f(n_1, n_2)$ is arbitrary. For example, $f(n_1, n_2)$ may be scaled such that 0 corresponds to the darkest level and 1 corresponds to the brightest. Alternatively, $f(n_1, n_2)$ may be scaled such that 0 corresponds to the darkest level and 255 corresponds to the brightest. What is the effect on the processed image $p(n_1, n_2)$ of the choice of scale for $f(n_1, n_2)$?

(b) The logarithmic operation has an undesirable behavior for the amplitude of $f(n_1, n_2)$ close to zero. One system proposed to eliminate this undesirable behavior is shown below.

$$f(n_1, n_2) \longrightarrow \boxed{(\cdot)^a} \longrightarrow \boxed{\begin{array}{c} \text{Highpass} \\ \text{filtering} \end{array}} \longrightarrow \boxed{(\cdot)^{1/a}} \longrightarrow p'(n_1, n_2)$$

In the figure, a is a positive real constant. What is a reasonable choice of the constant a in order for this to approximate the homomorphic system?

(c) In the system in (b), what effect does the choice of scale for $f(n_1, n_2)$ have on the processed image $p'(n_1, n_2)$?

(d) One sometimes-cited advantage of the homomorphic system is the property that the processed image $p(n_1, n_2)$ always has nonnegative amplitude. Does the system in (b) have this property?

8.9. The system in Figure 8.12 modifies the local contrast as a function of the local luminance mean and modifies the local luminance mean through some nonlinearity.

With proper choice of the parameters, the system can also be used as a linear shift-invariant highpass filter. How should the parameters in the system be chosen?

8.10. One limitation of highpass filtering for contrast enhancement is the accentuation of background noise.

(a) Discuss why the noise accentuation problem is typically much more severe in low local contrast regions, such as areas of uniform background, than in high local contrast regions, such as edges.

(b) Design an adaptive highpass filter that reduces the noise visibility problem discussed in (a).

8.11. In reducing the effect of cloud cover in images taken from an airplane, we can exploit the following two properties:

Property 1. Regions covered by cloud are typically brighter than regions not covered by cloud.

Property 2. The contrast in regions covered by cloud is typically lower than the contrast in regions not covered by cloud.

One reasonable approach to reducing the effect of cloud cover is the adaptive system (System 1) shown in Figure 8.12. An alternative approach that can be used to reduce the effect of cloud cover is the following adaptive system (System 2):

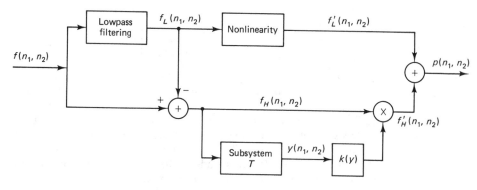

Figure P8.11

(a) Determine a reasonable Subsystem T, and sketch a reasonable $k(y)$ as a function of y and a reasonable f'_L as a function of f_L. Label the axes clearly.

(b) What is one major disadvantage of System 2 in its performance compared with System 1?

8.12. Determine and sketch the frequency response of each of the filters in Figure 8.14. In sketching the frequency response $H(\omega_1, \omega_2)$, you need to sketch only $H(\omega_1, \omega_2)|_{\omega_1 = 0}$.

8.13. For each of the following input sequences, determine the result of median filtering with window sizes of (i) 3×3 points and (ii) 5×5 points.

(a) $\delta(n_1, n_2)$

(b) $u(n_1, n_2)$

(c) $u_T(n_1)$

(d) $u(n_1, n_2) - u(n_1 - 2, n_2)$

8.14. Repeated application of a 1-D median filter to a 1-D sequence eventually leads to what is called a root signal, which is invariant under further applications of the 1-D median filter. Consider the 1-D sequence $x(n)$ shown below.

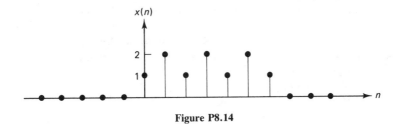

Figure P8.14

(a) Determine the root signal of $x(n)$, using a 3-point median filter.

(b) Is it possible to recover $x(n)$ from the root signal without additional information about $x(n)$?

(c) A median filter without additional specification refers to a nonrecursive median filter. It is possible to define a recursive median filter. Let $y(n)$ denote the output of a recursive 3-point median filter. The output $y(n)$ is given by

$$y(n) = \text{Median} [y(n - 1), x(n), x(n + 1)].$$

Determine the output $y(n)$ when a recursive 3-point median filter is applied to $x(n)$ above.

(d) Applying a 1-D recursive median filter to a 1-D sequence $x(n)$ is known to result in the root signal of $x(n)$ without repeated application of the filter. Are your results in (a) and (c) consistent with this?

8.15. A useful property of a median filter is its tendency to preserve step discontinuities. An $N \times N$-point median filter, however, distorts the unit step sequence $u(n_1, n_2)$. One method that preserves step discontinuities better than an $N \times N$-point median filter is to use a separable $N \times N$-point median filter. In this method, a 1-D N-point median filter is applied along the horizontal direction and then along the vertical direction. Another method is to use a star-shaped window for the median filter. The star-shaped 3×3-point window is shown below.

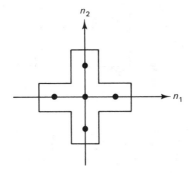

Figure P8.15

For this filter, the output $y(n_1, n_2)$ for an input $x(n_1, n_2)$ is determined by

$$y(n_1, n_2) = \text{Median} [x(n_1, n_2), x(n_1, n_2 + 1), x(n_1, n_2 - 1), x(n_1 - 1, n_2), x(n_1 + 1, n_2)].$$

(a) Does the star-shaped $N \times N$-point median filter distort $u(n_1, n_2)$?

(b) Which method preserves step discontinuities better, a separable $N \times N$-point median filter, or a star-shaped $N \times N$-point median filter? Consider steps with various orientations.

(c) Suppose the input is a uniform background degraded by wideband random noise. Which method reduces more background noise, a separable $N \times N$-point median filter or a star-shaped $N \times N$-point median filter?

8.16. In gradient-based edge detection algorithms, a gradient is approximated by a difference. One component of the gradient is $\partial f(x, y)/\partial x$. Suppose $\partial f(x, y)/\partial x$ is approximated by

(a) $f(n_1, n_2) - f(n_1 - 1, n_2)$

(b) $f(n_1 + 1, n_2) - f(n_1, n_2)$

(c) $f(n_1 + 1, n_2 + 1) - f(n_1 - 1, n_2 + 1) + 2[f(n_1 + 1, n_2) - f(n_1 - 1, n_2)]$
$+ f(n_1 + 1, n_2 - 1) - f(n_1 - 1, n_2 - 1)$

The differencing operation in each of the above three cases can be viewed as convolution of $f(n_1, n_2)$ with some impulse response of a filter $h(n_1, n_2)$. Determine $h(n_1, n_2)$ for each case.

8.17. Consider an image $f(n_1, n_2)$ of 7×7 pixels shown in Figure P8.17a. The numbers in the boxes represent the pixel intensities.

60	60	62	65	68	70	70
60	60	62	65	68	70	70
70	70	72	75	78	80	80
100	100	102	105	108	110	110
130	130	132	135	138	140	140
140	140	142	145	148	150	150
140	140	142	145	148	150	150

	40.79	44.72	46.65	44.72	40.79	
	160.20	161.25	161.79	161.25	160.20	
	240.13	240.83	241.20	240.83	240.13	
	160.20	161.25	161.79	161.25	160.20	
	40.79	44.72	46.65	44.72	40.79	

Figure P8.17a   **Figure P8.17b**

(a) Show that when the Sobel edge detector is used, the result of discrete approximation of $|\nabla f(x, y)|$ is given by Figure P8.17b. We will denote this result by $|\nabla f(n_1, n_2)|$. At pixel locations at the image boundary, we cannot compute $|\nabla f(n_1, n_2)|$ based on the Sobel edge detector.

(b) Suppose we choose a threshold of 100. Determine the candidate edge points.

(c) From your result in (b), edge thinning may be necessary to avoid wide strips of edges. Suppose we decide that any point among the candidate edge points is a true edge point if it is a local maximum of $|\nabla f(n_1, n_2)|$ in either the horizontal or the vertical direction. On the basis of this assumption, determine the edge points.

(d) By looking at $f(n_1, n_2)$ and using your own judgment, determine edge points of $f(n_1, n_2)$.

(e) Suppose we impose the following additional constraints in (c):

(i) If $|\nabla f(n_1, n_2)|$ has a local maximum at (n_1', n_2') in the horizontal direction but not in the vertical direction, (n_1', n_2') will be an edge point when $|\nabla_x f(n_1, n_2)|$ is significantly larger than $|\nabla_y f(n_1, n_2)|$. The functions $\nabla_x f(n_1, n_2)$ and $\nabla_y f(n_1, n_2)$ are the horizontal and vertical components, respectively, of $\nabla f(n_1, n_2)$.

(ii) The same constraint described in (i) with the horizontal direction x and vertical direction y interchanged.

Solve (c) with these two constraints imposed.

(f) Compare your results in (d) and (e).

8.18. One way to approximate the Laplacian $\nabla^2 f(x, y)$ in the discrete domain is to convolve $f(n_1, n_2)$ with $h(n_1, n_2)$, shown below.

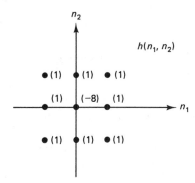

Figure P8.18

Show that the above approximation is reasonable.

8.19. In this problem, we illustrate that a reasonable generalization of $f''(x)$ to a 2-D function $f(x, y)$ for the purpose of edge detection is the Laplacian $\nabla^2 f(x, y)$. Consider an "ideal" 2-D edge, shown in Figure P8.19a. Given that the edge is in the direction perpendicular to the u-axis and that the point $(0, 0)$ lies on the edge contour, as shown in the figure, it is possible to model the function $f(x, y)$ in the neighborhood of $(0, 0)$ by

$$f(x, y) = g(u)|_{u = x\cos\theta + y\sin\theta}$$

where $g(u)$ is shown in Figure P8.19b. Note that $f(x, y)$ is constant in the direction of the edge line. In this case, the 2-D problem of locating the edge of $f(x, y)$ in the neighborhood of $(0, 0)$ can be viewed as a 1-D problem of locating the edge in $g(u)$.

(a) For (x, y) on the u-axis, show that

$$|g'(u)|_{u = x\cos\theta + y\sin\theta} = |\nabla f(x, y)|.$$

This is consistent with the generalization of $|f'(x)|$ to $|\nabla f(x, y)|$ in gradient-based methods for edge detection.

(b) For (x, y) on the u-axis, show that

$$g''(u)|_{u = x\cos\theta + y\sin\theta} = \nabla^2 f(x, y) = \frac{\partial^2 f(x, y)}{\partial x^2} + \frac{\partial^2 f(x, y)}{\partial y^2}.$$

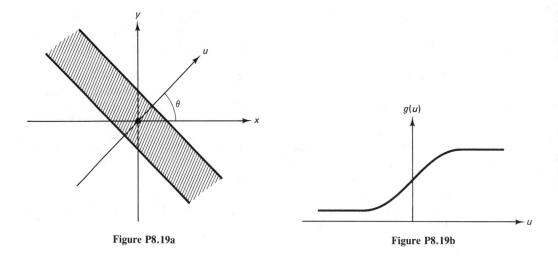

Figure P8.19a   Figure P8.19b

This is consistent with the generalization of $f''(x)$ to $\nabla^2 f(x, y)$ in Laplacian-based methods for edge detection.

(c) The result in (b) is not valid for 2-D edges with sharp corners. Discuss how this could effect the performance of Laplacian-based edge detection methods. You may want to look at some edge maps obtained by Laplacian-based edge detection methods.

8.20. Let $f(x, y)$ denote a 2-D function that is sampled on a Cartesian grid. The samples of $f(x, y)$ are shown in the following figure.

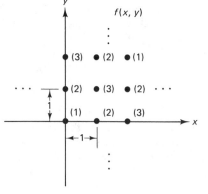

Figure P8.20

The values in parentheses in the figure represent the amplitudes of $f(x, y)$ evaluated at the spatial points corresponding to the filled-in dots. Suppose we wish to estimate $f(x, y)$ by using bilinear interpolation.

(a) Determine $f(x, y)$ for $0 \le x \le 2$, $0 \le y \le 1$.

(b) Are $\partial f(x, y)/\partial x$ and $\partial f(x, y)/\partial y$ well defined for all $0 \le x \le 2$ and $0 \le y \le 1$?

8.21. In image interpolation, $f(x, y)$ is approximated by

$$\hat{f}(x, y) = \sum_{i=1}^{N} S_i \phi_i(x, y)$$

where $\phi_i(x, y)$ is a set of predetermined functions. One method of estimating S_i from samples of $f(x, y)$ is by minimizing

$$\text{Error} = \sum_{(n_1, n_2) \in \psi} (f(x, y) - \sum_{i=1}^{N} S_i \, \phi_i(x, y))^2 \big|_{x = n_1 T_1, y = n_2 T_2}$$

where ψ denotes the pixels over which $f(x, y)$ is approximated.

(a) Determine the set of linear equations for S_i that results from solving the above minimization problem.

(b) Suppose $N = 3$, $\phi_1(x, y) = 1$, $\phi_2(x, y) = x$, $\phi_3(x, y) = xy$, $f(0, 0) = 1$, $f(0, 1) = f(1, 0) = 2$, and $f(1, 1) = 3$. Determine $\hat{f}(x, y)$, the interpolated result obtained by solving the above minimization problem.

8.22. Let $f(x, y, t)$ denote an image intensity as a function of two spatial variables (x, y) and a time variable t. Suppose $f(x, y, t)$ is generated from one image frame by translational motion with uniform velocities of v_x and v_y along the horizontal and vertical direction, respectively. Suppose $f(x, y, 0)$ is given by

$$f(x, y, 0) = x + y + 3xy$$

(a) Determine $f(x, y, t)$.

(b) For the $f(x, y, t)$ obtained in (a), show that

$$v_x \frac{\partial f(x, y, t)}{\partial x} + v_y \frac{\partial f(x, y, t)}{\partial y} + \frac{\partial f(x, y, t)}{\partial t} = 0.$$

8.23. The motion estimation methods we discussed in Section 8.4.2 are based on the assumption of translational motion given by (8.29):

$$f(x, y, t_0) = f(x - d_x, y - d_y, t_{-1})$$

If the overall illumination varies over time, a better model is given by

$$f(x, y, t_0) = \alpha f(x - d_x, y - d_y, t_{-1})$$

where α is some unknown constant. Discuss how we would develop region matching methods based on this new signal model.

8.24. In motion estimation methods based on the spatio-temporal constraint equation, we solve two linear equations for the velocity vector (v_x, v_y). The two equations can be expressed as

$$W\mathbf{v} = \boldsymbol{\gamma}$$

where $\mathbf{v} = [v_x, v_y]^T$ and where W and $\boldsymbol{\gamma}$ are given by (8.43). Let λ_1 and λ_2 denote the eigenvalues of W, and let $\boldsymbol{\alpha}_1$ and $\boldsymbol{\alpha}_2$ denote the corresponding orthonormal eigenvectors.

(a) Suppose $f(x, y, t)$ is constant. Show that $\lambda_1 = \lambda_2 = 0$, and that a reasonable estimate of $\mathbf{v}$ is $\mathbf{0}$.

(b) Consider $f(x, y, t)$ which does not depend on x and which results from uniform translational motion. Let $\lambda_1 \geq \lambda_2$. Show that $\lambda_2 = 0$, and that a reasonable estimate of $\mathbf{v}$ is $\mathbf{v} = [\boldsymbol{\alpha}_1^T \boldsymbol{\gamma}] \boldsymbol{\alpha}_1 / \lambda_1$.

9

Image Restoration

9.0 INTRODUCTION

In image restoration, an image has been degraded in some manner and the objective is to reduce or eliminate the degradation. Simple and heuristic enhancement algorithms for reducing degradation were discussed in Chapter 8. In this chapter, we study image restoration algorithms. Restoration algorithms are typically more mathematical and complex than enhancement algorithms. In addition, they are designed to exploit detailed characteristics of the signal and degradation.

A typical environment for an image restoration system is shown in Figure 9.1. If the digitizer and display were ideal, the output image intensity $f'(x, y)$ would be identical to the input $f(x, y)$ without any restoration. In practice, many different types of degradation may be present in the digitizer and display. With an image restoration system we attempt to deal with the degradation, so that the output $f'(x, y)$ will be as close as possible to the input $f(x, y)$.

For the purposes of studying image restoration, we will assume that all the degradation occurs before the image restoration system is employed, as shown in Figure 9.2. This will allow us to consider the image restoration problem entirely in the discrete-space domain (dotted line in Figure 9.2). We can consider $f(n_1, n_2)$ as the original digital image, $g(n_1, n_2)$ as the degraded digital image, and $p(n_1, n_2)$ as the processed digital image. The objective of image restoration is to make the processed image $p(n_1, n_2)$ be as close as possible to $f(n_1, n_2)$. It is not always reasonable to assume that all the degradation occurs before the restoration system is employed. One such example is additive random noise degradation in a display. In this case, it is more reasonable to process an image in anticipation of future degradation. However, many different types of degradation, such as blurring in the digitizer or display, can be modeled as occurring before the restoration system is applied. In this chapter, we assume that the original image $f(n_1, n_2)$ is degraded and an image restoration system attempts to restore $f(n_1, n_2)$ from the degraded image $g(n_1, n_2)$, as shown in Figure 9.2.

The development of an image restoration system depends on the type of degradation. Algorithms that attempt to reduce additive random noise are dif-

Figure 9.1 Typical environment for image restoration.

ferent from those that attempt to reduce blurring. The types of degradation we consider in this chapter are additive random noise, blurring, and signal-dependent noise such as multiplicative noise. These types of degradation are chosen because they often occur in practice and they have been extensively discussed in the literature. In addition to providing specific restoration systems for the degradation treated in this chapter, the general approaches used in the development of these systems apply to the reduction of other types of degradation.

Examples illustrating the performance of various algorithms are given throughout the chapter. These examples are included only for illustrative purposes and should not be used for comparing the performance of different algorithms. The performance of an image processing algorithm depends on many factors, such as the objective of the processing and the type of image used. One or two examples do not adequately demonstrate the performance of an algorithm.

In Section 9.1, we discuss how to obtain information on image degradation. Accurate knowledge about the degradation is essential in developing successful restoration algorithms. In Section 9.2, we discuss the problem of restoring an image degraded by additive random noise. Section 9.3 treats the problem of restoring an image degraded by blurring. Section 9.4 treats the problem of restoring an image degraded by both blurring and additive random noise, and more generally the problem of reducing an image degraded by more than one type of degradation. In Section 9.5, we develop restoration algorithms for reducing signal-dependent noise. In Section 9.6, we discuss temporal domain processing for image restoration. In Section 9.7, we describe how an image restoration problem can be phrased by using matrix notation and how tools of linear algebra may be brought to the solution of many image restoration problems.

9.1 DEGRADATION ESTIMATION

Since image restoration algorithms are designed to exploit characteristics of a signal and its degradation, accurate knowledge of the degradation is essential to devel-

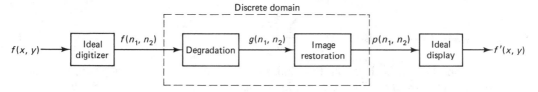

Figure 9.2 Image restoration based on the assumption that all the degradation occurs before image restoration. This allows us to consider the image restoration problem in the discrete-space domain.

oping a successful image restoration algorithm. There are two approaches to obtaining information about degradation. One approach is to gather information from the degraded image itself. If we can identify regions in the image where the image intensity is approximately uniform, for instance the sky, it may be possible to estimate the power spectrum or the probability density function of the random background noise from the intensity fluctuations in the uniform background regions. As another example, if an image is blurred and we can identify a region in the degraded image where the original undegraded signal is known, we may be able to estimate the blurring function $b(n_1, n_2)$. Let us denote the known original undegraded signal in a particular region of the image as $f(n_1, n_2)$ and the degraded image in the same region as $g(n_1, n_2)$. Then $g(n_1, n_2)$ is approximately related to $f(n_1, n_2)$ by

$$g(n_1, n_2) = f(n_1, n_2) * b(n_1, n_2). \tag{9.1}$$

Since $f(n_1, n_2)$ and $g(n_1, n_2)$ are assumed known, $b(n_1, n_2)$ can be estimated from (9.1). If $f(n_1, n_2)$ is the impulse $\delta(n_1, n_2)$, $b(n_1, n_2)$ is given by $g(n_1, n_2)$. This may be the case for a star in the night sky.

Another approach to obtaining knowledge about degradation is by studying the mechanism that caused the degradation. For example, consider an analog[*] image $f(x, y)$ blurred by a planar motion of the imaging system during exposure. Assuming that there is no degradation in the imaging system except the motion blur, we can express the degraded image $g(x, y)$ as

$$g(x, y) = \frac{1}{T} \int_{t=-T/2}^{T/2} f(x - x_0(t), y - y_0(t))\, dt \tag{9.2}$$

where $x_0(t)$ and $y_0(t)$ represent the horizontal and vertical translations of $f(x, y)$ at time t relative to the imaging system, and T is the duration of exposure. In the Fourier transform domain, (9.2) can be expressed as

$$\begin{aligned}
G(\Omega_x, \Omega_y) &= \int_{x=-\infty}^{\infty} \int_{y=-\infty}^{\infty} g(x, y) e^{-j\Omega_x x} e^{-j\Omega_y y}\, dx\, dy \\
&= \int_{x=-\infty}^{\infty} \int_{y=-\infty}^{\infty} \left[\frac{1}{T} \int_{t=-T/2}^{T/2} f(x - x_0(t), y - y_0(t))\, dt \right] \\
&\qquad \cdot e^{-j\Omega_x x} e^{-j\Omega_y y}\, dx\, dy \tag{9.3}
\end{aligned}$$

where $G(\Omega_x, \Omega_y)$ is the Fourier transform of $g(x, y)$. Simplifying (9.3), we obtain

$$G(\Omega_x, \Omega_y) = F(\Omega_x, \Omega_y) B(\Omega_x, \Omega_y) \tag{9.4a}$$

where

$$B(\Omega_x, \Omega_y) = \frac{1}{T} \int_{t=-T/2}^{T/2} e^{-j\Omega_x x_0(t)} e^{-j\Omega_y y_0(t)}\, dt. \tag{9.4b}$$

From (9.4), it is clear that the planar motion blur can be viewed as convolution of $f(x, y)$ with $b(x, y)$ whose Fourier transform $B(\Omega_x, \Omega_y)$ is given by (9.4b). The

*It is simpler to derive the effect of image degradation due to motion blur in the analog domain and then discretize the result than to derive it in the digital domain.

function $b(x, y)$ is sometimes referred to as the *blurring function*, since $b(x, y)$ typically is of lowpass character and blurs the image. It is also referred to as the *point spread function*, since it spreads an impulse. When there is no motion and thus $x_0(t) = 0$ and $y_0(t) = 0$, $B(\Omega_x, \Omega_y)$ is 1 and $g(x, y)$ is $f(x, y)$. If there is linear motion along the x direction so that $x_0(t) = kt$ and $y_0(t) = 0$, $B(\Omega_x, \Omega_y)$ in (9.4) reduces to

$$B(\Omega_x, \Omega_y) = \frac{\sin \frac{\Omega_x}{2} kT}{\frac{\Omega_x}{2} kT}. \tag{9.5}$$

A discrete image $g(n_1, n_2)$ may be approximately modeled by

$$g(n_1, n_2) = f(n_1, n_2) * b(n_1, n_2) \tag{9.6}$$

where $B(\omega_1, \omega_2)$, the discrete-space Fourier transform of $b(n_1, n_2)$, is the aliased version of $B(\Omega_x, \Omega_y)$ in (9.4b). Other instances in which the degradation may be estimated from its mechanism include film grain noise, blurring due to diffraction-limited optics, and speckle noise.

9.2 REDUCTION OF ADDITIVE RANDOM NOISE

The model of an image degraded by additive random noise is given by

$$g(n_1, n_2) = f(n_1, n_2) + v(n_1, n_2) \tag{9.7}$$

where $v(n_1, n_2)$ represents the signal-independent additive random noise. Examples of additive random noise degradation include electronic circuit noise, and in some cases amplitude quantization noise. In this section, we discuss a number of algorithms that have been proposed for reducing additive random noise in images.

9.2.1 Wiener Filtering

One of the first methods developed to reduce additive random noise in images is based on Wiener filtering, which was discussed in Section 6.1.4. If we assume that $f(n_1, n_2)$ and $v(n_1, n_2)$ are samples of zero-mean stationary random processes that are linearly independent of each other and that their power spectra $P_f(\omega_1, \omega_2)$ and $P_v(\omega_1, \omega_2)$ are known, the optimal linear minimum mean square error estimate of $f(n_1, n_2)$ is obtained by filtering $g(n_1, n_2)$ with a Wiener filter whose frequency response $H(\omega_1, \omega_2)$ is given by

$$H(\omega_1, \omega_2) = \frac{P_f(\omega_1, \omega_2)}{P_f(\omega_1, \omega_2) + P_v(\omega_1, \omega_2)}. \tag{9.8}$$

If we impose the additional constraint that $f(n_1, n_2)$ and $v(n_1, n_2)$ are samples of Gaussian random processes, then the Wiener filter in (9.8) is the optimal minimum mean square error estimator of the signal among both linear and nonlinear estimators. The Wiener filter was first considered for image restoration in the early 1960s. It has influenced the development of many other image restoration systems.

The Wiener filter in (9.8) was derived under the assumption that $f(n_1, n_2)$ and $v(n_1, n_2)$ are samples of zero-mean processes. If $f(n_1, n_2)$ has a mean of m_f and $v(n_1, n_2)$ has a mean of m_v, then m_f and m_v are first subtracted from the degraded image $g(n_1, n_2)$. The resulting signal $g(n_1, n_2) - (m_f + m_v)$ is next filtered by the Wiener filter. The signal mean m_f is then added to the filtered signal. This is shown in Figure 9.3. The treatment of nonzero means, as shown in Figure 9.3, minimizes the mean square error between $f(n_1, n_2)$ and $p(n_1, n_2)$ for Gaussian random processes $f(n_1, n_2)$ and $v(n_1, n_2)$. It also ensures that $p(n_1, n_2)$ will be an unbiased estimate of $f(n_1, n_2)$. If m_v is zero, m_f will be identical to the mean of $g(n_1, n_2)$. In this case, m_f can be estimated from $g(n_1, n_2)$.

The Wiener filter in (9.8) is a zero-phase filter. Since the power spectra $P_f(\omega_1, \omega_2)$ and $P_v(\omega_1, \omega_2)$ are real and nonnegative, $H(\omega_1, \omega_2)$ is also real and nonnegative. Therefore, the Wiener filter affects the spectral magnitude but not the phase. The Wiener filter preserves the high SNR frequency components while attenuating the low SNR frequency components. If we let $P_v(\omega_1, \omega_2)$ approach 0, $H(\omega_1, \omega_2)$ will approach 1, indicating that the filter tends to preserve the high SNR frequency components. If we let $P_v(\omega_1, \omega_2)$ approach ∞, $H(\omega_1, \omega_2)$ will approach 0, indicating that the filter tends to attenuate the low SNR frequency components.

The Wiener filter is based on the assumption that the power spectra $P_f(\omega_1, \omega_2)$ and $P_v(\omega_1, \omega_2)$ are known or can be estimated. In typical problems, estimating the noise power spectrum $P_v(\omega_1, \omega_2)$ is relatively easy with the methods discussed in Section 6.2. Estimating the image power spectrum $P_f(\omega_1, \omega_2)$ is not a simple matter. One method used is to average $|F(\omega_1, \omega_2)|^2$ for many different images $f(n_1, n_2)$. This is analogous to the periodogram averaging method for spectral estimation discussed in Section 6.2.1. Another method is to model $P_f(\omega_1, \omega_2)$ with a simple function such as

$$R_f(n_1, n_2) = \rho^{\sqrt{n_1^2 + n_2^2}} \tag{9.9a}$$

$$P_f(\omega_1, \omega_2) = F[R_f(n_1, n_2)] \tag{9.9b}$$

for some constant $0 < \rho < 1$. The parameter ρ is estimated from the degraded image $g(n_1, n_2)$.

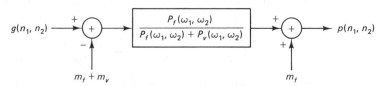

Figure 9.3 Noncausal Wiener filter for linear minimum mean square error estimation of $f(n_1, n_2)$ from $g(n_1, n_2) = f(n_1, n_2) + v(n_1, n_2)$.

The Wiener filter is generally implemented in the frequency domain by

$$p(n_1, n_2) = \text{IDFT} \left[G(k_1, k_2)H(k_1, k_2) \right]. \tag{9.10}$$

The sequences $G(k_1, k_2)$ and $H(k_1, k_2)$ represent the discrete Fourier transforms (DFTs) of $g(n_1, n_2)$ and $h(n_1, n_2)$. In (9.10), the size of the DFT and inverse DFT is at least $(N + M - 1) \times (N + M - 1)$ when the image size is $N \times N$ and the filter size is $M \times M$. As was discussed in Section 3.2.2, if the DFT size is less than $(N + M - 1) \times (N + M - 1)$, then the IDFT $[G(k_1, k_2)H(k_1, k_2)]$ will not be identical to $g(n_1, n_2) * h(n_1, n_2)$ near the boundaries of the processed image $p(n_1, n_2)$ because of aliasing effects. In most cases, however, the effective size of $h(n_1, n_2)$ is small, and adequate results can be obtained with a DFT and inverse DFT of size $N \times N$. One way to obtain $H(k_1, k_2)$ is to sample the Wiener filter frequency response $H(\omega_1, \omega_2)$ by

$$H(k_1, k_2) = H(\omega_1, \omega_2)|_{\omega_1 = 2\pi k_1/L, \omega_2 = 2\pi k_2/L} \tag{9.11}$$

where the size of the DFT and inverse DFT is $L \times L$.

The Wiener filter is typically a lowpass filter. As we discussed in Section 1.4.2, the energy of a typical image is concentrated in low-frequency regions. Since random background noise is generally wideband, the Wiener filter is lowpass in character. This is illustrated in Figure 9.4. Figure 9.4(a) shows an example of a $P_f(\omega_1, \omega_2)$ which decreases in amplitude as ω_1 and ω_2 increase. Figure 9.4(b) shows an example of a $P_v(\omega_1, \omega_2)$ that is constant independent of ω_1 and ω_2. The resultant Wiener filter $H(\omega_1, \omega_2)$ given by (9.8) is lowpass in character, as shown in Figure 9.4(c).

Throughout this chapter, we will rely on a subjective comparison of original, degraded, and processed images by a human observer to illustrate the performance of each image restoration algorithm. In addition when the information is available, we will provide the normalized mean square error (NMSE) between the original image $f(n_1, n_2)$ and the degraded image $g(n_1, n_2)$, and that between the original image $f(n_1, n_2)$ and the processed image $p(n_1, n_2)$. The NMSE between $f(n_1, n_2)$ and $p(n_1, n_2)$ is defined by

$$\text{NMSE} \left[f(n_1, n_2), p(n_1, n_2) \right] = 100 \times \frac{\text{Var} \left[f(n_1, n_2) - p(n_1, n_2) \right]}{\text{Var} \left[f(n_1, n_2) \right]} \% \tag{9.12}$$

where Var $[\cdot]$ is the variance. Using the variance ensures that the NMSE will not be affected by adding a bias to $p(n_1, n_2)$. The measure NMSE $[f(n_1, n_2), g(n_1, n_2)]$ is similarly defined. The SNR improvement due to processing is defined by

$$\text{SNR improvement} = 10 \log_{10} \frac{\text{NMSE} \left[f(n_1, n_2), g(n_1, n_2) \right]}{\text{NMSE} \left[f(n_1, n_2), p(n_1, n_2) \right]} \text{ dB}. \tag{9.13}$$

A human observing two images affected by the same type of degradation will generally judge the one with the smaller NMSE to be closer to the original. A very small NMSE generally can be taken to mean that the image is very close to the original. It is important to note, however, that the NMSE is just one of many possible objective measures and can be misleading. When images with different

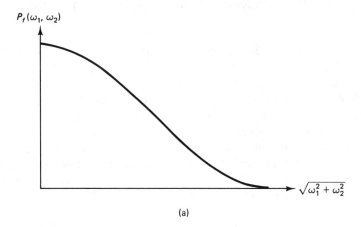

(a)

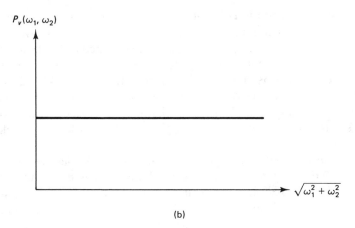

(b)

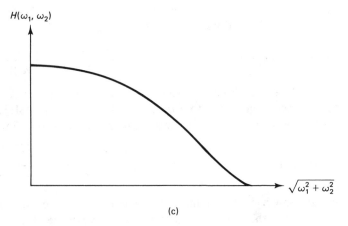

(c)

Figure 9.4 Illustration that the frequency response of a noncausal Wiener filter is typically lowpass in character.

types of degradation are compared, the one with the smallest NMSE will not necessarily seem closest to the original. As a result, the NMSE and SNR improvements are stated for reference only and should not be used in literally comparing the performance of one algorithm with another.

Figure 9.5 illustrates the performance of a Wiener filter for image restoration. Figure 9.5(a) shows an original image of 512×512 pixels, and Figure 9.5(b) shows the image degraded by zero-mean white Gaussian noise at an SNR of 7 dB. The SNR was defined in Chapter 8 as

$$\text{SNR in dB} = 10 \log_{10} \frac{\text{Var}\ [f(n_1, n_2)]}{\text{Var}\ [v(n_1, n_2)]}. \tag{9.14}$$

Figure 9.5(c) shows the result of the Wiener filter applied to the degraded image. In the Wiener filter, $P_v(\omega_1, \omega_2)$ was assumed given and $P_f(\omega_1, \omega_2)$ was estimated by averaging $|F(\omega_1, \omega_2)|^2$ for ten different images. For white noise degradation, $P_v(\omega_1, \omega_2)$ is constant independent of (ω_1, ω_2). The processed image has an SNR improvement of 7.4 dB. As Figure 9.5 shows, Wiener filtering clearly reduces the background noise. This is also evidenced by the SNR improvement. However, it also blurs the image significantly. Many variations of Wiener filtering have been proposed to improve its performance. Some of these variations will be discussed in the next section.

9.2.2 Variations of Wiener Filtering

The Wiener filter discussed in Section 9.2.1 was derived by minimizing the mean square error between the original and processed signals. The mean square error is not, however, the criterion used by a human observer in judging how close a processed image is to the original. Since the objective criterion consistent with human judgment is not known, many ad hoc variations have been proposed. One variation is power spectrum filtering. In this method, the filter used has the frequency response $H(\omega_1, \omega_2)$ given by

$$H(\omega_1, \omega_2) = \left(\frac{P_f(\omega_1, \omega_2)}{P_f(\omega_1, \omega_2) + P_v(\omega_1, \omega_2)} \right)^{1/2}. \tag{9.15}$$

The function $H(\omega_1, \omega_2)$ in (9.15) is the square root of the frequency response of the Wiener filter. If $f(n_1, n_2)$ and $v(n_1, n_2)$ are samples of stationary random processes linearly independent of each other, the output of the filter will have the same power spectrum as the original signal power spectrum $P_f(\omega_1, \omega_2)$. The method is thus known as *power spectrum filtering*. To show this,

$$P_p(\omega_1, \omega_2) = |H(\omega_1, \omega_2)|^2 P_g(\omega_1, \omega_2)$$
$$= |H(\omega_1, \omega_2)|^2 (P_f(\omega_1, \omega_2) + P_v(\omega_1, \omega_2)). \tag{9.16}$$

From (9.15) and (9.16),

$$P_p(\omega_1, \omega_2) = P_f(\omega_1, \omega_2). \tag{9.17}$$

(a)

(b)   (c)

Figure 9.5 (a) Original image of 512 × 512 pixels; (b) degraded image at SNR of 7 dB, with NMSE of 19.7%; (c) processed image by Wiener filtering, with NMSE of 3.6% and SNR improvement of 7.4 dB.

Several variations of Wiener filtering that have been proposed for image restoration can be expressed by the following $H(\omega_1, \omega_2)$:

$$H(\omega_1, \omega_2) = \left(\frac{P_f(\omega_1, \omega_2)}{P_f(\omega_1, \omega_2) + \alpha P_v(\omega_1, \omega_2)} \right)^{\beta} \tag{9.18}$$

where α and β are some constants. When $\alpha = 1$ and $\beta = 1$, $H(\omega_1, \omega_2)$ reduces

Figure 9.6 Image in Figure 9.5(a) processed by power spectrum filtering. The processed image has NMSE of 4.3% and SNR improvement of 6.6 dB.

to Wiener filtering. When $\alpha = 1$ and $\beta = \frac{1}{2}$, $H(\omega_1, \omega_2)$ reduces to power spectrum filtering. When α is a parameter and $\beta = 1$, the result is called a parametric Wiener filter. Since $H(\omega_1, \omega_2)$ in (9.18) is a straightforward generalization of a Wiener filter, all the comments made in Section 9.2.1 apply to this class of filters. Specifically, they are zero-phase filters, and tend to preserve high SNR frequency components. The power spectra $P_f(\omega_1, \omega_2)$ and $P_v(\omega_1, \omega_2)$ are assumed known, and such filters are typically implemented by using the DFT and inverse DFT. In addition, these filters are typically lowpass. They reduce noise but blur the image significantly. The performance of the power spectrum filtering is shown in Figure 9.6. The original and degraded images used are those shown in Figure 9.5. The SNR improvement is 6.6 dB.

9.2.3 Adaptive Image Processing*

One reason why the Wiener filter and its variations blur the image significantly is that a fixed filter is used throughout the entire image. The Wiener filter was developed under the assumption that the characteristics of the signal and noise do not change over different regions of the image. This has resulted in a space-invariant filter. In a typical image, image characteristics differ considerably from one region to another. For example, walls and skies have approximately uniform background intensities, whereas buildings and trees have large, detailed variations in intensity. Degradations may also vary from one region to another. It is reasonable, then, to adapt the processing to the changing characteristics of the image and degradation. The idea of adapting the processing to the local characteristics

---

*Some parts of this section were previously published in J. S. Lim, "Image Enhancement," in *Digital Image Processing Techniques*" (ed. M. P. Ekstrom). New York: Academic Press, 1984, Chapter 1.

of an image is useful not only in image restoration but in many other image processing applications, including image enhancement discussed in Chapter 8.

Two approaches to adaptive image processing have been developed. In one approach, called *pixel-by-pixel processing*, the processing is adapted at each pixel. At each pixel, the processing method is determined based on the local characteristics of the image, degradation, and any other relevant information in the neighborhood region centered around that pixel. Since each pixel is processed differently, this approach is highly adaptive and does not suffer from artificial intensity discontinuities in the processed image. However, the approach can be computationally expensive and is typically implemented only in the spatial domain.

In another approach, called *subimage-by-subimage* or *block-by-block processing*, an image is divided into many subimages, and each subimage is processed separately and then combined with the others. The size of the subimage is typically between 8×8 and 32×32 pixels. For each subimage, a space-invariant operation appropriate to the subimage is chosen on the basis of the local characteristics of the image, degradation, and any other relevant information in the subimage region. Because the processing that is applied to a subimage is a space-invariant operation, there is generally more flexibility in implementation than exists in pixel-by-pixel processing. For example, a lowpass filter may be implemented in either the spatial domain or the frequency domain. In addition, subimage-by-subimage processing is generally computationally less expensive than pixel-by-pixel processing, since the type of processing to be performed must be determined only once for the whole subimage. Because the type of processing changes abruptly as we move from one subimage to the next, artificial intensity discontinuities may appear along the boundaries of adjacent subimages in the processed image. This is called the *blocking effect*. In some applications, such as restoration of an image in a high SNR environment, the blocking effect may not be visible and need not be considered. In other applications, such as low bit-rate transform coding, the blocking effect can be quite pronounced and may be the most objectionable feature in the processed image.

The blocking effect can be reduced in some cases by lowpass filtering the processed image in only the subimage boundary regions. Another method of reducing the blocking effect is overlapping the subimages. In this method, to obtain a subimage, a window $w_{ij}(n_1, n_2)$ is applied to $g(n_1, n_2)$, the image to be processed. The window $w_{ij}(n_1, n_2)$ is chosen to satisfy two conditions. The first condition can be expressed as

$$\sum_i \sum_j w_{ij}(n_1, n_2) = 1 \quad \text{for all } (n_1, n_2) \text{ of interest.} \quad (9.19)$$

This condition guarantees that simple addition of the unprocessed subimages will result in the original image. The second condition requires $w_{ij}(n_1, n_2)$ to be a smooth function that falls close to zero near the window boundaries. This tends to reduce possible discontinuities or degradation that may appear at the subimage boundaries in the processed image.

One of the many ways to find a smooth 2-D window function that satisfies both of these conditions is to form a separable 2-D window from two 1-D windows that satisfy similar conditions:

$$w_{ij}(n_1, n_2) = w_i(n_1)w_j(n_2). \tag{9.20}$$

Two such window functions are the 2-D separable triangular and Hamming windows overlapped with their neighboring windows by half the window duration in each dimension. The 2-D separable triangular window is shown in Figure 9.7. In processing a subimage, the window used to obtain the subimage has to be taken into account.

There are many variations to pixel-by-pixel and subimage-by-subimage processing. For example, a filter may be designed only once every 8×8 to 32×32 pixels but implemented in a pixel-by-pixel processing manner.

A general adaptive image processing system is shown in Figure 9.8. The processing to be done at each pixel or subimage is adapted to the local characteristics of the image, the degradation, and any other relevant information. Knowledge of these characteristics can be obtained from two sources. One is some prior knowledge of the problem at hand. We may know, for example, what class of images to expect in a given application, or we may be able to deduce the degradation characteristics from our knowledge of the degrading source. Another source of information is the image to be processed. By such measured features as local variance, it may be possible to determine the presence of significant high-frequency details.

Determining what type of processing to use depends on a number of factors. These include the type of knowledge we have about the image and how this knowledge can be exploited in estimating the parameters of a processing method, such

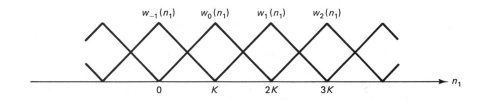

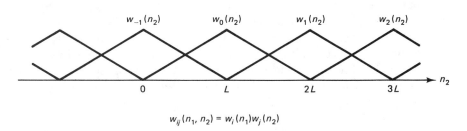

$$w_{ij}(n_1, n_2) = w_i(n_1)w_j(n_2)$$

Figure 9.7 Example of two-dimensional separable triangular window.

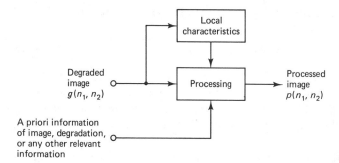

Figure 9.8 General adaptive image processing system.

as the lowpass filter cutoff frequency. Without a specific application context, only general statements can be made. In general, the more the available knowledge is used, the higher the resulting performance will be. If the available information is inaccurate, however, the system's performance may be degraded. In general, more sophisticated rules for adaptation are associated with subimage-by-subimage processing, while simple rules are associated with pixel-by-pixel processing for computational reasons.

When an adaptive image processing method is applied to the problem of restoring an image degraded by additive random noise, it is possible to reduce background noise without significant image blurring. In the next four sections, we discuss a few representative adaptive image restoration systems chosen from among the many proposed in the literature.

9.2.4 The Adaptive Wiener Filter

Most adaptive restoration algorithms for reducing additive noise in an image can be represented by the system in Figure 9.9. From the degraded image and prior knowledge, some measure of the local details of the noise-free image is determined. One such measure is the local variance. A space-variant* filter $h(n_1, n_2)$ which is a function of the local image details and of additional prior knowledge is then determined.

The space-variant filter is then applied to the degraded image in the local region from which the space-variant filter was designed. When the noise is wideband, the space-variant $h(n_1, n_2)$ is lowpass in character. In low-detail image regions such as uniform intensity regions, where noise is more visible than in high-detail regions, a large amount (low cutoff frequency) of lowpass filtering is performed to reduce as much noise as possible. Since little signal variation is present in low-detail regions, even a large amount of lowpass filtering does not significantly affect the signal component. In high-detail image regions such as edges, where a large signal component is present, only a small amount of lowpass filtering is

---

*For a space-variant filter, the filter coefficients change as a function of (n_1, n_2). For notational simplicity, we denote the filter coefficients by $h(n_1, n_2)$. It should be noted, however, that $h(n_1, n_2)$ changes as we process different parts of an image.

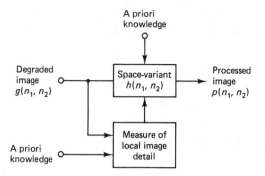

Figure 9.9 Typical adaptive image restoration system for additive noise reduction.

performed so as not to distort (blur) the signal component. This does not reduce much noise, but the same noise is less visible in the high-detail than in the low-detail regions.

A number of different algorithms can be developed, depending on which specific measure is used to represent local image details, how the space-variant $h(n_1, n_2)$ is determined as a function of the local image details, and what prior knowledge is available. One of the many possibilities is to adaptively design and implement the Wiener filter discussed in Section 9.2.1. As Figure 9.3 shows, the Wiener filter requires knowledge of the signal mean m_f, noise mean m_v, signal power spectrum $P_f(\omega_1, \omega_2)$, and noise power spectrum $P_v(\omega_1, \omega_2)$. Instead of assuming a fixed m_f, m_v, $P_f(\omega_1, \omega_2)$, and $P_v(\omega_1, \omega_2)$ for the entire image, they can be estimated locally. This approach will result in a space-variant Wiener filter. Even within this approach, many variations are possible, depending on how m_f, m_v, $P_f(\omega_1, \omega_2)$, and $P_v(\omega_1, \omega_2)$ are estimated locally and how the resulting space-variant Wiener filter is implemented. We will develop one specific algorithm to illustrate this approach.

We first assume that the additive noise $v(n_1, n_2)$ is zero mean and white with variance of σ_v^2. Its power spectrum $P_v(\omega_1, \omega_2)$ is then given by

$$P_v(\omega_1, \omega_2) = \sigma_v^2. \tag{9.21}$$

Consider a small local region in which the signal $f(n_1, n_2)$ is assumed stationary. Within the local region, the signal $f(n_1, n_2)$ is modeled by

$$f(n_1, n_2) = m_f + \sigma_f w(n_1, n_2) \tag{9.22}$$

where m_f and σ_f are the local mean and standard deviation of $f(n_1, n_2)$, and $w(n_1, n_2)$ is zero-mean white noise with unit variance.* There is some empirical evidence that (9.22) is a reasonable model for a typical image [Trussell and Hunt; Kuan et al. (1985)].

In (9.22), the signal $f(n_1, n_2)$ is modeled by a sum of a space-variant local

---

*The notation $w(n_1, n_2)$ is used to represent both a window function and white noise. Specifically which is meant will be clear from the context.

mean m_f and white noise with a space-variant local variance σ_v^2. Within the local region, then, the Wiener filter $H(\omega_1, \omega_2)$ is given by

$$H(\omega_1, \omega_2) = \frac{P_f(\omega_1, \omega_2)}{P_f(\omega_1, \omega_2) + P_v(\omega_1, \omega_2)}$$

$$= \frac{\sigma_f^2}{\sigma_f^2 + \sigma_v^2}.$$

(9.23)

From (9.23), $h(n_1, n_2)$ is a scaled impulse response given by

$$h(n_1, n_2) = \frac{\sigma_f^2}{\sigma_f^2 + \sigma_v^2} \delta(n_1, n_2).$$

(9.24)

From (9.24) and Figure 9.3, the processed image $p(n_1, n_2)$ within the local region can be expressed as

$$p(n_1, n_2) = m_f + (g(n_1, n_2) - m_f) * \frac{\sigma_f^2}{\sigma_f^2 + \sigma_v^2} \delta(n_1, n_2)$$

$$= m_f + \frac{\sigma_f^2}{\sigma_f^2 + \sigma_v^2} (g(n_1, n_2) - m_f).$$

(9.25)

If we assume that m_f and σ_f^2 are updated at each pixel,

$$p(n_1, n_2) = m_f(n_1, n_2) + \frac{\sigma_f^2(n_1, n_2)}{\sigma_f^2(n_1, n_2) + \sigma_v^2} (g(n_1, n_2) - m_f(n_1, n_2)).$$

(9.26)

Equation (9.26) is the essence of the algorithm developed by [Lee (1980)].

The algorithm based on (9.26) can be viewed as a special case of a two-channel process. In the two-channel process, the image to be processed is divided into two components, the local mean $m_f(n_1, n_2)$ and the local contrast $g(n_1, n_2) - m_f(n_1, n_2)$. The local mean and local contrast are modified separately and then the results are combined. In the case of (9.26), the local mean remains unmodified while the local contrast is scaled according to the relative amplitudes of σ_f^2 and σ_v^2. If σ_f^2 is much larger than σ_v^2, the local contrast of $g(n_1, n_2)$ is assumed to be primarily due to $f(n_1, n_2)$ and the local contrast of $g(n_1, n_2)$ is not attenuated. In this case, $p(n_1, n_2)$ is approximately the same as $g(n_1, n_2)$; little processing is done in such regions. If σ_f^2 is much smaller than σ_v^2, the local contrast of $g(n_1, n_2)$ is assumed to be primarily due to $v(n_1, n_2)$ and the local contrast of $g(n_1, n_2)$ is significantly attenuated. In this case, $p(n_1, n_2)$ is approximately m_f; $g(n_1, n_2)$ is smoothed significantly. Another example of the two-channel process is the adaptive algorithm developed in Section 8.1.4 to reduce the effect of cloud cover in an image taken from an airplane.

Noting that m_f is identical to m_g when m_v is zero, we can estimate $m_f(n_1, n_2)$ in (9.26) from $g(n_1, n_2)$ by

$$\hat{m}_f(n_1, n_2) = \frac{1}{(2M + 1)^2} \sum_{k_1 = n_1 - M}^{n_1 + M} \sum_{k_2 = n_2 - M}^{n_2 + M} g(k_1, k_2)$$

(9.27)

where $(2M + 1)^2$ is the number of pixels in the local region used in the estimation. Within the local region where $\sigma_f^2(n_1, n_2)$ can be assumed to be space invariant, substituting $\hat{m}_f(n_1, n_2)$ in (9.27) for $m_f(n_1, n_2)$ in (9.26) leads to

$$p(n_1, n_2) = g(n_1, n_2) * h(n_1, n_2) \tag{9.28a}$$

where

$$h(n_1, n_2) = \begin{cases} \dfrac{\sigma_f^2 + \dfrac{\sigma_v^2}{(2M + 1)^2}}{\sigma_f^2 + \sigma_v^2}, & n_1 = n_2 = 0 \\[4mm] \dfrac{\dfrac{\sigma_v^2}{(2M + 1)^2}}{\sigma_f^2 + \sigma_v^2}, & \begin{array}{l} -M \le n_1 \le M, \ -M \le n_2 \le M, \\ \text{except } n_1 = n_2 = 0 \end{array} \\[4mm] 0, & \text{otherwise.} \end{cases} \tag{9.28b}$$

The filters $h(n_1, n_2)$ when $\sigma_f^2 \gg \sigma_v^2$, $\sigma_f^2 = \sigma_v^2$, and $\sigma_f^2 \ll \sigma_v^2$ are shown in Figure 9.10 for $M = 1$. From Figure 9.10, as σ_f^2 decreases relative to σ_v^2, more noise smoothing is performed. To measure the local signal detail in the system in Figure 9.9, the algorithm developed uses the signal variance σ_f^2. The specific method used to design the space-variant $h(n_1, n_2)$ is given by (9.28b). Since designing the space-variant $h(n_1, n_2)$ is simple and the resulting filter $h(n_1, n_2)$ is typically a small FIR filter (of size 3×3, 5×5, or 7×7), pixel-by-pixel processing is often employed.

Since $\sigma_g^2 = \sigma_f^2 + \sigma_v^2$, σ_f^2 may be estimated from $g(n_1, n_2)$ by

$$\hat{\sigma}_f^2(n_1, n_2) = \begin{cases} \hat{\sigma}_g^2(n_1, n_2) - \sigma_v^2, & \text{if } \hat{\sigma}_g^2(n_1, n_2) > \sigma_v^2 \\ 0, & \text{otherwise} \end{cases} \tag{9.29a}$$

where

$$\hat{\sigma}_g^2(n_1, n_2) = \frac{1}{(2M + 1)^2} \sum_{k_1 = n_1 - M}^{n_1 + M} \sum_{k_2 = n_2 - M}^{n_2 + M} (g(k_1, k_2) - \hat{m}_f(n_1, n_2))^2. \tag{9.29b}$$

The local mean estimate $\hat{m}_f(n_1, n_2)$ can be obtained from (9.27), and σ_v^2 is assumed known.

Figure 9.11 illustrates the performance of this algorithm. Figure 9.11(a) shows the processed image. The original and degraded images used are those shown in Figures 9.5(a) and (b). The degradation used to obtain Figure 9.5(b) is additive white Gaussian noise. The SNR improvement is 7.1 dB. The processed image was obtained by using (9.27), (9.28), and (9.29) with $M = 2$. From the processed image, a significant amount of noise has been reduced without noticeable blurring. If a nonadaptive Wiener filter had been used, this amount of noise reduction would have been accompanied by noticeable image blurring. Figure 9.11(b) shows the result of a nonadaptive Wiener filter. Figure 9.11(b) is the same as Figure 9.5(c), which is repeated for convenience.

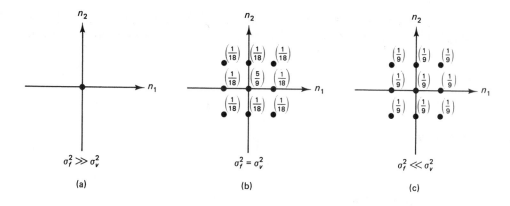

Figure 9.10 Impulse response of a space-variant image restoration filter as a function of σ_f^2 and σ_v^2. When the signal variance of σ_f^2 is much larger than the noise variance of σ_v^2, the filter is close to $\delta(n_1, n_2)$. As σ_v^2 increases relative to σ_f^2, $h(n_1, n_2)$ approaches a rectangular window.

9.2.5 Adaptive Restoration Based on the Noise Visibility Function

The adaptive restoration algorithm discussed in Section 9.2.4 was developed without a quantitative measure of how visible a given noise is to a human viewer. A quantitative measure of this type, if available, can be used in developing an image restoration system. A function representing such a measure is called the *noise*

Figure 9.11 Performance illustration of an adaptive Wiener filtering method. The degraded image used is the image in Figure 9.5(b). (a) Processed image by adaptive filtering, with NMSE of 3.8% and SNR improvement of 7.1 dB; (b) processed image by space-invariant Wiener filter, with NMSE of 3.6% and SNR improvement of 7.4 dB.

visibility function. The visibility of noise depends on the type of noise and also on the type of signal to which it has been added. The same level of white noise and colored noise will generally have different effects on the human observer. A high-detail region in an image tends to mask noise more than a low-detail region.

There are many ways to define and measure the noise visibility function. We will discuss one used by [Anderson and Netravali] in developing an image restoration system. We will assume white noise for the degrading background noise, although this approach applies to other types of noise as well. Let $M(n_1, n_2)$ denote some measure of local image detail of the original image $f(n_1, n_2)$. The function $M(n_1, n_2)$ is called the masking function, since a high-detail (high M) region masks a given level of noise better than a low-detail region. The noise visibility function $V(M)$ is defined to represent the relative visibility of a given level of noise at the masking level M. Specifically, suppose noise with variance of σ_1^2 at $M = M_1$ has been judged by a human viewer to be as visible as noise with variance of σ_2^2 at $M = M_2$. The function $V(M)$ is defined by

$$\sigma_1^2 V(M_1) = \sigma_2^2 V(M_2). \tag{9.30}$$

The higher the noise visibility $V(M)$ at $M = M_1$, the lower the level of noise σ_1^2 required to reach equal visibility to a fixed level of noise σ_2^2 at a fixed masking level M_2. Since the same level of noise is more visible in the low-detail region (small M), $V(M)$ is expected to decrease as M increases.

The noise visibility function $V(M)$ can be measured within a scaling factor, at least in theory, by using (9.30). Suppose we add noise with variance σ_1^2 to a local image region with a masking level of M_1. We can ask a human viewer to compare the noise visibility in the local region with that in another image region where M is M_2 and the noise level used is σ^2. We allow the viewer to vary σ^2 so that the noise in the two regions is equally visible, and denote the value of σ^2 chosen as σ_2^2. We will refer to this psychophysical experiment as a noise visibility matching experiment. From the σ_1^2 used in the experiment and the σ_2^2 chosen by the human viewer, we can determine $V(M_2)/V(M_1)$ by σ_1^2/σ_2^2.

Equation (9.30) is based on various assumptions. In (9.30), for example, $V(M)$ is assumed to depend only on M. Therefore, the masking function $M(n_1, n_2)$ has to be chosen such that the same level of noise is equally visible in all image regions with the same M. One choice of M proposed in [Anderson and Netravali] is

$$M(n_1, n_2) = \sum_{k_1 = n_1 - L}^{n_1 + L} \sum_{k_2 = n_2 - L}^{n_2 + L} 0.35^{\sqrt{(k_1 - n_1)^2 + (k_2 - n_2)^2}} (|f(k_1 + 1, k_2) \tag{9.31}$$
$$- f(k_1 - 1, k_2)| + |f(k_1, k_2 + 1) - f(k_1, k_2 - 1)|)$$

where $f(n_1, n_2)$ is the noise-free image and $(2L + 1) \times (2L + 1)$ is the size of the local region used in measuring the masking level M at (n_1, n_2). In (9.31), $M(n_1, n_2)$ increases as the horizontal and vertical slopes of $f(n_1, n_2)$ increase and the contributions that the horizontal and vertical slopes make on $M(n_1, n_2)$ decrease exponentially as the Euclidean distance between (n_1, n_2) and the point at which the slopes are measured increases. In (9.30), the equal noise visibility is assumed to hold as we scale σ_1^2 and σ_2^2 by the same scaling factor. This assumption is probably valid only over a small range of the scaling factor.

In addition to the assumptions made in (9.30), which may be only approximately valid, there are practical difficulties in measuring $V(M)$ by using (9.30). In a typical image, the number of pixels with a given M may be small, particularly for a large M. In such cases, it will be difficult to measure $V(M)$ by using the noise visibility matching experiment. Despite these difficulties, $V(M)$ has been approximately measured based on (9.30) and (9.31) by means of the noise visibility matching experiment. The result is shown in Figure 9.12. As expected, $V(M)$ decreases as M increases for a wide range of M.

The noise visibility function can be used in a number of ways to develop image restoration algorithms. We will develop one restoration algorithm, which can be viewed as a special case of the adaptive system shown in Figure 9.9. In this algorithm, the space-variant filter $h(n_1, n_2)$ has a Gaussian shape given by

$$h(n_1, n_2) = ke^{-(n_1^2 + n_2^2)/2\sigma^2}w(n_1, n_2) \tag{9.32}$$

where k and σ^2 are determined adaptively and $w(n_1, n_2)$ is a rectangular window that limits the region of support size of $h(n_1, n_2)$. To determine k and σ^2, one constraint imposed is

$$\sum_{n_1=-\infty}^{\infty} \sum_{n_2=-\infty}^{\infty} h(n_1, n_2) = 1. \tag{9.33}$$

Another constraint requires the noise in the processed image to be equally visible

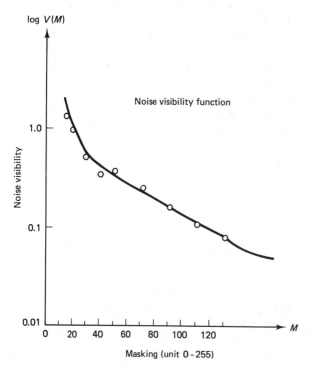

Figure 9.12 Noise visibility function $V(M)$. After [Anderson and Netravali].

Image Restoration   Chap. 9

throughout the entire image. To impose this constraint, note from elementary stochastic processes theory (see Chapter 6) that when the degrading noise $v(n_1, n_2)$ is white with variance of σ_v^2, the noise in the processed image is colored with variance of σ_p^2 where

$$\sigma_p^2 = \sigma_v^2 \sum_{n_1 = -\infty}^{\infty} \sum_{n_2 = -\infty}^{\infty} |h(n_1, n_2)|^2. \qquad (9.34)$$

If we choose $h(n_1, n_2)$ in each local region such that σ_p^2 satisfies

$$\sigma_p^2 V(M) = \text{constant}, \qquad (9.35)$$

the level of noise that remains in the processed image will be equally visible to the extent that $V(M)$ accurately reflects the definition in (9.30) and $V(M)$ is approximately the same for the white and colored noise. The constant in (9.35) is chosen such that some balance between noise reduction and signal distortion (blurring) is reached. If the constant chosen is too large, little background noise will be reduced. If the constant chosen is too small, noise will be reduced, but significant signal distortion will occur. At each pixel, the space-variant $h(n_1, n_2)$ can be determined from (9.32), (9.33), (9.34), and (9.35). Since the filter parameters k and σ^2 depend only on M in this algorithm, k and σ^2 can be precomputed and stored in a table as a function of M. To restore an image, $M(n_1, n_2)$ of the noise-free image $f(n_1, n_2)$ is estimated from the degraded image and $k(n_1, n_2)$ and $\sigma^2(n_1, n_2)$ can be determined from the precomputed table. At each pixel (n_1, n_2), the space-variant $h(n_1, n_2)$ can be determined from (9.32) by using the k and σ^2 that have been determined.

The algorithm developed above is based on the concept that the noise in the processed image is equally visible throughout the image, independent of local image detail. The amount of blurring that will result in the processed image, however, is not explicitly controlled. Fortunately, in the high-detail regions, where it is desirable to have as little signal blurring as possible, M is large, $V(M)$ is small, the noise level σ_p^2 that remains in the processed image is relatively large, and relatively little blurring occurs.

Figure 9.13 illustrates the performance of this algorithm. Figure 9.13(a) shows an original image of 512×512 pixels. Figure 9.13(b) shows the image degraded by white Gaussian noise at an SNR of 7 dB. Figure 9.13(c) shows the processed image, with an SNR improvement of 7.7 dB. The processed image was obtained by adapting the filter at each pixel and determining the masking function $M(n_1, n_2)$ from the original noise-free image.

Despite the various assumptions and approximations made in this algorithm's development, significant noise reduction is achieved with little noticeable signal blurring. In practice, the original noise-free image is not available for use in estimating $M(n_1, n_2)$. If $M(n_1, n_2)$ is obtained from the noisy image, the performance of the algorithm deteriorates. Figure 9.13(d) shows the image processed by the algorithm when $M(n_1, n_2)$ is obtained from the degraded image. The image has an SNR improvement of 4.5 dB. This algorithm is just one example of exploiting the noise visibility function $V(M)$. There are many other possible definitions of $V(M)$ and many other ways to exploit $V(M)$ in the development of image restoration algorithms.

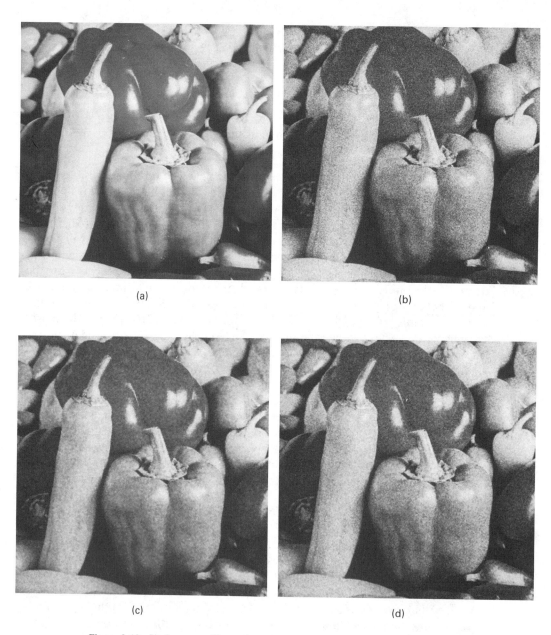

(a)

(b)

(c)

(d)

Figure 9.13 Performance illustration of an adaptive image restoration algorithm based on the noise visibility function. (a) Original image of 512 × 512 pixels; (b) image degraded by white noise, with SNR of 7 dB and NMSE of 19.8%; (c) processed image using the noise visibility function obtained from the original image, with NMSE of 3.4% and SNR improvement of 7.7 dB; (d) processed image using the noise visibility function obtained from the degraded image, with NMSE of 7.0% and SNR improvement of 4.5 dB.

9.2.6 Short-Space Spectral Subtraction

The method discussed in this section is a straightforward extension of the method developed to reduce additive random noise in speech [Lim]. Since the design and implementation of the space-variant filter in this method is computationally expensive, subimage-by-subimage processing is used.

Applying a window $w(n_1, n_2)$ to the degraded image $g(n_1, n_2)$, we obtain

$$g(n_1, n_2)w(n_1, n_2) = f(n_1, n_2)w(n_1, n_2) + v(n_1, n_2)w(n_1, n_2). \quad (9.36)$$

Rewriting (9.36), we have

$$g_w(n_1, n_2) = f_w(n_1, n_2) + v_w(n_1, n_2). \quad (9.37)$$

The window is chosen such that the subimage $g_w(n_1, n_2)$ can be assumed to be approximately stationary. With $G_w(\omega_1, \omega_2)$, $F_w(\omega_1, \omega_2)$, and $V_w(\omega_1, \omega_2)$ denoting the Fourier transforms of $g_w(n_1, n_2)$, $f_w(n_1, n_2)$, and $v_w(n_1, n_2)$, respectively, from (9.37)

$$|G_w(\omega_1, \omega_2)|^2 = |F_w(\omega_1, \omega_2)|^2 + |V_w(\omega_1, \omega_2)|^2 + F_w(\omega_1, \omega_2)V_w^*(\omega_1, \omega_2)$$
$$+ F_w^*(\omega_1, \omega_2)V_w(\omega_1, \omega_2). \quad (9.38)$$

The functions $V_w^*(\omega_1, \omega_2)$ and $F_w^*(\omega_1, \omega_2)$ are complex conjugates of $V_w(\omega_1, \omega_2)$ and $F_w(\omega_1, \omega_2)$. Rewriting (9.38), we obtain

$$|F_w(\omega_1, \omega_2)|^2 = |G_w(\omega_1, \omega_2)|^2 - |V_w(\omega_1, \omega_2)|^2 - F_w(\omega_1, \omega_2)V_w^*(\omega_1, \omega_2)$$
$$- F_w^*(\omega_1, \omega_2)V_w(\omega_1, \omega_2). \quad (9.39)$$

In the spectral subtraction method, $|F_w(\omega_1, \omega_2)|$ is estimated based on (9.39). From the degraded subimage $g_w(n_1, n_2)$, $|G_w(\omega_1, \omega_2)|^2$ can be obtained directly. The terms $|V_w(\omega_1, \omega_2)|^2$, $F_w(\omega_1, \omega_2)V_w^*(\omega_1, \omega_2)$, and $F_w^*(\omega_1, \omega_2)V_w(\omega_1, \omega_2)$ cannot be obtained exactly, and are approximated by $E[|V_w(\omega_1, \omega_2)|^2]$, $E[F_w(\omega_1, \omega_2)V_w^*(\omega_1, \omega_2)]$, and $E[F_w^*(\omega_1, \omega_2)V_w(\omega_1, \omega_2)]$. For $v_w(n_1, n_2)$ which is zero mean and uncorrelated with $f(n_1, n_2)$, $E[F_w(\omega_1, \omega_2)V_w^*(\omega_1, \omega_2)]$ and $E[F_w^*(\omega_1, \omega_2)V_w(\omega_1, \omega_2)]$ are zero and an estimate of $|F_w(\omega_1, \omega_2)|^2$ is suggested from (9.39) as

$$|\hat{F}_w(\omega_1, \omega_2)|^2 = |G_w(\omega_1, \omega_2)|^2 - E[|V_w(\omega_1, \omega_2)|^2] \quad (9.40)$$

where $E[|V_w(\omega_1, \omega_2)|^2]$ is obtained from the known or measured properties of $v(n_1, n_2)$. The estimate $|\hat{F}_w(\omega_1, \omega_2)|$ in (9.40) is not guaranteed to be nonnegative. To ensure that $|\hat{F}_w(\omega_1, \omega_2)|$ will be nonnegative, $|\hat{F}_w(\omega_1, \omega_2)|$ is assumed zero if $|G_w(\omega_1, \omega_2)|^2$ is less than $E[|V_w(\omega_1, \omega_2)|^2]$.

Given an estimate of $|F_w(\omega_1, \omega_2)|$, there are many different ways to estimate $f_w(n_1, n_2)$. One method which is frequently used and is consistent with the idea that minimum mean square error filters, such as Wiener filters, are zero phase is to approximate $\theta_{f_w}(\omega_1, \omega_2)$ by $\theta_{g_w}(\omega_1, \omega_2)$, so that

$$\hat{F}_w(\omega_1, \omega_2) = |\hat{F}_w(\omega_1, \omega_2)|e^{j\theta_{g_w}(\omega_1, \omega_2)} \quad (9.41a)$$

and

$$\hat{f}_w(n_1, n_2) = F^{-1}[\hat{F}_w(\omega_1, \omega_2)]. \quad (9.41b)$$

From (9.40), it is clear that $|F_w(\omega_1, \omega_2)|$ is estimated by subtracting a bias term $E[|V_w(\omega_1, \omega_2)|^2]$ from $|G_w(\omega_1, \omega_2)|^2$. This leads to the term *spectral subtraction*. Since $|V_w(\omega_1, \omega_2)|^2$ can be viewed as a periodogram, as discussed in Section 6.2.1, and since the variance of a periodogram is known to be very large, subtraction of $E[|V_w(\omega_1, \omega_2)|^2]$ does not reduce the background noise sufficiently. To reduce more noise at the expense of more signal distortion, $\alpha E[|V_w(\omega_1, \omega_2)|^2]$ with $\alpha > 1$ is often subtracted. In this case, the estimate $\hat{F}_w(\omega_1, \omega_2)$ is obtained from

$$\hat{F}_w(\omega_1, \omega_2) = \begin{cases} [|G_w(\omega_1, \omega_2)|^2 - \alpha E[|V_w(\omega_1, \omega_2)|^2]]^{1/2} e^{j\theta_{gw}(\omega_1,\omega_2)}, \\ \qquad\qquad |G_w(\omega_1, \omega_2)|^2 > \alpha E[|V_w(\omega_1, \omega_2)|^2] \\ 0, \quad \text{otherwise} \end{cases} \qquad (9.42)$$

where α is a parameter that controls the level of noise reduction. Since (9.42) is used for each subimage and the processed subimages are combined to form the whole image, the method is called *short-space* spectral subtraction. When a speech signal is processed in a segment-by-segment manner, the method is called short-time speech processing.

The spectral subtraction method can be viewed as enhancing the SNR. Since the same function is subtracted from $|G_w(\omega_1, \omega_2)|^2$, in both high-detail and low-detail regions, the subtraction has a relatively small effect in high-detail regions where $|G_w(\omega_1, \omega_2)|^2$ is relatively large. In low-detail regions, where $|G_w(\omega_1, \omega_2)|^2$ is relatively small and consists of primarily the noise component, the subtraction eliminates a relatively large portion of $|G_w(\omega_1, \omega_2)|^2$.

Figure 9.14 illustrates the performance of the short-space spectral subtraction method. Figure 9.14(a) shows an original image of 256 × 256 pixels. Figure 9.14(b) shows the image degraded by white Gaussian noise at an SNR of 10 dB. Figure 9.14(c) shows the processed image with $\alpha = 2$ in (9.42). The subimage size used is 32 × 32 pixels, and a separable overlapped triangular window was used in the subimage-by-subimage processing method.

9.2.7 Edge-Sensitive Adaptive Image Restoration

In the previous three sections, we have discussed adaptive image restoration algorithms that adapt to the local characteristics of an image. Within the local region, the image is typically assumed to be a sample of a stationary random process. One problem with this assumption arises in edge regions, where the signal cannot be adequately modeled even locally as a sample of a stationary random process. A filter based on this assumption will preserve edges by leaving a significant amount of noise near them. Although noise is less visible near edge regions than near uniform intensity regions, removing more noise near edges can be beneficial.

One approach to reducing more noise near edges without additional edge blurring is to model the image more accurately (by considering an edge as a deterministic component, for example) and to develop an image restoration algorithm based on the new model. However, modeling an image accurately is a difficult task, and an image restoration algorithm based on a detailed and accurate image model is likely to be quite complex. Another approach is to first detect edges using the edge detection algorithms discussed in Section 8.3 and then to use the

(a)

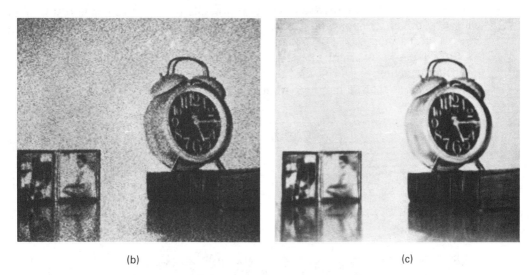

(b)   (c)

Figure 9.14 Performance illustration of the short-space spectral subtraction method. (a) Original image of 256 × 256 pixels; (b) image degraded by white noise, with SNR of 10 dB; (c) processed image by short-space spectral subtraction.

detected edges in designing and implementing an adaptive filter. Edges, for example, can be used in defining boundaries of local image regions over which the image is assumed stationary. When a space-variant filter $h(n_1, n_2)$ is designed, the region of support of $h(n_1, n_2)$ can be chosen such that $h(n_1, n_2)$ does not cover pixels that lie in more than one region where the regions are bounded by edges. This approach, however, requires explicit determination of edges, and detection of edges in the presence of noise is not a simple task.

Another approach is to use a cascade of 1-D adaptive filters without changing the image model or the basic principle used in the development of a 2-D image restoration system. If one of the 1-D filters is oriented in the same direction as the edge, the edge can be avoided and the image can be filtered along it. Other 1-D filters that cross the edge in the local region will perform little processing and will preserve the edge.

Let $T_i[\cdot]$ $(1 \le i \le N)$ represent a 1-D filter that is obtained in the same way as a 2-D adaptive filter but determined from a local 1-D region oriented in the ith direction. For practical reasons, N is typically chosen to be four, and the four directions are chosen to have angles of $0°$, $45°$, $90°$, and $135°$. The degraded image is filtered by a cascade of the four 1-D filters as if the image were a 1-D signal for each of the filters. The processed image $p(n_1, n_2)$ is

$$p(n_1, n_2) = T_4[T_3[T_2[T_1[g(n_1, n_2)]]]]. \tag{9.43}$$

The operators $T_i[\cdot]$ are space variant. Because the image is filtered sequentially, the signal and noise characteristics change after each filter and have to be updated for subsequent filtering.

To illustrate this approach of 1-D processing for adaptive image restoration, let us consider its application to the adaptive Wiener filter developed in Section 9.2.4. Equations (9.26), (9.27), and (9.29) define the restoration algorithm. Consider a 1-D space-variant filter oriented in the horizontal direction and designed and implemented by using the same principle as the 2-D restoration algorithm. The equations that define the output of the 1-D filter $p_1(n_1, n_2)$ are

$$p_1(n_1, n_2) = m_f(n_1, n_2) + \frac{\sigma_f^2(n_1, n_2)}{\sigma_f^2(n_1, n_2) + \sigma_v^2(n_1, n_2)}(g(n_1, n_2) - m_f(n_1, n_2)), \tag{9.44}$$

$$\hat{m}_f(n_1, n_2) = \frac{1}{(2M+1)} \sum_{k_1=n_1-M}^{n_1+M} g(k_1, n_2), \tag{9.45}$$

$$\text{and} \quad \hat{\sigma}_f^2(n_1, n_2) = \begin{cases} \hat{\sigma}_g^2(n_1, n_2) - \sigma_v^2(n_1, n_2), & \text{if } \hat{\sigma}_g^2(n_1, n_2) > \sigma_v^2(n_1, n_2) \\ 0, & \text{otherwise} \end{cases} \tag{9.46a}$$

$$\text{where} \quad \hat{\sigma}_g^2(n_1, n_2) = \frac{1}{(2M+1)} \sum_{k_1=n_1-M}^{n_1+M} (g(k_1, n_2) - \hat{m}_f(k_1, n_2))^2. \tag{9.46b}$$

Equations (9.44), (9.45), and (9.46) correspond to (9.26), (9.27), and (9.29). The second 1-D filter, say the vertical filter, is obtained in a similar way and applied to $p_1(n_1, n_2)$. The noise term $\sigma_v^2(n_1, n_2)$ should be updated, since the filter has reduced the noise power. The reduced noise power can be computed from the space-variant 1-D filter and $\sigma_v^2(n_1, n_2)$. Extension to the two remaining 1-D filters along the diagonal directions is similar.

The resulting system is a cascade of 1-D filters capable of adapting to the edge orientation in the image. A sharp edge inclined at a large angle to the filtering direction will remain practically intact, while the noise near the edge will be reduced by the filter oriented in directions close to the direction of the edge. The computational requirement of this approach is typically less than that of the corresponding 2-D restoration algorithm. In addition, this approach appears to improve the performance of some 2-D adaptive restoration algorithms.

Figure 9.15 illustrates the performance of this 1-D approach. Figure 9.15(a) shows an original image of 256×256 pixels. Figure 9.15(b) shows the image degraded by white Gaussian noise at an SNR of 6 dB. Figure 9.15(c) shows the image processed by the 2-D adaptive Wiener filter discussed in Section 9.2.4. The processed image has an SNR improvement of 6.79 dB. Figure 9.15(d) shows the image processed by a cascade of four 1-D space-variant filters designed on the basis of (9.44), (9.45), and (9.46). The processed image has an SNR improvement of 7.28 dB. Figure 9.16 shows enlarged segments of the images in Figure 9.15. Comparison of Figures 9.16(c) and (d) shows that less noise is present in the latter image.

In this and the previous three sections, we have discussed several adaptive image restoration algorithms. As is evident from the discussion, it is possible to develop many additional restoration algorithms. Adaptive restoration algorithms require considerably more computation than nonadaptive algorithms, but often perform better than nonadaptive restoration algorithms.

9.3 REDUCTION OF IMAGE BLURRING

An image degraded by blurring can be modeled as

$$g(n_1, n_2) = f(n_1, n_2) * b(n_1, n_2). \tag{9.47}$$

In (9.47), the degraded image $g(n_1, n_2)$ is the result of convolving the original image $f(n_1, n_2)$ with an impulse response $b(n_1, n_2)$. The sequence $b(n_1, n_2)$ is called the point spread function or the blurring function. The degradation that may be modeled by convolutional noise includes blurring due to lens misfocus, motion, and atmospheric turbulance.

The problem of reducing image blur can be classified into two categories. One is the deconvolution problem, in which $b(n_1, n_2)$ is assumed to be known. The second is the blind deconvolution problem, where $b(n_1, n_2)$ is not known and must be estimated from the available information. In Section 9.3.1, we discuss inverse filtering, a standard approach to solving the deconvolution problem. In Section 9.3.2, we discuss algorithms developed to solve the blind deconvolution problem.

9.3.1 Inverse Filtering

When the blurring function $b(n_1, n_2)$ is assumed known, one approach to deblurring is inverse filtering. From (9.47),

$$G(\omega_1, \omega_2) = F(\omega_1, \omega_2)B(\omega_1, \omega_2). \tag{9.48}$$

From (9.48),

$$F(\omega_1, \omega_2) = \frac{G(\omega_1, \omega_2)}{B(\omega_1, \omega_2)}. \tag{9.49}$$

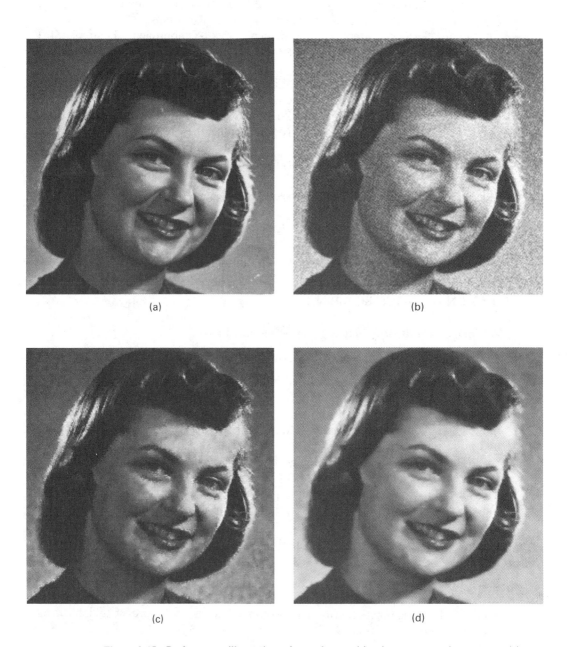

(a)

(b)

(c)

(d)

Figure 9.15 Performance illustration of an edge-sensitive image restoration system. (a) Original image of 256 × 256 pixels; (b) image degraded by white noise, with SNR of 6 dB and NMSE of 25.1%; (c) processed image by 2-D adaptive filtering, with NMSE of 5.2% and SNR improvement of 6.8 dB; (d) processed image by 1-D adaptive filtering, with NMSE of 4.7% and SNR improvement of 7.3 dB.

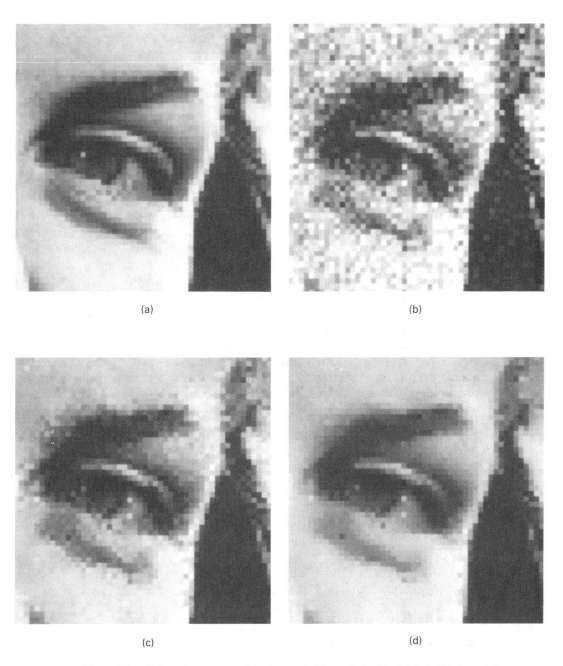

(a)

(b)

(c)

(d)

Figure 9.16 Enlarged segments of the images in Figure 9.15. (a) Original; (b) degraded; (c) processed by 2-D adaptive filtering; (d) processed by 1-D adaptive filtering.

From (9.49), a system that recovers $f(n_1, n_2)$ from $g(n_1, n_2)$ is an inverse filter, shown in Figure 9.17.

The inverse filter in Figure 9.17 tends to be very sensitive to noise. When $B(\omega_1, \omega_2)$ is very small, $1/B(\omega_1, \omega_2)$ is very large, and small noise in the frequency regions where $1/B(\omega_1, \omega_2)$ is very large may be greatly emphasized. One method of lessening the noise sensitivity problem is to limit the frequency response $1/B(\omega_1, \omega_2)$ to some threshold γ as follows:

$$
H(\omega_1, \omega_2) = \begin{cases} \dfrac{1}{B(\omega_1, \omega_2)}, & \text{if } \dfrac{1}{|B(\omega_1, \omega_2)|} < \gamma \\[2mm] \gamma \dfrac{|B(\omega_1, \omega_2)|}{B(\omega_1, \omega_2)}, & \text{otherwise.} \end{cases} \tag{9.50}
$$

The inverse filter $1/B(\omega_1, \omega_2)$ and its variation in (9.50) can be implemented in a variety of ways. We may design a filter whose frequency response approximates the desired one, using the filter design techniques discussed in Chapters 4 and 5, and then convolve the blurred image with the designed filter. Alternatively, we can implement the system, using DFTs and inverse DFTs in a manner analogous to the Wiener filter implementation discussed in Section 9.2.1.

Another method of implementing the inverse filter is to use an iterative procedure where an estimate of the signal $f(n_1, n_2)$ is updated in each iteration [Schafer et al.]. Let $\hat{f}_k(n_1, n_2)$ denote the signal estimate after k iterations. If $\hat{f}_k(n_1, n_2)$ is a good estimate of $f(n_1, n_2)$, $\hat{f}_k(n_1, n_2) * b(n_1, n_2)$ will be close to $g(n_1, n_2)$. The signal estimate after $k + 1$ iterations, $\hat{f}_{k+1}(n_1, n_2)$, is obtained by adding to $\hat{f}_k(n_1, n_2)$ a correction term which consists of a scaling constant λ multiplied to the difference between $g(n_1, n_2)$ and $\hat{f}_k(n_1, n_2) * b(n_1, n_2)$. Using $\lambda g(n_1, n_2)$ as the initial estimate $\hat{f}_0(n_1, n_2)$, the iterative procedure is

$$
\hat{f}_0(n_1, n_2) = \lambda g(n_1, n_2) \tag{9.51a}
$$

$$
\hat{f}_{k+1}(n_1, n_2) = \hat{f}_k(n_1, n_2) + \lambda(g(n_1, n_2) - \hat{f}_k(n_1, n_2) * b(n_1, n_2)) \tag{9.51b}
$$

where λ is a parameter which is positive and which can be used to control the convergence behavior of the iterative procedure.

To see how (9.51) is related to the inverse filter, we express (9.51) in the frequency domain:

$$
\hat{F}_0(\omega_1, \omega_2) = \lambda G(\omega_1, \omega_2), \tag{9.52a}
$$

$$
\hat{F}_{k+1}(\omega_1, \omega_2) = \hat{F}_k(\omega_1, \omega_2) + \lambda(G(\omega_1, \omega_2) - \hat{F}_k(\omega_1, \omega_2)B(\omega_1, \omega_2)). \tag{9.52b}
$$

Solving for $\hat{F}_k(\omega_1, \omega_2)$ recursively, using (9.52), we obtain

$$
\hat{F}_k(\omega_1, \omega_2) = \lambda G(\omega_1, \omega_2)[1 + (1 - \lambda B(\omega_1, \omega_2)) + \cdots + (1 - \lambda B(\omega_1, \omega_2))^k]
$$

$$
= \frac{G(\omega_1, \omega_2)}{B(\omega_1, \omega_2)}[1 - (1 - \lambda B(\omega_1, \omega_2))^{k+1}]. \tag{9.53}
$$

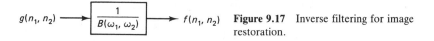

$g(n_1, n_2)$ → $\boxed{\dfrac{1}{B(\omega_1, \omega_2)}}$ → $f(n_1, n_2)$ **Figure 9.17**   Inverse filtering for image restoration.

From (9.53), as k approaches ∞, $\hat{F}_k(\omega_1, \omega_2)$ approaches $G(\omega_1, \omega_2)/B(\omega_1, \omega_2)$, which is the result of inverse filtering provided that

$$|1 - \lambda B(\omega_1, \omega_2)| < 1. \tag{9.54}$$

To the extent that λ which satisfies (9.54) can be found, (9.51) can be used in performing inverse filtering. One advantage of the iterative procedure is that it can be stopped after a finite number of iterations. The result after a finite number of iterations is not in general the same as that of inverse filtering, but it is less sensitive to the presence of noise in some cases.

Figure 9.18 illustrates the performance of inverse filtering. Figure 9.18(a) shows an original image of 512×512 pixels. Figure 9.18(b) shows the original image blurred by a Gaussian-shaped point spread function. The resulting image size is larger than 512×512 pixels, but is windowed by a 512×512-point rectangular window. The model of the degraded image in this case is

$$g(n_1, n_2) = [f(n_1, n_2) * b(n_1, n_2)]w(n_1, n_2). \tag{9.55}$$

Figure 9.18(c) shows the image processed by inverse filtering. The processed image $p(n_1, n_2)$ is computed by

$$p(n_1, n_2) = \text{IDFT}[G(k_1, k_2)H(k_1, k_2)] \tag{9.56}$$

where $G(k_1, k_2)$ is the DFT of $g(n_1, n_2)$ and $H(k_1, k_2)$ is obtained by

$$H(k_1, k_2) = \left.\frac{1}{B(\omega_1, \omega_2)}\right|_{\omega_1 = 2\pi k_1/N, \omega_2 = 2\pi k_2/N} \tag{9.57}$$

The DFT and inverse DFT sizes used are 512×512. In the absence of noise and very small $B(\omega_1, \omega_2)$, inverse filtering works quite well despite the fact that $g(n_1, n_2)$ in (9.55) is affected by windowing.

9.3.2 Algorithms for Blind Deconvolution

If the blurring function $b(n_1, n_2)$ is not accurately known, $b(n_1, n_2)$ must be estimated prior to inverse filtering. Since we are attempting to deconvolve $g(n_1, n_2)$ without detailed knowledge of $b(n_1, n_2)$, this is called the *blind deconvolution problem*.

If we know nothing about $f(n_1, n_2)$ or $b(n_1, n_2)$, it is not possible to solve the blind deconvolution problem. The problem is analogous to determining two numbers from their sum when nothing is known about either of the two numbers. To solve the blind deconvolution problem, therefore, something must be known about $f(n_1, n_2)$, $b(n_1, n_2)$, or both. Blind deconvolution algorithms differ depending on what is assumed known and how that knowledge is exploited.

(a)

(b)   (c)

Figure 9.18   (a) Original image of 512 × 512 pixels; (b) image blurred by a Gaussian-shaped point spread function; (c) result of inverse filtering.

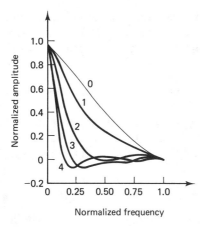

Figure 9.19 Modulation transfer function for a thin circular lens, as a function of the misfocus level. A higher number corresponds to a higher level of misfocus. After [Hopkins].

Suppose $f(n_1, n_2)$ and $b(n_1, n_2)$ are finite extent sequences with nonfactorable* z-transforms $F(z_1, z_2)$ and $B(z_1, z_2)$. Then we can recover $f(n_1, n_2)$ within translation and a scale factor from $g(n_1, n_2) = f(n_1, n_2) * b(n_1, n_2)$, using a polynomial factorization algorithm. Specifically, $G(z_1, z_2)$, the z-transform of $g(n_1, n_2)$, is given by $F(z_1, z_2)B(z_1, z_2)$. Since we assume that $f(n_1, n_2)$ and $b(n_1, n_2)$ are finite-extent sequences, $G(z_1, z_2)$ is a finite-order 2-D polynomial in z_1^{-1} and z_2^{-1}. In addition, we assume that $F(z_1, z_2)$ and $B(z_1, z_2)$ are nonfactorable, and therefore the only nontrivial factors of $G(z_1, z_2)$ are $F(z_1, z_2)$ and $B(z_1, z_2)$. Polynomial factorization algorithms that determine nontrivial factors of $G(z_1, z_2)$ are available [Izraelevitz and Lim] and may be used in determining $F(z_1, z_2)$ or $f(n_1, n_2)$ within translation and a scale factor. Unfortunately, this approach to solving the blind deconvolution problem has serious practical difficulties. Polynomial factorization algorithms developed to date are very expensive computationally. In addition, the algorithms are extremely sensitive to any deviation from the assumption that $G(z_1, z_2) = F(z_1, z_2)B(z_1, z_2)$ or $g(n_1, n_2) = f(n_1, n_2) * b(n_1, n_2)$. In practice, the convolutional model of $g(n_1, n_2) = f(n_1, n_2) * b(n_1, n_2)$ is not exact due to the presence of some background noise or due to approximations made in the modeling process.

One practical blind deconvolution algorithm is based on the assumption that $|B(\omega_1, \omega_2)|$ is a smooth function. This assumption is approximately valid in some applications. When an image is blurred by a thin circular lens, the modulation transfer function† $|H(\Omega_x, \Omega_y)|$ is a fairly smooth circularly symmetric lowpass filter shown in Figure 9.19. When a long-exposure image is blurred by atmospheric turbulence, the blurring function $b(x, y)$ and its Fourier transform $B(\Omega_x, \Omega_y)$ are approximately Gaussian-shaped [Goodman (1968)]. When an image is blurred by

---

*As we discussed in Chapter 2, a finite-extent 2-D sequence usually has a nonfactorable z-transform.

†The frequency response of an optical lens is called the *optical transfer function*. The amplitude function and phase responses of the optical transfer function are called the *modulation transfer function* and *phase transfer function*, respectively.

horizontal motion, $B(\Omega_x, \Omega_y)$ is a sinc function [Equation (9.5)] and $|B(\Omega_x, \Omega_y)|$ is smooth except in the regions where $B(\Omega_x, \Omega_y)$ crosses zero.

To estimate $|B(\omega_1, \omega_2)|$ under the assumption that $|B(\omega_1, \omega_2)|$ is smooth, we first note that

$$|G(\omega_1, \omega_2)| = |F(\omega_1, \omega_2)| \cdot |B(\omega_1, \omega_2)|. \tag{9.58}$$

Examples of $|F(\omega_1, \omega_2)|$, $|B(\omega_1, \omega_2)|$, and $|G(\omega_1, \omega_2)|$ are shown in Figures 9.20(a), (b), and (c). The function $|F(\omega_1, \omega_2)|$ can be viewed as a sum of two components, a smooth function, denoted by $|F(\omega_1, \omega_2)|_L$, and a rapidly varying function, denoted by $|F(\omega_1, \omega_2)|_H$:

$$|F(\omega_1, \omega_2)| = |F(\omega_1, \omega_2)|_L + |F(\omega_1, \omega_2)|_H. \tag{9.59}$$

The functions $|F(\omega_1, \omega_2)|_L$ and $|F(\omega_1, \omega_2)|_H$ for $|F(\omega_1, \omega_2)|$ in Figure 9.20(a) are shown in Figures 9.20(d) and (e). From (9.58) and (9.59),

$$|F(\omega_1, \omega_2)|_L \cdot |B(\omega_1, \omega_2)| + |F(\omega_1, \omega_2)|_H \cdot |B(\omega_1, \omega_2)| = |G(\omega_1, \omega_2)|. \tag{9.60}$$

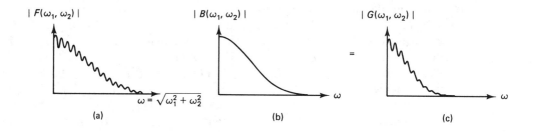

(a)   (b)   (c)

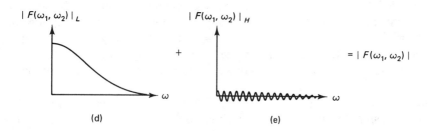

(d)   (e)

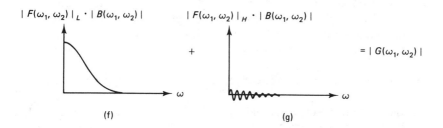

(f)   (g)

Figure 9.20 Development of a blind deconvolution method.

The functions $|F(\omega_1, \omega_2)|_L \cdot |B(\omega_1, \omega_2)|$ and $|F(\omega_1, \omega_2)|_H \cdot |B(\omega_1, \omega_2)|$ for this particular example are shown in Figures 9.20(f) and (g). Suppose we apply a smoothing operator S to (9.60). Assuming the smoothing operation S is linear, we have

$$S[|F(\omega_1, \omega_2)|_L \cdot |B(\omega_1, \omega_2)|] + S[|F(\omega_1, \omega_2)|_H \cdot |B(\omega_1, \omega_2)|] = S[|G(\omega_1, \omega_2)|]. \tag{9.61}$$

Since both $|F(\omega_1, \omega_2)|_L$ and $|B(\omega_1, \omega_2)|$ are smooth functions, smoothing will not affect $|F(\omega_1, \omega_2)|_L \cdot |B(\omega_1, \omega_2)|$ significantly. From Figure 9.20(g), however, smoothing will reduce $|F(\omega_1, \omega_2)|_H \cdot |B(\omega_1, \omega_2)|$ significantly. Based on this observation, (9.61) reduces to

$$|F(\omega_1, \omega_2)|_L \cdot |B(\omega_1, \omega_2)| \approx S[|G(\omega_1, \omega_2)|]. \tag{9.62}$$

From (9.62),

$$|B(\omega_1, \omega_2)| \approx \frac{S[|G(\omega_1, \omega_2)|]}{|F(\omega_1, \omega_2)|_L}. \tag{9.63}$$

Equation (9.63) is the basis for estimating $|B(\omega_1, \omega_2)|$. The numerator term $S[|G(\omega_1, \omega_2)|]$ can be determined from the blurred image $g(n_1, n_2)$. The denominator term $|F(\omega_1, \omega_2)|_L$ is estimated from the empirical observation that $|F(\omega_1, \omega_2)|_L$ is approximately the same for similar classes of images. Differences in the image details affect $|F(\omega_1, \omega_2)|_H$, but they do not appear to significantly affect $|F(\omega_1, \omega_2)|_L$. Based on this observation,

$$|F(\omega_1, \omega_2)|_L \approx |F'(\omega_1, \omega_2)|_L \tag{9.64}$$

where $|F'(\omega_1, \omega_2)|$ is obtained from an original undegraded image that is similar in content to $f(n_1, n_2)$. From (9.63) and (9.64),

$$|\hat{B}(\omega_1, \omega_2)| = \frac{S[|G(\omega_1, \omega_2)|]}{|F'(\omega_1, \omega_2)|_L}. \tag{9.65}$$

Despite various assumptions and heuristics used in the derivation of (9.65), a reasonable estimate of $|B(\omega_1, \omega_2)|$ can be obtained from (9.65). It is possible to derive an expression for the phase $\theta_b(\omega_1, \omega_2)$ in a manner analogous to the derivation of $|\hat{B}(\omega_1, \omega_2)|$ in (9.65). This approach has not yet been successful, however, partly because an image's details affect the phase function, and the phase of one image appears to contain little information about the phase of another image, even when the two images have similar contents. In the absence of a good method to estimate $\theta_b(\omega_1, \omega_2)$, $\theta_b(\omega_1, \omega_2)$ is assumed to be zero. Once $B(\omega_1, \omega_2)$ is estimated, inverse filtering or its variation, discussed in Section 9.3.1, can be used to reduce blurring.

Figure 9.21 illustrates the performance of a blind deconvolution method. Figure 9.21(a) shows a blurred image of the "Titanic" obtained with an underwater camera. Figure 9.21(b) shows the processed image. Since the blurring in this example is real (not introduced synthetically), the original unblurred image is not available. In processing the image, the blurring function $B(\omega_1, \omega_2)$ was estimated using a variation of the blind deconvolution method discussed above. The inverse filtering operation was implemented using the iterative procedure of (9.51). Com-

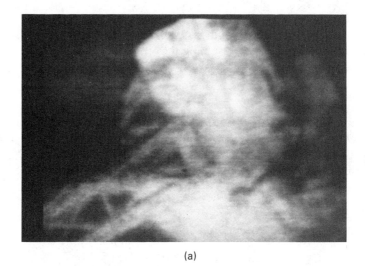

(a)

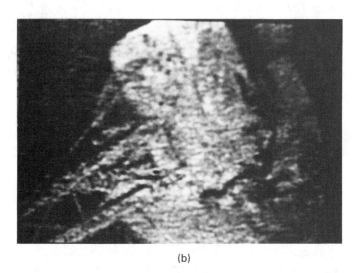

(b)

Figure 9.21 Performance illustration of a blind deconvolution method. Courtesy of Jules Jaffe. (a) Original blurred image of 512×340 pixels; (b) processed image.

parison of the unprocessed and processed images shows that blurring has been reduced significantly. However, the background noise present has been accentuated. Reduction of both blurring and background noise will be discussed in Section 9.4.

Another blind deconvolution method can be developed by assuming that the effective size (region of support) of $b(n_1, n_2)$ is much smaller than $f(n_1, n_2)$. In this method, the degraded image $g(n_1, n_2)$ is segmented into subimages $g_{ij}(n_1, n_2)$ by using nonoverlapping rectangular windows. The window size is chosen such that it is much larger than the effective size of $b(n_1, n_2)$ but much smaller than the size of $f(n_1, n_2)$. The subimage $g_{ij}(n_1, n_2)$ is then assumed to be

$$g_{ij}(n_1, n_2) \approx f_{ij}(n_1, n_2) * b(n_1, n_2) \tag{9.66}$$

where $f_{ij}(n_1, n_2)$ is the original image segmented by the same window used to obtain $g_{ij}(n_1, n_2)$. Equation (9.66) is not accurate near the boundaries of the subimage, but is a good representation away from the subimage boundaries. From (9.66),

$$|G_{ij}(\omega_1, \omega_2)| \approx |F_{ij}(\omega_1, \omega_2)| \cdot |B(\omega_1, \omega_2)|. \tag{9.67}$$

By summing both sides of (9.67) over all subimages and rewriting the expression, we obtain

$$|B(\omega_1, \omega_2)| \approx \frac{\sum_i \sum_j |G_{ij}(\omega_1, \omega_2)|}{\sum_i \sum_j |F_{ij}(\omega_1, \omega_2)|}. \tag{9.68}$$

Equation (9.68) is the basis for estimating $|B(\omega_1, \omega_2)|$. The numerator term $\Sigma_i \Sigma_j$ $|G_{ij}(\omega_1, \omega_2)|$ is obtained from $g(n_1, n_2)$. The term $\Sigma_i \Sigma_j |F_{ij}(\omega_1, \omega_2)|$ can be estimated from the empirical observation that

$$\sum_i \sum_j |F_{ij}(\omega_1, \omega_2)| \approx \sum_i \sum_i |F'_{ij}(\omega_1, \omega_2)| \tag{9.69}$$

where $F'_{ij}(\omega_1, \omega_2)$ is obtained from an undegraded image that is similar in content to $f(n_1, n_2)$. From (9.68) and (9.69), $|B(\omega_1, \omega_2)|$ is estimated by

$$|\hat{B}(\omega_1, \omega_2)| = \frac{\sum_i \sum_j |G_{ij}(\omega_1, \omega_2)|}{\sum_i \sum_j |F'_{ij}(\omega_1, \omega_2)|}. \tag{9.70}$$

The performance of the blind deconvolution method based on (9.70) is similar to the method based on (9.65).

9.4 REDUCTION OF BLURRING AND ADDITIVE RANDOM NOISE

In the previous two sections, we developed image restoration algorithms to reduce blurring and additive random noise separately. In practice, an image may be degraded by both blurring and additive noise:

$$g(n_1, n_2) = f(n_1, n_2) * b(n_1, n_2) + v(n_1, n_2). \tag{9.71}$$

This is shown in Figure 9.22.

From (9.71), one reasonable approach to restoring $f(n_1, n_2)$ is to apply a noise reduction system to estimate $r(n_1, n_2) = f(n_1, n_2) * b(n_1, n_2)$ from $g(n_1, n_2)$ and then to apply a deblurring system to estimate $f(n_1, n_2)$ from the estimated $r(n_1, n_2)$, as shown in Figure 9.23.

The approach of reducing one degradation at a time allows us to develop a restoration algorithm for each type of degradation and simply combine them when

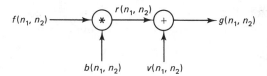

Figure 9.22 Model of image degradation by blurring and additive random noise.

more than one type of degradation is present in the degraded image. In addition, it is an optimal approach in some cases. For example, suppose $f(n_1, n_2)$ and $v(n_1, n_2)$ are samples of zero-mean stationary random processes that are linearly independent from each other. In addition, suppose $b(n_1, n_2)$ is known. Then the optimal linear estimator that minimizes $E[(f(n_1, n_2) - \hat{f}(n_1, n_2))^2]$ is an LSI system with a frequency response $H(\omega_1, \omega_2)$ given by

$$H(\omega_1, \omega_2) = \frac{P_f(\omega_1, \omega_2)B^*(\omega_1, \omega_2)}{P_f(\omega_1, \omega_2)|B(\omega_1, \omega_2)|^2 + P_v(\omega_1, \omega_2)}. \tag{9.72}$$

Equation (9.72) can be derived in a manner analogous to the derivation of the Wiener filter in Section 6.1.4. Equation (9.72) can be expressed as

$$\begin{aligned} H(\omega_1, \omega_2) &= \frac{P_f(\omega_1, \omega_2)|B(\omega_1, \omega_2)|^2}{P_f(\omega_1, \omega_2)|B(\omega_1, \omega_2)|^2 + P_v(\omega_1, \omega_2)} \frac{1}{B(\omega_1, \omega_2)} \\ &= \frac{P_r(\omega_1, \omega_2)}{P_r(\omega_1, \omega_2) + P_v(\omega_1, \omega_2)} \frac{1}{B(\omega_1, \omega_2)}. \end{aligned} \tag{9.73}$$

The expression $P_r(\omega_1, \omega_2)/(P_r(\omega_1, \omega_2) + P_v(\omega_1, \omega_2))$ is a noise reduction system by Wiener filtering. The system estimates $r(n_1, n_2) = f(n_1, n_2) * b(n_1, n_2)$ from $g(n_1, n_2)$. The expression $1/B(\omega_1, \omega_2)$ is the inverse filter, which estimates $f(n_1, n_2)$ from the estimated $\hat{r}(n_1, n_2)$. The overall system is thus a cascade of a noise reduction system and a deblurring system. This is shown in Figure 9.24.

Figure 9.25 illustrates the performance of the image restoration system when an image is degraded by blurring and additive noise. Figure 9.25(a) shows an original image of 512×512 pixels. Figure 9.25(b) shows the original image blurred by a Gaussian-shaped point spread function and then degraded by white Gaussian noise at an SNR of 25 dB. Figure 9.25(c) shows the image processed by the noise reduction system discussed in Section 9.2.4 followed by inverse filtering. The result of inverse filtering alone without noise reduction is shown in Figure 9.25(d). It is clear that inverse filtering is sensitive to the presence of noise, as was discussed in Section 9.3.1.

In processing an image, it is important to keep intermediate results with high accuracy to avoid a possible harmful effect of quantization. A small amount of

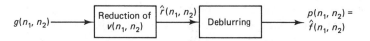

Figure 9.23 Reduction of blurring and additive random noise by cascade of a noise reduction system and a deblurring system.

$$g(n_1, n_2) \longrightarrow \boxed{\dfrac{P_r(\omega_1, \omega_2)}{P_r(\omega_1, \omega_2) + P_v(\omega_1, \omega_2)}} \xrightarrow{\hat{r}(n_1, n_2)} \boxed{\dfrac{1}{B(\omega_1, \omega_2)}} \longrightarrow \begin{array}{l} p(n_1, n_2) = \\ \hat{f}(n_1, n_2) \end{array}$$

Figure 9.24 Cascade of a Wiener filter for reduction of random noise and an inverse filter for reduction of blurring.

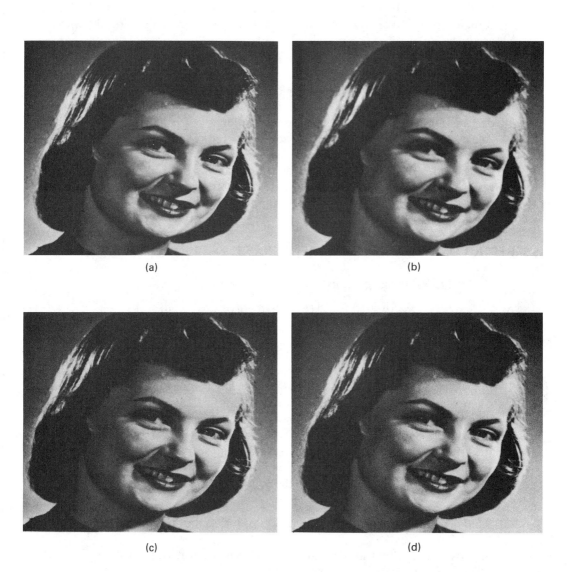

(a)

(b)

(c)

(d)

Figure 9.25 (a) Original image of 512 × 512 pixels; (b) image degraded by blurring and additive random noise with SNR of 25 dB; (c) image processed by an adaptive Wiener filter for random noise reduction and an inverse filter for reduction of blurring; (d) image processed by an inverse filter alone.

Sec. 9.4   Reduction of Blurring and Additive Random Noise   **561**

quantization noise may not be visible by itself but can be amplified by subsequent processing. For example, suppose we process an image degraded by blurring and additive random noise with a noise reduction system followed by a deblurring system. If the output of the noise reduction system is quantized to 8 bits/pixel, the quantization noise will not be visible at this stage. However, the subsequent deblurring system, which is typically a highpass filter, can amplify the quantization noise, and the noise may become clearly visible in the final result. Since the effect on the final result due to quantization in the intermediate results often is not straightforward to analyze, it is worthwhile to keep intermediate results with high accuracy.

In this section, we have discussed the problem of restoring an image degraded by two specific types of degradation. The idea of reducing one type of degradation at a time may be applicable to other types of degradation. Specifically, when an image is degraded by degradation 1, followed by degradation 2, and followed by degradation 3, one approach to consider is to reduce degradation 3 first, degradation 2 next, and finally degradation 1. Once the overall system consisting of subsystems is developed, it be made more computationally efficient by rearranging the subsystems. Such an approach, although not always optimal, often simplifies the restoration problem and is in some cases an optimal approach in that it leads to the same solution as the approach that treats all the degradations simultaneously.

9.5 REDUCTION OF SIGNAL-DEPENDENT NOISE

Any degraded image $g(n_1, n_2)$ can be expressed as

$$g(n_1, n_2) = D[f(n_1, n_2)] = f(n_1, n_2) + d(n_1, n_2) \qquad (9.74a)$$

where
$$d(n_1, n_2) = g(n_1, n_2) - f(n_1, n_2) \qquad (9.74b)$$

and $D[\cdot]$ is a degradation operator applied to $f(n_1, n_2)$. If $d(n_1, n_2)$ is not a function of the signal $f(n_1, n_2)$, $d(n_1, n_2)$ is called *additive signal-independent noise*. If $d(n_1, n_2)$ is a function of $f(n_1, n_2)$, $d(n_1, n_2)$ is called *additive signal-dependent noise*, often referred to as just *signal-dependent noise*. Examples of signal-dependent noise are speckle noise, film grain noise, and quantization noise. One approach to reducing signal-dependent noise is to transform $g(n_1, n_2)$ into a domain where the noise becomes additive signal-independent noise and then to reduce the signal-independent noise. Another approach is to reduce it directly in the signal domain. These approaches are discussed in the following two sections.

9.5.1 Transformation to Additive Signal-Independent Noise

Suppose we can find an operator T such that when it is applied to $g(n_1, n_2)$ in (9.74a), $T[g(n_1, n_2)]$ can be expressed as

$$\begin{aligned} T[g(n_1, n_2)] &= T[f(n_1, n_2) + d(n_1, n_2)] \\ &= T_1[f(n_1, n_2)] + v(n_1, n_2) \end{aligned} \qquad (9.75)$$

where $T_1[\cdot]$ is an operator that may be different from $T[\cdot]$ and $v(n_1, n_2)$ is an additive signal-independent noise. One approach to restoring $f(n_1, n_2)$ from $g(n_1, n_2)$ is first to estimate $T_1[f(n_1, n_2)]$ by reducing the additive signal-independent noise $v(n_1, n_2)$ and then estimate $f(n_1, n_2)$ from the estimated $T_1[f(n_1, n_2)]$. This approach exploits the fact that reducing additive signal-independent noise is generally much simpler than reducing signal-dependent noise, and a number of algorithms have already been developed to reduce additive signal-independent noise.

To illustrate this approach, let us consider the problem of reducing multiplicative noise. An example of multiplicative noise is the speckle effect [Dainty], which is commonly observed in images generated with highly coherent laser light, such as infrared radar images. The degraded image due to multiplicative noise, $g(n_1, n_2)$, can be expressed as

$$g(n_1, n_2) = f(n_1, n_2)v(n_1, n_2) \tag{9.76}$$

where $v(n_1, n_2)$ is random noise that is not a function of $f(n_1, n_2)$. Since $g(n_1, n_2)$ and $f(n_1, n_2)$ represent image intensities and are therefore nonnegative, $v(n_1, n_2)$ is also nonnegative. By applying the logarithmic operation to (9.76), we obtain

$$T[g(n_1, n_2)] = \log g(n_1, n_2) = \log f(n_1, n_2) + \log v(n_1, n_2). \tag{9.77}$$

If we denote $\log g(n_1, n_2)$ by $g'(n_1, n_2)$ and denote $\log f(n_1, n_2)$ and $\log v(n_1, n_2)$ similarly, (9.77) becomes

$$g'(n_1, n_2) = f'(n_1, n_2) + v'(n_1, n_2). \tag{9.78}$$

The multiplicative noise $v(n_1, n_2)$ has now been transformed to additive noise $v'(n_1, n_2)$ and image restoration algorithms developed for reducing additive signal-independent noise may be applied to reduce $v'(n_1, n_2)$. The resulting image is exponentiated to compensate for the logarithmic operation. The overall system is shown in Figure 9.26.

Figure 9.27 illustrates the performance of this image restoration algorithm in reducing multiplicative noise. The noise $v(n_1, n_2)$ is white noise generated by using the probability density function

$$p_{v(n_1, n_2)}(v_0) = ke^{-(v_0 - \sigma_1)^2/2\sigma_2^2}u(v_0) \tag{9.79}$$

where $u(v_0)$ is the unit step function, σ_1 and σ_2 are some constants, and k is a scale factor that ensures that the integral of the probability density function is 1. Figure 9.27(a) shows an original image of 512×512 pixels. Figure 9.27(b) shows the image degraded by multiplicative noise $v(n_1, n_2)$ obtained from (9.79) with $\sigma_1 = 1$, and $\sigma_2 = .1$. Figure 9.27(c) shows the processed image, using the system in Figure 9.26. The processed image has an SNR improvement of 5.4 dB. The restoration

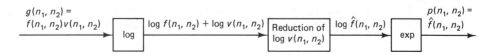

Figure 9.26 System for reduction of multiplicative noise.

(a)

(b)   (c)

Figure 9.27   Performance illustration of the multiplicative noise reduction system in Figure 9.26. (a) Original image of 512 × 512 pixels; (b) image degraded by multiplicative noise with NMSE of 5.1%; (c) processed image, with NMSE of 1.5% and SNR improvement of 5.4 dB.

system used for reducing additive noise is the adaptive Wiener filter, discussed in Section 9.2.4.

Image blurring, discussed in Section 9.3, is also signal-dependent noise. The inverse filter developed to reduce image blurring can also be viewed as transformation of image blurring to an additive signal-independent component, reduction of the additive component, and then transformation back to the signal domain. Specifically, from the model of degradation due to blurring,

$$G(\omega_1, \omega_2) = F(\omega_1, \omega_2)B(\omega_1, \omega_2). \qquad (9.80)$$

Applying a complex logarithmic operation to (9.80), we obtain

$$\log G(\omega_1, \omega_2) = \log F(\omega_1, \omega_2) + \log B(\omega_1, \omega_2). \qquad (9.81)$$

Subtracting the additive component $\log B(\omega_1, \omega_2)$ from $\log G(\omega_1, \omega_2)$ and exponentiating the result are equivalent to inverse filtering. Another example of the transformation of a signal-dependent noise to an additive signal-independent noise for its reduction is the decorrelation of quantization noise in image coding, which is discussed in Chapter 10.

9.5.2 Reduction of Signal-Dependent Noise in the Signal Domain

A major advantage of the approach discussed in the previous section is its simplicity. The approach, however, is based on the assumption that a domain can be found in which the signal-dependent noise becomes an additive signal-independent component. Such a domain may not exist for some types of signal-dependent noise. Even when such a domain can be found, the image restoration problem will be solved in the new domain. This may cause some degradation in performance. To see this, suppose an algorithm has been developed to reduce the signal-independent noise $v(n_1, n_2)$ in

$$g(n_1, n_2) = f(n_1, n_2) + v(n_1, n_2) \qquad (9.82)$$

by attempting to minimize

$$\text{Error} = E[(f(n_1, n_2) - \hat{f}(n_1, n_2))^2]. \qquad (9.83)$$

If the same algorithm is used to reduce the signal-independent noise $v(n_1, n_2)$ in

$$T[g(n_1, n_2)] = T_1[f(n_1, n_2)] + v(n_1, n_2) \qquad (9.84)$$

it will attempt to reduce

$$\text{Error} = E[(T_1[f(n_1, n_2)] - \hat{T}_1[f(n_1, n_2)])^2]. \qquad (9.85)$$

The error expressions in (9.83) and (9.85) are not the same.

An alternative approach is to reduce the signal-dependent noise by using the error criterion in (9.83) directly. As was briefly discussed in Section 6.1.5, the optimal solution that minimizes the Error in (9.83) is given by

$$\hat{f}(n_1, n_2) = E[f(n_1, n_2)|g(n_1, n_2)], \qquad (9.86)$$

which is the conditional expectation of $f(n_1, n_2)$ given $g(n_1, n_2)$. Even though (9.86) appears simple, its evaluation is typically very difficult. To simplify the solution, the estimator can be assumed to be linear:

$$\hat{f}(n_1, n_2) = \sum_{k_1} \sum_{k_2} g(k_1, k_2)h(n_1, n_2; k_1, k_2) + c(n_1, n_2) \qquad (9.87)$$

where $h(n_1, n_2; k_1, k_2)$ and $c(n_1, n_2)$ are chosen to minimize $E[(f(n_1, n_2) - \hat{f}(n_1, n_2))^2]$. The solution to the linear optimization problem can be written compactly by using matrix notation. As will be discussed in Section 9.7, the general solution is still very computationally expensive and requires such first- and second-order statistics as the cross covariance function of the signal $f(n_1, n_2)$ and the signal-dependent noise $d(n_1, n_2)$, which may be very difficult to estimate in practice. If we make some assumptions, however, the solution may be simplified considerably.

As an example, suppose we assume that the signal $f(n_1, n_2)$ and signal-dependent noise $d(n_1, n_2)$ are samples of stationary random processes. Then the solution to the linear minimum mean square error estimation of $f(n_1, n_2)$ is

$$\hat{f}(n_1, n_2) = m_f + (g(n_1, n_2) - m_g) * h(n_1, n_2), \qquad (9.88)$$

$$H(\omega_1, \omega_2) = F[h(n_1, n_2)] = \frac{P_{fg}(\omega_1, \omega_2)}{P_g(\omega_1, \omega_2)}, \qquad (9.89)$$

where $m_f = E[f(n_1, n_2)]$, $m_g = E[g(n_1, n_2)]$, $P_{fg}(\omega_1, \omega_2)$ is the cross power spectrum of $f(n_1, n_2)$ and the degraded signal $g(n_1, n_2)$ and $P_g(\omega_1, \omega_2)$ is the power spectrum of $g(n_1, n_2)$. This is the Wiener filter solution discussed in Section 6.1.4. When $d(n_1, n_2)$ is signal independent, (9.89) simplifies to the Wiener filter discussed in Section 9.2. Since the signal $f(n_1, n_2)$ and the signal-dependent noise $d(n_1, n_2)$ cannot be assumed stationary in an image restoration problem, the filter in (9.88) and (9.89) may be implemented locally in an adaptive manner to the extent that $P_{fg}(\omega_1, \omega_2)$ and $P_g(\omega_1, \omega_2)$ can be estimated locally.

In another example where the solution to the linear minimum mean square error signal estimation problem is greatly simplified, the signal $f(n_1, n_2)$ is modeled as

$$f(n_1, n_2) = m_f(n_1, n_2) + \sigma_f(n_1, n_2)w(n_1, n_2) \qquad (9.90)$$

where $m_f(n_1, n_2)$ is $E[f(n_1, n_2)]$, $\sigma_f^2(n_1, n_2)$ is the variance of $f(n_1, n_2)$ and $w(n_1, n_2)$ is zero-mean white noise with unit variance. This is the same model used in the image restoration algorithm developed in Section 9.2.4. For a certain class of signal-dependent noise which includes multiplicative noise and Poisson noise, the signal model (9.90) leads to a particularly simple algorithm. To illustrate this, we again consider the problem of reducing multiplicative noise.

Consider a degraded image $g(n_1, n_2)$ given by

$$g(n_1, n_2) = f(n_1, n_2)v(n_1, n_2) \qquad (9.91)$$

where $v(n_1, n_2)$ is a stationary white noise with mean of m_v and variance of σ_v^2. From (9.90) and (9.91),

$$g(n_1, n_2) = (m_f(n_1, n_2) + \sigma_f(n_1, n_2)w(n_1, n_2))v(n_1, n_2). \qquad (9.92)$$

Since both $w(n_1, n_2)$ and $v(n_1, n_2)$ are white, all the relevant information about $f(n_1, n_2)$ at a particular (n_1', n_2') is contained in $g(n_1', n_2')$. Therefore, we can consider the problem of estimating $f(n_1, n_2)$ at each pixel as estimating one random variable f from one observation g given by

$$g = fv = (m_f + \sigma_f w)v. \tag{9.93}$$

Assuming a linear estimator, we obtain

$$\hat{f} = ag + b \tag{9.94}$$

where a and b can be simply determined. From the elementary estimation theory, $E[(f - \hat{f})^2]$ is minimized by imposing the orthogonality principle, which imposes the following two conditions when applied to the above estimation problem:

$$E[f - \hat{f}] = 0 \tag{9.95a}$$

and

$$E[(f - \hat{f})g] = 0. \tag{9.95b}$$

From (9.94) and (9.95),

$$E[f - ag - b] = 0 \tag{9.96a}$$

and

$$E[(f - ag - b)g] = 0. \tag{9.96b}$$

Solving the two linear equations in (9.96) for a and b, we obtain

$$a = \frac{E[fg] - E[f]E[g]}{\sigma_g^2} \tag{9.97a}$$

and

$$b = E[f] - E[g]a \tag{9.97b}$$

where σ_g^2 is the variance of g. From (9.93), (9.94), and (9.97),

$$\hat{f} = m_f + (g - m_g)a \tag{9.98a}$$

where

$$a = \frac{m_v \sigma_f^2}{\sigma_g^2}. \tag{9.98b}$$

From (9.93) and (9.98) and after some algebra (see Problem 9.18),

$$\hat{f} = \frac{m_g}{m_v} + (g - m_g)\frac{\sigma_g^2 m_v^2 - m_g^2 \sigma_v^2}{m_v \sigma_g^2 (\sigma_v^2 + m_v^2)}. \tag{9.99}$$

Since (9.99) can be used for estimating $f(n_1, n_2)$ at every pixel,

$$\hat{f}(n_1, n_2) = \frac{m_g(n_1, n_2)}{m_v} + (g(n_1, n_2) - m_g(n_1, n_2))\frac{\sigma_g^2(n_1, n_2)m_v^2 - m_g^2(n_1, n_2)\sigma_v^2}{m_v \sigma_g^2(n_1, n_2)(\sigma_v^2 + m_v^2)}.$$

$$\tag{9.100}$$

The noise statistics σ_v^2 and m_v are assumed known. To the extent that the signal model is valid, and $m_g(n_1, n_2)$ and $\sigma_g^2(n_1, n_2)$ can be estimated from the local neighborhood at each pixel from $g(n_1, n_2)$, $f(n_1, n_2)$ may be estimated from (9.100). This approach can be used to develop algorithms for reducing other types of signal-dependent noise, including Poisson noise [Kuan et al. (1985)].

Figure 9.28 Image in Figure 9.27(b) processed by a multiplicative noise reduction system based on (9.100). The processed image has NMSE of 2.8% and SNR improvement of 2.5 dB.

Figure 9.28 shows the result of applying (9.100) to reduce multiplicative noise. The degraded image used is the one in Figure 9.27(b). The processed image shown in the figure has an SNR improvement of 2.5 dB.

9.6 TEMPORAL FILTERING FOR IMAGE RESTORATION

In the previous sections, we discussed spatial filtering algorithms for image restoration. In applications such as motion pictures, a sequence of images correlated in the temporal dimension is available, and this temporal correlation may be exploited through temporal filtering. One major advantage of temporal filtering over spatial filtering is its potential of reducing degradation without signal distortion. In this section, we discuss temporal filtering algorithms.

9.6.1 Frame Averaging

The simplest method of temporal filtering is frame averaging, which is very effective in processing a sequence of image frames in which the image does not change from frame to frame but the degradation does. There are many different types of frame averaging. The simplest and most common form of frame averaging is estimating an image $f(n_1, n_2)$ from a sequence of N degraded image frames $g_i(n_1, n_2)$ for $1 \leq i \leq N$ by

$$\hat{f}(n_1, n_2) = \frac{1}{N} \sum_{i=1}^{N} g_i(n_1, n_2). \tag{9.101}$$

Another example of frame averaging is

$$\hat{f}(n_1, n_2) = \left(\prod_{i=1}^{N} g_i(n_1, n_2) \right)^{1/N}. \tag{9.102}$$

The specific type of temporal averaging that is best in a given application depends on various factors, including the error criterion used and the assumptions made about the image degradation.

Suppose a sequence of degraded images $g_i(n_1, n_2)$ can be expressed as

$$g_i(n_1, n_2) = f(n_1, n_2) + v_i(n_1, n_2), \qquad 1 \le i \le N \qquad (9.103)$$

where $v_i(n_1, n_2)$ is zero-mean stationary white Gaussian noise with variance of σ_v^2 and $v_i(n_1, n_2)$ is independent from $v_j(n_1, n_2)$ for $i \ne j$. If we assume that $f(n_1, n_2)$ is nonrandom, the maximum likelihood (ML) estimate of $f(n_1, n_2)$ that maximizes $p_{g_i(n_1,n_2)|f(n_1,n_2)}(g_i'(n_1, n_2)|f'(n_1, n_2))$ is given by (see Problem 9.19)

$$\hat{f}(n_1, n_2) = \frac{1}{N} \sum_{i=1}^{N} g_i(n_1, n_2). \qquad (9.104)$$

From (9.103) and (9.104),

$$\hat{f}(n_1, n_2) = f(n_1, n_2) + \frac{1}{N} \sum_{i=1}^{N} v_i(n_1, n_2). \qquad (9.105)$$

From (9.105), the degradation in the frame-averaged image remains a zero-mean stationary white Gaussian noise with variance of σ_v^2/N, which represents a reduction of noise variance by a factor of N in comparison with $v_i(n_1, n_2)$. Since the noise characteristics in this example remain the same while the variance is reduced, the processed image will not suffer from any blurring despite the noise reduction. Frame averaging in (9.104) is often used in reducing noise that may occur in the digitization of a still image using an imaging device such as a vidicon camera.

As another example, suppose $g_i(n_1, n_2)$ can be expressed as

$$g_i(n_1, n_2) = f(n_1, n_2)v_i(n_1, n_2) \qquad (9.106)$$

where $v_i(n_1, n_2)$ for each i and (n_1, n_2) is an independent sample derived from the following probability density function:

$$p_v(v_0) = \lambda e^{-\lambda v_0} u(v_0). \qquad (9.107)$$

where $u(\cdot)$ is the unit step function. The speckle noise can be modeled in some cases by the multiplicative noise in (9.106) and (9.107) with $\lambda = 1$ [Dainty]. If we denote $g_i(n_1, n_2)$ at a particular i and (n_1, n_2) by g and denote the corresponding $f(n_1, n_2)$ and $v_i(n_1, n_2)$ by f and v, from (9.106) and (9.107)

$$p_{g|f}(g_0|f_0) = \frac{\lambda}{f_0} e^{-\lambda g_0/f_0} u(g_0). \qquad (9.108)$$

From (9.106) and (9.108), if we assume that $f(n_1, n_2)$ is nonrandom, the ML estimate of $f(n_1, n_2)$ is given by (see Problem 9.20)

$$\hat{f}(n_1, n_2) = \frac{\lambda}{N} \sum_{i=1}^{N} g_i(n_1, n_2). \qquad (9.109)$$

From (9.106) and (9.109),

$$\hat{f}(n_1, n_2) = f(n_1, n_2) \frac{\lambda}{N} \sum_{i=1}^{N} v_i(n_1, n_2). \qquad (9.110)$$

From (9.110), the type of degradation in the processed image remains as multiplicative noise in this example.

The ML estimation method used in the above two examples is typically much simpler than the minimum mean square error (MMSE) estimation method or the maximum a posteriori (MAP) estimation method. In the ML estimation method (see Section 6.1.5), we assume that no prior knowledge of the parameters (or signal) estimated is available, so a useful solution is generally obtained when the number of parameters estimated is less than the number of observations. In (9.104) and (9.110), for example, if $N = 1$, so that the number of parameters estimated is the same as the number of observations, $\hat{f}(n_1, n_2)$ is just the degraded signal or a scaled version of the degraded signal. This is one reason why we have relied on the MMSE criterion rather than the ML criterion in the development of spatial filtering algorithms in previous sections. The MMSE criterion combines some prior knowledge of the parameters to be estimated with the observations. To the extent that the prior knowledge available is accurate and significant, a useful estimation procedure can be developed even when the number of parameters estimated is comparable to the number of observations.

Figures 9.29 and 9.30 illustrate the performance of frame averaging. Figure 9.29(a) shows an original image of 512×512 pixels. Figure 9.29(b) shows an image degraded by zero-mean stationary white Gaussian noise at an SNR of 2 dB. Figure 9.29(c) shows the result of averaging eight degraded image frames using (9.104). Figure 9.30(a) shows the original image in Figure 9.29(a) degraded by the multiplicative noise of (9.106) and (9.107) with $\lambda = 1$. Figure 9.30(b) shows the result of averaging eight degraded image frames using (9.110).

In addition to temporal filtering, spatial filtering can be performed. For example, the processed image in Figure 9.30(b) can be modeled as an original image degraded by multiplicative noise whose characteristics can be derived from the noise characteristics in the single degraded frame. If we apply the method described in Section 9.5.2 to reduce the multiplicative noise in the image in Figure 9.30(b), we obtain the image shown in Figure 9.30(c). In this example, separable processing, where temporal and spatial filtering are performed separately, has been used. It is possible to design and implement a more general 3-D spatio-temporal filter, but this approach is typically much more computationally expensive than separable processing, without offering significant performance improvement.

Although frame averaging in some form is typically very simple and effective, precise signal registration is essential for its success. In such applications as motion pictures and television, the image may change from one frame to the next due to object movement, camera movement, and the like. To apply some form of frame averaging to such image frames, we need to estimate the movement of the image from one frame to the next. Motion-compensated image restoration is discussed in the next section.

9.6.2 Motion-Compensated Image Restoration

In motion-compensated image restoration, typically the motion parameters are estimated first; then the image frames are filtered along estimated motion trajec-

(a)

(b)   (c)

Figure 9.29 Performance illustration of frame averaging. (a) Original image of 512×512 pixels; (b) image degraded by white noise, with SNR of 2 dB and NMSE of 65.2%; (c) processed image by frame averaging (8 frames). The processed image has NMSE of 8.5% and SNR improvement of 8.8 dB.

(a)

(b)   (c)

Figure 9.30 Performance illustration of frame averaging. (a) Image in Figure 9.29(a) degraded by multiplicative noise, with NMSE of 28.8%; (b) processed image by frame averaging (8 frames), with NMSE of 4.7% and SNR improvement of 7.9 dB; (c) image processed by applying a spatial filter to the image in (b). The processed image has NMSE of 4.3% and SNR improvement of 8.3 dB.

tories. The motion estimation algorithms discussed in Section 8.4.2 can be used to estimate motion trajectories. Since the algorithms in Section 8.4.2 were developed under the assumption that noise-free image frames are available, it may be necessary to take degradation into account prior to motion estimation. Some motion estimation algorithms, such as the spatio-temporal constraint motion estimation algorithm with polynomial interpolation of signals discussed in Section 8.4.2, tend to be less sensitive to the presence of noise. For other algorithms, some simple form of degradation reduction prior to motion estimation can improve performance.

The type of filtering performed along the motion trajectories depends on the type of degradation. The motion-compensated image restoration considered in the literature has been primarily used for the reduction of additive random noise. For additive random noise reduction, the image intensities in the different frames along the motion trajectories are lowpass filtered. Both FIR and IIR filters have been considered. Since temporal filtering involves storage of frames, only low-order FIR and IIR filters are generally used. An example of a three-point FIR filter applied along the motion trajectories is illustrated in Figure 9.31. The middle frame $g(n_1, n_2, t_0)$ in the figure is the current frame being processed. The processed image $\hat{f}(n_1, n_2, t_0)$ is obtained from $g(n_1, n_2, t_0)$, the past frame $g(n_1, n_2, t_{-1})$, and the next frame $g(n_1, n_2, t_1)$. At each pixel in the current frame, two velocities (or displacements) are computed, one from $g(n_1, n_2, t_{-1})$ and $g(n_1, n_2, t_0)$ and the other from $g(n_1, n_2, t_0)$ and $g(n_1, n_2, t_1)$. The two displacements are used in identifying the spatial positions corresponding to times t_{-1} and t_1, as shown in Figure 9.31. The three image intensities at the corresponding three pixels are then averaged to obtain $\hat{f}(n_1, n_2, t_0)$ at the pixel. Since the corresponding spatial positions at times t_{-1} and t_1 do not generally fall on the sampling grid, spatial interpolation is required to determine $g(x, y, t_{-1})$ and $g(x, y, t_1)$ at the desired spatial locations.

Figure 9.32 illustrates the performance of a three-point temporal FIR filter applied along the motion trajectories. Figure 9.32(a) shows the current frame degraded by zero-mean white Gaussian noise. Figure 9.32(b) shows the result of motion-compensated frame averaging. A sequence of three degraded frames (past,

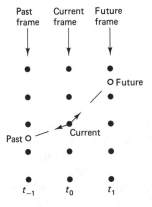

Figure 9.31 Motion-compensated image restoration.

(a)

(b)

(c)

Figure 9.32 Performance illustration of motion-compensated image restoration. (a) Degraded current frame; (b) processed image by motion-compensated frame averaging; (c) processed image by frame averaging without motion compensation. The blurring in this image shows the amount of motion present in the sequence of three image frames used.

current, and future) were used in the processing. The motion estimation algorithm used is the spatio-temporal constraint method with polynomial signal interpolation discussed in Section 8.4.2. After the motion parameters are estimated, the spatial interpolation required for temporal filtering is performed by using the truncated ideal interpolation filter. Figure 9.32(c) shows the result of frame averaging without motion compensation. The blurring in this image shows the amount of motion present in the sequence of three image frames used.

9.7 ADDITIONAL COMMENTS

In this chapter, we used monochrome images to illustrate the performance of various image restoration systems. Most of our discussions also apply to the restoration of color images. As we discussed in Chapter 7, a color image can be represented by three monochrome images. To restore a color image, we can process each of the three monochrome images separately and combine the results. Alternatively, we can restore a vector of three monochrome images jointly, exploiting our knowledge of the signal and degradation within each monochrome image and the correlation among them. The general principles and basic approaches we discussed in this chapter to develop various image restoration systems for a monochrome image also apply to a vector of monochrome images.

In this chapter, we developed image restoration algorithms, using notations typically employed in developing digital signal processing theories. Image restoration algorithms can also be developed by using matrix notation. Major advantages of the matrix notation include the compact form in which a restoration algorithm can be represented and the availability of extensive results of linear algebra and numerical analysis.

Consider an image $f(n_1, n_2)$ of $N \times N$ pixels. The N^2 elements in $f(n_1, n_2)$ can be represented by an $N^2 \times 1$-column vector $\mathbf{f}$. There are many ways to order the N^2 elements in forming $\mathbf{f}$. One method is to order the elements from bottom to top and left to right; that is,

$$\mathbf{f} = [f(0,0), f(0,1), \ldots, f(0, N-1), f(1,0), f(1,1), \ldots, f(1, N-1), \ldots,$$
$$f(N-1, 0), f(N-1, 1), \ldots, f(N-1, N-1)]^T.$$

$$(9.111)$$

Consider an image degradation model given by

$$g(n_1, n_2) = f(n_1, n_2) * b(n_1, n_2) + v(n_1, n_2). \tag{9.112}$$

Equation (9.112) can be expressed as

$$\mathbf{g} = B\mathbf{f} + \mathbf{v} \tag{9.113}$$

with a proper choice of $\mathbf{g}$, B, $\mathbf{f}$, and $\mathbf{v}$. For example, suppose that $f(n_1, n_2)$ and $b(n_1, n_2)$ are 2×2-point sequences, zero outside $0 \le n_1 \le 1$, $0 \le n_2 \le 1$, and that $g(n_1, n_2)$ and $v(n_1, n_2)$ are 3×3-point sequences, zero outside $0 \le n_1 \le 2$, $0 \le n_2 \le 2$. Then one set of $\mathbf{f}$, B, $\mathbf{v}$, and $\mathbf{g}$ is

$$\mathbf{f} = [f(0, 0), f(0, 1), f(1, 0), f(1, 1)]^T \tag{9.114a}$$

$$\mathbf{v} = [v(0, 0), v(0, 1), v(0, 2), v(1, 0), v(1, 1), v(1, 2), v(2, 0), v(2, 1), v(2, 2)]^T$$

$$(9.114b)$$

$$\mathbf{g} = [g(0, 0), g(0, 1), g(0, 2), g(1, 0), g(1, 1), g(1, 2), g(2, 0), g(2, 1), g(2, 2)]^T$$

$$\tag{9.114c}$$

$$B = \begin{bmatrix} b(0, 0) & 0 & 0 & 0 \\ b(0, 1) & b(0, 0) & 0 & 0 \\ 0 & b(0, 1) & 0 & 0 \\ b(1, 0) & 0 & b(0, 0) & 0 \\ b(1, 1) & b(1, 0) & b(0, 1) & b(0, 0) \\ 0 & b(1, 1) & 0 & b(0, 1) \\ 0 & 0 & b(1, 0) & 0 \\ 0 & 0 & b(1, 1) & b(1, 0) \\ 0 & 0 & 0 & b(1, 1) \end{bmatrix} \tag{9.114d}$$

The expression in (9.113) is quite general. For example, we can represent a space-variant blur by properly choosing the elements in the matrix B.

If we estimate $\mathbf{f}$ by a linear estimator that minimizes

$$\text{Error} = E[(\mathbf{f} - \hat{\mathbf{f}})^T (\mathbf{f} - \hat{\mathbf{f}})] \tag{9.115}$$

where $\hat{\mathbf{f}}$ is the estimate of $\mathbf{f}$, the solution is given by [Sage and Melsa]

$$\hat{\mathbf{f}} = E[\mathbf{f}] + \Phi_{fg}\Phi_g^{-1}[\mathbf{g} - E[\mathbf{g}]] \tag{9.116a}$$

where

$$\Phi_{fg} = E[(\mathbf{f} - E[\mathbf{f}])(\mathbf{g} - E[\mathbf{g}])^T] \tag{9.116b}$$

and

$$\Phi_g = E[(\mathbf{g} - E[\mathbf{g}])(\mathbf{g} - E[\mathbf{g}])^T]. \tag{9.116c}$$

The solution in (9.116) is written in a very compact form.

The advantages of matrix notation come at a price. In a typical image restoration problem, $\mathbf{f}$ may consist of a quarter-million elements, and matrices such as B, Φ_{fg}, or Φ_g may consist of a quarter-million × a quarter-million elements. Therefore, some approximations or simplifying assumptions have to be made before the solution can be used in practice. The process of making appropriate approximations or assumptions may sometimes become quite involved.

If the same error criterion is used and the same assumptions are made in solving a given image restoration problem, the results will of course be the same regardless of which notations are employed. However, they can provide different insights into the solution and can contribute to a deeper understanding of the image restoration problem. An extensive discussion of image restoration based on matrix representation can be found in [Andrews and Hunt].

REFERENCES

For a book exclusively on image restoration, see [Andrews and Hunt]. Most books on image processing [see references in Chapter 7] have at least one chapter devoted to image restoration. For a survey article, see [Andrews].

For models of image blur, see [Hopkins; Goodman (1968, 1985)] on blur due to optical systems, and [Hufnagle and Stanley] on blur due to atmospheric turbulence. For models of film grain noise, see [Mees; Falconer; Froelich et al.].

For models of speckle noise, see [Dainty; Goodman (1976, 1985)].

For readings on Wiener filtering for image restoration, see [Helstrom; Slepian; Horner; Hunt]. For variations of Wiener filtering, see [Cannon et al.]. The Wiener filter we discussed is a noncausal Wiener filter. For a "causal" Wiener filter in two dimensions, see [Ekstrom]. A Kalman filter adapts to changing signal and noise parameters and can be viewed as a generalization of a Wiener filter. For Kalman filtering in two dimensions, which was not discussed in this chapter, see [Woods and Radewan; Woods and Ingle].

For simple image models involving nonstationary mean and variance, see [Hunt and Cannon; Trussell and Hunt; Kuan et al. (1985)]. For adaptive restoration of images degraded by blurring and additive noise, see [Anderson and Netravali; Trussell and Hunt; Lee (1980); Lim; Frost et al.; Rajala and de Figueiredo; Abramatic and Silverman]. For edge-sensitive adaptive image restoration, see [Nagao and Matsuyama; Lee (1981); Chin and Yeh; Chan and Lim; Bovik et al.]. For image restoration based on a feasible solution that satisfies all known constraints on the solution, see [Youla and Webb; Sezan and Stark; Trussell and Civanlar; Civanlar and Trussell].

For inverse filtering, see [McGlamery; Sondhi; Robbins and Huang]. For implementation of an inverse filter using an iterative procedure, see [Schafer et al.]. For blind deconvolution methods, see [Stockham et al.; Cannon; Pohlig et al.].

For reduction of signal-dependent noise such as multiplicative noise, see [Walkup and Choens; Froehlich et al.; Lim and Nawab; Kuan et al. (1985, 1987)].

For temporal filtering for image restoration, see [Dennis; Dubois and Sabri]. For color image restoration, see [Mancill].

J. F. Abramatic and L. M. Silverman, Nonlinear restoration of noisy images, *IEEE Trans. Patt. Anal. Mach. Intell.*, Vol. PAMI-4, March 1982, pp. 141–149.

G. L. Anderson and A. N. Netravali, Image restoration based on a subjective criterion, *IEEE Trans. Syst.*, *Man, Cybern.*, Vol. SMC-6, December 1976, pp. 845–853.

H. C. Andrews, Digital image restoration: a survey, *Computer*, Vol. 7, May 1974, pp. 36–45.

H. C. Andrews and B. R. Hunt, *Digital Image Restoration.* Englewood Cliffs, NJ: Prentice-Hall, 1977.

A. C. Bovik, T. S. Huang, and D. C. Munson, Jr., Edge-sensitive image restoration using order-constrained least squares method, *IEEE Trans. Acoust., Speech Sig. Proc.*, Vol. ASSP-33, October 1985, pp. 1253–1263.

T. M. Cannon, Blind deconvolution of spatially invariant image blurs with phase, *IEEE Trans. Acoust., Speech Sig. Proc.*, Vol. ASSP-24, February 1976, pp. 58–63.

T. M. Cannon, H. J. Trussell, and B. R. Hunt, Comparison of different image restoration methods, *Appl. Opt.*, Vol. 17, November 1978, pp. 3384–3390.

P. Chan and J. S. Lim, One-dimensional processing for adaptive image restoration, *IEEE Trans. Acoust., Speech Sig. Proc.*, Vol. ASSP-33, February 1985, pp. 117–126.

R. T. Chin and C. L. Yeh, Quantitative evaluation of some edge-preserving noise-smoothing techniques, *Computer Vision*, *Graphics, and Image Processing*, Vol. 23, 1983, pp. 67–91.

M. R. Civanlar and H. J. Trussell, Digital signal restoration using fuzzy sets, *IEEE Trans. Acoust., Speech and Sig. Proc.*, Vol. ASSP-34, August 1986, pp. 919–936.

J. C. Dainty, *The Statistics of Speckle Patterns*, in *Progress in Optics*, ed. E. Wolf, Vol. XIV. Amsterdam: North-Holland, 1976.

T. J. Dennis, Nonlinear temporal filter for television picture noise reduction, *IEE Proc.*, Vol. 127, Pt. G, No. 2, April 1980, pp. 52–56.

E. Dubois and S. Sabri, Noise reduction in image sequences using motion-compensated temporal filtering, *IEEE Trans. on Commun.*, Vol. COM-32, July 1984, pp. 826–831.

M. P. Ekstrom, Realizable Wiener filtering in two dimensions, *IEEE Trans. Acoust., Speech Sig. Proc.*, Vol. ASSP-30, February 1982, pp. 31–40.

D. G. Falconer, Image enhancement and film-grain noise, *Optica Acta*, Vol. 17, September 1970, pp. 693–705.

B. R. Friden, Image enhancement and restoration, in *Picture Processing and Digital Filtering*, ed. T. S. Huang. New York: Springer-Verlag, 1975.

G. K. Froehlich, J. F. Walkup, and T. F. Krile, Estimation in signal-dependent film-grain noise, *Appl. Opt.*, Vol. 20, October 1981, pp. 3619–3626.

V. S. Frost, J. A. Stiles, K. S. Shanmugam, J. C. Holtzman, and S. A. Smith, An adaptive filter for smoothing noisy radar images, *Proc. IEEE*, Vol. 69, January 1981, pp. 133–135.

J. W. Goodman, *Introduction to Fourier Optics*. New York: McGraw-Hill, 1968.

J. W. Goodman, Some fundamental properties of speckle, *J. Opt. Soc. Amer.*, Vol. 66, November 1976, pp. 1145–1150.

J. W. Goodman, *Statistical Optics*. New York: Wiley, 1985.

C. W. Helstrom, Image restoration by the method of least squares, *J. Opt. Soc. Amer.*, Vol. 57, March 1967, pp. 297–303.

H. H. Hopkins, The frequency response of a defocused optical system, *Proc. Roy. Soc.*, Vol. 231, 1955, pp. 91–103.

J. L. Horner, Optical spatial filtering with the least mean-square-error filter, *J. Opt. Soc. Amer.*, Vol. 50, May 1969, pp. 553–558.

R. E. Hufnagle and N. R. Stanley, Modulation transfer function associated with image transmission through turbulent media, *J. Opt. Soc. Amer.*, Vol. 54, January 1964, pp. 52–61.

B. R. Hunt, The application of constrained least squares estimation to image restoration by digital computer, *IEEE Trans. Comput.*, Vol. C-22, September 1973, pp. 805–812.

B. R. Hunt and T. M. Cannon, Nonstationary assumptions for Gaussian models of images, *IEEE Trans. Syst., Man, Cybern.*, Vol. SMC-6, December 1976, pp. 876–882.

D. Izraelevitz and J. S. Lim, A new direct algorithm for image reconstruction from Fourier transform magnitude, *IEEE Trans. Acoust., Speech Sig. Proc.*, Vol. ASSP-35, April 1987, pp. 511–519.

D. T. Kuan, A. A. Sawchuk, T. C. Strand, and P. Chavel, Adaptive noise smoothing filter for images with signal-dependent noise, *IEEE Trans. Patt. Ana. Mach. Int.*, Vol. PAMI-7, March 1985, pp. 165–177.

D. T. Kuan, A. A. Sawchuk, T. C. Strand, and P. Chavel, Adaptive restoration of images with speckle, *IEEE Trans. Acoust., Speech Sig. Proc.*, Vol. ASSP-35, March 1987, pp. 373–383.

J. S. Lee, Digital image enhancement and noise filtering by use of local statistics, *IEEE Trans. Patt. Ana. Mach. Int.*, Vol. PAMI-2, March 1980, pp. 165–168.

J. S. Lee, Refined filtering of image noise using local statistics, *Comput. Graphics Image Proc.*, Vol. 15, 1981, pp. 380–389.

J. S. Lim, Image restoration by short space spectral subtraction, *IEEE Trans. Acoust., Speech Sig. Proc.*, Vol. ASSP-28, April 1980, pp. 191–197.

J. S. Lim and H. Nawab, Techniques for speckle noise removal, *Opt. Eng.*, Vol. 20, May/June 1981, pp. 472–480.

C. E. Mancill, Digital color image restoration, University of Southern California, Image Processing Institute, *USCIPI Report 630*, August 1975.

B. L. McGlamery, Restoration of turbulence-degraded images, *J. Opt. Soc. Amer.*, Vol. 57, March 1967, pp. 293–297.

C. E. K. Mees, ed., *The Theory of the Photographic Process.* New York: McMillan Company, 1954.

M. Nagao and T. Matsuyama, Edge preserving smoothing, *Computer Graphics Image Proc.*, Vol. 9, 1979, pp. 394–407.

S. C. Pohlig et al., New techniques for blind deconvolution, *Optical Eng.*, Vol. 20, March/April 1981, pp. 281–284.

S. A. Rajala and R. J. P. de Figueiredo, Adaptive nonlinear image restoration by a modified Kalman filtering approach, *IEEE Trans. Acoust., Speech Sig. Proc.*, Vol. ASSP-29, October 1981, pp. 1033–1042.

G. M. Robbins and T. S. Huang, Inverse filtering for linear shift-variant imaging systems, *Proc. IEEE*, Vol. 60, July 1972, pp. 862–872.

A. P. Sage and J. L. Melsa, *Application Theory with Applications to Communications and Control.* New York: McGraw-Hill, 1971.

R. W. Schafer, R. M. Mersereau, and M. A. Richards, Constrained iterative restoration algorithms, *Proc. IEEE*, Vol. 69, April 1981, pp. 432–450.

M. I. Sezan and H. Stark, Image restoration by the method of convex projections: part 2—applications and numerical results, *IEEE Trans. Med. Imaging*, Vol. MI-1, October 1982, pp. 95–101.

D. Slepian, Linear least-squares filtering of distorted images, *J. Opt. Soc. Amer.*, Vol. 57, July 1967, pp. 918–922.

M. M. Sondhi, Image restoration: The removal of spatially invariant degradations, *Proc. IEEE*, Vol. 60, July 1972, pp. 842–853.

T. G. Stockham Jr., T. M. Cannon, and R. B. Ingebretsen, Blind deconvolution through digital signal processing, *Proc. IEEE*, Vol. 63, April 1975, pp. 678–692.

H. J. Trussell and M. R. Civanlar, The feasible solution in signal restoration, *IEEE Trans. Acoust., Speech Sig. Proc.*, Vol. ASSP-32, April 1984, pp. 201–212.

H. J. Trussell and B. R. Hunt, Sectioned methods in image processing, *IEEE Trans. Acoust., Speech Sig. Proc.*, Vol. ASSP-26, April 1978, pp. 157–164.

J. F. Walkup and R. C. Choens, Image processing in signal-dependent noise, *Opt. Eng.*, Vol. 13, May/June 1974, pp. 258–266.

J. W. Woods and V. K. Ingle, Kalman filtering in two dimensions: Further results, *IEEE Trans. Acoust., Speech Sig. Proc.*, Vol. ASSP-29, April 1981, pp. 188–197.

J. W. Woods and C. H. Radewan, Kalman filtering in two dimensions, *IEEE Trans. Inform. Theory*, Vol. IT-23, July 1977, pp. 473–482.

D. C. Youla and H. Webb, Image restoration by the method of convex projections: part 1—theory, *IEEE Trans. Med. Imaging*, Vol. MI-1, October 1982, pp. 81–94.

PROBLEMS

9.1. Consider an image restoration problem in which we wish to develop a restoration system to compensate for blurring in an image display system. One model for this restoration problem is shown in the following figure.

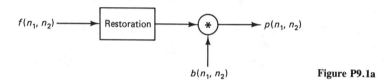

Figure P9.1a

In the figure, $f(n_1, n_2)$ is the original image, $b(n_1, n_2)$ is the blurring function, and $p(n_1, n_2)$ is the processed image. The restoration system in this problem is placed prior to the image degradation, since the display system displays the processed image. The restoration problem we discussed in this chapter is based on a model in which the degradation occurs before the restoration system is applied, as shown in the following figure.

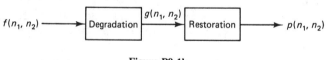

Figure P9.1b

Discuss how the model in Figure P9.1b can be used in solving the restoration problem in Figure 9.1a.

9.2. Let $g(n_1, n_2)$ denote the response of an LSI system when the input $f(n_1, n_2)$ is a line impulse $\delta_T(n_1)$. From $g(n_1, n_2)$, what can we say about the impulse response $h(n_1, n_2)$ and the frequency response $H(\omega_1, \omega_2)$ of the LSI system?

9.3. In developing image restoration systems, it is necessary to determine the characteristics of image degradation. In this problem, we consider the degradation due to speckle noise that occurs in infrared radar images. Let $g(n_1, n_2)$ denote the original noise-free image $f(n_1, n_2)$ degraded by speckle noise. Suppose the degraded image is sampled coarsely enough so that the degradation at any pixel can be assumed to be independent from all other pixels. According to a model derived theoretically and verified experimentally, $g(n_1, n_2)$ at each pixel (n_1, n_2) is an independent sample from the following density:

$$p_{g(n_1,n_2)}(g) = \frac{1}{f(n_1, n_2)} e^{-g/f(n_1,n_2)} u(g)$$

where

$$u(g) = \begin{cases} 1, & \text{for } g \geq 0 \\ 0, & \text{otherwise.} \end{cases}$$

From the above probability density function, speckle noise can be modeled as multiplicative noise $v(n_1, n_2)$ such that

$$g(n_1, n_2) = f(n_1, n_2)v(n_1, n_2).$$

Show that $p_v(v_0)$, the probability density function from which $v(n_1, n_2)$ is obtained, is given by

$$p_v(v_0) = e^{-v_0} u(v_0).$$

9.4. In Section 9.1, we derived a model for image restoration when an analog image $f(x, y)$ is blurred by a planar motion of an imaging system during the exposure time. Assume that the exposure time is for $0 \leq t \leq T$.

(a) Suppose the planar motion during the exposure time has a constant horizontal velocity v_x. Determine $b(x, y)$, the point spread function of the blurring. Does your answer make sense?

(b) Suppose the planar motion during the exposure time has a horizontal velocity $v_x(t)$ given by

$$v_x(t) = 2t, \quad 0 \leq t \leq T.$$

Determine the point spread function $b(x, y)$ and its Fourier transform $B(\Omega_x, \Omega_y)$.

9.5. At the peripheral level of the human visual system, the image intensity appears to be subjected to some form of nonlinearity, such as a logarithmic operation. This is evidenced in part by the approximate validity of Weber's law. Weber's law states that the just-noticeable difference in image density (logarithm of image intensity) is independent of the density. Since the image density domain is, then, more central to the human visual system than is the intensity domain, it appears that processing an image in the density domain may be more effective than processing an image in the intensity domain. Although the above reasoning sounds logical and has been used in the development of image restoration and enhancement systems, in many instances processing an image in the intensity domain rather than in the density domain makes more sense. Give one such example.

9.6. Let $g(n_1, n_2)$ denote a degraded image that can be expressed as

$$g(n_1, n_2) = f(n_1, n_2) + v(n_1, n_2)$$

where $f(n_1, n_2)$ is a signal and $v(n_1, n_2)$ is a background noise. In Wiener filtering, we assume that $P_f(\omega_1, \omega_2)$ is given. One method of estimating $P_f(\omega_1, \omega_2)$ is to model the correlation function $R_f(n_1, n_2)$ by

$$R_f(n_1, n_2) = \rho_1^{|n_1|} \rho_2^{|n_2|}$$

and then obtain $P_f(\omega_1, \omega_2)$ by

$$\hat{P}_f(\omega_1, \omega_2) = F[R_f(n_1, n_2)].$$

Assuming $v(n_1, n_2)$ is a zero-mean white noise with variance of σ_v^2 independent of $f(n_1, n_2)$, develop a method of estimating ρ_1 and ρ_2 from $g(n_1, n_2)$.

9.7. Many different measures may be used in quantifying the degree of closeness between an original image $f(n_1, n_2)$ and a processed image $p(n_1, n_2)$. In this problem we compare two measures. One is the normalized mean square error (NMSE), given by

$$\text{NMSE} = 100 \times \frac{\text{Var} \left[f(n_1, n_2) - p(n_1, n_2) \right]}{\text{Var} \left[f(n_1, n_2) \right]} \%$$

where Var [·] is the variance. Another measure is the scale-invariant normalized mean square error (SINMSE), given by

$$\text{SINMSE} = 100 \times \frac{\text{Var} \left[f(n_1, n_2) - ap(n_1, n_2) \right]}{\text{Var} \left[f(n_1, n_2) \right]} \%$$

where a is chosen to minimize the SINMSE.

(a) Show that the constant a is given by

$$a = \frac{E[f(n_1, n_2)p(n_1, n_2)] - E[f(n_1, n_2)]E[p(n_1, n_2)]}{\text{Var} \left[p(n_1, n_2) \right]}$$

where $E[\cdot]$ is the expectation.

(b) What is one major advantage of the SINMSE in comparison with the NMSE?

(c) The NMSE is an objective measure and does not always represent how close $p(n_1, n_2)$ is to $f(n_1, n_2)$ as judged by a human observer. Give an example where $p(n_1, n_2)$ with a larger NMSE will be judged by a human observer to be closer to $f(n_1, n_2)$ than $p(n_1, n_2)$ with a smaller NMSE.

9.8. One class of filters considered for reducing background noise in images has a frequency response $H(\omega_1, \omega_2)$ given by

$$H(\omega_1, \omega_2) = \left[\frac{P_f(\omega_1, \omega_2)}{P_f(\omega_1, \omega_2) + P_v(\omega_1, \omega_2)} \right]^\beta$$

where $P_f(\omega_1, \omega_2)$ is the signal power spectrum and $P_v(\omega_1, \omega_2)$ is the noise power spectrum. If $\beta = 1$, the filter is a Wiener filter. If $\beta = \frac{1}{2}$, the filter is a power spectrum filter. Suppose $P_f(\omega_1, \omega_2)$ has a lowpass character and its amplitude decreases as ω_1 and ω_2 increase, while $P_v(\omega_1, \omega_2)$ is approximately constant independent of ω_1 and ω_2.

(a) For a given $P_f(\omega_1, \omega_2)$ and $P_v(\omega_1, \omega_2)$, which filter reduces more noise, the Wiener filter or the power spectrum filter?

(b) For a given $P_f(\omega_1, \omega_2)$ and $P_v(\omega_1, \omega_2)$, which filter blurs the image more, the Wiener filter or the power spectrum filter?

9.9. In block-by-block processing, a window $w(n_1, n_2)$ is applied to $g(n_1, n_2)$, the image to be processed. To reduce the blocking effect, $w(n_1, n_2)$ may be chosen to satisfy two conditions. One condition is

$$\sum_i \sum_j w_{ij}(n_1, n_2) = 1 \quad \text{for all } (n_1, n_2) \text{ of interest.}$$

Another condition is that $w_{ij}(n_1, n_2)$ must be a smooth function that falls close to zero near the window boundaries. Show that both of these conditions are satisfied if we form a separable 2-D window from two 1-D windows that satisfy similar conditions.

9.10. Consider the adaptive image restoration system sketched in the following figure.

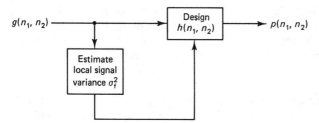

Figure P9.10

In this system, $h(n_1, n_2)$ is designed at each pixel based on the local signal variance σ_f^2 estimated from the degraded image $g(n_1, n_2)$. The filter $h(n_1, n_2)$ is assumed to have the form

$$h(n_1, n_2) = k_1 e^{-k_2(n_1^2 + n_2^2)} w(n_1, n_2)$$

where $w(n_1, n_2)$ is a 5×5-point rectangular window. The two parameters k_1 and k_2 are real and positive, and completely specify $h(n_1, n_2)$. We require that

$$\sum_{n1=-\infty}^{\infty} \sum_{n2=-\infty}^{\infty} h(n_1, n_2) = 1.$$

(a) Sketch $h(n_1, n_2)$ for a very small k_2 (close to 0).
(b) Sketch $h(n_1, n_2)$ for a very large k_2 (close to ∞).
(c) Sketch one reasonable choice of k_2 as a function of σ_f^2.
(d) Denote your choice in (c) as $k_2(\sigma_f^2)$. Determine k_1.
(e) The image restoration system discussed in this problem can exploit the observation that a particular noise is typically less visible to human viewers in image regions of high detail, such as edge regions, than in image regions of low detail, such as uniform background regions. The system, however, cannot exploit the observation that a particular noise is typically less visible to human viewers in bright areas than in dark areas. How would you modify the image restoration system so that this additional piece of information can be exploited?

9.11. Consider a degraded image $g(n_1, n_2)$ that can be expressed as

$$g(n_1, n_2) = f(n_1, n_2) + v(n_1, n_2)$$

where $f(n_1, n_2)$ is an undegraded original image and $v(n_1, n_2)$ is a signal-independent zero-mean white noise with variance σ_v^2. To reduce $v(n_1, n_2)$, we filter $g(n_1, n_2)$ with a linear shift-invariant filter $h(n_1, n_2)$. The processed image $p(n_1, n_2)$ can be expressed as

$$p(n_1, n_2) = f'(n_1, n_2) + v'(n_1, n_2)$$

where $f'(n_1, n_2)$ is the filtered signal and $v'(n_1, n_2)$ is the filtered noise.
(a) Show that Var $[v'(n_1, n_2)]$, the variance of $v'(n_1, n_2)$, is given by

$$\text{Var}\,[v'(n_1, n_2)] = \sigma_v^2 \sum_{n1=-\infty}^{\infty} \sum_{n2=-\infty}^{\infty} h^2(n_1, n_2).$$

(b) Determine $P_v'(\omega_1, \omega_2)$, the power spectrum of $v'(n_1, n_2)$. Is $v'(n_1, n_2)$ white noise?
9.12. In Section 9.2.5, we discussed a specific image restoration algorithm that exploits the noise visibility function to reduce additive background noise. In this problem, we consider some additional details associated with the algorithm. Let $V(M)$ denote the

noise visibility function, where M represents the masking level. The masking level $M(n_1, n_2)$ is a measure of the local spatial detail of an image, and we assume in this problem that $M(n_1, n_2)$ can be obtained from the degraded image. At each pixel, we design $h(n_1, n_2)$, which has a Gaussian shape given by

$$h(n_1, n_2) = ke^{-(n_1^2 + n_2^2)/2\sigma^2}w(n_1, n_2)$$

where k and σ are constants and $w(n_1, n_2)$ is a 5×5-point rectangular window. The design of the space-variant $h(n_1, n_2)$, then, involves the determination of k and σ at each pixel. Suppose we compute the two constants k and σ by imposing the following two constraints:

$$\sum_{n1=-\infty}^{\infty} \sum_{n2=-\infty}^{\infty} h(n_1, n_2) = 1$$

and

$$\sigma_v^2 \sum_{n1=-\infty}^{\infty} \sum_{n2=-\infty}^{\infty} |h(n_1, n_2)|^2 \, V(M) = a$$

where σ_v^2 is the variance of the additive noise present in the degraded image and a controls the relative amount of noise reduction and signal blurring. We assume that both σ_v^2 and a are known. Develop an efficient method to compute k and σ.

9.13. In Section 9.3.2, we discussed two specific methods of solving the blind deconvolution problem. There are many variations to the methods we developed. In this problem, we consider one of them. Let $g(n_1, n_2)$ denote the degraded image, which can be expressed as

$$g(n_1, n_2) = f(n_1, n_2) * b(n_1, n_2) \tag{1}$$

where $f(n_1, n_2)$ is an original image and $b(n_1, n_2)$ is the blurring point spread function. We assume that the effective size of $b(n_1, n_2)$ is much smaller than $f(n_1, n_2)$. Under this assumption,

$$g_{ij}(n_1, n_2) \approx f_{ij}(n_1, n_2) * b(n_1, n_2) \tag{2}$$

where $g_{ij}(n_1, n_2)$ is a subimage obtained by applying a nonoverlapping rectangular window to $g(n_1, n_2)$, and $f_{ij}(n_1, n_2)$ is similarly obtained from $f(n_1, n_2)$. From (2),

$$|G_{ij}(\omega_1, \omega_2)| \approx |F_{ij}(\omega_1, \omega_2)| \cdot |B(\omega_1, \omega_2)|. \tag{3}$$

One blind deconvolution method we developed in Section 9.3.2 is based on (3). From (3), we can also obtain

$$\log |G_{ij}(\omega_1, \omega_2)| \approx \log |F_{ij}(\omega_1, \omega_2)| + \log |B(\omega_1, \omega_2)|. \tag{4}$$

Develop a blind deconvolution method based on (4) in a manner analogous to the development of a blind deconvolution method based on (3).

9.14. In Section 9.3.2, we discussed blind deconvolution methods in which $|B(\omega_1, \omega_2)|$, the magnitude response of the unknown blurring, is estimated. In this problem, we consider the problem of estimating $\theta_b(\omega_1, \omega_2)$, the phase response of the unknown blurring. Let $g(n_1, n_2)$ denote the degraded image, which can be expressed as

$$g(n_1, n_2) = f(n_1, n_2) * b(n_1, n_2) \tag{1}$$

where $f(n_1, n_2)$ is an original image and $b(n_1, n_2)$ is the blurring point spread function. Assuming the effective size of $b(n_1, n_2)$ is much smaller than $f(n_1, n_2)$, we obtain

$$g_{ij}(n_1, n_2) \approx f_{ij}(n_1, n_2) * b(n_1, n_2) \tag{2}$$

where $g_{ij}(n_1, n_2)$ is a subimage obtained by applying a nonoverlapping rectangular window to $g(n_1, n_2)$ and $f_{ij}(n_1, n_2)$ is similarly obtained from $f(n_1, n_2)$.

(a) From (2), determine an approximate relationship among $\theta_{g_{ij}}(\omega_1, \omega_2)$, $\theta_{f_{ij}}(\omega_1, \omega_2)$ and $\theta_b(\omega_1, \omega_2)$. The functions $\theta_{g_{ij}}(\omega_1, \omega_2)$, $\theta_{f_{ij}}(\omega_1, \omega_2)$, and $\theta_b(\omega_1, \omega_2)$ are the phase functions of $g_{ij}(n_1, n_2)$, $f_{ij}(n_1, n_2)$ and $b(n_1, n_2)$, respectively. Clearly define the phase function you use. For example, can the phase function you use be the principal value of the phase function?

(b) Develop a method of estimating $\theta_b(\omega_1, \omega_2)$ in a manner analogous to estimating $|B(\omega_1, \omega_2)|$, discussed in Section 9.3.2. Clearly state the assumptions you make.

(c) Discuss if the assumptions you made in (b) are reasonable.

9.15. Consider a degraded image $g(n_1, n_2)$ which can be expressed as

$$g(n_1, n_2) = f(n_1, n_2) * b(n_1, n_2) + v(n_1, n_2)$$

where $f(n_1, n_2)$ is an original image, $b(n_1, n_2)$ is a known point spread function, and $v(n_1, n_2)$ is background noise. Assuming that $f(n_1, n_2)$ and $v(n_1, n_2)$ are samples of zero-mean stationary random processes that are linearly independent of each other, show that the optimal linear estimator that minimizes $E[(f(n_1, n_2) - \hat{f}(n_1, n_2))^2]$ is a linear shift-invariant system with its frequency response $H(\omega_1, \omega_2)$ given by

$$H(\omega_1, \omega_2) = \frac{P_f(\omega_1, \omega_2)B^*(\omega_1, \omega_2)}{P_f(\omega_1, \omega_2)|B(\omega_1, \omega_2)|^2 + P_v(\omega_1, \omega_2)}$$

where $P_f(\omega_1, \omega_2)$ and $P_v(\omega_1, \omega_2)$ denote the power spectrum of $f(n_1, n_2)$ and $v(n_1, n_2)$, respectively. You may wish to refer to the derivation of the noncausal Wiener filter in Section 6.1.4, where $b(n_1, n_2)$ is assumed to be $\delta(n_1, n_2)$.

9.16. Let $f(n_1, n_2)$ denote an original signal. Let $g(n_1, n_2)$ denote $f(n_1, n_2)$ degraded by two different additive random noises $v(n_1, n_2)$ and $w(n_1, n_2)$, as shown in the following figure.

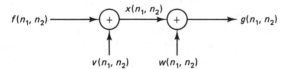

Figure P9.16a

Assume that $f(n_1, n_2)$, $v(n_1, n_2)$, and $w(n_1, n_2)$ represent zero-mean stationary random processes that are independent of each other and have known power spectra $P_f(\omega_1, \omega_2)$, $P_v(\omega_1, \omega_2)$, and $P_w(\omega_1, \omega_2)$, respectively. We wish to estimate $f(n_1, n_2)$ using a linear estimator that minimizes

$$E[(f(n_1, n_2) - \hat{f}(n_1, n_2))^2]$$

where $\hat{f}(n_1, n_2)$ is the estimate of $f(n_1, n_2)$.

(a) One approach is to model $g(n_1, n_2)$ as

$$g(n_1, n_2) = f(n_1, n_2) + r(n_1, n_2)$$

where

$$r(n_1, n_2) = v(n_1, n_2) + w(n_1, n_2),$$

and then design one linear shift-invariant filter $H(\omega_1, \omega_2)$ that estimates $f(n_1, n_2)$ from $g(n_1, n_2)$ directly. Determine $H(\omega_1, \omega_2)$.

(b) Another approach is to estimate $x(n_1, n_2) = f(n_1, n_2) + v(n_1, n_2)$ from $g(n_1, n_2)$ and then estimate $f(n_1, n_2)$ from the estimated $x(n_1, n_2)$, as shown in the following figure.

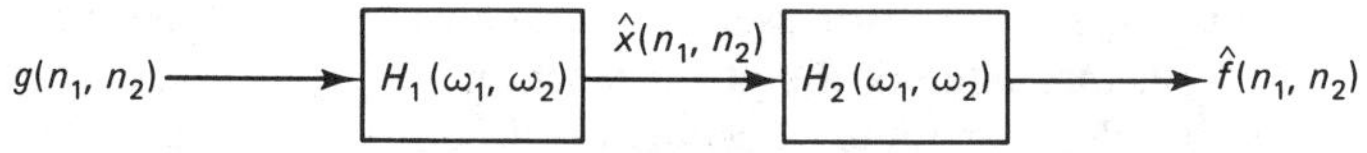

$$g(n_1, n_2) \longrightarrow \boxed{H_1(\omega_1, \omega_2)} \xrightarrow{\hat{x}(n_1, n_2)} \boxed{H_2(\omega_1, \omega_2)} \longrightarrow \hat{f}(n_1, n_2)$$

Figure P9.16b

We design $H_1(\omega_1, \omega_2)$ by minimizing $E[(x(n_1, n_2) - \hat{x}(n_1, n_2))^2]$. We then assume that $\hat{x}(n_1, n_2)$ is the true $x(n_1, n_2)$ and design $H_2(\omega_1, \omega_2)$ by minimizing $E[(f(n_1, n_2) - \hat{f}(n_1, n_2))^2]$. Determine $H_1(\omega_1, \omega_2)$ and $H_2(\omega_1, \omega_2)$.

(c) Compare the estimated signal $\hat{f}(n_1, n_2)$ from each of the two approaches discussed in (a) and (b).

9.17. Let $g(n_1, n_2)$ denote a degraded image that can be expressed as

$$g(n_1, n_2) = [f(n_1, n_2)]^{v(n_1, n_2)}$$

where $f(n_1, n_2)$ is an original image and $v(n_1, n_2)$ is random noise that is independent of $f(n_1, n_2)$. We assume that $f(n_1, n_2)$ and $v(n_1, n_2)$ satisfy

$$f(n_1, n_2) > 1 \quad \text{for all } (n_1, n_2)$$

and

$$v(n_1, n_2) > 0 \quad \text{for all } (n_1, n_2).$$

One approach to estimating $f(n_1, n_2)$ from $g(n_1, n_2)$ is to transform $g(n_1, n_2)$ into a domain where the degradation due to $v(n_1, n_2)$ becomes additive and apply an additive noise reduction system in the new domain. Develop a method based on this approach to estimate $f(n_1, n_2)$ from $g(n_1, n_2)$.

9.18. Consider $g = fv$, where f is a random variable with mean of m_f and variance of σ_f^2 and v is a random variable with mean of m_v and variance of σ_v^2. The two random variables f and v are statistically independent. We denote the mean and variance of g by m_g and σ_g^2. From Equations (9.94) and (9.97), one method of estimating f from the observation g is

$$\hat{f} = ag + b$$

where

$$a = \frac{E[fg] - E[f]E[g]}{\sigma_g^2}$$

and

$$b = E[f] - E[g]a.$$

(a) Show that $a = m_v \sigma_f^2 / \sigma_g^2$ and $b = m_f$. The result is Equation (9.98).

(b) Using the result in (a), show that

$$a = \frac{\sigma_g^2 m_v^2 - m_g^2 \sigma_v^2}{m_v \sigma_g^2 (\sigma_v^2 + m_v^2)}$$

and $b = \dfrac{m_g}{m_v}$.

The result is Equation (9.99).

9.19. Consider a sequence of degraded images $g_i(n_1, n_2)$ that can be expressed as

$$g_i(n_1, n_2) = f(n_1, n_2) + v_i(n_1, n_2), \qquad 1 \leq i \leq N$$

where $v_i(n_1, n_2)$ is zero-mean stationary white Gaussian noise with variance of σ_v^2 and $v_i(n_1, n_2)$ is independent from $v_j(n_1, n_2)$ for $i \neq j$. We assume that $f(n_1, n_2)$ is nonrandom and wish to estimate $f(n_1, n_2)$ using the maximum likelihood (ML) method.

(a) Explain why the ML estimate of $f(n_1, n_2)$ at a particular pixel (n_1, n_2) is affected by $g_i(n_1, n_2)$ at only that pixel.

(b) From (a), we can consider one pixel at a time, and therefore we can consider the problem of estimating one nonrandom variable f from N observations g_i given by

$$g_i = f + v_i, \qquad 1 \leq i \leq N$$

where v_i is N independent samples obtained from the zero-mean Gaussian density function with variance of σ_v^2. Determine $p_{g_1, g_2, \ldots, g_N | f}(g_1', g_2', \ldots, g_N' | f)$, the probability density function of g_i for $1 \leq i \leq N$ conditioned on f.

(c) Show that the ML estimate of f obtained by maximizing $p_{g_1, g_2, \ldots, g_N | f}(g_1', g_2', \ldots, g_N' | f)$ obtained in (b) is given by

$$\hat{f}_{\text{ML}} = \frac{1}{N} \sum_{i=1}^{N} g_i.$$

Since the above result applies to each pixel,

$$\hat{f}_{\text{ML}}(n_1, n_2) = \frac{1}{N} \sum_{i=1}^{N} g_i(n_1, n_2).$$

9.20. Consider a sequence of degraded images $g_i(n_1, n_2)$ that can be expressed as

$$g_i(n_1, n_2) = f(n_1, n_2) v_i(n_1, n_2), \qquad 1 \leq i \leq N$$

where $v_i(n_1, n_2)$ for each i and (n_1, n_2) is an independent sample derived from the following probability density function:

$$p_v(v_0) = \lambda e^{-\lambda v_0} u(v_0)$$

where $u(v_0)$ is the unit step function. We assume that $f(n_1, n_2)$ is nonrandom, and we wish to estimate $f(n_1, n_2)$ using the maximum likelihood (ML) method. Following the steps analogous to those in Problem 9.19, show that the ML estimate of $f(n_1, n_2)$ is given by

$$\hat{f}_{\text{ML}}(n_1, n_2) = \frac{\lambda}{N} \sum_{i=1}^{N} g_i(n_1, n_2).$$

9.21. Consider a sequence of degraded images $g_i(n_1, n_2)$ that can be expressed as

$$g_i(n_1, n_2) = f(n_1, n_2) + v_i(n_1, n_2)$$

where i is the time variable, $v_i(n_1, n_2)$ is zero-mean stationary white Gaussian noise with variance of σ_v^2, and $v_i(n_1, n_2)$ is independent from $v_j(n_1, n_2)$ for $i \neq j$. We assume that $g_i(n_1, n_2)$ has been observed for a long time. To reduce the noise $v_i(n_1, n_2)$, we consider performing frame averaging by

$$p_i(n_1, n_2) = \frac{1}{N} \sum_{l=i-N+1}^{i} g_l(n_1, n_2)$$

where $p_i(n_1, n_2)$ is the processed image at time index i.

(a) Determine the signal-to-noise ratio (SNR) improvement in dB achieved by the above frame averaging method. The SNR improvement in dB is defined as

$$10 \log \frac{\sigma_v^2}{\mathrm{Var}\,[p(n_1, n_2) - f(n_1, n_2)]}\ dB$$

where Var $[\cdot]$ is the variance.

(b) The processing method used above requires storage of N frames. One method of reducing the storage requirement is to obtain the processed image $p_i(n_1, n_2)$ by

$$p_i(n_1, n_2) = ap_{i-1}(n_1, n_2) + (1 - a)g_i(n_1, n_2)$$

where a is a real constant between 0 and 1. This method requires the storage of only one frame $p_{i-1}(n_1, n_2)$ in addition to the current observed frame $g_i(n_1, n_2)$. Determine the value of a required to achieve an SNR improvement equal to that in (a).

(c) The two methods considered in this problem are based on the assumption that $f(n_1, n_2)$ does not change from frame to frame. Suppose this assumption is valid only over some small finite number of frames. Which of the two methods will perform better in this environment?

10

Image Coding

10.0 INTRODUCTION

A major objective of image coding is to represent an image with as few bits as possible while preserving the level of quality and intelligibility required for the given application. Image coding has two major application areas. One is the reduction of channel bandwidth required for image transmission systems. Examples of this application include digital television, video conferencing, and facsimile. The other application is reduction of storage requirements. Examples of this application include reduction in the storage of image data from space programs and of video data in digital VCRs.

The levels of image quality and intelligibility required vary widely, depending on the application. In such applications as storage of image data from space programs and images of objects with historical value that no longer exist, regeneration of the original digital data may be very expensive or even impossible. In such applications, we may want to preserve all the information in the original digital data for possible future use. Image coding techniques which do not destroy any information and which allow exact reconstruction of the original digital data are said to be *information-preserving*. In applications such as digital television, it is not necessary for the coder to be information-preserving. In such applications, high quality is very important, but some information in the original data may be destroyed, so long as the decoded video on the TV m nitor is acceptable to human viewers. In applications such as remotely piloted vehicles (RPVs), image intelligibility is essential, but we may be able to sacrifice a significant degree of quality. The more quality and intelligibility we can sacrifice, the lower will be the required bit rate.

Image coding is related to image enhancement and restoration. If we can enhance the visual appearance of the reconstructed image or if we can reduce the degradation that results from the image coding algorithm (quantization noise being an example), we may be able to reduce the number of bits required to represent an image at a given level of quality and intelligibility, or conversely to hold the number of bits steady while improving the image quality and intelligibility.

A typical environment for image coding is shown in Figure 10.1. The digital image is encoded by an image coder, also referred to as a *source coder*. The output of the image coder is a string of bits that represents the source image. The channel coder transforms the string of bits to a form suitable for transmission over a communication channel through some form of modulation. The modulated signal is then transmitted over a communication channel. The communication channel typically introduces some noise, and provision for error correction is usually made in the channel coder to compensate for this channel noise. At the receiver, the received signal is demodulated and transformed back into a string of bits by a channel decoder. The image decoder reconstructs the image from the string of bits. For human viewing, the reconstructed image is typically displayed on a CRT monitor or printed out. In contrast to the communication environment in Figure 10.1, no communication channel is involved in applications of image coding for storage reduction. In storage reduction applications, the string of bits from the image coder is stored in proper format on a recording medium, ready for future retrieval. In this chapter, we are concerned primarily with the image coder and decoder.

The image coder in Figure 10.1 consists of three elements, shown in Figure 10.2. The first and most important element is the transformation of the image to the most suitable domain for quantization and codeword assignment. In essence, this element determines what specifically is coded. Image coding algorithms are classified into three broad categories, depending on what aspect of an image they code. In one class of algorithms, called *waveform coders*, the image intensity itself or some very simple variation of image intensity, such as the difference between two consecutive pixel intensities, is coded. In another class of algorithms, *transform coders*, an image is transformed to another domain such as the Fourier transform or cosine transform domain, which is significantly different from the intensity domain, and the transform coefficients are coded. In the third class of algorithms, *image model coders*, an image or some portion of an image is modeled and the model parameters are coded. The image is then synthesized from the coded model parameters. The performance and complexity of coding algorithms vary greatly, depending on what specifically is coded.

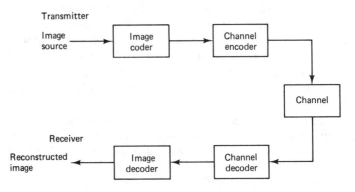

Figure 10.1 Typical environment for image coding.

Figure 10.2 Three major components in image coding.

The second element in the image coder is quantization. To represent an image with a finite number of bits, image intensities, transform coefficients, or model parameters must be quantized. Quantization involves assignment of the quantization (reconstruction) levels and decision boundaries. The third element in the image coder is assignment of codewords, the strings of bits that represent the quantization levels.

Each of these three elements attempts to exploit the redundancy present in the image source and the limitations of the display device and the human visual system. As a result, the elements are closely interrelated. For example, we will see that if the transformation element in the coder decorrelates the data sufficiently, the potential gain from vector quantization over scalar quantization is reduced. If the reconstruction levels in the quantizer are chosen such that each level is used with equal probability, the potential gain from variable-length codewords over fixed-length codewords is reduced. Despite this interrelation, we will study the elements separately for convenience. We will, however, point out the interrelationships at various points in our discussion.

In Section 10.1, we discuss quantization. Section 10.2 concerns codeword assignment. In the following three sections, we discuss the problem of what to code. Section 10.3 covers waveform coding. Section 10.4 covers transform coding and Section 10.5 covers image model coding. In Section 10.6, we discuss the problem of coding a sequence of image frames and color images. We also discuss the effects of channel errors.

Throughout this chapter, examples illustrating the performance of the various algorithms are given. These examples are intended for illustrative purposes only and are not intended for comparison of the performances among different algorithms. As we pointed out in Chapters 8 and 9, the performance of an image processing algorithm depends on many factors, such as the application for which the algorithm was developed and the class of images used. One or two examples cannot adequately demonstrate the performance of an algorithm.

10.1 QUANTIZATION

10.1.1 Scalar Quantization

Let f denote a continuous scalar quantity that may represent a pixel intensity, transform coefficient, or image model parameter. To represent f with a finite number of bits, only a finite number of reconstruction or quantization levels can be used. We will assume that a total of L levels are used to represent f. The process of assigning a specific f to one of L levels is called amplitude quantization, or quantization for short. If each scalar is quantized independently, the procedure is called *scalar quantization*. If two or more scalars are quantized jointly, the procedure is called *vector quantization* or *block quantization*. Vector quantization is discussed in Section 10.1.2.

Let $\hat{f}$ denote an f that has been quantized. We can express $\hat{f}$ as

$$\hat{f} = Q(f) = r_i, \qquad d_{i-1} < f \le d_i \tag{10.1}$$

where Q represents the quantization operation, r_i for $1 \le i \le L$ denotes L reconstruction levels, and d_i for $0 \le i \le L$ denotes $L + 1$ decision boundaries or decision levels. From (10.1), if f falls between d_{i-1} and d_i, it is mapped to the reconstruction level r_i. If we have determined reconstruction and decision levels, quantization of f is a deterministic process.

We can also express $\hat{f}$ in (10.1) as

$$\hat{f} = Q(f) = f + e_Q \tag{10.2}$$

where e_Q is the quantization error given by

$$e_Q = \hat{f} - f. \tag{10.3}$$

The quantization error e_Q is also called *quantization noise*. The quantity e_Q^2 can be viewed as a special case of a distortion measure $d(f, \hat{f})$, which is a measure of distance or dissimilarity between f and $\hat{f}$. Other examples of $d(f, \hat{f})$ include $|\hat{f} - f|$ and $\left| |\hat{f}|^p - |f|^p \right|$. The reconstruction and decision levels are often determined by minimizing some error criterion based on $d(f, \hat{f})$, such as the average distortion D given by

$$D = E[d(f, \hat{f})] = \int_{f_0 = -\infty}^{\infty} d(f_0, \hat{f}) p_f(f_0) \, df_0. \tag{10.4}$$

The most straightforward method of quantization is uniform quantization, in which the reconstruction and decision levels are uniformly spaced. Specifically, for a uniform quantizer,

$$d_i - d_{i-1} = \Delta, \qquad 1 \le i \le L \tag{10.5a}$$

$$r_i = \frac{d_i + d_{i-1}}{2}, \qquad 1 \le i \le L \tag{10.5b}$$

where Δ is the step size equal to the spacing between two consecutive reconstruction levels or two consecutive decision levels. An example of a uniform quantizer when $L = 4$ and f is assumed to be between 0 and 1 is shown in Figure 10.3.

The quantization noise e_Q is typically signal dependent. For example, the quantization noise e_Q for the uniform quantizer in Figure 10.3 is sketched in Figure 10.4. From Figure 10.4, e_Q is a function of f and therefore is signal dependent. It is possible to decorrelate the quantization noise e_Q for the uniform quantizer by a method known as *dithering* or *Roberts's pseudonoise technique*. As will be discussed in Section 10.3, decorrelation of quantization noise can be useful in improving the performance of an image coding system. It changes the characteristics of the degradation in the coded image. In addition, the decorrelated quantization noise may be reduced by the image restoration algorithms discussed in Chapter 9.

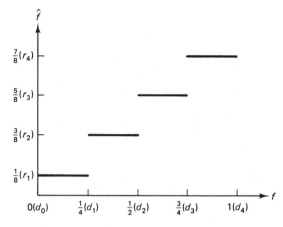

Figure 10.3 Example of uniform quantizer. The number of reconstruction levels is 4, f is assumed to be between 0 and 1, and $\hat{f}$ is the result of quantizing f. The reconstruction levels and decision boundaries are denoted by r_i and d_i, respectively.

Although uniform quantization is quite straightforward and appears to be a natural approach, it may not be optimal. Suppose f is much more likely to be in one particular region than in others. It is reasonable to assign more reconstruction levels to that region. Consider the example in Figure 10.3. If f rarely falls between d_0 and d_1, the reconstruction level r_1 is rarely used. Rearranging the reconstruction levels r_1, r_2, r_3, and r_4 so that they all lie between d_1 and d_4 makes more sense. Quantization in which reconstruction and decision levels do not have even spacing is called nonuniform quantization.

The optimum determination of r_i and d_i depends on the error criterion used. One frequently used criterion is the minimum mean square error (MMSE) criterion. Suppose we assume that f is a random variable with a probability density function $p_f(f_0)$. Using the MMSE criterion, we determine r_k and d_k by minimizing the average distortion D given by

$$D = E[d(\hat{f} - f)] = E[e_Q^2] = E[(\hat{f} - f)^2] \tag{10.6}$$

$$= \int_{f_0 = -\infty}^{\infty} p_f(f_0)(\hat{f} - f_0)^2 \, df_0.$$

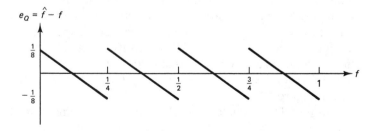

Figure 10.4 Illustration of signal dependence of quantization noise.

Noting that $\hat{f}$ is one of the L reconstruction levels obtained by (10.1), we can write (10.6) as

$$D = \sum_{i=1}^{L} \int_{f_0 = d_{i-1}}^{d_i} p_f(f_0)(r_i - f_0)^2 \, df_0. \tag{10.7}$$

To minimize D,

$$\frac{\partial D}{\partial r_k} = 0, \qquad 1 \le k \le L$$

$$\frac{\partial D}{\partial d_k} = 0, \qquad 1 \le k \le L - 1 \tag{10.8}$$

$$d_0 = -\infty$$

$$d_L = \infty.$$

From (10.7) and (10.8),

$$r_k = \frac{\displaystyle\int_{f_0 = d_{k-1}}^{d_k} f_0 p_f(f_0) \, df_0}{\displaystyle\int_{f_0 = d_{k-1}}^{d_k} p_f(f_0) \, df_0}, \qquad 1 \le k \le L \tag{10.9a}$$

$$d_k = \frac{r_k + r_{k+1}}{2}, \qquad 1 \le k \le L - 1 \tag{10.9b}$$

$$d_0 = -\infty \tag{10.9c}$$

$$d_L = \infty. \tag{10.9d}$$

The first set of equations in (10.9) states that a reconstruction level r_k is the centroid of $p_f(f_0)$ over the interval $d_{k-1} \le f_0 \le d_k$. The remaining set of equations states that the decision level d_k except d_0 and d_L is the middle point between two reconstruction levels r_k and r_{k+1}. Equation (10.9) is a necessary set of equations for the optimal solution. For a certain class of probability density functions, including uniform, Gaussian, and Laplacian densities, (10.9) is also sufficient.

Solving (10.9) is a nonlinear problem. The nonlinear problem has been solved for some specific probability density functions. The solutions when $p_f(f_0)$ is uniform, Gaussian, and Laplacian are tabulated in Table 10.1. A quantizer based on the MMSE criterion is often referred to as a Lloyd-Max quantizer [Lloyd; Max]. From Table 10.1, the uniform quantizer is the optimal MMSE quantizer when $p_f(f_0)$ is a uniform probability density function. For other densities, the optimal solution is a nonuniform quantizer. For example, the optimal reconstruction and decision levels for the Gaussian $p_f(f_0)$ with variance of 1 when $L = 4$ are shown in Figure 10.5.

It is useful to evaluate the performance improvement that the optimal MMSE quantizer gives over the simpler uniform quantizer. As an example, consider a Gaussian $p_f(f_0)$ with mean of 0 and variance of 1. The average distortion D in

TABLE 10.1 PLACEMENT OF RECONSTRUCTION AND DECISION LEVELS FOR LLOYD-MAX QUANTIZER. FOR UNIFORM PDF, $p_r(f_0)$ IS ASSUMED UNIFORM BETWEEN -1 AND 1. THE GAUSSIAN PDF IS ASSUMED TO HAVE MEAN OF 0 AND VARIANCE OF 1. FOR THE LAPLACIAN PDF,

$$p_f(f_0) = \frac{\sqrt{2}}{2\sigma} e^{-\frac{\sqrt{2}|f_0|}{\sigma}} \text{ with } \sigma = 1.$$

Bits	Uniform r_i	Uniform d_i	Gaussian r_i	Gaussian d_i	Laplacian r_i	Laplacian d_i
1	-0.5000	-1.0000	-0.7979	$-\infty$	-0.7071	$-\infty$
	0.5000	0.0000	0.7979	0.0000	0.7071	0.0000
		1.0000		∞		∞
2	-0.7500	-1.0000	-1.5104	$-\infty$	-1.8340	$-\infty$
	-0.2500	-0.5000	-0.4528	-0.9816	-0.4198	-1.1269
	0.2500	0.0000	0.4528	0.0000	0.4198	0.0000
	0.7500	0.5000	1.5104	0.9816	1.8340	1.1269
		1.0000		∞		∞
3	-0.8750	-1.0000	-2.1519	$-\infty$	-3.0867	$-\infty$
	-0.6250	-0.7500	-1.3439	-1.7479	-1.6725	-2.3796
	-0.3750	-0.5000	-0.7560	-1.0500	-0.8330	-1.2527
	-0.1250	-0.2500	-0.2451	-0.5005	-0.2334	-0.5332
	0.1250	0.0000	0.2451	0.0000	0.2334	0.0000
	0.3750	0.2500	0.7560	0.5005	0.8330	0.5332
	0.6250	0.5000	1.3439	1.0500	1.6725	1.2527
	0.8750	0.7500	2.1519	1.7479	3.0867	2.3769
		1.0000		∞		∞
4	-0.9375	-1.0000	-2.7326	$-\infty$	-4.4311	$-\infty$
	-0.8125	-0.8750	-2.0690	-2.4008	-3.0169	-3.7240
	-0.6875	-0.7500	-1.6180	-1.8435	-2.1773	-2.5971
	-0.5625	-0.6250	-1.2562	-1.4371	-1.5778	-1.8776
	-0.4375	-0.5000	-0.9423	-1.0993	-1.1110	-1.3444
	-0.3125	-0.3750	-0.6568	-0.7995	-0.7287	-0.9198
	-0.1875	-0.2500	-0.3880	-0.5224	-0.4048	-0.5667
	-0.0625	-0.1250	-0.1284	-0.2582	-0.1240	-0.2664
	0.0625	0.0000	0.1284	0.0000	0.1240	0.0000
	0.1875	0.1250	0.3880	0.2582	0.4048	0.2644
	0.3125	0.2500	0.6568	0.5224	0.7287	0.5667
	0.4375	0.3750	0.9423	0.7995	1.1110	0.9198
	0.5625	0.5000	1.2562	1.0993	1.5778	1.3444
	0.6875	0.6250	1.6180	1.4371	2.1773	1.8776
	0.8125	0.7500	2.0690	1.8435	3.0169	2.5971
	0.9375	0.8750	2.7326	2.4008	4.4311	3.7240
		1.0000		∞		∞

(10.6) as a function of the number of reconstruction levels L is shown in Figure 10.6 for the optimal MMSE quantizer (solid line). The average distortion D as a function of L is also shown in Figure 10.6 for the uniform quantizer* (dotted line), in which the reconstruction levels r_i are chosen to be symmetric with respect to

*The definition of uniform quantization by (10.5) has been extended in this case to account for a Gaussian random variable f whose value can range between $-\infty$ and ∞.

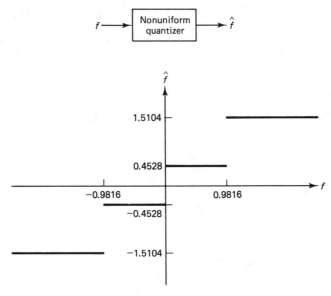

Figure 10.5 Example of a Lloyd-Max quantizer. The number of reconstruction levels is 4, and the probability density function for f is Gaussian with mean of 0 and variance of 1.

the origin, the minimum and maximum decision boundaries are assumed to be $-\infty$ and ∞, respectively, and the reconstruction step size Δ is chosen to minimize the average distortion D. From Figure 10.6, if codewords of uniform length are used to represent the reconstruction levels, the saving in the number of bits is in the range of $0 \sim \frac{1}{2}$ bit for L between 2 (1 bit) and 128 (7 bits). This example is based

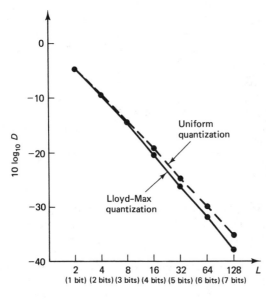

Figure 10.6 Comparison of average distortion $D = E[(\hat{f} - f)^2]$ as a function of L, the number of reconstruction levels, for a uniform quantizer (dotted line) and the Lloyd-Max quantizer (solid line). The vertical axis is $10 \log_{10} D$. The probability density function is assumed to be Gaussian with variance of 1.

Image Coding   Chap. 10

on the assumption that $p_f(f_0)$ is Gaussian. Similar analyses can, of course, be performed for other probability density functions. The more the density function deviates from being uniform, the higher will be the gain from nonuniform quantization over uniform quantization.

The notion that the uniform quantizer is the optimal MMSE quantizer when $p_f(f_0)$ is uniform suggests another approach. Specifically, we can map f to g by a nonlinearity in such a way that $p_g(g_0)$ is uniform, quantize g with a uniform quantizer, and then perform the inverse nonlinearity. This method is illustrated in Figure 10.7. The nonlinearity is called *companding*. From the elementary probability theory, one choice of the nonlinearity or companding $C[\cdot]$ that results in a uniform $p_g(g_0)$ is given by (see Problem 10.4)

$$g = C[f] = \int_{x=-\infty}^{f} p_f(x)\,dx - \tfrac{1}{2}. \qquad (10.10)$$

The resulting $p_g(g_0)$ is uniform in the interval $-\tfrac{1}{2} \le g \le \tfrac{1}{2}$. Although (10.10) is considerably easier to solve than the nonlinear equations in (10.9), the system in Figure 10.7 minimizes D' given by

$$D' = E[(\hat{g} - g)^2]. \qquad (10.11)$$

The distortion D' in (10.11) is not the same as D in (10.6).

In this section, we have considered the problem of quantizing a scalar f. In image coding, we have to quantize many scalars. One approach is to quantize each scalar independently. This approach is called scalar quantization of a vector source. Suppose we have N scalars f_i for $1 \le i \le N$ and each scalar f_i is quantized to L_i reconstruction levels. If L_i can be expressed as a power of 2 and each reconstruction level is coded with an equal number of bits (i.e., with a uniform-length codeword, which will be discussed in Section 10.2), L_i will be related to the required number of bits B_i by

$$L_i = 2^{B_i} \qquad (10.12a)$$

or
$$B_i = \log_2 L_i. \qquad (10.12b)$$

The total number of bits B required in coding N scalars is

$$B = \sum_{i=1}^{N} B_i. \qquad (10.13)$$

From (10.12) and (10.13), the total number of reconstruction levels L is

$$L = \prod_{i=1}^{N} L_i = 2^B. \qquad (10.14)$$

Figure 10.7 Nonuniform quantization by companding.

It is important to note from (10.13) and (10.14) that the total number of bits B is the sum of B_i, while the total number of reconstruction levels L is the product of L_i.

If we have a fixed number of bits B to code all N scalars using scalar quantization of a vector source, B has to be divided among N scalars. The optimal strategy for this bit allocation depends on the error criterion used and the probability density functions of the scalars. The optimal strategy typically involves assigning more bits to scalars with larger variance and fewer bits to scalars with smaller variance. As an example, suppose we minimize the mean square error $\sum_{i=1}^{N} E[(\hat{f}_i - f_i)^2]$ with respect to B_i for $1 \leq i \leq N$, where $\hat{f}_i$ is the result of quantizing f_i. If the probability density functions are the same for all the scalars except for their variance and we use the same quantization method such as the Lloyd-Max quantizer for each of the scalars, an approximate solution to the bit allocation problem is given by [Huang and Schultheiss; Segall]:

$$B_i = \frac{B}{N} + \frac{1}{2} \log_2 \frac{\sigma_i^2}{\left[\prod_{j=1}^{N} \sigma_j^2 \right]^{1/N}}, \quad 1 \leq i \leq N \quad (10.15)$$

where σ_i^2 is the variance of the ith scalar f_i. From (10.15),

$$L_i = 2^{B_i} = 2^{B/N} \frac{\sigma_i}{\left(\prod_{j=1}^{N} \sigma_i \right)^{1/N}}, \quad 1 \leq i \leq N. \quad (10.16)$$

From (10.16), the number of reconstruction levels for f_i is proportional to σ_i, the standard deviation of f_i. Although (10.15) is an approximate solution obtained under some specific assumptions, it is useful as a reference in other bit allocation problems. We note that B_i in (10.15) can be negative and is not in general an integer. In scalar quantization, B_i is a nonnegative integer. This constraint has to be imposed in solving a bit allocation problem in practice.

10.1.2 Vector Quantization

In the previous section, we discussed scalar quantization of a scalar source and a vector source. An alternate approach to coding a vector source is to divide scalars into blocks, view each block as a unit, and then jointly quantize the scalars in the unit. This is called vector quantization (VQ) or block quantization.

Let $\mathbf{f} = [f_1, f_2, \ldots, f_N]^T$ denote an N-dimensional vector that consists of N real-valued, continuous-amplitude scalars f_i. In vector quantization, $\mathbf{f}$ is mapped to another N-dimensional vector $\mathbf{r} = [r_1, r_2, \ldots, r_N]^T$. Unlike $\mathbf{f}$, whose elements have continuous amplitudes, the vector $\mathbf{r}$ is chosen from L possible reconstruction or quantization levels. Let $\hat{\mathbf{f}}$ denote $\mathbf{f}$ that has been quantized. We can express $\hat{\mathbf{f}}$ as

$$\hat{\mathbf{f}} = \text{VQ}(\mathbf{f}) = \mathbf{r}_i, \quad \mathbf{f} \in C_i \quad (10.17)$$

where VQ represents the vector quantization operation, $\mathbf{r}_i$ for $1 \leq i \leq L$ denotes L reconstruction levels, and C_i is called the ith cell. If $\mathbf{f}$ is in the cell C_i, $\mathbf{f}$ is mapped to $\mathbf{r}_i$. An example of vector quantization when $N = 2$ and $L = 9$ is shown in Figure 10.8. The filled-in dots in the figure are reconstruction levels, and the solid lines are cell boundaries. Note that a cell can have an arbitrary size and shape in vector quantization. This is in sharp contrast with scalar quantization, where a cell (the region between two consecutive decision levels) can have an arbitrary size but its shape is fixed. Vector quantization exploits this added flexibility.

As in the scalar case, we can define a distortion measure $d(\mathbf{f}, \hat{\mathbf{f}})$, which is a measure of dissimilarity between $\mathbf{f}$ and $\hat{\mathbf{f}}$. An example of $d(\mathbf{f}, \hat{\mathbf{f}})$ is $\mathbf{e}_Q^T \mathbf{e}_Q$, where the quantization noise $\mathbf{e}_Q$ is defined by

$$\mathbf{e}_Q = \hat{\mathbf{f}} - \mathbf{f} = VQ(\mathbf{f}) - \mathbf{f}. \tag{10.18}$$

The reconstruction levels $\mathbf{r}_i$ and the boundaries of cells C_i are determined by minimizing some error criterion, such as the average distortion measure D given by

$$D = E[d(\mathbf{f}, \hat{\mathbf{f}})]. \tag{10.19}$$

If $d(\mathbf{f}, \hat{\mathbf{f}})$ is $\mathbf{e}_Q^T \mathbf{e}_Q$, from (10.18) and (10.19),

$$\begin{aligned} D = E[\mathbf{e}_Q^T \mathbf{e}_Q] &= E[(\hat{\mathbf{f}} - \mathbf{f})^T(\hat{\mathbf{f}} - \mathbf{f})] \\ &= \int \int_{-\infty}^{\infty} \cdots \int (\hat{\mathbf{f}} - \mathbf{f}_0)^T(\hat{\mathbf{f}} - \mathbf{f}_0) p_{\mathbf{f}}(\mathbf{f}_0) \, d\mathbf{f}_0 \tag{10.20} \\ &= \sum_{i=1}^{L} \int \int \int_{\mathbf{f}_0 \in C_i} \cdots \int (\mathbf{r}_i - \mathbf{f}_0)^T(\mathbf{r}_i - \mathbf{f}_0) p_{\mathbf{f}}(\mathbf{f}_0) \, d\mathbf{f}_0. \end{aligned}$$

The average distortion in (10.20) is the mean square error (MSE) and is a generalization of (10.7).

A major advantage of vector quantization is its performance improvement over scalar quantization of a vector source. Vector quantization can lower the average distortion D with the number of reconstruction levels held constant or can reduce the required number of reconstruction levels when D is held constant. There are various ways that vector quantization can improve its performance over scalar quantization. The most significant way is to exploit the statistical dependence among the scalars in the block.

To illustrate that vector quantization can exploit statistical dependence, let us consider two examples. In the first example, we consider the exploitation of linear dependence (correlation). Consider two random variables f_1 and f_2 with a joint probability density function (pdf) $p_{f_1, f_2}(f_1', f_2')$ shown in Figure 10.9(a). The pdf is uniform with amplitude of $1/2a^2$ in the shaded region and zero in the unshaded region. The two marginal pdfs $p_{f_1}(f_1')$ and $p_{f_2}(f_2')$ are also shown in the figure. Since $E[f_1 f_2] \neq E[f_1]E[f_2], f_1$ and f_2 are correlated or linearly dependent. Suppose we quantize f_1 and f_2 separately, using scalar quantization and the MMSE criterion.

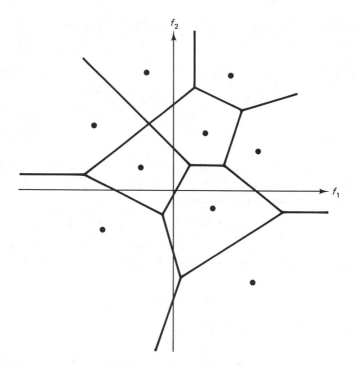

Figure 10.8 Example of vector quantization. The number of scalars in the vector is 2, and the number of reconstruction levels is 9.

Since each of the two scalars has a uniform pdf, the optimal scalar quantizer is a uniform quantizer. If we allow two reconstruction levels for each scalar, the optimal reconstruction levels for each scalar are $a/2$ and $-a/2$. The resulting four (2×2) reconstruction levels in this case are the four filled-in dots shown in Figure 10.9(b). Clearly, two of the four reconstruction levels are wasted. With vector quantization, we can use only the reconstruction levels shown in Figure 10.9(c). This example shows that vector quantization can reduce the number of reconstruction levels without sacrificing the MSE. We can eliminate the linear dependence between f_1 and f_2 in this example by rotating the pdf clockwise 45°. The result of this invertible linear coordinate transformation is shown in Figure 10.10. In the new coordinate system, g_1 and g_2 are uncorrelated, since $E[g_1 g_2] = E[g_1]E[g_2]$. In this new coordinate system, it is possible to place the two reconstruction levels at the filled-in dots shown in the figure by scalar quantization of the two scalars, and the advantage of vector quantization disappears. Eliminating the linear dependence reduces the advantage of vector quantization in this example. This is consistent with the notion that vector quantization can exploit linear dependence of the scalars in the vector.

Vector quantization can also exploit nonlinear dependence. This is illustrated in the second example. Consider two random variables f_1 and f_2 whose joint pdf

$p_{f_1,f_2}(f_1', f_2')$ is shown in Figure 10.11(a). The pdf is again uniform with amplitude of $1/(8a^2)$ in the shaded region and 0 outside the shaded region. The marginal density functions $p_{f_1}(f_1')$ and $p_{f_2}(f_2')$ are also shown in Figure 10.11(a). From the joint pdf, $E[f_1 f_2] = E[f_1]E[f_2]$ and therefore f_1 and f_2 are linearly independent. However, $p_{f_1,f_2}(f_1', f_2') \neq p_{f_1}(f_1')p_{f_2}(f_2')$ and therefore f_1 and f_2 are statistically dependent. When the random variables are linearly independent but statistically

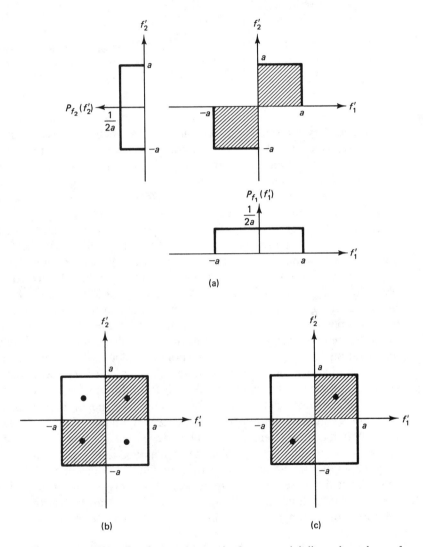

Figure 10.9 Illustration that vector quantization can exploit linear dependence of scalars in the vector. (a) Probability density function $p_{f_1,f_2}(f_1', f_2')$; (b) reconstruction levels (filled-in dots) in scalar quantization; (c) reconstruction levels (filled-in dots) in vector quantization.

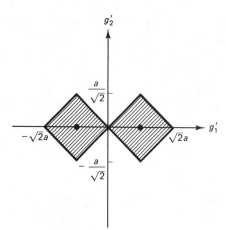

Figure 10.10 Result of eliminating linear dependence of the two scalars f_1 and f_2 in Figure 10.9 by linear transformation of f_1 and f_2.

dependent, they are said to be nonlinearly dependent.* If we quantize f_1 and f_2 separately, using the MSE criterion, and allow two reconstruction levels for each of f_1 and f_2, the optimal reconstruction levels for each scalar are $-a$ and a. The resulting reconstruction levels in this case are the four filled-in dots shown in Figure 10.11(b). The average distortion $D = E[\mathbf{e}_Q^T \mathbf{e}_Q]$ in this example is $a^2/2$. Using vector quantization, we can choose the four reconstruction levels as shown in Figure 10.11(c). The average distortion $D = E[\mathbf{e}_Q^T \mathbf{e}_Q]$ in this example is $5a^2/12$. This example shows that using vector quantization can reduce the MSE without increasing the number of reconstruction levels. We can eliminate the nonlinear dependence between f_1 and f_2 in this example by a simple invertible nonlinear operation. The result of one such nonlinear operation is shown in Figure 10.12. Since $p_{g_1,g_2}(g_1', g_2') = p_{g_1}(g_1')p_{g_2}(g_2')$, g_1 and g_2 are statistically independent. In this case, it is possible to place the two reconstruction levels at the filled-in dots shown in the figure by scalar quantization of the two scalars, and the advantage of vector quantization disappears. Eliminating nonlinear dependence reduces the advantage of vector quantization in this example. This is consistent with the notion that vector quantization can exploit nonlinear dependence of the scalars in the vector.

Linear transformation can always eliminate linear dependence. We can sometimes eliminate nonlinear dependence with an invertible nonlinear operation. If we eliminate the linear or nonlinear dependence through some invertible operation prior to quantization, then the advantage that vector quantization derives from exploiting linear or nonlinear dependence will disappear. This brings into sharp focus the close relationship between the transformation stage (what to code) and the quantization stage in image coding. If the transformation stage eliminates the linear or nonlinear dependence among the scalars to be coded, the potential

---

*This definition differs slightly from that in [Makhoul et al.]. In [Makhoul et al.], nonlinear dependence is defined as the statistical dependence that cannot be eliminated by a linear operator. With this definition, a linear operator cannot eliminate nonlinear dependence. With the definition in the text, a linear operator can eliminate nonlinear dependence in some special cases.

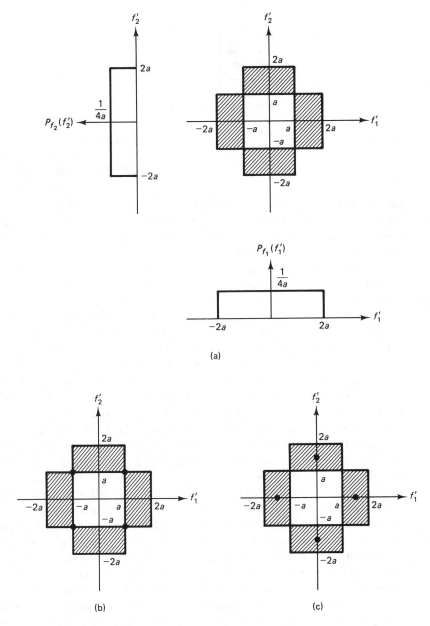

Figure 10.11 Illustration that vector quantization can exploit nonlinear dependence of scalars in the vector. (a) Probability density function $p_{f_1, f_2}(f'_1, f'_2)$; (b) reconstruction levels (solid dots) in scalar quantization; (c) reconstruction levels (solid dots) in vector quantization.

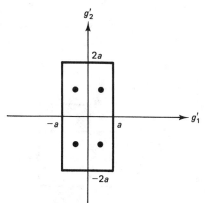

Figure 10.12 Result of eliminating nonlinear dependence of the two scalars f_1 and f_2 in Figure 10.11.

performance improvement of vector quantization over scalar quantization in the quantization stage diminishes and vector quantization becomes less attractive. This partially explains why the performance improvement by vector quantization is more pronounced for waveform coders than for transform coders. The scalars used in waveform coders, such as image intensities, tend to be considerably more correlated than the scalars used in transform coders, such as DCT coefficients. This is discussed further in Sections 10.3 and 10.4.

In addition to exploiting statistical dependence, vector quantization can also exploit the increase in dimensionality. To illustrate this point, let us consider two random variables f_1 and f_2 with a joint pdf that is uniform over a square region with area A. Clearly, f_1 and f_2 are statistically independent. Suppose we have a very large number of reconstruction levels L and therefore the cell size is much smaller than the square over which the pdf is nonzero. We first consider scalar quantization of f_1 and f_2. Since the pdfs for f_1 and f_2 are uniform, uniform quantization is optimal under the MMSE criterion. Uniform quantization of f_1 and f_2 individually results in the reconstruction levels and cells shown in Figure 10.13(a). The cell shape is a square with length a for the scalar quantization case. In vector quantization of f_1 and f_2, we can choose the reconstruction levels and cells as shown in Figure 10.13(b). The cell shape is a hexagon. After some algebra, it can be shown [Makhoul et al.] that the MSE for vector quantization is about 4% lower than for the scalar quantization at the same number of reconstruction levels. It can also be shown that the number of reconstruction levels required in vector quantization is about 2% less than that required in scalar quantization at the same MSE. This performance improvement is typically much smaller than that achieved by vector quantization when statistical dependence is exploited. However, the improvement becomes more significant as the dimension size (the number of scalars in the vector) increases. Note that this additional improvement has been achieved even when the scalars in the block are statistically independent.

The improvement in performance by vector quantization allows us to code a scalar with less than one bit in some cases. If we code each scalar independently and allow at least two reconstruction levels per scalar (using one reconstruction level is equivalent to not coding it), the minimum bit rate possible is one bit per scalar. With vector quantization, it is possible to allow each scalar to have two or more reconstruction levels when viewed individually, with a bit rate less than one bit per scalar when viewed jointly. To illustrate this point, we can go back to the example in Figure 10.9. In scalar quantization [Figure 10.9(b)], assigning two reconstruction levels per scalar requires a total of four reconstruction levels for two scalars, and therefore the bit rate is one bit per scalar. In vector quantization [Figure 10.9(c)], we allow two reconstruction levels for each scalar when each scalar is viewed individually, and we still have only two reconstruction levels for two scalars. The bit rate in this case is $\frac{1}{2}$ bit/scalar. If we decided to code pixel intensities and each pixel intensity must be represented by at least two reconstruction levels, vector quantization is one approach to achieve a bit rate of less than 1 bit/pixel.

The potential performance improvement by vector quantization over scalar quantization comes at a price. The costs associated with vector quantization in terms of computational and storage requirements are far greater than with scalar

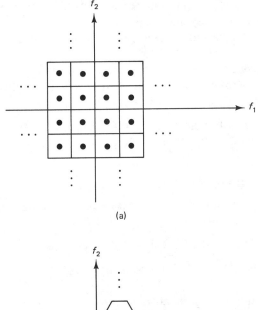

(a)

(b)

Figure 10.13 Illustration that vector quantization can exploit the dimensionality increase. In this case, the mean square error due to vector quantization is approximately 4% less than that due to scalar quantization. (a) Scalar quantization of f_1 and f_2; (b) vector quantization of f_1 and f_2.

quantization. Most of the computational and storage requirements are for the design and storage of a codebook and the need to search the codebook to identify the reconstruction level to be assigned to a vector **f**. This is discussed next.

Codebook design and the K-means algorithm. In vector quantization, we need to determine the reconstruction levels r_i and corresponding cells C_i. A list of reconstruction levels is called a *reconstruction codebook* or a *codebook* for short. If there are L reconstruction levels in the list, the list is said to be an L-level codebook. The codebook is needed at the transmitter to quantize a source vector to one of L reconstruction levels and at the receiver to determine the reconstruction level from the received codeword. The same codebook should, of course, be known to both the transmitter and receiver through prior agreement.

Designing a codebook for vector quantization is a very difficult problem. Unlike the scalar quantization case, in vector quantization the reconstruction levels are vectors, and the cell boundaries are no longer points. The difficulty of designing a codebook is one reason why vector quantization was not seriously considered in such applications as speech and image coding until the late 1970s. The optimal determination of r_i and C_i depends on the error criterion used. The often used MSE criterion can be expressed as average distortion $D = E[d(\mathbf{f}, \hat{\mathbf{f}})]$ with $d(\mathbf{f}, \hat{\mathbf{f}}) = (\hat{\mathbf{f}} - \mathbf{f})^T(\hat{\mathbf{f}} - \mathbf{f})$.

Optimal design of a codebook is a highly nonlinear problem. Attempts to solve this problem typically exploit the following two necessary conditions for its solution.

Condition 1. For a vector **f** to be quantized to one of the reconstruction levels, the optimal quantizer must choose the reconstruction level r_i which has the smallest distortion between **f** and r_i:

$$\text{VQ}(\mathbf{f}) = \mathbf{r}_i \quad \text{if and only if} \quad d(\mathbf{f}, \mathbf{r}_i) \leq d(\mathbf{f}, \mathbf{r}_j), \quad j \neq i, \quad 1 \leq j \leq L. \quad (10.21)$$

Condition 2. Each reconstruction level r_i must minimize the average distortion D in C_i:

$$\text{Minimize } E[d(\mathbf{f}, \mathbf{r}_i)| \mathbf{f} \in C_i] \text{ with respect to } \mathbf{r}_i. \quad (10.22)$$

The level r_i that satisfies (10.22) is called the *centroid* of C_i.

If (10.21) is not satisfied, the average distortion D can be reduced by imposing (10.21). If (10.22) is not satisfied, D can be reduced by imposing (10.22). The above two conditions are thus necessary for the optimal solution.

Condition 1 specifies a rule for quantizing **f** without explicit use of C_i. In other words, the distortion measure $d(\mathbf{f}, \hat{\mathbf{f}})$ together with all the reconstruction levels r_i for $1 \leq i \leq L$ specifies all the cells C_i for $1 \leq i \leq L$. Condition 2 shows a way to determine r_i from C_i and $d(\mathbf{f}, \hat{\mathbf{f}})$. The two conditions show that given $d(\mathbf{f}, \hat{\mathbf{f}})$, reconstruction levels and cells are not independent. In fact, reconstruction levels specify cells and cells specify reconstruction levels. Therefore, only the reconstruction levels alone are sufficient in the codebook; explicit information about cells does not have to be stored.

The two necessary conditions suggest an iterative procedure for designing an optimal codebook. Suppose we begin with an initial estimate of r_i. Given r_i and the distortion measure $d(\mathbf{f}, \hat{\mathbf{f}})$, we can determine C_i, at least in theory, by determining the corresponding r_i for all possible values of $\mathbf{f}$, using Condition 1 in (10.21). Given an estimate of C_i, we can determine r_i by computing the centroid of C_i using Condition 2 in (10.22). The r_i obtained is a new estimate of the reconstruction levels, and the procedure can be continued. This iterative procedure has two practical difficulties. First, it requires the determination of r_i for all possible $\mathbf{f}$. Second, $p_f(\mathbf{f}_0)$ which is required to compute the centroid of C_i is not given in practice. Instead, we have training vectors that are representative of the data to be coded. A modification of the above iterative procedure which takes into account these two practical difficulties is the K-means algorithm. In our case $K = L$. The K-means algorithm was discovered by Lloyd in 1957 for the scalar case ($N = 1$). It was described by [Forgy] in 1965 for the vector case. The algorithm is also called the LBG algorithm, since Y. Linde, A. Buzo, and R. M. Gray [Linde et al.] have shown that the algorithm can be used with many different distance measures and have popularized its use in the application of vector quantization to speech and image coding.

To describe the K-means algorithm, let us suppose that we have M training vectors denoted by $\mathbf{f}_i$ for $1 \leq i \leq M$. Since we estimate L reconstruction levels from M training vectors, we assume that $M \gg L$. In typical applications, M is on the order of $10L$ to $50L$ or more. The reconstruction levels r_i are determined by minimizing the average distortion, defined by

$$D = \frac{1}{M} \sum_{i=1}^{M} d(\mathbf{f}_i, \hat{\mathbf{f}}_i). \tag{10.23}$$

In the K-means algorithm, we begin with an initial estimate of r_i for $1 \leq i \leq L$. We then classify the M training vectors into L different groups or clusters corresponding to each reconstruction level, using (10.21). This can be done by comparing a training vector with each of the reconstruction levels and choosing the level that results in the smallest distortion. Note that we quantize only the given training vectors, not all possible vectors of $\mathbf{f}$. A new reconstruction level is determined from the vectors in each cluster. For notational convenience, suppose $\mathbf{f}_i$ for $1 \leq i \leq M_1$ are M_1 training vectors quantized to the first reconstruction level r_1. The new estimate of r_1 is obtained by minimizing $\sum_{i=1}^{M_1} d(\mathbf{f}_i, r_1)/M_1$. If the $d(\mathbf{f}, \hat{\mathbf{f}})$ used is the squared error $(\hat{\mathbf{f}} - \mathbf{f})^T (\hat{\mathbf{f}} - \mathbf{f})$, then the new estimate of r_1 is just the average of M_1 vectors $\mathbf{f}_i$, $1 \leq i \leq M_1$. A new estimate of all other reconstruction levels r_i for $2 \leq i \leq L$ is similarly obtained. This is a modification of (10.22), made so as to be consistent with D in (10.23). This completes one iteration of the procedure, which can be stopped when the average distortion D does not change significantly between two consecutive iterations. This iterative algorithm, the K-means algorithm, is shown in Figure 10.14. Due to the clustering involved, this algorithm is known as the *clustering algorithm* in the pattern recognition literature.

The K-means algorithm has been shown to converge to a local minimum of D in (10.23) [Linde, Buzo, and Gray]. To determine r_i with D closer to the global

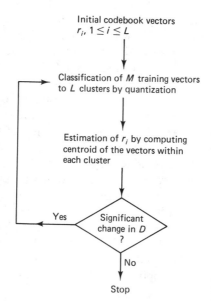

Initial codebook vectors
$r_i, 1 \le i \le L$

Classification of M training vectors
to L clusters by quantization

Estimation of r_i by computing
centroid of the vectors within
each cluster

Significant
change in D
?

Yes

No

Stop

Codebook designed: List of r_i

Figure 10.14 Codebook design by the K-means algorithm for vector quantization.

minimum, we can repeat the algorithm with different initial estimates of $\mathbf{r}_i$ and choose the set that results in the smallest average distortion D.

The most computationally expensive part of the K-means algorithm is the quantization of training vectors in each iteration. For each of M training vectors, the distortion measure must be evaluated L times (once for each reconstruction level), so ML evaluations of the distortion measure are necessary in each iteration. If we assume that there are N scalars in the vector, R bits are used per scalar, and a uniform-length codeword is assigned to each reconstruction level, then L is related to N and R by

$$L = 2^B = 2^{NR} \qquad (10.24)$$

where B is the total number of bits used per vector. If we further assume that the distortion measure used is the squared error $\mathbf{e}_Q^T \mathbf{e}_Q$, one evaluation of the distortion measure requires N arithmetic operations (N multiplications and N additions). The number of arithmetic operations required in the training vector quantization step of each iteration is given by

$$\text{Number of arithmetic operations} = NML = NM \cdot 2^{NR}. \qquad (10.25)$$

From (10.25), the computational cost grows exponentially with N (the number of scalars in a vector) and R (number of bits per scalar). When $N = 10$, $R = 2$, and $M = 10L = 10 \cdot 2^{NR} = 10 \cdot 2^{20}$, the number of arithmetic operations given by (10.25) is 100 trillion per iteration.*

*The choice of $NR = 20$ was made to illustrate how fast the number can grow when it depends exponentially on NR. In typical applications of vector quantization, NR chosen is significantly less than 20.

In addition to the computational cost, there is also storage cost. Assuming that each scalar requires one unit of memory, storage of the training vectors requires MN memory units and storage of the reconstruction levels requires LN memory units. Therefore,

$$\text{Total number of memory units required} = (M + L)N = (M + 2^{NR})N. \qquad (10.26)$$

Since M is typically much greater than L, memory requirements are dominated by the storage of training vectors. When $N = 10$, $R = 2$, and $M = 10L = 10 \cdot 2^{20}$, the number of memory units given by (10.26) is on the order of 100 million. Since the number in (10.26) again grows exponentially with N and R, both computational and storage requirements dictate that vector quantization will mainly be useful for a small number of scalars in the vector and a small number of bits per scalar.

So far we have discussed the computational and storage requirements of designing a codebook. Once the codebook is designed, it must be stored at both the transmitter and the receiver. Since the training vectors are no longer needed once the codebook is designed, only the reconstruction levels need to be stored. The amount of storage required is still very large. In this case,

$$\text{Number of memory units required in a codebook} = NL = N \cdot 2^{NR}. \qquad (10.27)$$

When $N = 10$ and $R = 2$, the number in (10.27) is on the order of 10 million.

For each vector $\mathbf{f}$ to be quantized, the distortion measure $d(\mathbf{f}, \mathbf{r}_i)$ has to be computed for each of the L reconstruction levels at the transmitter. Therefore, for each vector,

$$\text{Number of arithmetic operations} = NL = N \cdot 2^{NR}. \qquad (10.28)$$

When $N = 10$ and $R = 2$, the number in (10.28) is on the order of 10 million. From (10.27) and (10.28), both the number of memory units required in the codebook and the number of arithmetic operations required in quantizing one vector $\mathbf{f}$ grow exponentially with N (scalars per vector) and R (bits per scalar). Note that the above arithmetic operations are required at the transmitter. Fortunately, the receiver requires only simple table look-up operations.

Tree codebook and binary search. The major computations involved in codebook design by the K-means algorithm are in the quantization of the training vectors. Quantization of vectors is also required when the codebook is used at the transmitter. When a codebook is designed by the K-means algorithm, quantization of a vector during the design and the data transmission necessitates evaluating a distortion measure between the vector and each of the L reconstruction levels. This is called a *full search* and is responsible for the exponential dependence of the number of computations involved on the number of scalars in the vector and the number of bits per scalar. Various methods have been developed to eliminate this exponential dependence [Makhoul et al.]. They achieve a reduction in computations by modifying the codebook, by sacrificing performance in achieved average distortion, and/or by increasing storage requirements. We will describe one such method, which results in what is termed a *tree codebook*.

The basic idea behind the tree codebook is to divide the N-dimensional space of $\mathbf{f}$ into two regions using the K-means algorithm with $K = 2$, then divide each of these two regions into two more regions, again using the K-means algorithm, and continue the process. Specifically, assuming that L can be expressed as a power of 2, we first design a codebook with two reconstruction levels $\mathbf{r}_1$ and $\mathbf{r}_2$ using the K-means algorithm. We then classify all the training vectors into two clusters, one cluster corresponding to $\mathbf{r}_1$ and the other to $\mathbf{r}_2$. Each of the two clusters is treated independently, and a codebook with two reconstruction levels is designed for each cluster. This process is repeated until we have a total of L reconstruction levels at the last stage. This is shown in Figure 10.15 for the case where $L = 8$. By this process, the tree codebook is designed.

We first consider the computational and storage requirements in designing the codebook. We again assume that one evaluation of the distortion measure requires N arithmetic operations. Since there are $\log_2 L$ stages and the distortion measure is evaluated only twice for each of M training vectors for each stage and for each iteration of the K-means algorithm,

$$\text{Total number of arithmetic operations/iteration} = 2NM \log_2 L. \qquad (10.29)$$

When this number is compared with the corresponding number in (10.25), for the full search case, the reduction in the number of computations is seen to be by a factor of $L/(2 \log_2 L)$ or $2^{NR}/(2NR)$. When $N = 10$ and $R = 2$, the reduction in computation is by a factor of 26,000. The storage required in the design of the tree codebook is slightly more than in the case of K-means algorithm, since the storage requirement is dominated in both cases by the need to store all the training vectors.

We will now consider the computation involved in quantizing a vector $\mathbf{f}$ using the tree codebook. In the first stage, we compute the distortion measure between $\mathbf{f}$ and the two reconstruction levels $\mathbf{r}_1$ and $\mathbf{r}_2$ in Figure 10.15. Suppose $d(\mathbf{f}, \mathbf{r}_2)$ is smaller than $d(\mathbf{f}, \mathbf{r}_1)$, so we choose $\mathbf{r}_2$. In the second stage, we compute the distortion measure between $\mathbf{f}$ and the two reconstruction levels $\mathbf{r}_5$ and $\mathbf{r}_6$ in Figure 10.15 and choose the reconstruction level that results in the smaller distortion

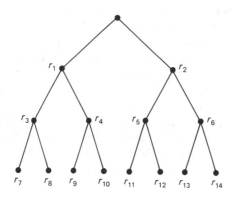

Figure 10.15 Example of a tree code-book.

measure. Suppose $\mathbf{r}_5$ is chosen. In the third stage, we compare $\mathbf{f}$ to $\mathbf{r}_{11}$ and $\mathbf{r}_{12}$ in Figure 10.15. This procedure is continued until we reach the last stage; the reconstruction level chosen at the last stage is the reconstruction level of $\mathbf{f}$. If L = 8 and $\mathbf{r}_{12}$ has been chosen in the third stage in the above example, $\mathbf{r}_{12}$ will be the reconstruction level of $\mathbf{f}$. In this procedure, we simply follow the tree and perform a search between two reconstruction levels at each node of the tree. Since the search is made between two levels at a time, it is said to be a *binary search*. Since $\log_2 L$ stages are involved and the distortion measure is evaluated twice at each stage, the number of arithmetic operations required to quantize $\mathbf{f}$ using the tree codebook is given by

$$\text{Number of arithmetic operations} = 2N \log_2 L = 2N^2 R. \qquad (10.30)$$

From (10.30), the computational cost does not increase exponentially with N and R. When the number in (10.30) is compared with the corresponding number in (10.28) for the full search case, the computational cost is reduced by a factor of $2^{NR}/(2NR)$. When $N = 10$ and $R = 2$, the reduction is by a factor of 26,000.

The reduction in the number of computations has a cost. The codebook used at the transmitter must store all the intermediate reconstruction levels as well as the final reconstruction levels, because the intermediate levels are used in the search. The codebook size is, therefore, increased by a factor of two over the codebook designed by the K-means algorithm. In addition, the tree codebook's performance in terms of the average distortion achieved is reduced slightly in typical applications as compared to the codebook designed by the K-means algorithm. In many cases, however, the enormous computational advantage more than compensates for the twofold increase in storage requirements and the slight decrease in performance.

The binary search discussed above is a special case of a more general class of methods known as tree-searched vector quantization. It is possible, for example, to divide each node into more than two branches. In addition, we can terminate a particular node earlier in the design of the codebook when only a very small number of training vectors are assigned to the node.

In this section, we have discussed vector quantization. The advantage of vector quantization over scalar quantization lies in its potential to improve performance. The amount of performance improvement possible depends on various factors, for instance, the degree of statistical dependence among the scalars in the vector. This performance improvement, however, comes at a price in computation and storage requirements. Whether or not the performance improvement justifies the additional cost depends on the application. Vector quantization is likely to be useful in low bit rate applications, where any performance improvement is important and the additional cost due to vector quantization is not too high due to the low bit rate. Vector quantization can also be useful in such applications as broadcasting, where the number of receivers is much larger than the number of transmitters and the high cost of a transmitter can be tolerated. The receiver in vector quantization has to store the codebook, but it requires only simple table look-up operations to reconstruct a quantized vector.

10.2 CODEWORD ASSIGNMENT

10.2.1 Uniform-Length Codeword Assignment

In Section 10.1, we discussed the problem of quantizing a scalar or a vector source. As the result of quantization, we obtain a specific reconstruction level. To transmit to the receiver which of the L possible reconstruction levels has been selected, we need to assign a specific codeword (a string of 0s and 1s) to each of the L reconstruction levels. Upon receiving the codeword, the receiver can identify the reconstruction level by looking up the appropriate entry in the codebook. For the receiver to be able to uniquely identify the reconstruction level, each reconstruction level must be assigned a different codeword. In addition, since more than one reconstruction level may be transmitted in sequence, the codewords have to be designed so that they can be identified when received sequentially. A code having these characteristics is said to be *uniquely decodable*. When $L = 4$, assigning 00 to r_1, 01 to r_2, 10 to r_3, and 11 to r_4 results in a uniquely decodable code. A code constructed by assigning 0 to r_1, 1 to r_2, 10 to r_3, and 11 to r_4 is not uniquely decodable. When 100 is received, for example, it could be taken to represent either $r_3 r_1$ or $r_2 r_1 r_1$.

It is convenient to think of the result of quantizing a scalar or a vector as a message that has L different possibilities a_i, $1 \leq i \leq L$, with each possibility corresponding to a reconstruction level. The simplest method of selecting codewords is to use codewords of uniform length. In this method each possibility of the message is coded by a codeword that has the same length as all the other possibilities of the message. An example of uniform-length codeword selection for $L = 8$ is shown in Table 10.2. The length of each codeword in this example is $\log_2 L = \log_2 8 = 3$ bits. We will refer to the number of bits required to code a message as the bit rate. The bit rate in our example is 3 bits/message. If we code more than one message, the average bit rate is defined as the total number

TABLE 10.2 AN EXAMPLE OF UNIFORM-LENGTH CODEWORD SELECTION FOR A MESSAGE WITH EIGHT POSSIBILITIES.

Message	Codeword
a_1	0 0 0
a_2	0 0 1
a_3	0 1 0
a_4	0 1 1
a_5	1 0 0
a_6	1 0 1
a_7	1 1 0
a_8	1 1 1

of bits required divided by the number of messages. For uniform-length codeword assignment, the average bit rate is the same as the bit rate.

10.2.2 Entropy and Variable-Length Codeword Assignment

Uniform-length codeword assignment, although simple, is not in general optimal in terms of the required average bit rate. Suppose some message possibilities are more likely to be sent than others. Then by assigning shorter codewords to the more probable message possibilities and longer codewords to the less probable message possibilities, we may be able to reduce the average bit rate.

Codewords whose lengths are different for different message possibilities are called *variable-length codewords*. When the codeword is designed based on the statistical occurrence of different message possibilities, the design method is called *statistical coding*. To discuss the problem of designing codewords such that the average bit rate is minimized, we define the entropy H by

$$H = -\sum_{i=1}^{L} P_i \log_2 P_i \tag{10.31}$$

where P_i is the probability that the message will be a_i. Since $\sum_{i=1}^{L} P_i = 1$, it can be shown that

$$0 \leq H \leq \log_2 L. \tag{10.32}$$

The entropy H can be interpreted as the average amount of information that a message contains. Suppose $L = 2$. If $P_1 = 1$ and $P_2 = 0$, the entropy H is 0, and is the minimum possible for $L = 2$. In this case, the message is a_1 with probability of 1; that is, the message contains no new information. At the other extreme, suppose $P_1 = P_2 = \frac{1}{2}$. The entropy H is 1 and is the maximum possible for $L = 2$. In this case, the two message possibilities a_1 and a_2 are equally likely. Receiving the message clearly adds new information.

From the information theory, the entropy H in (10.31) is the theoretically minimum possible average bit rate required in coding a message. This result, although it does not specify a method to design codewords, is very useful. Suppose the average bit rate using the codewords we have designed is the same as the entropy. We then know that these codewords are optimal, and we do not have to search any further. For example, suppose L can be expressed as a power of 2 and each message possibility a_i is equally probable so that $P_i = 1/L$ for $1 \leq i \leq L$. From (10.32), the entropy H for this case is $\log_2 L$. Since uniform-length codeword assignment results in an average bit rate of $\log_2 L$ bits/message, we can conclude that it is an optimal method to design codewords in this case. The entropy also provides a standard against which the performance of a codeword design method can be measured. If the average bit rate achieved by a codeword design method is close to the entropy, for example, the method is very efficient.

If we code each message separately, it is not in general possible to design codewords that result in an average bit rate given by the entropy. For example, suppose $L = 2$, $P_1 = \frac{1}{8}$, and $P_2 = \frac{7}{8}$. Even though the entropy H of this message is 0.544, it is not possible to design codewords that result in an average bit rate of less than 1 bit/message.

One optimal codeword design method which is simple to use, which is uniquely decodable, and which results in the lowest possible average bit rate is *Huffman coding*. An example of Huffman coding is shown in Figure 10.16. In the example, $L = 6$ with the probability for each message possibility noted at each node, as shown in the figure. In the first step of Huffman coding, we select the two message possibilities that have the two lowest probabilities. In our example, the two message possibilities are a_4 and a_6. We combine the probabilities of these two message possibilities and form a new node with the combined probability. We assign 0 to one of the two branches formed from this node and 1 to the other branch. Reversing 0 and 1 affects the specific codewords assigned, but does not affect the average bit rate. We now consider the two message possibilities a_4 and a_6 as one message possibility a_7 with a combined probability of $\frac{1}{16}$. We now select, from among the five messages a_1, a_2, a_3, a_5, and a_7, the two message possibilities with the lowest two probabilities. The two message possibilities chosen this time are a_3 and a_7. They are again combined as one message, and 0 is assigned to one branch and 1 is assigned to the other. We can continue this process until we are left with one message possibility with probability 1. To determine the specific codeword assigned to each message possibility, we begin with the last node with probability 1, follow the branches that lead to the message possibility of interest, and combine the 0s and 1s on the branches. For example, the message possibility a_4 has the codeword 1110. The codewords obtained in this way are also shown in

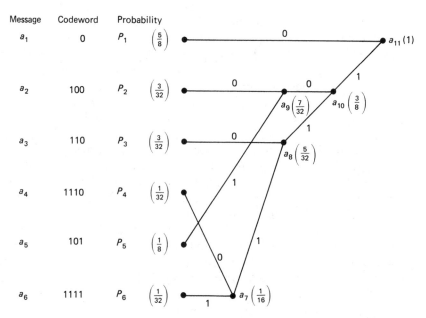

Figure 10.16 Illustration of codeword generation in Huffman coding. Message possibilities with higher probabilities are assigned with shorter codewords.

the figure. Message possibilities with higher probability can clearly be seen to be assigned shorter codewords, and message possibilities with lower probability, longer codewords.

To compare the performance of Huffman coding with the entropy H and uniform-length codeword assignment for the above example, we compute the average bit rate achieved by uniform-length codeword assignment, variable-length codeword assignment by Huffman coding, and the entropy. The results are as follows.

- Uniform-length codeword: 3 bits/message
- Huffman coding: $\frac{5}{8} \cdot 1 + \frac{3}{32} \cdot 3 + \frac{3}{32} \cdot 3 + \frac{1}{32} \cdot 4 + \frac{1}{8} \cdot 3 + \frac{1}{32} \cdot 4 = \frac{29}{16}$ bits/message $\simeq 1.813$ bits/message
- Entropy: $-(\frac{5}{8} \log_2 \frac{5}{8} + \frac{3}{32} \log_2 \frac{3}{32} + \frac{3}{32} \log_2 \frac{3}{32} + \frac{1}{32} \log_2 \frac{1}{32} + \frac{1}{8} \log_2 \frac{1}{8} + \frac{1}{32} \log_2 \frac{1}{32}) \simeq 1.752$ bits/message

In the above example, Huffman coding results in an average bit rate that is close to the entropy and that is lower by more than 1 bit/message than the rate associated with uniform-length codeword assignment.

The potential reduction in the average bit rate when variable-length codeword assignment is used instead of uniform-length codeword assignment comes with cost. The actual bit rate is variable. When message possibilities with low probability of occurrence are coded one after another, the actual bit rate may be considerably higher than the average bit rate. In the opposite situation, the actual bit rate may be considerably lower than the average bit rate. To transmit messages with variable-length codewords through a fixed bit rate system, a buffer that can store the messages when the actual bit rate increases must be maintained. Maintaining a buffer imposes additional complexity on the coding system and also involves delay.

If we are allowed to collect as many messages as we wish and then assign a codeword jointly, we can design codewords with an average bit rate per message arbitrarily close to the entropy H. For this reason, Huffman coding is also called an entropy coding method. Consider the case when $L = 2$, $P_1 = \frac{1}{8}$, and $P_2 = \frac{7}{8}$. The entropy H in this case is 0.544 bit. If we assign a codeword to one message at a time, the average bit rate is 1 bit/message. Suppose we wait until we have two messages and then assign codewords jointly. Table 10.3 shows the codewords

TABLE 10.3 HUFFMAN CODES FOR CODING TWO MESSAGES JOINTLY. THE TWO MESSAGES ARE INDEPENDENT AND EACH MESSAGE HAS TWO POSSIBILITIES WITH PROBABILITIES OF $\frac{1}{8}$ and $\frac{7}{8}$.

Joint message	Probability	Huffman Code
$a_1 a_1$	$\frac{1}{64}$	0 0 0
$a_1 a_2$	$\frac{7}{64}$	0 0 1
$a_2 a_1$	$\frac{7}{64}$	0 1
$a_2 a_2$	$\frac{49}{64}$	1

obtained by Huffman coding. The average bit rate in this case is 0.680 bit/message. As we increase the number of messages that are coded jointly, the average bit rate obtained by Huffman coding approaches the entropy. In practice, we cannot generally wait for a large number of messages, since doing so would involve delay and would require a large additional codebook. The average bit rate achieved by Huffman coding, therefore, will be typically higher than the entropy.

10.2.3 Joint Optimization of Quantization and Codeword Assignment

In Section 10.1, we discussed the quantization problem. So far in Section 10.2, we discussed the problem of assigning codewords to reconstruction levels. Although these topics have been discussed separately, they are closely related. For example, suppose we quantize a scalar f with a nonuniform probability density function using a uniform quantizer. The reconstruction levels in this case are not equally probable, and variable-length codeword assignment has the potential to significantly reduce the average bit rate from that associated with uniform-length codeword assignment. On the other hand, if reconstruction levels are chosen in the quantization step such that they have approximately equal probabilities of occurrence, variable-length codeword assignment will not offer a significant performance advantage. Clearly, what is done in the quantization stage affects what can be done in the codeword assignment stage.

The close relationship between the quantization stage and the codeword assignment stage suggests that optimizing each stage separately may not be optimal in solving the overall problem. In Section 10.1, we considered the problem of minimizing the average distortion given a fixed number of reconstruction levels L, or alternatively the problem of minimizing the number of reconstruction levels at a given average distortion. In practice, our interest is often not in minimizing the number of reconstruction levels but in minimizing the number of bits. If we require uniform-length codeword assignment, then the number of bits will specify the number of reconstruction levels and the two problems can be viewed as equivalent. If variable-length codewords are used, however, a smaller number of reconstruction levels will not necessarily imply fewer bits. For example, four reconstruction levels with unequal probabilities of occurrence can have a lower entropy than two reconstruction levels with equal probabilities of occurrence. Minimizing the number of reconstruction levels at a given average distortion and then minimizing the average bit rate by designing codewords optimally for the given reconstruction levels will not in general minimize the average bit rate at a given average distortion. Minimizing the average bit rate at a given average distortion by joint optimization of the quantization stage and codeword assignment stage is a highly nonlinear problem, and approximate solutions are available only for some special cases [Makhoul et al.].

In Sections 10.1 and 10.2, we discussed various approaches to quantization and codeword assignment. Our main concern in the discussion has been minimizing the average bit rate at a given average distortion. In practice, a number of factors, such as the computational and storage requirements and acceptable amount of delay, must be considered. Furthermore, our discussion in Sections 10.1 and 10.2 has been based on the assumption that the necessary statistics such as $p_f(f_0)$ and the distortion measure $d(f, \hat{f})$ are known. In practice, the necessary statistics must be estimated, and the distortion measure consistent with a given application is not generally known. Inaccuracies in the estimated statistics or assumed distortion measure will, of course, affect the results. For these reasons, the theoretical results of the various methods of quantization and codeword assignment should be considered only as guidelines in deciding which method to employ in a given image coding application.

10.3 WAVEFORM CODING

In Sections 10.1 and 10.2, we discussed quantization and codeword assignment, which are the second and third of the three stages in image coding. We now discuss the first stage, transforming an image into the most suitable domain for quantization and codeword assignment. This part of the process determines what specifically is coded. Image coding algorithms are classified broadly into three categories, depending on what aspect of an image is coded. In this section, we will discuss one class of algorithms, waveform coders. In Sections 10.4 and 10.5, we discuss transform coders and image model coders, respectively, the other two classes of algorithms.

In waveform coding, we code the image intensity itself or some simple variation of image intensity such as the difference between two consecutive pixel intensities. One major advantage of waveform coding over transform coding and image model coding is its simplicity. Since the waveform itself is coded, the coders are very simple both conceptually and computationally. Waveform coding techniques attempt to reconstruct the waveform faithfully without detailed exploitation of information specific to a particular class of signals, and can therefore be used in coding many different classes of signals, such as images and speech. In addition, the bit rate reduction that can be achieved by waveform coders is comparable to that of transform coders in applications such as digital television, where high quality images are required. In applications such as video conferencing and remotely piloted vehicles, where bit rate reduction is important and some sacrifice in quality can be tolerated, waveform coders do not generally perform as well as transform coders.

In principle, we can use any of the quantization and codeword assignment methods discussed in Sections 10.1 and 10.2 in waveform coding. However, scalar quantization and uniform-length codeword assignment have been predominantly used because of their simplicity. In all our discussions in this section, we will assume that scalar quantization and uniform-length codeword assignment are used unless specified otherwise.

In this section and throughout this chapter, we will use examples to illustrate the performance of different image coding algorithms. In addition, when the information is available, we will provide the NMSE (normalized mean square error) and the SNR (signal to noise ratio) defined as

$$\text{NMSE in } \% = 100\left(\frac{\text{Var}\,[\hat{f}(n_1, n_2) - f(n_1, n_2)]}{\text{Var}\,[f(n_1, n_2)]}\right) \% \quad (10.33a)$$

$$\text{SNR in dB} = 10 \log \left(\frac{\text{NMSE in } \%}{100}\right)^{-1} \text{dB} \quad (10.33b)$$

where $f(n_1, n_2)$ is the original image, and $\hat{f}(n_1, n_2)$ is the coded image. The NMSE measure in (10.33a) is identical to the NMSE used in Chapter 9. The use of variance ensures that adding a bias to $\hat{f}(n_1, n_2)$ will not affect the NMSE.

10.3.1 Pulse Code Modulation (PCM)

The simplest waveform coding method is the basic pulse code modulation (PCM) system, in which the image intensity $f(n_1, n_2)$ is quantized by a uniform quantizer. The basic PCM system is shown in Figure 10.17. The image intensity $f(n_1, n_2)$ that has been quantized is denoted by $\hat{f}(n_1, n_2)$ in the figure. The PCM system can be used not only to code image intensities, but also to code transform coefficients and image model parameters. However, it was first used, and is still extensively used, in coding waveforms. Therefore, a PCM system without additional specification is regarded as a waveform coder. This also applies to other waveform coders, such as delta modulation (DM) and differential pulse code modulation (DPCM), which will be discussed later.

The basic PCM system in Figure 10.17 is typically used in obtaining an original digital image $f(n_1, n_2)$ from an analog image in most digital image processing applications. The spatial resolution of a digital image $f(n_1, n_2)$ is primarily determined by its size (number of pixels). The size of $f(n_1, n_2)$ is chosen on the basis of the resolution desired in a given application. A digital image of 1024 × 1024 pixels has resolution comparable to that of 35 mm film. A digital image of 512 × 512 pixels has resolution comparable to that of a TV frame (two fields). Digital images of 256 × 256 pixels and 128 × 128 pixels are also used in such applications as video telephone. As we decrease an image's size, its resolution is decreased, and details begin to disappear. This was illustrated in Figure 1.11. The bit rate used in the original monochrome digital image in most applications is 8 bits/pixel. Except where a very accurate representation of the original analog image is required, such as in medical image processing, the basic PCM system at 8 bits/pixel preserves sufficient quality and intelligibility for most applications. A digital image with a bit rate of 8 bits/pixel will be considered an original image in our discussions of monochrome image coding.

The bit rate used in our discussion is expressed in bits/pixel. It is important to note that the measure bits/pixel can be misleading. For example, if we obtain a digital image by sampling the analog image at a rate much higher than can be

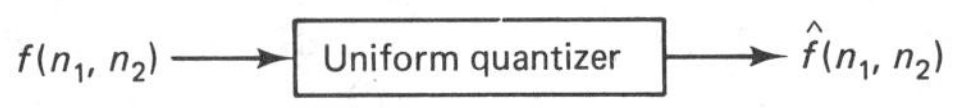

$f(n_1, n_2) \longrightarrow \boxed{\text{Uniform quantizer}} \longrightarrow \hat{f}(n_1, n_2)$

Figure 10.17 Basic pulse code modulation system.

resolved by a human visual system, the number of bits/pixel can be reduced without any visible resolution decrease by simply undersampling the digital image. A more meaningful measure is the number of bits/frame in single frame image coding or the number of bits/sec in the case of image sequence coding. For convenience, however, the bit rate measure we use is bits/pixel. In addition, we will specify an image frame size in single frame image coding and, when applicable, both the size and the number of frames/sec in image sequence coding.

If the input signal fluctuates sufficiently and if the quantization step size is small enough in the basic PCM system, the quantization noise can be modeled approximately as additive random noise. For typical images the quantization noise begins to be visible as random noise at about 5 to 6 bits/pixel. As we lower the bit rate below 3 to 4 bits/pixel, the signal-dependent quantization noise becomes visible as false contours, due to a luminance jump in the reconstructed image in regions where the original image intensity varies slowly. This is illustrated in Figure 10.18. Examples of images coded by the basic PCM system at different bit rates were previously shown in Figure 1.10. Several simple ways to improve the performance of the basic PCM system have been developed. These are discussed next.

Pulse code modulation with nonuniform quantization. A simple way to improve the performance of the basic PCM system is to use nonuniform quantization. Image intensities are not generally distributed equally in the available dynamic range. In such cases, performance can be improved by nonuniform quantization, as was discussed in Section 10.1. One method of performing nonuniform quantization is to apply a nonlinearity to $f(n_1, n_2)$ prior to uniform quantization and then apply the inverse nonlinearity after uniform quantization. The nonlinearity can be chosen such that the output is approximately equally probable throughout the available dynamic range. Even though nonuniform quantization can improve the performance of a basic PCM system, the amount of improvement is not very large for typical images. As was discussed in Section 10.1.1 (see Figure 10.6), nonuniform quantization can reduce the mean square error by less than 3 dB for a Gaussian-shaped image histogram for bit rates up to 7 bits/pixel. When the image histogram deviates considerably from a uniform histogram, however, nonuniform quantization can improve the performance of a basic PCM system considerably. For example, suppose the intensities of an image are concentrated in

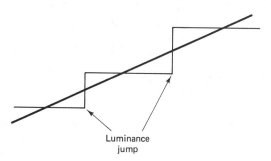

Luminance
jump

Figure 10.18 Illustration of luminance jumps that cause false contours in PCM image coding.

a few very narrow intensity ranges. Uniform quantization in this case will place many reconstruction levels which are wasted. Through nonuniform quantization, the reconstruction levels can be placed in the intensity regions where they are needed.

Roberts's pseudonoise technique. Another approach to improving the performance of a PCM system is to remove the signal dependence of the quantization noise, which appears as false contours at low bit rates. In this approach [Roberts], we transform the signal-dependent quantization noise to less annoying signal-independent random noise. Roberts's pseudonoise technique, also known as *dithering*, is a method that removes the signal dependence of the quantization noise. The method is shown in Figure 10.19. In this method, a known random noise $w(n_1, n_2)$ is added to the original image $f(n_1, n_2)$ before quantization at the transmitter and then the same $w(n_1, n_2)$ is subtracted at the receiver. Since $w(n_1, n_2)$ is known at both the transmitter and the receiver prior to image transmission, it does not have to be transmitted. A white noise sequence with a uniform pdf $p_w(w_0)$ given by

$$p_w(w_0) = \begin{cases} \dfrac{1}{\Delta}, & -\dfrac{\Delta}{2} \le w_0 \le \dfrac{\Delta}{2} \\ 0, & \text{otherwise} \end{cases} \tag{10.34}$$

where Δ is the quantizer step size can be used as the random noise $w(n_1, n_2)$. With $p_w(w_0)$ in (10.34), it can be shown that the reconstructed image $\hat{f}(n_1, n_2)$ can be modeled approximately as the original image $f(n_1, n_2)$ degraded by signal-independent additive random noise which is white and has the same uniform pdf as (10.34).

Simply removing the signal dependence of the quantization noise alone significantly improves the performance of a PCM system. In addition, it allows us to apply any of the additive random noise reduction systems discussed in Chapter 9 to reduce the signal-independent quantization noise. A system that includes

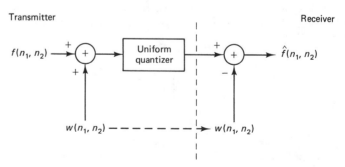

Figure 10.19 Decorrelation of quantization noise by Roberts's pseudonoise technique.

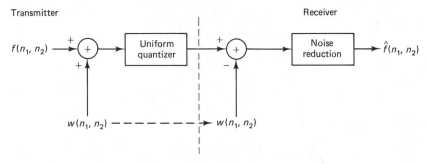

Figure 10.20 Reduction of quantization noise in PCM image coding.

quantization noise reduction in addition to Roberts's noise technique is shown in Figure 10.20. The noise reduction system is attached only at the receiver. This type of system may be useful in applications such as spacecrafts and remotely piloted vehicles, where it is desirable to have a simple transmitter and where some complexity in the receiver can be tolerated.

To more clearly illustrate the concept of eliminating the signal dependence of quantization noise, let us consider an example involving a 1-D signal. Figure 10.21(a) shows a segment of noise-free voiced speech. Figure 10.21(b) shows the speech waveform reconstructed by a PCM system with uniform quantization at 2 bits/sample. The effect of signal-dependent quantization noise is quite visible in the staircase shape of the waveform. Figure 10.21(c) shows the result of adding and then subtracting dither noise in a PCM system with a 2 bits/sample uniform quantizer. The result can be modeled by the original signal degraded by signal-independent additive random noise. Figure 10.21(d) shows the result obtained by applying a noise reduction system [Lim and Oppenheim] to the waveform in Figure 10.21(c).

Figure 10.22 shows an example that involves an image. Figure 10.22(a) shows an original image of 512×512 pixels at 8 bits/pixel. Figure 10.22(b) shows the result of a PCM system with a 2 bits/pixel uniform quantizer. Figure 10.22(c) shows the result of Roberts's pseudonoise technique. Although the two images in Figures 10.22(b) and (c) have essentially the same NMSE, the image in Figure 10.22(c) is more visually pleasing. Figure 10.22(d) shows the result of applying the adaptive Wiener filter discussed in Section 9.2.4 to the image in Figure 10.22(c).

The process of removing the signal dependence of the quantization noise and subsequently reducing it by an image restoration algorithm can be applied to any system that has a uniform quantizer as a component. For example, consider the PCM system with a nonuniform quantizer shown in Figure 10.23(a). Roberts's pseudonoise can be added prior to the uniform quantizer and subtracted after it. This is shown in Figure 10.23(b). The signal $\hat{g}(n_1, n_2)$ can be viewed as $g(n_1, n_2)$ degraded by additive random noise independent of $g(n_1, n_2)$. If noise reduction is desired, a noise reduction system can be applied to $\hat{g}(n_1, n_2)$, as shown in Figure 10.23(b).

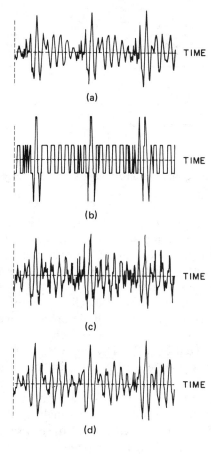

(a)

TIME

(b)

TIME

(c)

TIME

(d)

TIME

Figure 10.21 Example of quantization noise reduction in PCM speech coding. (a) Segment of noise-free voiced speech; (b) PCM-coded speech at 2 bits/sample; (c) PCM-coded speech at 2 bits/sample by Roberts's pseudonoise technique; (d) PCM-coded speech at 2 bits/sample with quantization noise reduction.

10.3.2 Delta Modulation (DM)

In the PCM system, the image intensity is coded by scalar quantization, and the correlation among pixel intensities is not exploited. One way to exploit some of the correlation, still using scalar quantization, is delta modulation (DM). In the DM system, the difference between two consecutive pixel intensities is coded by a one-bit (two reconstruction levels) quantizer. Although the dynamic range of the difference signal is doubled as a result of differencing, the variance of the difference signal is significantly reduced due to the strong correlation typically present in the intensities of two pixels that are spatially close.

In discussing DM, it is useful to assume that the pixels in an image have been arranged in some sequential manner so that $f(n_1, n_2)$ can be expressed as a 1-D signal $f(n)$. If $f(n)$ is obtained by reading one row of $f(n_1, n_2)$ and then reading the next row, it will preserve some of the spatial correlation present in $f(n_1, n_2)$. A DM system is shown in Figure 10.24. In the figure, $\hat{f}(n)$ represents $f(n)$ reconstructed by DM. To code $f(n)$, the most recently reconstructed $\hat{f}(n-1)$ is subtracted from $f(n)$. The difference signal $e(n) = f(n) - \hat{f}(n-1)$ is quantized

Figure 10.22 Example of quantization noise reduction in PCM image coding. (a) Original image of 512 × 512 pixels; (b) PCM-coded image at 2 bits/pixel; (c) PCM-coded image at 2 bits/pixel by Roberts's pseudonoise technique; (d) PCM-coded image at 2 bits/pixel with quantization noise reduction.

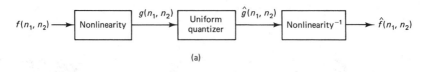

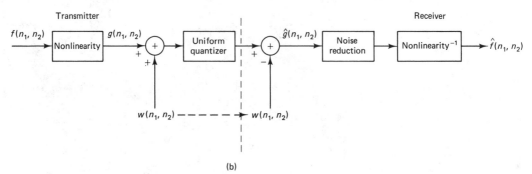

Figure 10.23 Quantization noise reduction in a PCM system with nonuniform quantizer. (a) PCM system; (b) PCM system with quantization noise reduction.

to $\Delta/2$ if $e(n)$ is positive and $-\Delta/2$ otherwise, where Δ is the step size. The difference signal $e(n)$ quantized by the one-bit quantizer is denoted by $\hat{e}(n)$, and $\hat{e}(n)$ is sent at the transmitter. At the receiver, $\hat{e}(n)$ is added to $\hat{f}(n-1)$ to obtain $\hat{f}(n)$. The reconstructed $\hat{f}(n)$ is also needed at the transmitter, since $\hat{f}(n)$ is used in coding $f(n+1)$. The dotted line in the figure represents a delay and indicates that $\hat{f}(n)$ is computed so that it can be used in coding the next sample $f(n+1)$.

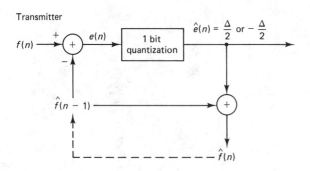

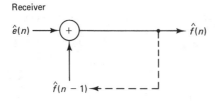

Figure 10.24 Delta modulation system.

The equations that govern the DM system in Figure 10.24 are

$$e(n) = f(n) - \hat{f}(n - 1) \tag{10.35a}$$

$$\hat{e}(n) = \begin{cases} \dfrac{\Delta}{2}, & e(n) > 0 \\[2mm] -\dfrac{\Delta}{2}, & \text{otherwise} \end{cases} \tag{10.35b}$$

$$\hat{f}(n) = \hat{f}(n - 1) + \hat{e}(n). \tag{10.35c}$$

From (10.35a) and (10.35c), the quantization noise $e_Q(n)$ is given by

$$e_Q(n) = \hat{f}(n) - f(n) = \hat{e}(n) - e(n). \tag{10.36}$$

In DM, $e(n)$ quantized is the difference between $f(n)$ and $\hat{f}(n - 1)$. If $f(n - 1)$ is used instead of $\hat{f}(n - 1)$, the transmitter does not have to include the receiver structure to compute $\hat{f}(n - 1)$. However, the quantization noise can accumulate from one pixel to the next. This is illustrated in Figure 10.25. The solid staircase ($\hat{f}_1(n)$) is the result of DM when $\hat{f}(n - 1)$ is used. The dotted staircase ($\hat{f}_2(n)$) is the result when $f(n - 1)$ is used. The reconstructed signal can be significantly different from $f(n)$ in this case, since the receiver does not have $f(n - 1)$ and therefore uses $\hat{f}(n - 1)$ in reconstructing $\hat{f}(n)$, while the error $e(n)$ quantized at the transmitter is based on $f(n - 1)$.

An important design parameter in DM is the step size Δ. Consider a signal reconstructed by DM, as shown in Figure 10.26. In the region where the signal varies slowly, the reconstructed signal varies rapidly around the original signal. This is called *granular noise*. A large Δ results in a correspondingly large amount of granular noise, so a small Δ is desired. When the signal increases or decreases rapidly, however, it may take many pixels before $\hat{f}(n)$ can catch up with $f(n)$ using a small Δ. The reconstructed $\hat{f}(n)$ will appear blurred in such regions. This is called *slope overload distortion*. A small Δ results in a correspondingly large degree of slope overload distortion, so a large Δ is desired. Clearly, reduction of the granular noise and reduction of the slope overload distortion are conflicting re-

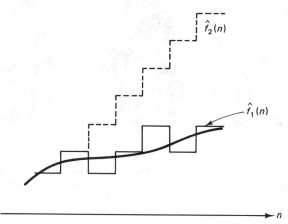

Figure 10.25 Illustration that quantization noise accumulates in delta modulation when $f(n - 1)$ is used in predicting $f(n)$ instead of $\hat{f}(n - 1)$. The solid staircase denoted by $\hat{f}_1(n)$ is the reconstruction when $\hat{f}(n - 1)$ is used. The dotted staircase denoted by $\hat{f}_2(n)$ is the reconstruction when $f(n - 1)$ is used.

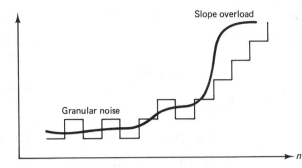

Figure 10.26 Granular noise and slope-overload distortion in delta modulation.

quirements, and the step size Δ is chosen through some compromise between the two requirements.

Figure 10.27 illustrates the performance of a DM system. Figures 10.27(a) and (b) show the results of DM with step sizes of $\Delta = 8\%$ and 15%, respectively, of the overall dynamic range of $f(n_1, n_2)$. The original image used is the 512 × 512-pixel image in Figure 10.22(a). When Δ is small [Figure 10.27(a)], the granular noise is reduced, but the slope overload distortion problem is severe and the resulting image appears blurred. As we increase Δ [Figure 10.27(b)], the slope overload distortion is reduced, but the graininess in the regions where the signal varies slowly is more pronounced.

(a)   (b)

Figure 10.27 Example of delta-modulation (DM)-coded image. The original image used is the image in Figure 10.22(a). (a) DM-coded image with $\Delta = 8\%$ of the overall dynamic range. NMSE = 14.8%, SNR = 8.3 dB; (b) DM-coded image with $\Delta = 15\%$, NMSE = 9.7%, SNR = 10.1 dB.

Figure 10.28 DM-coded image at 2 bits/pixel. The original image used is the image in Figure 10.22(a). NMSE = 2.4%, SNR = 16.2 dB.

To obtain good quality image reconstruction using DM without significant graininess or slope overload distortion, 3–4 bits/pixel is typically required. A bit rate higher than 1 bit/pixel can be obtained in DM by oversampling the original analog signal relative to the sampling rate used in obtaining $f(n_1, n_2)$. Oversampling reduces the slope of the digital signal $f(n)$ so a smaller Δ can be used without increasing the slope overload distortion. An example of an image coded by DM at 2 bits/pixel is shown in Figure 10.28. To obtain the image in Figure 10.28, the size of the original digital image in Figure 10.22(a) was increased by a factor of two by interpolating the original digital image by a factor of two along the horizontal direction. The interpolated digital image was coded by DM with $\Delta = 12\%$ of the dynamic range of the image and the reconstructed image was undersampled by a factor of two along the horizontal direction. The size of the resulting image is the same as the image in Figure 10.27, but the bit rate in this case is 2 bits/pixel.

10.3.3 Differential Pulse Code Modulation

Differential pulse code modulation (DPCM) can be viewed as a generalization of DM. In DM, the difference signal $e(n) = f(n) - \hat{f}(n - 1)$ is quantized. The most recently coded $\hat{f}(n - 1)$ can be viewed as a prediction of $f(n)$ and $e(n)$ can be viewed as the error between $f(n)$ and a prediction of $f(n)$. In DCPM, a prediction of the current pixel intensity is obtained from more than one previously coded pixel intensity. In DM, only one bit is used to code $e(n)$. In DPCM, more than one bit can be used in coding the error.

A DPCM system is shown in Figure 10.29. To code the current pixel intensity $f(n_1, n_2), f(n_1, n_2)$ is predicted from previously reconstructed pixel intensities. The predicted value is denoted by $f'(n_1, n_2)$. In the figure, we have assumed that $\hat{f}(n_1 - 1, n_2), \hat{f}(n_1, n_2 - 1), \hat{f}(n_1 - 1, n_2 - 1), \ldots$ were reconstructed prior to coding $f(n_1, n_2)$. We are attempting to reduce the variance of

$e(n_1, n_2) = f(n_1, n_2) - f'(n_1, n_2)$ in this system by using previously coded pixels in the prediction of $f(n_1, n_2)$. The prediction error $e(n_1, n_2)$ is quantized by a PCM system with a uniform or nonuniform quantizer. The quantized $e(n_1, n_2)$, $\hat{e}(n_1, n_2)$, is then transmitted. At the receiver, $\hat{e}(n_1, n_2)$ is combined with $f'(n_1, n_2)$, the prediction of $f(n_1, n_2)$. Since the previously reconstructed pixel values and the specific way to predict $f(n_1, n_2)$ from the previously reconstructed pixel values are known at the transmitter and the receiver, the same $f'(n_1, n_2)$ is obtained at the transmitter and the receiver. The reconstructed $\hat{f}(n_1, n_2)$ is also needed at the transmitter, since it is used in coding the pixel intensities which have not yet been coded. The dotted line in the figure indicates that $\hat{f}(n_1, n_2)$ is computed to code these pixel intensities. As in DM, previously reconstructed values are used to avoid propagating the quantization noise.

The equations that govern the DPCM system in Figure 10.29 are

$$e(n_1, n_2) = f(n_1, n_2) - f'(n_1, n_2) \tag{10.37a}$$

$$\hat{e}(n_1, n_2) = Q[e(n_1, n_2)] \tag{10.37b}$$

$$\hat{f}(n_1, n_2) = f'(n_1, n_2) + \hat{e}(n_1, n_2) \tag{10.37c}$$

where $Q[e(n_1, n_2)]$ is the quantization of $e(n_1, n_2)$ by a PCM system. From (10.37a) and (10.37c), the quantization noise $e_Q(n_1, n_2)$ is given by

$$e_Q(n_1, n_2) = \hat{f}(n_1, n_2) - f(n_1, n_2) = \hat{e}(n_1, n_2) - e(n_1, n_2). \tag{10.38}$$

The DPCM system in (10.37) can also be viewed as a generalization of PCM. Specifically, DPCM reduces to PCM when $f'(n_1, n_2)$ is set to zero.

In DPCM, $f(n_1, n_2)$ is predicted by linearly combining the previously reconstructed values:

$$f'(n_1, n_2) = \sum_{(k_1, k_2) \in R_a} \sum a(k_1, k_2)\hat{f}(n_1 - k_1, n_2 - k_2) \tag{10.39}$$

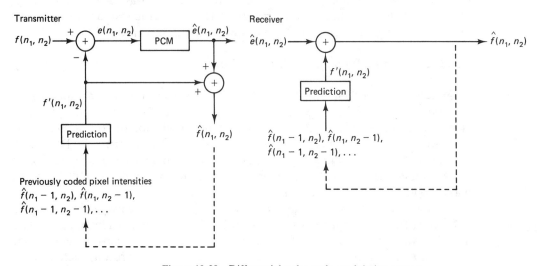

Figure 10.29 Differential pulse code modulation system.

where R_a is the region of (k_1, k_2) over which $a(k_1, k_2)$ is nonzero. Typically, $f'(n_1, n_2)$ is obtained by linearly combining $\hat{f}(n_1 - 1, n_2), \hat{f}(n_1, n_2 - 1)$, and $\hat{f}(n_1 - 1, n_2 - 1)$. Since the prediction of $f(n_1, n_2)$ is made in order to reduce the variance of $e(n_1, n_2)$, it is reasonable to estimate $a(k_1, k_2)$ by minimizing

$$E[e^2(n_1, n_2)] = E[(f(n_1, n_2) - \sum_{(k_1, k_2) \in R_a} \sum a(k_1, k_2)\hat{f}(n_1 - k_1, n_2 - k_2))^2]. \tag{10.40}$$

Since $\hat{f}(n_1, n_2)$ is a function of $a(k_1, k_2)$ and depends on the specific quantizer used, solving (10.40) is a nonlinear problem. Since $\hat{f}(n_1, n_2)$ is the quantized version of $f(n_1, n_2)$, and is therefore a reasonable representation of $f(n_1, n_2)$, the prediction coefficients $a(k_1, k_2)$ are estimated by minimizing

$$E\left[(f(n_1, n_2) - \sum_{(k_1, k_2) \in R_a} \sum a(k_1, k_2)f(n_1 - k_1, n_2 - k_2))^2\right]. \tag{10.41}$$

Since the function in (10.41) minimized is a quadratic form of $a(k_1, k_2)$, solving (10.41) involves solving a linear set of equations in the form of

$$R_f(l_1, l_2) = \sum_{(k_1, k_2) \in R_a} \sum a(k_1, k_2)R_f(l_1 - k_1, l_2 - k_2) \tag{10.42}$$

where $f(n_1, n_2)$ is assumed to be a stationary random process with the correlation function $R_f(n_1, n_2)$. The linear equations in (10.42) are the same as those used in the estimation of the autoregressive model parameters discussed in Chapters 5 and 6.

Figure 10.30 illustrates the performance of a DPCM system. Figure 10.30 shows the result of a DPCM system at 3 bits/pixel. The original image used is the image in Figure 10.22(a). The PCM system used in Figure 10.30 is a nonuniform quantizer. The prediction coefficients $a(k_1, k_2)$ used to generate the example are

Figure 10.30 Example of differential pulse code modulation (DPCM)-coded image at 3 bits/pixel. Original image used is the image in Figure 10.22(a), NMSE = 2.2%, SNR = 16.6 dB.

$a(1, 0) = a(0, 1) = 0.95$ and $a(1, 1) = -0.95$. At 3 bits/pixel, the result of a DPCM system is typically an image of good quality.

Since a PCM system is a component of a DPCM system, it is possible to use Roberts's pseudonoise technique in a DPCM system. However, the error signal $e(n_1, n_2)$ that is quantized in a DPCM system typically varies rapidly from pixel to pixel, and the reconstructed image suffers less from the false contour problem in a DPCM system than it does in a PCM system. For this reason, Roberts's pseudonoise technique, which is very useful in PCM systems, is not as useful in DPCM systems. In addition, reducing quantization noise by using an image restoration system is less useful in a DPCM system. Both the error sequence $e(n_1, n_2)$ and the quantization noise $e_Q(n_1, n_2)$ have broad-band spectra, and reducing $e_Q(n_1, n_2)$ in $\hat{e}(n_1, n_2) = e(n_1, n_2) + e_Q(n_1, n_2)$ is not very effective.

Since the prediction of $f(n_1, n_2)$ from neighborhood pixels is more difficult in regions such as edges where the local contrast is relatively high, the error signal $e(n_1, n_2)$ is likely to be large in such regions. A given level of noise is known to be less visible in high-contrast than in low-contrast regions. This knowledge can be exploited [Netravali (1977)] in determining the reconstruction levels of $e(n_1, n_2)$ in a DPCM system, since the amplitude of $e(n_1, n_2)$ can be related to the amount of local contrast.

10.3.4 Two-Channel Coders

A two-channel coder [Schreiber et al.; Troxel et al.; Schreiber and Buckley] is an example of two-channel processing, which was discussed in Sections 8.1.4 and 9.2.4. In a two-channel coder, an image $f(n_1, n_2)$ is divided into two components, the lows and highs. The lows component $f_L(n_1, n_2)$ consists primarily of low-frequency components and represents the local luminance mean of $f(n_1, n_2)$. The highs component $f_H(n_1, n_2)$ consists primarily of high-frequency components and represents the local contrast of $f(n_1, n_2)$. Since the lows component is a lowpass filtered version of $f(n_1, n_2)$, it can be severely undersampled depending on the lowpass filter used. The highs component can be coarsely quantized, since it does not include the local luminance mean information, and since regions with large amplitude of $f_H(n_1, n_2)$ have high local contrast, so a given level of noise will be less visible.

A two-channel coder is shown in Figure 10.31. The original image $f(n_1, n_2)$ is lowpass filtered by an FIR filter. The lows component $f_L(n_1, n_2)$ is subsampled, typically by a factor of 8×8. This subsampled lows component $f_{LS}(n_1, n_2)$ is quantized by a PCM system and the result is transmitted. Although $f_{LS}(n_1, n_2)$ is accurately represented, typically by 8 to 10 bits/pixel, its contribution to the overall bit rate is on the order of 0.1 to 0.2 bit/pixel due to severe undersampling. An estimate of the original lows component $f_L(n_1, n_2)$ is obtained by interpolating $\hat{f}_{LS}(n_1, n_2)$ and is denoted by $\hat{f}_L(n_1, n_2)$. The highs component $f_H(n_1, n_2)$ is obtained by subtracting $\hat{f}_L(n_1, n_2)$ from $f(n_1, n_2)$, and is quantized by a PCM system which may incorporate nonuniform quantization and Roberts's pseudonoise technique. We can choose the reconstruction levels in the nonuniform quantizer exploiting the notion that the same level of noise is less visible in regions with high local

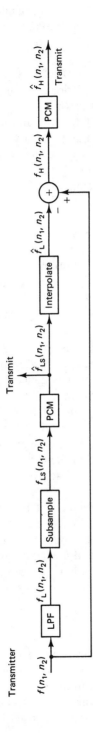

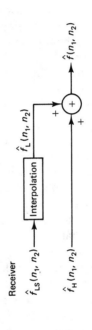

Figure 10.31 Two-channel image coder.

contrast. Allowing 3 bits/pixel to code $f_H(n_1, n_2)$ results in good quality for typical images. At the receiver, $\hat{f}_L(n_1, n_2)$ is obtained by interpolating the received $\hat{f}_{LS}(n_1, n_2)$. The result is combined with the received $\hat{f}_H(n_1, n_2)$ to obtain the reconstructed image $\hat{f}(n_1, n_2)$.

A two-channel coder is similar to a DPCM system. The lows component $\hat{f}_L(n_1, n_2)$ can be viewed as the DPCM prediction $f'(n_1, n_2)$, and the highs component $f_H(n_1, n_2)$ can be viewed as the DPCM error $e(n_1, n_2) = f(n_1, n_2) - f'(n_1, n_2)$. The difference lies in the way in which $\hat{f}_L(n_1, n_2)$ and $f'(n_1, n_2)$ are obtained. In a two-channel coder, $\hat{f}_L(n_1, n_2)$ is obtained directly from $f(n_1, n_2)$. Since the receiver does not have $f(n_1, n_2)$, $\hat{f}_L(n_1, n_2)$ has to be transmitted. In a DPCM system, $f'(n_1, n_2)$ is obtained from previously reconstructed pixel intensities, so it does not have to be transmitted. Transmission of $\hat{f}_L(n_1, n_2)$, although increasing the bit rate, offers some advantages. In a DPCM system, $f'(n_1, n_2)$ is obtained recursively from previously reconstructed pixel intensities, so what is done in coding the current pixel intensity affects all the pixels coded in the future. Therefore, any channel error in the transmission will affect not only the current pixel intensity, but also all future pixel intensities. In addition, modifying the difference signal $e(n_1, n_2)$, perhaps to enhance the image, is difficult, since modifying $e(n_1, n_2)$ at the current pixel intensity affects all future pixel intensities. The effect of the channel error or modification of the local contrast $f_H(n_1, n_2)$ is much more localized in a two-channel coder.

A two-channel coder can be viewed as a special case of a subband image coder. In a two-channel coder, an image is divided into two channels, that is, the lowpass and highpass, and each component is coded by a coder that is adapted specifically to that channel. We can, of course, divide an image into many bands (typically 16 bands in image coding) by using bandpass filters and code each band with a coder specifically adapted to it. A coding method in which a signal is divided into many channels and each channel is then coded with its own coder is called *subband signal coding*. The subband signal coding method was first applied to speech coding [Crochiere, et al.] and then later to image coding [Vetterli; Woods and O'Neil].

Figure 10.32 illustrates the performance of a two-channel coder. Figure 10.32(a) shows an original digital image of 512×512 pixels at 8 bits/pixel. Figure 10.32(b) shows the result of a two-channel coder at $3\frac{1}{8}$ bits/pixel. The subsampling rate used to generate $f_{LS}(n_1, n_2)$ is 1:64 and $f_{LS}(n_1, n_2)$ is quantized at 8 bits/pixel. Therefore, the bit rate used for coding the lows component $f_L(n_1, n_2)$ is $\frac{1}{8}$ bit/pixel. The highs component $f_H(n_1, n_2)$ is quantized at 3 bits/pixel.

10.3.5 Pyramid Coding

A pyramid is a data structure that provides successively condensed information of an image. A pyramid is useful in a variety of image processing applications, including image coding and analysis. The multigrid approach for motion estimation that we discussed in Section 8.4.2 is an example of pyramid image processing. We will refer to a coding method that exploits the data structure of a pyramid as *pyramid coding*.

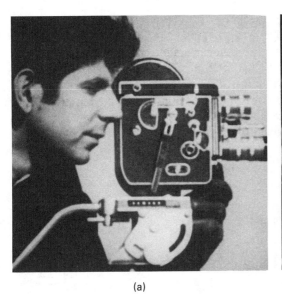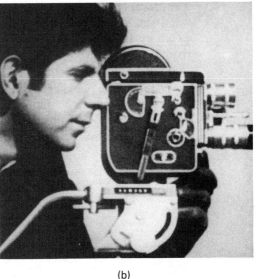

(a)   (b)

Figure 10.32 Example of image coding by a two-channel coder. (a) Original image of 512 × 512 pixels; (b) coded image at 3⅛ bits/pixel. NMSE = 1.0%, SNR = 19.8 dB.

It is possible to develop many image representations [Rosenfeld] that can be viewed as pyramids. In this section, we discuss one particular representation developed by [Burt and Adelson]. This pyramid representation consists of an original image and successively lower resolution (blurred) images and can be used for image coding.

Let $f_0(n_1, n_2)$ denote an original image of $N \times N$ pixels where $N = 2^M + 1$, for example, 129×129, 257×257, 513×513, and so forth. It is straightforward to generate an image of $(2^M + 1) \times (2^M + 1)$ pixels from an image of $2^M \times 2^M$ pixels, for example, by simply repeating the last column once and the last row once. We assume a square image for simplicity. We will refer to $f_0(n_1, n_2)$ as the base level image of the pyramid. The image at one level above the base is obtained by lowpass filtering $f_0(n_1, n_2)$ and then subsampling the result. Suppose we filter $f_0(n_1, n_2)$ with a lowpass filter $h_0(n_1, n_2)$ and denote the result by $f_0^L(n_1, n_2)$ so that

$$f_0^L(n_1, n_2) = L[f_0(n_1, n_2)] = f_0(n_1, n_2) * h_0(n_1, n_2) \qquad (10.43)$$

where $L[\cdot]$ is the lowpass filtering operation. Since $f_0^L(n_1, n_2)$ has a lower spatial resolution than $f_0(n_1, n_2)$ due to lowpass filtering, we can subsample $f_0^L(n_1, n_2)$. We denote the result of the subsampling operation by $f_1(n_1, n_2)$. The image $f_1(n_1, n_2)$ is smaller in size than $f_0(n_1, n_2)$ due to subsampling and is the image at one level above the base of the pyramid. We will refer to $f_1(n_1, n_2)$ as the first-level image of the pyramid. The second-level image, $f_2(n_1, n_2)$, is obtained by lowpass filtering the first-level image $f_1(n_1, n_2)$ and then subsampling the result. This procedure can be repeated to generate higher level images $f_3(n_1, n_2)$, $f_4(n_1, n_2)$, and so forth.

The process of generating $f_{i+1}(n_1, n_2)$ from $f_i(n_1, n_2)$ is shown in Figure 10.33. We will assume that the Kth-level image $f_K(n_1, n_2)$ is at the top of the pyramid. As we go up the pyramid, the image size becomes smaller and the image has lower spatial resolution. The images $f_i(n_1, n_2)$ for $0 \le i \le K$ can be viewed as multi-resolution images and form the pyramid.

Depending on what lowpass filter is used and how the lowpass filtered image is subsampled, many variations are possible. In the *Gaussian pyramid representation* [Burt and Adelson], the lowpass filter used is a separable filter with 5×5-point impulse response $h(n_1, n_2)$ given by

$$h(n_1, n_2) = h(n_1)h(n_2) \tag{10.44a}$$

where

$$h(n) = \begin{cases} a, & n = 0 \\ \dfrac{1}{4}, & n = \pm 1 \\ \dfrac{1}{4} - \dfrac{a}{2}, & n = \pm 2. \end{cases} \tag{10.44b}$$

The constant a in (10.44b) is a free parameter and is chosen typically between 0.3 and 0.6. The sequence $h(n)$ is sketched in Figure 10.34 for $a = 0.3, 0.4, 0.5,$ and 0.6. When $a = 0.4$, $h(n)$ has approximately Gaussian shape and thus the name "Gaussian pyramid." The choice of $h(n_1, n_2)$ in (10.44) ensures that $h(n_1, n_2)$ is zero phase, and the filter passes the DC component unaffected ($H(0, 0) = 1$, $\Sigma_{n_1} \Sigma_{n_2} h(n_1, n_2) = 1$). In addition, separability of $h(n_1, n_2)$ reduces computations involved in the filtering operation. The image $f_0^L(n_1, n_2)$ obtained from $f_0(n_1, n_2)$ $* h(n_1, n_2)$ is then subsampled by a factor of 4, that is, a factor of 2 along n_1 and a factor of 2 along n_2. The subsampled image $f_1(n_1, n_2)$ is given by

$$f_1(n_1, n_2) = \begin{cases} f_0^L(2n_1, 2n_2), & 0 \le n_1 \le 2^{M-1}, \quad 0 \le n_2 \le 2^{M-1} \\ 0, & \text{otherwise.} \end{cases} \tag{10.45}$$

The size of $f_1(n_1, n_2)$ is $(2^{M-1} + 1) \times (2^{M-1} + 1)$ pixels, approximately one quarter of the size of $f_0(n_1, n_2)$. From (10.45), $f_0^L(n_1, n_2)$ has to be computed for only even values of n_1 and n_2 to obtain $f_1(n_1, n_2)$. Higher-level images are generated by repeatedly applying the lowpass filtering and subsampling operations. A one-dimensional graphical representation of the above process is shown in Figure 10.35. An example of the Gaussian pyramid representation for an image of 513×513 pixels is shown in Figure 10.36.

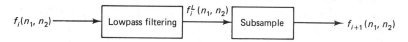

Figure 10.33 Process of generating the $i + 1$th-level image $f_{i+1}(n_1, n_2)$ from the ith-level image $f_i(n_1, n_2)$ in Gaussian pyramid image representation.

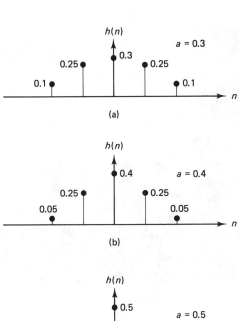

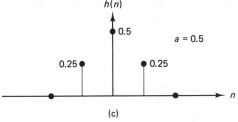

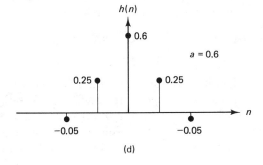

Figure 10.34 Impulse response $h(n)$ as a function of the free parameter "a." The 2-D lowpass filter $h(n_1, n_2)$ used in the Gaussian pyramid image representation is obtained from $h(n)$ by $h(n_1, n_2) = h(n_1)h(n_2)$.

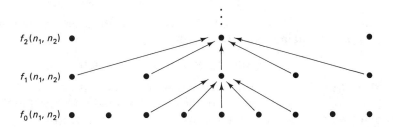

Figure 10.35 One-dimensional graphical representation of the Gaussian pyramid generation.

Figure 10.36 Example of the Gaussian pyramid representation for image of 513×513 pixels with $K = 4$.

The Gaussian pyramid representation can be used in developing an approach to image coding. To code the original image $f_0(n_1, n_2)$, we code $f_1(n_1, n_2)$ and the difference between $f_0(n_1, n_2)$ and a prediction of $f_0(n_1, n_2)$ from $f_1(n_1, n_2)$. Suppose we predict $f_0(n_1, n_2)$ by interpolating $f_1(n_1, n_2)$. Denoting the interpolated image by $f_1^I(n_1, n_2)$, we find that the error signal $e_0(n_1, n_2)$ coded is

$$
\begin{aligned}
e_0(n_1, n_2) &= f_0(n_1, n_2) - I[f_1(n_1, n_2)] \\
&= f_0(n_1, n_2) - f_1^I(n_1, n_2)
\end{aligned}
\tag{10.46}
$$

where $I[\cdot]$ is the spatial interpolation operation. The interpolation process expands the support size of $f_1(n_1, n_2)$, and the support size of $f_1^I(n_1, n_2)$ is the same as $f_0(n_1, n_2)$. One advantage of coding $f_1(n_1, n_2)$ and $e_0(n_1, n_2)$ rather than $f_0(n_1, n_2)$ is that the coder used can be adapted to the characteristics of $f_1(n_1, n_2)$ and $e_0(n_1, n_2)$. If we do not quantize $f_1(n_1, n_2)$ and $e_0(n_1, n_2)$, from (10.46) $f_0(n_1, n_2)$ can be recovered exactly by

$$
f_0(n_1, n_2) = I[f_1(n_1, n_2)] + e_0(n_1, n_2).
\tag{10.47}
$$

In image coding, $f_1(n_1, n_2)$ and $e_0(n_1, n_2)$ are quantized and the reconstructed image $\hat{f}_0(n_1, n_2)$ is obtained from (10.47) by

$$
\hat{f}_0(n_1, n_2) = I[\hat{f}_1(n_1, n_2)] + \hat{e}_0(n_1, n_2)
\tag{10.48}
$$

where $\hat{f}_0(n_1, n_2)$ and $\hat{e}_0(n_1, n_2)$ are quantized versions of $f_0(n_1, n_2)$ and $e_0(n_1, n_2)$. If we stop here, the structure of the coding method is identical to the two-channel coder we discussed in the previous section. The image $f_1(n_1, n_2)$ can be viewed as the subsampled lows component $f_{LS}(n_1, n_2)$ and $e_0(n_1, n_2)$ can be viewed as the highs component $f_H(n_1, n_2)$ in the system in Figure 10.31.

The idea that an image can be decomposed into two components with very different characteristics can also be applied to coding $f_1(n_1, n_2)$, the first-level image in the Gaussian pyramid. Instead of coding $f_1(n_1, n_2)$, we can code $f_2(n_1, n_2)$ and $e_1(n_1, n_2)$ given by

$$e_1(n_1, n_2) = f_1(n_1, n_2) - I[f_2(n_1, n_2)]. \tag{10.49}$$

This procedure, of course, can be repeated. Instead of coding $f_i(n_1, n_2)$, we can code $f_{i+1}(n_1, n_2)$ and $e_i(n_1, n_2)$ given by

$$e_i(n_1, n_2) = f_i(n_1, n_2) - I[f_{i+1}(n_1, n_2)]. \tag{10.50}$$

If we do not quantize $f_{i+1}(n_1, n_2)$ and $e_i(n_1, n_2)$, from (10.50) $f_i(n_1, n_2)$ can be recovered exactly from $f_{i+1}(n_1, n_2)$ and $e_i(n_1, n_2)$ by

$$f_i(n_1, n_2) = I[f_{i+1}(n_1, n_2)] + e_i(n_1, n_2). \tag{10.51}$$

We can repeat the above process until we reach the top level of the pyramid. This is shown in Figure 10.37. Instead of coding $f_0(n_1, n_2)$, we code $e_i(n_1, n_2)$ for $0 \le i \le K - 1$ and $f_K(n_1, n_2)$. An example of $e_i(n_1, n_2)$ for $0 \le i \le K - 1$ and $f_K(n_1, n_2)$ for an original image $f_0(n_1, n_2)$ of 513×513 pixels for $K = 4$ is shown in Figure 10.38. If $e_i(n_1, n_2)$ for $0 \le i \le K - 1$ and $f_k(n_1, n_2)$ are not quantized, then $f_0(n_1, n_2)$ can be reconstructed exactly from them by recursively solving (10.51) for $i = K - 1, K - 2, \ldots, 0$. Note that (10.51) is valid independent of the specific choice of the interpolation operation $I[\cdot]$. Equation (10.51) can be used for reconstructing $f_0(n_1, n_2)$ from quantized $e_i(n_1, n_2)$ for $0 \le i \le K - 1$ and $f_K(n_1, n_2)$.

The images $f_K(n_1, n_2)$ and $e_i(n_1, n_2)$ for $0 \le i \le K - 1$ form a pyramid called the *Laplacian pyramid*, where $e_i(n_1, n_2)$ is the ith-level image of the pyramid and $f_K(n_1, n_2)$ is the top level image of the pyramid. From (10.50),

$$e_0(n_1, n_2) = f_0(n_1, n_2) - I[f_1(n_1, n_2)]. \tag{10.52}$$

From Figure 10.33, $f_1(n_1, n_2)$ is the result of subsampling $f_0(n_1, n_2) * h(n_1, n_2)$. Approximating the interpolation operation $I[\cdot]$ as the reversal of the subsampling operation,

$$\begin{aligned} e_0(n_1, n_2) &\simeq f_0(n_1, n_2) - f_0(n_1, n_2) * h(n_1, n_2) \\ &= f_0(n_1, n_2) * (\delta(n_1, n_2) - h(n_1, n_2)). \end{aligned} \tag{10.53}$$

Since $h(n_1, n_2)$ has lowpass character, $e_0(n_1, n_2)$ has a highpass character. Now consider $e_1(n_1, n_2)$, the first-level image of the Laplacian pyramid. Following the steps similar to those that led to (10.53) and making a few additional approximations, we obtain

$$I[e_1(n_1, n_2)] \simeq f_0(n_1, n_2) * h_1(n_1, n_2) \tag{10.54}$$

where

$$h_1(n_1, n_2) = h(n_1, n_2) - h(n_1, n_2) * h(n_1, n_2). \tag{10.55}$$

From (10.54), the result of interpolating $e_1(n_1, n_2)$ such that its size is the same as $f_0(n_1, n_2)$ is approximately the same as the result of filtering $f_0(n_1, n_2)$ with $h_1(n_1, n_2)$.

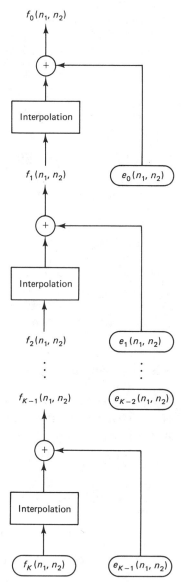

Figure 10.37 Laplacian pyramid generation. The base image $f_0(n_1, n_2)$ can be reconstructed from $e_i(n_1, n_2)$ for $0 \leq i \leq K - 1$ and $f_K(n_1, n_2)$.

Since $h(n_1, n_2)$ is a lowpass filter, $h_1(n_1, n_2)$ in (10.56) is a bandpass filter. If we continue the above analysis further, we can argue that the result of repetitive interpolation of $e_i(n_1, n_2)$ for $1 \leq i \leq K - 1$ is approximately the same as the result of filtering $f_0(n_1, n_2)$ with a different bandpass filter. As we increase i from 1 to $K - 1$, the frequency response of the bandpass filter has a successively smaller effective bandwidth with successively lower passband frequencies. If $h(n_1, n_2)$ has approximately Gaussian shape, then so will $h(n_1, n_2) * h(n_1, n_2)$. For a Gaussian-shaped $h(n_1, n_2)$, from (10.56) the bandpass filter has an impulse response that is

Figure 10.38 Example of the Laplacian pyramid image representation with $K = 4$. The original image used is the 513 $\times$ 513-pixel image $f_0(n_1, n_2)$ in Figure 10.36. $e_i(n_1, n_2)$ for $0 \leq i \leq 3$ and $f_4(n_1, n_2)$.

the difference of the two Gaussian functions. The difference of two Gaussians can be modeled [Marr] approximately by the Laplacian of a Gaussian, hence the name "Laplacian pyramid."

From the above discussion, the pyramid coding method we discussed can be viewed as an example of subband image coding. As we have stated briefly, in subband image coding, an image is divided into different frequency bands and each band is coded with its own coder. In the pyramid coding method we discussed, the bandpass filtering operation is performed implicitly and the bandpass filters are obtained heuristically. In a typical subband image coder, the bandpass filters are designed more theoretically [Vetterli; Woods and O'Neil].

Figure 10.39 illustrates the performance of an image coding system in which $f_K(n_1, n_2)$ and $e_i(n_1, n_2)$ for $0 \leq i \leq K-1$ are coded with coders adapted to the signal characteristics. Qualitatively, higher-level images have more variance and more bits/pixel are assigned. Fortunately, however, they are smaller in size. Figure 10.39 shows an image coded at $\frac{1}{2}$ bit/pixel. The original image used is the 513 $\times$ 513-pixel image $f_0(n_1, n_2)$ in Figure 10.36. The bit rate of less than 1 bit/pixel was possible in this example by entropy coding and by exploiting the observation that most pixels of the 513 $\times$ 513-pixel image $e_0(n_1, n_2)$ are quantized to zero.

One major advantage of the pyramid-based coding method we discussed above is its suitability for progressive data transmission. By first sending the top-level image $f_K(n_1, n_2)$ and interpolating it at the receiver, we have a very blurred image. We then transmit $e_{K-1}(n_1, n_2)$ to reconstruct $f_{K-1}(n_1, n_2)$, which has a higher spatial resolution than $f_K(n_1, n_2)$. As we repeat the process, the reconstructed image at the receiver will have successively higher spatial resolution. In some applications, it may be possible to stop the transmission before we fully

Figure 10.39 Example of the Laplacian pyramid image coding with $K = 4$ at $\frac{1}{2}$ bit/pixel. The original image used is the 513×513-pixel image $f_0(n_1, n_2)$ in Figure 10.36.

recover the base level image $f_0(n_1, n_2)$. For example, we may be able to judge from a blurred image that the image is not what we want. Fortunately, the images are transmitted from the top to the base of the pyramid. The size of images increases by approximately a factor of four as we go down each level of the pyramid.

In addition to image coding, the Laplacian pyramid can also be used in other applications. For example, as we discussed above, the result of repetitive interpolation of $e_i(n_1, n_2)$ such that its size is the same as that of $f_0(n_1, n_2)$ can be viewed as approximately the result of filtering $f_0(n_1, n_2)$ with the Laplacian of a Gaussian. As we discussed in Section 8.3.3, zero-crossing points of the result of filtering $f_0(n_1, n_2)$ with the Laplacian of a Gaussian are the edge points in the edge detection method by Marr and Hildreth.

10.3.6 Adaptive Coding and Vector Quantization

The waveform coding techniques discussed in previous sections can be modified to adapt to changing local image characteristics. In a PCM system, the reconstruction levels can be chosen adaptively. In a DM system, the step size Δ can be chosen adaptively. In regions where the intensity varies slowly, for example, Δ can be chosen to be small to reduce granular noise. In regions where the intensity increases or decreases rapidly, Δ can be chosen to be large to reduce the slope overload distortion problem. In a DPCM system, the prediction coefficients and the reconstruction levels can be chosen adaptively. Reconstruction levels can also be chosen adaptively in a two-channel coder and a pyramid coder. The number of bits assigned to each pixel can also be chosen adaptively in all the waveform coders we discussed. In regions where the quantized signal varies very slowly, for example, we may want to assign a smaller number of bits/pixel. It is also possible to have a fixed number of bits/frame, while the bit rate varies at a pixel level.

In adaptive coding, the parameters in the coder are adapted based on some local measure, such as local image contrast. If the local measure used can be obtained from previously coded pixel intensities, then it does not have to be transmitted. If the local measure used is obtained directly from $f(n_1, n_2)$, it must be transmitted, since the receiver does not have access to the original image. Adaptive coding certainly adds complexity to an image coder, but can often significantly improve its performance.

PCM systems do not exploit the statistical dependence of neighborhood pixel intensities. One way to exploit the statistical dependence is to use methods such as DM, DPCM, and two-channel coders, where the difference between $f(n_1, n_2)$ and a prediction of $f(n_1, n_2)$ is coded. An alternate way is to use vector quantization. As we discussed in Section 10.2, vector quantization can exploit the statistical dependence of the parameters coded. Vector quantization has been considered in coding the waveform $f(n_1, n_2)$. The blocks used consist of neighborhood pixel intensities of a small size, typically 2×2, 3×3, and 4×4. Vector quantization has been primarily applied in low bit rate (below 1 bit/pixel) applications, since the computational and storage costs increase rapidly with the block size and the bit rate. Intelligible images can be reconstructed with some sacrifice in quality at bit rates below 1 bit/pixel, which is not possible with DM, DPCM, or two-channel coders with scalar quantization and uniform-length codeword assignment. For waveform coders, vector quantization is an effective way to code an image at a bit rate lower than 1 bit/pixel. Figure 10.40 illustrates the performance of vector quantization when applied to coding the waveform $f(n_1, n_2)$. Figure 10.40(a) shows an original image of 512×512 pixels. Figure 10.40(b) shows the

(a)   (b)

Figure 10.40 Example of an image coded by vector quantization. Courtesy of William Equitz. (a) Original image of 512×512 pixels; (b) coded image by vector quantization at $\frac{1}{2}$ bit/pixel. The block size used is 4×4 pixels and the codebook is designed by using a variation of the K-means algorithm. NMSE = 2.7%, SNR = 15.7 dB.

result of vector quantization at $\frac{1}{2}$ bit/pixel. The block size used is 4×4 pixels, and the codebook in this example was designed by using a variation of the K-means algorithm discussed in Section 10.1.2. The training data used are all the blocks of four different images.

10.4 TRANSFORM IMAGE CODING

In transform image coding, an image is transformed to a domain significantly different from the image intensity domain, and the transform coefficients are then coded. In low bit rate applications (below 1 or 2 bits/pixel) such as video conferencing, transform coding techniques with scalar quantization typically perform significantly better than waveform coding techniques with scalar quantization. They are, however, more expensive computationally.

Transform coding techniques attempt to reduce the correlation that exists among image pixel intensities more fully than do waveform coding techniques. When the correlation is reduced, redundant information does not have to be coded repeatedly. Transform coding techniques also exploit the observation that for typical images a large amount of energy is concentrated in a small fraction of the transform coefficients. This is called the *energy compaction property*. Because of this property, it is possible to code only a fraction of the transform coefficients without seriously affecting the image. This allows us to code images at bit rates below 1 bit/pixel with a relatively small sacrifice in image quality and intelligibility.

In coding transform coefficients, we can use any of the quantization methods discussed in Section 10.1. However, the method that has been used most frequently is scalar quantization, because of its simplicity. In this section, we will assume scalar quantization unless otherwise specified.

10.4.1 Transforms

A schematic diagram of a transform image coder is shown in Figure 10.41. At the transmitter, the image $f(n_1, n_2)$ is transformed and the transform coefficients $T_f(k_1, k_2)$ are quantized. The quantized $\hat{T}_f(k_1, k_2)$ are then coded. At the receiver, the codewords are decoded and the resulting quantized transform coefficients $\hat{T}_f(k_1, k_2)$ are inverse transformed to obtain the reconstructed image $\hat{f}(n_1, n_2)$.

Several properties are desirable in the transform. Since we compute the transform at the transmitter and the inverse transform at the receiver, it is important to have efficient ways to compute them. Transforms which have been considered for image coding are linear transforms that can be expressed as

$$T_f(k_1, k_2) = \sum_{n_1=0}^{N_1-1} \sum_{n_2=0}^{N_2-1} f(n_1, n_2) a(n_1, n_2; k_1, k_2) \tag{10.56a}$$

$$f(n_1, n_2) = \sum_{k_1=0}^{N_1-1} \sum_{k_2=0}^{N_2-1} T_f(k_1, k_2) b(n_1, n_2; k_1, k_2) \tag{10.56b}$$

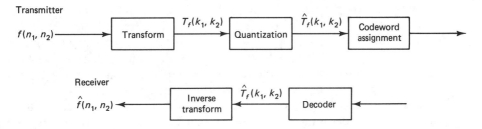

Figure 10.41 Transform image coder.

where $f(n_1, n_2)$ is an $N_1 \times N_2$-point sequence, $T_f(k_1, k_2)$ is also an $N_1 \times N_2$-point sequence and represents the transform coefficients, and $a(n_1, n_2; k_1, k_2)$ and $b(n_1, n_2; k_1, k_2)$ are the basis functions that satisfy (10.56). From (10.56), we can deduce that $f(n_1, n_2)$ is a linear combination of the basis functions $b(n_1, n_2; k_1, k_2)$ and that the transform coefficients $T_f(k_1, k_2)$ are the amplitudes of the basis functions in the linear combination. When the basis functions have some form of sinusoidal behavior, the transform coefficients can be interpreted as amplitudes of generalized spectral components. From computational considerations, the basis functions used in most transform image coders are separable, so (10.56) can be expressed as

$$T_f(k_1, k_2) = \sum_{n_1=0}^{N_1-1} \sum_{n_2=0}^{N_2-1} f(n_1, n_2) a_R(n_1; k_1) a_C(n_2; k_2), \tag{10.57a}$$

$$f(n_1, n_2) = \sum_{k_1=0}^{N_1-1} \sum_{k_2=0}^{N_2-1} T_f(k_1, k_2) b_R(n_1; k_1) b_C(n_2; k_2). \tag{10.57b}$$

An example of a transform in the form of (10.57) is the discrete Fourier transform (DFT) discussed in Section 3.2;

$$F(k_1, k_2) = \sum_{n_1=0}^{N_1-1} \sum_{n_2=0}^{N_2-1} f(n_1, n_2) e^{-j(2\pi/N_1)k_1n_1} e^{-j(2\pi/N_2)k_2n_2} \tag{10.58a}$$

$$f(n_1, n_2) = \frac{1}{N_1N_2} \sum_{k_1=0}^{N_1-1} \sum_{k_2=0}^{N_2-1} F(k_1, k_2) e^{j(2\pi/N_1)k_1n_1} e^{j(2\pi/N_2)k_2n_2}. \tag{10.58b}$$

When the basis functions are separable, the transform and inverse transform can be computed by row-column decomposition, in which 1-D transforms are computed row by row and then column by column. Row-column decomposition for computing a 2-D DFT was discussed in Section 3.4.1. Row-column decomposition can reduce the number of arithmetic operations by orders of magnitude compared to direct computation. In the case of a 512×512-point DFT computation, row-column decomposition alone reduces the number of arithmetic operations by a factor of several hundred (see Section 3.4.1). For transforms such as the DFT, additional computational savings can be obtained by exploiting the sinusoidal behavior of the basis functions.

Another desirable property of a transform is reduction of correlation among transform coefficients. We will refer to this property as the *correlation reduction property*. By proper choice of the basis function $b(n_1, n_2; k_1, k_2)$, the correlation among the coefficients can be reduced. Another desirable property related to the correlation reduction property is energy compaction. Compaction of most energy into a fraction of the transform coefficients allows us to discard many coefficients without seriously affecting the image. Note that correlation reduction contributes to energy compaction but is not by itself sufficient for energy compaction. For example, an amplitude of a random white noise signal is uncorrelated with any other amplitude of the signal, but the signal does not have the energy compaction property. Fortunately, typical images have been observed to have the energy compaction property, when the correlation among the pixel intensities is reduced by a transform operation.

An issue closely related to the energy compaction property is the preservation of some measure of energy in the transform domain. In other words, a transform coefficient with little energy should contribute only a small amount to the signal energy. The DFT has such a property. Specifically, Parseval's theorem states that

$$\sum_{n_1=0}^{N_1-1} \sum_{n_2=0}^{N_2-1} |f(n_1, n_2)|^2 = \frac{1}{N_1 N_2} \sum_{k_1=0}^{N_1-1} \sum_{k_2=0}^{N_2-1} |F(k_1, k_2)|^2. \tag{10.59}$$

From (10.59), a DFT coefficient with small magnitude contributes only a small amount to the signal energy. Discarding DFT coefficients with small magnitudes, therefore, does not seriously distort the signal. All the linear transforms* that have been considered for image coding have properties similar to (10.59).

Many different transforms have been considered for transform image coding. They differ in efficiency in energy compaction and in computational requirements. The Karhunen-Loève (KL) transform is the best of all linear transforms for energy compaction. In the KL transform, the basis functions $a(n_1, n_2; k_1, k_2)$ are obtained within a scaling factor by solving

$$\lambda(k_1, k_2)a(n_1, n_2; k_1, k_2) = \sum_{l_1=0}^{N_1-1} \sum_{l_2=0}^{N_2-1} K_f(n_1, n_2; l_1, l_2)a(l_1, l_2; k_1, k_2) \tag{10.60}$$

where

$$K_f(n_1, n_2; l_1, l_2) = E[(f(n_1, n_2) - E[f(n_1, n_2)])(f(l_1, l_2) - E[f(l_1, l_2)])]. \tag{10.61}$$

In (10.60), it is assumed that $f(n_1, n_2)$ is a random process with a known covariance function $K_f(n_1, n_2; l_1, l_2)$. To the extent that this assumption is valid, the coefficients of KL transforms are completely uncorrelated and the KL transform has the best energy compaction property. Specifically, suppose we order the transform coefficients from the largest expected squared magnitude to the smallest. If we

---

*The energy preservation property holds for all unitary transforms. The forward and inverse unitary transforms can be expressed as $\mathbf{F} = Q\mathbf{f}$ and $\mathbf{f} = Q^H\mathbf{F}$, respectively, where H represents the Hermitian (conjugate transpose) operation, $\mathbf{f}$ is a column vector (see Section 9.7) obtained from $f(n_1, n_2)$, and $\mathbf{F}$ is a column vector obtained from $T_f(k_1, k_2)$. From the forward and inverse transforms, signal energy = $\mathbf{f}^H\mathbf{f} = (Q^H\mathbf{F})^H(Q^H\mathbf{F}) = \mathbf{F}^H Q\, Q^H\mathbf{F} = \mathbf{F}^H\mathbf{F}$
The DFT is a unitary transform to within a scale factor.

choose the first M coefficients for any M and compute the energy in the signal corresponding to the M transform coefficients, the energy is largest on the average for the KL transform among all linear transforms.

The KL transform is interesting theoretically, but it has serious practical difficulties. The basis function $a(n_1, n_2; k_1, k_2)$ used in the KL transform depends on $K_f(n_1, n_2; l_1, l_2)$. Different classes of images are likely to have different $K_f(n_1, n_2; l_1, l_2)$, and for each different $K_f(n_1, n_2; l_1, l_2)$, the implicit equation in (10.60) must be solved to obtain the basis function $a(n_1, n_2; k_1, k_2)$. In addition, for a general $K_f(n_1, n_2; l_1, l_2)$, there is no computationally efficient algorithm to compute the transform coefficients. Computing the transform of an $N_1 \times N_2$-point sequence without a computationally efficient algorithm requires on the order of $N_1^2 N_2^2$ multiplications and $N_1^2 N_2^2$ additions. Furthermore, the KL transform is based on the assumption that $f(n_1, n_2)$ is a random process and its covariance function $K_f(n_1, n_2; l_1, l_2)$ is known. In practice, the random process assumption for an image may not be valid. In addition, $K_f(n_1, n_2; l_1, l_2)$ is not known and must be estimated. If the estimate is not accurate, the optimal theoretical properties of the KL transform no longer hold. Due to these practical difficulties, the KL transform is seldom used in practice for image coding.

One transform which has a fixed set of basis functions, an efficient algorithm for its computation and good energy compaction for typical images is the DFT. Fast Fourier transform (FFT) algorithms can be used to compute the DFT and the inverse DFT efficiently. In addition, for typical images, a major portion of the signal energy is concentrated in the low-frequency regions, as discussed in Section 1.4.2. The DFT also plays an important role in digital signal processing. For these reasons, when transform image coding techniques were first developed in the late 1960's, the DFT was the first that was considered.

It is possible to improve the energy compaction property of the DFT without sacrificing other qualities such as the existence of a computationally efficient algorithm. The discrete cosine transform (DCT) has this improved characteristic. The DCT [Ahmed et al.] is closely related to the DFT and is the most widely used transform in transform image coding. To illustrate the relationship between the DCT and the DFT, and the improvement in energy compaction that the DCT offers over the DFT, we will consider the N-point 1-D sequence shown in Figure 10.42(a). In this example, $N = 4$. The DFT $F(k)$ consists of the discrete Fourier series (DFS) coefficients $\tilde{F}(k)$ for $0 \leq k \leq N - 1$. The DFS coefficients $\tilde{F}(k)$ are obtained from the periodic sequence $\tilde{f}(n)$ shown in Figure 10.42(b). From Figure 10.42(b), $\tilde{f}(n)$ has sharp artificial discontinuities which arise from joining the beginning and the end parts of $f(n)$. The discontinuities contribute energy to high frequencies and reduce the DFT's efficiency in energy compaction. To eliminate the artificial discontinuities, we create a $2N$-point sequence $g(n)$ from $f(n)$ as shown in Figure 10.42(c). The sequence $g(n)$ is symmetric with respect to the point halfway between $n = N - 1$ and $n = N$, and the first N points of $g(n)$ are the same as $f(n)$. If we now compute the $2N$-point DFT of $g(n)$, the DFT $G(k)$ is the same as the DFS coefficients $\tilde{G}(k)$ obtained from the periodic sequence $\tilde{g}(n)$ shown in Figure 10.42(d). The DCT is defined in terms of $G(k)$. Clearly, the sequence $\tilde{g}(n)$ does not have artificial discontinuities and the DCT's efficiency in energy

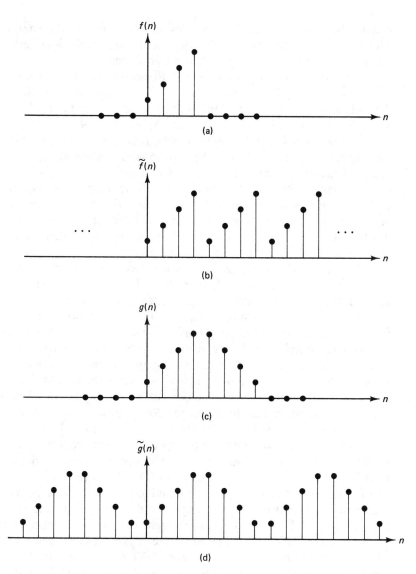

Figure 10.42 Qualitative explanation of better energy compaction property of the discrete cosine transform compared with the discrete Fourier transform.

compaction is improved. Even though the DFT size involved has doubled, the computational requirements are essentially the same as those of an N-point DFT because of the symmetry in $g(n)$. The DCT was discussed in detail in Section 3.3.

In addition to the DFT and the DCT, many other transforms, including the Haar, Hadamard, discrete sine, and slant transforms, have been considered for transform image coding. When the energy compaction property and computational requirements are considered together, the DCT is generally regarded as the best choice. Transforms such as the Haar and Hadamard require fewer computations than the DCT, but their energy compaction properties are not as good as that of the DCT for typical images. Since the DCT is the most widely used transform, we will use it in our subsequent discussion unless specified otherwise. Many of our discussions, however, apply to other transforms as well.

10.4.2 Implementation Considerations and Examples

Subimage-by-subimage coding. In transform image coding, an image is divided into many subimages or blocks, and each block is coded separately. By coding one subimage at a time, the coder can be made adaptive to local image characteristics. For example, the method of choosing quantization and bit allocation methods may differ between uniform background regions and edge regions. Subimage-by-subimage coding also reduces storage and computational requirements. Since one subimage is processed at a time, it is not necessary to store the entire image. To see the reduction in computational requirements, consider an image $f(n_1, n_2)$ of $N \times N$ pixels where N can be expressed as a power of 2. We divide $f(n_1, n_2)$ into many subimages. The size of each subimage is $M \times M$ pixels, where M can also be expressed as a power of 2. The number of subimages in the image is N^2/M^2. If we assume that the number of arithmetic operations involved in computing an $M \times M$-point transform is $M^2 \log_2 M^2$, the total number of arithmetic operations required in computing all the transforms in an image is $(N^2/M^2)M^2 \log_2 M^2 = N^2 \log_2 M^2$. Computing the transform of one $N \times N$-point image requires $N^2 \log_2 N^2$ arithmetic operations. When $N = 512$ and $M = 8$, the computation required in subimage-by-subimage coding is $\log_2 M^2/\log_2 N^2 = \frac{1}{3}$ of the computation required in transforming the entire image. Furthermore, in a multiple processor environment, transforms of subimages can be computed in parallel.

Although a smaller size subimage is more efficient computationally and allows us to design a coder more adaptive to local image characteristics, the subimage size cannot be reduced indefinitely. As we divide an image into smaller segments, transform coding exploits less of the correlation present among image pixel intensities. As the subimage size decreases, the correlation among neighboring subimages increases. However, each subimage is coded independently, and the increasing correlation among neighboring subimages is not exploited. This diminishes the performance of transform image coding and imposes a lower limit on the subimage size. Subimage sizes typically used in transform image coding are 8×8 and 16×16. Note that the size used in transform coding is much larger than that used in vector quantization. The block size used in vector quantization is typically 2×2, 3×3, or 4×4, due to the exponential dependence of the computational and memory requirements on the block size. In this respect, scalar quantization of transform coefficients has the potential to exploit the correlation present among pixel intensities better than vector quantization of the waveform.

Zonal and threshold coding. Transform image coding exploits the energy compaction property. Only a small fraction of the transform coefficients is typically coded. Two approaches used to determine which transform coefficients to code are zonal coding and threshold coding. In zonal coding, only the coefficients within a specified region are coded. Many factors, such as the transform used and the available number of bits, affect the zone shape and size. Two typical zone shapes used in DCT coding are shown in Figure 10.43. The coefficients in the shaded regions are coded, while those in the unshaded regions are discarded and assumed to be zero. The zone shapes used are consistent with the observation that most of the energy in typical images is concentrated in the low-frequency regions.

In threshold coding, transform coefficients are compared with a threshold, and those above the threshold are coded. Strictly from the energy compaction point of view, threshold coding is preferable to zonal coding. In zonal coding, some transform coefficients with small magnitudes may be coded while those with large magnitudes are discarded, since zones are prespecified. In threshold coding, only the coefficients with large magnitudes are selected. However, the locations of transforms coded are not known in advance, and information on the locations must be encoded in threshold coding. A typical method of coding coefficient locations is run-length coding [Jayant and Noll]. In one version of run-length coding, all the coefficients are ordered in some manner, and only the locations in which one coefficient is coded while its immediate neighbor coefficient is discarded are represented.

Threshold coding is an adaptive method. The choice of which transform coefficients are coded depends on the local image characteristics. For a fixed threshold, the number of coefficients coded varies from one subimage to the next. If we wish to develop a fixed bit rate system, in which the number of bits per subimage or per image remains the same, a control mechanism must be established to modify the threshold for different subimages and to adjust the bit allocation among different subimages and different coefficients within each subimage.

Bit allocation. Given a fixed number of bits for coding a subimage, we have to divide these bits among the transform coefficients coded within the subimage. As discussed in Section 10.1, it is beneficial to allocate more bits to a coefficient with a larger expected variance. The expected variance of each trans-

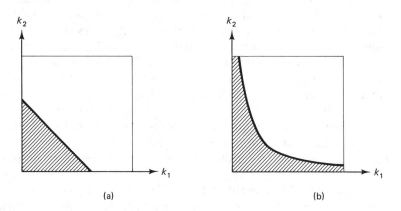

(a)   (b)

Figure 10.43 Examples of zones used in discrete cosine transform zonal coding.

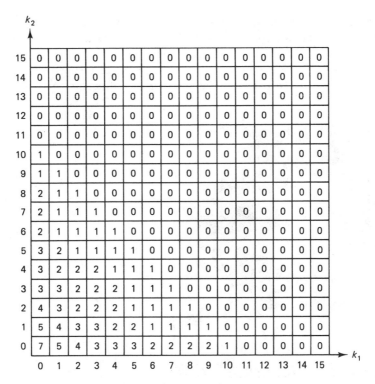

Figure 10.44 Example of bit allocation map at $\frac{1}{2}$ bit/pixel for zonal discrete cosine transform image coding. Block size $= 16 \times 16$ pixels.

form coefficient for a typical subimage varies greatly among transform coefficients. For the DCT, the expected variance is much larger for low-frequency coefficients than for high-frequency coefficients. An example of bit allocation for DCT coefficients of size 16×16 at a bit rate of $\frac{1}{2}$ bit/pixel is shown in Figure 10.44. The number in each square in the figure represents the number of bits allocated for each coefficient in zonal coding.

The choice of reconstruction levels in the quantization step depends on the number of bits allocated and on the expected variance of the coefficient. To eliminate this dependence on the expected variance, each coefficient is typically normalized by the expected standard deviation, and only one set of quantization levels is used for a given number of bits.

Degradations due to transform coding. Quantization noise in the reconstructed image manifests itself in transform image coding differently from waveform image coding. In general, the effect of quantization noise is less localized in transform image coding. Quantization of one transform coefficient affects all the image intensities within the subimage.

Several types of image degradation result from quantization noise in transform image coding. One type is loss of spatial resolution. In the DCT coding of images, the discarded transform coefficients are typically high-frequency components. The result is loss of detail in the image. An example of this type of degradation is shown in Figure 10.45. Figure 10.45(a) shows an original image of 512 × 512 pixels. Figures 10.45(b) and (c) show images reconstructed by retaining 14% and 8% of the DCT coefficients in each subimage, respectively. The subimage size used is 16 × 16 pixels. The transform coefficients retained are not quantized,

(a)

(b)

(c)

Figure 10.45 Illustration of spatial resolution loss due to discarding discrete cosine transform (DCT) coefficients.(a) Original image of 512 × 512 pixels. (b) reconstructed image with 14% of DCT coefficients retained. NMSE = 0.8%, SNR = 21.1 dB. (c) reconstructed image with 8% of DCT coefficients retained. NMSE = 1.2%, SNR = 19.3 dB.

Figure 10.46 Illustration of graininess increase due to quantization of DCT coefficients. A 2-bit/pixel uniform quantizer was used to quantize each DCT coefficient retained to reconstruct the image in Figure 10.45(b).

and are selected from a zone of triangular shape shown in Figure 10.43(a). From Figure 10.45, it is clear that the reconstructed image appears more blurry as we retain a smaller number of coefficients. It is also clear that an image reconstructed from only a small fraction of the transform coefficients looks quite good, illustrating the energy compaction property.

Another type of degradation results from quantization of the retained transform coefficients. The degradation in this case typically appears as graininess in the image. Figure 10.46 shows the result of coarse quantization of transform coefficients. This example is obtained by using a 2-bit uniform quantizer for each retained coefficient to reconstruct the image in Figure 10.45(b).

A third type of degradation arises from subimage-by-subimage coding. Since each subimage is coded independently, the pixels at the subimage boundaries may have artificial intensity discontinuities. This is known as the *blocking effect*, and is more pronounced as the bit rate decreases. An image with a visible blocking effect is shown in Figure 10.47. A DCT with zonal coding, a subimage of 16 × 16 pixels, and a bit rate of 0.15 bit/pixel were used to generate the image in Figure 10.47.

Examples. To design a transform coder at a given bit rate, different types of image degradation due to quantization have to be carefully balanced by a proper choice of various design parameters. As was discussed, these parameters include the transform used, subimage size, selection of which coefficients will be retained, bit allocation, and selection of quantization levels. If one type of degradation dominates other types of degradation, the performance of the coder can usually be improved by decreasing the dominant degradation at the expense of some increase in other types of degradation.

Figure 10.48 shows examples of transform image coding. Figure 10.48(a)

Figure 10.47 DCT-coded image with visible blocking effect.

and (b) show the results of DCT image coding at 1 bit/pixel and $\frac{1}{2}$ bit/pixel, respectively. The original image is the 512×512-pixel image shown in Figure 10.45(a). In both examples, the subimage size used is 16×16 pixels, and adaptive zonal coding with the zone shape shown in Figure 10.43(b) and the zone size adapted to the local image characteristics has been used.

(a)

(b)

Figure 10.48 Example of DCT image coding. (a) DCT-coded image at 1 bit/pixel. NMSE = 0.8%, SNR = 20.7 dB. (b) DCT-coded image at $\frac{1}{2}$ bit/pixel. NMSE = 0.9%, SNR = 20.2 dB.

10.4.3 Reduction of Blocking Effect

When the bit rate is sufficiently low, the blocking effect, which results from independent coding of each subimage, becomes highly visible. Reconstructed images exhibiting blocking effects can be very unpleasant visually, and blocking effects that are clearly visible often become the dominant degradation.

Two general approaches to reducing the blocking effect have been considered. In one approach, the blocking effect is dealt with at the source. An example of this approach is the overlap method, which modifies the image segmentation process. A typical segmentation procedure divides an image into mutually exclusive regions. In the overlap method, the subimages are obtained with a slight overlap around the perimeter of each subimage. The pixels at the perimeter are coded in two or more regions. In reconstructing the image, a pixel that is coded more than once can be assigned an intensity that is the average of the coded values. Thus, abrupt boundary discontinuities caused by coding are reduced because the reconstructed subimages are woven together. An example of the overlap method is shown in Figure 10.49. In the figure, a 5×5-pixel image is divided into four 3×3-pixel subimages by using a one-pixel overlap scheme. The shaded area indicates pixels that are coded more than once. The overlap method reduces blocking effects well. However, some pixels are coded more than once, and this increases the number of pixels coded. The increase is about 13% when an image of 256×256 pixels is divided into 16×16-pixel subimages with a one-pixel overlap. This increase shows why overlap of two or more pixels is not very useful. It also shows a difference between image coding and other image processing applications such as image restoration in dealing with blocking effects. As was discussed in Section 9.2.3, a blocking effect can occur in any subimage-by-subimage processing environment. In image restoration, the cost of overlapping subimages is primarily an increase in the number of computations. An overlap of 50% of the subimage size is common in subimage-by-subimage restoration. In image coding, however, the cost of overlapping subimages is an increase in the number of computations and, more seriously, a potential increase in the required bit rate. An overlap of more than one pixel is thus seldom considered in DCT image coding.

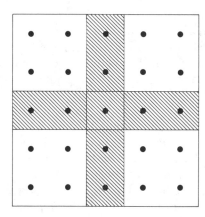

Figure 10.49 Example of one-pixel overlap in the overlap method for reducing blocking effect.

Another example of the general approach to deal with the blocking effect at the source is to use a different transform. In one class of transforms called lapped orthogonal transforms [Cassereau et al.], the subimages used are overlapped. However, the total number of transform coefficients remains the same as the number of pixels in the image. This is accomplished by representing a subimage with a smaller number of transform coefficients than the subimage size. Even though a subimage cannot be reconstructed exactly from the transform coefficients of the subimage, the entire image can be reconstructed exactly from the transform coefficients of all the subimages. Overlapping subimages reduces the blocking effect. Unlike the previous method where the DCT is used, the total number of lapped orthogonal transform coefficients necessary for exact reconstruction remains the same as the image size.

Another approach to reducing the blocking effect is to filter the image at the decoder after the blocking effect has occurred. In this approach, the coding procedure is not changed. For example, subimages remain mutually exclusive. The type of filter used is lowpass, since boundary discontinuities caused by segmentation are similar to sharp edges, which contribute to high-frequency components. In addition, the filter is typically applied only at or near the subimage boundaries to avoid unnecessary image blurring. Since the segmentation procedure used at the encoder is known at the decoder, locations of the artificial image discontinuities are known. Unlike the overlap method, the filtering method does not increase the number of pixels coded. In a limited study, the filtering method was shown to be more effective than the overlap method using the DCT in reducing the blocking effect at a given bit rate [Reeve and Lim]. How the filtering method compares in performance with the overlap method using the lapped orthogonal transforms has not yet been studied. Figure 10.50 shows an example of reduction of the blocking effect by the filtering method. Figure 10.50(a) shows an image with visible blocking effect. Figure 10.50(b) shows the image processed by applying a 3×3-point FIR filter with lowpass characteristics to only the pixels at the subimage boundaries in the image in Figure 10.50(a).

10.4.4 Hybrid Transform Coding

Transform coding methods perform very well in low bit rate applications, while waveform coding methods are very simple to implement. A hybrid transform/waveform coder, or hybrid coder for short, combines waveform and transform coding methods to achieve performance better than waveform-coding methods at low bit rates, while requiring simpler implementation than true 2-D transform coders.

In a hybrid coder, an image $f(n_1, n_2)$ is transformed by a 1-D transform, such as a 1-D DCT, along each row (or column). The result $T_f(k_1, n_2)$ is then coded by a waveform coder, such as DPCM, along each column (or row). This is illustrated in Figure 10.51. The 1-D transform decorrelates each row of data well. The remaining correlation is reduced further by DPCM. Due to the transform, the correlation in the data is reduced more than it would have been by waveform coding alone. Since a 1-D transform is used, the implementation issues such as

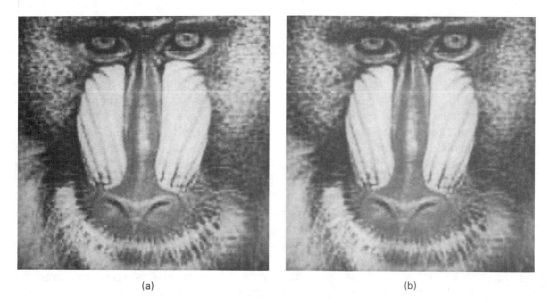

Figure 10.50 Example of blocking effect reduction using a filtering method. (a) Image of 512 × 512 pixels with visible blocking effect. The image is coded by a zonal DCT coder at 0.2 bit/pixel. (b) Image in (a) filtered to reduce the blocking effect. The filter used is a 3 × 3-point $h(n_1, n_2)$ with $h(0, 0) = \frac{1}{5}$ and $h(n_1, n_2) = \frac{1}{10}$ at the remaining eight points.

selection of the zone shape and size in zonal coding are simpler than those with a 2-D transform coder. Hybrid coding of a single image frame has not been used extensively in practice, perhaps because the method does not reduce the correlation in the data as much as a 2-D transform coder and the complexity in a 2-D transform coder implementation is not much higher than a hybrid coder. As will be discussed in Section 10.6, however, hybrid coding is useful in interframe image coding.

10.4.5 Adaptive Coding and Vector Quantization

Transform coding techniques can be made adaptive to the local characteristics within each subimage. In zonal coding, for example, the shape and size of the

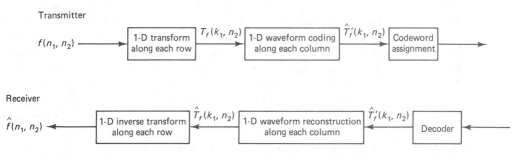

Figure 10.51 Hybrid transform/waveform coder.

zone can be adapted. In addition, the number of bits allocated to each subimage and each coefficient within a subimage can be adapted. For example, we may wish to code fewer coefficients and allocate fewer bits in uniform background regions than in edge regions.

A coding method may be adapted continuously based on some measure such as the variance of pixel intensities within a subimage. If the local measure can be obtained from previously coded subimages, it does not have to be transmitted. An alternative to continuous adaptation is to classify all subimages into a small number of groups and design a specific coding method for each group. An advantage of this approach is the small number of bits necessary to code the group that a given subimage belongs to. If the number of bits/frame is fixed in adaptive coding, a control mechanism that allocates the appropriate number of bits to each subimage is required. For a variable bit rate coder, a buffer must be maintained to accommodate the variations in local bit rates.

Adaptive coding significantly improves the performance of a transform coder while adding to its complexity. Transform coding methods are used extensively in low bit rate applications where reducing the bit rate is important, even if it involves additional complexity in the coder. Many transform coders used in practice are adaptive.

In transform coding, the pixel intensities in a block are transformed jointly. In this respect, transform coding is similar to vector quantization of a waveform. The transform coefficients, however, are typically quantized by scalar quantization. It is possible to use vector quantization to quantize transform coefficients. However, the additional complexity associated with vector quantization may not justify the performance improvement that vector quantization can offer. Transform coefficients are not correlated much with each other, while correlation among scalars is a major element exploited by vector quantization.

10.5 IMAGE MODEL CODING

In image model coding, an image or some portion of an image is modeled and the model parameters are used for image synthesis. At the transmitter, the model parameters are estimated from analyzing the image, and at the receiver the image is synthesized from the estimated and quantized model parameters. An image model coder is shown in Figure 10.52. It can be viewed as an analysis/synthesis system. Image model coders have the potential to synthesize intelligible images at a bit rate substantially lower than that necessary for waveform or transform coders. However, they are still at the research stage, and much needs to be done before their use becomes feasible in very low bit rate applications, such as video telephones for the deaf. Developing a simple model capable of synthesizing intelligible images is not an easy problem. In addition, estimating model parameters and synthesizing an image from them are likely to be very expensive computationally.

Image model coding exploits the notion that synthesizing intelligible images does not require detailed reproduction of image intensities. For example, back-

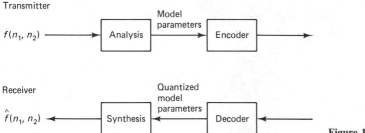

Transmitter

$f(n_1, n_2)$ ⟶ Analysis ⟶ Model parameters ⟶ Encoder ⟶

Receiver

$\hat{f}(n_1, n_2)$ ⟵ Synthesis ⟵ Quantized model parameters ⟵ Decoder ⟵

Figure 10.52 Image model coder.

ground regions of an image, such as grass, sky, and wall, may not be essential to image intelligibility, and we may be able to replace them with similar backgrounds that can be synthesized with a simple model. As another example, a cartoonist can draw a very intelligible image with a small number of simple lines. In image model coding, we attempt to retain the features of an image essential to its intelligibility and grossly approximate nonessential features, using simple image models. This approach contrasts sharply with waveform and transform coding, in which we attempt to accurately reconstruct the image intensity $f(n_1, n_2)$. In waveform and transform coding, the difference between $f(n_1, n_2)$ and the reconstructed $\hat{f}(n_1, n_2)$ comes from the quantization of the parameters, and $f(n_1, n_2)$ can be reconstructed exactly if the parameters used are not quantized. In image model coding, the difference between $f(n_1, n_2)$ and the synthesized $\hat{f}(n_1, n_2)$ is due both to quantization of the model parameters and to modeling error. It is generally not possible to exactly reconstruct $f(n_1, n_2)$ from the model parameters, even if we do not quantize them. The number of parameters involved is likely to be much smaller in image model coding than in waveform and transform coding, indicating the image model coding's potential for application to the development of very low bit rate systems.

A typical image consists of various regions with different characteristics. It is convenient to model different classes of regions with different models. Regions such as grass, water, sky, and wall have order or repetitive structure analogous to the texture of cloth. We will refer to such regions as texture regions [Brodatz]. There are two basic approaches to texture modeling. One is to use some basic elementary pattern and repeat it according to some deterministic or probabilistic rule. Another approach is to model a texture region as a random field with some given statistics. Studies indicate that when two textures have similar second-order statistics, they will usually appear to human viewers to have similar textures. Many models can be used to model texture as a random field with some given second-order statistics [Pratt et al.]. Figure 10.53 shows an example of a texture region's synthesis by a random process model. The figure shows an original image of 512 × 512 pixels with a region of 96 × 128 pixels replaced by texture synthesized by using a first-order Markov model [Cross and Jain; Cohen and Cooper] with 3 unknown model parameters. Even though more than 10,000 pixel intensities have been synthesized by using only 3 parameters, the synthesized texture blends well with the rest of the image. To use an approach of this type, some method of segmenting an image into regions of similar texture must also be developed.

Figure 10.53 Original image of 512 × 512 pixels with a region (upper right) of 96 × 128 pixels replaced with synthesized texture. First-order Markov model was used to synthesize the texture.

Adequately modeling image regions containing objects as purely textured regions may not be possible. Another approach to modeling such regions is based on the idea that a small number of well-placed contours can retain a great deal of intelligibility about the objects they represent. Using this approach, we can represent an image region with a set of contours, and, if we wish, fill in the regions bounded by contours with some texture. One method of determining the appropriate contours is to map the image to a bi-level image and look for the locations where the change between the two levels occurs. One advantage of this approach lies in the possible use of the bi-level image itself for image coding. Various methods of mapping a gray scale image to a bi-level image that retains important features of the original gray scale image have been considered [Covell]. We will discuss one of these methods.

A gray scale image can be mapped to a bi-level image following a hierarchy of decision rules. In this method [Covell], pixels that are mapped by a higher-level decision rule are not tested by lower-level decision rules. All pixels are mapped to white or black by the time they reach the lowest decision level. At the highest level, the pixel intensity is compared to high and low clipping levels. In many images, most regions where edge information is of interest have middle-range intensities. Depending on the clipping levels chosen, a significant percentage of the bi-level image may be generated at this level. At the next decision level, two parallel dynamic thresholds that take into account both the pixel intensity and the difference between the pixel intensity and the local luminance mean are used. A pixel of high intensity can be mapped to black only if it is significantly lower than the local luminance mean. Similarly, a pixel of low intensity can be mapped to white only if it is significantly above the local luminance mean. At the next decision level, an edge detection algorithm is applied to the pixels that have not yet been mapped to black or white. Based on the result of the edge detection algorithm, the pixel intensity, and the difference between the pixel intensity and local luminance mean, pixels with clear indication of edge presence are mapped to black. The pixels not mapped after this stage are rather nondescript. These pixels are mapped to black or white in some reasonable fashion. For example, in a multiple frame environment, they may be mapped to the same level as the pixel in the same location on the previous frame. The result is a bi-level image. The bi-level image thus obtained may be further processed by a median or similar filter to eliminate some isolated rapid changes that may be due to noise. Figure 10.54 shows an example of a bi-level image obtained from a gray level image of 512 × 512 pixels.

We can code the bi-level image directly, exploiting the notion that the level does not vary rapidly from one pixel to the next. In the spirit of image model coding, we may wish to model the contours (pixels where transitions between the two levels occur) with a set of such basic elements as straight lines, polynomials, circles, and ellipses which can be represented by a small number of parameters. The choice of the basic elements and the estimation of their parameters from the contours will depend on the class of objects that the contours represent. For example, the choice of the basic elements for contours representing human faces will be different from the choice for contours representing houses. Modeling a specific class of images such as human faces for very low bit rate applications is similar to modeling human speech. By detailed modeling of human speech, a particular class of acoustic signals, very low bit rate systems with reasonable intelligibility have been developed [Rabiner and Schafer]. Speech is, of course, very different from images in its method of generation at the source and its perception by humans. Whether or not the approach of modeling a particular class of objects in an image will result in a successful system in practice remains to be seen.

Figure 10.54 Example of bi-level image constructed from a gray level image.

10.6 INTERFRAME IMAGE CODING, COLOR IMAGE CODING, AND CHANNEL ERROR EFFECTS

10.6.1 Interframe Image Coding

We have discussed so far the problem of coding a single image frame. Coding a single frame exploits only the spatial correlation and is called *intraframe coding*. Applications such as motion pictures and television present a sequence of frames. A typical scene often does not change rapidly from one frame to the next, and the high correlation between neighboring frames can be exploited in image coding. Exploiting the temporal correlation as well as spatial correlation to code a sequence of images is called *interframe coding*.

Interframe coding requires storage of the frames used in the coding process. If we use N previous frames in coding the current frame, then these previous N frames have to be stored. In addition, there may be some inherent delay if all N frames are needed at the same time in coding the current frame, as in transform coding. This limits the number of frames that can be used in interframe coding. In most cases, only one or two additional frames are used in coding the current frame.

One approach to interframe coding is to simply extend 2-D intraframe coding methods to 3-D interframe coding. In DPCM waveform coding, for example, we can predict the current pixel intensity being coded as a linear combination of previously coded neighborhood pixel intensities in both the current and the previous frames. Transform coding methods can also be extended in a straightforward manner. A strong temporal domain correlation will manifest itself as an energy concentration in the low temporal-frequency regions, and transform coefficients in the high temporal-frequency regions may be discarded without seriously distorting the image frame intensities. Some studies [Pratt] indicate that in typical scenes, the bit rate can be reduced by a factor of five without sacrificing intelligibility or quality if a 3-D DCT with a subimage size of $16 \times 16 \times 16$ pixels is used instead of a 2-D DCT with a subimage size of 16×16 pixels. In practice, the storage requirements and delay involved make using sixteen frames difficult in most cases.

Hybrid transform/waveform coding can also be extended to interframe coding. We can compute a 2-D transform for each frame and then apply a waveform coder such as DPCM along the temporal dimension, as shown in Figure 10.55. The sequence $f(n_1, n_2, n_3)$ in the figure represents the intensities of image frames, with n_1 and n_2 denoting two spatial variables and n_3 denoting the time variable. At each n_3, a 2-D transform is computed with respect to variables n_1 and n_2, and the result is denoted by $T_f(k_1, k_2, n_3)$. At each k_1 and k_2, a waveform coder is used to quantize $T_f(k_1, k_2, n_3)$. The result is $\hat{T}_f'(k_1, k_2, n_3)$. The process is reversed at the decoder. The 2-D transform used is typically the DCT, and the waveform coder used is typically DPCM. Hybrid coders have several advantages over transform coders in interframe coding. When we are restricted to using only a small number of frames, which is typically the case in practice, the advantage of transform coding along the temporal dimension over waveform coding in terms of correlation reduction and energy compaction diminishes significantly. In hybrid coding, 2-D

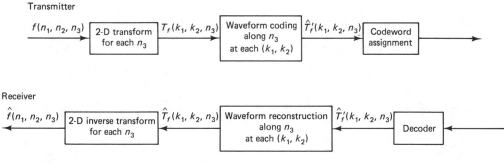

Figure 10.55 Interframe hybrid coder.

transforms are first computed. Since many coefficients are discarded at this point, waveform coding is applied to only a fraction of transform coefficients. In transform coding, however, all the transform coefficients have to be computed first. In addition, transform coding imposes an inherent delay, because computing any one transform coefficient requires all the frames involved. In hybrid coding, the current frame is predicted from one or two previously coded frames and significant delay is not necessary.

Interframe hybrid coding can be viewed as an example of dimension-dependent processing, which was discussed in Section 6.2.2. In dimension-dependent processing, a different processing method is applied to each different dimension, chosen on the basis of the dimension's particular characteristics. Many data points are available along the two spatial dimensions, and transform coding, which exploits the correlation among the data points, is used. Along the temporal dimension, only a few data points are typically available, and waveform coding is used.

The frame replenishment method which is related to DPCM codes the difference between the current frame and the previously coded frame. Let $f(n_1, n_2, n_3)$ represent the current frame and $\hat{f}(n_1, n_2, n_3 - 1)$ represent the previously coded frame. In the simplest form of the frame replenishment method, $f(n_1, n_2, n_3)$ is predicted as $\hat{f}(n_1, n_2, n_3 - 1)$, and $e(n_1, n_2, n_3) = f(n_1, n_2, n_3) - \hat{f}(n_1, n_2, n_3 - 1)$ is quantized. Since $|e(n_1, n_2, n_3)|$ is typically very small except in the small regions where there is motion, only $e(n_1, n_2, n_3)$ with magnitude above a certain threshold is coded along with its spatial location. At the decoder, the quantized $e(n_1, n_2, n_3)$ is combined with $\hat{f}(n_1, n_2, n_3 - 1)$ to reconstruct the current frame $f(n_1, n_2, n_3)$. Since the number of pixels at which $e(n_1, n_2, n_3)$ is retained depends on the local frames involved, a buffer must be established to smooth out the higher-than-average data rates in frames with large motion and the lower-than-average data rates in frames with little motion. An example of the frame replenishment method applied to bi-level images is shown in Figure 10.56. Figure 10.56(a) shows a sequence of sixteen original image frames. The size of each frame is 128×128 pixels. Each of the sixteen frames is mapped to a bi-level image by using the method described in Section 10.5, and the resulting bi-level images are coded by using the frame replenishment method. The result at a bit rate of 0.08 bit/pixel is shown in Figure 10.56(b). When a frame rate of 15 frames/sec is used, the resulting bit rate is approximately 20 kbits/sec. We note that it is very difficult to

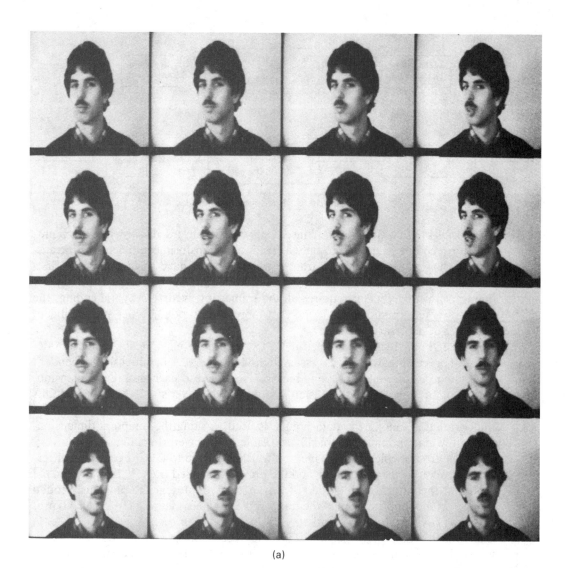

(a)

Figure 10.56 Example of the frame replenishment method applied to bi-level image coding. (a) 16 original frames. The size of each frame is 128 × 128 pixels. The sequence of the frames is from left to right (first row), right to left (second row), left to right (third row), and right to left (fourth row). (b) Reconstructed frames at 0.08 bit/pixel by application of the frame replenishment method to the bi-level images obtained from the gray level images in (a). The method used to generate bi-level images is the same as that used in Figure 10.54.

visualize from Figure 10.56 how the frames will appear when displayed in a sequence. For example, a random fluctuation of intensity from one frame to the next in the same local region may not be annoying when the frames are viewed one frame at a time. However, when the frames are displayed as a video sequence, the random fluctuations may appear as an annoying flicker in the local region. In

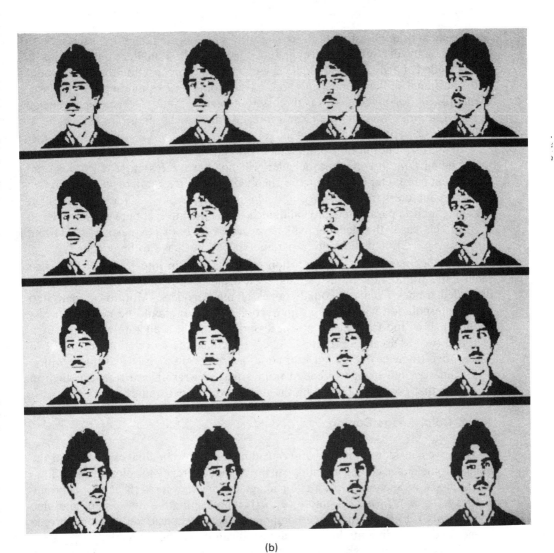

(b)

Figure 10.56 (continued)

addition, degradations such as the dirty window effect and crawling, which are due to correlation between frames, may not be that apparent when one frame at a time is viewed.

One way to improve the performance of this frame replenishment method is to predict the current frame $f(n_1, n_2, n_3)$ by using motion estimation algorithms. Specifically, we can form $e(n_1, n_2, n_3)$ by $e(n_1, n_2, n_3) = f(n_1, n_2, n_3) - \hat{f}(n_1 - d_x, n_2 - d_y, n_3 - 1)$, where d_x and d_y are the horizontal and vertical displacements, which are functions of pixel location. To the extent that the intensity change between the current frame and the previously coded frame is due to motion that can be modeled as translational motion at least at the local level, and that the displacements can be estimated accurately, $e(n_1, n_2, n_3)$ obtained by using motion compensation will have smaller magnitude. As a result, a smaller number of pixels will be coded at a given threshold level with motion compensation than they would with the frame replenishment method. The problem of estimating the displacements (or motion velocities) was discussed in Section 8.4.2. If the displacements d_x and d_y are not integers, the closest integers may be chosen or $\hat{f}(n_1, n_2, n_3 - 1)$ will have to be spatially interpolated.

If we estimate motion parameters from $f(n_1, n_2, n_3)$ and $\hat{f}(n_1, n_2, n_3 - 1)$ in encoding $f(n_1, n_2, n_3)$, the decoder will not have access to $f(n_1, n_2, n_3)$, so the motion parameter will have to be encoded. As an alternative, we can estimate motion parameters from $\hat{f}(n_1, n_2, n_3 - 1)$ and $\hat{f}(n_1, n_2, n_3 - 2)$ and assume that the same motion parameters are valid during the time interval between $n_3 - 1$ and n_3. In this case, the decoder will have access to both $\hat{f}(n_1, n_2, n_3 - 1)$ and $\hat{f}(n_1, n_2, n_3 - 2)$, and the motion parameters will not have to be encoded.

Another method of using motion compensation in interframe coding is to reduce the frame rate by discarding some frames and then reconstructing the discarded frames from the coded frames at the decoder. Motion-compensated frame interpolation was discussed in Section 8.4.3. It should be noted that discarding half of the frames will not necessarily result in a bit rate reduction by a factor of two. Discarding frames is likely to reduce the correlation between two consecutive frames to be coded. In addition, replacing a discarded frame with a new frame created from two coded frames may not result in a frame as high in quality as that produced by coding the discarded original frame.

10.6.2 Color Image Coding

Thus far we considered the problem of coding monochrome images. Many of the monochrome image coding methods can be extended directly to color image coding. As discussed in Section 7.1.4, a color image can be viewed as three monochrome images $f_R(n_1, n_2)$, $f_G(n_1, n_2)$, and $f_B(n_1, n_2)$, representing the red, green, and blue components. Each of the three components can be considered a monochrome image, and the coding methods we discussed can be applied directly.

The three components $f_R(n_1, n_2)$, $f_G(n_1, n_2)$, and $f_B(n_1, n_2)$ are highly correlated with each other, and coding each component separately is not very efficient. One approach that exploits the correlation is to transform the three components to another set of three components with less correlation. One such set is $f_Y(n_1, n_2)$, $f_I(n_1, n_2)$, and $f_Q(n_1, n_2)$, where $f_Y(n_1, n_2)$ is the luminance component and $f_I(n_1, n_2)$ and $f_Q(n_1, n_2)$ are the two chrominance components. The linear 3×3 matrix transformation between R, G, and B components and Y, I, and Q components was discussed in Section 7.1.4. The approach to transform the R, G, and B components to the Y, I, and Q components and then code the Y, I, and Q components is illustrated in Figure 10.57.

One advantage of coding Y, I, and Q components rather than R, G, and B components is that most high-frequency components of a color image are concentrated in the Y component. Therefore, the chrominance components I and Q can be undersampled by a factor of 2×2 to 4×4 in waveform coding without seriously

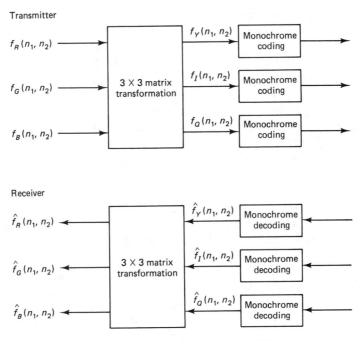

Figure 10.57 Color image coding in the *YIQ* space.

affecting the high-frequency details of a color image. In transform coding, a smaller number of coefficients needs to be coded for the *I* and *Q* components than for the *Y* component. Typically, the total number of bits assigned to both the *I* and *Q* components is only approximately half the number of bits used to code the *Y* component; adding color does not increase the bit rate by a factor of three. In addition, the aesthetics of color lead to a more visually pleasant reconstructed image and tend to hide degradations in the image. A color image coded at a given bit rate typically looks better than a black-and-white image obtained by coding only the *Y* component of the color image at the same total bit rate.

Two examples of color image coding are shown in Figures 10.58 and 10.59 (see color insert). Figure 10.58(a) shows an original color image of 512 × 512 pixels at 24 bits/pixel, eight bits for each of the *R*, *G*, and *B* components. Figure 10.58(b) shows the reconstructed image at 0.4 bit/pixel. Figure 10.59 is another example of a color image coded at 0.4 bit/pixel. The coding scheme used is the scene adaptive coder [Chen and Pratt]. It is an adaptive DCT coder with a sub-image size of 16 × 16 pixels, threshold coding and run-length coding for the transform coefficient locations, and scalar quantization and Huffman coding for the transform coefficient amplitudes.

10.6.3 Channel Error Effects

A communication channel may introduce some errors when transmitting an image. One frequently occurring case, often used to study the effect of channel error, is the reversal of a bit from 0 to 1 or from 1 to 0 with some given probability. The effect of a reversed bit on a reconstructed image will vary depending on what the bit represents. For example, if the reversed bit is part of a location representation, an incorrect location will be decoded. In our discussion, we will assume that the reversed bit represents the amplitude of a waveform or a transform coefficient.

In PCM coding, a bit reversal error affects only the particular pixel whose intensity is represented by the bit in error. An example of an image reconstructed with bit reversal error with error probability of 0.5% is shown in Figure 10.60. The bit reversals appear as noise of the impulsive type. This noise can be reduced by such methods as median filtering and out-range pixel smoothing, discussed in Section 8.2.3.

In DPCM image coding, the effect of a bit reversal error is not limited to one pixel intensity. Instead, the error propagates until a pixel intensity is reinitialized at some later time. When a bit reversal occurs in the error signal at a certain pixel, the pixel reconstructed at the transmitter is different from the pixel reconstructed at the receiver. Since reconstructed pixel intensities are used recursively in DM or DPCM encoding and since the encoder does not know that an error has occurred, the error affects all subsequent pixel intensities from that point on until reinitialization at the encoder and decoder. This error propagation is one reason why the predicted value is typically multiplied by a leakage factor of less than one in forming the error signal to be quantized in DM or DPCM. Figure 10.61 shows an example of a reconstructed image with bit reversal error probabilities of 0.5%. The images were obtained by applying a 1-D DM coder along each row and the pixel intensity was reinitialized at the beginning of each row. The effect of channel error manifests itself as noise of a streaking type in these examples. Such noise can be reduced by applying a 1-D median filter along the vertical direction.

A bit reversal in transform image coding affects one particular coefficient.

Figure 10.60 Illustration of channel error effect in pulse code modulation image coding. Reconstructed image at 8 bits/pixel with error probability of 0.5%.

Figure 10.61 Illustration of channel error effect in delta modulation image coding. Reconstructed image at 1 bit/pixel with error probability of 0.5%.

However, each transform coefficient affects all the pixel intensities within the subimage. Figure 10.62 shows an example of an image coded at $\frac{1}{2}$ bit/pixel with bit reversal error probability of 0.5%. An adaptive zonal DCT coder was used in the example.

10.7 ADDITIONAL COMMENTS

In this chapter, we discussed a number of image coding methods. We now briefly summarize their relative advantages. One major advantage of waveform coders over transform coders is their simplicity. Since the waveform itself or some simple

Figure 10.62 Illustration of channel error effect in discrete cosine transform image coding. Reconstructed image at $\frac{1}{2}$ bit/pixel with error probability of 0.5%.

variation is coded, the coders are very simple both conceptually and computationally. In addition, in applications such as digital television where high quality images are required, waveform coders can perform as well as transform coders at a given bit rate.

Transform coders are typically more expensive computationally than waveform coders, since an image must be transformed at the transmitter and inverse transformed at the receiver. In low bit rate applications such as video telephones and remotely piloted vehicles, where some quality can be sacrificed, transform coders generally perform better than waveform coders.

Image model coders are likely to be the most expensive computationally, since the model parameters of an image are estimated at the transmitter and an image is synthesized at the receiver. However, they have the potential to be useful in very low bit rate applications, such as video telephones for the deaf, where intelligibility is the major concern and a significant amount of quality may be sacrificed to reduce the required bit rate. Image model coders are still at the research stage, and much needs to be done before they can be considered for practical applications.

Vector quantization improves the performance of a given coding algorithm over scalar quantization in the required bit rate. The amount of performance improvement possible depends on various factors, such as the degree of statistical dependence among scalars in the vector. Vector quantization generally improves the performance of waveform coders more than that of transform coders. The parameters coded in waveform coders typically have more correlation among them than those coded in transform coders. The performance improvement by vector quantization comes at a price. Vector quantization adds considerable complexity in the computational and storage requirements. However, it is important to note that most of the computational complexity is at the transmitter. In applications such as broadcasting, where the number of receivers is much larger than the number of transmitters, the cost of additional complexity due to vector quantization may not be significant.

Adaptive image coding methods generally perform considerably better in the required bit rate than nonadaptive methods. Even though adaptive coding adds additional complexity to a coder, the performance improvement often justifies the cost.

Interframe coding is quite useful in applications such as television and motion pictures where the sequence of frames has significant temporal correlation. Interframe coding, however, requires storage of the frames required in the coding process and may involve some inherent delay.

In general, the reduction in the required bit rate is accompanied by an increase in computational and memory costs. The required bit rate is directly related to the cost of using a communication channel (or cost of storage medium in the case an image is coded to reduce the storage cost). The computational and memory costs are directly related to the hardware cost of the transmitter (encoder) and receiver (decoder). The relative importance of reducing the channel bandwidth and hardware cost varies drastically, depending on the application, and often dictates which coding algorithm to choose. In such applications as digital television, many receivers share a channel. Reducing the hardware cost of receivers is very important, while reducing the channel cost is less significant. In applications such as video telephone, however, reducing the channel cost is very important.

The relative hardware cost of the transmitter and receiver also plays an important role in deciding which coding method to choose. In applications such as digital television, there are many more receivers than transmitters. Vector quantization, which adds considerable computational complexity to the transmitter but not to the receiver, may be a viable approach. In applications such as remotely piloted vehicles, we may have many more transmitters than receivers. It is important to reduce the cost of transmitters by using simple coding methods. At the receiver, however, we may be able to tolerate some complexity such as a quantization noise reduction system.

From the above discussion, it is clear that a large number of factors affect the choice of an image coding system in practice. Development of an image coding algorithm suitable for a given application is a challenging process, often requiring many iterations and considerable human interaction. We hope that the fundamentals of digital image coding discussed in this chapter can serve as a guide to this difficult process.

10.8 CONCLUDING REMARKS

In Chapters 7, 8, 9, and 10, we discussed fundamentals of digital image processing. We presented a number of different image processing algorithms and discussed their capabilities and limitations. It should be noted that the objective in this book has not been to provide off-the-shelf algorithms that can be used in specific applications. Instead, the objective has been to study fundamentals and basic ideas that led to the development of various algorithms. Our discussion should be used as a guide to developing the image processing algorithm most suitable to a given application. The factors that have to be considered in developing an image processing algorithm differ considerably in different applications, and thus one cannot usually expect to find an off-the-shelf image processing algorithm ideally suited to a particular application.

An important step in developing an image processing system in a practical application environment is to identify clearly the overall objective. In applications in which images are processed for human viewers, the properties of human visual perception have to be taken into account. In applications in which an image is processed to improve machine performance, the characteristics of the machine are important considerations. Thus, approaches to developing an image processing system vary considerably, depending on the overall objective in a given application.

Another important step is to identify the constraints imposed in a given application environment. In some applications, system performance is the overriding factor, and any reasonable cost may be justified, whereas in other applications, cost may be an important factor. In some applications, real-time image processing is necessary, whereas in others the real time aspect may not be as critical. Clearly, the approach to developing an image processing system is influenced by the constraints imposed by the application context.

A third important step is to gather information about the images to be processed. This information can be exploited in developing an image processing system. For example, if the class of images to be processed consists mostly of buildings, which have a large number of horizontal and vertical lines, this information can be used in designing an adaptive lowpass filter to reduce background noise.

Given the overall objective, the constraints imposed, and information about the class of images to be processed, a reasonable approach to developing an image processing algorithm is to determine if one or more existing methods, such as the ones discussed in this book, are applicable to the given problem. In general, significant developmental work is needed to adapt existing methods to a given problem. If existing methods do not apply, or if the performance achieved by existing methods does not meet the requirements of the problem, new approaches have to be developed. Digital image processing has a relatively short history, and there is considerable room for new approaches and methods. We hope that the fundamentals of digital image processing covered in this book will form a foundation for additional reading on the topic, application of theoretical results to real-world problems, and advancement of the field through research and development.

REFERENCES

For books on image coding, see [Pratt; Clarke; Netravali and Haskell]. Most books on image processing (see Chapter 7 References) also treat image coding. For special journal issues on image coding, see [Cutler; Aaron et al.; Habibi (1977); Netravali (1980); Netravali and Prasada (1985)]. For review articles on the topic, see [Huang; Schreiber (1967); Habibi and Robinson; Netravali and Limb; Jain; Musmann et al.]. For a review article on digital television coding, see [Kretz and Nasse]. For a review of video conferencing systems, see [Sabri and Prasada].

For overviews of information theory, quantization, and codeword assignment, see [Gallagher; Abramson; Jayant and Noll]. For tutorial articles on vector quantization, see [Gray; Makhoul et al.]. For Shannon's original work, see [Shannon]. On Huffman coding, see [Huffman].

On digitization of images, see [Nielsen et al.]. On delta modulation, see [Steele]. On DPCM image coding, see [Habibi (1971); Netravali (1977)]. On two-channel coders, see [Schreiber et al.; Troxel et al.; Schreiber and Buckley]. On subband signal coding, see [Crochiere et al.; Vetterli; Woods and O'Neil]. On pyramid coding, see [Burt and Adelson; Rosenfeld].

For readings on transform image coding, see [Habibi and Wintz; Wintz; Ahmed and Rao]. On hybrid coding, see [Habibi (1974, 1981)]. On reduction of blocking effects, see [Reeve and Lim; Ramamurthi and Gersho; Cassereau et al.].

For readings on texture, see [Haralick; Wechsler; Pratt et al.; Cross and Jain; VanGool et al.]. On very low bit rate image coding, see [Letellier et al.; Pearson and Robinson]. On adaptive image coding, see [Habibi (1977); Netravali and Prasada (1977); Habibi (1981); Chen and Pratt].

For readings on interframe coding, see [Haskell et al.; Netravali and Robbins; Jain and Jain]. On color image coding, see [Limb et al.]. For readings on graphics coding, which was not discussed in this chapter, see [Netravali (1980)].

M. R. Aaron et al., ed., Special Issue on Signal Processing for Digital Communications, *IEEE Trans. Commun. Tech.*, Vol. COM-19, part 1, Dec. 1971.

A. Abramson, ed., *Information Theory and Coding.* New York: McGraw-Hill, 1983.

N. Ahmed and K. R. Rao, *Orthogonal Transforms for Digital Signal Processing.* New York: Springer Verlag, 1975.

N. Ahmed, T. Natarajan, and K. R. Rao, Discrete cosine transform, *IEEE Trans. Computers*, Vol. C-23, Jan. 1974, pp. 90–93.

P. Brodatz, *Textures.* New York: Dover, 1966.

P. J. Burt and E. H. Adelson, The Laplacian pyramid as a compact image code, *IEEE Trans. Commun.*, Vol. COM-31, April 1983, pp. 532–540.

P. M. Cassereau, D. H. Staelin, and G. deJager, Encoding of images based on a lapped orthogonal transform, to be published, IEEE Trans. on Communications, 1989.

W. H. Chen and W. K. Pratt, Scene adaptive coder, IEEE Trans. Commun., Vol. COM-32, March 1984, pp. 225–232.

R. J. Clarke, *Transform Coding of Images.* London: Academic Press, 1985.

F. S. Cohen and D. B. Cooper, Simple parallel hierarchical and relaxation algorithms for segmenting noncausal Markovian random fields, *IEEE Trans. Patt. Anal. Mach. Intell.*, Vol. PAMI-9, March 1987, pp. 195–219.

M. Covell, Low data-rate video conferencing, master's thesis, Dept. of EECS, MIT, Dec. 1985.

R. E. Crochiere, A. A. Webber, and J. L. Flanagan, Digital coding of speech in subbands, *Bell Syst. Tech. J.*, Vol. 55, Oct. 1976, pp. 1069–1085.

G. R. Cross and A. K. Jain, Markov random field texture models, *IEEE Trans. Patt. Anal. Mach. Intell.*, Vol. PAMI-5, Jan. 1983, pp. 25–39.

C. C. Cutler, ed., Special Issue on Redundancy Reduction, *Proc. IEEE*, Vol. 55, March 1967.

E. W. Forgy, Cluster analysis of multivariate data: efficiency vs. interpretability of classifications, abstract, *Biometrics*, Vol. 21, 1965, pp. 768–769.

R. G. Gallager, *Information Theory and Reliable Communication.* New York: Wiley, 1968.

R. M. Gray, Vector quantization, *IEEE ASSP Mag.*, April 1984, pp. 4–29.

A. Habibi, Comparison on nth-order DPCM encoder with linear transformations and block quantization techniques, *IEEE Trans. Commun. Technol.*, Vol. COM-19, Dec. 1971, pp. 948–956.

A. Habibi, Hybrid coding of pictorial data, *IEEE Trans. Comm. Tech.*, Vol. COM-22, May 1974, pp. 614–624.

A. Habibi, ed., Special issue of Image Bandwidth Compression, *IEEE Trans. on Communications*, Vol. COM-25, November, 1977.

A. Habibi, Survey of adaptive image coding techniques, *IEEE Trans. Commun.*, Vol. COM-25, Nov. 1977, pp. 1275–1284.

A. Habibi, An adaptive strategy for hybrid image coding, *IEEE Trans. Commun.*, Vol. COM-29, Dec. 1981, pp. 1736–1740.

A. Habibi and G. S. Robinson, A survey of digital picture coding, *Computer 7*, May 1974, pp. 22–34.

A. Habibi and P. A. Wintz, Image coding by linear transformation and block quantization, *IEEE Trans. Comm. Tech.*, Vol. COM-19, Feb. 1971, pp. 50–62.

R. M. Haralick, Statistical and structural approaches to texture, *Proc. IEEE*, Vol. 67, May 1979, pp. 786–804.

B. G. Haskell, F. W. Mounts, and J. C. Candy, Interframe coding of videotelephone pictures, *Proc. IEEE*, Vol. 60, July 1972, pp. 792–800.

T. S. Huang, PCM picture transmission, *IEEE Spectrum*, Vol. 12, Dec. 1965, pp. 57–63.

J. J. Y. Huang and P. M. Schultheiss, Block quantization of correlated Gaussian random variables, *IEEE Trans. Comm. Syst.*, Sept. 1963, pp. 289–296.

D. A. Huffman, A method for the construction of minimum redundancy codes, *Proc. IRE*, Vol. 40, Sept. 1952, pp. 1098–1101.

A. K. Jain, Image data compression: A review, *Proc. IEEE*, Vol. 69, March 1981, pp. 349–389.

J. R. Jain and A. K. Jain, Displacement measurement and its application in interframe image coding, *IEEE Trans. Commun.*, Vol. COM-29, Dec. 1981, pp. 1799–1808.

N. S. Jayant and P. Noll, *Digital Coding of Waveforms*, Englewood Cliffs, NJ: Prentice-Hall, 1984.

F. Kretz and D. Nasse, Digital television: transmission and coding, *Proc. IEEE*, Vol. 73, April 1985, pp. 575–591.

P. Letellier, M. Nadler, and J. F. Abramatic, The telesign project, *Proc. IEEE*, Vol. 73, April 1985, pp. 813–827.

J. S. Lim and A. V. Oppenheim, Reduction of quantization noise in PCM speech coding, *IEEE Trans. Acoust. Speech Sig. Proc.*, Vol. ASSP-28, Feb. 1980, pp. 107–110.

J. O. Limb, C. B. Rubinstein, and J. E. Thomson, Digital coding of color video signals—A review, *IEEE Trans. Commun.*, Vol. COM-25, Nov. 1977, pp. 1349–1385.

Y. Linde, A. Buzo, and R. M. Gray, An algorithm for vector quantizer design, *IEEE Trans. Commun.*, Vol. COM-28, Jan. 1980, pp. 84–95.

S. P. Lloyd, Least squares quantization in PCM, *IEEE Trans. Inform. Theory*, Vol. IT-28, March 1982, pp. 129–137.

J. Makhoul, S. Roucos, and H. Gish, Vector quantization in speech coding, *Proc. IEEE*, Vol. 73, Nov. 1985, pp. 1551–1588.

J. Max, Quantizing for minimum distortion, *IEEE Trans. Inform. Theory*, Vol. IT-6, March 1960, pp. 7–12.

D. Marr, *Vision, A Computational Investigation into the Human Representation and Processing of Visual Information.* New York: W. H. Freeman and Company, 1982.

H. G. Musmann, P. Pirsch, and H. J. Grallert, Advances in picture coding, *Proc. IEEE*, Vol. 73, April 1985, pp. 523–548.

A. N. Netravali, On quantizers for DPCM coding of picture signals, *IEEE Trans. Inform. Theory*, Vol. IT-23, May 1977, pp. 360–370.

A. N. Netravali, ed., Special Issue on Digital Encoding of Graphics, *Proc. IEEE*, Vol. 68, July 1980.

A. N. Netravali and B. G. Haskell, *Digital Pictures: Representation and Compression.* New York: Plenum Press, 1988.

A. N. Netravali and L. O. Limb, Picture coding: A review, *Proc. IEEE*, Vol. 68, March 1980, pp. 366–406.

A. N. Netravali and B. Prasada, Adaptive quantization of picture signals using spatial masking, *Proc. IEEE*, Vol. 65, April 1977, pp. 536–548.

A. N. Netravali and B. Prasada, eds., Special Issue on Visual Communications Systems, Proc. IEEE, April 1985.

A. N. Netravali and J. D. Robbins, Motion compensated television coding: Part I, *Bell Syst. Tech. J.*, Vol. 58, March 1979, pp. 631–670.

L. Nielsen, K. J. Astrom, and E. I. Jury, Optimal digitization of 2-D images, *IEEE Trans. Acoust. Speech Sig. Proc.*, Vol. ASSP-32, Dec. 1984, pp. 1247–1249.

D. E. Pearson and J. A. Robinson, Visual communication at very low data rates, *Proc. IEEE*, Vol. 73, April 1985, pp. 795–812.

W. K. Pratt, *Image Transmission Techniques*. New York: Academic Press, 1979.

W. K. Pratt, O. D. Faugeras, and A. Gagalowicz, Applications of stochastic texture field models to image processing, *Proc. IEEE*, Vol. 69, May 1981, pp. 542–551.

L. R. Rabiner and R. W. Schafer, *Digital Processing of Speech Signals*. Englewood Cliffs, NJ: Prentice-Hall, 1978.

B. Ramamurthi and A. Gersho, Nonlinear space-variant postprocessing of block coded images, *IEEE Trans. Acoust. Speech Sig. Proc.*, Vol. ASSP-34, Oct. 1986, pp. 1258–1267.

H. C. Reeve III and J. S. Lim, Reduction of blocking effects in image coding, *J. of Opt. Eng.*, Vol. 23, Jan./Feb. 1984, pp. 34–37.

L. G. Roberts, Picture coding using pseudo-random noise, *IRE Trans. Inf. Theory*, Vol. IT-8, Feb. 1962, pp. 145–154.

A. Rosenfeld, ed., *Multiresolution Image Processing and Analysis*. Berlin: Springer Verlag, 1984.

S. Sabri and B. Prasada, Video Conferencing Systems, *Proc. IEEE*, Vol. 73, April 1985, pp. 671–688.

W. F. Schreiber, Picture coding, *Proc. IEEE*, Vol. 55, March 1967, pp. 320–330.

W. F. Schreiber, C. F. Knapp, and N. D. Kay, Synthetic highs: An experimental TV bandwidth reduction system, *J. Soc. Motion Pict. Telev. Eng.*, Vol. 68, Aug. 1959, pp. 525–537.

W. F. Schreiber and R. R. Buckley, A two-channel picture coding system: II—adaptive companding and color coding, *IEEE Trans. Commun.*, Vol. COM-29, Dec. 1981, pp. 1849–1858.

A. Segall, Bit allocation and encoding for vector sources, *IEEE Trans. on Information Theory*, Vol. IT-22, March 1976, pp. 162–169.

C. E. Shannon, A mathematical theory of communication Parts I and II, *Bell Syst. Tech. J.*, Vol. 27, July 1948, pp. 379–423; pp. 623–656.

R. Steele, *Delta Modulation Systems*. New York: Wiley, 1975.

D. E. Troxel et al., A two-channel picture coding system: I—Real-time implementation, *IEEE Trans. Commun.*, Vol. COM-29, Dec. 1981, pp. 1841–1848.

L. VanGool, P. Dewaele, and A. Oosterlinck, Survey Texture Analysis Anno 1983, *Computer Vision, Graphics, and Image Processing*, Vol. 29: 1985, pp. 336–357.

M. Vetterli, Multi-dimensional sub-band coding: some theory and algorithms, *Signal Processing*, Vol. 6, April 1984, pp. 97–112.

H. Wechsler, Texture analysis—a survey, *Sig. Proc.*, Vol. 2, July 1980, pp. 271–282.

P. A. Wintz, Transform picture coding, *Proc. IEEE*, Vol. 60, July 1972, pp. 809–820.

J. W. Woods and S. D. O'Neil, Subband coding of images, *IEEE Trans. Acoust. Speech Sig. Proc.*, Vol. ASSP-34, Oct. 1986, pp. 1278–1288.

PROBLEMS

10.1. Let f denote a scalar that lies in the range of $0 \le f \le 5$. We wish to quantize f to six reconstruction levels using a uniform quantizer.

    (a) Determine the reconstruction levels and decision boundaries.

    (b) Let $\hat{f}$ denote the result of quantization. Show that $e_Q = \hat{f} - f$ is signal dependent.

10.2. Let f be a random variable with a probability density function $p_f(f_0)$. We wish to quantize f. One way to choose the reconstruction levels r_k, $1 \le k \le L$, and the decision boundaries d_k, $0 \le k \le L$, is to minimize the average distortion D given by

$$D = E[(\hat{f} - f)^2]$$

where $\hat{f}$ is the result of quantization. Show that a necessary set of conditions that r_k and d_k have to satisfy is given by Equation (10.9).

10.3. Let f be a random variable with a Gaussian probability density function with mean of 3 and standard deviation of 5. We wish to quantize f to four reconstruction levels by minimizing $E[(\hat{f} - f)^2]$ where $\hat{f}$ is the result of quantization. Determine the reconstruction levels and decision boundaries.

10.4. Consider a random variable f with a probability density function $p_f(f_0)$. We wish to map f to g by a nonlinearity such that $p_g(g_0)$, the probability density function of g, is uniform.

    (a) Show that one choice of the nonlinearity or companding $C[\cdot]$ such that $p_g(g_0)$ will be uniform between $-\frac{1}{2} \le g_0 \le \frac{1}{2}$ and zero otherwise is given by

$$g = C[f] = \int_{x=-\infty}^{f} p_f(x)\, dx - \tfrac{1}{2}.$$

    (b) Show that another choice of the nonlinearity $C[\cdot]$ such that $p_g(g_0)$ will be uniform between $-\frac{1}{2} \le g_0 \le \frac{1}{2}$ and zero otherwise is given by

$$g = C[f] = \int_{x=f}^{\infty} p_f(x)\, dx - \tfrac{1}{2}.$$

10.5. Let f denote a random variable with a probability density function $p_f(f_0)$ given by

$$p_f(f_0) = \begin{cases} e^{-f_0}, & f_0 \ge 0 \\ 0, & \text{otherwise.} \end{cases}$$

Determine a nonlinearity $C[\cdot]$ such that g defined by $g = C[f]$ will have a uniform probability density function $p_g(g_0)$ given by

$$p_g(g_0) = \begin{cases} 1, & 0 \le g_0 \le 1 \\ 0, & \text{otherwise.} \end{cases}$$

10.6. We wish to quantize, with a total of 7 bits, two scalars f_1 and f_2, which are Gaussian random variables. The variances of f_1 and f_2 are 1 and 4, respectively. Determine how many (integer number) bits we should allocate to each of the f_1 and f_2, under the assumption that scalar quantization is used and that we wish to minimize $E[(\hat{f}_1 - f_1)^2 + (\hat{f}_2 - f_2)^2]$. The variables $\hat{f}_1$ and $\hat{f}_2$ are the results of quantizing f_1 and f_2, respectively.

10.7. Consider two random variables $\mathbf{f} = [f_1, f_2]^T$ with a joint probability density function $p_{f_1,f_2}(f'_1, f'_2)$ given by

$$p_{f_1,f_2}(f'_1, f'_2) = \begin{cases} \frac{1}{4}, & \text{in the shaded region in the following figure} \\ 0, & \text{otherwise.} \end{cases}$$

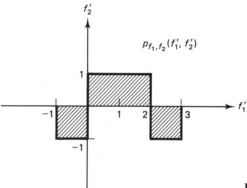

Figure P10.7

(a) We wish to quantize f_1 and f_2 individually, using scalar quantization. We wish to minimize the mean square error $E[(\hat{\mathbf{f}} - \mathbf{f})^T (\hat{\mathbf{f}} - \mathbf{f})]$. If we are given a total of eight reconstruction levels for the vector $\mathbf{f}$, where should we place them?

(b) If we are allowed to use vector quantization in quantizing $\mathbf{f}$, how many reconstruction levels are required to have the same mean square error $E[(\hat{\mathbf{f}} - \mathbf{f})^T (\hat{\mathbf{f}} - \mathbf{f})]$ as in (a)? Determine the reconstruction levels in this case.

10.8. Let $\mathbf{f} = (f_1, f_2)^T$ denote a vector that we wish to quantize. We quantize $\mathbf{f}$ by minimizing $E[(\hat{\mathbf{f}} - \mathbf{f})^T (\hat{\mathbf{f}} - \mathbf{f})]$, where $\hat{\mathbf{f}}$ is the result of quantizing $\mathbf{f}$. Suppose the reconstruction levels $\mathbf{r}_i$, $1 \le i \le 3$, are the filled-in dots shown in the following figure.

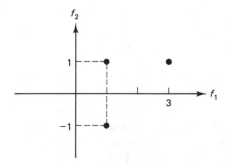

Figure P10.8

Determine the cells C_i, $1 \le i \le 3$. Note that if **f** is in C_i, it is quantized to $\mathbf{r}_i$.

10.9. Let $\mathbf{f} = (f_1, f_2)^T$ denote a vector to be quantized by using vector quantization. Suppose the four training vectors used for the codebook design are represented by the filled-in dots in the following figure.

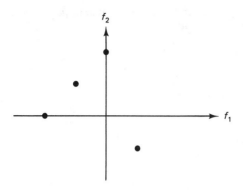

Figure P10.9

Design the codebook using the K-means algorithm with $K = 2$ (two reconstruction levels) with the average distortion D given by

$$D = E[(\hat{\mathbf{f}} - \mathbf{f})^T(\hat{\mathbf{f}} - \mathbf{f})]$$

where $\hat{\mathbf{f}}$ is the result of quantizing **f**.

10.10. Consider a message with L different possibilities. The probability of each possibility is denoted by P_i, $1 \le i \le L$. Let H denote the entropy.
(a) Show that $0 \le H \le \log_2 L$.
(b) For $L = 4$, determine one possible set of P_i such that $H = 0$.
(c) Answer (b) with $H = \log_2 L$.
(d) Answer (b) with $H = \frac{1}{2} \log_2 L$.

10.11. Suppose we wish to code the intensity of a pixel. Let f denote the pixel intensity. The five reconstruction levels for f are denoted by r_i, $1 \le i \le 5$. The probability that f will be quantized to r_i is P_i, and it is given in the following table.

r_i	P_i
20	$\frac{1}{6}$
68	$\frac{1}{24}$
95	$\frac{1}{2}$
150	$\frac{1}{24}$
220	$\frac{1}{4}$

(a) Suppose we wish to minimize the average bit rate in coding f. What is the average number of bits required? Design a set of codewords that will achieve this average bit rate. Assume scalar quantization.
(b) Determine the entropy H.
(c) Suppose we have many pixel intensities that have the same statistics as f. Discuss

a method that can be used in coding the pixel intensities with the required average bit rate being as close as possible to the entropy.

(d) What codebook size is needed if the method you developed in (c) is used?

10.12. Suppose we have an image intensity f, which can be viewed as a random variable whose probability density function $p_f(f_0)$ is shown in the following figure.

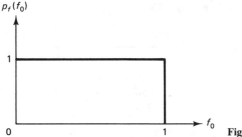

Figure P10.12

Given the number of reconstruction levels, the reconstruction levels and decision boundaries are chosen by minimizing

$$\text{Error} = E[(\hat{f} - f)^2]$$

where $E[\cdot]$ denotes expectation and $\hat{f}$ denotes the reconstructed intensity. We wish to minimize the Error defined above and at the same time keep the average number of bits required to code f below 2.45 bits. Determine the number of reconstruction levels, the specific values of the reconstruction levels and decision boundaries, and the codewords assigned to each reconstruction level.

10.13. Consider an image intensity f which can be modeled as a sample obtained from the probability density function sketched below:

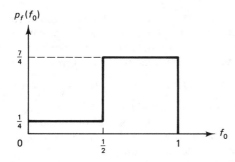

Figure P10.13

Suppose four reconstruction levels are assigned to quantize the intensity f. The reconstruction levels are obtained by using a uniform quantizer.

(a) Determine the codeword to be assigned to each of the four reconstruction levels such that the average number of bits in coding f is minimized. Specify what the reconstruction level is for each codeword.

(b) For your codeword assignment in (a), determine the average number of bits required to represent f.

(c) What is a major disadvantage of variable-length bit assignment relative to uniform-length bit assignment? Assume that there are no communication channel errors.

10.14. Let f denote a pixel value that we wish to code. The four reconstruction levels for f are denoted by r_i, $1 \leq i \leq 4$. The true probability that f will be quantized to r_i is P_i. In practice, the true probability P_i may not be available and we may have to use P_i', an estimate of P_i. The true probability P_i and the estimated probability P_i' are shown in the following table.

r_i	P_i	P_i'
50	$\frac{1}{2}$	$\frac{1}{3}$
100	$\frac{1}{4}$	$\frac{1}{3}$
150	$\frac{1}{8}$	$\frac{1}{6}$
200	$\frac{1}{8}$	$\frac{1}{6}$

If more than one pixel is coded, we will code each one separately.
(a) What is the minimum average bit rate that could have been achieved if the true probability P_i had been known?
(b) What is the actual average bit rate achieved? Assume that the codewords were designed to minimize the average bit rate under the assumption that P_i' was accurate.

10.15. Let $f(n_1, n_2)$ denote a zero-mean stationary random process with a correlation function $R_f(n_1, n_2)$ given by

$$R_f(n_1, n_2) = \rho^{|n_1|}\rho^{|n_2|}, \quad \text{where } 0 \leq \rho \leq 1.$$

(a) In a PCM system, $f(n_1, n_2)$ is quantized directly. Determine σ_f^2, the variance of $f(n_1, n_2)$.
(b) In a DM system, we quantize the prediction error $e_1(n_1, n_2)$ given by

$$e_1(n_1, n_2) = f(n_1, n_2) - a\,\hat{f}(n_1 - 1, n_2)$$

where a is a constant and $\hat{f}(n_1 - 1, n_2)$ is the quantized intensity of the previous pixel coded. Determine a reasonable choice of a. With your choice of a, determine $E[e_1^2(n_1, n_2)]$. Clearly state any assumptions you make.
(c) In a DPCM system, we quantize the prediction error $e_2(n_1, n_2)$ given by

$$e_2(n_1, n_2) = f(n_1, n_2) - a\hat{f}(n_1 - 1, n_2) - b\hat{f}(n_1, n_2 - 1) - c\hat{f}(n_1 - 1, n_2 - 1)$$

where a, b, and c are constants, and $\hat{f}(n_1 - 1, n_2)$, $\hat{f}(n_1, n_2 - 1)$, and $\hat{f}(n_1 - 1, n_2 - 1)$ are the quantized intensities of previously coded pixels. Determine a reasonable choice of a, b, and c. With your choice of a, b, and c, determine $E[e_2^2(n_1, n_2)]$. Clearly state any assumptions you make.
(d) Compare σ_f^2, $E[e_1^2(n_1, n_2)]$ and $E[e_2^2(n_1, n_2)]$, obtained from (a), (b), and (c). Based on this comparison, which of the following three expressions would you code if you wished to minimize the average bit rate at a given level of distortion: $f(n_1, n_2)$, $e_1(n_1, n_2)$, or $e_2(n_1, n_2)$?
(e) State significant advantages of quantizing $f(n_1, n_2)$ over quantizing $e_1(n_1, n_2)$ and $e_2(n_1, n_2)$.

10.16. Consider the following two different implementations of a two-channel image coder. In Figure P10.16, $f(n_1, n_2)$ is the image, $f_L(n_1, n_2)$ is the lows component, and

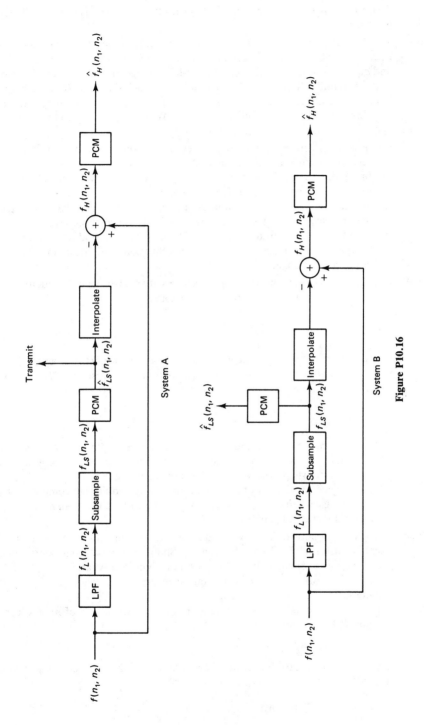

Figure P10.16

$f_H(n_1, n_2)$ is the highs component. How do the two systems differ? Which system is preferable?

10.17. We wish to code a 257×257-pixel image using the Laplacian pyramid coding method discussed in Section 10.3.5. The images we quantize are $f_3(n_1, n_2)$, $e_2(n_1, n_2)$, $e_1(n_1, n_2)$, and $e_0(n_1, n_2)$, where $f_3(n_1, n_2)$ is the top-level image and $e_0(n_1, n_2)$ is the base image in the pyramid. Suppose we design the coders such that 2 bits/pixel, 1 bit/pixel, 0.5 bit/pixel are allocated to quantize $f_3(n_1, n_2)$, $e_2(n_1, n_2)$, $e_1(n_1, n_2)$, and $e_0(n_1, n_2)$, respectively. What is the bit rate (bits/pixel) of the resulting coding system?

10.18. One way to develop an adaptive DM system is to choose the step size Δ adaptively.
(a) In regions where the intensity varies slowly, should we decrease or increase Δ relative to some average value of Δ?
(b) In regions where the intensity varies quickly, should we decrease or increase Δ relative to some average value of Δ?
(c) To modify Δ according to your results in (a) and (b), we have to determine if the intensity varies quickly or slowly, and this information has to be available to both the transmitter and receiver. Develop one method that modifies Δ adaptively without transmission of extra bits that contain information about the rate of intensity change.

10.19. In transform coding, the class of transforms that have been considered for image coding are linear transforms that can be expressed as

$$T_f(k_1, k_2) = \sum_{n1=0}^{N_1-1} \sum_{n2=0}^{N_2-1} f(n_1, n_2)a(n_1, n_1; k_1, k_2) \tag{1}$$

where $f(n_1, n_2)$ is an image, $a(n_1, n_2; k_1, k_2)$ are basis functions, and $T_f(k_1, k_2)$ are transform coefficients. If $a(n_1, n_2; k_1, k_2)$ is separable so that $T_f(k_1, k_2)$ can be expressed as

$$T_f(k_1, k_2) = \sum_{n1=0}^{N_1-1} \sum_{n2=0}^{N_2-1} f(n_1, n_2)a_R(n_1; k_1)a_c(n_2; k_2), \tag{2}$$

the required number of computations can be reduced significantly.
(a) Determine the number of arithmetic operations required in computing $T_f(k_1, k_2)$ for the case when $a(n_1, n_2; k_1, k_2)$ is not separable.
(b) Determine the number of arithmetic operations required in computing $T_f(k_1, k_2)$ for the case when $a(n_1, n_2; k_1, k_2)$ is separable and the row-column decomposition method is used in computing $T_f(k_1, k_2)$.

10.20. Let $f(n_1, n_2)$ denote an $N_1 \times N_2$-point sequence, and let $C_f(k_1, k_2)$ denote the discrete cosine transform (DCT) of $f(n_1, n_2)$. Show that discarding some coefficients $C_f(k_1, k_2)$ with small amplitude does not significantly affect $\sum_{n1} \sum_{n2} (\hat{f}(n_1, n_2) - f(n_1, n_2))^2$, where $\hat{f}(n_1, n_2)$ is the image reconstructed from $C_f(k_1, k_2)$ with small amplitude coefficients set to zero.

10.21. Let $f(n_1, n_2)$ denote a stationary random process with mean of m_f and correlation function $R_f(n_1, n_2)$ given by

$$R_f(n_1, n_2) = \rho^{|n_1| + |n_2|} + m_f^2, \qquad \text{where } 0 < \rho < 1.$$

Let $f_w(n_1, n_2)$ denote the result of applying a 2×2-point rectangular window to $f(n_1, n_2)$ so that

$$f_w(n_1, n_2) = \begin{cases} f(n_1, n_2), & 0 \le n_1 \le 1, 0 \le n_2 \le 1 \\ 0, & \text{otherwise.} \end{cases}$$

Let $F_w(k_1, k_2)$ denote the 2×2-point DFT of $f_w(n_1, n_2)$.

(a) In transform coding, we exploit the property that a properly chosen transform tends to decorrelate the transform coefficients. This allows us to avoid repeatedly coding the redundant information. Illustrate this correlation reduction property by comparing the correlation of $f_w(n_1, n_2)$ and the correlation of $F_w(k_1, k_2)$. Note that there is an arbitrary scaling factor associated with the DFT definition, and proper normalization should be made in the comparison.

(b) A property related to the correlation reduction property is the energy compaction property. The energy compaction property allows us to discard some coefficients with small average amplitudes. Illustrate the energy compaction property by comparing $f_w(n_1, n_2)$ and $F_w(k_1, k_2)$.

(c) Answer (a) with $m_f = 0$.

(d) Answer (b) with $m_f = 0$.

10.22. Consider a 4×4-pixel subimage $f(n_1, n_2)$. The variances of the 4×4-point DCT coefficients of $f(n_1, n_2)$ are shown in the following figure.

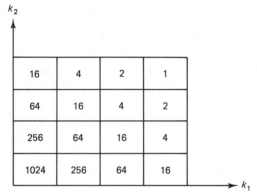

Figure P10.22

For a 2 bits/pixel DCT coder, determine a good bit allocation map that can be used in quantizing the DCT coefficients.

10.23. In a subband image coder, we divide an image into many bands (typically 16 bands) by using bandpass filters and then code each band with a coder specifically adapted to it. Explain how we can interpret the discrete Fourier transform coder as a subband coder with a large number of bands.

10.24. The notion that we can discard (set to zero) variables with low amplitude without creating a large error between the original and reconstructed value of the variable is applicable not only to transform coefficients but also to image intensities. For example, if the image intensity is small (close to zero), setting it to zero does not create a large error between the original and reconstructed intensities. Zonal coding and threshold coding are two methods of discarding variables with low amplitudes.

(a) Why is zonal coding useful for transform coding but not for waveform coding?

(b) Why is threshold coding useful for transform coding but not for waveform coding?

10.25. In some applications, detailed knowledge of an image may be available and may be exploited in developing a very low bit rate image coding system. Consider a video telephone application. In one system that has been proposed, it is assumed that the primary changes that occur in the images will be in the eye and mouth regions. Suppose we have stored at both the transmitter and the receiver the overall face and many possible shapes of the left eye, the right eye, and the mouth of the video telephone user. At the transmitter, an image frame is analyzed, and the stored eye shape and mouth shape closest to those in the current frame are identified. The identification numbers are transmitted. At the receiver, the stored images of the eyes and mouth are used to create the current frame.

(a) Suppose 100 different images are stored for each of the right eye, the left eye, and the mouth. What is the bit rate/sec of this system? Assume a frame rate of 30 frames/sec.

(b) What type of performance would you expect from such a system?

10.26. Let an $N_1 \times N_2 \times N_3$-point sequence $f(n_1, n_2, n_3)$ denote a sequence of frames, where n_1 and n_2 represent the spatial variables and n_3 represents the time variable. Let $F(k_1, k_2, k_3)$ denote the $N_1 \times N_2 \times N_3$-point discrete Fourier transform (DFT) of $f(n_1, n_2, n_3)$.

(a) Suppose $f(n_1, n_2, n_3)$ does not depend on n_3. What are the characteristics of $F(k_1, k_2, k_3)$?

(b) Suppose $f(n_1, n_2, n_3) = f(n_1 - n_3, n_2 - n_3, 0)$. What are the characteristics of $F(k_1, k_2, k_3)$?

(c) Discuss how the results in (a) and (b) may be used in the 3-D transform coding of $f(n_1, n_2, n_3)$.

10.27. Let $f(n_1, n_2, n_3)$ for $0 \leq n_3 \leq 4$ denote five consecutive image frames in a motion picture. The variables (n_1, n_2) are spatial variables. Due to the high temporal correlation, we have decided to transmit only $f(n_1, n_2, 0)$, $f(n_1, n_2, 4)$, and the displacement vectors $d_x(n_1, n_2)$ and $d_y(n_1, n_2)$ that represent the translational motion in the five frames. We assume that the motion present can be modeled approximately by uniform velocity translation in a small local spatio-temporal region. At the receiver, the frames $f(n_1, n_2, 1)$, $f(n_1, n_2, 2)$, and $f(n_1, n_2, 3)$ are created by interpolation. Since $d_x(n_1, n_2)$ and $d_y(n_1, n_2)$ do not typically vary much spatially, the hope is that coding $d_x(n_1, n_2)$ and $d_y(n_1, n_2)$ will be easier than coding the three frames.

(a) One method of determining $d_x(n_1, n_2)$ and $d_y(n_1, n_2)$ is to use only $f(n_1, n_2, 0)$ and $f(n_1, n_2, 4)$. An alternate method is to use all five frames. Which method would lead to a better reconstruction of the three frames $f(n_1, n_2, 1)$, $f(n_1, n_2, 2)$, and $f(n_1, n_2, 3)$?

(b) Determine one reasonable error criterion that could be used to estimate $d_x(n_1, n_2)$ and $d_y(n_1, n_2)$ from all five frames. Assume that we use a region-matching method for motion estimation.

10.28. Let $f(n_1, n_2)$ denote an image intensity. When $f(n_1, n_2)$ is transmitted over a communication channel, the channel may introduce some errors. We assume that the effect of channel error is a bit reversal from 0 to 1 or from 1 to 0 with probability P.

(a) Suppose we code $f(n_1, n_2)$ with a PCM system with 8 bits/pixel. When $P = 10^{-3}$, determine the expected percentage of pixels affected by channel errors.

(b) Suppose we code $f(n_1, n_2)$ with a PCM system with 4 bits/pixel. When $P = 10^{-3}$, determine the expected percentage of pixels affected by channel errors.

(c) Based on the results of (a) and (b), for a given coding method and a given P, does the channel error affect more pixels in a high bit rate system or in a low bit rate system?

(d) Suppose we code $f(n_1, n_2)$ with a DCT image coder with an average bit rate of 1 bit/pixel. When $P = 10^{-3}$, determine the expected percentage of pixels affected by the channel error. Assume that a subimage size of 8×8 pixels is used.

Index